BIOLOGICAL RESPONSE SIGNATURES

Indicator Patterns Using Aquatic Communities

BIOLOGICAL RESPONSE SIGNATURES

Indicator Patterns Using Aquatic Communities

Thomas P. Simon

CRC PRESS

Boca Raton London New York Washington, D.C.

Cover Illustrations: Rainbow Darter, Southern Redbelly Dace, and Rock Bass (© Joseph R. Tomelleri). Euglenoid (*Trachelomonas armata* (Ehr.) Stein) courtesy of Philadelphia Academy of Natural Sciences.

Library of Congress Cataloging-in-Publication Data

Biological response signatures : indicator patterns using aquatic communities / edited by Thomas P. Simon.
p. ; cm.
Includes bibliographical references (p.).
ISBN 0-8493-0905-0
1. Indicators (Biology) 2. Water quality biological assessment. 3. Aquatic ecology. I. Simon, Thomas P.
QH541.15.I5 B56 2002
577.6'028'7—dc21 2002276541
CIP

Direct all inquiries to CRC Press LLC, 2000 N.W. Corporate Blvd., Boca Raton, Florida 33431.

International Standard Book Number 0-8493-0905-0
Library of Congress Card Number 2002276541
Printed in the United States of America 1 2 3 4 5 6 7 8 9 0
Printed on acid-free paper

Foreword

Biological indicators have become primary measures of the conditions of our water resources. The integration of biological assessments and criteria into water quality management programs worldwide is impressive. Whether freshwater or coastal, running waters or wetlands, biological indicators have transcended the difficulties in communicating their results by the adoption and refinement of indices of biotic integrity (IBIs). IBIs have been developed for fish, macroinvertebrates, periphyton/diatoms, macrophytes, and even birds and terrestrial ecosystems. The key remaining challenge for such biological indicators is diagnostic — to demonstrate clearly the causes and effects needed to take actions to protect and restore water resources. *Biological Response Signatures* by Dr. Thomas Simon helps to meet this challenge by presenting an unprecedented compilation of technical approaches and case studies that allow the reader to better understand biological response signatures and stressor identification and how they can be applied successfully in other programs.

The United States Environmental Protection Agency (USEPA) has been particularly interested in identifying specific stressors that cause impairment to aquatic communities. It published a *Stressor Identification Guidance Document* with a logical approach to evaluating evidence and identifying the main stressors causing biological impairments. In fact, several issues surrounding the total maximum daily load (TMDL) regulation and guidance directly affect biological indicators. Most recently, USEPA has decided that a TMDL does not need to be developed if a pollutant cannot be identified as the stressor causing biological impairments. Therefore, although a TMDL may not be required, it is imperative to determine the real cause of impairment to the biological community, whether the stressor is chemical contamination, nutrient enrichment, poor habitat quality, or hydrologic alteration.

Biological Response Signatures takes this discipline to the next level, just as Dr. Simon's last effort, *Assessing the Sustainability and Biological Integrity of Water Resources Using Fish Communities,* has become the standard reference for fish community assessment. From the conceptual framework to the case studies, this book provides those key elements to support better diagnostic evaluations of the stressors to biological communities.

Wayne S. Davis
United States Environmental Protection Agency
Washington, D.C.

Preface

The primary purpose of this book is to further the technical knowledge of biological indicators necessary to assess the causes and sources of environmental effects. In an effort to do this, I have built from two previous texts I was involved with and attempt to supply the reader with new information and relevant summaries. Although this book is not comprehensive, I have encouraged the chapter authors to include relevant ideas and information. This book does not review the status of biocriteria nor their underpinnings (see Karr, 1981; Karr et al., 1986; and Karr and Chu, 1999). Those interested should consult Davis and Simon (1995) for information on application and background. This book does not attempt to review the status of multimetric indices for all biological indicator assemblages (see Simon, 1999, for additional information on fish assemblage indicator development).

Rarely are environmental impacts the results of single chemical or even single industrial sources; rather, they are complex mixtures and interactions of contaminants. The use of environmental assessment procedures within monitoring frameworks demands some relevancy to the decisions based on biological criteria that management agencies make. These biological criteria standards are the basis for environmental indicators that provide direct measures of environmental quality.

The use of biological criteria in monitoring and assessment programs resulted from the impressive degree of precision that can be achieved with the index of biotic integrity (IBI). The IBI was originally developed by James R. Karr (1981) to evaluate midwestern stream fish assemblages. This single index has been further developed to represent a full range of issues that explain biological integrity (Fausch et al., 1984, 1990; Angermeier and Karr, 1986; Angermeier and Schlosser, 1987; Karr et al., 1986; Simon, 1999).

The IBI is considered a multimetric index; it represents a family of indices adapted for use in organismal indicator groups besides fish. Simon (2000) indicated that the IBI is one single type of biocriterion. I suggested that our tool boxes contain a wide range of biological indicators including diversity indices (e.g., Shannon-Weiner diversity index), univariate indices (e.g., species richness indices, Hilsenhoff biotic index, index of well-being), the widely adapted multimetric indices of biological integrity for a variety of indicator assemblages (e.g., Karr, 1981; Karr et al., 1986; Simon and Lyons, 1995; Simon, 1999; Simon et al., 2001), and indices of sustainability (e.g., tailwaters index, reservoir fishery assessment index; included in Simon, 1999) in order to make accurate assessments.

This book was begun to evaluate what is known about patterns in multimetric indices relating to known point or non-point source impacts. It focuses on the IBI, but I attempted to include viewpoints related to other univariate and multivariate approaches as well. Current research on environmental assessment patterns has not kept pace with the prognosis of environmental health and condition (USEPA, 2000). This book is state-of-the-art and describes the results of years of biological indicator development; it is the first to address patterns in multimetric indices based on site assessment. It attempts to evaluate the differences in biological integrity between natural and altered landscapes and discusses the types of organismal indicator groups used for assessments of diatoms, aquatic macrophytes, aquatic invertebrates, mussels, fish, amphibians, and birds.

The 26 chapters in this book are designed to build on the foundation established by *Biological Assessment and Criteria: Tools for Water Resource Planning and Decision Making* (Davis and Simon, 1995). That book described the foundational concepts and background necessary to utilize fully the ideas presented in this volume. It was used by many resource agencies, researchers, and in college classrooms to teach the next generation of environmental scientists and biologists a wide range of issues ranging from environmental assessment to natural resource decision making and ecosystem management.

This book does not repeat information contained in *Assessing the Sustainability and Biological Integrity of Water Resources Using Fish Communities* which highlighted a number of inconsistencies, including the lack of information on framing points such as zoogeographic implications to developing reference conditions, differences between biological integrity of altered environments and natural habitats, inaccurate or misinformation in guild descriptions and classifications that are the premises behind the various metrics, and application of the index to areas other than small warmwater streams. I have divided this book into five sections: (1) conceptual framework, (2) contaminant patterns in ecosystems, (3) method advancement, (4) land use modification patterns and effects of non-point sources, and (5) case studies.

The conceptual framework section (Chapters 1 through 4) discusses ideas behind biological response signatures. Following are a series of chapters that describe ramifications of the environmental relevancy of biological criteria, biological stress responses, use of multiple indicators to diagnose biological response patterns, and setting restoration and ecological recovery endpoints using biological criteria. The use of biological response signatures and advancing the efforts to understand changes in community response to anthropogenic disturbance will probably require major efforts in future years.

The majority of regulatory agencies are in the process of developing response indicators based on reference conditions or reference sites within regional frameworks and calibrated multimetric indices. Changes in assemblage indicators must be able to separate natural variability from human disturbance gradients. Cairns describes the effects of environmental stress and the unique abilities of biological organisms to detect and diagnose these changes. The state of Ohio has been a model for other states in the area of understanding and evaluating processes and how they might work within a regulatory framework. Sources of impairments are presented using a series of case studies that reflect a variety of stressor responses. The last chapter in this section describes how biological criteria and indicators can be used to diagnose the degree of impact and determine appropriate restoration goals and options. These ecological recovery endpoints can be used to determine when restoration as a result of total maximum daily load (TMDL) and national pollutant discharge and elimination (NPDES) permit limits are in compliance or when, after an oil or major spill, the environment has fully recovered.

The contaminant patterns in the ecosystems section (Chapters 5 through 8) describes patterns from specific effects on different organism groups. Contaminated dredge spoil effects on wetland plant communities, the effects of depth of fines on aquatic ecosystem health, pesticide effects on assemblages, and metal contamination of macroinvertebrate assemblages are highlighted. These chapters are paired with case studies in Sections IV and V that describe specific applications. Limited information on large-scale cutting edge issues such as sediment quantity effects and new generation pesticides is available. Likewise, the extensive degradation of wetland plant assemblages and metal effects on trophic dynamics in aquatic systems result from sediment and water degradation that affects thousands of stream miles.

The method advancement section (Chapters 9 through 12) discusses assessment approaches and evaluates specific IBI metric relationships with human disturbance. The assessment of point source impacts has classically been monitored using a simple upstream versus downstream approach. This assessment approach is not necessarily effective for testing metric response and determining whether the IBI reveals noise or natural variability. The use of the new traveling zone (T-zone) approach was applied to assessments on the Ohio River. Pioneer species have classically been used to evaluate headwater streams, but do they have a wider application?

Rankin and Simon evaluate patterns in Ohio streams and rivers and specifically assess four case studies. Thoma and Simon evaluate preliminary patterns observed in the Great Lakes and correlations between omnivores and nutrient stimulation. The need for USEPA and the U.S. Department of Agriculture to establish protective nutrient criteria caused interest in evaluating specific patterns between fish assemblage structure and function and nutrient stimulation. The future of biocriteria and biological integrity assessment is presented in the chapter by Wiley et al. They show how using a model of stream integrity based on geographic information system technology and an extensive database from Michigan revealed predicted versus observed effects in streams in the lower peninsula.

The land use modification patterns and effects of non-point sources section (Chapters 13 through 18) represent a variety of non-point source impact types from urbanization to mining, pesticide application, and land use changes as a result of confined disposal facilities (CDFs). Wang and Lyons review the resultant impacts of urbanization and how they may impact biological integrity. The relevant question becomes whether urban streams can attain the same levels of biological integrity they formerly had as natural streams.

Wilhelm et al. describe the recovery of seral soils after impacts and management of confined disposal facilities. Their studies in the Great Lakes show that limited recovery of wetland assemblages occurred in these CDFs. Carlisle et al. evaluated the effects of aerial deposition and former land use changes in the Cuyahoga National Park as a result of heavy metals, polyaromatic hydrocarbons (PAHs), and mercury. They evaluated the ability of benthic macroinvetebrate assemblages to detect changes and effects that could assist in the recovery of these systems.

Hard rock mining impacts in the western United States greatly impacted large expanses of land. Mebane and Fore describe issues surrounding hard rock mining impacts in Idaho and Colorado streams using macroinvertebrate assemblages. Lydy et al. conducted a case study in Kansas to evaluate patterns in IBI metrics with pesticide residues in fish tissue and sediment. Simon and Exl evaluated patterns in silviculture impacts on stream communities related to water quality, habitat changes, and biological indicators.

The last section of the book (Chapters 20 through 26) deals with case studies. Impacts that result from confined animal feedlots, iron and steel manufacturing, acid mine leachate and acid rain, thermal discharge, and agriculture, urbanization, and coal mining are described. Many chapters in this section are based on multiple indicator groups and provide a wealth of information on responses. Several chapters include information on point source discharge, including two chapters on iron and steel manufacturing in southern Lake Michigan and discharge effects on fish and thermal effects on fish and macroinvertebrate assemblages in the Ohio River drainage. The testing and patterns in diatom assemblages in the Appalachian Mountains were evaluated, while multiple indicators in this same region were evaluated for response signatures.

This book is the beginning of further work using aquatic assemblages as environmental indicators of biological integrity. It is my hope that environmental managers, biologists, hydrologists, and others using this book will benefit from the experiences of the authors who are at the forefront of this field. Although this book puts into perspective how much is known about response patterns and environmental assessment, additional work is required to answer remaining questions. Historically, many disciplines worked in isolation — concerned only with their own indicators. We now know that many groups observe similar patterns and responses. It is with great pleasure that we continue our attempts to provide an accurate prognosis on the condition of our environmental resources in an effort to restore, protect, and enhance the biological integrity of our nation's surface waters.

REFERENCES

Angermeier, P.L. and J.R. Karr. 1986. Applying an index of biotic integrity based on stream fish communities: considerations in sampling and interpretation, *North American Journal of Fisheries Management*, 6, 418–429.

Angermeier, P.L. and I.J. Schlosser. 1987. Assessing biological integrity of the fish community in a small Illinois stream, *North American Journal of Fisheries Management*, 7, 331–338.

Davis, W.S. and T.P. Simon, Eds. 1995. *Biological Assessment and Criteria: Tools for Water Resources Planning and Decision Making*. Lewis Publishers, Boca Raton, FL.

Fausch, K.D., J.R. Karr, and P.R. Yant. 1984. Regional application of an index of biotic integrity based on stream fish communities, *Transactions of the American Fisheries Society*, 113, 39–55.

Fausch, K.D., J. Lyons, J.R. Karr, and P.L. Angermeier. 1990. Fish communities as indicators of environmental degradation, in S.M. Adams (Ed.), *Biological Indicators of Stress in Fish*. American Fisheries Society Symposium 8, Bethesda, MD, 123–144.

Karr, J.R. 1981. Assessment of biotic integrity using fish communities, *Fisheries*, 6, 21–27.

Karr, J.R. and E.W. Chu. 1999. *Restoring Life in Running Waters: Better Biological Monitoring*. Island Press, Washington, D.C.

Karr, J.R., K.D. Fausch, P.L. Angermeier, P.R. Yant, and I.J. Schlosser. 1986. *Assessing Biological Integrity in Running Waters: A Method and Its Rationale*. Illinois Natural History Survey Special Publication 5, Champaign, IL.

Simon, T.P., Ed. 1999. *Assessing the Sustainability and Biological Integrity of Water Resources Using Fish Communities.* CRC Press, Boca Raton, FL.

Simon, T.P. 2000. The use of biological criteria as a tool for water resource management, *Environmental Science and Policy,* 3, S43-S50.

Simon, T.P. and J. Lyons. 1995. Application of the index of biotic integrity to evaluate water resource integrity in freshwater ecosystems, in Davis, W.S. and T.P. Simon, Eds. *Biological Assessment and Criteria: Tools for Water Resource Planning and Decision Making.* Lewis Publishers, Boca Raton, FL.

Simon, T.P., P.M. Stewart, and P.L. Rothrock. 2001. Development of multimetric indices of biotic integrity for riverine and palustrine wetland plant communities along southern Lake Michigan, *Aquatic Ecosystem Health and Management*, 4, 293–309.

U.S. Environmental Protection Agency. 2000. *Stressor Identification Guidance Document*. EPA 822/B-00/025. USEPA, Office of Water, Washington, D.C.

Acknowledgments

I gratefully acknowledge the chapter authors for their efforts and dedication at completing the task at hand. Chapters affiliated with the U.S. Environmental Protection Agency, U.S. Geological Survey, and U.S. Fish and Wildlife Service were part of the official USEPA, USGS, and USFWS peer review process, and those reviewers are acknowledged.

I want to thank my wife and children for their support. I especially want to thank my wife who provided enormous strength for this project. Without her encouragement, support, and handling the additional load of our daily family activities, I would never have been able to complete this task.

I wrote portions of this book and edited it in my private capacity. No official support or endorsement by the U.S. Environmental Protection Agency, the U.S. Fish and Wildlife Service, or any other agency of the federal government is intended or should be inferred.

Contributors

Holly H. Bennett
Department of Biological and Environmental Sciences
Troy State University
Troy, Alabama

Sandra A. Bryce
Dynamac Corporation
Corvallis, Oregon

Jason T. Butcher
U.S. Forest Service
Superior National Forest
Duluth, Minnesota

John Cairns, Jr.
Virginia Polytechnic Institute and State University
Department of Biology
Blacksburg, Virginia

Daren M. Carlisle
U.S. National Park Service
Omaha, Nebraska

William H. Clements
Colorado State University
Department of Biology
Fort Collins, Colorado

Jeffrey S. DeShon
Ohio Environmental Protection Agency
Division of Surface Water
Groveport, Ohio

Ronda L. Dufour
Dufour Consulting
Indianapolis, Indiana

Erich B. Emery
Ohio River Valley Water Sanitation Commission
Cincinnati, Ohio

Joseph A. Exl
U.S. Fish and Wildlife Service
Bloomington, Indiana

Wayne C. Faatz
Indiana Department of Natural Resources
Indianapolis, Indiana

Leska S. Fore
Statistical Design, Inc.
Seattle, Washington

James R. Gammon
Department of Biological Science
DePauw University
Greencastle, Indiana

Eric L. Garza
U.S. Geological Survey
Lake Michigan Ecological Research Station
Porter, Indiana

Mary G. Henry
U.S. Fish and Wildlife Service
Washington, D.C.

Robert M. Hughes
Dynamac Corporation
Corvallis, Oregon

Krzysztof M. Jop
Marion, Massachusetts

Peter M. Kiffney
Northwest Fisheries Science Center
Seattle, Washington

Michael J. Lydy
Department of Zoology
Southern Illinois University
Carbondale, Illinois

John Lyons
Fish and Habitat Research Section
Wisconsin Department of Natural Resources
Monona, Wisconsin

Frank H. McCormick
U.S. Environmental Protection Agency
Cincinnati, Ohio

Joan S. Martin
Huron River Watershed Council
Ann Arbor, Michigan

Christopher A. Mebane
Idaho Department of Environmental Quality
Boise, Idaho

Charles C. Morris
Department of Biological and Environmental Sciences
Troy State University
Troy, Alabama

Steven A. Newhouse
Indiana Department of Environmental Management
Indianapolis, Indiana

Edward T. Rankin
Ohio Environmental Protection Agency
Division of Surface Water
Groveport, Ohio

Christy D. Robinson
Department of Biological and Environmental Sciences
Troy State University
Troy, Alabama

James A. Sawyer, IV
Department of Biological and Environmental Sciences
Troy State University
Troy, Alabama

Paul W. Seelbach
Institute for Fisheries Research
Ann Arbor, Michigan

Thomas P. Simon
U.S. Fish and Wildlife Service
Bloomington, Indiana

Scott A. Sobiech
U.S. Fish and Wildlife Service
Carlsbad, California

Daniel W. Sparks
U.S. Fish and Wildlife Service
Bloomington, Indiana

Paul M. Stewart
Department of Biologcal and Environmental Sciences
Troy State University
Troy, Alabama

Roger F. Thomas
Ohio Environmental Protection Agency
Division of Surface Water
Twinsburg, Ohio

Jeffrey A. Thomas
Ohio River Valley Water Sanitation Commission
Cincinnati, Ohio

Lizhu Wang
Wisconsin Department of Natural Resources
Fish and Habitat Research Section
Monona, Wisconsin

Kevin Wehrly
Institute for Fisheries Research
Ann Arbor, Michigan

Gerould S. Wilhelm
Conservation Design Forum, Inc.
Elmhurst, Illinois

Michael J. Wiley
The University of Michigan
School of Natural Resources and Environment
Ann Arbor, Michigan

Chris O. Yoder
Midwest Biodiversity Institute
Columbus, Ohio

Reviewers

Loren L. Bahls
Hannaea
Helena, Montana

William H. Clements
Colorado State University
Department of Biology
Fort Collins, Colorado

Darla Donald
Virginia Polytechnic University
Department of Biology
Blacksburg, Virginia

Ronda L. Dufour
Dufour Consulting
Indianapolis, Indiana

Leska S. Fore
Statistical Design, Inc
Seattle, Washington

James R. Gammon
Department of Biological Science
DePauw University
Greencastle, Indiana

Mary G. Henry
U.S. Fish and Wildlife Service
Washington, D.C.

Alan Herlihy
Oregon State University
Department of Fish and Wildlife
Western Ecology Division
Corvallis, Oregon

Robert M. Hughes
Dynamac Corporation
Corvallis, Oregon

Phil Kaufmann
U.S. Environmental Protection Agency
Western Ecology Division
Corvallis, Oregon

Tim Kubiak
U.S. Fish and Wildlife Service
Washington, D.C.

D. Phil Larsen
U.S. Environmental Protection Agency
Corvallis, Oregon

John Lyons
Fish and Habitat Research Section
Wisconsin Department of Natural Resources
Monona, Wisconsin

Frank H. McCormick
U.S. Environmental Protection Agency
Cincinnati, Ohio

Steven A. Newhouse
Indiana Department of Environmental Management
Indianapolis, Indiana

James T. Oris
Miami University
Department of Zoology
Oxford, Ohio

Edward T. Rankin
Ohio Environmental Protection Agency
Division of Surface Water
Groveport, Ohio

Thomas P. Simon
U.S. Fish and Wildlife Service
Bloomington, Indiana

Scott A. Sobiech
U.S. Fish and Wildlife Service
Carlsbad, California

Paul M. Stewart
Department of Biological and Environmental Sciences
Troy State University
Troy, Alabama

John Stoddard
U.S. Environmental Protection Agency
Western Ecology Division
Corvallis, Oregon

Roger F. Thoma
Ohio Environmental Protection Agency
Division of Surface Water
Twinsburg, Ohio

Ian Waite
U.S. Geological Survey
Portland, Oregon

Thomas Whittier
Dynamac Corporation
Corvallis, Oregon

Gerould S. Wilhelm
Conservation Design Forum, Inc.
Elmhurst, Illinois

Douglas Wilcox
U.S. Geological Survey
Biological Resources Division
Ann Arbor, Michigan

Contents

Section I

Conceptual Framework

1 Biological Response Signatures: Toward the Detection of Cause-and-Effect and Diagnosis in Environmental Disturbance

Thomas P. Simon

CONTENTS

1.1 INTRODUCTION

Yoder and Rankin (1995a) were the first to coin the term, "biological response signatures." The term is defined as discernable patterns in the response of aquatic community attributes, so that the information is able to discriminate between different stressor types. Unique combinations of biological community characteristics that aid in distinguishing one impact type over another are detected in the biological community data and respond with discrete signatures. In their paper, which described the effects of select environmental disturbances using biological indicators, Yoder and Rankin were able to segregate various impacts into nine categories of disturbance. These response signatures were considered the mechanisms that would assist environmental managers in diagnosing and providing a prognosis on cause and effect. However, as with many tools the practitioners wanted more resolution and the ability to determine chemical-specific impacts.

Suter (1993) critically evaluated the concepts behind ecological health and the index of biotic integrity (IBI). He stated that his paper "does not attack the concept [IBI] but rather the much more limited belief that the best way to use… biosurvey data is to create an index of heterogenous variables [multimetric approach] and claim that it represents ecosystem health." He outlined ten criticisms that serve as the foundation of any good environmental indicator (Herricks and Schaefer, 1985). Among Suter's criticisms of the IBI are ambiguity, eclipsing, arbitrary variance, unreality,

0-8493-0905-0/03/$0.00+$1.50

post hoc justification, unitary response scales, lack of diagnostic results, disconnection from testing and modeling, nonsense results, and improper analogy to other indices. Karr (1993) responded to many of the issues Suter raised, as did Simon and Lyons (1995). However, one question that has remained about multimetric indices is whether they are relevant.

This project began with the need to determine whether the IBI was relevant, and whether the underlying assumption that metrics could be diagnostic tools to determine specific impacts could be used to identify cause and effect. However, as the project progressed, it became increasingly apparent that the relevancy issue should not be limited to only multimetric indices but expanded to include all biological criteria. Biological criteria are ecological benchmarks based on a variety of biological integrity response measures sensitive to human-induced modifications (Davis and Simon, 1995; Simon, 1999). The U.S. Environmental Protection Agency (USEPA) indicated in 1990 that biological criteria are "narrative and numerical expressions that describe the reference [least-impacted] biological integrity of aquatic communities inhabiting waters of a given designated aquatic life use."

Simon (2000) indicated that biological criteria include narrative and numerical expressions; thus in the broadest sense biological criteria can be based upon a variety of indices including diversity indices (Washington, 1984), univariate indices (e.g., Hilsenhoff biotic index [Hilsenhoff, 1982]), floristic quality index (Swink and Wilhelm, 1994), index of well-being (Gammon, 1976), numerous fisheries population and stock assessment indices (Nielsen and Johnson, 1983), and multimetric indices of biological integrity and sustainability (e.g., Karr, 1981; Karr et al., 1986; DeShon, 1995; Stewart et al., 1999; see Davis and Simon, 1995; Simon, 1999; Fore, Chapter 22). The diagnostic capability of multimetric indices has not been demonstrated on a widespread basis, although specific examples exist (Eagleson et al., 1990; Yoder and DeShon, Chapter 3). It is the intention of this book to explore whether biological criteria developed over the last two decades can address the issue of relevancy.

1.1.1 Independent Application and the Weight-of-Evidence Approach

USEPA's policy of independent application (IA) suggests that all environmental data be weighted equally for evaluation. IA is considered controversial since the biological data, which are direct measures of aquatic life-designated uses, can usually only affect management decisions unilaterally. Under the IA policy, biological data can only affect management decisions when both the whole effluent toxicity and water chemistry data indicate that no problem exists; thus, assessments can only cause more stringent environmental management decisions. The problems associated with habitat modification, loss in biological integrity to less desirable levels, and diffuse non-point source impairments are not easily detected and described under IA. However, the Clean Water Act's purpose is to protect and restore the chemical, physical, and biological integrity of the nation's streams and these issues must be addressed (Karr, 1995).

1.1.2 The Three-Legged Stool and Other Landscape Features

USEPA used the analogy of a three-legged stool to support different monitoring approaches; i.e., water quality parameters, whole effluent toxicity testing, and ambient biological surveys. Karr (1993) challenged this concept as too rigid and inadequate to address changing environmental conditions. Instead, he proposed the analogy of a tripod supporting a spotting scope. In order to see a distant object, such as a designated use, the three legs of the tripod must be adjusted to accommodate the terrain, which is the nature of the water resource problem. The concept of the three-legged stool suggests that all environmental data are equal, while Karr's tripod approach selects the best tool for the assessment.

USEPA suggested that it was important for biologists to outfit a "toolbox" of methods and indicators that could be used for evaluating biological integrity of the nation's surface waters. The use of multiple biological criteria approaches for assessing the anthropogenic disturbance gradients in the environment is still in its infancy. With further development of this process, it will become obvious where more research is needed. However, it is imperative that we not stop short of our intended goal but continue the development and application of new and refined tools.

1.2 PATTERNS IN ENVIRONMENTAL ASSESSMENT APPROACHES

Two of the regulated community's most frequently asked questions are:

1. Will biological criteria result in limits in their National Pollution Discharge and Elimination System (NPDES) permits (Polls, 1994; Reash, 1995)?
2. How will the states implement the criteria within the Clean Water Act and other federal and state legislation?

Many misconceptions about the development and implementation of biological criteria are based on misinformation (Seegert, 2000a). Once these misconceptions are discussed, the issues quickly become site specific, rather than issues surrounding the development of biocriteria (Seegert, 2000b). However, many of the issues have been addressed in recent literature (see Davis and Simon, 1995; Karr and Chu, 1999; Simon, 1999); thus, the remaining question becomes, how good is this tool? However, confounding the issue of the impacts of specific types of industry is the cultural basis for these impacts. It is necessary that as impacts are described and documented, society determine the levels of protection that it will accept and how much it values the aesthetic quality of the natural environment.

Yoder and Rankin (1995a) were among the first to evaluate the conditions of aquatic assemblages using interpreted biological information. Few biologists dispute the value of biological communities in demonstrating impairment due to any number of stressors in the environment. However, the capability to use the resultant community data to discriminate between different stressors is frequently questioned. Yoder and Rankin argued that the new multimetric indices were capable of determining the source of an impaired condition (see Yoder and DeShon, Chapter 3). They found that the response patterns of the various metrics and components of the indices when studied near predominant sources showed certain patterns (Table 1.1). They concluded that the inclusion of multiple indicator groups, detailed taxonomic resolution, standardized sampling procedures, and an adequate database would eliminate the problems associated with using a single indicator (i.e., seasonal variation, periodic absences of key indicator taxa). Most programs compensate for these potentially confounding circumstances by standardizing index sampling periods, using macro-scale (e.g., fish) and micro-scale (e.g., macroinvertebrates) indicators, and by recognizing that differences between indicators may be a result of signal differences to recovery (Simon et al., Chapter 4).

1.3 DECIPHERING PATTERNS IN "NOISE" VS. "SIGNAL"

The index of biotic integrity (IBI) has proven responsive to a wide variety of disturbances that affect fish assemblage stability and function (Karr, 1981; Karr et al., 1986; Davis and Simon, 1995; Karr and Chu, 1999; Simon, 1999). The use of multimetric indices to assess human disturbance of aquatic systems has often shown that repeat visits to the same site provided a range of index scores that were too variable (Yoder and Rankin, 1995b). However, the response of this "noise" is often a signal that can provide information on the health and status of the community.

Yoder and Rankin (1995b) noted that greater "noise" in the IBI was found in Ohio streams with low biological integrity. Similar results were observed on the Ohio River, where repeated sampling over 11 weeks showed that differences in stable instream cover and hard substrate habitat

TABLE 1.1
Characteristics of impact types including sources, characteristics, and aspects of multimetric indices affected. Nine impact types are used by Yoder and Rankin (1995) to describe biological response signatures in Ohio rivers and streams.

Type	Major Source	Characteristics	Biocriteria Effects
Complex Toxic	Major municipal WWTP; industrial point sources	These facilities comprise a significant portion of the summer base flow of the receiving stream and generally have one of the following characteristics: 1) serious instream chemical water quality impairments involving toxics; 2) recurrent whole effluent toxicity, fish kills, or severe sediment contamination involving toxics; or 3) this may include areas that have combined sewer overflows (CSOs) and/or urban areas located upstream from the point source.	Lowest quality for IBI, MIwb, darter species, percent round-bodied suckers, sensitive species, percent DELT anomalies, intolerant species, and density (less tolerant species).
Conventional Municipal/Industrial	Municipal WWTP's that discharge conventional substances	These facilities may or may not dominate stream flows and no serious or recurrent whole effluent toxicity is evident or small industrial discharges that may be toxic, but do not comprise a significant fraction of the summer base flow; other influences, i.e., CSOs and urban runoff may be present upstream from the point sources.	High incidence of extreme outliers for darter species, number of species, percent carnivores, percent simple lithophils, density (minus tolerants) and biomass. Extreme range for DELT anomalies >10% observed within or in close proximity to WWTP mixing zones.
Combined Sewer Overflows/Urban	Impacts from CSOs and urban runoff within cities and Metropolitan areas that are in direct proximity to sampling sites	Areas include both free flowing and impounded areas upstream from the major WWTP discharges. Minor point sources may also be present in some areas.	Moderate decline in IBI and MIwb, loss of darter, intolerant, and sensitive species, decline in percent round-bodied suckers, number of species and few DELT anomalies.
Channelization	Areas impacted by extensive, large-scale channel modification projects	Little or no habitat recovery has occurred and some minor point source influences may be present.	Low or even lower metric values for percent round-bodied suckers, intolerant species, sunfish species (lowest), percent top carnivores, percent simple lithophils, and biomass. Exhibits the highest maximums and outliers for density (including tolerants), however, did not indicate toxic impacts (e.g., DELT anomalies very low).
Agricultural Nonpoint	Areas that are principally impacted from row-crop agriculture	Dominant land use in the Corn Belt. Some minor point source and localized habitat influences may be present.	Metric and index values indicative of good and exceptional performance. May also show fair and poor scores under extended low flows due to water withdrawal, higher effluent loads, and more intensive land use and riparian impacts.

TABLE 1.1 (CONTINUED)
Characteristics of impact types including sources, characteristics, and aspects of multimetric indices affected. Nine impact types are used by Yoder and Rankin (1995) to describe biological response signatures in Ohio rivers and streams.

Type	Major Source	Characteristics	Biocriteria Effects
Flow Alteration	Controlled releases	Sites affected by flow alteration include controlled releases downstream from major reservoirs or areas affected by water withdrawals as the predominant impact	Good to exceptional performance of IBI and MIwb, little effect on number of sensitive, darter, and number of species.
Impoundment	Navigation dams, low-head dams, flood control and water supply reservoirs	River segments that have been artificially impounded by low-head dams or flood control and water supply reservoirs	Good performance of IBI and MIwb, loss of sensitive and intolerant species, including darters and round-bodied suckers, decline in number of species. Low numbers of omnivores and tolerant species.
Combined Sewer Overflow/Urban with Toxics	Same as CSO/Urban Conventional	A significant presence of toxics usually associated with municipal CSO systems with significant pretreatment programs and sources of industrial contributions to the sewer system	Metric values consistently show lowest quality for IBI, MIwb; darter species, percent round-bodied suckers, sensitive species, percent DELT anomalies, intolerant species, percent tolerant species, and density (less tolerant species).
Livestock Access	Sites directly impacted by livestock operations	Animals have unrestricted access to adjacent streams	Declines in caddisflies, declines in percent tolerant taxa.

produced less noise in the IBI than soft substrates and poor instream habitat (Simon and Sanders, 1999). Karr et al. (1985a) showed how changes in chlorine and ammonia levels from wastewater treatment facilities caused changes in IBI scores. Likewise, Yoder and Rankin (1995a, b) showed how changes in water resource management decisions can affect the biological integrity of watersheds in Ohio. Processes that improved combined sewer overflow (CSO) and domestic effluent treatment in the Scioto River downstream of Columbus improved IBI scores; likewise, contaminants in the Ottawa River near Lima, Ohio, showed the effects of lowering IBI scores for miles downstream of oil refineries.

The ability to protect biological resources is dependent on our ability to detect differences between natural and human-induced variation in biological condition (Karr and Chu 1999). To determine changes that result from human disturbance, sampling and analysis should concentrate on multiple sites within the same environmental setting across a range of conditions from least-impacted to severely disturbed as a result of human disturbance (Emery and Thomas, Chapter 9).

The development of response signatures is dependent on sampling a variety of disturbance intensities for only a single human activity; thus, a changing biological response is similar to a dose response curve. This approach is difficult to construct; however, if successful, it would then produce biological response signatures for that particular activity (Karr et al. 1986; Yoder and Rankin 1995a). Knowledge of such biological response signatures would give researchers a diagnostic tool for watersheds influenced by unknown or multiple human activities; however, the complex nature of chemical contaminants from outfalls and divergent land use practices makes it virtually impossible to separate single human actions. Although it is often important to diagnosis-specific contaminants causing impacts, it is often not feasible or necessary, as the biological indicators can be used to classify impacts into specific classes of human disturbance.

For example, non-point source impacts from logging and agriculture cause different response signatures. Logging impacts may include warming of the stream, removal of riparian vegetation, and increased sedimentation, while agriculture may increase nutrients to the stream, increasing channelization and loss of instream cover. Changes in fish assemblage structure and function would be detected through agricultural changes in the percentage of tolerant species, increase in relative abundance, loss of sensitive species, and decline of simple lithophilic spawning species. Sedimentation impacts may have the biggest effects on logged streams with the loss of sensitive benthic habitat specialists, increase in species richness due to invading cool and warm water species, decline in the percentage of specialized insectivores, and decline in species' relative abundance. Both of these human disturbance types would be different from contaminants or toxicity changes from point source discharges (Table 1.1).

Diverse human activities generally interact to affect watersheds; however, this may enable sites to be grouped and placed on a gradient according to activities and their effects (Rossano 1995; Karr and Chu 1999). For example, since industrial effluents are more toxic than domestic effluents and both are more serious than low-head dams, weirs, or levees, a dichotomous flowchart of human disturbance threats could be produced that groups sites into categories of biological conditions across a gradient of human disturbance. Sometimes a single variable can capture and integrate multiple sources of influence. For example, relatively simple descriptors that act as surrogates for human disturbance can explain these biological differences; e.g., percent impervious area. Alternatively, sites can be grouped in qualitative disturbance categories.

Patterson (1996) classified northern Rocky Mountain stream sites into four categories of human activity: (1) little or no human influence on the watershed, (2) light recreational use (hiking, backpacking), (3) heavy recreational use (major trailheads, camping areas), and (4) urbanization, grazing, agriculture, or wastewater discharge. Light recreational use did not alter B-IBIs compared to undisturbed watersheds; however, heavy recreational use significantly altered invertebrates but not to the same extent as intensive urbanization and agriculture.

Thorne and Williams (1997) classified sites in South America, Africa, and Southeast Asia according to a pollution gradient based on six measures of chemical pollution. Biological condition based on individual metrics such as total taxa richness (families) and mayfly, stonefly, and caddisfly richness declined as pollution increased. The three tropical regions' biological responses were similar, and paralleled those observed in temperate regions even though the fauna were very different. Thus, environmental biologists should recognize that patterns may not necessarily be species specific but rather trophic or niche specific. This puts value on the function and the structure of the community.

Data collected over a number of years at the same site can also reveal biological responses as human activities change. Regardless of how a range of human influences is selected among study sites, sampling at sites with different intensities and types of human activity is essential to detect and understand biological responses to human influence.

1.4 FUTURE DIRECTIONS

Although the IBI enables the analysis of anthropogenic impact at index or submetric levels, selection of certain metrics to show impacts was not always diagnostic for select disturbances. As our knowledge base continues to expand, one area of research may be the development of impact-specific multimetric indices or the development of specific metrics that will assist in diagnosing the effects from specific contaminants or environmental disturbances.

Fore (Chapter 19) completed such an analysis for evaluating hard mining areas in Colorado; however, the impacts may be similar to the toxicity reduction evaluations (TREs) used in toxicity test procedures when toxicity is observed. The TRE approach takes a sample of effluent and changes it by filtration, aeration, chelation, and other manipulations to reveal the cause of toxicity. When a biological community sample is collected and an impact is detected from the IBI or other biocriteria approach, then specific ecological disturbance evaluations (EDEs) could be conducted so that the biologist can extract data that might be a response signature for the suspected cause of impact. This would include additional sampling stratified across disturbance gradient (i.e., spatial scales), and multiple collections (i.e., temporal scales).

The USEPA developed a guidance document on stressor identification that outlines the steps necessary to determine causes and effects of specific impacts (USEPA, 2000). The process is an iterative approach that resembles a risk management method for determining cause. The method is based on a strength-of-evidence approach and evaluation using Koch's postulate, which combines different lines of evidence to provide compelling evidence for causation. The approach was originally developed for pathogen-induced diseases (Yerushalmy and Palmer, 1959; Hackney and Kinn, 1979) and ecological effects (Adams, 1963; Woodman and Cowling, 1987; Suter, 1990, 1993) and had been recommended for ecological risk assessment (USEPA, 1998). Koch's postulate infers that:

1. The injury, dysfunction or other effect of the pathogen or toxicant must be regularly associated with exposure to the pathogen or toxicant in association with any contributing causal factors.
2. The pathogen, toxicant, or a specific indicator of exposure must be found in the affected organism.
3. The effects must be seen when healthy organisms are exposed to the pathogen or toxicant under controlled conditions, and any contributory factors should contribute in the same way during the controlled exposures.
4. The pathogen, toxicant, or a specific indicator of exposure must be found in the experimentally affected organism.

The power of Koch's postulate is how the four types of evidence are combined. For example, the requirement of regular association cannot be determined in the field because it usually cannot be controlled in such a manner as to establish whether the stressor works alone or in combination with other correlated causes. In addition, field associations cannot document the temporal sequence of cause and effect. The second and fourth postulates suggest that the field observations must correspond to experimental exposures. This suggests that the exposure and field correspondence are unlikely to be coincidental. Thus, each cause can be evaluated separately or in combination with other stressors. Yoder and DeShon (Chapter 3) show how a biological stressor gradient can be used to identify the cause of environmental impacts using a logical argument progression.

One promising tool is the Michigan Rivers Inventory model (Seelbach et al., Chapter 12), which uses a modeling approach to evaluate stream impacts. The changes in the stream can be evaluated using various geographic information system (GIS) layers that detect changes in river morphology, chemistry, and groundwater contributions. The assemblage indicator can then be modeled and sampled to compare predicted and observed differences. Significant differences from the control population can be assumed to result from anthropogenic disturbance.

1.5 CONCLUSIONS

The science of environmental assessment has changed dramatically over the last century (Davis, 1995), with improvements in methods, theory, and detection. Unfortunately, over this same period, we have seen rampant degradation and loss of significant amounts of the landscape. Although in some areas the biological integrity can reflect pristine or pre-Columbian landscapes, most often we must manage our environmental resources using least-impacted or best-that-remain sites to establish the reference condition.

The increase in complex effluent and sediment toxicity, pesticide toxicity, and the synergistic, additive, and antagonistic effects of these complex contaminants confounds the ability of the scientist to determine the sources of effects. These biological response signatures are the bases for determining patterns in the multimetric biocriteria and signal-to-noise ratios for select impacts. This effort is the future research direction that must make use of data patterns from biological databases, dose-response curves, traveling zones that replace the conventional upstream versus downstream approaches, and models that evaluate predicted versus observed patterns in heterogeneous landscapes. The focus of this book is to change our monitoring objectives and increase our resolution to assess biological indicator patterns. Use of logical response patterns, multivariate analysis, and analysis of specific metric patterns will further our assessment discriminatory ability past the nine classes Ohio has devised.

ACKNOWLEDGMENTS

Without the many influences that have shaped the concepts discussed in this book, I would never have attempted to undertake this process. Many of the authors are "giants" who helped to make biomonitoring and biological integrity concepts and research issues relevant. I especially wish to thank Jim Karr, John Cairns, Wayne Davis, Chris Yoder, and Bob Hughes for their support, counsel, and ideas. The opinions expressed do not necessarily represent those of the U.S. Fish and Wildlife Service, and no official endorsement should be inferred.

REFERENCES

Adams, D.F. 1963. Recognition of the effects of fluorides on vegetation, *Journal of Air Pollution Control Association,* 13, 360–362.

Davis, W.S. and T.P. Simon (Eds.). 1995. *Biological Assessment and Criteria: Tools for Water Resource Planning and Decision Making,* Lewis Publishers, Boca Raton, FL.

DeShon, J.E. 1995. Development and application of the invertebrate community index (ICI). In W.S. Davis and T.P. Simon (Eds.). *Biological Assessment and Criteria: Tools for Water Resource Planning and Decision Making,* Lewis Publishers, Boca Raton, FL, 217–244.

Eagleson, K.W., D.L. Lenat, L.W. Rusley, and R.B. Winborne. 1990. Comparison of measured instream biologcal responses and responses predicted using the *Ceriodaphnia dubia* chronic toxicity test, *Environmental Toxicology and Chemistry,* 9, 1019–1028.

Emery, E.B. and J.A. Thomas. 2002. A method for assessing outfall effects on Great River fish populations: the traveling zone approach, Chapter 9, this volume.

Fore, L.S. 2002a Response of diatom assemblages to human disturbance: development and testing of a multimetric index for the mid-Atlantic Region (USA), Chapter 22, this volume.

Fore, L.S. 2002b. Biologcal assessment of mining disturbance on stream invertebrates in mineralized areas of Colorado, Chapter 19, this volume.

Gammon, J.R. 1976. The fish populations of the middle 340 km of the Wabash River, Purdue University Water Resources Research Center Technical Report 86, Lafayette, IN.

Hackney, J.D. and W.S. Kinn. 1979. Koch's postulates updated: a potentially useful application to laboratory research and policy analysis in environmental toxicology, *American Review in Respiration and Disease,* 1119, 849–852.

Herricks, E.E. and D.J. Schaefer. 1985. Can we optimize biomonitoring? *Environmental Management,* 9, 487–492.

Hilsenhoff, W.L. 1982. Using a biotic index to evaluate water quality of streams. Technical Bulletin Number 132. Wisconsin Department of Natural Resources, Madison, WI.

Karr, J.R. 1981. Assessment of biological integrity using fish communities, *Fisheries,* 6(6), 21–27.

Karr, J.R. 1993. Defining and assessing ecological integrity: beyond water quality, *Environmental Toxicology and Chemistry,* 12, 1521–1531.

Karr, J.R. 1995. Protecting aquatic ecosystems: clean water is not enough, in W.S. Davis and T.P. Simon (Eds.), *Biological Assessment and Criteria: Tools for Water Resource Planning and Decision Making,* Lewis Publishers, Boca Raton, FL, 7–14.

Karr, J.R. and E.W. Chu. 1999. *Restoring Life in Running Waters: Better Biological Monitoring,* Island Press, Washington, D.C.

Karr, J.R., R.C. Heidinger, and E.H. Helmer. 1995a. Sensitivity of the index of biotic integrity to changes in chlorine and ammonia levels from wastewater treatment facilities, *Journal of the Water Pollution Control Federation,* 57, 912–915.

Karr, J.R., K.D. Fausch, P.L. Angermeier, P.R. Yant, and I.J. Schlosser. 1986. *Assessing the Biological Integrity in Running Waters: A Method and Its Rationale,* Illinois Natural History Survey, Special Publication 5, Champaign, IL.

Nielsen, L. and D. Johnson. 1983. *Fisheries Techniques.* American Fisheries Society, Bethesda, MD.

Patterson, A.J. 1996. The effect of recreation on biotic integrity of small streams in Grand Teton National Park. M.S. thesis, University of Washington, Seattle, WA.

Polls, I. 1994. How people in the regulated community view biological integrity, *Journal North American Benthological Society,* 13, 598–604.

Reash, R.J. 1995. Biocriteria: a regulated industry perspective, in W.S. Davis and T.P. Simon (Eds.). *Biological Assessment and Criteria: Tools for Water Resource Planning and Decision Making*, Lewis Publishers, Boca Raton, FL, 153–166.

Rossano, E.M. 1995. Development of an index of biological integrity for Japanese streams (IBI-J), M.S. thesis, University of Washington, Seattle, WA.

Seegert, G. 2000a. The development, use, and misuse of biocriteria with an emphasis on the index of biotic integrity, *Environmental Science and Policy,* 3, S51–S58.

Seegert, G. 2000b. Considerations regarding development of index of biotic integrity metrics for large rivers, *Environmental Science and Policy,* 3, S99-S106.

Simon, T.P. 1999. *Assessing the Sustainability and Biological Integrity of Water Resources Using Fish Communities.* CRC Press, Boca Raton, FL.

Simon, T.P. 2000. The use of biological criteria as a tool for water resource management, *Environmental Science and Policy,* 3, S43-S49.

Simon, T.P. and J. Lyons. 1995. Application of the index of biotic integrity to evaluate water resource integrity in freshwater ecosystems, in W.S. Davis and T.P. Simon (Eds.). *Biological Assessment and Criteria: Tools for Water Resource Planning and Decision Making*. Lewis Publishers, Boca Raton, FL, 245–262.

Simon, T.P. and R.E. Sanders. 1999. Applying an index of biotic integrity based on Great River fish communities: considerations in sampling and interpretations, in T.P. Simon (Ed.). *Assessing the Sustainability and Biological Integrity of Water Resources using Fish Communities.* CRC Press, Boca Raton, FL, 475–505.

Simon, T.P., E.T. Rankin, R.L. Dufour, and S.A. Newhouse. 2002. Using biological criteria for establishing restoration and ecological recovery endpoints, Chapter 4, this volume.

Stewart, P.M., R. Scribiallo, and T.P. Simon. 1999. The use of aquatic macrophytes in monitoring and in assessment of biological integrity, in A. Gerhardt (Ed.). *Biomonitoring of Polluted Waters — Reviews on Actual Topics.* Environmental Science Forum 96. Trans Tech Publications, Limited, Uetikon-Zuerch, Switzerland, 275–302.

Suter, G.W., II. 1990. Use of biomarkers in ecological risk assessment, in J.F. McCarthy and L.L. Shugart (Eds.). *Biomarkers of Environmental Contamination,* Lewis Publishers, Ann Arbor, MI, 419–426.

Suter, G.W., II. 1993. A critique of ecosystem health concepts and indexes, *Environmental Toxicology and Chemistry,* 12, 1533–1539.

Swink, W.R. and G. Wilhelm. 1994. *Plants of the Chicago Region.* 4th ed. Indiana Academy of Science, Indianapolis, IN.

Thorne, R. St. J. and W.P. Williams. 1997. The response of benthic invertebrates to pollution in developing countries: a multimetric system of bioassessments, *Freshwater Biology,* 37, 671–686.

U.S. Environmental Protection Agency (USEPA). 1990. *Biological Criteria: National Program Guidance for State Managers.* EPA 440-4-90-010. USEPA, Office of Water, Washington, D.C.

U.S. Environmental Protection Agency (USEPA). 1998. *Guidelines for Ecological Risk Assessment.* EPA 822-R-98-008. USEPA, Office of Research and Development. Risk Assessment Forum, Washington, D.C.

U.S. Environmental Protection Agency (USEPA). 2000. *Stressor Identification Guidance Document*. EPA 822/B-00/025. USEPA, Office of Research and Development, Washington, D.C.

Washington, H.G. 1984. Diversity, biotic and similarity indices: a review with special relevance to aquatic ecosystems, *Water Research,* 18, 653–694.

Woodman, J.N. and E.B. Cowling. 1987. Airborne chemicals and forest health, *Environmental Science*, 21, 120–126.

Yerushalmy, J. and C.E. Palmer. 1959. On the methodology of investigations of etiologic factors in chronic disease, *Journal of Chronic Disease,* 10, 27–40.

Yoder, C.O. and J.E. DeShon. 2002. Using biological response signatures within a framework of multiple indicators to assess and diagnose causes and sources of impairments to aquatic assemblages in selected Ohio rivers and streams, Chapter 3, this volume.

Yoder, C.O. and E.T. Rankin. 1995a. Biological response signatures and the area of degradation value: new tools for interpreting multimetric data, in W.S. Davis and T.P. Simon (Eds.). *Biological Assessment and Criteria: Tools for Water Resource Planning and Decision Making,* Lewis Publishers, Boca Raton, FL, 263–286.

Yoder, C.O. and E.T. Rankin. 1995b. Biological criteria program development and implementation in Ohio, in W.S. Davis and T.P. Simon (Eds.). *Biological Assessment and Criteria: Tools for Water Resource Planning and Decision Making,* Lewis Publishers, Boca Raton, FL, 109–144.

2 Biotic Community Response to Stress

John Cairns, Jr.

CONTENTS

2.1 INTRODUCTION

Since the development of the saprobian index nearly a century ago (Kolkwitz and Marsson, 1908, 1909), practitioners of bioassessment have accumulated information on the effects of anthropogenic stress on aquatic communities. While every metric supplies some information, no single metric presents all the information needed. The amount of information obtained per unit effort can vary widely from one metric to the next. Metrics that never vary, vary without reference to any observable impact, or are difficult to measure well or with consistency have been discarded in favor of more practical metrics with more discriminatory power. Investigators have occasionally been humbled by looking for effects in one metric and missing devastating effects in another.

The search has been broadened to include metrics that range more widely over differing attributes of a community. With this more inclusive net, screening for impact is more effective in systems that are largely unfamiliar. The standards by which impact is judged have been customized to differing ecoregions and types of ecosystems. These formulaic assessments of biological conditions are useful tools for making the everyday decisions required in environmental management. Because of this usefulness, these tools, which largely originated in streams, have been developed and applied to other ecosystems, e.g., coastal freshwater wetlands (Simon, 1998; Simon and Stewart, 1998), small lakes (Schulz et al., 1999), estuaries (Weisberg et al., 1997), and forests (Canterbury et al., 2000). Experience has broadened the use of these assessment tools and the accumulating body of knowledge fits into the larger ideas about stress ecology.

0-8493-0905-0/03/$0.00+$1.50

2.2 STRESS ECOLOGY

2.2.1 DEFINITIONS

Environmental stress is an action, agent, or condition that impairs the structure or function of a biological system. The responses to an environmental stress are either structural (e.g., describing the number and kinds of biotic components) or functional (e.g., describing performance or flux). Many early studies of community response to pollution stress were carried out by systematists or taxonomists. Structural changes involving alterations in species composition, trophic ratios, and indicator species were far more common than functional changes, such as nutrient cycling and energy flow.

In addition to differences in type (i.e., structure or function), responses to environmental stress also differ by hierarchical scale. Environmental stress can individually affect biological communities at the levels of cells, organisms, populations, communities, ecosystems, and landscapes. Examples of environmental stresses include floods, fires, droughts, hurricanes, volcanic activity, climate change, land use change, introduction of exotic species, physical change of attributes such as temperature, substrate, or hydrology, and chemical changes such as pollution or nutrient enrichment. Responses to environmental stresses include DNA shearing, mortality, reduced recruitment, reduced diversity, and loss of energy or nutrients, among countless others. Clearly, environmental stress can be either natural or anthropogenic. Both types of environmental stresses can be characterized on the basis of their spatial and temporal extents and patterns, their intensity, and their novelty (e.g., Kelly and Harwell, 1989; Cairns, 2001).

2.2.2 AN ENVIRONMENTAL GENERAL STRESS SYNDROME

Systems ecologists and others outlined the general ways in which ecosystems may respond to various types of environmental stress (Table 2.1; Barrett et al., 1976; Odum, 1969, 1985; Rapport et al., 1985; Schindler, 1987; Costanza and Mageau, 1999). While some of these hypothesized responses are based on observation, others are based on thermodynamics; the expectations for a persistent system are defined from these principles. One criticism of multimetric indices has been that they are tautological (Suter, 1993), i.e., the characteristics of pristine and damaged systems are used to define a scoring system which is then used to define pristine and damaged systems. By looking at *a priori* assumptions about the changes in ecosystems under stress, this circle can be broken. Metrics that are consistent with general theories of stress ecology, as well as successful at discriminating obviously damaged from obviously intact systems, are more defensible.

Many metrics included in common indices focus on one prediction of response to stress at the level of community structure. According to Item 13 in Table 2.1, species diversity decreases, dominance increases, and functional redundancy declines. These responses are particularly useful for assessing environmental damage, since changes in community composition often precede other changes, especially changes in ecosystem level function (Schindler, 1987; Niederlehner and Cairns, 1994).

There is support for several other of the metrics suggested from the framework of stress ecology. Cattaneo et al. (1998) found that the mean sizes of organisms in a lake chronically impacted by metals decreased in both individual species and in three groups of organisms spanning three kingdoms and many trophic levels. Havens and Carlson (1998) examined plankton communities in a synoptic survey of lakes across a pH range of 7.3 to 4.2. As pH decreased, food web complexity, functional redundancy (which they call complementarity), and species numbers decreased. In a comprehensive evaluation, Schindler (1987) found that accumulated data from whole lake experiments supported hypothesized changes in proportion of r-strategists, life spans of organisms, species diversity, relative openness of the ecosystem, and relative sensitivity of structure versus function on comparable temporal scales. Schindler (1990) found small, unclear or contradictory responses

TABLE 2.1
Responses Expected in Stressed Ecosystems

Energetics
1. Community respiration increases
2. P/R becomes unbalanced
3. Maintenance cost increase; P/B and R/B ratios increase
4. Importance of auxiliary energy increases
5. Exported or unused primary production increases
Nutrient Cycling
6. Nutrient turnover increases
7. Horizontal transport increases; vertical cycling of nutrients decreases
8. Nutrient loss increases
Community Structure
9. Proportion of r-strategist increases
10. Size of organisms decreases
11. Life spans of organisms or parts decrease
12. Food chains shorten
13. Species diversity decreases; dominance increases; redundance declines
General System Level Trends
14. Ecosystem becomes more open
15. Autogenic successional trends reverse
16. Efficiency of resource use decreases
17. Parasitism increases; mutualism decreases
18. Functional properties more robust than structural properties

Source: Modified from Odum, E.P. 1985. Trends expected in stressed ecosystems, *BioScience*, 35, 419–422. With permission.

in several other categories, including community respiration, P/R ratios, transport of C and N, and average sizes of organisms.

2.2.3 What Other Characteristics are Desirable in a Metric?

Cairns et al. (1993) and Suter (1989), among many others, compiled lists of characteristics thought to be desirable in a metric used for assessment of ecosystem condition (Table 2.2). The ideal characteristics for a metric vary with the purpose for which the information is collected, for example, the ideal metrics for assessing current condition will differ from those that are best for distinguishing trends over time, diagnosing existing damage, or monitoring for early warning. In addition, some desirable characteristics may be mutually exclusive. For example, it is not easy to conceive of a metric that could simultaneously be diagnostic of a particular environmental stress while remaining broadly applicable, i.e., responding to many environmental stresses.

Metrics capable of predicting serious damage are seldom socially relevant because, by definition, they occur earlier in the progression of impact. For example, changes in community composition of small, rapidly reproducing and widely dispersed species such as phytoplankton were consistently among the earliest responses to various stresses in whole lake experiments (Schindler, 1987), but these species have no obvious value to most observers. Changes in phytoplankton communities are unlikely to trigger remedial action unless they are linked to metrics with more social relevance, like changes in populations of commercially or recreationally important fish species.

TABLE 2.2
A Compilation of Desirable Characteristics in Metrics Used to Assess Ecosystems

1. Condition
2. Biologically relevant
3. Socially relevant
4. Sensitive to environmental stress
5. Broadly applicable to many environmental stresses and ecosystems
6. Diagnostic of damage from a particular environmental stress
7. Measurable
8. Interpretable
9. Cost effective
10. Integrative, i.e., related to the responses of many unmeasured metrics
11. Historical data are available for the metric to aid in measurement and interpretation
12. Anticipatory, i.e., observable before serious harm occurs
13. Nondestructive
14. Continuity
15. Of appropriate scale for the management problem addressed
16. Lack of redundancy
17. Timely

It is important to have a good match between the type of information gathered and the type of decision to be made. Too often, the communication of goals between those charged with measuring biological condition and those who must make policy decisions is imperfect. Those collecting the scientific evidence used in probabilistic risk assessments must be thoroughly acquainted with the manner in which the information they collect will be used. Those making policy decisions must become more literate in the complex determination of biotic community responses to stress and be familiar with the uncertainties inherent in each type of data.

`In addition to structural and functional response types, unique responses occur at many distinct spatial and temporal scales and levels of biological organization (e.g., cells, tissues, organs, organisms, populations, communities, ecosystems, landscapes, and biosphere). Some attributes at higher levels of biological organization are not present at lower levels, for example, energy flow and nutrient spiraling are properties of ecosystems, but not of organisms. Other attributes are present in some form at many levels, for example, one can measure the diversity of phenotypes at population level, the diversity of species at community level, and the diversity of habitat patches at landscape level. It is somewhat paradoxical that the same action, agent, or condition can adversely affect one biological system at the same time it does not affect or may benefit others. These discrepancies can be hierarchical. The same flood that benefits an ecosystem can be devastating to many individuals or an entire species. Because environmental management goals include objectives at population, community, ecosystem, regional, and global levels, assessments must encompass these same levels.

2.2.4 Hierarchy

Awareness of scale provides two contrasting approaches to studying environmental stress. Top-down methods start with observed damage to a biological system of interest, and investigations move down through hierarchical levels. Component structures and functions are examined in order to diagnose the causative agent and plan remedial actions. At the outset, the damage has already been done, so the relevance of the changes is known. However, the causative agent and the chain of events leading to unacceptable damage are not known. Bottom-up methods start with an environmental stress. The effects of that stress on biological systems are determined through designed experiments. Because experiments on small and quick biological systems at lower scales are generally less expensive, they are common. In bottom-up assessments, the causative agent is known at the outset, but the importance of ultimate changes at any ecologically relevant higher scale is unknown.

There are substantive difficulties in extrapolating from laboratory systems, even those using naturally assembled communities, to communities existing in ecosystems. Biotic community response to stress consists of a multidimensional array of breakpoints and thresholds. Responses to environmental stress are often non-linear and often display radiance that appears counterintuitive. Mayer et al. (1987) demonstrated the difficulties involved in extrapolating from one species to another. Species that are closely related taxonomically often display different physiological responses to stress. The difficulty is exacerbated when one attempts to extrapolate from one biological level to another, such as from a single species to a biotic community (Smith and Cairns, 1993).

2.3 NEW CHALLENGES: NEW DEMANDS

2.3.1 More Ambitious Goals in Environmental Management

As successes in environmental protection have mounted, goals have become more ambitious. Individuals are not satisfied to be free of obvious disease; they want to be healthy and function optimally. The same type of goal inflation applies to ecosystems. The maximum amount of services is needed from ecosystems (e.g., Costanza et al., 1997), while the ecosystem continues to be self-maintaining. Aquatic systems must provide storage and distribution of freshwater, flood control, purification of water through decomposition of wastes, regeneration of nutrients, and removal of sediments. Ecosystems must also provide areas for recreation and aesthetic satisfaction. If these services are diminished, problems abound.

Humans exhibit homeostasis, or feedback loops, which tend to bring such things as temperature, blood chemistry, respiratory rate, pulse rate, and other physiological attributes to a nominative state. Many other species of animals exhibit this characteristic. However, aggregations of species (communities), including both plants and animals, clearly have no single nominative state. Instead, they may exhibit homeorhesis, which means variability within limits (e.g., Odum, 1997). Humans stressed beyond tolerance limits suffer severely or die, while communities that suffer deleterious effects often reach a new equilibrium state quite unlike the previous state. As is the case with human health, the gradient between optimal functioning and collapse may be quite lengthy in environmental systems; however, in some cases, an abrupt transition may occur from functioning to collapse.

2.3.2 Industrial Ecology and Natural Capital

As historian McNeill (2000) notes, the massive changes wrought in the physical world have created something new. Humans have refashioned the air, water, soil, and biosphere to a degree unprecedented in human history. Of course, human beings have been reshaping the planet and affecting biotic communities for millennia; however, unquestionably, the 20th century witnessed biotic transformations of a scale and variety unprecedented in human history. The human population increased slowly to one billion people in 1804, then rapidly to six billion in 1999, and is expected to increase by another billion people every 12 to 15 years. In addition, increased affluence for people across the globe increases the environmental stress on natural systems by increasing the per capita human use of natural systems. If the lifestyle of every person living in 1996 was raised to that of a typical North American, another two Earths would be needed to provide the surface area required to produce the resources used and assimilate the wastes produced (Rees, 1996). It seems likely that virtually every ecological system on the planet will eventually be simultaneously subjected to a multiplicity of environmental stresses.

With the expansion of the impact of the human population, viewing industrial and ecological areas as totally separate entities seems less and less possible on a finite planet. Tibbs (1992) views the relationship as a gradient with industry at one end and nature at the other. However, along this continuum, starting with single material ecosystems (Tibbs views industries as a type of ecosystem), the transition is to interindustrial ecosystems, then to hybrid industrial ecosystems, to bioengineered ecosystems, to modified natural ecosystems, to reclaimed natural ecosystems, and finally to pristine natural ecosystems. Natural ecosystems are characterized by provision of a wide variety of services, high integrity, self-maintenance, and great resilience to natural perturbations. Industrial systems are specialists; they provide fewer services.

Tibbs (1992) views this gradient from industrial to natural systems as a co-evolution of environmentalism and industrialism that would lead to a single interactive system and envisions a gradient in industrial environmental management strategies. Basic regulatory compliance would give way to partial recycling initiatives, followed by development of management tools, highly developed closed-loop recycling, significant changes in products and packaging, full integration of environmentalism into the corporate culture, and the development of synergistic ecosystems in which industrial and ecological systems benefit from and are dependent upon each other's functions.

Hybrid industrial ecosystems clearly are not comparable to pristine natural ecosystems; yet, by changing industrial practice to emphasize closed cycles for material and energy rather than one-way flow, they resemble intact, functioning ecosystems and may be thought of as having a degree of ecological integrity not commonly associated with industries. Each point along the gradient from natural to industrial systems represents a different ecological state and will have attributes unique to that particular system. Assessing biotic community response to stress in the field of industrial ecology requires the determination of the level of stress that jeopardizes the functioning of the system and the degree of resilience that permits it to rebound once the stress is diminished or eliminated.

The move to industrial ecology is not a license to further damage natural systems. Instead, it is a call to protect the biological integrity of those systems while restoring as many ecological functions to damaged systems as possible. Industries that play by the same material and energy flow rules as ecosystems will export less material for those ecosystems to deal with. Initially at least, the ecological portions of hybrid systems will have to be constructed by using colonizing organisms tolerant of the levels of biotic stress existing at the time the co-evolutionary relationship begins. Determining stress in a dynamic, co-evolving system will require continual reevaluation of the criteria used to measure biotic community response.

The assessment process will include the redefinition of a community. Distinguishing natural successional changes, which do not impair integrity or indicate that it has been damaged, from those that impair integrity (e.g., invasion of exotics) or signal that ecological integrity has been impaired or damaged will be difficult. Clearly, this will require a shift in the ways a community's response is measured on the part of both the industrialist and ecologist. However, if the hybrid systems function as Tibbs hopes they will, there will be less biotic community stress near industrial sites than exists now, and these systems will serve both as buffer zones and early warning zones of industrial malfunction vis-a-vis the environment.

Hawken et al. (1999) regard natural capital (i.e., old growth forests, gene pools, and all the other parts of the biosphere that provide ecosystem services) as the basis for all other forms of capital, including economic, industrial, and human. They present persuasive evidence that natural capitalism can link corporate profits with environmental sensitivity. They assert that the next industrial revolution will favor those industries and economic systems that practice natural capitalism. In their view, natural capital should not be diminished and ideally should accumulate. This paradigm cuts the Gordian knot with regard to assessing biotic community response to stress. If natural capital decreases, stress is excessive. If it remains constant, there is a good balance; if it is increasing, biotic community stress is decreasing. The great attraction of natural capitalism is that it presents a single unifying theme upon which to base all measurements of biotic community response to stress.

While it is impossible to do justice to the ideas of industrial ecology and natural capital in this space, both paradigms focus in somewhat different ways upon the protection of ecosystem services. This protection cannot occur unless biotic community stress is reduced over large spatial and temporal scales. The effects of stress on the delivery of ecosystem services are extremely important endpoints in the measurement of biotic community stress. Since ecosystems deliver a multiplicity of services and all these services may cycle seasonally, such measurements are both challenging and critically important.

2.3.3 Cross-Disciplinary Collaboration

Effective environmental management requires an enormous variety of skills. It is not enough to be a biologist, sociologist, urban planner, geographer, hydrologist, teacher, legal scholar, or politician, even though all their skills are necessary. Yet professionals in different disciplines belong to different tribal cultures, each with its own rites of passage, language, literature, and meeting sites. They are typically housed in different buildings on campuses (Cairns, 1993). However, the barriers between disciplines are decreasing (Cairns, 1998). Each group still emphasizes certain components as important or unimportant, has special requirements for minimal literacy in its respective field, and imposes different meanings on common terms such as restoration. Association with those in other disciplines produces a form of culture shock which may lead to a feeling of insecurity or worse, a feeling that those in other disciplines should pay more attention to the researcher's own discipline.

Most of the modern problems of human society are so multidimensional that their resolution requires more than any single discipline can offer. The relatively unexplored areas between disciplines can engender research of exceptional promise and enrich the cultures of all engaged professionals. Participation in cross-disciplinary research too early in a professional career can have damaging results if the tenure and promotion committee has members with strong beliefs about disciplinary purity. Still, journals such as *Ecological Economics* celebrate the commingling of disciplines, and most universities and college campuses have interdisciplinary centers. Nevertheless, desirable linkages among disciplines do not develop accidentally; they need much administrative encouragement to survive. Fortunately, such efforts often attract extramural funding otherwise unavailable, which is of interest to both participants and academic administrators.

In the final analysis, it is important to understand how different sources of information are combined to produce the final judgment. This combination is especially difficult in situations involving a mixture of sciences, even though the mixture is a necessary condition for most societal decisions. Estimations of probability present special difficulties in making judgments, particularly when attitudes differ about the value of the precautionary principle that contends that protective action is necessary if there is a threat of severe or irreversible environmental damage, even if the threat is uncertain (Sandin, 1999). Furthermore, estimates of probability are influenced by the way questions are framed and the temporal and spatial scales involved.

2.4 CONCLUSIONS

Scientists have an insidious desire to select a single indicator of stress, especially scientists who naturally think that the organism or community or response that they study is the most important and valuable one. However, all metrics are not equal in the decision-making process, and each decision requires that unique properties of the community be incorporated into the study design and that knowledge of the community and its overall structure and function be incorporated into the analytical process.

By choosing a suite of powerful metrics covering a wide range of attributes of communities and ecosystems, the ability to characterize anthropogenic impact has improved. However, to avoid the tautology of using the characteristics of pristine and damaged systems to define a scoring system, which is then used to define pristine and damaged systems, metrics must also be relevant ecologically or socially. They must be strongly linked to theory or to practical concerns in order to make them the most effective management tools possible. It is easier to make a case for the ecological relevance of metrics that are consistent with current concepts of a general ecosystem stress response. These metrics are theoretically based and have been observed to vary in predictable ways over many stresses and ecosystem types. Alternately, a metric that is closely related to a characteristic of obvious value to shareholders is also convincing. These socially relevant metrics are now evolving beyond regulatory limits to the preservation of ecosystem services and natural capital. Clearly, if natural capital is decreasing, there is too much stress. Thus, natural capital provides a simple and unifying concept upon which to base biotic community response to stress.

ACKNOWLEDGMENTS

I am indebted to Eva Call for transcribing the dictation of the first draft of this manuscript and for subsequent revisions. Darla Donald provided her usual skilled editorial assistance. I greatly appreciate the comments of B.R. Niederlehner, Alan Heath, Peter Leigh, and David Orvos on an early draft of this manuscript. The Cairns Foundation paid the cost of preparing and processing it.

REFERENCES

Barrett, G.W., G.M Van Dyne, and E.P. Odum. 1976. Stress ecology, *BioScience*, 26, 192–194.

Cairns, J., Jr. 1993. The intellectual electric fence, *Annals of Earth*, XI(3), 17–18.

Cairns, J., Jr. 1998. The diminished charge on the intellectual electric fence, *The Social Contract*, IX(3), 145–151.

Cairns, J. Jr. 2001. Stress, environmental, in S. Levin (Ed.), *Encyclopedia of Biodiversity*, Academic Press, New York, 515–522.

Cairns, J., Jr., P.V. McCormick, and B.R. Niederlehner. 1993. A proposed framework for developing indicators of ecosystem health, *Hydrobiologia*, 263(1), 1–44.

Canterbury, E.G., E.T. Martin, R.D. Petit, J.L. Petit, and F.D. Bradford. 2000. Bird communities and habitat as ecological indicators of forest condition in regional monitoring, *Conservation Biology*, 14(2), 544–558.

Cattaneo, A., A. Asioli, P. Comoli, and M. Manca. 1998. Organisms' response in a chronically polluted lake supports hypothesized link between stress and size, *Limnology and Oceanography*, 43(8), 1938–1943.

Costanza, R. and M. Mageau. 1999. What is a healthy ecosystem? *Aquatic Ecology*, 33(1), 105–115.

Costanza, R., R. D'Arge, R. deGroot, S. Farber, M. Grasso, B. Hannon, K. Limburg, S. Naeem, R.V. O'Neill, J. Paruelo, R.G. Raskin, P. Sutton, and M. van den Belt. 1997. The value of the world's ecosystem services and natural capital, *Nature*, 387(6630), 253–260.

Havens, K.E. and R.E. Carlson. 1998. Functional complementarity in plankton communities along a gradient of acid stress. *Environmental Pollution*, 101(2), 427–436.

Hawken, P., A. Lovins, and H. Lovins. 1999. *Natural Capitalism: Creating the Next Industrial Revolution.* Little, Brown and Company Publishers, New York. 378 pp.

Kelly, J.R. and M.A. Harwell. 1989. Indicators of ecosystem response and recovery, in S.A. Levin, M.A. Harwell, J.R. Kelly, and K.D. Kimball (Eds.), *Ecotoxicology: Problems and Approaches*, Springer-Verlag, New York, 9–35.

Kolkwitz, R. and M. Marsson. 1908. Okologie der pflanzlichen Saprobien, *Berichte der Deutschen Botanischen Gesellschaft*, 26, 505–519.

Kolkwitz, R. and M. Marsson. 1909. Okologie der tierischen Saprobien, *Bietrage sur Lehre von der biologische Gewasserbeuteilung. Internationale Revue der gesamten Hydrobiologie und Hydrographie*, 2, 126–152.

Mayer, F.L. Jr., C.H. Deans, and A.G. Smith. 1987. *Inter-taxa Correlations for Toxicity to Aquatic Organisms.* USEPA EPA/600/X-87/332, National Technical Information Service, Springfield, VA.

McNeill, J.R. 2000. *Something New Under the Sun: An Environmental History of the Twentieth-Century World.* W. W. Norton & Co., New York.

Niederlehner, B.R. and J. Cairns, Jr. 1994. Consistency and sensitivity of community level endpoints in microcosm tests, *Journal of Aquatic Ecosystem Health*, 3, 93–99.

Odum, E.P. 1969. The strategy of ecosystem development, *Science*, 164, 262–270.

Odum, E.P. 1985. Trends expected in stressed ecosystems, *BioScience*, 35, 419–422.

Odum, E.P. 1997. *Ecology: A Bridge between Science and Society.* Sinauer Associates, Sunderland, MA.

Rapport, D.L., H.A. Regier, and T.C. Hutchinson. 1985. Ecosystem behavior under stress, *American Naturalist*, 125, 617–640.

Rees, W.E. 1996. Revisiting carrying capacity: area-based indicators of sustainability, *Population and Environment*, 17, 195–214.

Sandin, P. 1999. Dimensions of the precautionary principle, *Human Ecological Risk Assessment*, 55, 889–907.

Schindler, D.W. 1987. Detecting ecosystem responses to anthropogenic stress, *Canadian Journal of Fisheries and Aquatic Science*, 44(1), 6–25.

Schindler, D.W. 1990. Experimental perturbations of whole lakes as tests of hypotheses concerning ecosystem structure and function, *Oikos*, 57, 25–41.

Schulz, E.J., M.V. Hoyer, and D.E. Canfield, Jr. 1999. An index of biotic integrity: a test with limnological and fish data from sixty Florida lakes, *Transactions of the American Fisheries Society*, 128(4), 564–577.

Simon, T.P. 1998. Modification of an index of biotic integrity and development of reference condition expectations on dunal, palustrine wetland fish communities along the southern shore of Lake Michigan, *Aquatic Ecosystem Health and Management*, 1(1), 49–62.

Simon, T.P. and P.M. Stewart. 1998. Application of an index of biotic integrity for dunal, palustrine wetlands: emphasis on assessment of nonpoint source landfill effects on the Grand Calumet Lagoons, *Aquatic Ecosystem Health and Management*, 1, 63–74.

Smith, E.P. and J. Cairns, Jr. 1993. Extrapolation methods for setting ecological standards for water quality, *Ecotoxicology*, 2, 203–219.

Suter, G.W., II. 1989. Ecological endpoints, in W. Warren-Hicks, B.R. Parkhurst, and S.S. Baker (Eds.), *Ecological Assessment of Hazardous Waste Sites: A Field and Laboratory Reference*, EPA/600/3-89/013. National Technical Information Service, Springfield, VA, 2–1 to 2–26.

Suter, G.W., II. 1993. A critique of ecosystem health concepts and indices, *Environmental Toxicology and Chemistry*, 12, 1533–1539.

Tibbs, H.B.C. 1992. Industrial ecology: an environmental agenda for industry, *Whole Earth Review, Winter*, 4–19.

Weisberg, J.B., J.A. Ranasinghe, D.M. Dauer, L.C. Schaffner, R.J. Diaz, and J.B. Frithsen. 1997. An estuarine benthic index of biotic integrity (B-IBI) for Chesapeake Bay, *Estuaries*, 20 (1), 149–158.

3 Using Biological Response Signatures within a Framework of Multiple Indicators to Assess and Diagnose Causes and Sources of Impairments to Aquatic Assemblages in Selected Ohio Rivers and Streams

Chris O. Yoder and Jeffrey E. DeShon

CONTENTS

0-8493-0905-0/03/$0.00+$1.50

3.1 INTRODUCTION

The use and value of biological data based on assessments of indigenous aquatic assemblages have seen unprecedented interest and growth in the past 20 years (Davis and Simon, 1995). Much of this is owed to the emphasis placed on biological assessments and criteria by the U.S. Environmental Protection Agency (USEPA) and several states for evaluating the condition of surface waters as they reflect the goals of the Clean Water Act (CWA) and state water quality standards. Recent texts such as Davis and Simon (1995), Rosenberg and Resh (1999), and Simon (1999) among many others illustrate the methods and procedures used to develop consistent and reliable bioassessment approaches. However, significant skepticism remains about the diagnostic value and utility of the biological assessments and criteria despite this recent interest (Houck, 1999).

Commonly used approaches to managing causes and sources of water pollution in the USA, including both mandatory and voluntary controls, continue to be predominated by a focus on the measurement and control of pollutants in general and selected chemicals in particular (Karr, 1995). This emphasis continues in water quality management programs despite established evidence that factors such as habitat modification, introduction of invasive alien species, and modification of ecosystem processes at the watershed scale are not only widespread, but pose greater risk to the long term well-being of water resources than individual pollutants (Loeb and Spacie, 1994; Karr and Chu, 1998).

The total maximum daily load (TMDL) approach to managing water quality is one such example where the principal focus remains on substances defined as pollutants in the CWA, even though Section 502 of the act defines pollution as the human-made alteration of the chemical, physical, biological, and radiological quality of the water. In addition, some states, following the guidance in federal water quality regulations (40 CFR, Part 131), define designated uses in more specific terms and at least four have formally adopted biological criteria as direct measurements of protection and restoration efforts. Several other states use biological data to directly assess degradation and impairment.

Houck (1999) surmises that USEPA's emphasis on pollutants in the TMDL process partially reflects the perception that causes and sources of biologically measured impairments cannot be adequately defined; hence its emphasis on the more easily measured and managed pollutant paradigm. Certainly, this skepticism has been long standing and is embraced in past guidance documents (USEPA, 1985) and in the literature (Suter, 1993).

However, a significant concern with a pollutant-only focus is that it leaves the aforementioned water resource and ecosystem process problems unanswered, and lacks sufficient accuracy so that significant pollutant-caused problems are sometimes overlooked or underrated. This narrow focus also constrains water quality management to the point where significant impairments to fundamental ecosystem processes will be ignored, improperly understood, or deemed irretrievable. Given the significant risks inherent to limiting water quality management in such a manner, it is important that the perceived barriers to the interpretation and diagnosis of biological impairments be overcome. It is the purpose of this chapter to show how biological data can be used to support the determination of causes and sources of pollution associated with biological impairments and used within an adequate monitoring and assessment and multiple indicators framework to better guide water quality management.

Deficiencies in implementing such an integrated process are at issue in the shortcomings of the TMDL approach to water quality management. A review of the scientific adequacy of TMDL approaches used to date recommended that the program should encompass all stressors, both pollutants and pollution that determine the condition of a waterbody (NRC, 2001). Thus, not only are tools and indicators that can reveal the sequence and order of these types of impacts needed, but processes by which assessments are conducted so as to accurately identify which factors limit attainment of designated uses are required. Certainly, biological assessment as part of an adequate approach to monitoring and assessment (Yoder, 1998) plays a vital role in making this process work.

Some attempts to develop improved methods, criteria, and procedures for characterizing and diagnosing biological impairments have recently been made. Yoder and Rankin (1995a) developed the concept of biological response signatures. The signatures are combinations of biological community data that consistently indicate different types of impacts. Yoder and Rankin found discernable patterns in aggregated biological assemblage data in the form of individual metric responses to different types of stressors, particularly complex toxic stressors. Significant diagnostic power in the data and aggregations of the data into metrics and index values were amply demonstrated with diverse and robust datasets. The technique can help prioritize the applications of other assessment tools to more accurately direct water quality management to address problems that are truly limiting and which are restorable. Eagleson et al. (1990) demonstrated this use of bioassessment data to characterize different types of impacts in validating effluent toxicity testing. USEPA (2000) recently published a stressor identification and evaluation process that involves the informed usage of multiple chemical, physical, and biological parameters in an iterative, diagnostic process. Other recent studies used multivariate and correlation analysis to demonstrate relationships between ambient chemical, physical, and biological data (Norton et al., 2000; Majumder et al., 2001).

We present here a process for using biological data within a disciplinary framework of activity, stressor, exposure, and response indicators to characterize and quantify the extent and severity of impairments and an interpretive process for determining the associated causes and sources of those impairments. This is accomplished by combining the concepts of the biological response signatures developed by Yoder and Rankin (1995a) within a hierarchical process where chemical, physical, and biological indicators from sources of potential stress and the ambient environment are linked to form a rationale for diagnosis. The indicator hierarchy used in this process was originally developed by USEPA (1995) and has been described further by Yoder and Rankin (1998). Essential elements of this process include the systematic and consistent use of adequate monitoring and assessments (Yoder, 1998), adequately calibrated and robust biological criteria (Yoder and Rankin, 1995b), and tiered aquatic life use designations within state water quality standards. These are the essential pieces of a monitoring and assessment information architecture in which water quality management is guided by the results observed in the environment rather than administrative and prescriptive approaches alone.

Case examples from six Ohio rivers and streams are used to illustrate the practical application of these methods and procedures in establishing the extent and severity of impairments to aquatic life designated uses, diagnosing associated causes and sources of impairments, and tracking changes through time. The results of this process support the impaired waters listing processes (CWA Sections 303e and 305b) and the site-specific management of administrative processes including water quality standards (designated uses, criteria), NPDES permitting, and TMDL development.

3.2 METHODS AND PROCEDURES

Chemical, physical, and biological data from six Ohio river and stream segments were accessed to illustrate the use of biological response signatures and their integration with chemical, physical, and activity indicators to diagnose impairments to aquatic life. The six segments chosen were the Ottawa River mainstem, lower Cuyahoga River mainstem, middle Scioto River mainstem, lower reach of Paint Creek, Dicks Creek, and the Rocky Fork of the Mohican River (Figure 3.1). Ohio EPA conducted comprehensive monitoring of several years' duration in each river and stream and dealt with the many associated water quality management issues and challenges. Thus, we used the existing knowledge of the issues and impacts in combination with the feedback provided by the biological responses in each area to illustrate the efficacy of the approach.

3.2.1 Biological and Water Quality Assessments

The analysis of the case examples follows the process used by Ohio EPA in producing biological and water quality assessments. Ohio EPA relies on an integrated and hierarchical approach in assessing the status of aquatic life uses and determining associated causes and sources of threats or impairments. This is accomplished by completing a biological and water quality survey, or

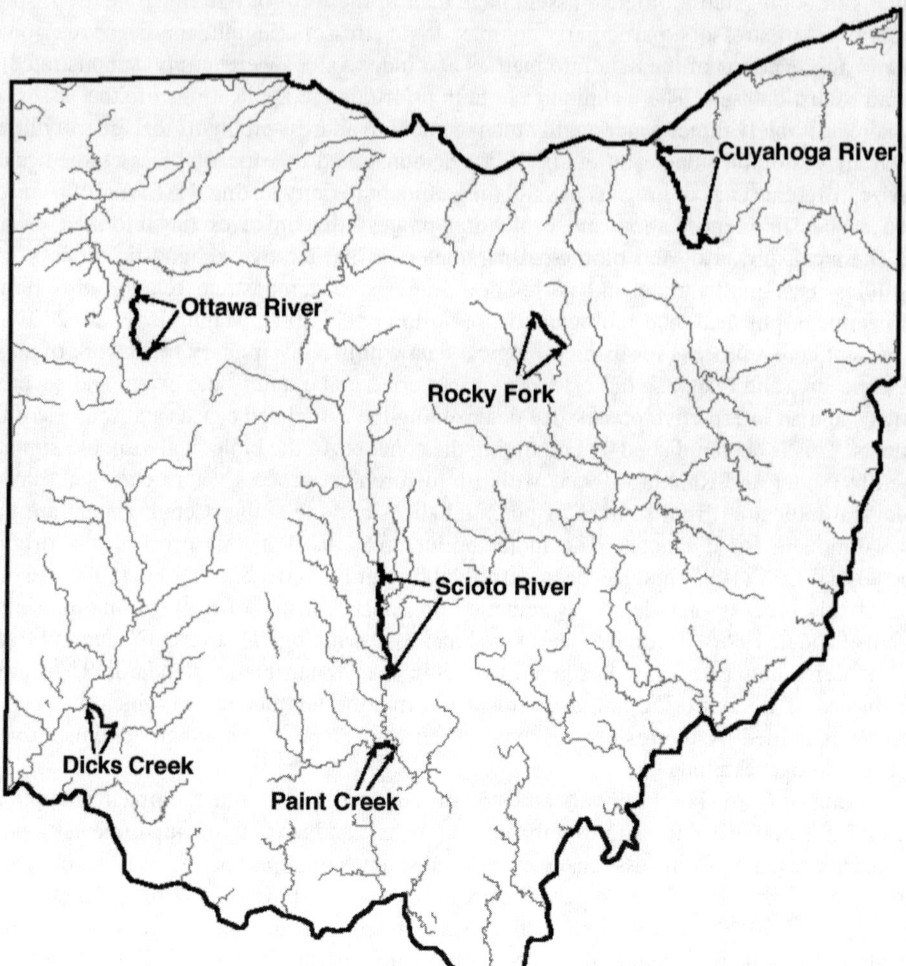

FIGURE 3.1 Locations of rivers and streams and approximate locations of study areas used as case examples.

"biosurvey," which is an interdisciplinary monitoring and assessment effort coordinated on a waterbody-specific or watershed scale. This may involve a relatively simple setting focusing on one or two streams, one or two principal stressors, and a handful of sampling sites to a much more complex effort including mainstem reaches, entire subwatersheds, multiple and overlapping stressors, and tens of sites. The case examples highlighted here correspond to the more complex efforts. Ohio EPA annually conducts biosurveys in three or four subwatershed areas and numerous segment- or site-specific study areas for an aggregate total of 500 to 600 sampling sites statewide.

Each biosurvey has three principal objectives: (1) to determine the extent that use designations assigned in the water quality standards (WQS) are either attained or not attained; (2) to determine whether use designations assigned to a given water body are appropriate and attainable; and (3) to determine whether changes in key ambient biological, chemical, or physical indicators have occurred over time, particularly before and after the implementation of point source pollution controls or best management practices.

The data gathered by a biosurvey are processed, evaluated, and synthesized in a biological and water quality report. Each biological and water quality assessment contains a summary of major findings, a site-by-site description of attainment status, recommendations for revisions to WQS, future monitoring needs, and recommendations for actions that may be needed to resolve impairments of designated uses. While the principal focus of a biosurvey is on the status of aquatic life uses, the status of other uses such as recreation and water supply and human health concerns are also addressed. The findings and conclusions of a biological and water quality study may factor into regulatory actions taken by Ohio EPA (e.g., NPDES permits, director's orders, the WQS [Ohio Administrative Code 3745–1]), and are eventually incorporated into water quality permit support documents (WQPSDs), state water quality management plans, the Ohio nonpoint source assessment, the Ohio water resource inventory (305[b] report), and more recently, the development of total maximum daily loads (TMDLs). The use of assessed data and information comprised of multiple chemical, physical, biological, and administrative indicators, each used within its most appropriate role, serves as feedback on the effectiveness of water quality management programs.

3.2.2 Hierarchy of Surface Water Indicators

Key to implementing an adequate monitoring and assessment approach is the use of cost effective, robust indicators comprised of biological, chemical, and physical measures which ensure that all relevant pollution sources are assessed objectively and on the basis of environmental results (Yoder, 1998; Yoder and Rankin, 1998). Ohio EPA relies on a sequential approach in linking the results of water quality management activities with such measures (1999a). This integrated approach is outlined in Figure 3.2 and includes a hierarchical continuum from administrative to true environmental indicators. The six levels of indicators include:

1. Management actions (e.g., permitting, enforcement, grants)
2. Responses to management actions (e.g., treatment works upgrades, pollution prevention, best management practices)
3. Changes in human activity outputs (e.g., reduced/increased pollutant loadings, land use changes)
4. Changes in ambient conditions (e.g., chemical/physical water quality, instream habitat quality)
5. Changes in uptake and/or assimilation (e.g., tissue contamination, biomarkers, wasteload allocation variables)
6. Changes in health, ecology, or other effects (e.g., ecological condition)

The results of administrative activities (Levels 1 and 2) can be linked to efforts to improve water quality (Levels 3 through 5), which should translate into environmental results (Level 6).

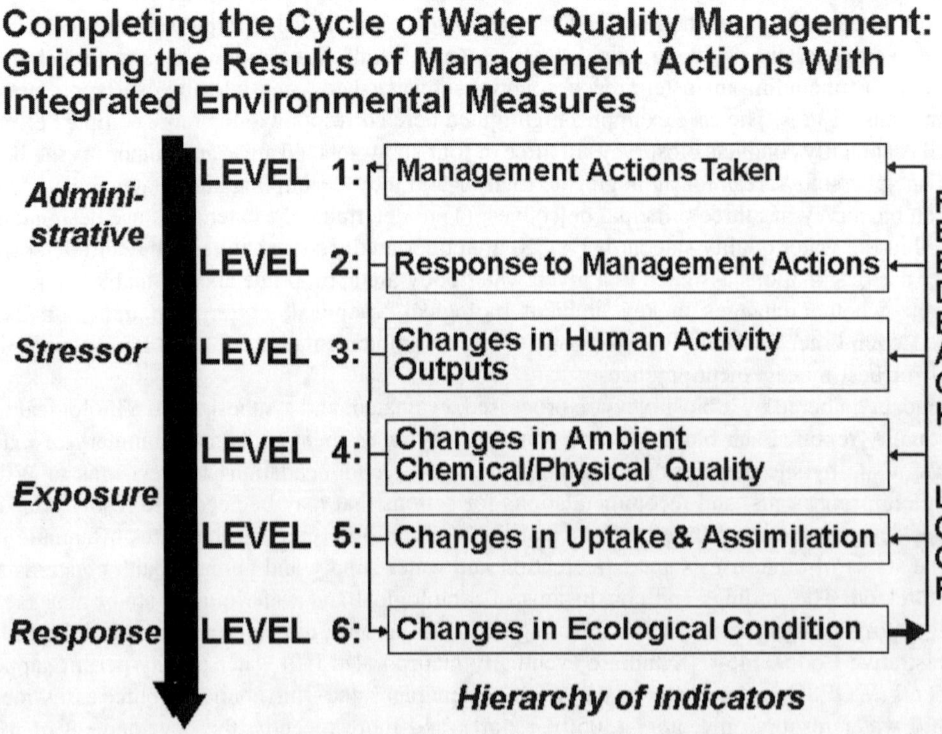

FIGURE 3.2 A hierarchy of surface water indicators from administrative through true environmental indicators. Patterned after those developed by U.S. EPA (1995) for evaluating the effectiveness of water quality management program efforts.

Superimposed on this hierarchy are the concepts of stressor, exposure, and response indicators. *Stressor* indicators include activities that have the potential to degrade the aquatic environment, e.g., pollutant discharges, land use effects, and habitat modifications. *Exposure* indicators measure the initial effects of stressors and can include concentrations of toxic chemicals, bioassay endpoints, tissue residues, and biomarkers, each of which provides evidence of biological exposure to a stressor-caused agent. *Response* indicators are generally composite measures of the cumulative effects of stress and exposure and include the more direct measures of community and population response that are represented here by the biological indices, metrics, and other assemblage data attributes that comprise Ohio's biological criteria.

This framework for using multiple indicators represents the essential technical assessment process for watershed-based management approaches (Yoder, 1998). One important condition, however, is to use the different indicators *within* the roles that are most appropriate for each. When indicators are used outside of their most appropriate role, problems with the accuracy of ambient assessments become most evident (Yoder and Rankin, 1998).

3.2.3 Water Quality Standards: Designated Aquatic Life Uses

WQS are essential cornerstones of water quality management. They consist of designated uses and chemical, physical, and biological criteria designed to represent measurable properties of the environment that are consistent with the characteristics and level of protection specified by a designated use. Use designations consist of two broad categories, aquatic life and non-aquatic life uses. In applications of WQS to flowing waters, the aquatic life use criteria frequently result in the most stringent protection and restoration requirements; hence their common focus in state water quality management programs. Also, an emphasis on protecting for aquatic life generally results in water quality suitable for all uses.

The Ohio standards employ a tiered system of refined aquatic life use classifications which is different from the "one-size-fits-all" approach of general uses. Tiered uses are based on the reality that reference aquatic assemblages vary locally and regionally, and management goals for these resources should be stratified accordingly. The Ohio standards also offer an opportunity to stratify water quality management goals and end-points, thus reducing the risks of underprotection or overprotection inherent to a general use approach. The Ohio WQS designate five principal aquatic life uses:

1. **Warmwater Habitat (WWH)** — Defines typical warmwater assemblages of aquatic organisms for Ohio rivers and streams. Biocriteria are stratified by ecoregion and site-type.* This use represents the principal restoration target for most water resource management efforts in Ohio.
2. **Exceptional Warmwater Habitat (EWH)** — Reserved for waters that support unusual and exceptional assemblages of aquatic organisms characterized by a high diversity of species, particularly those that are highly intolerant and/or rare, threatened, endangered, or have special status (i.e., declining species). Biocriteria are set uniformly across ecoregions and stratified by site type. This designation represents a protection goal for water resource management efforts dealing with Ohio's best water resources.
3. **Coldwater Habitat (CWH)** — Intended for waters that support assemblages of cold water organisms and/or are stocked with salmonids with the intent of providing a put-and-take fishery on a year-round basis, which was further sanctioned by the Ohio DNR's Division of Wildlife. This use is complemented by the Seasonal Salmonid Habitat (SSH) use that applies to Lake Erie tributaries supporting periodic "runs" of salmonids during the spring, summer, and/or fall. No numeric biocriteria have been developed specifically for coldwater streams; such streams are expected to meet WWH biocriteria.
4. **Modified Warmwater Habitat (MWH)** — Applies to streams and rivers that have been subjected to extensive, maintained, and essentially permanent hydromodifications such that the biocriteria for WWH use are not attainable; where restoration to a CWA goal use has been ruled out via use attainability analysis. Representative aquatic assemblages are predominated by species that are tolerant to low dissolved oxygen, siltation, nutrient enrichment, and poor quality habitat. Biocriteria are stratified by ecoregion and site type.
5. **Limited Resource Water (LRW)** — Applies to small streams (usually <3 mi. drainage area*) and other water courses that have been irretrievably altered to the extent that no appreciable assemblage of aquatic life can be supported. Such waterways generally include small streams in extensively urbanized areas and those in watersheds with extensive drainage modifications or other irretrievably altered waterways. These streams are expected to support poor quality biological assemblages at a minimum.

Chemical, physical, and/or biological criteria are generally assigned to each use designation in accordance with the broad goals and objectives cited in each definition. The system of use designations employed in the Ohio WQS constitutes a tiered approach in that each designation provides varying and graduated levels of protection. This is best illustrated by the biological criteria that are stratified across the state by ecoregion, site type, and designated use (Figures 3.3 and 3.4). This stratification has also been applied to Ohio's chemical water quality criteria including parameters

* A site type distinguishes between headwaters (<20 sq. mi. drainage area), wadeable (sampled with wading methods), and boat sites (sampled with boat-mounted methods) for fish community assessment purposes.

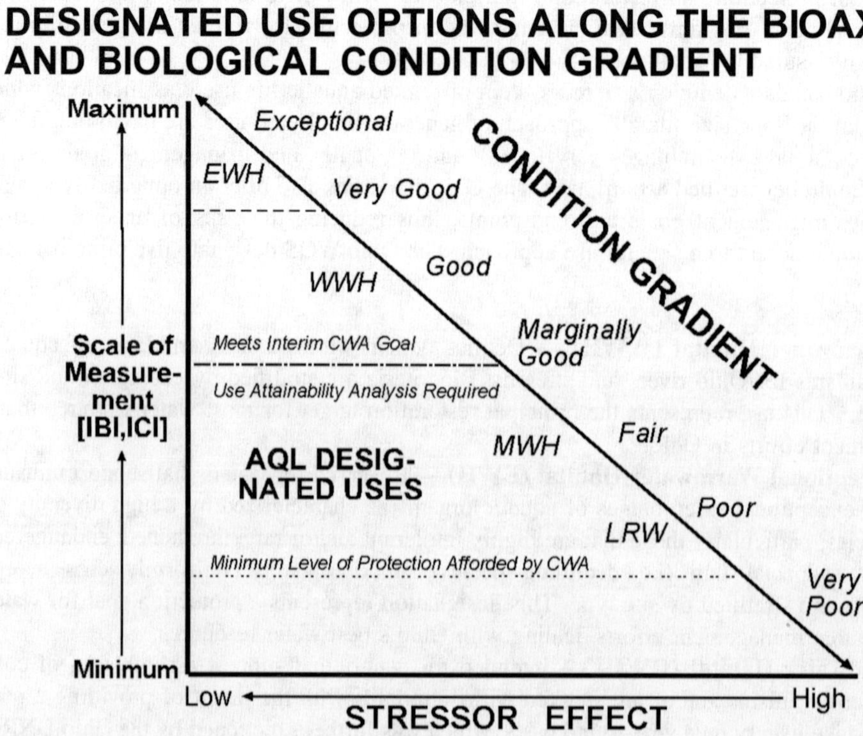

FIGURE 3.3 Relationship between tiered aquatic life uses in the Ohio WQS and narrative evaluations of biological community condition and how they correspond to a continuum of stressor effects (*x* axis) and measured biological index values (*y* axis). Thresholds for tiered uses and correspondence to Clean Water Act minimum goals are shown.

such as dissolved oxygen, ammonia–nitrogen, selected heavy metals, and temperature. For other parameters, such as less commonly occurring metals and xenobiotic compounds, the database to construct an equally graduated set of criteria has been lacking, and the same water quality criteria may apply to multiple use designations. Among the six study areas, the WWH use applies to all of them; the EWH and MWH uses apply to one area each.

3.2.4 Determining Aquatic Life Use Attainment Status

Use attainment status describes the degree to which environmental indicators exceed or fall below a water quality criterion. Aquatic life use attainment status in Ohio is based on biological criteria (Ohio Administrative Code 3745–1–07; Figure 3.4). These are most applicable to ambient assessments and apply to rivers and streams outside allowable discharge mixing zones.

Numerical biological criteria are based on multimetric biological parameters that include the index of biotic integrity (IBI) and the modified index of well-being (MIwb) based on attributes of the fish community, and the invertebrate community index (ICI) based on attributes of the macroinvertebrate community. The IBI and ICI are multimetric indices patterned after an original IBI described by Karr (1981) and Fausch et al. (1984). The IBI was modified for application to Ohio rivers and streams in accordance with guidance prescribed by Karr et al. (1986). The ICI was developed by Ohio EPA (1987a) and further described by DeShon (1995); it offers a macroinvertebrate assemblage analog to the fish IBI. The MIwb is a measure of fish community abundance and diversity using numbers and biomass information. It is a modification of the original index of well-being originally developed by Gammon (1976) and further described by Gammon et al. (1981). Numerical endpoints for each index are stratified by ecoregion, use

Ohio Biological Criteria: Adopted May 1990 (OAC 3745-1-07; Table 7-14)

Huron Erie Lake Plain (HELP)

Use	Size	IBI	MIwb	ICI
WWH	H	28	NA	34
	W	32	7.3	34
	B	34	8.6	34
MWH-C	H	20	NA	22
	W	22	5.6	22
	B	20	5.7	22
MWH-I	B	30	5.7	NA

Erie Ontario Lake Plain (EOLP)

Use	Size	IBI	MIwb	ICI
WWH	H	40	NA	34
	W	38	7.9	34
	B	40	8.7	34
MWH-C	H	24	NA	22
	W	24	6.2	22
	B	24	5.8	22
MWH-I	B	30	6.6	NA

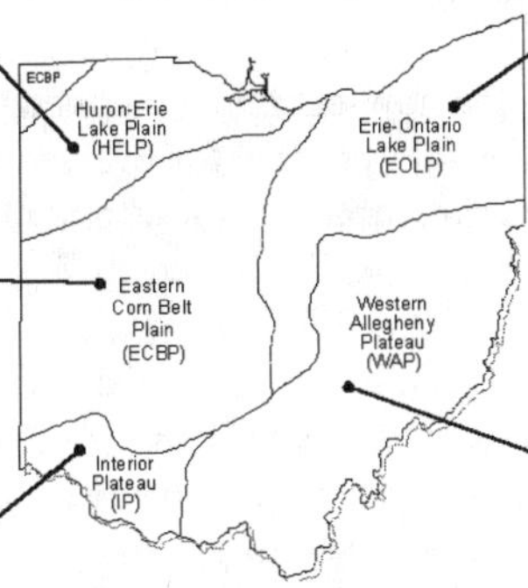

Eastern Corn Belt Plains (ECBP)

Use	Size	IBI	MIwb	ICI
WWH	H	40	NA	36
	W	40	8.3	36
	B	42	8.5	36
MWH-C	H	24	NA	22
	W	24	6.2	22
	B	24	5.8	22
MWH-I	B	30	6.6	NA

Western Allegheny Plateau (WAP)

Use	Size	IBI	MIwb	ICI
WWH	H	44	NA	34
	W	44	8.4	34
	B	40	8.6	34
MWH-C	H	24	NA	22
	W	24	6.2	22
	B	24	5.8	22
MWH-A	H	24	NA	30
	W	24	5.5	30
	B	24	5.5	30
MWH-I	B	30	6.6	NA

Interior Plateau (IP)

Use	Size	IBI	MIwb	ICI
WWH	H	40	NA	30
	W	40	8.1	30
	B	38	8.7	30
MWH-C	H	24	NA	22
	W	24	6.2	22
	B	24	5.8	22
MWH-I	B	30	6.6	NA

Statewide Exceptional Criteria

Use	Size	IBI	MIwb	ICI
EWH	H	50	NA	46
	W	50	9.4	46
	B	48	9.6	46

FIGURE 3.4 Biological criteria in the Ohio WQS for Warmwater Habitat (WWH), Exceptional Warmwater Habitat (EWH), and Modified Warmwater Habitat (MWH) use designations arranged by ecoregion, biological index, use designation, site type for fish assemblages, and ecoregion.

designation, site type, and stream or river size (Figure 3.4) and their derivation has been extensively described elsewhere (Ohio EPA, 1987a and b, 1989a and b; DeShon, 1995; Yoder and Rankin, 1995b).

3.2.5 Causal Associations

The identification of impairment in rivers and streams is straightforward. Numerical biological criteria are the principal arbiters of aquatic life use attainment and impairment. The rationale for using biological criteria in the principal arbiter role within a weight-of-evidence framework has been extensively discussed elsewhere (Karr et al., 1986; Karr, 1991; Ohio EPA, 1987b; Miner and Borton, 1991; Yoder, 1995).

Describing the causes and sources associated with observed impairments revealed by the biological criteria and linking this with pollution sources involve interpretation of multiple lines of evidence including (but not limited to) water chemistry data, sediment chemistry data, habitat data, effluent quality data, whole effluent toxicity results, land use data, and biological response signatures (Yoder and Rankin, 1995a). Thus, the assignment of principal causes and sources of impairment in this framework represents the association of impairments (based on response indicators, i.e., biological criteria) with stressor and exposure indicators that are reinforced by previous experience with a stratum of different situations and impacts.

The reliability of the interpretation of associated causes and sources is increased where many such associations have been previously identified. These are further validated when management efforts directed at abating the identified causes and sources are followed by expected changes in ecological response, some of which are evident in the results presented here. This process is similar to making medical diagnoses by relying on multiple lines of evidence concerning symptoms of patient health. Such diagnoses incorporate the results of clinical observations and controlled research that experimentally or statistically link symptoms and test results to specific diseases, pathologies, and responses to treatments. Thus, a doctor relies on clinical experience and observation combined with the results of research in interpreting symptoms (i.e., multiple lines of evidence based on patient examinations) to establish a diagnosis, principal causes and sources of the malady, a prognosis, and a strategy for treating the condition.

As in medical science, where the ultimate arbiter of success is the eventual recovery and well-being of the patient, the ultimate measure of success in water quality management is the restoration of impaired designated uses. While the application of the metaphor of ecosystem health has been criticized (Suter, 1993), here we refer to a process for evaluating biological condition and causes and sources associated with observed impairments, not whether human health and ecosystem health are equivalent concepts.

A critically important aspect of the delineation of associated causes and sources is the consistent custody of data and information from genesis through final analysis, i.e., those who plan, conduct, and create the original data also oversee the analysis and conclusions made about those data. This is a critical and frequently overlooked component of the assessment process. The value of observations made in the field is of equal importance to the analyses performed later on the same data. We view this consistency, conduct, and custody of watershed assessment data and information as an essential component of quality assurance/quality control (QA/QC) and quality management plans (QMPs).

3.3 ANALYSIS OF RESULTS

The data and information used here were obtained by biological and water quality surveys conducted by Ohio EPA between 1981 and 2000 in the six study areas as follows: Ottawa River (1985, 1987, 1989, 1991, and 1996), Cuyahoga River (1984, 1985, 1986, 1987, 1988, 1991, 1996, and 2000), Scioto River (1981, 1986, 1988, 1991, and 1996), Paint Creek (1985, 1992, and 1997), Dicks Creek (1987, 1995, and 2000), and Rocky Fork (1993 and 1998). All data were collected in accordance with the agency's standard procedures for sampling, survey design, and data processing. Two biological and water quality reports were completed for each study area except Paint Creek, and all data years to date were utilized in the assessments. In the years when reports were not completed, the data and information were assessed for 305b reporting, 303d listing, and WQS revision purposes at a minimum. In addition, water quality permit support documents (PSDs) were prepared for the major NPDES permitted entities in each study area. In all cases, the interdisciplinary process of determining the extent and severity of impairments and associated causes and sources was completed for all except year 2000 surveys which are in progress.

3.3.1 Descriptions of Study Areas and Stressors

The six study areas were chosen based on the existence of multiple years of data, breadth of knowledge about the environmental setting, and impacts sufficient to demonstrate the concept of biological response signatures and the hierarchy of surface water indicators. The extent and trends of biological response in each area varied in accordance with the ability of water quality management to effectively limit and reduce pollution from all sources. Each study area was the subject of a use attainability analysis and with only one exception (lower Scioto River, EWH), the recommended uses were adopted into Ohio WQS.

3.3.1.1 Ottawa River

The Ottawa River is located in northwest Ohio. It is 53 miles long and drains approximately 365 square miles (Lake Erie drainage basin). The mainstem flows through the city of Lima and smaller rural communities, and land use is predominantly agricultural. The river and its tributaries are impacted by a combination of point and non-point sources of pollution. Point sources include large municipal and industrial discharges, combined sewer overflows (CSOs) in Lima, and inadequately or untreated sewage discharges from unsewered areas in the watershed. The Lima wastewater treatment plant (WWTP) discharges an average of 18.5 million gallons per day (MGD) of treated wastewater and is located just upstream from two large industries, the Clark Oil (formerly BP Oil) Refinery and PCS Nitrogen Ohio (formerly Arcadian LP). The effluent flow from these entities enters the Ottawa River within a 0.8 mile reach and comprises the vast majority of the flow in the mainstem during dry months.

Significant improvements have been made to the wastewater treatment facilities since the late 1970s. Ongoing improvements at the Lima WWTP, Clark Oil, and PCS Nitrogen Ohio, as well as new wastewater treatment facilities at the Shawnee and American Bath WWTPs, have further decreased the loadings of conventional pollutants to the mainstem since 1991.

The Ottawa River is impacted by other sources including leachate from the old Bath Township landfill, the former BP Oil L5 landfill, runoff from refining and chemical manufacturing facilities, and runoff or spills from petroleum transfer and tank farm facilities. The Clark Oil facility ranked in the top five entities in the Ohio toxic release inventory during the past decade. This site has served as a refinery for more than 100 years and the extent of site and receiving stream contamination by legacy pollutants is significant. Known causes of impairment to the Ottawa River and tributary streams include organic enrichment, low dissolved oxygen, unknown toxicity, habitat alterations, nutrients, ammonia, metals, oil and grease, and chlorine (Ohio EPA, 1998). Our analysis encompassed a 25-mile segment of the river that is directly impacted by the Lima CSOs, the Lima WWTP, Clark Oil, and PCS Nitrogen Ohio and includes zones of immediate impact and recovery. Physical habitat in the mainstem is good to excellent and the study area is designated WWH.

3.3.1.2 Cuyahoga River

The Cuyahoga River, located in northeast Ohio, drains 813 square miles and extends nearly 100 miles before emptying into Lake Erie at Cleveland. The Akron and Northeast Ohio Regional Sewer District (NEORSD) Southerly WWTPs are the two largest point sources to the lower mainstem, discharging averages of 75 and 120 MGD, respectively. These discharges constitute most of the flow in the mainstem during dry weather. The city of Akron has 41 combined sewer overflows and the NEORSD in Cleveland has 74 of its permitted 135 combined sewer overflow outfalls within the lower Cuyahoga watershed. While most Akron CSOs are located in the Little Cuyahoga River drainage (just upstream from the study area), two discharge directly to the Cuyahoga mainstem between the confluence of the Little Cuyahoga River and the Akron WWTP. Combined sewers in Cleveland are mostly located in the Mill Creek and Big Creek watersheds and the navigation channel just downstream from the study area. Sewer overflows, pollutant spills, and unauthorized discharges are chronic problems in the NEORSD service area and the greater Cleveland area in general (Ohio EPA, 1999b). Our analyses encompassed a 35-mile segment impacted by the Akron and NEORSD Southerly WWTPs and some of the CSOs. Physical habitat ranges from good to excellent and the study area is designated WWH.

3.3.1.3 Scioto River

The Scioto River is located in central and south central Ohio, extending nearly 130 miles and draining 6500 square miles (Ohio River drainage). In the middle portion of the mainstem, the city of Columbus WWTPs and CSOs are the principal pollution sources. The Jackson Pike and Columbus Southerly WWTPs discharge an average of 75 and 125 MGD, respectively, of treated wastewater to the mainstem. Due to a combination of surface water withdrawals and regulated flows to a major tributary upstream from Columbus, these discharges comprise 90 to 95% of the flow in the mainstem during low flow periods. Treatment processes at both WWTPs have undergone extensive upgrades since the early 1980s. Our analyses encompassed a 40-mile segment impacted by the Columbus

CSOs and WWTPs. Habitat in the middle section of the mainstem ranges from good to excellent and is designated WWH. There is a long-standing proposal to upgrade the lower 8 miles of this segment to EWH based on attainment of the biocriteria for that use (*circa* 1991).

3.3.1.4 Paint Creek

Paint Creek is a major tributary of the Scioto River in south central Ohio (Ohio River drainage). Only the lower 5 miles are included in our analyses. Mead Paper operates a kraft pulp mill that discharges approximately 33.5 MGD of treated wastewater at RM 2.54. Upgrades to the treatment process and changes in the paper manufacturing process improved effluent quality since the mid to late 1980s and early 1990s. The effluent comprises a majority of the flow during dry weather. Physical habitat quality in the lower mainstem is good to excellent and the study area is designated WWH.

3.3.1.5 Dicks Creek

Dicks Creek is a small tributary of the lower Great Miami River mainstem in southwest Ohio. It is about 10.5 miles long and drains approximately 48 square miles (Ohio River drainage). AK Steel discharges to three locations on Dicks Creek (outfalls 002, 003, and 015) at RM 2.92, 3.80, and 4.15, respectively, and one to the North Branch of Dicks Creek (outfall 004) at RM 0.22. The plant produces flat rolled steel and intermediate products of pig iron and coke in addition to steel finishing and coating. Pollutants also come from runoff and leachate from landfills and the plant property. Our analyses included the 5 miles of the mainstem downstream from the North Branch. Portions of this segment have been channelized and managed for flood control purposes while the lower 2 miles offer good to excellent habitat. The channelized reach is designated MWH and the areas outside the direct influence of flood control maintenance are designated WWH.

3.3.1.6 Rocky Fork of the Mohican River

The Rocky Fork of the Mohican River is in northeast Ohio. It is 25 miles long and drains about 87 square miles (Ohio River drainage). Major impacts to the mainstem include discharges from the Armco steel making facility, the Mansfield WWTP, and runoff from steel scrap yards adjacent to Armco. The Mansfield WWTP discharges an average of 10.5 MGD of treated wastewater to the mainstem. Its treatment process was improved in the late 1980s. The Armco facility uses scrap steel to manufacture carbon, silicon, and stainless steel coils and sheets. Four outfalls discharge to Rocky Fork. Changes to the manufacturing process in late 1993 (after the 1993 sampling) significantly improved effluent quality. The shearing of scrap metal and the use of cutting oils and lubricants at the scrap yards are likely sources of contamination via runoff. Our analyses included the lower 16 miles of the mainstem. Physical habitat is generally good, but some areas in Mansfield have been locally channelized. However, habitat throughout Rocky Fork is adequate to support warmwater aquatic assemblages and this is reflected by the WWH use.

3.4 DISCUSSION

3.4.1 Synthesis of Results: Associated Causes and Sources of Impairment

In the six study areas, the predominant stressors range from toxic to a mix of toxic and conventional* impacts to predominantly conventional pollution. The predominant impact types changed over time

* Toxic stressors include a wide variety of toxic substances and frequently produce acute toxicity and long-term contamination of the aquatic environment. They are difficult to control via conventional wastewater treatment. Conventional stressors include treated and untreated municipal sewage and associated pollutants such as BOD, ammonia, nutrients, and common heavy metals; they are amenable to commonly available treatment technology.

because the NPDES permitting process prompted pollution abatement, albeit incomplete in some instances. Significant habitat modification is a prevalent issue in only one of the study areas — Dicks Creek. However, the impaired aquatic life uses noted in all six study areas has been blamed on habitat deficiencies. We examined some of the key response signatures identified by Yoder and Rankin (1995a) to determine the extent of toxic responses in each study area and the extent of any changes observed over time.

As described in Yoder and Rankin (1995a), the macroinvertebrate assemblage "signature" of a complex toxic impact is best characterized by three aspects of the data; the invertebrate community index (ICI) score, the number of qualitative EPT taxa, and the proportion of individuals comprised by the midge genus *Cricotopus*. The combined response of these three components to the complex toxic impact type was described as follows, based on the assessment of the Ohio EPA statewide macroinvertebrate database performed by Yoder and Rankin, 1995a); ICI $\leq$18, qualitative EPT $\leq$4, and percent *Cricotopus* $\geq$5.0.

The concurrence of all three was strongly indicative of complex toxic impacts as opposed to the other impact types tested: conventional municipal/industrial, CSOs/urban, channelization, agricultural row cropping, flow alteration, CSOs/urban with toxics, and unrestricted livestock access. The complex toxic impact type was characterized by Yoder and Rankin as the combination and interaction of major WWTP and industrial point sources (including indirect discharges to WWTPs) that comprise a significant portion of the summer–fall base flow of the receiving stream and where one or more of the following had been recorded: serious instream chemical water quality impairment by toxic parameters, recurrent whole effluent toxicity, fish kills, and/or extreme contamination of bottom sediments by toxic parameters. In addition to these three biological parameters, the proportion of toxic tolerant taxa and proportion of organic tolerant taxa (as originally defined by Yoder and Rankin, 1995a) were used in our analysis. The criteria for threshold responses are found in Table 3.1.

The values for individual macroinvertebrate and fish assemblage attributes and the number of toxic signatures expressed at each site for each sampling year in each study area appear in Appendix Tables 1 through 3. We included the frequency of occurrence of toxic response signatures in these analyses in an attempt to determine the strength of a toxic response exhibited by the macroinvertebrate and fish assemblages. This consisted of the number of accumulated toxic response signatures (based on the thresholds in Table 3.1) in each sample, which provided the opportunity to analyze the spatial and temporal extent over which a particular stream or river reach exhibits a toxic response.

For macroinvertebrates, >75% occurrence of toxic signatures was indicative of a strong toxic response, indicating that three of four attributes exceeded individual toxic response thresholds. For the fish assemblage, this was set at 85%, which means that six of seven attributes exceeded toxic response thresholds. Only the reaches most directly impacted by the major sources of impact in each study area were included. Upstream sites and far field reaches below the documented recovery points were not included in the graphical analyses; however, these sites are included in the appendix tables. Graphical analyses consisting of box-and-whisker plots of the indices, metrics, and aggregate attributes were used to visually reveal changes in a study area over time and differences between study areas.

3.4.2 Case Study Responses

3.4.2.1 Ottawa River Responses

The responses of fish and macroinvertebrate metrics and indices in the Ottawa River were strongly toxic for at least 25% of the samples in all years except 1996 (Figures 3.5 and 3.6) and were the strongest among the six study areas. All the fish assemblage responses were strongly toxic and the 75th percentile aggregate frequency of toxic responses was greater than 85% in all years except 1996. There was a lessening of the aggregate toxic response in 1996 and for the IBI, MIwb, fish numbers, and proportion of round-bodied suckers attributes (Figure 3.5). The percent DELT anomalies remained elevated above the toxic response threshold for the majority of samples, as did the proportion of highly tolerant species.

Macroinvertebrate responses were strongly toxic through 1991, but a trend toward lessening the frequency of these responses was more evident in 1996 and was most apparent in the ICI, qualitative EPT taxa, and percent toxic tolerant taxa (Figure 3.6). The percent *Cricotopus* response remained strongly toxic through 1996. The response of the percent organic tolerant taxa was not indicative of a shift toward a more conventional type of impact that can emerge as the more severe

TABLE 3.1
Criteria Used to Determine the Extent of a Response Signature Exhibited by the Macroinvertebrate and Fish Assemblages in the Six Study Areas

Macroinvertebrates	
Invertebrate Community Index (ICI)	≤18
Qualitative EPT Taxa	≤4
Percent *Cricotopus* sp.	≥5
Percent Toxic-Tolerant Taxa[a]	≥35
Percent Organic/Nutrient/DO Tolerant Taxa[b]	≥35
Fish — Wading/Headwater Sites	
Index of Biotic Integrity (IBI)	≤22
Modified Index of Well-Being (MIwb)	≤5.9
Percent DELT Anomalies	≥10
Percent Tolerant	≥70
Number of Intolerant Species[c]	<1
Density (Numbers less Tolerants)	<150
Number of Darter Species[d]	<1
Fish — Boat Sites	
Index of Biotic Integrity (IBI)	≤22
Modified Index of Well-Being (MIwb)	≤5.9
Percent DELT Anomalies	≥10
Percent Tolerant	≥70
Number of Intolerant Species	<1
Density (Numbers less Tolerants)	<150
Percent Round-Bodied Suckers	<5

[a] Toxic-tolerant taxa include *Cricotopus* sp., *Dicrotendipes simpsoni*, *Glyptotendipes (G.) barbipes*, *Polypedilum (P.) fallax* group, *Polypedilum (P.) illinoense*, and *Nanocladius (N.) distinctus*.

[b] Organic-tolerant taxa include *Oligochaeta*, *Glyptotendipes (G.)* sp. (not *G. barbipes*), *Chironomus (C.) decorus* group, *Chironomus (C.) riparius* group, *Dicrotendipes lucifer*, *Dicrotendipes neomodestus*, *Polypedilum (Tripodura) scalaenum* group, *Turbellaria*, *Physella* sp., *Simulium* sp.

[c] Sensitive species (intolerant plus moderately intolerant) at headwaters sites.

[d] Includes sculpins at headwaters sites.

Source: Yoder, C.O. and E.T. Rankin. 1995b. Biological criteria program development and implementaion in Ohio, in Davis, W.S. and T.P. Simon, Eds. *Biological Assessment and Criteria: Tools for Water Resource Planning and Decision Making.* Lewis Publishers, Boca Raton, FL, 109–144.

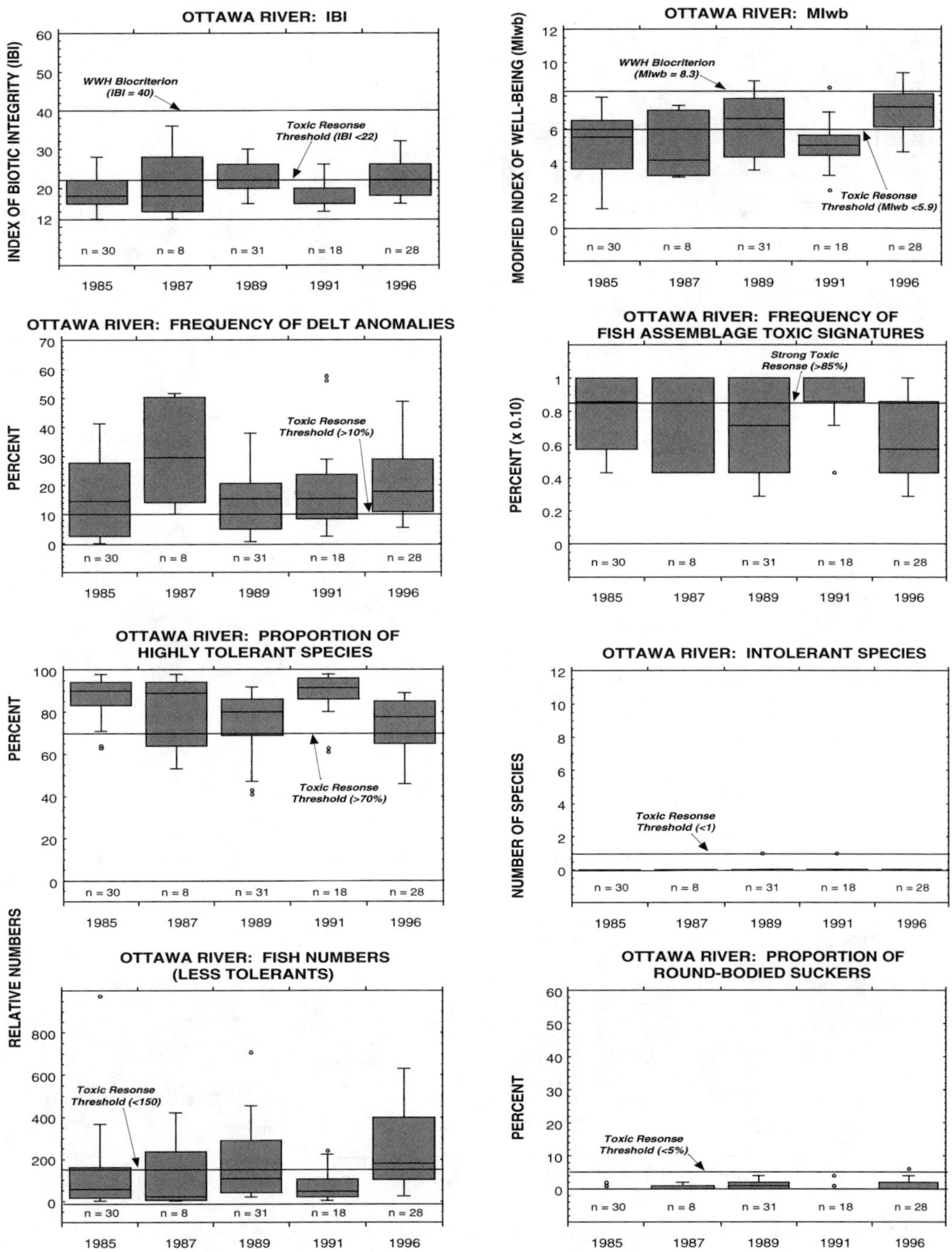

FIGURE 3.5 Results for fish assemblage metrics and indices and the frequency of toxic responses in the Ottawa River study area between 1985 and 1996.

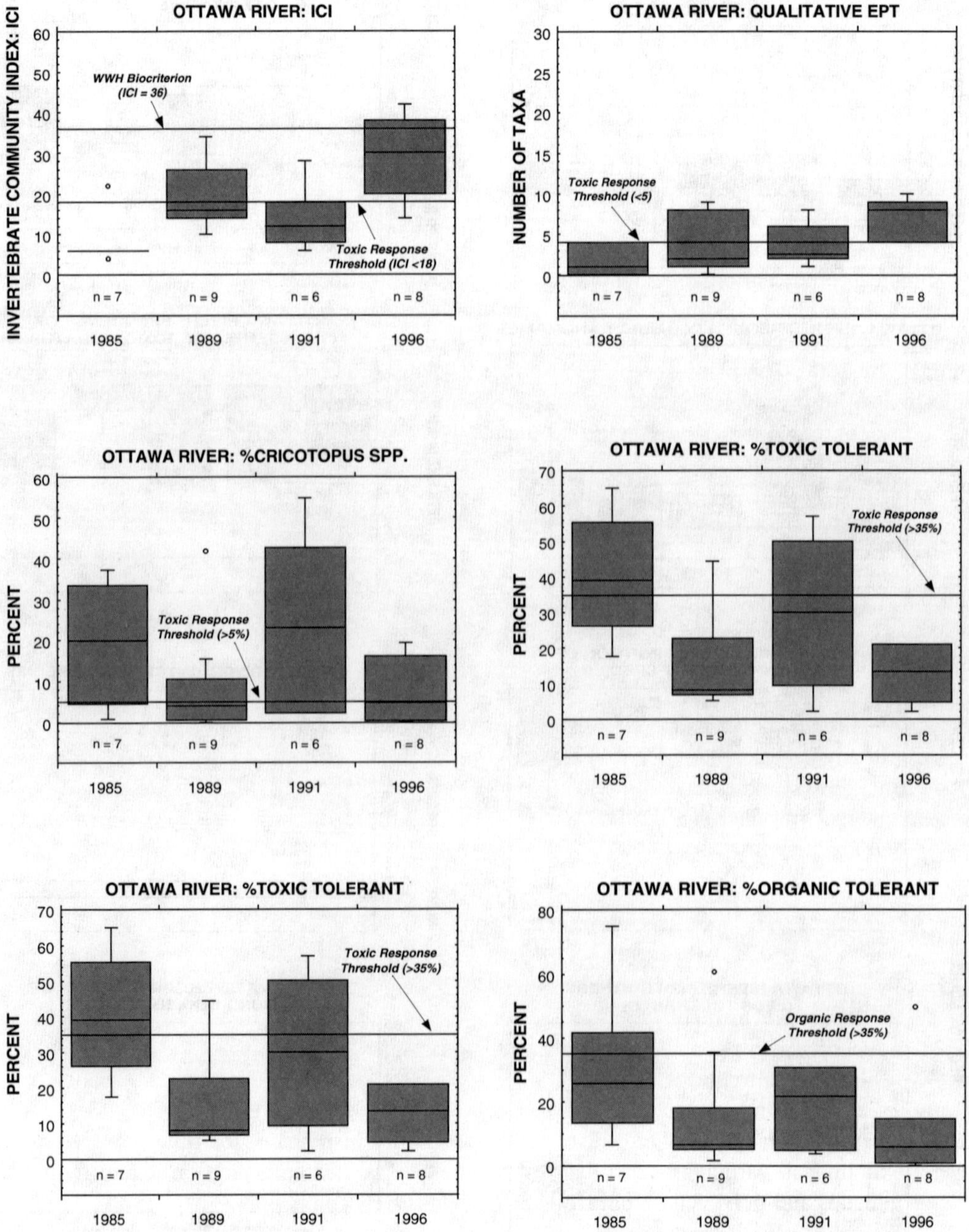

FIGURE 3.6 Results for macroinvertebrate assemblage metrics and indices and the frequency of toxic responses in the Ottawa River study area between 1985 and 1996.

toxic responses abate. Given the decline in the occurrence of this response signature, it seems that pollution abatement efforts are capturing this impact type as well.

3.4.2.2 Cuyahoga River Responses

Both the fish and macroinvertebrate responses showed strong and positive trends between 1984 and 2000 from a strongly toxic signature and severe impairment of the WWH use designation to a virtual absence of the toxic response and partial to full attainment of the WWH use in 2000 (Figures 3.7 and 3.8). The toxic response was strongest in 1984 in both groups and persisted through 1986. In 1984, the degradation was so severe (Ohio EPA, 1999b) that it masked the true extent of the toxic response in some of the metrics and the aggregate response by all measures. For example, only a single fish was captured at three sampling sites downstream from the Akron WWTP in 1984 which produced highly skewed responses for some of the assemblage attributes. In 1987, the strength of the toxic responses began to decrease and this continued through 1991 for both groups and through 1996 for fish assemblage.

The macroinvertebrate assemblage showed an earlier positive response and this was not replaced by an increased percent organic tolerant taxa response, which indicates a substantial improvement in overall assemblage condition and good control of the major point source stressors. In terms of the ICI, the recovery to WWH performance levels in all the impacted segments was virtually complete by 2000. The fish assemblage response has been slower and, while showing only initial indications of full recovery in 2000, exhibited a complete eradication of toxic signatures. The only metric that continued to perform at toxic response levels was the number of intolerant species, but this metric is also responsive to other, less severe stressors.

3.4.2.3 Scioto River Responses

The responses observed in the Scioto River between 1981 and 1996 were not strongly indicative of toxic stressors in any samples, although some metrics exhibited slight degrees of response by exceeding toxic response thresholds in a few of the samples (Figures 3.9 and 3.10). These diminished over time and were virtually absent in 1996. When present, they occurred in metrics that also respond to non-toxic stressors and two of the principal toxic signatures, DELT anomalies and percent *Cricotopus,* never exceeded the toxic response thresholds. Thus, the evidence of a toxic response in any sample was weak when evident at all. In addition, a strong signature of organic enrichment shown by the percent organic tolerant taxa was indicative of the influence of conventional municipal impacts on the study area. This is consistent with previous assessments (Ohio EPA, 1999c) and is corroborated by the positive response of the receiving stream to water quality management targeted at conventional pollution control.

3.4.2.4 Rocky Fork of the Mohican River Responses

Responses in 1993 were strongly toxic (Figures 3.11 and 3.12), but were localized to zones of immediate impact downstream from Armco and its neighboring scrap yards. The fish assemblage exhibited a particularly strong response in 1993. Impacts downstream from the Mansfield WWTP were typical of the conventional municipal impact type. In 1998, most of the toxic responses were greatly lessened, but not absent. Recovery was still incomplete (Ohio EPA, 2000a), based on the biological attributes analyzed here. Five years is probably insufficient for full recovery from a severe toxic impact, particularly one that results in extensive contamination of bottom sediments by heavy metals and polycyclic aromatic hydrocarbon (PAH) compounds.

3.4.2.5 Dicks Creek Responses

Strong toxic responses in the macroinvertebrate assemblage existed in Dicks Creek, particularly in 1995 (Figures 3.11 and 3.12). Evidence of a co-occurring response to organic enrichment was also apparent, unlike that observed in the other study areas. While the ICI and qualitative EPT taxa showed some improvement between 1987/1995 and 2000, the percent *Cricotopus* and percent toxic tolerant taxa remained elevated well above toxic response thresholds. The increased range in some of these attributes and the aggregate frequency of toxic response was evidence that the zone or extent of toxic impact lessened over time. Nevertheless, based on the macroinvertebrate responses, recovery is only in the initial stages and depends on continued pollution abatement at AK Steel.

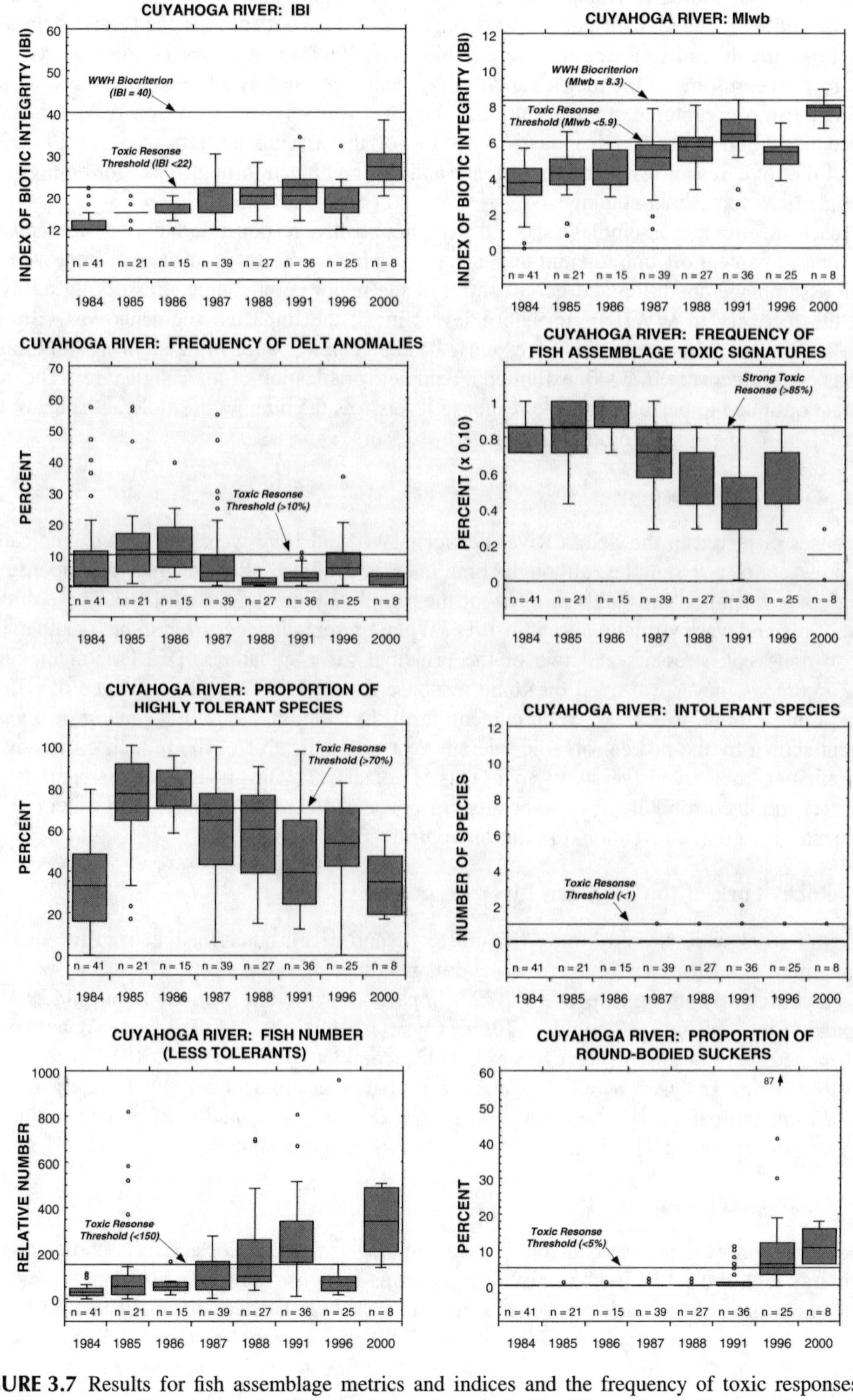

FIGURE 3.7 Results for fish assemblage metrics and indices and the frequency of toxic responses in the Cuyahoga River study area between 1984 and 2000.

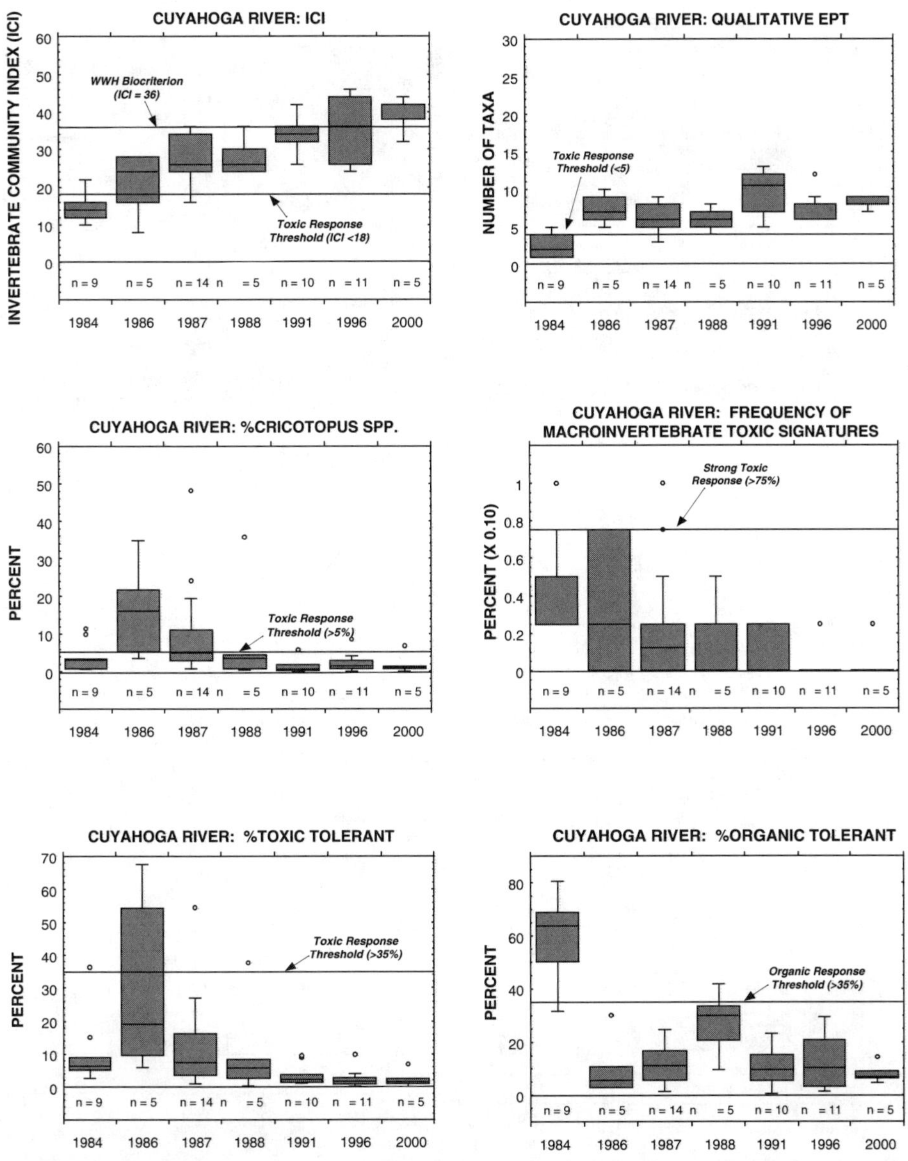

FIGURE 3.8 Results for macroinvertebrate assemblage metrics and indices and the frequency of toxic responses in the Cuyahoga River study area between 1984 and 2000.

The fish assemblage showed a markedly differing response, exhibiting fewer episodes of toxic response. Evidence of strong toxic responses by the fish community was periodic and indicated the effects of acute, episodic impacts, particularly in the 1987 and 1995 results (Ohio EPA, 2000b). Toxic response thresholds were exceeded sporadically and only in the most severely impacted reaches of the creek. The temporal manifestations of toxic signatures in the fish assemblage followed episodic releases of toxic substances (Ohio EPA 2000b). These results indicate the need to use at least two biological assemblages and repeated intervals of sampling where complex, poorly controlled, and episodic releases are potential problems.

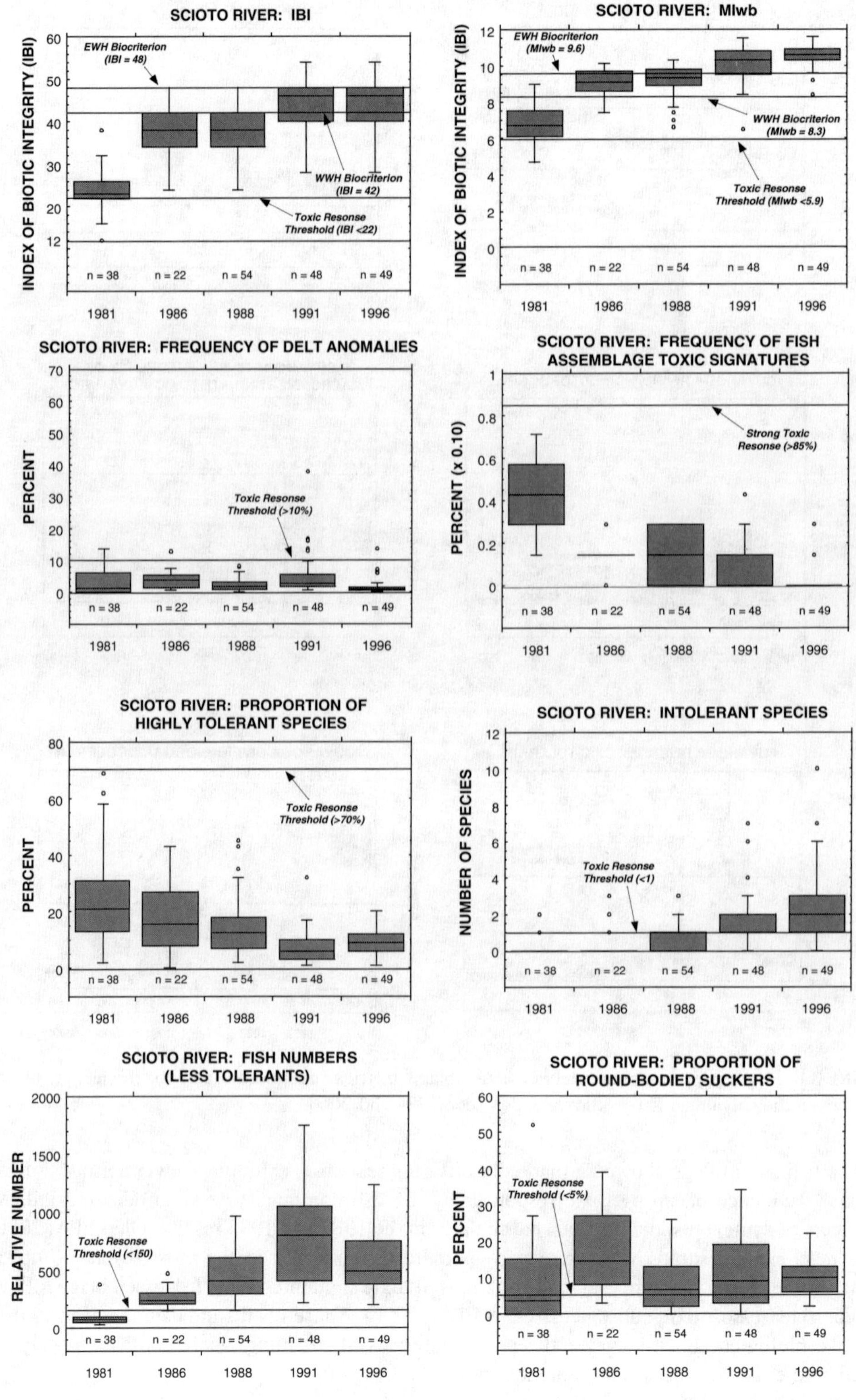

FIGURE 3.9 Results for fish assemblage metrics and indices and the frequency of toxic responses in the Scioto River study area between 1981 and 1996.

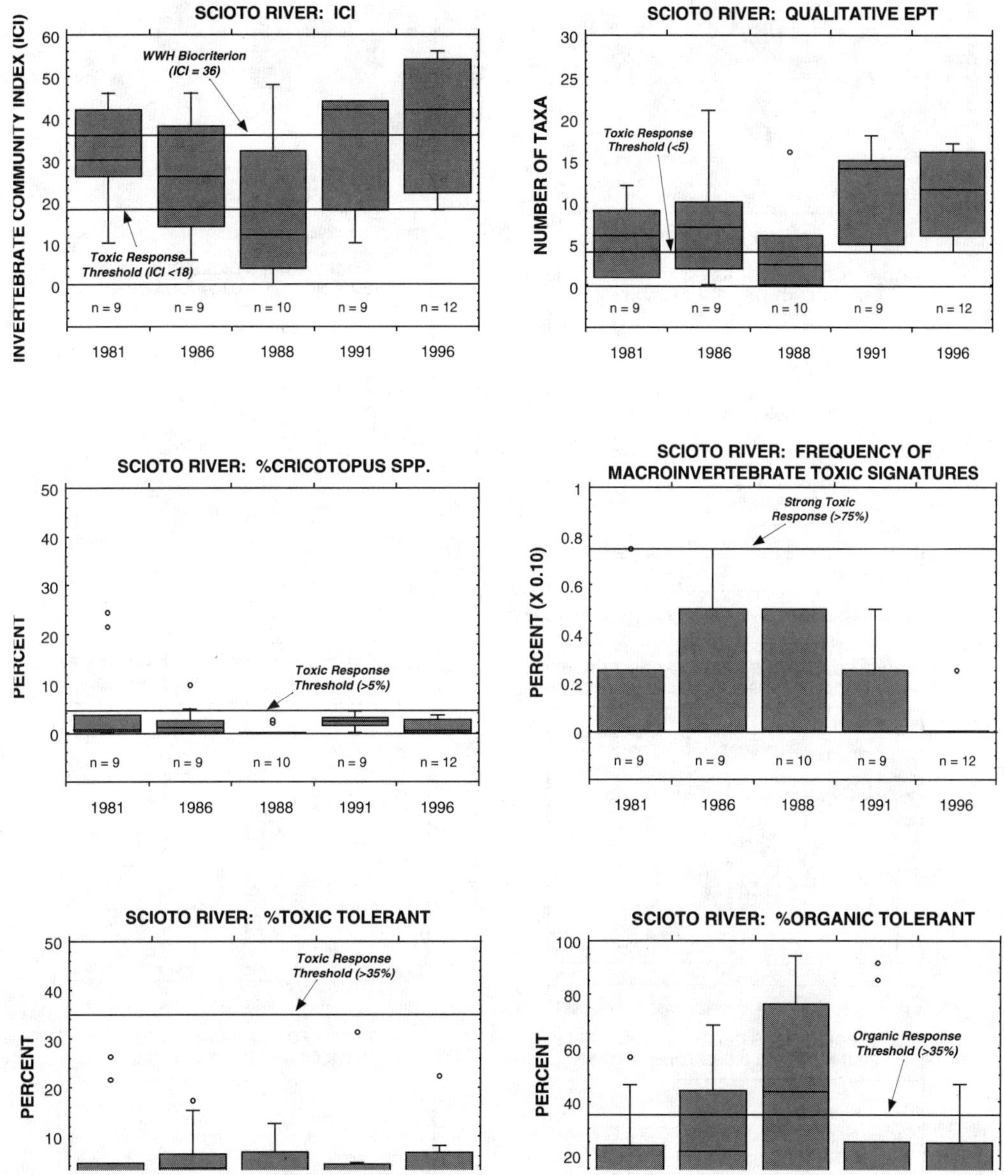

FIGURE 3.10 Results for macroinvertebrate assemblage metrics and indices and the frequency of toxic responses in the Scioto River study area between 1981 and 1996.

3.2.4.6 Paint Creek Responses

Paint Creek showed the least evidence of a toxic response and the strongest evidence of organic enrichment among the six study areas (Figures 3.11 and 3.12). The results also indicate dramatic improvements in biological condition between 1985 and 1997, with all deleterious response signatures virtually absent in 1997 and full attainment of the WWH biocriteria. This was the first case of full recovery downstream from a major industrial discharge in Ohio. However, the characteristics of the Mead Paper effluent are very similar to those of the conventional municipal impact type.

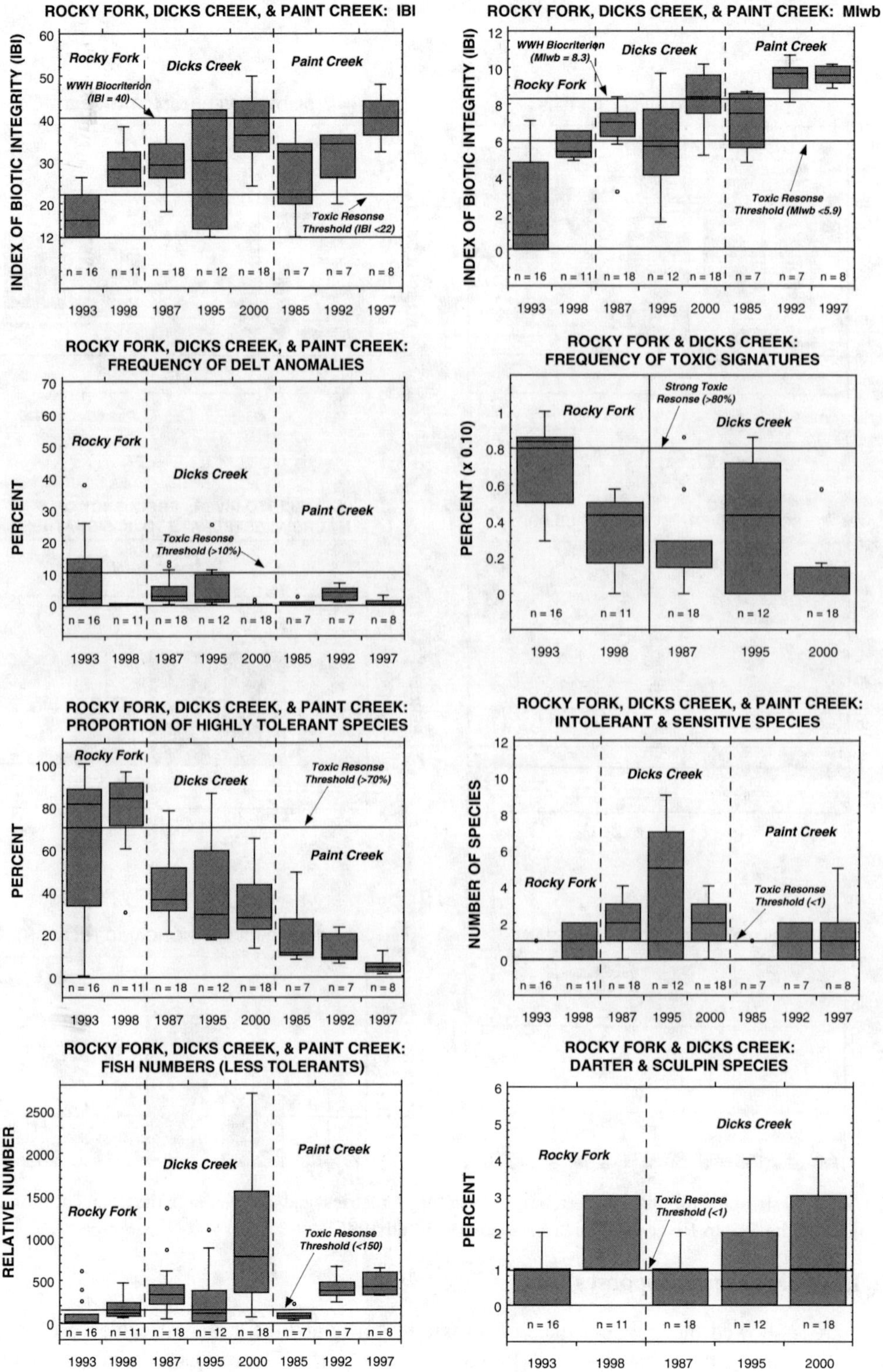

FIGURE 3.11 Results for fish assemblage metrics and indices and the frequency of toxic responses in the Rocky Fork, Dicks Creek, and Paint Creek study areas between 1985 and 2000.

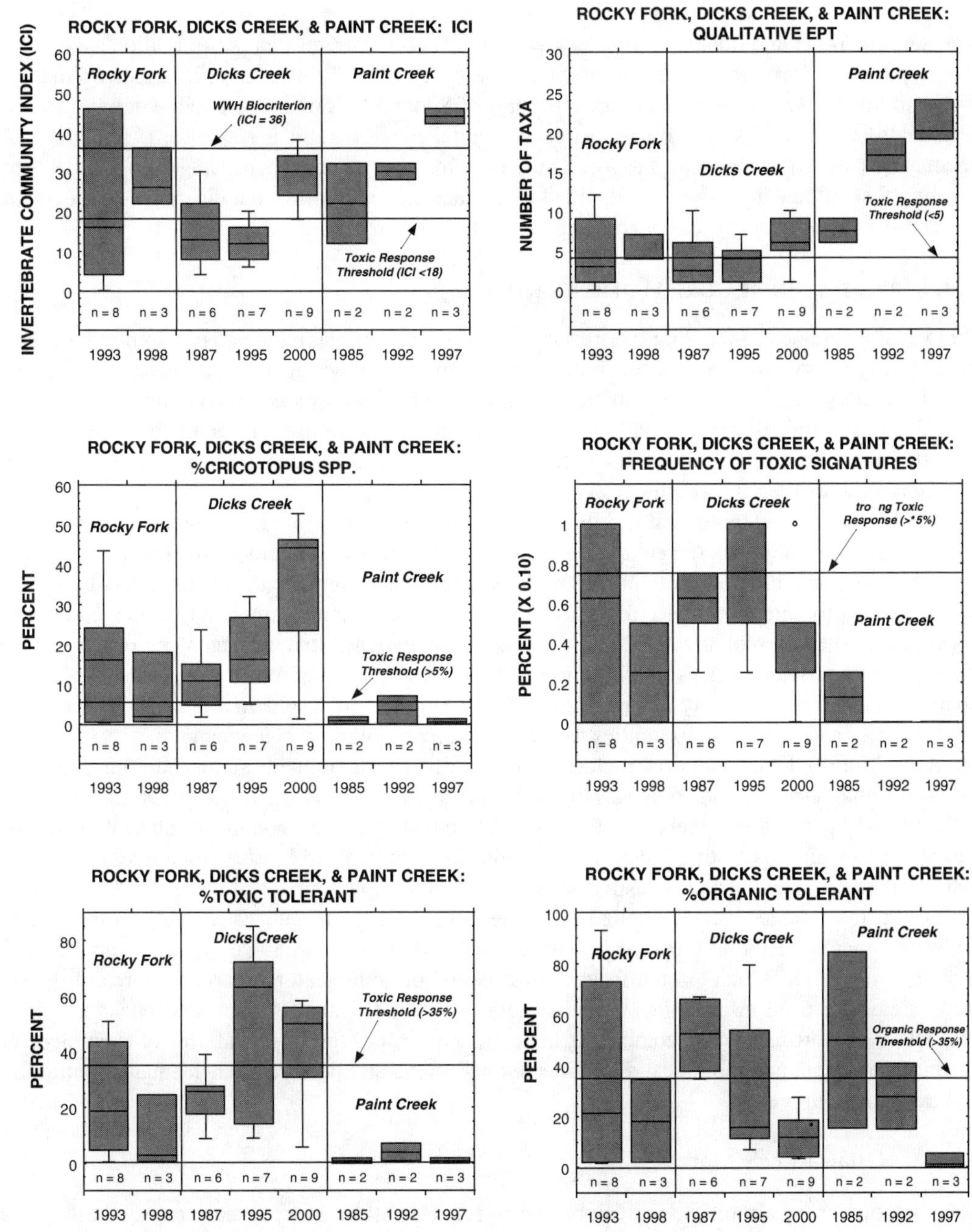

FIGURE 3.12 Results for macroinvertebrate assemblage metrics and indices and the frequency of toxic responses in the Rocky Fork, Dicks Creek, and Paint Creek study areas between 1985 and 2000.

3.2.4.7 Overall Observations

Nothing in our analyses changed the original observations of Yoder and Rankin (1995a) about the combinations of macroinvertebrate and fish assemblage attributes that most strongly indicate complex toxic impacts. The presence of toxic responses exhibited by the additional assemblage attributes examined here served to strengthen their observations. They were insufficient when considered alone, and in some cases were responsive to non-toxic impacts. In one instance, the number of intolerant fish species continued to perform within the toxic response threshold while the major signatures of Yoder and Rankin completely recovered. This illustrates that these additional attributes can respond to other stressors, some of which are "background" influences. In another situation, the toxic impact was so severe that key attributes of Yoder and Rankin were below toxic response thresholds due to a lack of sufficient organism numbers to exhibit a response, a situation also encountered by Eagleson et al. (1990). Thus, care must be taken when using this approach by assuring that sufficiently experienced and skilled practitioners perform the diagnosis — a demand society places on other professional disciplines for similar reasons.

3.4.3 Multiple Indicators Matrix Analysis

Biological assessment, including the use of biological response signatures, is an important aspect in assigning causal associations to observed impairments. However, the process of determining associated causes and sources of aquatic life impairments includes a corresponding assessment of the available chemical/physical data from ambient media and both point and non-point sources. This fulfills the role of biological assessment information as a response indicator in the indicator hierarchy described earlier (Figure 3.2).

While informative, biological results alone are insufficient to provide detailed cause and effect data on which pollution abatement options can be developed and implemented. However, linkages between indicators of human activity and biological response complete the essential feedback loop that is needed to determine the effectiveness of water quality management and steps required to improve it, whereas biological feedback alone might indicate that management responses are incomplete or misplaced. This confirmation of the success of the administrative aspects of water quality management is largely lacking in the United States. This results in a lack of confidence in the processes and loss of opportunities to further improve pollution abatement.

A multiple indicators matrix includes aquatic life use attainment status, multiple stressor, exposure, and response indicators arranged in an upstream-to-downstream direction for major mainstem segments and tributaries. Each representative indicator column is shaded, with the darkness of shading indicating increasingly serious departures from levels compatible with attainment of the designated use or a desired state consistent with attainment of the use. The data used to develop these tables were compiled from the latest biological and water quality assessments performed by Ohio EPA and the underlying documentation for the summary data that appeared in each report. Sufficient data existed for the construction of an indicators matrix for three of the six study areas. We used information from the Ottawa River, Cuyahoga River, and Dicks Creek to illustrate this approach. When completed, the matrix illustrates the roles and uses of each piece of chemical, physical, and biological data collected within the architecture of an adequate monitoring and assessment approach.

3.4.3.1 Ottawa River Matrix

The construction of a multiple indicators matrix for the Ottawa River mainstem was based on a 1996 study by Ohio EPA (1998). The accumulation of increasingly serious toxic responses and exposures reached a maximum in the segment directly impacted by the three major point sources (Figure 3.13). Evidence of toxic exposure appeared in the water column chemical results, sediment chemistry results, whole effluent toxicity results, frequency of DELT anomalies, fish tissue contaminants, and biochemical markers. These indicators pointed strongly to impacts of a toxic character and the biological response signatures provided corroborating feedback.

The designation of this segment of the Ottawa River as WWH was contested as unattainable due to habitat limitations. Clearly, the habitat data indicates the contrary as the QHEI scores are more than adequate to support the WWH designation (Rankin, 1989, 1995) as does the biological recovery to WWH in far-field reaches of the mainstem. Also, the biological response signatures strongly indicated toxicity, which was a fundamentally different response from what would occur

SEGMENT	DES. USE	RESPONSE INDICATORS				EXPOSURE INDICATORS						STRESSORS				
	Attain-ment Status	QHEI	IBI	MIwb	ICI	Water Chem.	Sedi-ment Chem.	Tox-icity	% DELT	Fish Tiss.	Bio-marker	# Dams/ Pools	Urban-Indust. Landuse	Cumulative Loads	Spills	CSO SSOs
Ottawa River mainstem - 1996																
Thayer Rd to Sugar St.	FULL-PART.	68	Fair-Good	Fair-Good	Good	Nitrates	Low	NA	Mod-High	Mer-cury	Low	Mod-e	Low	Low	Low	Low
Sugar St. to Lima WWTP	NON	"#	Poor to Fair	Poor to Fair	Poor to M.G.	CBOD TSS . .	As,Cr d ($u Ni()*	Mod-erate	High	Pesti-cides	BUN Nah %a-+	High	High	Mod-erate	Mod- e	High
Lima WWTP Ae ntown dam	NON	7/	Poor	Poor to Fair	Fair to Good	Amm. CBOD TSS D.O. Nitrates Phos Chrom. PAH Pesticid	As,Cr Cd,Cu Ni,Zn PAH	Mod-erate	Very High	Selen-ium Pest-icides	EROD Naph B(a)p BUN	Mod-e	High	High	High	High
Allentown dam to 1alida	PAR-T2AL	69	Poor -Fair	Fair-Good	Good -Ec.	TSS	Low	NA	High	Pesti-cides	Low	Low	Low	High	Low	Low
Kalida to mouth	FULL	69	Good	Good	Exc.	TSS	Low	NA	Very High	Pesti-cides	Low	Low	Low	High	Low	Low

FIGURE 3.13 A matrix of stressor, exposure, and response indicators for the Ottawa River mainstem based on data collected in 1996 (after Ohio EPA, 1998). The darkness of shading indicates the degree of severity in effect expressed by an indicator.

due to habitat alone (Yoder and Rankin, 1995a). In another instance, low dissolved oxygen was advanced as the primary factor limiting the biota and controlling the impairment observed in the 1980s and early 1990s. Again, the biological response signatures exhibited very strong responses to toxic chemical contamination and resultant aggregate toxicity compared to low dissolved oxygen (DO) alone.

Low DO is present in the Ottawa River as indicated by extensive data collected during Ohio EPA biosurveys (Ohio EPA, 1998). However, its full impact is masked by the more serious toxic effects evident in the biological response signatures and the cadre of stressor and exposure indicators in the matrix. This raises an important water quality management issue in that while the low DO may need to be addressed, the current toxicity problems render it potentially academic in terms of restoring the river mainstem. In addition, the application of water quality management via NPDES permitting in this and other areas addresses the stressors that contribute to DO problems. Thus, feedback on the performance of these abatement measures in terms of biological response is needed before changing the course of water quality management in this study area.

3.4.3.2 Cuyahoga River Matrix

The multiple indicators matrix for the Cuyahoga River (Figure 3.14) cites an accumulation of toxic responses and exposures in the segment assessed by our analyses (lower Cuyahoga mainstem). However, one of the exposure indicators, whole effluent toxicity, shows a historical change that corresponds to the lessening in toxic response signatures observed in 1996 and earlier. Other exposures have also been reduced from historical levels (Ohio EPA, 1999b) and this also corresponds to the reduced frequency of toxic response signatures. Like the Ottawa, the designation of WWH use for the lower Cuyahoga mainstem was contested as unattainable due to habitat limitations. Again, the habitat data indicates to the contrary as the QHEI scores are more than adequate

SEGMENT	DES. USE	RESPONSE INDICATORS				EXPOSURE INDICATORS					STRESSORS		
	Attain-ment Status	QHEI	IBI	MIwb	ICI	PAH Tumor Potential	Sediment Contam. Met. / Org.	Toxicity Past / Present	Fish tissue	Bio-mark.	Urban - Industrial Land Use	SPILLS (#s by county)	CSOs SSOs
Upper Cuyahoga River mainstem													
East Br. Res. to US 422	PAR-TIAL	52	F*-VG	F*-G	F*	Low	Low/Low	na/No	Low	na	Low	Low	None
Hiram Rapids to Kent	FULL	74	G-E	G-E	G-E	Low	Low-Mod/ Low	No/No	Low	High	Low-Mod	Mod	None
Kent to L. Cuy. River	PAR-TIAL	81	F*	F*-G	F*-G	Low-High	Low-Mod/ Low-High	No / No	Low	Very High	Mod	Mod - High	YES
Little Cuyahoga River *(lower section)*													
In Akron	NON	69.5	P*	F*	P*-F*	Low-High	Mod / Low	Yes / No	na	Very High	Very High	High	YES
Lower Cuyahoga River *(upper section)*													
L. Cuy. River to Boston	NON	75	F*-P*	F*	F*-G	Low-High	Low-Mod/ Low	Yes / No	High	Very High	High	High	YES
Tinkers Cr. to Estuary	NON	71	P*	VP*-F*	F*-VG	Low-High	Low-Mod/ Low	Yes/Yes	Mod	na	High	Very High	YES
Estuary	NON	55.5	P*	P*	G	Mod	Low / Mod	Yes / No	Mod	na	Very High	Very High	YES
Ship Channel	NON	33	P*	P*	P*	High	V. High / Mod	Yes/Yes	Mod	na	Very High	Very High	YES

FIGURE 3.14 A matrix of stressor, exposure, and response indicators for the Cuyahoga River mainstem and major tributaries based on data collected in 1996 (after Ohio EPA, 1999a). The darkness of shading indicates the degree of severity in effect expressed by an indicator.

to support the WWH use designation (Rankin, 1989, 1995) and the biological response signatures strongly indicate toxicity as opposed to habitat limitations. Finally, the emergence of the macroinvertebrate assemblage to fully attain the WWH biocriteria and the incremental improvement of the fish assemblage toward that goal are further evidence of WWH use attainability and the effectiveness of management driven by that goal.

3.4.3.3 Dicks Creek Matrix

The largest body of information from all indicator levels was collected in 1995 and used as part of the middle and lower Great Miami River assessment (Ohio EPA, 1997). This and other available information was used to construct a multiple indicators matrix (Figure 3.15; Ohio EPA, 2000b). This matrix is spatially more detailed than the previous examples and illustrates the level of detailed analysis that is possible using this approach.

The accumulation of indicators that show increasingly serious departures from their respective criteria or other benchmarks were most numerous downstream from the AK 003 outfall. The only exposure indicator that did not show at least one detection was sediment polychlorinated biphenyls (PCBs). In the matrix, this indicator did not become visible until downstream from the AK landfill tributary, which is where most of the PCB contamination emanates. Exceedences of selected indicators were evident at most of the other sampling locations including those outside the direct influence of the AK Steel discharges. In particular, RM 5.2 in Dicks Creek (upstream from our study area) exceeded several water quality criteria and showed highly elevated metals in sediment. RM 1.0 in the North Branch exceeded water quality criteria and revealed elevated metals, and detectable concentrations of polycyclic aromatic hydrocarbons (PAHs) in sediments.

These two sites are upstream from the direct influence of the AK Steel wastewater discharges, and impacts from other sources occur in the study area. However, while these signals also correspond to failures to meet the biocriteria for at least the ICI, and hence non-attainment of the applicable aquatic life use, the biological response signatures were not indicative of the toxic impacts observed downstream from AK Steel. The accumulation of toxic stressor and exposure indicators associated with toxic biological response signatures all occurred downstream from the AK Steel discharges and are the most severe where stressor and exposure indicators connected with the AK Steel discharges accumulate.

The co-occurrence of toxic chemical compounds and substances that are byproducts of the steel-making process are characteristic of the complex toxic impact types described by Yoder and Rankin (1995a). These responses are commensurate with the toxic response by the macroinvertebrate assemblage that strongly implicates the roles of selected AK outfalls in the failure to attain the applicable aquatic life use. There is little reason to assert that habitat alone plays a major role in the observed impairments, other than that already incorporated into the biological criteria for the MWH use designation.

3.4.3.4 Other Study Areas

Only one of the remaining three study areas exhibited a response characteristic of toxic impact. The Rocky Fork of the Mohican River was extensively impacted by heavy metals and PAH compounds from Armco Steel, an adjacent scrapyard, and a number of industries that no longer discharge directly to the mainstem. There was a strong negative correlation between the fish and macroinvertebrate indices and the concentration of heavy metals and PAHs in bottom sediments in 1993 (Ohio EPA, 2000a). By 1998, the severity of the biological response had lessened to the point that the strong toxic response was virtually diminished, which corresponded to upgrades and process changes at Armco Steel in late 1993. This was the first case in Ohio where toxic impacts associated with an industrial discharge were virtually abated.

The Scioto River and Paint Creek clearly exhibited non-toxic responses. The sequence of treatment upgrades and instream recovery of the Scioto River is well documented and this has been observed statewide *where toxic impacts and associated stressors are not prevalent.* Paint Creek is similar, yet while the source is industrial, the response of the biological assemblages indicates a virtual absence of toxicity. In the case of Mead Paper, the recovery of Paint Creek was due to extensive reductions in loadings of conventional substances such as suspended solids and BOD. Toxic chemicals occur in the Mead effluent, but they are in very low concentrations or result in little or no toxicity; no toxic response is exhibited by the assemblages.

3.4.4 Relevance to Water Quality Management

The relevance of biological response signatures and the indicator hierarchy to water quality management is significant and includes making defensible diagnoses of causes and sources associated with biological impairments. They provide the critical process and elements of a monitoring and assessment architecture for supporting administrative activities such as permitting, standards setting, planning, and TMDL development. Frequently, administrative actions are challenged as arbitrary, capricious, or based on little or no "real world" information. The availability of comprehensive, robust, and reliable information that links administrative efforts to control pollution to environmental results closes important gaps in that process.

This was at least partially demonstrated with the three multiple indicators matrices where in each case, habitat and conventional problems were initially blamed for the biological impairments, and the biological response signatures strongly indicated otherwise. The stress and exposure indicators that provided evidence of exceeding important chemical and physical thresholds and criteria that emanated from the major pollution sources that are subject to regulation and management

SEGMENT	DESIGNATED USE	STRESSOR INDICATORS					EXPOSURE INDICATORS						RESPONSE INDICATORS			
	Attainment Status[a]	Permit Exceedences	Spills/Releases	Fish Kills	Toxicity (WET)	Habitat Modifications	Water Column Chem.[b]	Water Column PAHs	Sediment Metals	Sediment PAHs	Sediment PCBs	Habitat (QHEI)	IBI[c]	MIwb	ICI[d]	BRS[e]
Dicks Creek & N. Branch - MWH Use Designation																
Dts. Moraine Materials (RM 5.2)	**[NON]**	—	0	—	—	—	15/8 [0>max.]	—	**HIGHLY** (As, Cr)	0 PAH det.	0 PCB det.	—	—	NA	**VP***	
N. Branch Ust. AK 004 (RM 1.0)	**NON**	—	0	—	—	Maintained Channel	5/4 [0>max.]	0 PAH	ELEVATED (Al, Ba, Cr)	6 PAH det. [5>LEL]	0 PCB det.	42.0	45	NA	**8***	ENRICHMENT
N. Branch Dst. AK 004 (RM 0.1)	**NON**	26	1	0	YES	Recovering Channel	13/7 [0>max.]	0 PAH	Non/Slight	0 PAH det.	0 PCB det.	52.5	48	NA	**VP***	**TOXIC**
Dst. N. Branch & AK 004 (RM 5.0/4.7)	**NON**	—	—	—	—	Maintained Channel	12/6 [0>max.]	0 PAH	ELEVATED (As, Cr)	0 PAH det.	0 PCB det.	44	43	NA	**6***	ENRICHMENT/ **TOXIC**
Dst. Shakers Cr. (RM 4.4/4.1)	**NON**	—	—	—	—	Recovering Channel	—	—	—	—	—	58.5	41	9.7	**P***	**TOXIC**
Dst. AK 015 (RM 3.9)	**[NON]**	1	1	Jun 29, 1995 (10K+)	NO	—	—	—	—	—	—	—	—	—	**8***	**TOXIC**
Dst. AK 003 (RM 3.0/3.7)	**NON**	8	5	Jul 26, 1995 (12K+)	YES/NO	Maintained	23/16 **[10>max;3**	17 PAH (36% det.)	**EXTREME** (Cr)	4 PAH det. [3>LEL]	0 PCB det.	40.0	30/**22***	5.8/**5.6***	**12***	**TOXIC**
Dst. AK 002 (RM 2.8)	**[NON]**	5	2	0	NO	—	—	—	—	—	—	—	—	—	**12***	**TOXIC**
Dst. Landfill Trib. (RM 2.6)	**NON**	0	0	0	—	Maintained Channel	12/10 **1>max.**	6 PAH (42% det.)	**HIGHLY** (Cr)	9 PAH det. [8>LEL]	**EXTREME**	52.0	34/**14***	7.7/**4.1***	**8***	**TOXIC**
Dicks Creek - WWH Use Designation																
Amanda Elem. (RM 2.4/1.7)	**NON**	—	0	0	—	Recovering Channel	—	—	—	—	—	62.5	28/**12***	**4.4*/2.1***	16*	**TOXIC**
Ust. Mouth (RM 0.4/0.2)	**NON**	—	—	—	—	Natural Channel	8/7 [0>max.]	1 PAH (1 of 3 det.)	**HIGHLY** (Cd)	9 PAH det. [8>LEL]	**EXTREME**	72.5	30*/**12***	6.9*/**1.5***	20*	ENRICHMENT

FOOTNOTES:

[a] Attainment status based on one organism group is parenthetically expressed.
[b] Includes conventional, nutrients, demand, and heavy metals parameters; results given as number of exceedences/parameters of chronic aquatic concentration (CAC; 30 day average criterion), criterion maximum concentration (CC, outside mixing zone maximum), and final acute value (FAV) and number of parameters with exceedences; exceedences of CMC and FAV are further highlighted.
[c] IBI and MIwb results given for individual sampling pass in area affected by the July 26, 1995 fish kill.
[d] Narrative evaluation used in lieu of ICI in flow limited situations or in the absenceof an ICI value (E = Exceptional; G = good; MG = Marginally Good; F = Fair; P = Poor; VP = Very Poor).
[e] Biological Response Signature (BRS) based on macroinvertebrate community composition and response.
[*] Significant departure from ecoregional biocriteria; poor and very poor results are in boldface type and underlined.
[ns] Nonsignificant departure from ecoregional biocriteria for WWH only (4 IBI or ICI units; 0.5 MIwb units); does not apply to MWH use.

FIGURE 3.15 Environmental indicators matrix for the Dicks Creek mainstem based on data and information compiled from Ohio EPA (1997). Darker shading indicates the severity of departure from criteria or the severity of the impact implicated by the results.

further supported these observations. The integration of biological assessment information can have the effect of redirecting water quality management toward the problems and stressors that are truly limiting, as opposed to those that might be sought based on an incomplete, administrative and stressor-driven process.

The indicators matrix approach employs data from a variety of sources and necessarily includes information that might not be regarded as sampling data. Furthermore, the data are sometimes reduced to qualitative indicators of severity that are not as amenable to quantitative correlation with the biological results. This does not make them any less valuable for interpretation and diagnostic purposes. One result of including them may be a more complete use of the biological data. For example, Norton et al. (2000) used only chemical/physical data in their analysis of stress responses by the biological data, much of which overlaps with the data used in our analyses. Yet one of the most important biological response attributes of this study and the one by Yoder and Rankin (1995a), percent DELT anomalies, was not a significant correlate in their results.

We believe the failure to detect such an important biological response signature was the result of an incomplete array of stress and exposure indicators, rather than the lack of response in the percent DELT metric. Thus, care must be taken to include as many stress and exposure variables as possible and then question potential omissions if biologically meaningful responses are not triggered by correlation with the initially selected set of stressor variables. This was recognized by Norton et al. (2000) and points to the need to continue to develop and refine efforts to increase the predictability of stress and exposure indicators and information. It also raises questions about how precise stress and exposure linkages to response must be in order to guide and confirm the adequacy and success of pollution abatement strategies. Do we need to pinpoint individual pollutants before water quality management can be developed and implemented or are general assignments of impact types adequate to guide this process? Certainly there are advantages if the latter is an adequate approach.

Biological assessment information, if used within an adequate monitoring and assessment architecture, provides information that is largely complementary to administrative processes and the role of chemical/physical indicators in this process (Table 3.2). The complementary use of stress, exposure, and response indicators strengthens water quality management and provides safeguards against unintentional bias that can lead to under- or over-protective actions. There is an urgent need to advance biological assessment and criteria beyond use as simple screening or pass/fail

TABLE 3.2
Strengths and Limitations of Chemical and Biological Assessments

Attribute	Chemical-Based	Bioassessment-Based
Expressed in WQS as	Parameter-specific criteria	Biological criteria
Representation of biointegrity	Surrogate measure	Direct measure
Principal focus	Pollutant focused	Resource focused
Breadth of coverage[a]	Partial	Complete
Operative direction	Bottom-up approach	Top-down approach
Effect properties	Individual effects	Cumulative effects
Indicator role	Stressor/exposure	Response
Best strength	Design criteria	Impact assessment criteria

[a] Based on the five factors that determine the integrity of water resources.

Source: National Research Council. 2001. *Assessing the TMDL Approach to Water Quality Management*. National Academy Press, Washington, D.C.

tools to include vital diagnoses and in communicating incremental changes in response to management actions. Certainly, the level of biological assessment demonstrated by this analysis is the minimum needed by state and other agencies to successfully support the growing demands of the WQS and TMDL processes.

ACKNOWLEDGMENTS

The database on which this analysis is based is the result of many years of dedication and long hours by Ohio EPA staff and interns of the Ecological Assessment Section (EAS) and district water quality groups, who are too numerous to mention individually. Key members of EAS (past and present) who contributed to the process of biological criteria and assessment development during the past two decades include Marc Smith, Ed Rankin, Dave Altfater, Jack Freda, Mike Bolton, Roger Thoma, Randy Sanders, Chuck McKnight, Chuck Boucher, Bob Miltner, Brian Alsdorf, Marty Knapp, Dennis Mishne, and Ed Moore. We also thank the management of Ohio EPA's Division of Surface Water for consistent support of adequate monitoring and assessment. This chapter is dedicated to the memory of the late Bernie Counts, who dedicated his career to the advancement of biological assessment.

REFERENCES

Davis, W.S. and T.P. Simon (Eds.). 1995. *Biological Assessment and Criteria: Tools for Water Resource Planning and Decision Making*. Lewis Publishers, Boca Raton, FL.

DeShon, J.E. 1995. Development and application of the invertebrate community index (ICI), in W.S. Davis and T. Simon (Eds.). *Biological Assessment and Criteria: Tools for Water Resource Planning and Decision Making*. Lewis Publishers, Boca Raton, FL, 217–244.

Eagleson, K.W., D.L. Lenat, L.W. Ausley, and F.B. Winborne. 1990. Comparison of measured instream biological responses with responses predicted using the *Ceriodaphnia dubia* chronic toxicity test, *Environmental Toxicology and Chemistry*, 9, 1091–1028.

Fausch, D.O., Karr, J.R. and P.R. Yant. 1984. Regional application of an index of biotic integrity based on stream fish communities, *Transactions of the American Fisheries Society*, 113, 39–55.

Gammon, J.R. 1976. The Fish Populations of the Middle 340 km of the Wabash River. Technical Report 86, Purdue University, Water Resources Research Center, West Lafayette, IN.

Gammon, J.R., A. Spacie, J.L. Hamelink, and R.L. Kaesler. 1981. Role of electrofishing in assessing environmental quality of the Wabash River, in J.M. Bates and C.I. Weber (Eds.), *Ecological Assessments of Effluent Impacts on Communities of Indigenous Aquatic Organisms*. ASTM STP 703, Philadelphia, PA, 307–324.

Houck, O.A. 1999. *The Clean Water Act TMDL Program: Law, Policy, and Implementation*. Environmental Law Institute, Washington, D.C.

Hughes, R.M., D.P. Larsen, and J.M. Omernik. 1986. Regional reference sites: a method for assessing stream pollution, *Environmental Management*, 10, 629–635.

Karr, J.R. 1981. Assessment of biotic integrity using fish communities, *Fisheries*, 6 (6), 21–27.

Karr, J.R. 1991. Biological integrity: a long-neglected aspect of water resource management, *Ecological Applications*, 1, 66–84.

Karr, J.R. 1995. Protecting aquatic ecosystems: clean water is not enough, in W.S. Davis and T.P. Simon (Eds.), *Biological Assessment and Criteria: Tools for Water Resource Planning and Decision Making*. Lewis Publishers, Boca Raton, FL, 7–13.

Karr, J.R., K.D. Fausch, P.L. Angermier, P.R. Yant, and I.J. Schlosser. 1986. *Assessing Biological Integrity in Running Waters: A Method and Its Rationale*. Illinois Natural History Survey Special Publication 5, Champaign, IL.

Karr, J.R. and E.W. Chu. 1998. *Restoring Life in Running Waters: Better Biological Monitoring*. Island Press, Washington, D.C.

Loeb, S.L. and A. Spacie. 1994. *Biological Monitoring of Aquatic Systems*. CRC Press, Boca Raton, FL.

Majumder, S., F. Fulk, and S. Courmier. 2001. Assessing the effects of environmental stressors on macroinvertebrate indicators in Ohio. 22nd Meeting of the Society of Environmental Toxicologists and Chemists, Baltimore, MD. Poster presentation.

Miner, R. and D. Borton. 1991. Considerations in the development and implementation of biocriteria, in *Water Quality Standards for the 21st Century*, U.S. Environmental Protection Agency, Office of Science and Technology, Washington, D.C.

National Research Council. 2001. *Assessing the TMDL Approach to Water Quality Management*. National Academy Press, Washington, D.C.

Norton, S.B., S.M. Courmier, M. Smith, and R.C. Jones. 2000. Can biological assessments discriminate among types of stress? A case study from the Eastern Corn Belt Plains Ecoregion, *Environmental Toxicology and Chemistry*, 19, 1113–1119.

Ohio Environmental Protection Agency. 1987a. *Biological Criteria for the Protection of Aquatic Life: Volume II. Users Manual for Biological Field Assessment of Ohio Surface Waters*. Ohio EPA, Division of Water Quality Monitoring and Assessment, Surface Water Section, Columbus, OH.

Ohio Environmental Protection Agency. 1987b. *Biological Criteria for the Protection of Aquatic Life: Volume I. The Role of Biological Data in Water Quality Assessment*. Ohio EPA, Division of Water Quality Monitoring and Assessment, Surface Water Section, Columbus, OH.

Ohio Environmental Protection Agency. 1989a. *Addendum to Biological Criteria for the Protection of Aquatic Life: Users Manual for Biological Field Assessment of Ohio Surface Waters*. Ohio EPA, Division of Water Quality Planning and Assessment, Surface Water Section, Columbus, OH.

Ohio Environmental Protection Agency. 1989b. *Biological Criteria for the Protection of Aquatic Life: Volume III. Standardized Biological Field Sampling and Laboratory Methods for Assessing Fish and Macroinvertebrate Communities*. Ohio EPA, Division of Water Quality Planning and Assessment, Columbus, OH.

Ohio Environmental Protection Agency. 1997. *Biological and Water Quality Study of the Middle and Lower Great Miami River and Selected Tributaries, 1995*. MAS/1996-12-8. Ohio EPA, Division of Surface Water, Monitoring and Assessment Section, Columbus, OH.

Ohio Environmental Protection Agency. 1998. *Biological and Water Quality Study of the Ottawa River Basin (1996)*. MAS/1997-12-6. Ohio EPA, Division of Surface Water, Monitoring and Assessment Section, Columbus, OH.

Ohio Environmental Protection Agency. 1999a. *Ohio EPA Five-Year Surface Water Monitoring Strategy: 2000–2004*. MAS/1999-7-2. Ohio EPA, Division of Surface Water, Monitoring and Assessment Section, Columbus, OH.

Ohio Environmental Protection Agency. 1999b. *Biological and Water Quality Study of the Cuyahoga River and Selected Tributaries, Volume 1*. MAS/1997-12-4. Ohio EPA, Division of Surface Water, Monitoring and Assessment Section, Columbus, OH.

Ohio Environmental Protection Agency. 1999c. *Biological and Water Quality Study of the Middle Scioto River and Alum Creek*. MAS/1997-12-12. Ohio EPA, Division of Surface Water, Monitoring and Assessment Section, Columbus, OH.

Ohio Environmental Protection Agency. 2000a. *1998 Biological and Water Quality Study of the Black Fork, Clear Fork, Rocky Fork and Jerome Fork of the Mohican River and Selected Tributaries*. Technical Report 1999-12-2. Ohio EPA, Division of Surface Water, Monitoring and Assessment Section, Columbus, OH.

Ohio Environmental Protection Agency.2000b. *Assessment of the Impacts of the AK Steel Middletown Facilities in the Dicks Creek Watershed and the Great Miami River mainstem: 1980–1998*. Technical Report EAS/2000-12-5. Ohio EPA, Division of Surface Water, Monitoring and Assessment Section, Columbus, OH.

Rankin, E.T. 1989. *The Qualitative Habitat Evaluation Index (QHEI): Rationale, Methods, and Application*. Ohio EPA, Division of Water Quality Planning and Assessment, Columbus, OH.

Rankin, E.T. 1995. Habitat indices in water resource quality assessments, in W.S. Davis and T.P. Simon (Eds.), *Biological Assessment and Criteria: Tools for Water Resource Planning and Decision Making*, Lewis Publishers, Boca Raton, FL, 181–208.

Rosenberg, D.M. and V.H. Resh (Eds.). 1999. *Freshwater Biomonitoring and Benthic Macroinvertebrates*. Chapman & Hall, New York, NY.

Simon, T.P. (Ed.). 1999. *Assessing the Sustainability and Biological Integrity of Water Resources Using Fish Communities*. CRC Press, Boca Raton, FL.

Suter, G.W., II. 1993. A critique of ecosystem health concepts and indexes, *Environmental Toxicology and Chemistry*, 12, 1533–1539.

U.S. Environmental Protection Agency. 1985. *Technical Support Document for Water Quality-Based Toxics Control*. EPA/44/4-85/03, USEPA, Office of Water, Washington, D.C.

U.S. Environmental Protection Agency. 1995. *A Conceptual Framework to Support Development and Use of Environmental Information in Decision-Making*. EPA 239-R-95-012.USEPA, Office of Policy, Planning, and Evaluation, Washington, D.C.

U.S. Environmental Protection Agency. 2000. *Stressor Identification Guidance Document*. EPA 822/B-00/025.USEPA, Office of Water, Washington, D.C.

Yoder, C.O. 1995. Policy issues and management applications of biological criteria, in W.S. Davis and T.P. Simon (Eds.). *Biological Assessment and Criteria: Tools for Water Resource Planning and Decision Making*. Lewis Publishers, Boca Raton, FL, 327–344.

Yoder, C.O. 1998. Important concepts and elements of an adequate state watershed monitoring and assessment program, in *Proceedings of the National Water Quality Monitoring Conference, National Conference on Monitoring: Critical Foundations to Protecting Our Waters*. U.S. Environmental Protection Agency, Washington, D.C, 615–628.

Yoder, C.O. and E.T. Rankin. 1995a. Biological response signatures and the area of degradation value: new tools for interpreting multi-metric data, in W.S. Davis and T.P. Simon (Eds.), *Biological Assessment and Criteria: Tools for Water Resource Planning and Decision Making*. Lewis Publishers, Boca Raton, FL, 263–286.

Yoder, C.O. and E.T. Rankin. 1995b. Biological criteria program development and implementation in Ohio, in W.S. Davis and T.P. Simon (Eds.), *Biological Assessment and Criteria: Tools for Water Resource Planning and Decision Making*. Lewis Publishers, Boca Raton, FL, 109–144.

Yoder, C.O. and E.T. Rankin. 1998. The role of biological indicators in a state water quality management process, *Journal of Environmental Monitoring and Assessment*, 51, 61–88.

APPENDIX TABLE 1
Expression of Toxic Response Signatures (Boldface) among Selected Metrics and Aggregations of Macroinvertebrate Assemblage Data in Selected Ohio Rivers and Streams Using Biological Response Signatures Described by Yoder and Rankin (1995). Organic/Nutrient/DO Response Signatures are Underlined.

Location RM (Source)	ICI	Qual. EPT	Percent Cricotopus	% Toxic Tolerant	% Organic/Nutrient/DO	#Toxic Signatures
Scioto River - 1996						
136.3	48	15	0.0	0.0	3.3	0
129.0 (CSO)	22	6	2.9	8.1	46.5	0
127.8	**18**	5	0.0	0.0	26.4	1/4
126.5 (Jackson Pike)	22	**4**	3.6	22.4	8.9	1/4
123.2	32	6	1.5	6.7	14.5	0
119.3	36	9	0.0	0.9	8.8	0
117.3 (Southerly STP)	46	11	0.5	0.5	6.7	0
116.3 (Picway EGS)	40	13	2.6	5.2	24.6	0
114.0	44	12	1.5	1.9	7.3	0
109.4	54	15	0.4	0.4	4.5	0
106.0	54	17	0.0	0.0	1.7	0
102.0	54	17	0.0	0.0	1.3	0
100.0	56	16	0.2	0.2	2.1	0
Scioto River - 1991						
136.3	42	12	0.2	0.2	3.6	0
133.4	22	10	0.0	14.3	40.7	0
129.0 (CSO)	**18**	**4**	4.2	31.4	15.3	2/4
127.8	**10**	**4**	0.0	2.5	85.2	2/4
126.5 (Jackson Pike)	**12**	5	2.4	4.7	91.7	1/4
119.3	42	16	3.1	4.1	25.0	0
117.3 (Southerly STP)	36	15	0.0	0.0	6.6	0
114.0	44	13	1.3	1.6	6.6	0
109.4	44	15	4.4	4.4	12.4	0
106.0	44	14	1.7	1.7	3.2	0
102.0	44	18	2.3	2.3	6.6	0
Scioto River - 1988						
133.0	**14**	**2**	0.0	24.6	67.5	2/4
131.1	**6**	**2**	0.0	3.0	93.2	2/4
129.0 (CSO)	**6**	**2**	0.0	12.6	76.3	2/4
127.7	**0**	**0**	0.0	9.5	89.4	2/4
126.5 (Jackson Pike)	**4**	**1**	0.0	0.7	94.3	2/4
125.5	**2**	**0**	0.0	6.8	72.6	2/4
119.3	**14**	**3**	1.9	1.9	46.6	2/4
117.3 (Southerly STP)	**10**	**0**	2.5	3.4	35.9	2/4
114.0	**16**	5	0.0	1.0	40.8	1/4
109.4	32	6	0.0	3.2	11.9	0
102.0	46	16	0.0	0.0	0.6	0
97.9 (CCA)	48	16	0.0	0.6	6.3	0
Scioto River - 1986						
129.0 (CSO)	**14**	7	4.8	17.4	36.0	1/4
125.5 (Jackson Pike)	**6**	**0**	**9.7**	15.4	68.3	**3/4**

APPENDIX TABLE 1 (CONTINUED)
Expression of Toxic Response Signatures (Boldface) among Selected Metrics and Aggregations of Macroinvertebrate Assemblage Data in Selected Ohio Rivers and Streams Using Biological Response Signatures Described by Yoder and Rankin (1995). Organic/Nutrient/DO Response Signatures are Underlined.

Location RM (Source)	ICI	Qual. EPT	Percent Cricotopus	% Toxic Tolerant	% Organic/Nutrient/DO	#Toxic Signatures
Scioto River - 1986						
119.1	**12**	**2**	2.1	6.4	<u>44.2</u>	2/4
117.3 (Southerly STP)	**16**	**2**	2.4	3.5	4.1	2/4
114.0	26	7	0.0	1.6	<u>49.2</u>	0
111.9	28	9	1.0	4.1	10.6	0
109.3	38	10	0.0	0.0	21.5	0
102.0	46	16	0.0	0.0	5.2	0
100.0	46	21	0.0	0.0	21.0	0
Scioto River - 1981						
136.7	46	5	0.5	0.5	2.6	0
133.0	30	**3**	4.7	5.5	11.1	1/4
131.1E	**16**	**2**	4.2	10.5	<u>53.2</u>	2/4
131.1W	22	**1**	3.6	8.3	34.2	1/4
129.0 (CSO)	26	**1**	0.6	2.4	11.6	1/4
124.5 (Jackson Pike)	**10**	**1**	**24.5**	26.3	<u>56.6</u>	**3/4**
117.3 (Southerly STP)	**10**	**1**	**21.7**	21.7	<u>46.3</u>	**3/4**
116.3E (Picway EGS)	30	5	3.3	4.6	23.7	0
116.3W	28	6	3.6	4.2	16.4	0
101.4E	42	9	0.0	1.7	2.4	0
101.4W	46	12	0.0	0.0	0.1	0
98.4E	42	9	0.2	0.5	6.8	0
98.4W	38	7	0.4	0.4	3.5	0
Paint Creek - 1997						
7.8	52	24	0.7	1.5	3.3	0
4.4	56	27	0.0	0.0	1.2	0
1.9 (Mead Paper)	42	20	0.0	0.0	5.4	0
1.3	46	24	1.4	1.7	0.1	0
0.2	44	19	0.5	0.9	1.1	0
Paint Creek - 1992						
8.9	58	29	0.7	0.7	1.7	0
7.8	56	29	1.3	1.7	0.5	0
5.2	56	25	0.0	0.2	0.5	0
1.9 (Mead Paper)	28	15	0.0	0.9	<u>40.8</u>	0
0.1	32	19	7.0	7.0	14.9	0
Paint Creek - 1985						
5.1	56	17	0.5	1.0	1.3	0
1.8 (Mead Paper)	**12**	9	0.0	0.0	<u>84.3</u>	1/4
0.1	32	6	1.7	1.7	15.3	0

APPENDIX TABLE 1 (CONTINUED)
Expression of Toxic Response Signatures (Boldface) among Selected Metrics and Aggregations of Macroinvertebrate Assemblage Data in Selected Ohio Rivers and Streams Using Biological Response Signatures Described by Yoder and Rankin (1995). Organic/ Nutrient/DO Response Signatures are Underlined.

Location RM (Source)	ICI	Qual. EPT	Percent Cricotopus	% Toxic Tolerant	% Organic/Nutrient/DO	#Toxic Signatures
Ottawa River - 1996						
45.9	48	16	0.3	0.3	5.8	0
44.5	40	13	0.3	1.2	4.9	0
43.4	40	13	0.1	0.1	0.8	0
41.2	42	13	0.2	0.5	2.7	0
40.1	20	8	**5.0**	5.0	<u>44.9</u>	1/4
39.6 (CSO)	34	12	0.9	5.2	8.7	0
38.6 (CSOs)	**6**	**2**	0.0	26.5	<u>62.1</u>	2/4
37.9 (CSOs)	28	8	0.3	5.8	14.8	0
37.7 (L-5 landfill)	**14**	**4**	0.0	18.6	<u>49.6</u>	2/4
37.4 (Lima STP)	42	8	3.3	4.7	5.2	0
37.0 (Refinery)	20	**4**	**16.2**	20.2	6.6	2/4
36.1 (PCS nitrogen)	22	**4**	**10.8**	21.0	0.8	2/4
34.5	38	9	**6.0**	8.3	1.8	1/4
32.6	32	8	**19.4**	21.0	9.3	1/4
28.8	32	10	2.2	2.2	0.0	0
Ottawa River - 1991						
45.9	36	15	0.0	4.2	<u>40.5</u>	0
37.8 (CSOs)	**6**	6	2.2	2.2	<u>94.6</u>	1/4
37.4 (Lima STP)	**8**	**1**	**54.9**	**57.0**	30.6	**4/4**
37.0 (Refinery)	**14**	**2**	**39.8**	**41.5**	19.5	**4/4**
36.1 (PCS nitrogen)	**10**	**2**	**42.7**	**50.1**	3.6	**4/4**
34.5	**18**	**3**	**6.3**	18.5	4.8	**3/4**
28.8	28	8	2.2	9.4	23.8	0
Ottawa River - 1989						
45.9	48	12	0.0	1.5	2.3	0
41.2	32	11	0.6	21.7	3.5	0
37.9 (CSOs)	**16**	**2**	**5.8**	22.6	<u>60.8</u>	**3/4**
37.4 (Lima STP)	**10**	**0**	**41.8**	**44.4**	<u>35.3</u>	**4/4**
37.0 (Refinery)	**12**	**1**	**15.4**	27.9	18.0	**3/4**
36.8	**14**	**1**	**10.6**	15.9	12.8	**3/4**
36.1A (PCS nitrogen)	**16**	**2**	3.1	6.8	6.5	2/4
36.1B (PCS nitrogen)	**14**	**2**	3.9	8.1	6.4	2/4
34.5	30	8	0.0	5.7	1.5	0
30.1	26	9	0.5	8.2	5.1	0
28.8	34	8	0.0	5.2	2.4	0
Ottawa River - 1985						
45.9	42	10	0.4	5.4	5.1	0
41.1	**16**	**4**	3.5	33.8	17.8	2/4
38.6 (CSOs)	**2**	**2**	0.0	13.1	<u>86.4</u>	2/4
37.8 (CSOs)	**6**	**4**	0.8	17.6	<u>75.0</u>	2/4
37.3 (Lima STP)	**6**	**0**	**19.9**	30.4	25.0	**3/4**
37.0 (Refinery)	**6**	**0**	**33.5**	**65.1**	13.4	**4/4**

APPENDIX TABLE 1 (CONTINUED)

Expression of Toxic Response Signatures (Boldface) among Selected Metrics and Aggregations of Macroinvertebrate Assemblage Data in Selected Ohio Rivers and Streams Using Biological Response Signatures Described by Yoder and Rankin (1995). Organic/Nutrient/DO Response Signatures are Underlined.

Location RM (Source)	ICI	Qual. EPT	Percent Cricotopus	% Toxic Tolerant	% Organic/Nutrient/DO	#Toxic Signatures
36.1 (PCS nitrogen)	**4**	**1**	**20.2**	**39.0**	<u>41.5</u>	**4/4**
34.5	**6**	**1**	**16.2**	**43.4**	31.0	**4/4**
32.6	**6**	**1**	**37.3**	**55.4**	25.8	**4/4**
28.8	22	**4**	4.4	26.2	6.5	1/4
Dicks Creek - 2000						
4.9 (AK Steel 004)	24	5	**44.9**	**50.7**	27.5	2/4
4.1 (AK Steel 015)	30	6	**46.2**	**55.9**	4.2	1/4
3.9 (AK Steel 005)	**18**	**1**	**52.8**	**58.3**	18.7	**4/4**
3.7 (AK Steel *003*)	20	**4**	**52.0**	**57.8**	18.3	**3/4**
2.8 (AK Steel 002)	30	6	**44.3**	**50.0**	10.4	2/4
2.6 (AK Steel 006)	26	8	**23.4**	30.9	22.3	1/4
1.7	34	9	**32.8**	**39.2**	3.3	2/4
0.9	38	9	**12.8**	13.3	11.8	1/4
0.2	34	10	1.3	5.7	3.4	0
Dicks Creek - 1995						
4.7 (AK Steel 004)	**6**	**4**	4.9	9.1	<u>79.5</u>	2/4
3.9 (AK Steel 005)	**8**	**1**	**32.0**	**63.2**	21.9	**4/4**
3.7 (AK Steel 003)	**12**	**0**	**16.2**	**70.3**	11.4	**4/4**
2.8 (AK Steel 002)	**12**	**4**	**15.8**	**72.3**	15.7	**4/4**
2.6 (AK Steel 006)	**8**	**3**	**21.9**	**85.0**	7.1	**4/4**
1.7	**16**	7	**26.8**	**59.8**	11.7	**3/4**
0.2	20	5	**10.6**	14.1	<u>53.8</u>	1/4
Dicks Creek - 1987						
4.7 (AK Steel 004)	30	10	**15.0**	17.8	34.8	1/4
4.1 (AK Steel 015)	**16**	**4**	4.7	8.9	<u>67.0</u>	2/4
3.6 (AK Steel 005/003)	**4**	**1**	**10.8**	24.8	<u>63.9</u>	**3/4**
2.7 (AK Steel 002)	**10**	**1**	**23.6**	26.8	<u>66.0</u>	**3/4**
1.7 (AK Steel 006)	**8**	**1**	**10.9**	27.7	<u>37.6</u>	**3/4**
0.2	22	6	1.7	**39.0**	<u>41.0</u>	1/4
Rocky Fork of the Mohican River - 1998						
16.3	46	9	2.3	2.3	14.0	0
15.8 (Scrap yards)	22	**4**	1.9	2.8	34.5	1/4
14.2 (Armco 002)	26	7	**17.9**	24.5	18.2	1/4
0.6	36	7	0.6	0.6	2.2	0
Rocky Fork of the Mohican River - 1993						
16.3	50	8	1.1	2.8	2.1	0
16.1	42	8	1.0	4.9	4.6	0
15.8 (Scrap yards)	28	**1**	**12.3**	17.5	25.6	2/4
14.2 (Armco 002)	**0**	**0**	0.5	0.5	<u>93.2</u>	2/4
13.3	**4**	**1**	**20.0**	20.3	<u>72.9</u>	**3/4**
11.5	**16**	5	**22.5**	**43.7**	21.7	**3/4**
11.1 (Mansfield STP)	**12**	**2**	**43.5**	**50.9**	21.0	**4/4**

APPENDIX TABLE 1 (CONTINUED)

Expression of Toxic Response Signatures (Boldface) among Selected Metrics and Aggregations of Macroinvertebrate Assemblage Data in Selected Ohio Rivers and Streams Using Biological Response Signatures Described by Yoder and Rankin (1995). Organic/Nutrient/DO Response Signatures are Underlined.

Location RM (Source)	ICI	Qual. EPT	Percent Cricotopus	% Toxic Tolerant	% Organic/Nutrient/DO	#Toxic Signatures
10.2	**16**	**4**	**24.2**	**39.1**	5.0	**4/4**
6.4	46	9	2.8	4.7	1.6	0
0.7	46	12	0.4	5.0	2.2	0
Cuyahoga River - 2000						
42.8	38	5	0.0	0.0	0.9	0
39.7 (L. Cuyahoga R.)	34	6	0.0	2.8	1.5	0
33.3 (Akron STP)	42	7	**6.9**	6.9	9.0	1/4
29.2	32	9	1.3	2.5	6.6	0
15.6	44	9	1.0	1.6	4.6	0
8.3 (Southerly STP)	38	8	0.7	1.0	14.4	0
7.1 (Big Cr.)	42	9	0.0	0.0	6.2	0
Cuyahoga River - 1996						
42.8	40	10	0.0	0.0	4.0	0
41.9 (L. Cuyahoga R.)	34	6	0.9	0.9	3.8	0
38.0	32	8	0.0	0.8	11.2	0
37.1 (Akron STP)	26	7	0.6	1.7	23.5	0
35.3	28	8	0.7	0.7	20.8	0
33.3	24	6	2.8	2.8	29.4	0
26.5	36	6	4.1	4.1	6.0	0
20.7	46	8	1.8	2.4	1.9	0
17.0	42	12	2.7	2.7	6.6	0
15.6	44	8	0.9	0.9	1.4	0
11.0	46	8	1.5	1.8	3.2	0
9.7 (Southerly STP)	36	6	0.8	1.5	17.4	0
8.3	36	9	0.0	0.3	10.2	0
7.1 (Big Cr.)	24	8	**8.6**	9.9	17.0	1/4
Cuyahoga River - 1991						
42.8	28	8	3.0	3.0	16.0	0
39.7 (L. Cuyahoga R.)	32	6	1.4	2.1	14.6	0
37.2 (Akron STP)	26	6	**5.7**	9.0	15.3	1/4
35.3	32	5	1.9	3.7	12.6	0
26.5	36	7	0.4	2.3	5.5	0
15.6	42	12	0.3	1.5	0.5	0
11.0	42	13	0.7	1.3	5.2	0
10.4 (Southerly STP)	36	10	0.3	1.6	7.2	0
8.5	34	8	0.0	2.5	9.6	0
8.2	32	11	0.9	1.3	23.2	0
7.3	34	13	0.4	2.1	21.6	0
7.1 (Big Cr.)	34	12	**5.9**	9.6	9.8	1/4
Cuyahoga River - 1988						
42.8	26	5	2.6	3.5	12.6	0
39.7 (L. Cuyahoga R.)	24	**4**	**5.7**	9.3	32.1	2/4

APPENDIX TABLE 1 (CONTINUED)
Expression of Toxic Response Signatures (Boldface) among Selected Metrics and Aggregations of Macroinvertebrate Assemblage Data in Selected Ohio Rivers and Streams Using Biological Response Signatures Described by Yoder and Rankin (1995). Organic/Nutrient/DO Response Signatures are Underlined.

Location RM (Source)	ICI	Qual. EPT	Percent Cricotopus	% Toxic Tolerant	% Organic/Nutrient/DO	#Toxic Signatures
35.3 (Akron STP)	24	**4**	3.5	8.4	33.5	1/4
29.0	26	5	4.4	5.8	29.9	0
15.6	36	8	0.7	2.6	9.6	0
11.4	30	7	0.4	0.4	20.7	0
7.1 (Big Cr.)	24	6	**35.8**	**37.6**	41.7	2/4
Cuyahoga River - 1987						
42.8	**12**	**4**	1.3	6.9	73.7	2/4
39.7 (L. Cuyahoga R.)	**22**	**4**	2.8	5.7	22.0	2/4
37.2 (Akron STP)	24	**3**	**11.1**	16.3	24.7	2/4
35.3	24	6	**19.4**	22.7	12.9	1/4
33.2	22	**4**	**24.2**	26.9	9.2	2/4
29.0	24	6	**8.5**	8.5	10.8	1/4
24.1	26	9	**9.5**	16.1	18.2	1/4
20.7	34	9	3.6	3.6	3.5	0
17.3	32	6	4.2	4.7	5.0	0
15.6	36	9	**5.2**	5.9	1.4	1/4
13.1	34	5	1.7	2.4	11.3	0
12.0	26	5	2.8	6.3	10.1	0
11.4	34	8	0.8	1.0	16.7	0
9.5 (Southerly STP)	26	6	1.5	2.0	19.6	0
7.3	20	7	4.7	9.3	13.3	0
7.1 (Big Cr.)	**16**	**3**	**48.3**	**54.6**	5.7	**4/4**
Cuyahoga River - 1986						
42.8	30	8	4.2	5.9	29.6	0
39.7 (L. Cuyahoga R.)	**20**	6	0.0	**49.3**	17.1	1/4
35.3 (Akron STP)	**8**	6	**21.7**	**54.4**	30.1	1/4
33.2	**16**	5	**34.9**	**67.6**	10.8	1/4
29.0	24	9	**16.1**	19.2	5.7	0
24.1	28	7	**5.3**	9.7	2.8	0
15.6	28	10	3.5	5.9	2.9	0
Cuyahoga River - 1984						
42.8	32	**4**	0.0	0.4	6.8	1/4
40.2 (L. Cuyahoga R.)	24	6	**12.5**	24.9	13.3	1/4
37.2 (Akron STP)	**14**	**1**	**10.0**	**36.4**	50.3	**4/4**
35.3	**10**	**1**	3.0	6.7	80.6	2/4
33.2	**10**	**1**	3.4	9.0	63.9	2/4
29.0	**16**	**2**	3.1	5.6	69.1	2/4
20.7	**20**	5	0.8	6.4	67.2	1/4
17.3	**16**	**2**	1.6	4.8	54.4	2/4
15.6	**22**	5	0.7	2.7	68.9	1/4
13.1	**12**	**2**	0.6	5.2	31.7	2/4
9.5 (Southerly STP)	**12**	**4**	**11.4**	15.2	32.7	**3/4**

APPENDIX TABLE 2
Expression of Toxic Response Signatures among Selected Metrics and Aggregations of Fish Assemblage Data in Wadeable Ohio Streams Using Biological Response Signatures Described by Yoder and Rankin (1995)

Location RM (Source)	IBI	Modified Iwb	%DELT Anomalies	Percent Tolerant	Intolerant Species[a]	Density[b] (No./300m)	Darter Sp.[c]	#Toxic Signatures[d]
Dicks Creek - 2000								
6.4	26	NA[e]	0.0	67	**0**	525	1	1/6
5.5	26	NA	0.0	**71**	**0**	686	1	2/6
5.0 (AK 004)	40	NA	0.0	13	3	1430	**0**	1/6
5.0 (AK 004)	40	NA	0.0	19	3	1906	1	0
4.2	34	**5.7**	1.4	23	**0**	**114**	**0**	4/7
4.2	36	7.5	0.0	19	**0**	500	1	1/7
3.9 (AK 005)	32	7.5	0.0	38	1	350	**0**	1/7
3.9 (AK 005)	32	8.4	0.0	28	1	1546	**0**	1/7
3.7 (AK 003)	36	8.4	0.0	45	1	404	3	0
3.7 (AK 003)	36	8.9	0.0	29	2	1022	1	0
2.8 (AK 002)	38	8.7	0.0	43	2	856	3	0
2.8 (AK 002)	46	9.6	0.0	27	3	2202	3	0
2.6 (AK 006)	24	**5.2**	0.0	65	**0**	**66**	**0**	4/7
2.6 (AK 006)	28	8.0	0.0	49	2	478	**0**	1/7
1.7	30	7.7	0.6	48	1	**126**	1	1/7
1,7	36	7.9	0.3	34	1	294	3	0
0.9	48	10.0	0.0	22	4	1672	4	0
0.9	44	10.2	0.0	24	3	2704	3	0
0.4	50	10.0	0.0	19	2	942	4	0
0.4	50	9.4	0.0	25	3	702	3	0
Dicks Creek - 1998								
6.3	36	NA	0.0	54	**0**	1220	2	1/6
5.0 (AK 004)	52	NA	0.0	21	5	1398	5	0
4.4	26	--	0.0	51	5	174	2	0
3.0 (AK 003)	28	--	0.0	40	3	221	1	0
2.6 (AK 006)	44	--	0.0	42	8	342	4	0
0.4	42	--	0.0	26	8	297	4	0
Dicks Creek - 1995								
5.0 (AK 004)	42	NA	0.9	17	4	376	2	0
5.0 (AK 004)	42	NA	1.3	17	7	222	2	0
4.4	42	9.7	1.4	18	8	878	2	0
4.4	40	9.7	1.0	20	6	1092	2	0
3.0 (AK 003)	30	**5.8**	9.4	51	5	**83**	**0**	4/7
3.0 (AK 003)	**20**	**5.6**	1.4	18	2	**104**	**0**	4/7
2.6 (AK 006)	34	7.7	3.0	46	6	**130**	**0**	2/7
2.6 (AK 006)	**14**	**4.1**	0.0	33	**0**	**16**	**0**	5/7
2.4	28	**4.4**	**11.0**	**72**	5	**40**	1	4/7
2.4	**12**	**2.1**	0.0	**86**	**0**	**2**	**0**	**6/7**
0.4	30	6.9	1.7	59	9	**118**	4	1/7
0.4	**12**	**1.5**	**25.0**	25	**0**	**5**	**0**	**6/7**

APPENDIX TABLE 2 (CONTINUED)
Expression of Toxic Response Signatures among Selected Metrics and Aggregations of Fish Assemblage Data in Wadeable Ohio Streams Using Biological Response Signatures Described by Yoder and Rankin (1995)

Location RM (Source)	IBI	Modified Iwb	%DELT Anomalies	Percent Tolerant	Intolerant Species[a]	Density[b] (No./300m)	Darter Sp.[c]	#Toxic Signatures[d]
Dicks Creek - 1987								
5.5	38	NA	0.1	6.0	**0**	2288	**0**	2/6
5.5	24	NA	0.1	**70**	**0**	820	1	2/6
4.6 (AK 004)	40	NA	0.0	30	2	509	**0**	1/6
4.6 (AK 004)	38	NA	1.3	23	2	220	**0**	1/6
4.6 (AK 004)	38	NA	1.4	20	2	854	1	0
4.2	26	6.6	3.4	62	3	222	**0**	1/7
4.2	34	7.0	1.8	42	2	312	1	0
4.2	26	**5.9**	1.3	58	3	**143**	1	2/7
3.4 (AK 003)	30	7.3	0.8	31	1	613	**0**	1/7
3.4 (AK 003)	32	8.4	0.3	31	2	1347	**0**	1/7
3.4 (AK 003)	30	7.3	0.5	30	1	444	**0**	1/7
2.7 (AK 002)	28	**5.8**	3.8	52	1	232	**0**	2/7
2.7 (AK 002)	24	7.3	2.4	51	2	368	**0**	1/7
2.7 (AK 002)	28	6.9	**11.0**	36	2	228	**0**	2/7
2.5 (AK 006)	24	6.4	**12.3**	51	2	186	**0**	2/7
2.5 (AK 006)	**22**	6.2	4.0	**78**	1	**93**	**0**	4/7
2.5 (AK 006)	**18**	**3.2**	**13.3**	42	**0**	**50**	**0**	**6/7**
0.2	28	7.6	2.6	34	4	414	**0**	1/7
0.2	34	7.9	7.7	36	3	366	2	0
0.2	32	7.5	5.6	36	4	351	1	0
Rocky Fork of the Mohican River - 1998								
16.4	40	NA	0.0	**79**	3	463	3	1/6
16.4	40	NA	0.2	**70**	2	327	2	1/6
15.8 (Scrap yards)	24	NA	0.3	**88**	**0**	**69**	1	3/6
15.8 (Scrap yards)	24	NA	0.0	**91**	1	**60**	1	2/6
14.3 (Armco 002)	34	NA	0.0	**71**	2	374	3	1/6
14.3 (Armco 002)	32	NA	0.3	60	2	462	3	0
11.6	32	6.5	0.0	**77**	2	236	3	1/7
11.6	24	**5.2**	0.4	**74**	1	**98**	2	3/7
11.2 (Mansfield)	28	**5.1**	0.0	**92**	**0**	**132**	2	4/7
11.2 (Mansfield)	24	**4.9**	0.0	**96**	**0**	**78**	1	4/7
10.2	30	**5.9**	0.0	**84**	2	**141**	3	3/7
10.2	28	**5.4**	0.3	**88**	1	**105**	3	3/7
0.6	38	6.5	0.0	30	2	**110**	3	1/7
Rocky Fork of the Mohican River - 1993								
16.4	32	NA	0.0	66	2	459	2	0
16.4	32	NA	0.2	61	2	291	2	0
15.8 (Scrap yards)	**16**	NA	1.3	**85**	1	**99**	1	3/6
15.8 (Scrap yards)	**20**	NA	2.4	**82**	**0**	**96**	1	4/6
14.6	**12**	NA	**15.0**	**80**	**0**	**7**	**0**	**6/6**
14.6	**20**	NA	5.3	**87**	**0**	**20**	**0**	**5/6**

APPENDIX TABLE 2 (CONTINUED)
Expression of Toxic Response Signatures among Selected Metrics and Aggregations of Fish Assemblage Data in Wadeable Ohio Streams Using Biological Response Signatures Described by Yoder and Rankin (1995)

Location RM (Source)	IBI	Modified Iwb	%DELT Anomalies	Percent Tolerant	Intolerant Species[a]	Density[b] (No./300m)	Darter Sp.[c]	#Toxic Signatures[d]
14.45 (Armco 002)	**12**	NA	0.0	**0**	**0**	**0**	**0**	**5/6**
14.45 (Armco 002)	**12**	NA	0.0	**100**	**0**	**0**	**0**	**5/6**
14.2	**12**	NA	0.0	**0**	**0**	**0**	**0**	**5/6**
14.2	**12**	NA	0.0	**0**	**0**	**0**	**0**	**5/6**
13.4	**12**	**1.6**	0.0	33	**0**	**2**	**0**	5/7
13.4	**12**	**3.3**	**37.5**	**75**	**0**	**2**	**0**	**7/7**
11.7	**22**	**4.8**	**25.5**	**78**	**0**	**54**	1	**6/7**
11.7	**16**	**2.9**	**14.4**	**94**	**0**	**9**	**0**	**7/7**
11.1 (Mansfield)	**22**	6.2	0.9	**88**	**0**	252	2	3/7
11.1 (Mansfield)	24	7.1	1.5	**88**	**0**	612	2	2/7
10.2	26	**5.7**	**11.5**	60	**0**	390	2	3/7
10.2	**16**	**4.6**	9.6	**91**	**0**	**38**	1	5/7
6.4	28	6.6	**11.3**	64	1	338	3	1/7
6.4	24	6.6	3.4	**75**	1	318	3	1/7
0.6	42	9.5	0.6	26	4	1386	5	0
0.6	40	8.7	0.0	26	3	662	5	0

[a] Sensitive species included at headwaters sites.

[b] Metric 12 of Ohio EPA modified IBI; numbers/300m less highly tolerant species.

[c] Includes sculpins at headwaters sites.

[d] Number of metrics or attributes that reflect a toxic response after Yoder and Rankin (1995); values in boldface type are within toxic response criteria for that metric or attribute.

[e] NA: MIwb not applicable at headwaters sites.

APPENDIX TABLE 3

Expression of Toxic Response Signatures among Selected Metrics and Aggregations of Fish Assemblage Data in Boatable Ohio Rivers Using the Biological Response Signatures Described by Yoder and Rankin (1995)

Location RM (Source)	IBI	Modified Iwb	%DELT Anomalies	Percent Tolerant	Intolerant Species	Density[a] (No./300m)	% Round-Bodied	#Toxic Signatures
Scioto River - 1996								
136.2 (Upstream)	48	10.4	0.4	4	6	791	15	0
136.2 (Upstream)	52	10.3	0.1	3	5	1248	21	0
133.3	46	9.3	0.2	7	5	741	9	0
133.3	50	9.8	0.0	12	4	1066	21	0
133.0	40	9.2	0.4	14	2	826	7	0
133.0	42	9.9	0.3	15	3	1264	14	0
131.8 (Impound.)	34	7.9	0.6	9	**0**[d]	1075	5	1/7
131.8 (Impound.)	36	8.6	0.3	12	1	1187	9	0
129.1 (CSO)	52	11.2	0.3	14	4	800	5	0
129.1 (CSO)	48	11.0	0.1	13	2	931	5	0
127.5	50	10.0	0.2	13	2	772	6	0
127.5	46	10.5	0.3	10	2	843	5	0
127.2 (JP WWTP)	34	8.4	6.1	3	1	200	**4**	1/7
127.2 (JP WWTP)	34	9.8	2.7	12	1	416	**4**	1/7
126.4	42	10.4	1.3	9	1	376	**3**	1/7
126.4	46	10.3	0.5	16	3	623	**2**	1/7
123.5	38	10.4	2.5	12	**0**	366	7	1/7
123.5	42	10.4	0.9	20	1	480	10	0
119.9	46	10.6	0.7	9	2	672	5	0
119.9	44	11.1	0.4	10	1	1050	6	0
118.3 (CS WWTP)	40	9.6	**13.6**	7	**0**	488	11	2/7
118.3 (CS WWTP)	36	9.2	6.7	15	1	306	10	0
118.3 (CS WWTP)	28	10.1	7.1	3	1	912	7	0
118.1	48	10.3	0.9	9	2	502	12	0
118.1	54	10.4	0.2	6	4	753	5	0
118.1	38	10.5	0.7	9	1	676	7	0
117.1	46	10.3	1.7	9	1	375	16	0
117.1	50	10.9	1.0	10	3	535	13	0
117.1	40	10.0	2.3	4	**0**	320	13	1/7
116.3 (Pic. EGS)	44	11.0	1.2	7	2	337	20	0
116.3 (Pic. EGS)	52	11.0	1.6	9	2	469	22	0
116.3 (Pic. EGS)	48	10.6	1.9	2	3	327	21	0
113.8	46	11.1	1.3	9	2	488	15	0
113.8	42	10.9	1.8	13	1	500	10	0
113.8	44	10.4	1.6	3	1	456	10	0
109.2	42	10.7	0.9	6	1	485	9	0
109.2	46	11.3	1.2	10	3	971	10	0
109.2	40	10.8	2.7	12	1	560	10	0
107.4	40	11.0	0.6	10	0	540	12	0
107.4	50	11.4	0.3	13	5	840	11	0
107.4	40	10.3	1.3	13	**0**	546	6	1/7
105.9	54	11.6	1.1	5	6	748	22	0
105.9	42	10.6	1.1	1	3	762	8	0
105.2	40	10.5	1.2	7	2	358	16	0
105.2	42	10.8	0.6	12	1	383	12	0

APPENDIX TABLE 3 (CONTINUED)
Expression of Toxic Response Signatures among Selected Metrics and Aggregations of Fish Assemblage Data in Boatable Ohio Rivers Using the Biological Response Signatures Described by Yoder and Rankin (1995)

Location RM (Source)	IBI	Modified Iwb	%DELT Anomalies	Percent Tolerant	Intolerant Species	Density[a] (No./300m)	% Round-Bodied	#Toxic Signatures
102.0	48	9.9	0.9	8	5	358	17	0
102.0	48	10.8	0.0	7	3	413	16	0
102.0	50	10.3	0.5	5	4	302	6	0
100.0	48	11.3	1.4	12	7	530	11	0
100.0	54	10.9	0.1	4	10	1235	**4**	1/7
100.0	40	10.0	0.8	3	3	376	6	0
99.7 (CCA)	44	10.4	0.0	6	4	322	15	0
99.7 (CCA)	50	10.6	0.0	7	2	409	8	0
98.9 (Circleville)	46	10.9	0.9	4	3	446	11	0
98.9 (Circleville)	48	10.7	0.7	6	2	520	17	0
97.9	50	11.0	0.0	13	6	480	8	0
97.9	52	10.9	0.8	8	7	780	6	0
Scioto River - 1991								
136.2	50	9.1	3.1	5	3	430	22	0
136.2	52	9.8	1.4	8	4	526	15	0
133.3	40	9.7	2.3	11	2	764	8	0
133.3	48	9.4	1.3	4	4	854	17	0
133.3	50	10.3	1.9	2	3	874	24	0
133.0	38	9.6	0.7	20	1	734	**3**	1/7
133.0	48	9.8	0.8	5	2	760	7	0
131.8	34	9.0	1.2	12	1	750	**2**	1/7
129.1 (CSO)	42	9.5	4.9	9	2	710	**3**	1/7
129.1 (CSO)	42	10.0	2.5	10	3	1234	**2**	1/7
129.1 (CSO)	46	10.4	1.4	11	2	1246	**2**	1/7
127.5	42	9.0	2.8	10	1	836	**0**	1/7
127.5	48	9.7	1.7	6	2	1246	**2**	1/7
127.5	44	9.4	0.6	8	1	1050	5	0
127.1 (JP WWTP)	30	8.7	**37.7**	1	1	220	**3**	2/7
127.1 (JP WWTP)	30	8.4	5.6	6	1	240	**3**	1/7
127.1 (JP WWTP)	30	8.5	9.3	5	**0**	290	**1**	2/7
126.4	32	9.1	**13.5**	13	1	378	**0**	2/7
126.4	38	9.3	1.9	17	1	514	**0**	1/7
126.4	44	9.6	2.3	6	1	708	**1**	1/7
123.5	30	9.0	**13.0**	17	1	284	**1**	2/7
123.5	46	10.4	1.8	6	1	758	**3**	1/7
123.5	44	10.5	1.3	8	**0**	958	6	1/7
119.9	32	9.2	**16.1**	15	**0**	306	5	2/7
119.9	46	10.8	2.7	7	2	716	5	0
119.9	46	10.5	0.7	7	1	908	8	0
118.4 (CS WWTP)	24	6.5	4.1	**84**	**0**	215	**0**	3/7
118.4 (CS WWTP)	40	8.9	1.1	32	**0**	985	9	1/7
118.4 (CS WWTP)	42	9.5	1.8	13	1	1369	9	0
118.1	46	10.2	**16.2**	13	2	408	31	1/7
118.1	48	10.8	1.6	7	3	1046	17	0
118.1	54	11.3	1.6	5	2	1194	34	0
117.1	36	9.5	**16.7**	13	**0**	418	10	2/7

APPENDIX TABLE 3 (CONTINUED)
Expression of Toxic Response Signatures among Selected Metrics and Aggregations of Fish Assemblage Data in Boatable Ohio Rivers Using the Biological Response Signatures Described by Yoder and Rankin (1995)

Location RM (Source)	IBI	Modified Iwb	%DELT Anomalies	Percent Tolerant	Intolerant Species	Density[a] (No./300m)	% Round-Bodied	#Toxic Signatures
117.1	44	10.5	3.3	8	1	644	22	0
117.1	50	10.6	2.0	8	**0**	958	19	1/7
116.3 (Pic. EGS)	36	10.4	9.7	6	1	382	12	0
116.3 (Pic. EGS)	50	10.7	2.3	4	3	746	12	0
116.3 (Pic. EGS)	46	10.9	4.0	3	1	868	14	0
113.8	38	10.4	**16.2**	13	2	312	13	1/7
113.8	44	9.9	6.6	4	2	668	9	0
113.8	54	10.7	0.7	3	4	1104	9	0
109.2	46	10.8	3.5	3	1	876	8	0
109.2	44	10.3	3.7	1	1	742	16	0
109.2	50	10.6	2.4	3	2	954	19	0
107.4	48	10.2	2.1	3	1	866	16	0
107.4	50	11.3	4.3	1	4	1492	25	0
107.4	54	11.4	1.9	1	2	1428	26	0
105.2	44	9.9	7.5	5	2	380	12	0
105.2	46	10.8	1.3	1	2	1156	16	0
105,2	54	11.5	0.7	2	4	1746	22	0
102.0	54	10.3	2.8	4	7	634	21	0
102.0	56	11.2	1.9	2	7	1476	26	0
102.0	56	10.8	1.4	1	6	1042	29	0
100.2	40	9.5	3.1	9	1	358	9	0
100.2	46	10.8	3.6	3	1	986	19	0
100.2	44	9.7	2.1	9	**0**	532	10	1/7
Scioto River - 1988								
133.0	38	9.5	1.2	17	2	478	**2**	1/7
133.0	40	10.1	0.9	25	1	704	**3**	1/7
131.8	36	9.1	0.9	20	**0**	694	**0**	2/7
131.8	32	8.4	0.3	18	**0**	1110	**1**	2/7
131.0	34	8.4	0.2	20	**0**	746	**0**	2/7
131.0	34	8.9	0.4	17	1	776	**0**	1/7
129.1 (CSO)	36	9.2	1.0	13	1	730	**1**	1/7
129.1 (CSO)	42	10.0	0.0	13	**0**	848	**2**	2/7
129.1 (CSO)	38	9.4	1.1	8	**0**	644	**2**	2/7
128.6	38	8.7	0.8	14	**0**	776	**1**	2/7
128.6	38	9.0	0.2	13	**0**	866	**0**	2/7
128.6	34	9.0	0.7	10	**0**	602	**3**	2/7
127.5	28	7.0	0.2	32	1	582	**0**	1/7
127.5	32	6.6	0.3	18	**0**	796	**0**	2/7
127.5	34	8.6	0.6	20	**0**	266	**4**	2/7
126.4 (JP WWTP)	24	7.4	0.8	45	**0**	**146**	**2**	2/7
126.4 (JP WWTP)	26	7.7	1.2	30	**0**	236	**1**	2/7
126.4 (JP WWTP)	34	8.4	2.2	23	**0**	280	**2**	2/7
123.5	30	7.8	0.0	43	**0**	154	**2**	2/7
123.5	32	8.6	1.7	16	**0**	846	**2**	2/7
123.5	32	8.6	0.0	15	**0**	378	**2**	2/7
119.9	34	8.9	2.5	35	1	236	**3**	1/7

APPENDIX TABLE 3 (CONTINUED)

Expression of Toxic Response Signatures among Selected Metrics and Aggregations of Fish Assemblage Data in Boatable Ohio Rivers Using the Biological Response Signatures Described by Yoder and Rankin (1995)

Location RM (Source)	IBI	Modified Iwb	%DELT Anomalies	Percent Tolerant	Intolerant Species	Density[a] (No./300m)	% Round-Bodied	#Toxic Signatures
119.9	32	9.5	2.4	25	**0**	356	**3**	2/7
119.9	40	10.1	2.2	20	1	430	**6**	1/7
118.3 (CS WWTP)	26	8.1	7.9	32	2	404	2	0
118.3 (CS WWTP)	38	9.2	3.3	15	2	872	**3**	1/7
118.3 (CS WWTP)	36	8.9	0.9	15	1	966	**1**	1/7
118.1	32	9.3	3.6	30	**0**	236	13	1/7
118.1	38	10.2	1.6	18	1	728	7	0
118.1	36	9.6	8.3	18	1	384	19	0
117.1	28	9.0	3.9	18	**0**	306	10	1/7
117.1	38	10.0	1.3	10	**0**	674	6	1/7
117.1	40	10.3	2.6	7	**0**	564	9	1/7
116.3 (Pic. EGS)	36	9.4	3.1	11	**0**	282	9	1/7
116.3 (Pic. EGS)	40	9.6	0.4	15	**0**	394	5	1/7
116.3 (Pic. EGS)	36	9.0	3.3	10	**0**	272	5	1/7
113.8	40	8.5	0.8	10	1	226	**3**	1/7
113.8	40	9.7	0.0	11	**0**	326	**4**	2/7
113.8	42	9.0	2.7	3	**0**	290	9	1/7
109.2	34	9.5	3.3	15	**0**	406	12	1/7
109.2	40	9.8	0.8	12	**0**	690	6	1/7
107.4	36	9.5	5.6	16	1	266	23	0
107.4	42	10.2	2.0	13	1	438	12	0
107.4	44	10.3	6.4	11	2	354	18	0
105.2	42	9.0	4.2	12	2	364	14	0
105.2	46	10.0	1.0	11	1	472	16	0
105.2	48	10.3	3.3	5	2	454	26	0
102.0	48	9.0	3.0	9	2	318	25	0
102.0	44	9.3	0.7	7	3	652	13	0
102.0	48	9.7	0.0	3	1	458	17	0
100.2	44	9.2	0.0	9	**0**	370	10	1/7
100.2	44	10.1	0.9	2	**0**	548	8	1/7
100.2	46	10.3	0.8	3	1	508	9	0
99.5 (CCA)	38	9.5	2.7	7	**0**	370	17	1/7
99.5 (CCA)	42	9.6	0.6	5	**0**	294	12	1/7
99.5 (CCA)	40	9.9	5.0	7	1	316	21	0
97.8	46	9.3	1.7	2	**0**	454	9	1/7
97.8	44	9.8	1.7	4	**0**	552	9	1/7
97.8	44	9.8	2.7	7	1	272	22	0
Scioto River - 1986								
133.0	40	8.5	0.0	10	1	276	**3**	1/7
133.0	38	7.5	0.0	15	**0**	288	**1**	2/7
133.0	32	8.8	0.0	27	1	334	**1**	1/7
131.8	32	7.5	0.0	13	**0**	494	**0**	2/7
131.8	30	7.9	0.5	28	**0**	290	**4**	2/7
131.8	30	8.0	0.0	22	**0**	458	**3**	2/7
131.0	36	7.8	0.0	13	**0**	196	**1**	2/7
131.0	32	8.2	1.2	8	**0**	714	**1**	2/7

APPENDIX TABLE 3 (CONTINUED)
Expression of Toxic Response Signatures among Selected Metrics and Aggregations of Fish Assemblage Data in Boatable Ohio Rivers Using the Biological Response Signatures Described by Yoder and Rankin (1995)

Location RM (Source)	IBI	Modified Iwb	%DELT Anomalies	Percent Tolerant	Intolerant Species	Density[a] (No./300m)	% Round-Bodied	#Toxic Signatures
131.0	34	8.3	0.0	10	**0**	408	**2**	2/7
129.1 (CSO)	38	10.0	1.4	10	**0**	394	5	2/7
129.1 (CSO)	34	8.9	1.2	15	**0**	470	**4**	2/7
126.4 (JP WWTP)	30	8.5	3.3	24	**0**	168	11	1/7
126.4 (JP WWTP)	24	7.6	5.4	43	**0**	**146**	6	1/7
123.5	36	9.0	5.0	39	**0**	212	9	1/7
123.5	36	8.7	1.5	26	**0**	202	8	1/7
119.9	28	7.4	1.2	33	**0**	210	**3**	2/7
119.9	28	7.9	4.4	31	**0**	206	**3**	2/7
118.1 (CS WWTP)	46	9.7	5.6	18	1	274	38	0
118.1 (CS WWTP)	44	9.8	**12.7**	16	1	236	37	1/7
117.1	32	9.0	6.8	21	**0**	210	15	1/7
117.1	38	10.1	7.5	14	**0**	378	12	1/7
113.5	38	8.6	5.2	27	**0**	236	16	1/7
113.5	34	8.2	1.2	27	**0**	240	9	1/7
109.2	42	8.8	3.5	11	**0**	302	23	1/7
109.2	42	9.9	3.6	7	1	358	27	0
105.2	44	9.4	3.7	6	2	208	23	0
105.2	42	9.3	5.6	7	**0**	300	36	1/7
102.0	48	9.7	4.0	8	3	228	39	0
102.0	42	9.2	0.6	0	**0**	310	30	1/7
100.2	42	9.4	3.1	11	**0**	172	20	1/7
100.2	40	9.4	1.3	3	**0**	193	14	1/7
Scioto River - 1981								
133.0	26	**5.5**	2.3	42	**0**	**50**	**2**	4/7
133.0	42	7.7	2.1	11	1	**84**	28	1/7
133.0	36	7.1	5.8	9	**0**	**98**	31	2/7
131.8	**20**	6.3	**20.6**	38	**0**	**42**	21	4/7
131.8	**16**	7.0	**13.7**	30	**0**	**80**	5	5/7
131.8	24	7.7	0.0	16	1	**132**	5	2/7
130.4	28	6.1	2.0	24	**0**	**74**	**0**	2/7
130.4	24	5.7	2.4	21	**0**	**66**	**0**	3/7
130.4	30	7.5	0.0	42	1	**122**	**3**	2/7
129.1 (CSO)	**16**	6.1	3.6	16	**0**	**92**	**2**	4/7
129.1 (CSO)	**22**	7.1	8.7	7	**0**	376	**1**	3/7
129.1 (CSO)	**22**	6.4	1.3	11	**0**	**136**	5	4/7
128.1	**22**	**5.8**	0.0	46	**0**	**56**	**0**	5/7
128.1	**20**	7.0	**13.3**	17	**0**	**96**	**0**	5/7
128.1	**16**	6.4	2.4	19	**0**	**68**	**0**	3/7
126.4 (JP WWTP)	24	**5.8**	0.0	69	1	**26**	**2**	3/7
126.4 (JP WWTP)	**20**	6.5	2.1	13	**0**	**82**	**0**	4/7
126.4 (JP WWTP)	**22**	6.6	0.0	36	**0**	**46**	**0**	4/7
122.9	24	6.3	0.0	26	**0**	**50**	**0**	3/7
122.9	26	6.8	3.2	29	**0**	**44**	6	2/7
122.9	26	**5.9**	0.0	27	**0**	**76**	**2**	4/7
119.9	**18**	**5.5**	0.0	31	**0**	**48**	**0**	5/7

APPENDIX TABLE 3 (CONTINUED)

Expression of Toxic Response Signatures among Selected Metrics and Aggregations of Fish Assemblage Data in Boatable Ohio Rivers Using the Biological Response Signatures Described by Yoder and Rankin (1995)

Location RM (Source)	IBI	Modified Iwb	%DELT Anomalies	Percent Tolerant	Intolerant Species	Density[a] (No./300m)	% Round-Bodied	#Toxic Signatures
119.9	**22**	**5.9**	1.9	13	**0**	**90**	**2**	5/7
119.9	24	6.6	0.0	4	**0**	**100**	**4**	3/7
118.1 (CS WWTP)	28	7.1	0.0	14	**0**	**88**	**0**	3/7
118.1 (CS WWTP)	24	8.0	6.3	12	**0**	**90**	14	2/7
118.1 (CS WWTP)	24	7.6	1.5	2	1	**128**	**2**	2/7
117.1	20	6.9	0.0	37	**0**	**80**	5	2/7
117.1	32	8.6	**11.5**	21	1	**96**	20	2/7
117.1	**16**	6.0	6.1	62	**0**	**50**	**0**	4/7
116.3 (Pic. EGS)	24	7.5	0.0	33	1	**62**	**2**	2/7
116.3 (Pic. EGS)	32	9.0	6.2	18	**0**	**128**	5	2/7
114.0	**16**	**5.6**	0.0	25	**0**	**42**	**0**	5/7
114.0	**22**	**5.3**	0.0	33	**0**	**44**	**3**	5/7
114.0	**12**	**4.7**	7.5	58	**0**	**34**	**3**	5/7
104.8	26	7.7	3.4	15	**0**	**100**	17	2/7
104.8	26	7.4	8.0	25	1	**130**	21	1/7
104.8	24	8.1	1.2	10	**0**	154	16	1/7
102.0	**22**	7.4	3.9	27	**0**	**74**	27	3/7
102.0	38	8.6	**13.5**	10	2	**142**	52	2/7
102.0	24	7.7	6.3	23	1	**74**	17	1/7
100.2	**22**	6.5	0.0	37	**0**	**34**	15	3/7
100.2	**22**	7.3	0.0	20	**0**	**78**	8	3/7
100.2	**22**	7.5	0.0	30	**0**	**70**	8	3/7
98.3	32	7.6	0.0	11	**0**	**68**	21	2/7
98.3	26	6.3	0.0	21	**0**	**46**	10	2/7
98.3	24	6.3	0.0	15	1	**44**	15	1/7
Ottawa River - 1996								
43.6	34	7.7	2.9	41	**0**	250	23	1/7
43.6	30	7.3	7.8	48	**0**	194	30	1/7
42.3	28	7.4	4.4	61	**0**	464	**3**	1/7
40.8	26	6.4	3.0	51	**0**	166	**0**	2/7
40.8	28	6.9	8.6	45	**0**	154	**4**	2/7
40.2 (CSOs)	**22**	**5.0**	5.1	69	**0**	**80**	**0**	5/7
39.9 (CSOs)	24	6.3	1.7	**80**	**0**	158	**0**	3/7
39.9 (CSOs)	26	**5.2**	**23.5**	**78**	**0**	**52**	**2**	**6/7**
38.9 (CSOs)	24	**5.1**	2.2	**88**	**0**	**78**	**0**	5/7
38.9 (CSOs)	22	**4.8**	5.0	**87**	**0**	**50**	**0**	5/7
38.5 (CSOs)	26	6.3	8.8	63	**0**	**138**	**0**	3/7
38.5 (CSOs)	26	7.2	8.6	63	**0**	174	**1**	2/7
37.7 (L5 landfill)	26	6.9	**10.2**	67	**0**	431	**0**	3/7
37.7 (L5 landfill)	32	8.3	**18.8**	50	**0**	538	**0**	3/7
37.66 (Lima)	26	9.0	6.6	59	**0**	1190	**0**	2/7
37.66 (Lima)	32	9.4	**11.1**	46	**0**	1440	**0**	3/7
37.4	**22**	7.5	9.7	69	**0**	334	**0**	2/7
37.4	**22**	6.0	**18.0**	69	**0**	**146**	**0**	5/7
37.11 (Refinery)	28	8.1	**20.8**	**76**	**0**	530	**1**	4/7
37.11 (Refinery)	28	7.3	**14.1**	**87**	**0**	400	**0**	4/7

APPENDIX TABLE 3 (CONTINUED)
Expression of Toxic Response Signatures among Selected Metrics and Aggregations of Fish Assemblage Data in Boatable Ohio Rivers Using the Biological Response Signatures Described by Yoder and Rankin (1995)

Location RM (Source)	IBI	Modified Iwb	%DELT Anomalies	Percent Tolerant	Intolerant Species	Density[a] (No./300m)	% Round-Bodied	#Toxic Signatures
37.0	**22**	7.5	5.3	**76**	**0**	348	**0**	4/7
37.0	26	7.5	**17.6**	62	**0**	628	**0**	3/7
36.87 (PCS-N)	**20**	7.3	**10.5**	**85**	**0**	180	**0**	5/7
36.87 (PCS-N)	24	6.5	**14.2**	**89**	**0**	180	**0**	4/7
36.7	26	6.3	**28.9**	**83**	**0**	**54**	**4**	5/7
36.7	26	8.1	**48.8**	65	**0**	**124**	6	3/7
36.0	**18**	**5.5**	**25.9**	**84**	**0**	**110**	**0**	**7/7**
36.0	**16**	**5.3**	**33.1**	**88**	**0**	**70**	**0**	**7/7**
34.7	**20**	**4.9**	**26.3**	**87**	**0**	**26**	**2**	**7/7**
34.7	**16**	**5.4**	**37.1**	**87**	**0**	**38**	**3**	**7/7**
32.8	**22**	7.0	**34.2**	**83**	**0**	**102**	**2**	**6/7**
32.8	**20**	7.3	**22.5**	**84**	**0**	182	**2**	5/7
32.4 (Shawnee)	**18**	6.1	**22.2**	**77**	**0**	**98**	**1**	**6/7**
32.4 (Shawnee)	**18**	6.4	**33.8**	**83**	**0**	**120**	**1**	**6/7**
28.9	**18**	**4.6**	**11.9**	**88**	**0**	**46**	**0**	**7/7**
28.9	**18**	6.4	**31.5**	**84**	**0**	**130**	**0**	**6/7**
25.5	28	7.5	**10.8**	**75**	**0**	214	**3**	4/7
25.5	26	8.6	**15.8**	59	**0**	336	**2**	3/7
21.0 (Elida)	**20**	7.5	9.9	**78**	**0**	230	**1**	4/7
21.0 (Elida)	**22**	8.1	**12.0**	59	**0**	390	**1**	4/7
16.2	**20**	8.4	4.1	**70**	**0**	603	**0**	4/7
13.0	26	8.7	4.3	55	1	466	**2**	1/7
5.6	32	9.7	**16.7**	30	2	466	15	1/7
5.6	32	9.7	9.1	26	1	754	8	0
3.8	38	9.2	**32.7**	24	2	318	26	1/7
3.8	36	9.3	**34.2**	31	2	428	20	1/7
1.2	36	10.1	**19.3**	19	1	376	34	1/7
1.2	36	9.7	9.2	24	1	750	9	0
Ottawa River - 1991								
37.7 (L5 landfill)	**20**	**5.2**	7.8	**92**	**0**	**96**	**0**	**6/7**
37.7 (L5 landfill)	26	7.0	2.6	**94**	**0**	224	**0**	3/7
37.7 (L5 landfill)	**20**	6.0	**16.7**	88	**0**	**148**	**1**	5/7
37.4 (Lima STP)	**22**	**5.5**	**25.6**	**83**	**0**	**56**	**0**	**7/7**
37.4 (Lima STP)	**18**	**5.8**	8.5	**91**	**0**	**94**	**0**	**6/7**
37.4 (Lima STP)	**20**	**3.8**	2.4	**98**	**0**	**26**	**0**	**6/7**
37.0 (Refinery)	26	8.5	**19.4**	61	**0**	240	**4**	3/7
37.0 (Refinery)	**16**	**5.5**	9.5	**87**	**0**	**144**	**0**	**6/7**
37.0 (Refinery)	**20**	**5.4**	7.2	**97**	**0**	**28**	**0**	**6/7**
36.7 (PCS-N)	**20**	**5.6**	**29.0**	**86**	**0**	**20**	**1**	**7/7**
36.7 (PCS-N)	**20**	**4.3**	**14.8**	**95**	**0**	**16**	**0**	**7/7**
36.7 (PCS-N)	**22**	**4.4**	**14.5**	**97**	**0**	**14**	**0**	**7/7**
34.7	**14**	**2.3**	**57.5**	**97**	**0**	**4**	**0**	**7/7**
34.7	**16**	**4.8**	**56.0**	63	**0**	**104**	**0**	**6/7**
34.7	**14**	**3.2**	**17.9**	**96**	**0**	**18**	**0**	**7/7**
28.9	**18**	**4.7**	**16.0**	**80**	1	**50**	**1**	**6/7**
28.9	**16**	**4.5**	**23.6**	**89**	**0**	**44**	**0**	**7/7**

APPENDIX TABLE 3 (CONTINUED)
Expression of Toxic Response Signatures among Selected Metrics and Aggregations of Fish Assemblage Data in Boatable Ohio Rivers Using the Biological Response Signatures Described by Yoder and Rankin (1995)

Location RM (Source)	IBI	Modified Iwb	%DELT Anomalies	Percent Tolerant	Intolerant Species	Density[a] (No./300m)	% Round-Bodied	#Toxic Signatures
28.9	**16**	**4.4**	**13.1**	**93**	**0**	**42**	**0**	**7/7**
1.2	30	8.8	1.2	43	**0**	492	43	0
1.2	32	9.1	1.4	22	1	1384	**0**	1/7
1.2	34	9.8	1.2	25	1	1066	**2**	1/7
Ottawa River - 1989								
37.7 (L5 landfill)	24	6.6	0.7	**86**	**0**	175	**0**	3/7
37.7 (L5 landfill)	30	7.1	4.0	**74**	**0**	295	**0**	3/7
37.7 (L5 landfill)	26	8.4	4.0	**78**	1	400	**0**	2/7
37.5	28	7.6	1.8	**86**	**0**	360	**0**	3/7
37.4 (Lima STP)	**20**	**4.3**	**12.6**	**92**	**0**	**40**	**1**	**7/7**
37.4 (Lima STP)	28	**5.3**	5.1	**87**	**0**	**104**	**1**	5/7
37.4 (Lima STP)	**22**	6.7	6.1	**86**	**0**	**122**	**0**	5/7
37.0 (Refinery)	**22**	7.8	**14.7**	69	**0**	180	**0**	4/7
37.0 (Refinery)	26	7.1	**19.1**	**78**	**0**	211	**1**	5/7
37.0 (Refinery)	24	7.8	**15.4**	**82**	**0**	**106**	**2**	5/7
36.7 (PCS-N)	**22**	**4.0**	**38.0**	**72**	**0**	**56**	**0**	**7/7**
36.7 (PCS-N)	26	7.5	**33.5**	62	**0**	182	**3**	3/7
36.7 (PCS-N)	**20**	6.6	**23.6**	**77**	**0**	**86**	**2**	**6/7**
36.0	**18**	**3.5**	**27.6**	**84**	**0**	**20**	**0**	**7/7**
36.0	**16**	**4.1**	**23.0**	**89**	**0**	**22**	**0**	**7/7**
36.0	**16**	**3.9**	**29.5**	**86**	**0**	**34**	**0**	**7/7**
34.7	20	**4.8**	**36.0**	**71**	**0**	**50**	**2**	**7/7**
34.7	22	**4.1**	**20.2**	**88**	**0**	**32**	**0**	**7/7**
34.7	**16**	**3.9**	**20.7**	**90**	**0**	**32**	**0**	**7/7**
32.4 (Shawnee)	22	6.8	**15.5**	69	**0**	**128**	**4**	5/7
32.4 (Shawnee)	20	**5.9**	**17.7**	**80**	**0**	**102**	**2**	**7/7**
32.4 (Shawnee)	18	6.1	**20.1**	**82**	**0**	**83**	**2**	**6/7**
28.9	**16**	**4.1**	**19.8**	**84**	**0**	**30**	**2**	**7/7**
28.9	22	**5.3**	**11.8**	**85**	**0**	**42**	**1**	**7/7**
28.9	18	**4.7**	**11.9**	**89**	**0**	**34**	**1**	**7/7**
25.5	28	6.8	**11.0**	69	**0**	202	**4**	3/7
25.5	30	8.8	**15.9**	41	**0**	366	**4**	3/7
25.5	30	8.7	**11.2**	47	**0**	456	**2**	3/7
19.0 (Elida)	24	7.8	3.6	61	**0**	290	**0**	2/7
19.0 (Elida)	26	8.9	5.1	57	**0**	370	**1**	2/7
19.0 (Elida)	28	8.3	1.4	43	**0**	706	**0**	2/7
1.2	32	8.5	0.7	51	**0**	370	**2**	2/7
1.2	36	9.1	0.6	46	1	434	**2**	1/7
1.2	38	9.5	0.0	41	1	436	6	0
Ottawa River - 1987								
43.6	36	6.7	0.0	32	**0**	212	26	1/7
43.6	34	7.5	0.6	47	**0**	186	36	1/7
43.6	34	7.7	4.1	36	**0**	284	41	1/7
41.0	26	7.0	3.0	58	**0**	188	**3**	1/7
38.5 (CSOs)	**16**	**4.4**	**18.6**	65	**0**	**30**	5	5/7

APPENDIX TABLE 3 (CONTINUED)
Expression of Toxic Response Signatures among Selected Metrics and Aggregations of Fish Assemblage Data in Boatable Ohio Rivers Using the Biological Response Signatures Described by Yoder and Rankin (1995)

Location RM (Source)	IBI	Modified Iwb	%DELT Anomalies	Percent Tolerant	Intolerant Species	Density[a] (No./300m)	% Round-Bodied	#Toxic Signatures
38.5 (CSOs)	24	**4.5**	2.1	**74**	**0**	**48**	**1**	5/7
38.5 (CSOs)	24	**5.8**	3.1	**78**	**0**	**96**	**1**	5/7
37.7 (L5 landfill)	**22**	6.3	**14.1**	64	**0**	150	**0**	4/7
37.7 (L5 landfill)	36	7.4	**10.1**	66	**0**	421	**1**	3/7
36.0	**16**	**3.2**	**50.3**	**91**	**0**	**13**	**0**	**7/7**
34.7	**12**	**3.2**	**51.6**	**90**	**0**	**6**	**0**	**7/7**
34.7	**14**	**3.1**	**40.9**	**98**	**0**	**2**	**0**	**7/7**
34.7	**16**	**3.8**	**32.5**	**94**	**0**	**10**	**0**	**7/7**
28.9	**20**	**4.4**	**26.6**	**88**	**0**	**36**	**0**	**7/7**
25.5	28	7.1	**20.1**	53	**0**	236	**2**	3/7
7.9	**16**	**5.8**	3.7	**70**	**0**	**60**	**3**	**6/7**
1.2	24	**5.9**	1.8	63	**0**	**69**	**2**	4/7
Ottawa River - 1985								
40.2 (CSOs)	28	6.0	0.0	67	**0**	**116**	**2**	3/7
40.2 (CSOs)	24	6.0	0.5	**85**	**0**	**128**	**0**	4/7
40.2 (CSOs)	26	6.3	0.2	**80**	**0**	178	**0**	3/7
38.9 (CSOs)	**22**	6.4	3.5	**82**	**0**	**138**	**0**	5/7
38.9 (CSOs)	26	6.2	8.7	**86**	**0**	224	**0**	3/7
38.9 (CSOs)	26	6.3	3.8	**84**	**0**	320	**0**	3/7
37.7 (L5 landfill)	**20**	**5.9**	**12.4**	**83**	**0**	155	**0**	**6/7**
37.7 (L5 landfill)	24	6.5	2.7	**86**	**0**	365	**0**	3/7
37.7 (L5 landfill)	28	7.5	3.2	**74**	**0**	975	**0**	3/7
37.4 (Lima STP)	**22**	**4.9**	2.5	**93**	**0**	**57**	**0**	**6/7**
37.4 (Lima STP)	**20**	6.9	6.8	**88**	**0**	160	**0**	4/7
37.4 (Lima STP)	24	7.3	**15.9**	**92**	**0**	**146**	**0**	5/7
37.0 (Refinery)	**18**	**3.8**	2.5	**94**	**0**	**14**	**0**	**6/7**
37.0 (Refinery)	**18**	**5.6**	**17.3**	**94**	**0**	**34**	**1**	**7/7**
37.0 (Refinery)	**16**	**5.8**	**29.0**	**89**	**0**	**57**	**1**	**7/7**
36.7 (PCS-N)	**20**	**3.2**	1.4	**97**	**0**	**10**	**1**	**6/7**
36.7 (PCS-N)	**16**	**3.2**	**73.2**	**94**	**0**	**16**	**0**	**7/7**
36.7 (PCS-N)	**20**	**4.0**	**21.5**	**94**	**0**	**20**	**0**	**7/7**
34.7	**12**	**1.2**	0.0	**90**	**0**	**2**	**0**	**6/7**
34.7	**14**	**2.4**	**35.7**	**98**	**0**	**2**	**0**	**7/7**
34.7	**16**	**3.3**	**37.4**	**91**	**0**	**10**	**0**	**7/7**
32.8 (Shawnee)	**16**	**2.7**	**13.2**	**92**	**0**	**6**	**0**	**7/7**
32.8 (Shawnee)	**16**	**3.1**	**27.7**	**94**	**0**	**18**	**0**	**7/7**
32.8 (Shawnee)	**18**	**4.6**	**20.0**	**79**	**0**	**52**	**0**	**7/7**
28.9	**14**	**3.6**	**20.5**	**95**	**0**	**12**	**0**	**7/7**
28.9	**18**	**4.6**	**26.4**	**93**	**0**	**48**	**0**	**7/7**
28.9	**20**	**4.7**	**33.4**	**92**	**0**	**36**	**0**	**7/7**
25.5	26	7.6	**10.1**	63	**0**	**136**	**2**	4/7
25.5	28	7.9	**41.3**	64	**0**	366	**1**	3/7
25.5	24	6.6	**29.7**	**71**	**0**	178	**0**	4/7
22.3 (Elida)	**16**	5.8	1.1	**82**	**0**	**68**	**0**	**6/7**
22.3 (Elida)	**16**	5.4	8.6	**84**	**0**	**106**	**0**	**6/7**
22.3 (Elida)	**16**	6.3	**17.2**	**82**	**0**	**140**	**0**	**6/7**

APPENDIX TABLE 3 (CONTINUED)
Expression of Toxic Response Signatures among Selected Metrics and Aggregations of Fish Assemblage Data in Boatable Ohio Rivers Using the Biological Response Signatures Described by Yoder and Rankin (1995)

Location RM (Source)	IBI	Modified Iwb	%DELT Anomalies	Percent Tolerant	Intolerant Species	Density[a] (No./300m)	% Round-Bodied	#Toxic Signatures
16.2	**20**	6.1	0.9	**85**	**0**	193	**0**	4/7
16.2	**22**	7.7	0.7	**86**	**0**	367	**0**	4/7
16.2	**18**	6.0	1.5	**90**	**0**	196	**0**	4/7
12.3	**20**	**5.2**	1.3	**73**	**0**	**86**	**0**	**6/7**
12.3	28	6.5	0.3	**83**	**0**	226	**0**	3/7
12.3	24	7.1	2.4	**77**	**0**	200	**0**	3/7
7.9	**20**	**5.4**	5.0	63	**0**	**74**	**1**	5/7
7.9	28	7.9	2.1	64	**0**	522	**0**	2/7
7.9	24	7.6	3.4	**75**	**0**	182	**1**	3/7
4.4	26	6.7	2.7	**70**	**0**	**114**	**1**	4/7
4.4	28	8.6	1.8	64	**0**	402	**3**	2/7
4.4	30	9.1	3.0	63	**0**	432	**2**	2/7
1.2	28	7.4	2.0	62	**0**	**94**	7	2/7
1.2	34	9.0	1.3	44	**0**	504	**3**	2/7
1.2	30	9.2	0.0	50	**0**	392	**4**	2/7
Paint Creek - 1997								
8.9	48	11.1	0.0	9	8	546	18	0
8.9	54	11.1	0.0	3	10	528	28	0
7.8	52	11.1	0.0	6	10	664	15	0
7.8	50	11.1	1.5	5	6	504	27	0
4.6	50	10.6	0.0	4	3	578	14	0
4.6	46	11.4	0.0	3	9	936	6	0
2.4 (Mead)	36	9.9	0.0	6	**0**	340	**0**	2/7
2.4 (Mead)	32	8.9	2.9	3	**0**	380	**0**	2/7
2.3	36	10.1	0.0	6	1	515	**4**	0
2.3	40	9.6	1.2	5	2	330	**3**	1/7
1.3	40	9.6	0.0	12	1	326	**2**	1/7
1.3	48	10.2	0.0	4	5	644	**4**	1/7
0.2	44	9.4	0.0	1	1	590	6	0
0.2	42	9.2	0.4	2	2	470	**3**	1/7
Paint Creek - 1992								
8.9	50	10.6	1.0	3	3	540	43	0
4.6	42	10.6	5.6	6	3	382	27	0
4.6	44	10.3	2.7	7	3	418	23	0
2.4 (Mead)	26	8.9	3.8	8	**0**	240	**0**	2/7
2.4 (Mead)	**20**	8.1	6.8	23	**0**	340	**0**	3/7
2.3	36	9.2	0.9	8	**0**	468	**3**	2/7
1.3	34	10.0	3.3	6	1	378	16	0
1.3	36	10.7	5.0	20	2	662	14	0
0.2	34	10.0	1.3	9	**0**	430	7	1/7
0.2	36	9.7	4.5	10	2	326	17	0
Paint Creek - 1985								
5.0	50	10.0	0.0	5	3	332	47	0
5.0	50	10.4	0.0	12	7	374	35	0

APPENDIX TABLE 3 (CONTINUED)
Expression of Toxic Response Signatures among Selected Metrics and Aggregations of Fish Assemblage Data in Boatable Ohio Rivers Using the Biological Response Signatures Described by Yoder and Rankin (1995)

Location RM (Source)	IBI	Modified Iwb	%DELT Anomalies	Percent Tolerant	Intolerant Species	Density[a] (No./300m)	% Round-Bodied	#Toxic Signatures
2.1 (Mead)	**12**	**4.8**	0.0	27	**0**	**32**	**0**	5/7
2.1 (Mead)	**20**	**5.6**	0.0	49	**0**	**52**	**0**	5/7
1.5	40	8.2	2.4	10	**0**	**74**	24	2/7
1.5	34	7.4	0.0	10	**0**	**88**	18	2/7
0.4	32	8.7	0.0	12	**0**	**114**	14	2/7
0.4	28	7.5	0.0	8	**0**	**68**	**0**	3/7
0.4	32	8.6	0.8	11	1	220	12	0
Cuyahoga River - 2000								
49.7	30	8.3	0.3	52	**0**	328	**4**	2/7
49.7	30	8.5	0.0	32	**0**	310	14	1/7
39.7 (ust. Akron)	32	8.4	0.9	31	**0**	314	14	1/7
39.7 (ust. Akron)	34	8.1	0.9	36	**0**	292	19	1/7
33.2	26	7.4	0.6	57	1	294	6	0
33.2	**20**	8.0	3.9	47	**0**	298	12	1/7
26.5	30	8.0	0.2	40	**0**	506	8	1/7
26.5	30	7.7	0.3	36	**0**	456	9	1/7
15.6	38	8.9	3.1	25	**0**	488	18	1/7
11.0 (Southerly)	24	6.7	2.9	33	**0**	**136**	15	2/7
8.0	28	8.0	3.4	19	**0**	382	16	1/7
7.2 (Big Creek)	26	7.5	4.0	17	**0**	208	6	1/7
Cuyahoga River - 1996								
48.7	24	7.1	7.0	34	**0**	172	**4**	2/7
48.7	28	7.1	0.0	38	**0**	**110**	**3**	3/7
48.0	24	6.9	3.4	31	**0**	**122**	**2**	3/7
48.0	24	6.4	3.2	41	**0**	**112**	**0**	3/7
46.0	28	6.7	2.2	33	**0**	214	**0**	2/7
42.0 (L. Cuya.)	24	7.0	3.7	33	**0**	220	19	1/7
38.6 (ust. Akron)	32	6.0	3.6	25	**0**	**82**	49	2/7
38.6 (ust. Akron)	**20**	6.1	4.3	46	1	**102**	26	2/7
37.4 (dst. Akron)	**22**	**4.7**	0.0	61	**0**	**70**	6	4/7
37.4 (dst. Akron)	**16**	**5.5**	**10.5**	**82**	**0**	**70**	11	5/7
37.2	24	**5.5**	4.9	46	**0**	**44**	41	4/7
37.2	**20**	**5.5**	9.2	57	1	**76**	30	3/7
35.3	**16**	**4.7**	7.2	**77**	**0**	**32**	6	5/7
42.0 (L. Cuya.)	28	6.7	1.5	35	**0**	192	25	1/7
35.3	24	6.5	1.6	48	**0**	**132**	19	2/7
33.3	**14**	**4.9**	5.2	49	**0**	**98**	**0**	5/7
33.3	**20**	**4.6**	1.4	**77**	**0**	**34**	**0**	**6/7**
27.2	24	**5.4**	1.1	63	**0**	**66**	6	3/7
27.2	**20**	**5.4**	3.7	**73**	**0**	**90**	9	5/7
22.0	**14**	**3.8**	0.0	**70**	**0**	**20**	**0**	**6/7**
22.0	**12**	**4.0**	5.0	**80**	**0**	**16**	**0**	**6/7**
16.5	**18**	**5.7**	**19.4**	42	**0**	**36**	10	5/7
16.5	**12**	**5.7**	**11.1**	50	**0**	**36**	6	5/7
15.9	**20**	**5.7**	**12.5**	43	**0**	**46**	**3**	**6/7**

APPENDIX TABLE 3 (CONTINUED)

Expression of Toxic Response Signatures among Selected Metrics and Aggregations of Fish Assemblage Data in Boatable Ohio Rivers Using the Biological Response Signatures Described by Yoder and Rankin (1995)

Location RM (Source)	IBI	Modified Iwb	%DELT Anomalies	Percent Tolerant	Intolerant Species	Density[a] (No./300m)	% Round-Bodied	#Toxic Signatures
15.9	24	7.0	7.7	37	**0**	**82**	12	2/7
11.5	**22**	7.0	5.0	53	**0**	**98**	15	3/7
11.5	**14**	**5.5**	5.7	28	**0**	**76**	9	4/7
10.4 (Southerly)	32	**4.5**	**20.0**	0	**0**	150	87	3/7
9.6	**16**	**5.0**	**34.8**	**70**	**0**	**28**	13	**6/7**
9.6	**18**	**4.4**	**10.5**	58	**0**	**32**	8	5/7
8.3	**16**	6.0	8.6	41	**0**	**96**	5	3/7
8.3	**18**	**5.1**	5.3	56	**0**	**50**	**2**	5/7
7.2 (Big Creek)	**14**	**4.0**	**10.4**	24	**0**	**146**	**3**	**6/7**
7.2 (Big Creek)	**22**	6.4	0.0	16	**0**	960	**2**	3/7
Cuyahoga River - 1991								
49.8	30	8.1	1.1	39	1	458	5	0
49.8	34	9.0	0.6	28	1	1102	**4**	1/7
49.8	42	8.9	1.1	15	1	938	6	0
42.6	30	7.9	1.8	33	**0**	806	**0**	2/7
42.6	36	7.9	2.1	23	1	977	6	0
42.6	32	8.7	3.3	28	**0**	691	6	1/7
38.6 (ust. Akron)	24	6.2	1.5	66	1	252	**3**	1/7
37.4 (ust. Akron)	**20**	**5.1**	2.5	**74**	**0**	310	**1**	4/7
37.4 (dst. Akron)	32	7.6	2.6	17	**0**	1250	**3**	2/7
37.4 (dst. Akron)	34	8.3	0.0	20	**0**	2270	6	1/7
37.2	**22**	6.3	2.6	64	1	**110**	11	2/7
37.2	24	6.6	1.8	67	2	**146**	10	1/7
37.2	26	7.9	0.8	41	1	808	8	0
35.3	**16**	**5.6**	1.1	66	**0**	163	**1**	4/7
35.3	**20**	6.4	3.4	**73**	1	164	6	2/7
35.3	26	7.6	1.2	52	**0**	515	5	1/7
38.6 (ust. Akron)	24	7.3	0.6	53	1	452	**4**	1/7
38.6 (ust. Akron)	26	7.1	0.5	65	1	468	**1**	1/7
26.7	**22**	7.0	1.5	**71**	**0**	198	**1**	4/7
26.7	**20**	6.4	0.0	68	**0**	**140**	**1**	4/7
26.7	24	7.6	0.4	51	**0**	670	**0**	2/7
15.9	24	7.2	3.8	64	1	182	**1**	1/7
15.9	26	7.2	0.5	54	**0**	374	**0**	2/7
15.9	24	7.5	2.6	23	**0**	444	**0**	2/7
11.5	**20**	6.0	2.4	56	**0**	**148**	**0**	4/7
11.5	**18**	6.8	2.3	52	**0**	228	**0**	3/7
11.5	**20**	7.0	4.4	46	**0**	342	**0**	3/7
10.5 (Southerly)	**20**	**2.5**	0.0	**97**	**0**	**10**	**0**	**6/7**
10.5 (Southerly)	**18**	**3.3**	3.1	**93**	**0**	**50**	**0**	**6/7**
10.5 (Southerly)	24	6.0	1.8	34	**0**	370	**0**	2/7
10.3	**14**	**5.8**	8.0	28	**0**	**122**	**0**	5/7
10.3	**20**	6.2	3.5	34	**0**	230	**0**	3/7
10.3	**20**	6.4	5.7	32	**0**	256	**1**	3/7
8.9	**18**	**5.8**	7.7	37	**0**	**146**	**0**	5/7
8.9	**18**	**5.4**	3.2	29	**0**	208	**0**	4/7

APPENDIX TABLE 3 (CONTINUED)
Expression of Toxic Response Signatures among Selected Metrics and Aggregations of Fish Assemblage Data in Boatable Ohio Rivers Using the Biological Response Signatures Described by Yoder and Rankin (1995)

Location RM (Source)	IBI	Modified Iwb	%DELT Anomalies	Percent Tolerant	Intolerant Species	Density[a] (No./300m)	% Round-Bodied	#Toxic Signatures
8.9	**22**	6.2	8.9	43	**0**	214	**1**	3/7
8.3	**14**	**5.3**	1.9	33	**0**	174	**0**	4/7
8.3	**20**	6.4	3.1	36	**0**	158	**0**	3/7
8.3	**18**	6.0	7.5	32	**0**	162	**0**	3/7
7.4	**20**	6.4	4.3	12	**0**	282	**0**	3/7
7.4	**16**	6.0	5.2	24	**0**	226	**0**	3/7
7.4	**20**	6.5	**10.6**	14	**0**	194	**0**	4/7
7.1 (Big Creek)	**16**	6.2	2.5	18	**0**	198	**0**	3/7
7.1 (Big Creek)	**20**	7.0	3.3	20	**0**	312	**0**	3/7
7.1 (Big Creek)	26	7.4	2.8	12	**0**	310	**1**	2/7
Cuyahoga River - 1988								
42.6	30	7.7	1.9	41	**0**	644	**0**	2/7
42.6	30	9.0	0.4	41	**0**	582	**0**	2/7
42.6	24	7.2	1.5	44	**0**	302	**0**	2/7
41.0 (L. Cuya.)	**22**	6.2	3.1	65	**0**	**134**	**0**	4/7
41.0 (L. Cuya.)	**22**	**5.5**	1.6	63	**0**	**138**	**0**	4/7
41.0 (L. Cuya.)	**22**	**5.9**	1.3	**70**	**0**	**98**	**0**	**6/7**
40.4	24	**5.4**	1.4	**77**	**0**	**90**	**0**	**6/7**
40.4	26	**5.2**	0.0	69	**0**	**108**	**0**	4/7
40.4	**20**	**4.6**	**11.7**	**84**	**0**	**38**	**0**	**7/7**
38.6 (ust. Akron)	**18**	**5.0**	1.0	**74**	**0**	**108**	**0**	**6/7**
38.6 (ust. Akron)	**20**	**5.8**	0.2	43	**0**	390	**0**	4/7
38.6 (ust. Akron)	**18**	**4.9**	1.3	**82**	**0**	**68**	**1**	**6/7**
36.6 (dst. Akron)	**20**	**4.1**	0.0	**90**	**0**	**72**	**1**	**6/7**
36.6 (dst. Akron)	**22**	**5.0**	0.0	**89**	**0**	**126**	**0**	**6/7**
36.6 (dst. Akron)	**20**	**4.2**	3.4	**89**	1	**52**	**2**	5/7
33.6	**18**	**3.3**	2.8	**89**	**0**	**46**	**1**	**6/7**
33.6	**20**	**4.9**	0.0	**85**	**0**	**136**	**0**	**6/7**
26.7	**22**	6.5	0.4	**71**	**0**	322	**0**	4/7
24.1	**22**	6.6	0.6	65	**0**	230	**0**	3/7
24.1	**18**	**5.8**	1.2	**85**	**0**	**78**	**0**	**6/7**
21.0	28	**5.5**	0.0	**76**	**0**	**92**	**0**	5/7
17.3	24	**5.7**	1.1	67	**0**	312	**0**	3/7
17.3	**22**	8.0	1.3	44	**0**	364	**0**	3/7
15.9	**22**	6.1	0.9	44	**0**	248	**0**	3/7
15.9	**18**	6.0	2.0	57	**0**	**86**	**0**	4/7
13.1	**22**	**5.7**	0.6	55	**0**	184	**0**	4/7
13.1	**22**	6.4	1.1	48	**0**	**98**	**0**	4/7
12.1	26	6.2	0.5	17	**0**	690	**0**	2/7
12.1	**14**	**4.9**	4.8	64	**0**	**70**	**0**	5/7
11.5	**22**	6.3	1.1	15	**0**	700	**0**	3/7
11.5	**20**	**5.5**	2.3	60	**0**	**70**	**0**	5/7
9.8 (Southerly)	24	6.8	0.4	49	**0**	484	**0**	2/7
9.8 (Southerly)	**20**	**5.7**	3.2	61	**0**	**92**	**0**	5/7
7.5	**16**	**5.0**	5.0	35	**0**	**82**	**0**	5/7

APPENDIX TABLE 3 (CONTINUED)
Expression of Toxic Response Signatures among Selected Metrics and Aggregations of Fish Assemblage Data in Boatable Ohio Rivers Using the Biological Response Signatures Described by Yoder and Rankin (1995)

Location RM (Source)	IBI	Modified Iwb	%DELT Anomalies	Percent Tolerant	Intolerant Species	Density[a] (No./300m)	% Round-Bodied	#Toxic Signatures
7.5	**20**	**4.8**	0.0	**73**	**0**	**74**	**0**	**6/7**
7.5	**18**	**4.6**	1.9	39	**0**	**39**	**0**	5/7
7.1 (Big Creek)	**14**	**5.4**	1.3	24	**0**	**116**	**0**	5/7
7.1 (Big Creek)	**22**	**5.3**	2.5	19	**0**	260	**0**	4/7
7.1 (Big Creek)	**14**	**5.7**	3.1	35	**0**	**84**	**0**	5/7
Cuyahoga River - 1987								
42.6	24	7.1	2.3	34	**0**	340	**2**	2/7
42.6	28	7.9	2.9	26	**0**	848	**0**	2/7
42.6	26	7.8	1.2	35	**0**	1246	**0**	2/7
40.4 (L. Cuya.)	**20**	**5.5**	0.0	63	**0**	**114**	**0**	5/7
40.4 (L. Cuya.)	**20**	**5.7**	8.4	54	**0**	214	**1**	4/7
40.4 (L. Cuya.)	**22**	6.1	**11.3**	66	**0**	212	**0**	4/7
38.6 (ust. Akron)	**22**	**5.6**	0.0	**72**	1	**124**	**0**	5/7
38.6 (ust. Akron)	**18**	6.0	5.4	59	1	182	**0**	2/7
38.6 (ust. Akron)	**20**	**5.9**	4.7	58	**0**	320	**0**	4/7
36.5 (dst. Akron)	**16**	**2.7**	**19.6**	**96**	**0**	**12**	**0**	7/7
36.5 (dst. Akron)	**14**	**4.6**	6.7	**71**	**0**	**70**	**2**	**6/7**
36.5 (dst. Akron)	**16**	**4.5**	2.1	**80**	**0**	**82**	**0**	**6/7**
33.6	**16**	**1.1**	**46.5**	**99**	**0**	**2**	**0**	**7/7**
33.6	**18**	**4.2**	**27.9**	**86**	1	**40**	**1**	**6/7**
33.6	**22**	**4.5**	0.0	**75**	**0**	**86**	**0**	**6/7**
26.7	**16**	**5.4**	1.7	**77**	**0**	**95**	**0**	**6/7**
26.7	24	7.0	0.0	62	**0**	260	**0**	2/7
26.7	**16**	6.6	5.2	68	**0**	198	**0**	3/7
24.1	**14**	**4.8**	9.7	**79**	**0**	**62**	**0**	**6/7**
24.1	**22**	**5.5**	0.0	**82**	**0**	**78**	**0**	**6/7**
24.1	24	6.7	0.3	61	**0**	246	**0**	2/7
21.6	**12**	**4.4**	**20.8**	**76**	**0**	**37**	**0**	**7/7**
21.6	**14**	**4.4**	4.4	**88**	**0**	**18**	**0**	**6/7**
21.6	**22**	6.6	0.0	49	**0**	**114**	**0**	4/7
17.3	**22**	7.8	3.1	43	**0**	150	**0**	3/7
17.3	**22**	6.6	2.1	41	**0**	172	**0**	3/7
17.3	**20**	6.3	2.8	47	**0**	154	**0**	3/7
15.9	**18**	**4.8**	2.4	54	**0**	**82**	**0**	5/7
15.9	**16**	**5.1**	1.6	65	**0**	**44**	**0**	5/7
15.9	**16**	6.0	5.4	29	**0**	184	**0**	3/7
13.1	**16**	**5.4**	5.7	57	**0**	**48**	**0**	5/7
13.1	24	**5.1**	1.5	69	**0**	**42**	**0**	5/7
13.1	**22**	**5.7**	1.1	24	**0**	274	**0**	4/7
12.2	**20**	**5.4**	4.1	47	**0**	**78**	**0**	5/7
12.2	**16**	**4.1**	**14.0**	**70**	**0**	**32**	**0**	**7/7**
12.2	**16**	**5.8**	3.0	39	**0**	162	**0**	4/7
11.5	**22**	**4.5**	0.0	46	**0**	**92**	**0**	5/7
11.5	**14**	**3.2**	**10.7**	**91**	**0**	**8**	**0**	**7/7**
11.5	**14**	**5.1**	7.6	64	**0**	**42**	**0**	5/7

APPENDIX TABLE 3 (CONTINUED)
Expression of Toxic Response Signatures among Selected Metrics and Aggregations of Fish Assemblage Data in Boatable Ohio Rivers Using the Biological Response Signatures Described by Yoder and Rankin (1995)

Location RM (Source)	IBI	Modified Iwb	%DELT Anomalies	Percent Tolerant	Intolerant Species	Density[a] (No./300m)	% Round-Bodied	#Toxic Signatures
9.8 (Southerly)	**22**	**4.6**	3.8	33	**0**	166	**0**	4/7
9.8 (Southerly)	**16**	**3.7**	6.3	**75**	**0**	**16**	**0**	**6/7**
9.8 (Southerly)	**22**	6.1	**24.5**	50	**0**	**48**	**0**	5/7
7.5	30	**5.5**	1.5	6	**0**	246	**0**	3/7
7.5	**12**	**1.8**	**54.5**	**95**	**0**	**2**	**0**	**7/7**
7.5	**18**	**3.1**	0.0	16	**0**	**82**	**0**	5/7
7.1 (Big Creek)	26	**5.0**	5.6	21	**0**	234	**0**	4/7
7.1 (Big Creek)	**12**	**3.2**	**30.0**	**75**	**0**	**10**	**0**	**7/7**
7.1 (Big Creek)	**14**	**5.1**	5.7	18	**0**	**142**	**0**	5/7
Cuyahoga River - 1986								
42.6	24	6.5	0.9	40	**0**	274	**0**	2/7
42.6	30	8.2	1.6	36	**0**	434	**1**	2/7
42.6	24	7.4	5.2	47	1	246	**0**	1/7
40.4 (L. Cuya.)	20	6.3	9.9	49	1	262	**2**	1/7
40.4 (L. Cuya.)	24	6.9	3.4	45	1	436	**0**	1/7
40.4 (L. Cuya.)	**20**	6.5	3.1	42	1	250	**0**	2/7
38.6 (ust. Akron)	**16**	6.4	4.0	57	1	**106**	**3**	3/7
38.6 (ust. Akron)	**20**	6.5	1.9	65	1	338	**1**	2/7
38.6 (ust. Akron)	**18**	**5.5**	1.4	**75**	1	**146**	**0**	5/7
36.5 (dst. Akron)	**16**	**2.9**	**21.6**	**95**	**0**	**16**	**0**	**7/7**
36.5 (dst. Akron)	**16**	**4.3**	5.9	**87**	**0**	**46**	**1**	**6/7**
36.5 (dst. Akron)	**16**	**3.4**	**15.6**	**88**	**0**	**50**	**1**	**7/7**
33.6	**14**	**4.1**	6.8	**84**	**0**	**38**	**0**	**6/7**
33.6	**18**	**3.6**	5.0	**93**	**0**	**30**	**0**	**6/7**
33.6	**16**	**3.6**	**24.6**	**93**	**0**	**38**	**0**	**7/7**
26.7	**18**	**3.8**	**39.3**	**79**	**0**	**41**	**0**	**7/7**
26.7	**16**	**5.6**	**10.9**	59	**0**	**73**	**0**	**6/7**
26.7	**14**	**4.3**	**15.4**	**71**	**0**	**70**	**0**	**7/7**
24.1	**12**	**5.0**	**11.8**	**75**	**0**	**76**	**0**	**7/7**
24.1	**20**	**4.3**	5.1	**87**	**0**	**30**	**0**	**6/7**
24.1	**16**	**5.9**	9.8	68	**0**	162	**0**	**6/7**
21.0	**20**	**4.6**	**18.6**	**72**	**0**	**52**	**0**	**7/7**
21.0	**20**	**5.7**	9.9	58	**0**	**50**	**0**	5/7
21.0	**16**	**5.5**	2.9	**71**	**0**	**77**	**0**	**6/7**
Cuyahoga River - 1985								
42.6	30	8.5	0.0	28	**0**	334	**1**	2/7
40.4 (L. Cuya.)	**18**	**5.7**	4.7	50	1	150	**1**	3/7
40.4 (L. Cuya.)	**20**	7.4	3.5	52	1	286	**1**	2/7
40.4 (L. Cuya.)	**16**	**5.3**	0.9	63	1	168	**0**	3/7
38.6 (ust. Akron)	**20**	**5.7**	0.9	68	1	190	**0**	3/7
38.6 (ust. Akron)	**22**	6.2	1.4	69	1	170	**0**	2/7
38.6 (ust. Akron)	**16**	**5.6**	1.4	**83**	1	**92**	**1**	5/7
36.5 (dst. Akron)	**16**	**0.7**	**46.2**	**100**	**0**	**0**	**0**	**7/7**
36.5 (dst. Akron)	**12**	**1.4**	**57.1**	**86**	**0**	**4**	**0**	**7/7**
36.5 (dst. Akron)	**16**	**3.5**	1.9	**92**	**0**	**54**	**0**	**6/7**

APPENDIX TABLE 3 (CONTINUED)

Expression of Toxic Response Signatures among Selected Metrics and Aggregations of Fish Assemblage Data in Boatable Ohio Rivers Using the Biological Response Signatures Described by Yoder and Rankin (1995)

Location RM (Source)	IBI	Modified Iwb	%DELT Anomalies	Percent Tolerant	Intolerant Species	Density[a] (No./300m)	% Round-Bodied	#Toxic Signatures
33.6	**16**	**3.0**	**55.8**	**96**	**0**	**6**	**0**	**7/7**
33.6	**16**	**4.7**	**16.6**	**77**	**0**	**46**	**1**	**7/7**
33.6	**16**	**4.9**	**11.1**	57	**0**	**86**	**1**	**6/7**
26.7	**16**	**3.3**	**15.2**	**87**	**0**	**12**	**0**	**7/7**
26.7	**16**	**4.2**	**16.4**	**74**	**0**	820	**0**	**6/7**
24.1	**16**	**3.6**	**13.8**	**94**	**0**	**16**	**0**	**7/7**
24.1	**14**	**5.9**	**10.3**	66	**0**	**100**	**0**	5/7
24.1	**14**	**4.1**	6.4	**90**	**0**	**28**	**0**	**6/7**
21.0	**16**	**4.7**	4.6	**81**	**0**	**92**	**0**	**6/7**
21.0	**16**	**4.2**	6.7	**72**	**0**	**42**	**0**	**6/7**
21.0	**20**	**3.6**	**22.2**	**93**	**0**	**24**	**0**	**7/7**
17.3	**16**	**3.8**	0.6	44	**0**	370	**0**	5/7
15.9	**18**	**5.0**	2.0	**91**	**0**	**72**	**0**	**6/7**
13.1	**20**	6.3	1.1	17	**0**	582	**0**	3/7
11.5	**18**	6.5	**11.4**	33	**0**	520	**0**	4/7
9.8 (Southerly)	**16**	**5.1**	**11.4**	**76**	**0**	**80**	**0**	**7/7**
7.5	**16**	**3.6**	**20.0**	64	**0**	**18**	**0**	**6/7**
7.1 (Big Creek)	**14**	**5.9**	3.9	23	**0**	**140**	**0**	5/7
Cuyahoga River - 1984								
48.7	28	7.0	3.4	31	**0**	210	**0**	2/7
48.7	26	7.9	5.2	29	**0**	**138**	**0**	3/7
47.6	34	7.3	6.8	15	**0**	150	**0**	2/7
47.6	26	7.1	**10.2**	44	**0**	**98**	**0**	4/7
47.6	26	7.3	7.4	43	**0**	**108**	**0**	3/7
47.2	**22**	7.1	**10.2**	49	**0**	**104**	**0**	5/7
47.2	**18**	**5.7**	8.2	52	**0**	**58**	**0**	5/7
47.2	**22**	6.4	**13.3**	63	**0**	**44**	**0**	5/7
45.5	34	7.1	0.0	4	**0**	266	**0**	2/7
45.5	30	6.9	0.0	9	**0**	198	**0**	2/7
45.5	30	6.3	0.7	6	**0**	252	**0**	2/7
45.0	36	7.8	2.0	13	**0**	262	**0**	2/7
45.0	24	6.5	2.0	8	**0**	**144**	**0**	3/7
45.0	32	6.9	2.7	9	**0**	210	**0**	2/7
42.6	26	7.9	1.3	18	**0**	258	**0**	2/7
42.6	32	7.1	1.4	21	**0**	328	**0**	2/7
42.6	**22**	7.1	5.7	34	**0**	230	**1**	3/7
40.4 (L. Cuya.)	**18**	6.2	1.1	32	**0**	**120**	**0**	4/7
40.4 (L. Cuya.)	24	6.1	0.0	35	1	**118**	**0**	2/7
40.4 (L. Cuya.)	24	7.2	4.4	36	1	186	**0**	1/7
36.5 (dst. Akron)	**12**	**0.3**	0.0	0	**0**	**2**	**0**	5/7
36.5 (dst. Akron)	**12**	**0.0**	0.0	0	**0**	**0**	**0**	5/7
36.5 (dst. Akron)	**12**	**0.0**	0.0	0	**0**	**0**	**0**	5/7
33.6	**12**	**1.2**	0.0	0	**0**	**12**	**0**	5/7
33.6	**12**	**0.3**	0.0	0	**0**	**2**	**0**	5/7
33.6	**12**	**0.3**	0.0	0	**0**	**2**	**0**	5/7
26.7	**12**	**0.0**	0.0	0	**0**	**0**	**0**	5/7

APPENDIX TABLE 3 (CONTINUED)
Expression of Toxic Response Signatures among Selected Metrics and Aggregations of Fish Assemblage Data in Boatable Ohio Rivers Using the Biological Response Signatures Described by Yoder and Rankin (1995)

Location RM (Source)	IBI	Modified Iwb	%DELT Anomalies	Percent Tolerant	Intolerant Species	Density[a] (No./300m)	% Round-Bodied	#Toxic Signatures
26.7	**12**	**2.6**	7.7	15	**0**	**22**	**0**	5/7
26.7	**12**	**0.3**	0.0	0	**0**	**2**	**0**	5/7
24.1	**12**	**2.0**	0.0	33	**0**	**4**	**0**	5/7
24.1	**12**	**3.7**	**28.6**	57	**0**	**12**	**0**	**6/7**
24.1	**12**	**3.5**	0.0	59	**0**	**14**	**0**	5/7
21.0	**16**	**3.7**	7.4	19	**0**	**44**	**0**	5/7
21.0	**12**	**5.6**	4.0	52	**0**	**24**	**0**	5/7
21.0	**16**	**3.6**	3.3	**79**	**0**	**26**	**0**	**6/7**
17.3	**14**	**4.6**	**12.2**	51	**0**	**40**	**0**	**6/7**
17.3	**14**	**4.5**	8.6	31	**0**	**44**	**0**	5/7
17.3	**20**	**5.1**	**12.1**	45	**0**	**36**	**0**	**6/7**
17.3	**12**	**3.5**	0.0	38	**0**	**26**	**0**	5/7
15.9	**14**	**4.3**	3.7	33	**0**	**33**	**0**	5/7
15.9	**12**	**3.7**	0.0	36	**0**	**36**	**0**	5/7
15.9	**12**	**3.6**	8.7	57	**0**	**20**	**0**	5/7
15.9	**14**	**4.5**	**46.7**	19	**0**	**44**	**0**	**6/7**
13.1	**12**	**3.8**	4.7	56	**0**	**62**	**0**	5/7
13.1	**16**	**4.5**	8.1	35	**0**	**48**	**0**	5/7
13.1	**22**	**4.0**	1.1	**79**	**0**	**38**	**0**	**6/7**
13.1	**14**	**4.7**	**20.9**	17	**0**	**110**	**0**	**6/7**
11.5	**18**	**4.4**	0.0	24	**0**	**50**	**0**	5/7
11.5	**12**	**3.9**	5.9	41	**0**	**20**	**0**	5/7
11.5	**12**	**4.2**	**40.0**	48	**0**	**26**	**0**	**6/7**
11.5	**14**	**3.4**	**36.0**	40	**0**	**30**	**0**	**6/7**
9.8 (Southerly)	**16**	**3.8**	6.5	35	**0**	**40**	**0**	5/7
9.8 (Southerly)	**12**	**4.5**	0.0	29	**0**	**34**	**0**	5/7
9.8 (Southerly)	**14**	**4.2**	5.9	53	**0**	**32**	**0**	5/7
9.8 (Southerly)	**12**	**3.4**	**57.1**	0	**0**	**14**	**0**	**6/7**
7.5	**14**	**5.6**	7.0	34	**0**	**102**	**0**	5/7
7.5	**14**	**3.4**	**20.0**	16	**0**	**42**	**0**	**6/7**
7.5	**14**	**5.0**	3.4	21	**0**	**92**	**0**	5/7
7.1 (Big Creek)	**14**	**3.0**	**17.6**	**75**	**0**	**22**	**0**	**7/7**
7.1 (Big Creek)	**14**	**3.8**	**11.1**	22	**0**	**56**	**0**	**6/7**
7.1 (Big Creek)	**16**	**4.5**	0.0	30	**0**	**60**	**0**	5/7

[a] Metric 12 of Ohio EPA modified IBI for boat sites; numbers/km less highly tolerant species.
[b] Number of metrics or attributes that reflect a toxic response after Yoder and Rankin (1995).
[c] Values in boldface type are within toxic response criteria for that metric or attribute.

4 Using Biological Criteria for Establishing Restoration and Ecological Recovery Endpoints

Thomas P. Simon, Edward T. Rankin, Ronda L. Dufour, and Steven A. Newhouse

CONTENTS

4.1 INTRODUCTION

An increasing awareness among restoration ecologists is the need to restore damaged ecosystems to some model of a non-impacted environment (Benke, 1990; Fausch et al., 1990). Often these models are to pristine or pre-Columbian conditions and are concerned strictly with habitat (Hughes, 1995). Many pre-Columbian models are unrealistic expectations for industrial areas or heavily impaired urban streams; however, they may be appropriate for rural and least-impacted stream corridors. Environmental accidents such as spills, channelization, stream modifications, or other anthropogenic stresses are situations where biological criteria expectations can be used for evaluating stream restoration options in any stream corridor setting (Rankin and Simon, Chapter 10; Wilhelm et al., Chapter 14).

The principal goal of the Clean Water Act is to restore and maintain the chemical, physical, and biological integrity of the nation's surface waters. Karr and Dudley (1981) defined biological integrity as "the ability of an aquatic ecosystem, to support and maintain a balanced, integrated,

0-8493-0905-0/03/$0.00+$1.50

adaptive community of organisms having a species composition, diversity, and functional organization comparable to that of the natural habitats of a region." The U.S. Environmental Protection Agency (USEPA) in 1990 defined biological criteria as "numeric values or narrative expressions that describe the designated aquatic life uses." Biological criteria are used to define biological integrity goals for aquatic life designated uses.

Simon and Stewart (1999) used biological criteria developed for the Indiana portion of southern Lake Michigan to evaluate the influence of a public land program's protection of native fish communities. Management options that included supplemental stocking programs by wildlife agencies, increase of exotic and non-indigenous species, and large scale loss of habitat caused a large discrepancy between public land management and biological integrity goals for Northwest Indiana.

This chapter focuses on the premises behind ecological recovery endpoints and how they can be developed and used to restore habitats and aquatic communities to former levels of biological integrity. The development of ecological recovery endpoints is explained and this chapter explores how data from a variety of places can be used to set expectations. In addition, we will discuss limiting factors of this approach and tentative blocks to implementation based on two midwestern case studies.

4.2 USES OF ECOLOGICAL RECOVERY ENDPOINTS

4.2.1 Restoration of Damaged Systems

The need for a realistic restoration plan to ensure that a biological community is returned to a natural system is imperative before efforts are initiated. However, no measures have been proposed to determine when these goals are achieved. Previous efforts focused on chemical criteria as benchmarks of restored habitat or attempted to improve physical habitat. These approaches assume that if the chemical and physical limitations are removed, the biological system should be able to restore itself. Numerous studies conducted in Great Lakes areas of concern have shown that this approach is not adequate (Hartig and Zarull, 1992; Karr and Chu, 1999). To improve the biological integrity of damaged systems and enable them to return to biologically stable levels, it is necessary to have finite goals and measurement indicator objectives.

Moyle and Marchetti (1999) developed biological integrity goals using multimetric indices explicitly for the purpose of restoring aquatic communities in the Sierra Nevada streams of California. No additional efforts have been proposed to evaluate restoration options of aquatic assemblages, with the exception of the Leading Creek or Meigs Mine 31 discharge and Grand Calumet River natural resource damage assessment (NRDA).

4.2.2 Natural Resource Damage Assessment Targets

The premise behind natural resource damage assessment (NRDA) is that natural resources at damaged sites would be restored to "baseline" conditions. *Baseline* is not defined explicitly within the regulation; however, it is presumed to include the restoration of communities to represent mostly native species typical of that region. This definition is consistent with biological integrity goals. As more disturbance systems are observed, it becomes increasingly difficult to know what predisturbance conditions would or should have been. The loss of biological integrity is not necessarily limited to already disturbed places. As reference sites are degraded and become increasingly rarer, the development of expectations has become progressively more important (Simon and Stewart, 1999; Reash, 1999).

The development of reference conditions for rare resource types, i.e., wetlands, large rivers, and Great Lakes nearshore, is difficult to determine; however, suitable reference conditions can be developed when evaluated at ecoregion or watershed scale (Hughes, 1995; Simon and Emery, 1995;

Simon and Sanders, 1999). Simon and Emery used biological community attributes from least-impacted sites on the upper Ohio River to develop a model of community expectations and formulated an index of biotic integrity.

4.3 COMPONENTS OF ECOLOGICAL RECOVERY ENDPOINTS

The variety of ecological recovery endpoints that can be chosen for post-impact monitoring depends on the amount of pre-impact data collected. Use of historical data, anecdotal information, and current monitoring data can be used to develop the endpoints. The importance of ecological recovery endpoints as a baseline for establishing when restoration objectives have been reached is illustrated in Figure 4.1. The process depends on obtaining biological data at approved monitoring stations that are representative of the impacted area. As these areas are sampled over time, trajectories of ecological integrity can be monitored and multimetric indices and biological criteria can be calculated and compared to recovery endpoints. For ecological recovery endpoints to be successfully implemented, it is necessary to evaluate multiple trophic levels of ecological organization. A minimum of two biological indicators should be monitored and assessed (Davis and Simon, 1989).

4.3.1 INDEX OF BIOTIC INTEGRITY GOALS

The index of biotic integrity (IBI) is an important monitoring endpoint for restoration since it is capable of determining whether the impacted area is similar to reference conditions from the same ecoregion. The IBI is a family of multimetric indices originally developed by Karr (1981) and expounded upon by Karr et al. (1986). The index was developed to explain patterns in midwestern stream fish communities. It has been greatly expanded to a variety of applications including fish (Simon and Lyons, 1995; Hughes and Oberdorff, 1999), macroinvertebrates (Kearns and Karr, 1994), and aquatic plant assemblages (Simon et al., 2001). The index is based on species composition, indicator species, species sensitivity and tolerance, trophic feeding dynamics, abundance, and condition of individuals. The power of IBI is that it evaluates a variety of ecological levels of organization including ecosystems, communities, populations, and individuals.

The IBI can be used as an ecological recovery endpoint since it can verify whether a site has been restored to similar ecoregion levels of biological integrity. Ecoregions are areas of similar geological history, land use, soil, and potential natural vegetation (Omernik, 1987). The assumed expectation is that similar areas should have similar biological communities and integrity. For example, historical or anecdotal data for a disturbed stream may not be available. Post-disturbance data collected from the site could be assumed to at least approximate an ecoregion average. Thus, ecoregion means can be compared to post-impact data and recovery trajectories can be monitored.

4.3.2 MACROINVERTEBRATE GOALS

Several macroinvertebrate multimetric indices have been developed to assess the structure and function of aquatic macroinvertebrate assemblages (Cummins and Wilzbach, 1985; Rosenberg and Resh, 1993; Resh and Jackson, 1993). These include a variety of indices. The three most commonly used are the invertebrate community index (ICI), macroinvertebrate index of biotic integrity (mIBI), and the rapid bioassessment protocols (RBPs).

Like the IBI, the macroinvertebrate indices are regionally calibrated and include species composition, functional feeding guilds, sensitivity and tolerance values, relative abundance, and univariate measures, e.g., Hilsenhoff biotic index, Ephemeroptera, Plecoptera, and Tricoptera ratios. Even without pre-impact data, the macroinvertebrate assemblage data can be used to monitor recovery at a site.

The invertebrate community index was developed by the Ohio Environmental Protection Agency to evaluate Ohio stream communities (DeShon, 1995). The index is based on the lowest levels of

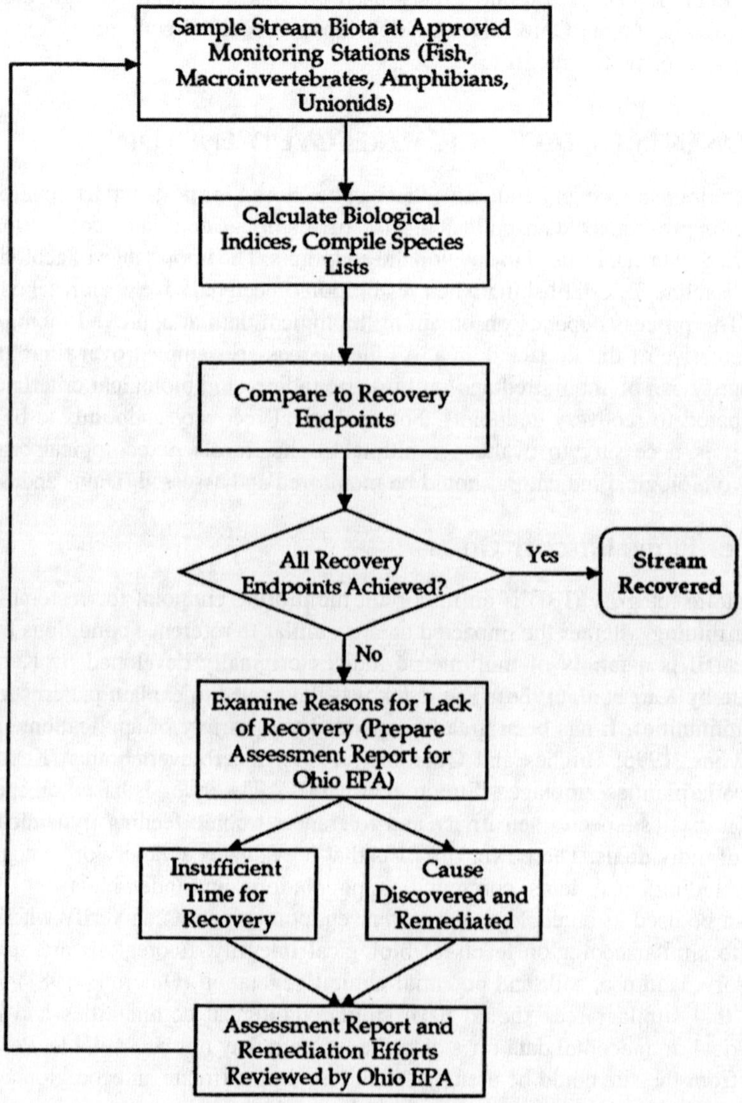

FIGURE 4.1 Process for assessing progress toward ecological recovery endpoints in stream segments.

taxonomic identifications, usually genus and species levels. The ICI scores range from 0 to 60 points. The scores are determined like the IBI from either ecoregion or statewide calibration of macroinvertebrate communities. This index has been used extensively in Ohio and has been modified for use in Indiana (Sobiech et al., 1994).

The macroinvertebrate index of biotic integrity (mIBI) was developed for use on Tennessee Valley Authority reservoirs (Kearns and Karr, 1994). It is modeled after the IBI and includes structure and function attributes of macroinvertebrate communities. The mIBI has not been used to the same extent that the RBPs and the ICI have been used.

The RBPs are perhaps the most widely used macroinvertebrate procedures (Plafkin et al., 1989; Barbour et al., 1997). The index is modified based on the level of taxonomic identification. The premise is that the riffle community within a stream is the most diverse portion of the site assemblage. Kick nets are used to sample a variety of habitats and dip nets are used to sample woody debris and other structures if riffles are limited. This method was specifically adapted for low-flow streams along the coastal plain of Delaware (Maxted et al., 1997), channelized streams of Florida (Barbour et al., 1996), ditches in Illinois (Barbour et al., unpublished), statewide criteria for Indiana (Simon et al., 2000), Cuyahoga River watershed in Ohio (Stewart et al., 1998), and northern lakes and forest ecoregion in Minnesota (Butcher, 2001).

The RBP provides a representative sample that is collected and identified following specific subsampling protocols either in the field to family level (Level II) or in the laboratory to the lowest taxonomic level (Level III). The expectations for the site are based on upstream controls or regional standards. Problems in using the RBPs are a function of the absence of riffle habitats or the lack of an ecoregion calibration. The RBP provides the recommendation that a specific application is modified regionally, but many applications have not been modified at all.

4.3.3 Univariate Biological Indicator Measures

Univariate biological criteria may include a variety of biological indices, such as the index of well-being (Gammon, 1976), floristic quality index (Swink and Wilhelm, 1994), or Hilsenhoff biotic index (Hilsenhoff, 1982). A variety of other measurement indices, such as diversity indices (e.g., Shannon-Weiner diversity) or population or stock indices (e.g., proportional stock densities, relative stock density, or relative weight indices) can be used as substitutes for these types of ecological recovery endpoints when the primary indicator may be a single species or aspect of the community that has been disturbed.

Univariate indices provide another dimension that the community based multimetric indices may not. For example, age structure, biomass, sensitivity to specific types of pollutants, and changes in weight with length are not measured by the multimetric indices but are important components of univariate indices. Univariate indices can also be used when pre-impact data is lacking or missing entirely by comparing the ecoregion or watershed mean against the post-discharge data.

4.3.4 Key Species Targets

Key species indicator taxa have long been important indicators of biological recovery. Some of the earliest biological response indicators were specific organisms or were based on simple indices, e.g., saprobic index (Cairns and Pratt, 1993; Davis, 1995). Indicator species constitute a species or species assemblage that has particular requirements with regard to a known set of physical or chemical variables such that changes in presence, numbers, morphology, physiology, or behavior of that species indicate that the given physical or chemical variables are outside its preferred limits. Johnson et al. (1993) reviewed sentinel species of benthic macroinvertebrates based on sensitivities to specific contaminants. The aquatic life criteria documents published by the USEPA review the literature on toxicities of specific contaminants to a variety of aquatic life and stages. Key taxa should be ideal indicators and exhibit the following attributes (Rosenberg and Wiens, 1976; Hellawell, 1986):

1. **Taxonomic soundness and easy recognition by the non-specialist**. Taxonomic uncertainties will complicate long-term monitoring and between-site interpretations.
2. **Cosmopolitan distribution**. Choice of a cosmopolitan species allows for comparative studies on regional, national, and international scales.
3. **Numerical abundance**. The numerical predominance of an indicator species allows for ease of sampling and for conclusions regarding quantitative distribution patterns.
4. **Low genetic and ecological variability**. Indicators should have narrow ecological demands.
5. **Large body size**. This factor facilitates sampling and sorting.
6. **Limited mobility and relatively long life history**. This factor allows ease of integration on spatial and temporal scales.

7. **Well known ecological characteristics**. Background physiological and autecological information should be widely available.
8. **Suitability for laboratory studies**. This may allow determination of cause and effect.

Properties of a good indicator may include responses to biochemical changes during life history, bioaccumulation, presence/absence or numerical predominance of organism populations or species assemblages, morphological deformities, and changes in community composition. Key indicator taxa may be representative of specific habitat types but should not be so rare that they would not be easily encountered during routine sampling.

4.4 CASE STUDIES: MODIFICATIONS TO ORIGINAL CONCEPTS

4.4.1 Leading Creek

The Leading Creek and Raccoon Creek watersheds and several of their tributaries received over 1.1 billion gallons of acid mine drainage effluent when the largest long-wall coal mine in the United States collapsed. Storage water had been pumped into an adjacent abandoned mine that collapsed and flooded the lower mine, Meigs 31. Water was pumped from the mine into several streams that drained into the Ohio River. Leading Creek suffered the most extensive damage from the Meigs 31 discharges. Four other tributaries of the Raccoon Creek watershed also received significant amounts of mine effluent. Essentially, the entire fish, aquatic insect, unionid mussel, and amphibian assemblages were eliminated from Parker Run, the stream that received mine effluent, downstream from the Meigs 31 discharge and from Leading Creek including 0.5 miles upstream from Parker Run. This upstream mortality was a result of the high volume of the discharge that created a reverse flow in the stream.

Leading Creek was designated a state resource water (SRW) in southeastern Ohio that achieved warmwater habitat (WWH) designation based on aquatic life criteria for the Allegheny Plateau ecoregion. The historical and pre-discharge data for Leading Creek showed that post-discharge biological integrity declined substantially after the release of the mine wastewater (Figure 4.2). Essential to enhancing the full recovery of Leading Creek was facilitating the upstream migration of species from downstream to upstream. However, a low-head dam at approximately RM 11.0 was a barrier to the migration of fish and unionid mussels glochidia. The removal of the old WPA dam facilitated access to reaches of Leading Creek upstream from the areas impacted by the discharge. These areas are necessary spawning habitats for the state-endangered silver lamprey, *Ichthyomyzon unicuspis*. Most of the best habitat in Leading Creek is located upstream from this dam. Ohio EPA (1994) suggests that stream recovery occurs more quickly when re-population occurs from the downstream reaches.

Historically, Leading Creek contained 49 fish species, 162 macroinvertebrate taxa (excluding unionids), and 10 species of unionid mussels (Ohio EPA, 1994). In addition, during the investigation of the fish kill, mudpuppies *Necturus maculosus,* were collected from multiple sites in middle Leading Creek (RM 9.1 to 15.6 or segment 3), where mostly dead individuals were collected during the summer discharge of untreated mine water were dead.

Post-discharge monitoring required that strict adherence to the standard operating procedures developed by Ohio EPA (Ohio EPA, 1989) be followed by company consultants, state and federal biologists, and others responsible for monitoring. As part of the court settlement, a recovery endpoints document specified the levels of recovery expected for fish, macroinvertebrate, unionid mussel, and amphibian communities that would meet previous designated uses (Figure 4.3). Cumulative frequency distribution plots of taxa or fish species versus number of samples were developed for Strongs Run, Robinson Run, Sugar Run, Leading Creek and Parker Run. These regression lines were developed to show the expected levels based on historical data collected pre-discharge

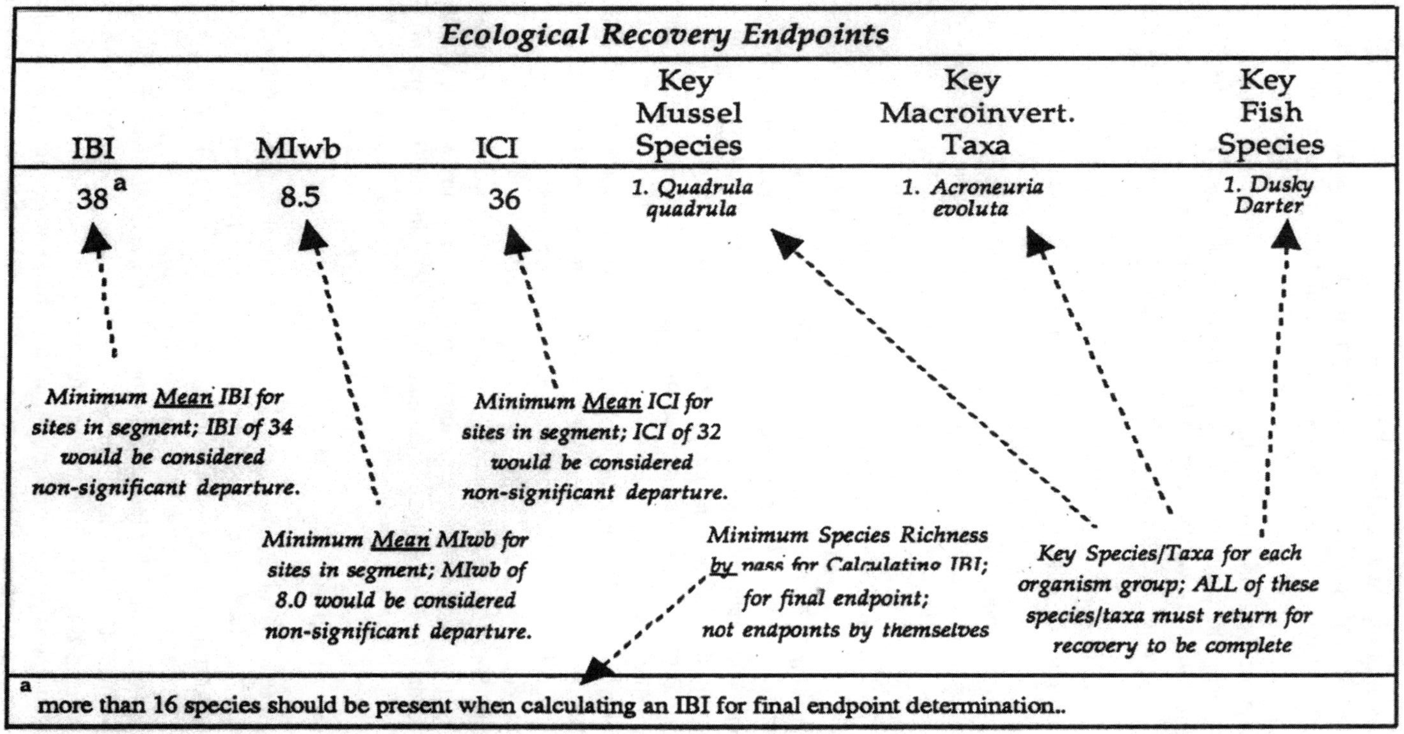

FIGURE 4.2 Example of ecological recovery endpoints of the type used for three segments of Leading Creek, Racoon Creek, Strongs Run, Sugar Run, Robinson Run, and Parker Run.

(Figure 4.4). Specific monitoring points were established by Ohio EPA for the company to monitor. However, several of the stations were subsequently changed to facilitate access and ensure greater accuracy in reaching the baseline. In addition, the Leading Creek watershed was divided into three segments that were used to evaluate longitudinal gradient changes in the river and specific key taxa were selected for each of the three reaches (Table 4.1).

The requirements that Ohio EPA established for sampling specified that a minimum number of species had to be collected before an IBI could be calculated for final endpoint determination. This was established so that if fewer species than expected were collected, an acceptable IBI showing that the site met designated uses (thus causing a false negative or type I error) could not be calculated. Segments where mudpuppies were collected after the spill had to show that mudpuppies were present in one of two segments of Leading Creek. Finally, key species were used to evaluate sensitive species recolonization and had to be present as sexually mature, reproducing adults and recruits.

Modifications of the recovery endpoints were required as implementation occurred for a variety of reasons. The consultants for the company knew that minimum numbers of individuals had to be collected to calculate an IBI, so they sampled excessively at a single site in order to collect the minimum number of individuals. This amount of effort in a small creek system was something the recovery endpoint developers had not considered as a rational approach. Interim IBI's were calculated by the consultants based on the less-than-sufficient numbers of species required in order to evaluate trends in recovery. The lack of a suitable assemblage structure caused numerous instances where the consultants indicated successful recovery shortly after the spill that were based on less-than-expected species richness and specimens less than 25 mm in total length. Problems with the key species were less significant. The most notable problem arose when only a single individual of the key species was collected. The consultants wanted the single fish to be counted as both a sexually mature adult and a new recruit. Neither scenario was considered valid and the controversy led to the development of specific sampling guidelines, e.g., a maximum number of minutes for sampling and multiple years of monitoring were necessary to show that mature adults and new recruits were coming into the site.

A second problem occurred when a very rare species, river redhorse, *Moxostoma carinatum*, was selected as a key species for the lower river. The river redhorse is a mollusk feeder that perhaps was previously collected as a transient species feeding on the mussel fauna near the mouth of Leading Creek. The company attempted to collect river redhorse in the spring, summer, and fall and never found any. Since the food base for the creek had not recovered and river redhorse was represented by only a single individual according to pre-discharge survey data, an allowance was made for special studies *in lieu* of the documentation of the redhorse. The special studies were designed to provide additional information on the biology and habitat requirements of the species.

4.4.2 Grand Calumet River

The Grand Calumet River is an area of concern in the Great Lakes. It is near Gary, in northwest Indiana (Figure 4.5). The river is the only area of concern with all 14 designated uses considered impaired by the International Joint Commission (IJC, 1989). Biological criteria developed from the Central Corn Belt Plain ecoregion have been applied (Simon, 1991). The area includes a set of criteria developed specifically for Indiana's portion of the Lake Michigan drainage.

Simon (1991) replaced the number of sucker species with an alternative metric (number of minnow species) developed for the Central Corn Belt Plain that was applied in headwater streams in the Lake Michigan drainage. An increasing number of minnow species was considered a good reflection of high biological integrity; however, this was based on the presence of native minnows that were not specified in the original calibration.

Numerous surveys were conducted between 1985 and 1999 to evaluate the biological integrity of the Grand Calumet River. An evaluation of exotic and non-indigenous species in the river was

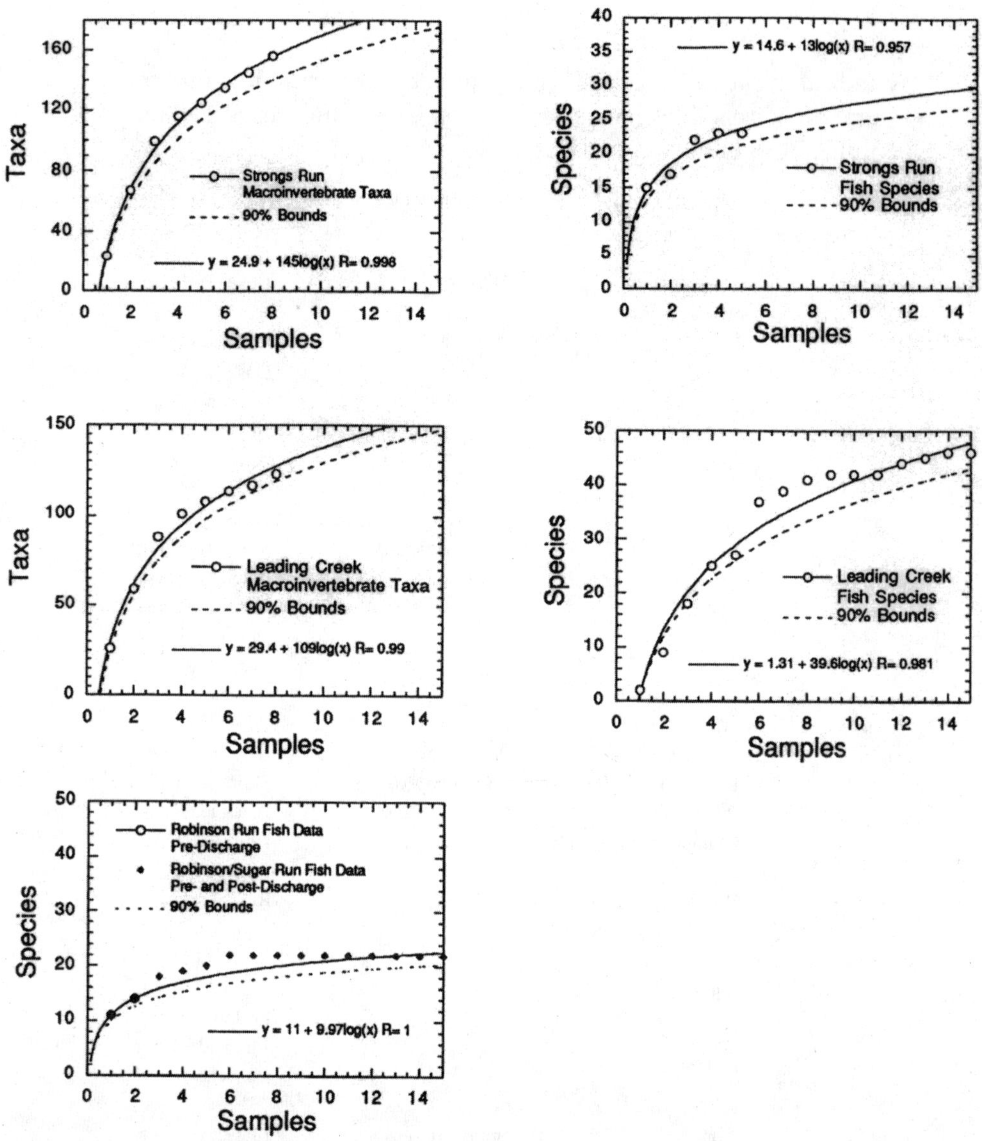

FIGURE 4.3 Cumulative frequency distribution plots of taxa (macroinvertebrates) or fish species vs. number of samples for Strongs Run, Robinson/Sugar Run, and Leading Creek/Parker Run. Dashed lines indicate the 90% of species represented by the regression line.

investigated; however, one incident specifically showed how misapplication of IBI criteria can cause problems in restoration goals. Simon et al. (Chapter 25) found that a significant number of exotic minnow species were present in the Grand Calumet River. Carp, *Cyprinus carpio*, goldfish *Carassius auratus*, and rudd *Scardinius erythrophthalmus,* are all exotic Asian species that have become established in the Grand Calumet River. In addition, golden shiner, *Notemigonus crysoleucas,* and bluntnose minnow *Pimephales notatus,* were increasing in abundance at these sites. The addition of these two species to the ubiquitous exotic species caused some sites to score well above the expected reference condition. Their presence at several sites actually caused the IBI metric for number of minnow species to achieve the highest score (5). Changes in the metric had to be made to reflect the need for native minnows at these sites.

TABLE 4.1
Key Indicator Species Used as Ecological Recovery Endpoints for the Restoration of Leading Creek after the Collapse of the Meigs 31 Mine in Southeastern Ohio

	Key Indicator Species		
Stream/Segment	**Mussel**	**Macroinvertebrate**	**Fish**
Leading Creek (Segment 1)	None	None	Channel shiner Longear sunfish Dusky darter Smallmouth buffalo Adult spotted bass Adult channel catfish
Leading Creek (Segment 2)	None	*Acroneuria evoluta* *Orconectes sanbornii* *Polycentropodidae* *Hydropsychidae*	Sand shiner Channel shiner Silverjaw minnow Longear sunfish Blackside darter Dusky darter Logperch Golden redhorse
Leading Creek (Segment 3)	Wabash pigtoe Fatmucket Pink heelsplitter Plain pocketbook Cylindrical papershell Giant floater White heelsplitter Fragile heelsplitter Mapleleaf Squawfoot	*Acroneuria evoluta* *Orconectes sanbornii* *Polycentropodidae* *Hydropsychidae*	Silver lamprey River redhorse Shorthead redhorse Black redhorse Northern hogsucker Sand shiner Redfin shiner Longnose gar Spotted bass Longear sunfish Fantail darter Blackside darter Logperch
Parker Run	None	*Acroneuria evoluta*	Blacknose dace Southern redbelly dace Silverjaw minnow Fantail darter
Raccoon Creek	Pink heelsplitter Fragile heelsplitter	*Cheumatopsyche* sp. *Stenacron* sp. *Stenonema* sp. *Isonychia* sp. *Caenis* sp.	Golden redhorse Northern hogsucker Spotted sucker Adult spotted bass Longear sunfish Warmouth Dusky darter Blackside darter Grass pickerel
Strongs Run	None	*Centroptilum* sp. *Stenacron* sp. *Stenonema femoratum*	Least brook lamprey Longear sunfish Southern redbelly dace

TABLE 4.1 (CONTINUED)
Key Indicator Species Used as Ecological Recovery Endpoints for the Restoration of Leading Creek after the Collapse of the Meigs 31 Mine in Southeastern Ohio

	Key Indicator Species		
Stream/Segment	**Mussel**	**Macroinvertebrate**	**Fish**
		Caenis sp. *Cheumatopsyche* sp.	Dusky darter
Sugar Run	None	*Centroptilum* sp. *Paraleptophlebia* sp. *Stenonema femoratum* *Hexagenia* sp. *Caenis* sp.	Least brook lamprey Redfin shiner Southern redbelly dace
Robinson Run	None	*Stenacron* sp. *Isonychia* sp. *Cheumatopysche* sp.	Least brook lamprey Redfin shiner Southern redbelly dace

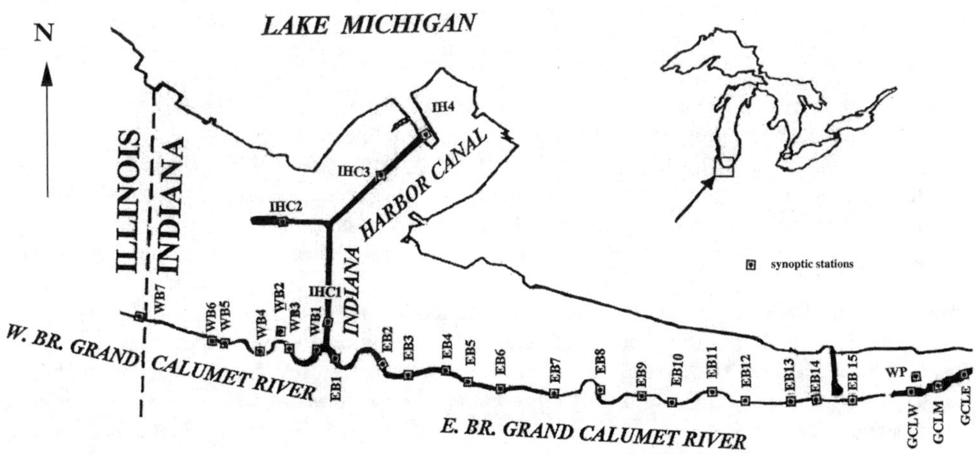

FIGURE 4.4 Grand Calumet River study area in northwest Indiana.

4.5 CONCLUSIONS

Attainable restoration goals and endpoints for restoring the biological integrity of water resources to degraded rivers and streams can be achieved by using ecological recovery endpoints as measures of success. The agency or primary responsible party needs to know to what level restoration should proceed. Restoration goals are often based solely on chemical and physical habitat information because of a lack of suitable biological information for the area under consideration. Ecological recovery endpoints are based on patterns in regionally calibrated multimetric indices for several biological indicators, key species concepts, univariate biological criteria (e.g., Hilsenhoff biotic indices, floristic quality index, index of well-being) and other stock assessment indices.

As restoration biologists attempt to restore watersheds that have had long histories of degraded water quality, it becomes increasingly more difficult to find pre-impact information. However, if historical data are not available, biologists can use ecological recovery endpoints for evaluating community structure and function, based on ecoregion means, presence and absence information, or pre-impact survey data.

Ecological recovery endpoints were suitable models for setting restoration goals after a large acid mine drainage impact occurred in several small creeks in southeastern Ohio. The endpoints were also suitable models for a natural resource damage assessment for the Grand Calumet River watershed in northwest Indiana — a Great Lakes area of concern. Simon (2000) suggested that in their broadest sense biological criteria are based on numerical and narrative biological information. The primary goal is to restore the chemical, physical, and biological integrity of water resources via direct measures of ecological integrity.

ACKNOWLEDGMENTS

The authors wish to express their gratitude to a number of colleagues whose insights and collaboration through the years assisted in the development of these ideas. We specifically thank Chris Yoder, Jim Karr, Jeff DeShon, Robert Hughes, Mike Barbour, Wayne S. Davis, and Michael Stewart. The opinions expressed in this manuscript do not necessarily reflect those of the agencies. No official endorsement should be inferred.

REFERENCES

Barbour, M.T., J. Gerritsen, G.E. Griffith, R. Frydenborg, E. McCarron, and J.S. White. 1996. A framework for biological criteria for Florida streams using benthic m macroinvertebrates, *Journal North American Benthological Society,* 15, 185–211.

Barbour, M.T., J. Gerritsen. Unpublished. Biological criteria development for streams and rivers of Illinois using benthic macroinvertebrates. Illinois Environmental Protection Agency, Springfield, IL.

Barbour, M.T., J. Gerritson, B.D. Snyder, and J.B. Stribling. 1997. Revision to Rapid Bioassessment Protocols for Use in Rivers and Streams: Periphyton, Benthic Macroinvertebrates, and Fish. EPA 841-D-97–002. U.S. Environmental Protection Agency, Office of Water, Washington, D.C.

Benke, A.C. 1990. A perspective on America's vanishing streams, *Journal North American Benthological Society,* 9, 77–88.

Butcher, J.T. 2001. The Development of Watershed Indicators in the Northern Lakes and Forests Ecoregion Using Benthic Macroinvertebrates. M.S. Thesis, Purdue University, West Lafayette, IN.

Cairns, J., Jr. and J.R. Pratt. 1993. A history of biological monitoring using benthic macroinvertebrates, in D.M. Rosenberg and V.H. Resh (Eds.). *Freshwater Biomonitoring and Benthic Macroinvertebrates.* Chapman & Hall, New York, 10–27.

Cummins, K.W. and M.A. Wilzbach. 1985. *Field Procedures for Analysis of Functional Feeding Groups of Stream Macroinvertebrates.* Contribution 1611. Appalachian Environmental Laboratory, University of Maryland, Frostburg, MD.

Davis, W.S. 1995. Biological assessment and criteria: building on the past, in W.S. Davis and T.P. Simon (Eds.). *Biological Assessment and Criteria: Tools for Water Resource Planning and Decision Making.* Lewis Publishers, Boca Raton, FL, 15–30.

Davis, W.S. and T.P. Simon. 1989. Sampling and data evaluation requirements for fish and macroinvertebrate communities, in T.P. Simon, L.L. Holst, and L.J. Shepard (Eds.). *Proceedings of the 1989 Midwest Pollution Control Biologists Meeting, 14–17 February 1989.* EPA 905–9–90–005. U.S. Environmental Protection Agency, Region 5, Environmental Sciences Division, Chicago, IL.

DeShon, J.E. 1995. Development and application of the Invertebrate Community Index (ICI), in W.S. Davis and T.P. Simon (Eds.). *Biological Assessment and Criteria: Tools for Water Resource Planning and Decision Making.* Lewis Publishers, Boca Raton, FL, 217–244.

Fausch, K.D., Lyons, J., Karr, J.R., and Angermeier, P.L. 1990. Fish communities as indicators of environmental degradation, *American Fisheries Society Symposium* 8, 123–136.

Gammon, J.R. 1976. *The Fish Populations of the Middle 340 km of the Wabash River.* Purdue University Water Resources Research Center Technical Report 86, Lafayette, IN.

Hartig, J.H. and M.A. Zarull (Eds.). 1992. *Under RAPs: Toward Grassroots Ecological Democracy in the Great Lakes Basin.* University of Michigan Press, Ann Arbor.

Hellawell, J.M. 1986. *Biological Indicators of Freshwater Pollution and Environmental Management.* Elsevier Press, London.

Hilsenhoff, W.L. 1982. Using a biotic index to evaluate water quality in streams. *Wisconsin Department of Natural Resources Technical Bulletin 132.*

Hughes, R.M. 1995. Defining acceptable biological status by comparing with reference conditions, in W.S. Davis and T.P. Simon (Eds.). *Biological Assessment and Criteria: Tools for Water Resource Planning and Decision Making.* Lewis Publishers, Boca Raton, FL, 31–48.

Hughes, R.M. and T. Oberdorff. 1999. Applications of IBI concepts and metrics to waters outside the United States and Canada, in T.P. Simon (Ed.). *Assessing the Sustainability and Biological Integrity of Water Resources Using Fish Communities.* CRC Press, Boca Raton, FL, 79–96.

Johnson, R.K., T. Wiederholm, and D.M. Rosenberg. 1993. Freshwater biomonitoring using individual organisms, populations, and species assemblages of benthic macroinvertebrates, in D.M. Rosenberg and V.H. Resh (Eds.). *Freshwater Biomonitoring and Benthic Macroinvertebrates.* Chapman & Hall, New York. 40–158.

Karr, J.R. 1981. Assessment of biotic integrity using fish communities. *Fisheries* 6: 21–27.

Karr, J.R., K.D. Fausch, P.L. Angermeier, P.R. Yant, and I.J. Schlosser. 1986. *Assessing Biological Integrity in Running Waters: A Method and its Rationale.* Illinois Natural History Survey Special Publication 5, Champaign, IL.

Karr, J.R. and E.W. Chu. 1999. *Restoring Life in Running Waters: Better Biological Monitoring.* Island Press, Washington, D.C.

Karr, J.R. and D.R. Dudley. 1981. Ecological perspective on water quality goals, *Environmental Management*, 5, 55–68.

Kerans, B.L. and J.R. Karr. 1994. Development and testing of a benthic index of biotic integrity (B-IBI) for rivers of the Tennessee Valley, *Ecological Applications,* 4 (4), 768–785.

Maxted, J.R., S.B. Weisberg, J.C. Chailou, R.A. Eskin, and F.W. Kutz. 1997. Ecological condition of dead-end canals of the Delaware and Maryland coastal bays, *Estuaries*, 20, 319–327.

Moyle, P.B. and M.P. Marchetti. 1999. Application of indices of biotic integrity to California streams and watersheds, in T.P. Simon (Ed.). *Assessing the Sustainability and Biological Integrity of Water Resources Using Fish Communities.* CRC Press, Boca Raton, FL. 367–382.

Ohio Environmental Protection Agency. 1989. *Biological Criteria for the Protection of Aquatic Life. Volume II. Users Manual for Biological Field Assessment of Ohio Surface Waters.* Ohio EPA, Division of Water Quality Monitoring and Assessment, Surface Water Section, Columbus, OH.

Ohio Environmental Protection Agency. 1994. *Ecological Recovery Endpoints for Streams Affected by the Meigs #31 Mine Discharges during July-September 1993.* Ohio EPA, Division of Surface Water, Ecological Assessment Section, Columbus, OH.

Omernik, J.M. 1987. Ecoregions of the conterminous United States, *Annals of the Association of American Geographers*, 77, 118–125.

Plafkin, J.L., M.T. Barbour, K.D. Porter, S.K. Gross, and R.M. Hughes. 1989. *Rapid Bioassessment Protocols for Use in Streams and Rivers. Benthic Macroinvertebrates and Fish.* EPA 440–4–89–001. Office of Water Regulation and Standards, U.S. Environmental Protection Agency, Washington, D.C.

Reash, R.J. 1999. Considerations for characterizing midwestern large river habitats, in T.P. Simon (Ed.), *Assessing the Sustainability and Biological Integrity of Water Resources Using Fish Communities.* CRC Press, Boca Raton, FL, 463–474.

Resh, V.H. and J.K. Jackson. 1993. Rapid assessment approaches to biomonitoring using benthic macroinvertebrates, in D.M. Rosenberg and V.H. Resh (Eds.), *Freshwater Biomonitoring and Benthic Macroinvertebrates*, Chapman & Hall, New York, 195–233.

Rosenberg, D.M. and V.H. Resh. 1993. *Freshwater Biomonitoring and Benthic Macroinvertebrates*, Chapman & Hall, New York.

Rosenberg, D.M. and A.P. Wiens. 1976. Community and species responses of *Chironomidae* (*Diptera*) to contamination of fresh waters by crude oil and petroleum products, with special reference to the Trail River, Northwest Territories, *Journal of Fisheries Research Board of Canada,* 33, 1955–1963.

Simon, T.P. 1991. *Development of Ecoregion Expectations for the Index of Biotic Integrity. I. Central Corn Belt Plain.* EPA 905–9–91–025. U.S. Environmental Protection Agency, Environmental Sciences Division, Chicago, IL.

Simon, T.P. and E.B. Emery. 1995. Modification and assessment of an index of biotic integrity to quantify water resource quality in Great Rivers, *Regulated Rivers Research and Management*, 11, 283–298.

Simon, T.P. and J. Lyons. 1995. Application of the Index of Biotic Integrity to Evaluate Water Resource Integrity in Freshwater Ecosystems, in W.S. Davis and T.P. Simon (Eds.), *Biological Assessment and Criteria: Tools for Water Resource Planning and Decision Making*. Lewis Publishers, Boca Raton, FL. 245–262.

Simon, T.P. and R.E. Sanders. 1999. Applying an index of biotic integrity based on Great River fish communities: considerations in sampling and interpretation, in T.P. Simon (Ed.), *Assessing the Sustainability and Biological Integrity of Water Resources Using Fish Communities*. CRC Press, Boca Raton, FL. 475–505.

Simon, T.P. and P.M. Stewart. 1999. Structure and function of Lake Michigan fish assemblages: with emphasis on the restoration of native fish communities, *Natural Areas Journal,* 19, 142–154.

Simon, T.P., P.M. Stewart, and P.L. Rothrock. 2001. Development of an index of biotic integrity for plant assemblages (P-IBI) in southern Lake Michigan, *Aquatic Ecosystem Health and Management,* 4, 293–309.

Simon, T.P., D.A. Sparks, R. Dufour, and S.A. Newhouse. 2000. *Baseline Determination for the Grand Calumet River and Indiana Harbor Canal.* U.S. Fish and Wildlife Service, Bloomington, IN.

Simon, T.P., S.A. Sobiech, D.A. Sparks, K. Jopp. 2002. Assessing the ecological integrity of the East Branch Grand Calumet River: response of four biological indicators, in T.P. Simon (Ed.). *Biological Response Signatures: Patterns in Biological Integrity for Assessment of Freshwater Aquatic Assemblages.* CRC Press, Boca Raton, FL.

Sobiech, S.A., T.P. Simon, and D.A. Sparks. 1994. *Preremedial Biological and Water Quality Assessment of the East Branch Grand Calumet River, Gary, IN.* U.S. Fish and Wildlife Service, Bloomington, IN.

Stewart, P.M., P. Hudson, J.T. Butcher, and R. Hesselberg. 1998. *Benthic Macroinvertebrate and Polycyclic Aromatic Hydrocarbon Inventory in Tributaries to the Cuyahoga River at the Cuyahoga Valley National Recreation Area.* U.S. Geological Survey, Great Lakes Science Center, Lake Michigan Ecological Research Station, Porter, IN.

Swink, F. and G. Wilhelm. 1994. *Plants of the Chicago Region*, 4th ed. Indiana Academy of Science, Indianapolis.

Section II

Contaminant Patterns in Ecosystems

5 Effects of Contaminated Dredge Spoils on Wetland Plant Communities: A Literature Review

Paul M. Stewart, Eric L. Garza, and Jason T. Butcher

CONTENTS

5.1 INTRODUCTION

Contaminated dredge spoil disposal is a national concern due to its scope and effects on biota, water quality, and the physical environment. In 1998, over $500 million was spent on dredging in the United States (Briuer, 1998). About 350 million tons of sediment are dredged each year, most of which is removed from the Gulf of Mexico drainage. New Orleans, Louisiana moves 60 million tons of river and harbor sediments each year, while about 4 million tons of sediments are moved from the Great Lakes drainage. About 20% of all dredged sediment is ocean-disposed and the remaining sediments are divided equally between freshwater and terrestrial disposal (R.M. Engler, U.S. Army Corps of Engineers, 2000, personal communication). The challenge for the 21st century is balancing navigation dredging needs with environmental protection (Briuer, 1998).

Dredging is done for several reasons, including harbor maintenance for industrial and recreational activities and protection of the public by removal and disposal of contaminants (Great Lakes Commission, 1999). Disposal of dredge material can have varied effects on both plant and animal communities. Dried dredge material often oxidizes, causing salinity (ion content) and pH to decrease drastically, making the sediment inhospitable to most plants and animals (Lee and Brandon, 1991). Contaminants may also leach out of disposed sediments, contaminating surrounding areas; the

0-8493-0905-0/03/$0.00+$1.50

kinetics of leachate dynamics depend on a variety of factors including time of sediment–water contact, sediment salinity, and sediment oxidation status (Brannon et al., 1990). Volatilization of contaminants (VOCs including PCBs and PAHs) is also a known transfer pathway (Semmler, 1990).

The impacts of navigational dredging and disposal of dredged material are controversial subjects (Allen and Hardy, 1980). Without question, according to Allen and Hardy, dredging can devastate fish and wildlife resources, but without definitive information the impacts are difficult to document. Dredging activities can increase uptake of contaminants into aquatic food chains and upland disposal can increase possible uptake of contaminants by waterfowl and other animals using the disposal area (Allen and Hardy, 1980).

Investigations on the effects of dredge spoils have been conducted on animals (Smith et al., 1995; Winger et al., 2000). Several earthworm species showed high sensitivity and were able to bioaccumulate various heavy metals, polychlorinated biphenyls (PCBs), and polyaromatic hydrocarbons (PAHs) from a variety of contaminated substances (U.S. Army Corps of Engineers, 1986). Laboratory experiments were done on marine and estuarine invertebrates and fish to determine the effects of suspended sediments (Peddicord, 1976). Many species were relatively insensitive to inert suspended solids and more sensitive species were easily killed at warmer temperatures and lower dissolved oxygen concentrations. Experiments with contaminated sediments suggest that they have much greater potential for adverse impact than uncontaminated sediments.

Few studies specifically examined the effects of dredge spoils on plant communities (U.S. Army Corps of Engineers, 1986; Gibson and Looney, 1994; Wilhelm et al., Chapter 14, this volume). Gibson and Looney did not discuss ecotoxicological effects of deposited sediments, but dealt with primary succession on deposited spoils. Lacking substantial literature specific to dredge-spoil placement, we incorporated literature on the effects of nutrient and toxin addition on plant communities. This chapter reviews available information and discusses the possible outcomes of nutrient and toxin enrichment.

Many factors influence plant communities, including present species composition (Wilson and Shure, 1993), herbivory (Bach, 1994, Ritchie et al., 1998), pathogens (van der Putten and Peters, 1997), climate fluctuations (Ritchie and Tilman, 1995), and fire (Whittle et al., 1997, Bowles and McBride, 1998). Anthropogenic disturbances such as habitat alteration (Wilcox, 1995; Poulson, 1999; Poulson and McClung, 1999), nutrient enrichment (Wilson and Shure, 1993; Van Oene et al., 1999), and contamination (Dinelli and Lombini, 1996; Panno et al., 1999; Stewart et al., 1999; Hart and Lovvorn, 2000) also impact plant assemblages and may affect higher trophic levels (Stewart et al., 1992). Anthropogenic and natural factors can simultaneously place selective pressure on plant communities (Ritchie and Tilman, 1995; Leach and Givnish, 1999). Aquatic communities respond to environmental degradation by loss of sensitive species, decline in species richness, trophic state changes (Ravera, 1983), and shifts in growth habit (Stewart et al., 1999). Contaminants more toxic to aquatic plants than animals include heavy metals, organics, alcohols, pesticides, surfactants, and effluents (Lewis, 1995).

5.2 CONTAMINANT AND HYDROLOGICAL EFFECTS ON PLANT COMMUNITIES

5.2.1 Nutrient Enrichment

Sediments tend to act as nutrient sinks and when moved, may continue to pose potential problems to plant communities. Nutrient materials from dredging operations may vary widely in content depending on the nature and origin of the sediments (Allen and Hardy, 1980). High nutrient concentrations, especially ammonia, are often released during disposal operations, as are lesser amounts of orthophosphates (Blom et al., 1976; Brannon et al., 1976; Schroeder et al., 1977). Nutrient addition has diverse effects on plant assemblages, such as increasing plant biomass (Boyer and Zedler, 1998; Hytönen and Kaunisto, 1999) and shifting the relative abundance of species

(Newman et al., 1996; Richardson et al., 1999). Nutrient enrichment may cause an alteration in plant successional dynamics that may persist long after the stress has ended (Milchunas and Lauenroth, 1995). Changes in plant community structure may not become evident until years after the initial nutrient introduction (Craft et al., 1995). Soil and sediment nutrients tend to be used more efficiently by plants in areas of high species diversity (Hooper and Vitousek, 1998).

Most plant species adapted to low nutrient conditions do not respond well to nutrient enrichment (Chapin et al., 1986). In nutrient-limited environments, low production species often dominate (Aerts and van der Piejl, 1993). The growth of each species will be limited by different nutrients (Bedford et al., 1999). Community dominance can shift by removing nutrient limitations on certain species, resulting in a change in species diversity (Tilman, 1987; Bedford et al., 1999). Additions of limiting nutrients may lead to an overall decrease in species diversity (Tilman, 1987; Goldberg and Miller, 1990).

5.2.2 Heavy Metal Enrichment

Except for the top layer, sediments should not normally contain high concentrations of contaminants. However, heavy metals from natural sources may be present and in some cases they can exceed water quality criteria (Gustafson, 1975). If anthropogenic sources such as iron and steel manufacturers and other industries are present, contaminant concentrations can far exceed natural levels.

When contaminated material is placed in a confined disposal facility (CDF), contaminants may be mobilized to form leachate that is transported to the site boundaries by seepage (Brannon et al., 1990). Subsurface drainage and seepage through dikes may reach adjacent surface and groundwater and act as a source of contamination (Schroeder, 2000). Contamination by heavy metals such as lead, zinc, and mercury, and metalloids such as aluminum, antimony, and arsenic is an increasing problem in surface waters (Yurukova and Kochev, 1994) and terrestrial environments (Siebe, 1995; Pichtel et al., 2000). Heavy metal bioaccumulation occurs in many plant species (Bosserman, 1985; Reimer and Duthie, 1993; St-Cyr and Campbell, 1994; Debusk et al., 1996; Bennett et al., 2000) and fungal species (Aruguete et al., 1998; Michelot et al., 1998). Although sediments can hold and immobilize small amounts of metals (Doyle and Otte, 1997), a greater proportion is generally absorbed by vegetation (Zhu and Sikora, 1995). Sources of metal enrichment include sewage runoff (Obarska-Pempkowiak and Klimkowska, 1999), agricultural runoff (Hill et al., 1997; Elowson, 1999), and atmospheric deposition (Jónsdóttir et al., 1995).

Metals such as zinc, copper, and iron are micronutrients essential for plant growth and development. Low contaminant concentrations can increase growth (Carbonell et al., 1998; He and Yang, 1999). Higher concentrations, however, may become toxic to cells (Steffens, 1990) and pose significant threats to plant communities (Gorham and Gordon, 1963; Mhatre and Chaphekar, 1985; Galbraith et al., 1995; Kapustka et al., 1995). Plant tolerance to different metals varies widely (Guilizzoni, 1991; Dahmani-Muller et al., 2000), and some species have the ability to adapt to heavy metal contamination (Rout et al., 1999). Plants evolved two main methods of tolerating toxic concentrations of heavy metals including avoidance and sequestration. Avoidance involves blocking metal uptake into cellular tissues; sequestration involves compartmentalization of the metal to mask its toxic effects, e.g., transport of metals into vacuoles (Tomsett et al., 1992).

Plant species take up heavy metals differentially, depending on their environment (Yurukova and Kochev, 1994) and relative location (Welsh and Denny, 1980; Abaychi and Al-Obaidy, 1987). Uptake rates can vary within the same genus, and trends regarding patterns of metal accumulation within plant tissues are sometimes inconsistent (Nissen and Lepp, 1997). Heavy metal uptake and accumulation in vegetation may vary throughout the year as well as among different species of plants and metals (Kufel and Kufel, 1985; Carbonell et al., 1998; Ge et al., 2000). Metal concentrations in plant tissues are influenced more by soil characteristics such as pH (Kapustka et al., 1995; Murray et al., 2000), percent organic carbon, percent clay, cation exchange capacity, and other mineral constituents than by concentrations of heavy metals in the soil (Tack and Verloo, 1996; Murray et al., 2000).

Arsenic has been relatively well researched by toxicologists; its speciation and toxicity are of great concern (Koch et al., 2000). Carbonell et al. (1998) demonstrated that arsenic concentrations in the roots and shoots of the emergent wetland plant, *Spartina alterniflora*, increased with increasing application. Inorganic arsenicals, arsenite, As(III), and arsenate, As(V), increased drymass production of exposed plants. Plants treated with arsenate had improved growth associated with increases in phosphorus availability compared to unexposed controls. Organic arsenicals such as monomethyl arsonic acid and dimethylarsinic acid caused the highest root sodium concentrations and higher leaf calcium concentrations, which indicate phytotoxicity. Similar results were found using arsenic compounds in other terrestrial environments (Pitten, 1999).

In greenhouse experiments, toxicity of heavy metals to individual plant species depended on metal type and concentration (Wang, 1987; Guilizzoni, 1991). In spring barley (*Hordeum vulgare*), toxicity of zinc was enhanced in the presence of copper, causing decreased growth (Luo and Rimmer, 1995). Toxicity of this nature is likely related to free-ion activity (Tomsett et al., 1992) and excessive uptake (Murray et al., 2000).

Terrestrial plant communities near a smelter in southwest Montana were severely impacted by emissions containing heavy metals and arsenic (Galbraith et al., 1995). The observed effects were the destruction of an evergreen forest, species impoverishment, increased weed dominance in grasslands, and reductions in diversity near the smelter. Species diversity was reduced by an average of 34% of its reference value.

5.2.3 Effects of Organic Compounds

Dredged material from harbors and other heavily industrialized areas may contain substantial amounts of contaminants (Allen and Hardy, 1980). They tend to be tightly bound to clay particles. They are not readily released into the water column but may be deposited into landfills. In addition to the effects of heavy metals and added nutrients, compounds such as polychlorinated biphenyls (PCBs) and other organics from industrial sources may pose a threat to plants if present in dredge materials. The amount of contaminants, such as petroleum hydrocarbons, pesticides, and PCBs can vary widely in material from maintenance dredging, depending on area sources (Allen and Hardy, 1980). Industrial harbors tend to be highly polluted, whereas interconnecting waterways may be relatively unpolluted.

PCBs affect some aquatic plants (Phillips, 1978), and PAHs readily accumulate in the membrane systems of organisms (Cooke and Dennis, 1983). Exposure of PAHs to light results in the formation of photoproducts that may be more phytotoxic than the parent compound (Duxbury et al., 1997; McConkey et al., 1997). A site in the Grand Calumet lagoons that had extremely high PAH concentrations contained fewer species and lacked floating plants at the time of sampling (Stewart et al., Chapter 21, this volume).

5.2.4 Changes in Hydrology

Confined disposal facilities may affect plant communities by causing alterations in hydrology by diking, mounding, and the formation of impermeable barriers. Hydrology is considered the driving force in maintenance of wetland plant communities (Rea and Ganf, 1994; Hunt et al., 1999). Anthropogenic disturbances in hydrology often negatively impact wetland structure and function (Wilcox, 1995; Kuhn et al., 1999). For example, woolgrass (*Scirpus cyperinus*) commonly invades disturbed wetlands by outcompeting other native plants for space and resources (Wilcox et al., 1985). Decreasing average water levels have been correlated with increases in willow abundance, while increasing average water levels have been correlated with co-dominance of sedge (*Carex* spp.) and joint grass (*Calamagrostris canadensis*) in Wisconsin wetlands (Ashworth, 1997). Dredge spoil disposal contributes to hydrologic disturbance by mounding and compression, thus affecting water levels and flow.

Elevation of an area relative to the water table can determine which plant species will dominate (Odland, 1997). Changes in wetland hydrology correlated with increased *Typha* biomass. In addition, a single or a few species also dominated a large area, which indicated past hydrologic disturbance (Wilcox, 1995). Fluctuations in water tables influenced seed germination rates in some species, among them the sedge, *Carex comosa* (Baskin et al., 1996).

In an Everglades wetland, rising water levels and decreased hydrologic fluctuation increased cattail (*Typha domingensis*) relative abundance (Newman et al., 1996, Newman et al., 1998, Dirk et al., 1999). Fire and nutrients also impacted this species (David, 1996). Phosphorus enrichment promoted a community shift from sawgrass (*Cladium jamaicense*), a plant adapted to nutrient-poor environments and low nutrient status, to competitive, high nutrient-status species such as cattail (Newman et al., 1996; Newman et al., 1998; Richardson et al., 1999).

Dirk et al. (1999) suggested that annual increases in water depth were related to increased bladderwort (*Utricularia* spp.) and water lily (*Nymphaea odorata*) abundance. In contrast, other species, such as the rush (*Rhyncospora tracyi*), decreased in abundance. The lowering of water tables caused by drought events may cause a shift in plant communities away from nutrient and stress-tolerant taxa, such as *Scirpus* spp., towards even more stress-tolerant, highly competitive taxa such as *Typha* spp.

5.3 EFFECTS ON SELECTED PLANT COMMUNITIES

5.3.1 Effects on Beach and Dune Communities

Beach nourishment projects are common and can serve as alternatives for the disposal of sandy material (Allen and Hardy, 1980). No information was found on the effects of CDF siting on beach and dune communities. However, adverse impacts may include increased turbidity, nutrient introduction, smothering of organisms, and contaminant addition. Potential nutrient enrichment effects on beach and dune plants are poorly understood. Sea rocket (*Cakile edentula*), when in a nutrient-rich environment, produced more lateral branches and fruit in the presence of adequate water resources, suggesting that fitness was increased by nutrient enrichment (Zhang, 1996). The response of western juniper (*Juniperus occidentalis*) to nitrogen fertilization included increased growth rate (Miller et al., 1991); this response may be mirrored by the two rare juniper species, ground juniper (*Juniperus communis*) and trailing juniper (*Juniperus horizontalis*), in the Great Lakes region. Legumes, such as *Lathyrus* spp., may be outcompeted by other non-nitrogen fixing plants in areas enriched by nitrogen (Theodose and Bowman, 1997). Some legumes, however, showed increased fitness when limitation by nutrients other than nitrogen was removed (Ritchie and Tilman, 1995).

Although the impact of metal enrichment on beach and dune communities is unclear, jack pine (*Pinus bankisana*) is known to bioaccumulate heavy metals in its leaves (Gratton et al., 2000). Some jack pines possess alleles that permit greater than average tolerance to heavy metal contamination (Xie and Knowles, 1992), potentially negating any adverse effects.

5.3.2 Effects on Wetland Communities

Disposal of dredged material in wetlands is becoming less frequent due to recognition of the inherent value of wetlands (Allen and Hardy, 1980). Wetlands are noted for their nutrient-absorbing properties (Johnston et al., 1990; Comín et al., 1997; Obarska-Pempkowiak and Klimkowska, 1999; Romero et al., 1999; Soto et al., 1999), and both phosphorus and nitrogen are incorporated into vegetative biomass (Zhu and Sikora, 1995). As nutrient-retentive ecosystems, wetlands benefit downstream watersheds by removing excess nutrients that can contribute to eutrophication (Johnston, 1991). Wetlands receive nutrients from an array of sources, including atmospheric deposition (Johnston, 1991; Jónsdóttir et al., 1995), surface runoff (Hill et al., 1997; Elowson, 1999), and ground water (Johnston, 1991).

Some wetland plants (e.g., woolgrass, *Scirpus lacustris,* and cattail, *Typha latifolia)*, tolerate high nitrogen concentrations and are used in the treatment of wastewater (Hubbard et al., 1999; Huang et al., 2000). Riparian wetlands are used to remove excess nutrients from flowing water (Hearne and Howard-Williams, 1988). Although wetlands have the capacity to absorb nutrients, excessive nutrient loading can negatively affect plant communities (Wilcox, 1995; Bedford et al., 1999). In general, patterns of excessive nutrient enrichment are positively correlated with low species diversity in many Midwest wetlands (Bedford et al., 1999). Nutrient limitation, particularly phosphorus limitation, has been suggested as essential to the maintenance of species diversity in some nutrient-limited wetlands (Dirk et al., 1997).

In a short-term fertilization experiment in a wet meadow, nitrogen addition increased sedge abundance, and nitrogen and phosphorus or phosphorus additions increased grass abundance; none of the nutrient enrichment regimes increased forb abundance (Theodose and Bowman, 1997). In another experiment, graminoid species increased in relative abundance when nitrogen or nitrogen and phosphorus treatments were applied (Bowman et al., 1993). In these separate studies, species diversity of nutrient-rich wet meadows and those that were intermediate in nutrient availability were negatively affected by nutrient enrichment. Diversity decreased as larger graminoids increased in abundance and shaded out smaller forbs (Bowman et al., 1993; Theodose and Bowman, 1997).

The competitive abilities of three wetland plant species, reed canary grass (*Phalaris arundinacea*), cattail (*Typha latifolia*), and common tussock sedge (*Carex stricta*), were observed in relation to nutrient concentrations and soil moisture. When grown with reed canary grass, the overall biomass of cattail and common tussock sedge was significantly less. With its faster growth, reed canary grass was able to effectively shade out the other species in both high and low nutrient trials (Wetzel and van der Valk, 1998).

In a nutrient-limited bog, nitrogen fertilization was used as a tool to determine the effects of fertilization on competition between two plant species tolerant of low-nutrient conditions. Addition of nitrogen, the limiting nutrient, spurred growth of round-leafed sundew (*Drosera rotundifolia*) and a moss (*Sphagnum fuscum*). While both increased their growth in response to nutrient addition, neither was able to out-compete the other species for sunlight (Svensson, 1995), suggesting that in some cases, nutrient addition may not have detrimental effects on plant communities tolerant to nutrient limitation. These effects were not examined over a long term nor in the presence of other, more invasive, species.

In a Minnesota wet meadow, increased stormwater runoff, agriculture, and urbanization were correlated with a reduction of native graminoid and herbaceous perennial abundance; this vegetation was replaced by annuals in recently cultivated sites or introduced perennials and floating aquatics in stormwater-impacted wetlands (Galatowitsch et al., 2000). Ditching in these areas had similar effects, although only in impacted landscapes. Nitrogen and phosphorus enrichment increased the rate of photosynthesis in wet meadow plants (Bowman et al., 1993).

Characteristics of fens include high water tables, accumulation of peat, and nutrient limitation (Aerts et al., 1999). These wetlands require sustained, high water tables to produce the carbonate precipitation associated with their unique vegetation (Almendinger and Leete, 1998). Fen-associated sedges (*Carex* sp.) remove nitrogen from sediment, although the amount of uptake and its effects on biomass production may depend on the availability of other nutrients such as phosphorus (Pérez-Corona and Verhoeven, 1996). In fens, although the response of some stress-tolerant sedges appeared independent of nutrient enrichment timing, other species such as jointed rush (*Juncus articulatus*) increased in abundance when nutrient enrichment occurred early in the growing season (Dirk et al., 1999).

In an overland-flow area constructed to purify runoff from a peat mine, nutrient addition caused a vegetation shift relative to a reference area that did not receive additional nutrient inflow. Biomass of graminoids such as sedges, herbs (*Menyanthes trifoliata*), and the root systems of plants in general were greater than in the reference area. Sedges and herbs dominated the nutrient enriched areas, whereas mosses (i.e., *Sphagnum* spp.) only maintained dominance in areas that did not receive additional nutrients (Huttunen et al., 1996).

In a constructed wetland enriched by ammonia at concentrations ranging from 20.5 to 82.4 mg/l NH_3-N, bulrush (*Scirpus acutus*) was the only plant that showed significant drymass production as a response to nutrient addition. Bulrush showed a peak drymass production at ammonia concentrations between 30 and 50 mg/l NH_3-N; at higher or lower concentrations, drymass production decreased. Within the constructed wetland enclosures, substantial differences in growth rates and biomass production of five wetland plants were ranked highest to lowest: common reed (*Phragmites australis*) > cattail (*Typha latifolia*) > bulrush (*Scirpus acutus*) > arrowhead (*Sagittaria latifolia*) > and common rush (*Juncus roemerianus*). Drymass production was highest in common reed, which nearly doubled that of the second highest plant. Common rush showed signs of severe stress and produced very little biomass during this study (Hill et al., 1997). Although nutrient limitation may not affect the survivorship of some rush species, survival may be negatively affected by increased plant density created by nutrient enrichment (Lentz, 1999). A constructed wetland planted with cattail and rush removed a significant amount of ammonia in flowthrough experiments, with the greatest removal occurring in summer (Sikora et al., 1995).

In a constructed marsh, addition of nitrogen spurred the growth of cordgrass (*Spartina foliosa*) and produced stems as tall as those in natural marshes, although cessation of fertilization caused stem height to shrink to pre-fertilization levels (Boyer and Zedler, 1998). Reeds in a constructed wetland significantly reduced the concentration of nitrogen and phosphorus in water by incorporation into plant biomass (Obarska-Pempkowiak and Klimkowska, 1999).

Willows (*Salix* spp.) represent another group that can remove excess nitrogen from agricultural runoff or sewage effluent (Elowson, 1999). They can also absorb heavy metals from water and soils in wetland habitats (Sander and Ericsson, 1998). Some species demonstrated the ability to tolerate soil lead concentrations up to 17,000 mg kg^{-1} (Eltrop et al., 1991). In a comparison of two sites, one with heavy metal contamination and one without, no difference was found between the concentrations of bioaccumulated metals in the stems of willows; however, willows from contaminated areas had higher metal concentrations in root tissues and lower transport rates of metal ions to stems (Landberg and Greger, 1996). *Salix viminalis* showed toxic effects on exposure to high concentrations of manganese (Tahvanainen and Rytkönen, 1999).

Aquatic macrophytes absorb heavy metals (Manny et al., 1991) and some show effects such as chlorosis and root brittleness when treated with high concentrations of mercury, which also affects biomass and leaf area index (Mhatre and Chaphekar, 1985). In a tailings pond near an abandoned mine, metal concentrations in plants strongly correlated with soil concentration (Chambers and Sidle, 1991); however, in an old antimony-mining area, antimony uptake by plants was inversely related to its concentration in the soil (Baroni et al., 2000). These studies suggest that plant metal uptake did not follow a consistent pattern and was affected by a variety of factors, not all of which were measured or understood.

In ponds and lakes near a metal smelter in Ontario, Canada, heavy metal contamination caused changes in aquatic plant community structure (Gorham and Gordan, 1963). A moss (*Leptodictyum riparium*) and needle spike rush (*Eleocharis acicularis*) were relatively tolerant of contamination in ponds within two miles of the smelter. Bladderwort (*Utricularia vulgaris*) and pondweed (*Potamogeton epihydrus*) were sensitive and only found at distances more than 24 km from the metal source. Overall, heavy metal pollution drastically reduced aquatic floral diversity.

Much information is available concerning nutrient and heavy metal addition to wetlands (Newman et al., 1996; Sander and Ericsson, 1998); however, no information dealing specifically with community shifts caused by nutrient and mineral enrichment by dredge spoils was found. Wetland areas may be particularly at risk of receiving nutrient- and metal-contaminated dredge-spoil runoff because wetlands are typically located in topographic depressions. Sedges (Huttunen et al., 1996), cattail (Newman et al., 1996; Richardson et al., 1999), and common reed (Hill et al., 1997) are common wetland plants that can quickly dominate an area after nutrient enrichment and may have the capacity to tolerate heavy metal enrichment (Lan et al., 1992; Carbonell et al., 1998). The competitive abilities of different wetland species varied based on local conditions and other species present (Keddy et al., 1994).

Species diversity may decrease rapidly with nutrient addition according to fertilization experiments undertaken in a wet meadow (Bowman et al., 1993; Theodose and Bowman, 1997). Species tolerant to nutrient limitation (e.g., round-leaved sundew (*Drosera rotundifolia*) and horned bladderwort (*Utricularia cornuta*)) may be at risk of competition from more aggressive, competitive species when faced with nutrient enrichment (Newman et al., 1996; Richardson et al., 1999), although this is uncertain (Svensson, 1995).

5.4 FUTURE RESEARCH NEEDS

When contaminated material is placed in a CDF, contaminants may be mobilized to form leachate that is transported to the site boundaries by seepage. Subsurface drainage and seepage through dikes may reach adjacent surface and groundwater and act as a source of contamination (Schroeder, 2000). More studies are needed regarding the affects of dredging activities on native plant communities. Little work has been published on the plant assemblages that colonize disposal facilities and the accumulation of contaminants in plant tissues (see Wilhelm et al., Chapter 14, this volume).

Autoecological studies of native plant species in North America are limited. More specifically, information on the effects of nutrient and heavy metal enrichment for the rare plant species is lacking. In gauging the effects of deposited dredge material on plant communities, aspects of species richness and diversity, dominance and abundance, proportion of exotic, indicator and endangered species, and other characteristics may be valuable (Stewart et al., 1999). Attention must be given to tolerant species and present species should be examined for signs of stress.

A floristic quality index (FQI) is a univariate metric that measures number of species and fidelity to natural areas, and may be a useful tool in assessing the overall health of a plant community (Swink and Wilhelm, 1994). Simon et al. (2001) presented the development of a multimetric index of biotic integrity for riverine and palustrine wetland plant communities. They evaluated over 20 characteristics of plant communities and 12 metrics in five categories were selected for inclusion into the index. Sites receiving the highest plant index scores were among those supported as the best sites in the study. Metrics included the coefficient of community as an alternative to a tolerance metric (Swink and Wilhelm, 1994). This plant index of biotic integrity (PIBI) concept needs further development and testing but shows potential as a multimetric tool for assessing aquatic plant assemblages. Using components of the PIBI, e.g., FQI, CC, and species richness, many plant communities located on or near CDFs scored poorly, suggesting negative impacts of the dredge material stored therein and the communities that developed on them (Wilhelm et al., Chapter 14, this volume).

5.5 CONCLUSIONS

Plant communities naturally shift over time with changing environmental conditions (Bowles and McBride, 1998; Huusela-Veistola and Vasarainen, 2000). Addition of toxins and nutrients and changes in hydrology may influence plant community structure (Wilson and Shure, 1993; Poulson, 1999). The storage and disposal of nutrient and metal contaminated dredge spoils may cause shifts in nearby plant communities. Shifts in species composition and diversity may not be observed for decades after nutrient enrichment (Findlay and Bourdages, 2000), causing any disturbance to remain undetected. Plant community shifts often have great amounts of inertia and are difficult to reverse (Milchunas and Lauenroth, 1995).

ACKNOWLEDGMENTS

We thank Douglas Wilcox for reviewing an early draft of this manuscript. Ann Zimmerman and Tanya Schwartz aided the literature search. Felicia Kirksey and Jean Sellars provided the scope of work and background information. The U.S. Army Corps of Engineers, Chicago District, provided funding for this project. This chapter is contribution 1154 of the USGS Great Lakes Science Center.

REFERENCES

Abaychi, J.K. and S.Z. Al-Obaidy. 1987. Concentrations of trace elements in aquatic vascular plants from Shatt Al Arab River, Iraq, *Journal of Biological Science and Research*, 18, 123–129.

Aerts, R. and M.J. van der Piejl. 1993. A simple model to explain the dominance of low-productive perennials in nutrient poor habitats, *Oikos*, 66, 144–147.

Aerts, R., J.T.A. Verhoeven, and D.F. Whigham. 1999. Plant-mediated controls on nutrient cycling in temperate fens and bogs, *Ecology*, 80, 2170–2181.

Allen, K.O. and J.W. Hardy. 1980. *Impacts of Navigational Dredging on Fish and Wildlife: A Literature Review*. FWS/OBS-80/07, Office of Biological Services, Fish and Wildlife Service, U.S. Department of the Interior, Washington, D.C.

Almendinger, J.E. and J.H. Leete. 1998. Regional and local hydrogeology of calcareous fens in the Minnesota river basin, USA, *Wetlands*, 18, 184–202.

Aruguete, D.M., J.H. Aldstadt, and G.M Mueller. 1998. Accumulation of several heavy metals and lanthanides in mushrooms (Agricales) from the Chicago region, *The Science of the Total Environment*, 224, 43–56.

Ashworth, S.M. 1997. Comparison between restored and references sedge meadow wetlands in south-central Wisconsin, *Wetlands*, 17, 518–527.

Bach, C.E. 1994. Effects of a specialist herbivore (*Altica subplicata*) on *Salix cordata* and sand dune succession, *Ecological Monographs*, 64, 423–445.

Baroni, F., A. Boscagli, G. Protano, and F. Riccobono. 2000. Antimony accumulation in *Achillea ageratum*, *Plantago lanceolata*, and *Silene vulgaris* growing in an old Sb-mining area, *Environmental Pollution*, 109, 347–352.

Baskin, C.C., E.W. Chester, and J.M. Baskin. 1996. Effect of flooding on annual dormancy cycles in buried seeds of two wetland *Carex* species, *Wetlands*, 16, 84–88.

Bedford, B.L., M.R. Walbridge, and A. Aldous. 1999. Patterns in nutrient availability and plant diversity of temperate North American wetlands, *Ecology*, 80, 2151–2169.

Bennett, J.P., E. Chiriboga, J. Coleman, and D.M. Waller. 2000. Heavy metals in wild rice from northern Wisconsin, *The Science of the Total Environment*, 246, 261–269.

Blom, B.E., T.S. Jenkins, D.C. Leggett, and R.P. Murrmann. 1976. Effects of sediment organic matter on migration of various chemical constituents during disposal of dredged material. U.S. Army Cold Regions Research and Engineering Laboratory, Hanover, NH. Published by U.S. Army Engineers Waterways Experimental Station, Vicksburg, MS.

Bosserman, R.W. 1985. Distribution of heavy metals in aquatic macrophytes from Okefenokee Swamp, in J. Salánki (Ed.), *Heavy Metals in Water Organisms: Symposia Biologica Hungarica*. Akadémiai Kiadó, Budapest, Hungary, 29, 31–39.

Bowles, M.L. and J.L. McBride. 1998. Vegetation composition, structure, and chronological change in a decadent Midwestern North American savanna remnant, *Natural Areas Journal*, 18, 14–27.

Bowman, W.D., T.A. Theodose, J.C. Schardt, and R.T. Conant. 1993. Constraints of nutrient availability on primary production in two alpine tundra communities, *Ecology*, 74, 2085–2097.

Boyer, K.E. and J.B. Zedler. 1998. Effects of nitrogen additions on the vertical structure of a constructed cordgrass marsh, *Ecological Applications*, 8, 692–705.

Brannon, J.M., R.M. Engler, J.R. Rose, and P.C. Hunt. 1976. Selective analytical partitioning of sediments to evaluate potential mobility of chemical constituents during dredging and disposal operations. Technical Report D-76–7. U.S. Army Engineers Waterways Experiment Station, Vicksburg, MS.

Brannon, J.M., T.E. Myers, and C.B. Price. 1990. Factors affecting the behavior of metals in leachate. Environmental Effects of Dredging Technical Notes EEDP-02–13, U.S. Army Engineers Waterways Experiment Station, Vicksburg, MS.

Briuer, E. 1998. Corps research will achieve efficient, environmentally sensitive dredging, *Dredging Research*, 1, 1–3.

Carbonell, A.A., M.A. Aarabi, R.D. DeLaune, R.P. Gambrell, and W.H. Patrick. 1998. Arsenic in wetland vegetation: availability, phytotoxicity, uptake and effects on plant growth and nutrition, *The Science of the Total Environment*, 217, 189–199.

Chambers, J.C. and R.C. Sidle. 1991. Fate of heavy metals in an abandoned lead-zinc tailings pond: I. Vegetation, *Journal of Environmental Quality*, 20, 745–751.

Chapin, F.S., P.M. Vitousek, and K. van Cleve. 1986. The nature of nutrient limitation in plant communities, *The American Naturalist*, 127, 48–58.

Comín, F.A., J.A. Romero, V. Astorga, and C. García. 1997. Nitrogen removal and cycling in restored wetlands used as filters of nutrients for agricultural runoff, *Water Science Technology*, 35, 255–261.

Cooke, M. and A.J. Dennis. 1983. *Polycyclic Aromatic Hydrocarbons: Formation, Metabolism, and Measurement*. Battelle, Columbus, Ohio.

Craft, C.B., J. Vymazal, and C.J. Richardson. 1995. Response of Everglades plant communities to nitrogen and phosphorus additions, *Wetlands*, 15, 258–271.

Dahmani-Muller, H., F. van Oort, B. Gélie, and M. Balabane. 2000. Strategies of heavy metal uptake by three plant species growing near a metal smelter, *Environmental Pollution*, 109, 231–238.

David, P.G. 1996. Changes in plant communities relative to hydrologic conditions in the Florida Everglades, *Wetlands*, 16, 15–23.

Debusk, T.A., R.B. Laughlin, and L.N. Schwartz. 1996. Retention and compartmentalization of lead and cadmium in wetland microcosms, *Water Research*, 30, 2707–2716.

Dinelli, E. and A. Lombini. 1996. Metal distributions in plants growing on copper mine spoils in Northern Appennines, Italy: the evaluation of seasonal variations, *Applied Geochemistry*, 11, 375–385.

Dirk, B., B. Verhagen, V. van Haesebroeck, and M. El-Kahloun. 1999. Phosphorus fertilization in a phosphorus-limited fen: effects of timing, *Applied Vegetation Science*, 2, 71–78.

Dirk, B., B. Verhagen, V. van Haesebroeck, and R.F. Verheyen. 1997. Nutrient limitation in species-rich lowland fens, *Journal of Vegetation Science*, 8, 415–424.

Doyle, M.O. and M.L. Otte. 1997. Organism-induced accumulation of iron, zinc and arsenic in wetland soils, *Environmental Pollution*, 96, 1–11.

Duxbury, C.L., D.G. Dixson, and B.M. Greenburg. 1997. Effects of simulated radiation on the bioaccumulation of polycyclic aromatic hydrocarbons by the duckweed *Lemna gibba*, *Environmental Toxicology and Chemistry*, 16, 1739–1748.

Elowson, S. 1999. Willow as a vegetation filter for cleaning of polluted drainage water from agricultural land, *Biomass and Bioenergy*, 16, 281–290.

Eltrop, L., G. Brown, O. Joachim, and K. Brinkmann. 1991. Lead tolerance of *Betula* and *Salix* in the mining area of Mechernich, Germany, *Plant and Soil*, 131, 275–285.

Findlay, C.S. and J. Bourdages. 2000. Response time of wetland biodiversity to road construction on adjacent lands, *Conservation Biology*, 14, 86–94.

Galatowisch, S.M., D.C. Whited, R. Lehtinen, J. Husveh, and K. Schik. 2000. The vegetation of wet meadows in relation to their land use, *Environmental Monitoring and Assessment*, 60, 121–144.

Galbraith, H., K. LeJeune, and J. Lipton. 1995. Metal and arsenic impacts to soils, vegetation communities and wildlife habitat in southwest Montana uplands contaminated by smelter emissions: I. Field evaluation, *Environmental Toxicology and Chemistry*, 14, 1895–1903.

Ge, Y., P. Murray, and W.H. Hendershot. 2000. Trace metal speciation and bioavailability in urban soils, *Environmental Pollution*, 107, 137–144.

Gibson, D.J. and P.B. Looney. 1994. Vegetation colonization of dredge spoil on Perdido Key, Florida, *Journal of Coastal Research*, 10, 133–143.

Goldberg, D.E. and T.E. Miller. 1990. Effects of different resource additions on species diversity in an annual plant community, *Ecology*, 71, 213–225.

Gorham, E. and A.G. Gordon. 1963. Some effects of smelter pollution upon aquatic vegetation near Sudbury, Ontario, *Canadian Journal of Botany*, 41, 371–378.

Gratton, W.S., K.K. Nkongolo, and G.A. Spiers. 2000. Heavy metal accumulation in soil and jack pine (*Pinus banksiana*) needles in Sudbury, Ontario, Canada, *Bulletin of Environmental Contamination and Toxicology*, 64, 550–557.

Great Lakes Commission. 1999. *Dredging and the Great Lakes*. Great Lakes Commission, Ann Arbor, MI.

Guilizzoni, P. 1991. The role of heavy metals and toxic materials in the physiological ecology of submersed macrophytes, *Aquatic Botany*, 41, 87–109.

Gustafson, J.F. 1975. Estimating pollution reality or fantasy, *World Dredging and Marine Construction*, Nov, 38–40.

Hart, E.A. and J.R. Lovvorn. 2000. Vegetation dynamics and primary production in saline, lacustrine wetlands of a Rocky Mountain basin, *Aquatic Botany*, 66, 21–39.

Hearne, J.W. and C. Howard-Williams. 1988. Modelling nitrate removal by riparian vegetation in a springfed stream: the influence of land-use practices, *Ecological Modelling*, 42, 179–198.

He, M. and J. Yang. 1999. Effects of different forms of antimony on rice during the period of germination and growth and antimony concentration in rice tissue, *The Science of the Total Environment*, 243/244, 149–155.

Hill, D.T., V.W.E. Payne, J.W. Rogers, and S.R. Kown. 1997. Ammonia effects on the biomass production of five constructed wetland plant species, *Bioresource Technology*, 62, 109–113.

Hooper, D.U. and P.M. Vitousek. 1998. Effects of plant composition and diversity on nutrient cycling, *Ecological Monographs*, 68, 121–150.

Huang, J., R.B. Reneau, and C. Hagedorn. 2000. Nitrogen removal in constructed wetlands employed to treat domestic wastewater, *Water Research*, 34, 2582–2588.

Hubbard, R.K., J.M. Ruter, G.L. Newton, and J.G. Davis. 1999. Nutrient uptake and growth response of six wetland/riparian plant species receiving swine lagoon effluent, *Transactions of the American Society of Agricultural Engineers*, 42, 1331–1341.

Hunt, R.J., J.F. Walker, and D.P. Krabbenhoft. 1999. Characterizing hydrology and the importance of ground-water discharge in natural and constructed wetlands, *Wetlands*, 19, 458–472.

Huttunen, A., K. Heikkinen, and R. Ihme. 1996. Nutrient retention in the vegetation of an overland flow treatment system in northern Finland, *Aquatic Botany*, 55, 61–73.

Huusela-Veistola, E. and A. Vasarainen. 2000. Plant succession in perennial grass strips and effects on the diversity of leafhoppers (Homoptera, Auchenorrhycha), *Agriculture, Ecosystems and Environment*, 80, 101–112.

Hytönen, J. and S. Kaunisto. 1999. Effect of fertilization on the biomass production of coppiced mixed birch and willow stands on a cut-away peatland, *Biomass and Bioenergy*, 17, 455–469.

Johnston, C.A., N.E. Detenbeck, and G.J. Niemi. 1990. The cumulative effect of wetlands on stream water quality and quantity: a landscape approach, *Biogeochemistry*, 10, 105–141.

Johnston, C.A. 1991. Sediment and nutrient retention by freshwater wetlands: effects on surface water quality, *Critical Reviews in Environmental Control*, 21, 491–565.

Jónsdóttir, I.J., T.V. Callaghan, and J.A. Lee. 1995. Fate of added nitrogen in a moss-edge Arctic community and effects of increased nitrogen deposition, *The Science of the Total Environment*, 160/161, 677–685.

Kapustka, L.A., J. Lipton, H. Galbraith, D. Cacela, and K. LeJeune. 1995. Metal and arsenic impacts to soils, vegetation communities and wildlife habitat in southwest Montana uplands contaminated by smelter emissions: II. Laboratory phytotoxicity studies, *Aquatic Toxicology and Chemistry*, 14, 1905–1912.

Keddy, P.A., L. Twolan-Strutt, and I.C. Wisheu. 1994. Competitive effect and response rankings in 20 wetland plants: are they consistent across three environments? *Journal of Ecology*, 82, 635–643.

Koch, I., L. Wang, C.A. Ollson, W.R. Cullen, and K.J. Reimer. 2000. The predominance of inorganic arsenic species in plants from Yellowknife, Northwest Territories, Canada, *Environmental Science and Technology*, 34, 22–26.

Kufel, I. and L. Kufel. 1985. Heavy metals and mineral nutrient budget in *Phragmites australis* and *Typha augustifolia*. in J. Salánki (Ed.), *Heavy Metals in Water Organisms: Symposia Biologica Hungarica*: Akadémiai Kiadó, Budapest, Hungary, 29, 61–66.

Kuhn, N.L., I.A. Mendelssohn, and D.J. Reed. 1999. Altered hydrology effects on Louisiana salt marsh function, *Wetlands*, 19, 617–626.

Lan, C., G. Chen, and M.H. Wong. 1992. Use of cattails in treating wastewater from a Pb/Zn mine, *Environmental Management*, 16, 75–80.

Landberg, T. and M. Greger. 1996. Differences in uptake and tolerance to heavy metals in *Salix* from unpolluted and polluted areas, *Applied Geochemistry*, 11, 175–180.

Leach, M.K. and T.J. Givnish. 1999. Gradients in the composition, structure, and diversity in remnant oak savannas in southern Wisconsin, *Ecological Monographs*, 69, 353–374.

Lee, C.R. and D.L. Brandon. 1991. *Long-term evaluation of plants and animals colonizing contaminated estuarine dredged material placed in an upland environment.* Technical note, U.S. Army Corps of Engineers Waterways Experiment Station, Vicksburg, MS.

Lentz, K.A. 1999. Effects of intraspecific competition and nutrient supply on the endangered northeastern bulrush, *Scirpus ancistrochaetus* Schuyler (Cyperaceae), *American Midland Naturalist*, 142, 47–54.

Lewis, M.A. 1995. Use of freshwater plants for phytotoxicity testing, *Environmental Pollution*, 87, 319–336.

Luo, Y. and D.L. Rimmer. 1995. Zinc-copper interaction affecting plant growth on a metal-contaminated soil, *Environmental Pollution*, 88, 79–83.

Manny, B.A., S.J. Nichols, and D.W. Schloesser. 1991. Heavy metals in aquatic macrophytes drifting in a large river, *Hydrobiologia*, 219, 333–344.

McConkey, B.J., C.L. Duxbury, D.G. Dixon, and B.M. Greenberg. 1997. Toxicity of a PAH photooxidation product to the bacteria *Photobacterium phosphoreum* and the duckweed *Lemna gibba*: effects of phenanthrene and its primary product, phenanthrenequinone, *Environmental Toxicology and Chemistry*, 16, 892–899.

Mhatre, G.N. and S.B. Chaphekar. 1985. The effect of mercury on some aquatic plants, *Environmental Pollution*, 39, 207–216.

Michelot, D., E. Siobud, J.-C. Doré, C. Veil, and F. Poirier. 1998. Update on metal content profiles in mushrooms — toxicological implications and tentative approach to the mechanisms of bioaccumulation, *Toxicon*, 36, 1997–2012.

Milchunas, D.G. and W.K. Lauenroth. 1995. Inertia in plant community structure: state changes after cessation of nutrient-enrichment stress, *Ecological Applications*, 5, 452–458.

Miller, P.M., L.E. Eddleman, and J.M. Miller. 1991. The response of juvenile and small adult western juniper (*Juniperus occidentalis*) to nitrate and ammonium fertilization, *Canadian Journal of Botany*, 69, 2344–2352.

Murray, P., Y. Ge, and W.H. Hendershot. 2000. Evaluating three trace metal contaminated sites: a field and laboratory investigation, *Environmental Pollution*, 107, 127–135.

Newman, S., J.B. Grace and J.W. Koebel. 1996. Effects of nutrients and hydroperiod on *Typha*, *Cladium*, and *Elocharis*: implications for Everglades restoration, *Ecological Applications*, 6, 774–783.

Newman, S., J. Schuette, J.B. Grace, K. Rutchey, T. Fontaine, K.R. Reddy, and M. Pietrucha. 1998. Factors influencing cattail abundance in the northern Everglades, *Aquatic Botany*, 60, 265–280.

Nissen, L.R. and N.W. Lepp. 1997. Baseline concentrations of copper and zinc in shoot tissues of a range of *Salix* species, *Biomass and Bioenergy*, 12, 115–120.

Obarska-Pempkowiak, H. and K. Klimkowska. 1999. Distribution of nutrients and heavy metals in a constructed wetland system, *Chemosphere*, 39, 303–312.

Odland, A. 1997. Development of vegetation in created wetlands in western Norway, *Aquatic Botany*, 59, 45–62.

Panno, S.V., V.A. Nuzzo, K. Cartwright, B.R. Hensel, and I.G. Krapac. 1999. Impact of urban development on the chemical composition of ground water in a fen-wetland complex, *Wetlands*, 19, 236–245.

Peddicord, R. 1976. *Biological Impacts of Suspensions of Dredged Material.* Proceedings of WODCON VII, San Francisco, CA.

Pérez-Corona, M.E. and J.T.A. Verhoeven. 1996. Effects of soil P status on growth and P and N uptake of *Carex* species from fens differing in P-availability, *Acta Botanica Neerlandica*, 45, 381–392.

Phillips, D.J. 1978. Use of biological indicator organisms to quantitate organochlorine pollutants in aquatic environments — a review, *Environmental Pollution*, 16, 167–229.

Pichtel, J., K. Kuroiwa and H.T. Sawyerr. 2000. Distribution of Pb, Cd and Ba in soils and plants of two contaminated sites, *Environmental Pollution*, 110, 171–178.

Pitten, F.-A., G. Müller, P. König, D. Schmidt, K. Thurow, and A. Kramer. 1999. Risk assessment of a former military base contaminated with organoarsenic-based warfare agents: uptake of arsenic by terrestrial plants, *The Science of the Total Environment*, 226, 237–245.

Poulson, T.L. 1999. Autogenic, allogenic, and individualistic mechanisms of dune succession at Miller, Indiana, *Natural Areas Journal*, 19(2), 172–176.

Poulson, T.L. and C. McClung. 1999. Anthropogenic effects on early dune succession at Miller, Indiana, *Natural Areas Journal*, 19, 177–179.

Ravera, O. 1983. Assessment of the trophic state of a body of water, *Annals of Limnology*, 19, 229–334.

Rea, N. and G.G. Ganf. 1994. The influence of water regime on the performance of aquatic plants, *Water Science and Technology*, 29, 127–132.

Reimer, P. and H.C. Duthie, 1993. Concentrations of zinc and chromium in aquatic macrophytes from the Sudbury and Muskoka regions of Ontario, Canada, *Environmental Pollution*, 79, 261–265.

Richardson, C.J., G.M. Ferrell, and P. Vaithiyanathan. 1999. Nutrient effects on stand structure, resorption efficiency, and secondary compounds in Everglades sawgrass, *Ecology*, 80, 2182–2192.

Ritchie, M.E., D. Tilman, and M.H. Knops. 1998. Herbivore effects on plant and nitrogen dynamics in oak savanna, *Ecology*, 79, 165–177.

Ritchie, M.E. and D. Tilman. 1995. Responses of legumes to herbivores and nutrients during succession on a nitrogen-poor soil, *Ecology*, 76, 2648–2655.

Romero, J.A., F.A. Comín, and C. García. 1999. Restored wetlands as filters to remove nitrogen, *Chemosphere*, 39, 323–332.

Rout, G.R., S. Samantaray, and P. Das. 1999. *In vitro* selection and biotechnical characterization of zinc and manganese adapted callus lines in *Brassica* spp, *Plant Science*, 137, 89–100.

Sander, M-L. and T. Ericsson. 1998. Vertical distributions of plant nutrients and heavy metals in *Salix viminalis* stems and their implications for sampling, *Biomass and Bioenergy*, 14, 57–66.

Schroeder, P.R. 2000. *Leachate screening considerations.* DOER Technical Notes Collection (ERDC TN-DOER-C16), U.S. Army Engineer Research and Development Center, Vicksburg, MS.

Schroeder, W.L., K.J. Williamson, R.T. Hudspeth, and D.H.K. Farness. 1977. *Dredging in Estuaries: Guide for Review of Environmental Impact Statements.* National Science Foundation, Washington, D.C., Oregon State University, Corvallis, OR.

Semmler, J.A. 1990. *PCB volatilization from dredged material, Indiana Harbor, Indiana.* Technical note, U.S. Army Corps of Engineers Waterways Experiment Station, Vicksburg, MS.

Siebe, C. 1995. Heavy metal availability to plants in soils irrigated with wastewater from Mexico City, *Water Science and Technology*, 32, 29–34.

Sikora, F.J., Z. Tong, L.L. Behrends, S.L. Steinberg and H.S. Coonrod. 1995. Ammonium removal in constructed wetlands with recirculating subsurface flow: removal rates and mechanisms, *Water Science and Technology*, 32, 193–202.

Simon, T.P., P.M. Stewart, and P.E. Rothrock. 2001. Development of multimetric indices of biotic integrity for riverine and palustrine wetland plant communities along southern Lake Michigan, *Aquatic Ecosystems Health and Management*, 4, 293–309.

Smith, V.J., R.J. Swindlehurst, P.A. Johnston, and A.D. Vethaak. 1995. Disturbance of host defense capability in the common shrimp, *Crangon crangon*, by exposure to harbour dredge spoils, *Aquatic Toxicology*, 32, 43–58.

Soto, F., M. García, E. de Luís, and E. Bécares. 1999. Role of *Scirpus lacustris* in bacterial and nutrient removal from wastewater, *Water Science and Technology*, 40(3), 241–247.

St-Cyr, L. and P.G.C. Campbell. 1994. Trace metals in submerged plants of the St. Lawrence River, *Canadian Journal of Botany*, 72, 429–439.

Steffens, J.C. 1990. Heavy metal stress and the phytochelatin response, in *Stress Responses in Plants: Adaptation and Acclimation Mechanisms*, Wiley-Liss, Inc., 377–394.

Stewart, A.J., G.J. Haynes, and M.I. Martinez. 1992. Fate and biological effects of contaminated vegetation in a Tennessee stream, *Environmental Toxicology and Chemistry*, 11, 653–664.

Stewart, P.M., J.T. Butcher, and T.P. Simon. 2002. Response signatures of four biological indicators to an iron and steel industrial landfill, Chapter 21, this volume.

Stewart, P.M., R.W. Scribailo, and T.P. Simon. 1999. The use of aquatic macrophytes in monitoring and in assessment of biological integrity, *Environmental Science Forum*, 96, 275–302.

Svensson, B.M. 1995. Competition between *Sphagnum fuscum* and *Drosera rotundifolia*: a case of ecosystem engineering, *Oikos*, 74, 205–212.

Swink, F. and G. Wilhelm. 1994. *Plants of the Chicago Region.* 4th ed. Indiana Academy of Science, Indianapolis, IN.

Tack, F.M. and M.G. Verloo. 1996. Metal contents in stinging nettle (*Urtica dioica* L.) as affected by soil characteristics, *The Science of the Total Environment*, 192, 31–39.

Tahvanainen, L. and V.-M. Rytkönen. 1999. Biomass production of *Salix viminalis* in southern Finland and the effect of soil properties and climate conditions on its production and survival, *Biomass and Bioenergy*, 16, 103–117.

Theodose, T.A. and W.D. Bowman. 1997. Nutrient availability, plant abundance, and species diversity in two alpine tundra communities, *Ecology*, 78, 1861–1872.

Tilman, D. 1987. Secondary succession and the pattern of plant dominance along experimental nitrogen gradients, *Ecological Monographs*, 57, 189–214.

Tomsett, A.B., A.K. Sewell, S.J. Jones, J.R. de Miranda, and D.A. Thurman. 1992. Metal-binding proteins and metal regulated gene expression in higher plants, in J.L. Wray (Ed.). *Society for Experimental Biology Series 49: Inducible Plant Proteins*, Cambridge University Press, Cambridge, UK, 1–24.

U.S. Army Corps of Engineers. 1986. Upland animal bioassays of dredged materials. Environmental Effects of Dredging Technical Notes EEDP-02–2, U.S. Army Engineer Waterways Experiment Station, Environmental Laboratory, Vicksburg, MS.

van der Putten, W.H. and B.A.M. Peters. 1997. How soil-borne pathogens may affect plant competition, *Ecology*, 78, 1785–1795.

van Oene, H., F. Berendse and C.G.F. de Kovel. 1999. Model analysis of the effects of historic CO_2 levels and nitrogen inputs on vegetation succession, *Ecological Applications*, 9, 920–935.

Wang, W. 1987. Toxicity of nickel to common duckweed (*Lemna minor*), *Environmental Toxicology and Chemistry*, 6, 961–967.

Welsh, R.P.H. and P. Denny. 1980. The uptake of lead and copper by aquatic macrophytes in two English lakes, *Journal of Ecology*, 68, 443–455.

Wetzel, P.R. and A.G. van der Valk. 1998. Effects of nutrient and soil moisture on competition between *Carex stricta*, *Phalaris arundinacea*, and *Typha latifolia*, *Plant Ecology*, 138, 179–190.

Whittle, C.A., L.C. Duchesne, and T. Needham. 1997. The impact of broadcast burning and fire severity on species composition and abundance of surface vegetation in a jack pine (*Pinus banksiana*) clear cut, *Forest Ecology and Management*, 94, 141–148.

Wilcox, D.A. 1995. Wetland and aquatic macrophytes as indicators of anthropogenic hydrologic disturbance, *Natural Areas Journal*, 15, 240–248.

Wilcox, D.A., N.B. Pavlovic, and M.L. Mueggler. 1985. Selected ecological characteristics of *Scirpus cyperinus* and its role as an invader of disturbed wetlands, *Wetlands*, 5, 87–97.

Wilhelm, G.S., P.M. Stewart, and T.P. Simon. Conservatism of confined disposal facilities based on the biological stability and integrity of plant communities: a case study in the Laurentian Great Lakes basin, Chapter 14, this volume.

Wilson, A.D. and D.J. Shure. 1993. Plant competition and nutrient limitation during early succession in the southern Appalachian Mountains, *American Midland Naturalist*, 129, 1–9.

Winger, P.V., P.J. Lasier, D.H. White, and J.T. Seginak. 2000. Effects of contaminants in dredge material from the lower Savannah River, *Archives of Environmental Contamination and Toxicology*, 38, 128–136.

Xie, C.Y. and P. Knowles. 1992. Associations between allozyme phenotypes and soil nutrients in a natural population of jack pine (*Pinus banksiana*), *Biochemical Systematics and Ecology*, 20, 179–185.

Yurukova, L. and K. Kochev. 1994. Heavy metal concentrations in freshwater macrophytes from the Aldomirovsko swamp in the Sofia District, Bulgaria, *Bulletin of Environmental Contamination and Toxicology*, 52, 627–632.

Zhang, J. 1996. Interactive effects of soil nutrients, moisture and sand burial on the development, physiology, biomass and fitness of *Cakile edentula*, *Annals of Botany*, 78, 591–598.

Zhu, T. and F.J. Sikora. 1995. Ammonium and nitrate removal in vegetated and unvegetated gravel bed microcosm wetlands, *Water Science and Technology*, 32, 219–228.

6 Effects of Sediment Quantity on the Health of Aquatic Ecosystems: A Case Study on Depth of Fines in Coastal Plain Streams in Alabama

Charles C. Morris, James A. Sawyer, IV, Holly H. Bennett, and Christy D. Robinson

CONTENTS

6.1 INTRODUCTION

Non-point source pollution results from nearly every type of human activity and land use, including urban and industrial storm water runoff, livestock and crop production, forestry, mining, construction, and hydrological modifications. Non-point source pollution affects more stream miles than point source pollution and is a significant reason U.S. rivers failed to meet the goals of the Clean Water Act (Wilkinson, 1987). The magnitude and cumulative effects of non-point source pollutants impart profound impacts on the health of aquatic ecosystems. Non-point sources generally cannot

0-8493-0905-0/03/$0.00+$1.50

be monitored at their points of origin because their sources are not readily identifiable. Furthermore, it is difficult to distinguish human-induced from naturally occurring non-point source pollution. Consequently, few definitive data quantitatively document the cause-and-effect relationships between non-point source pollutants and the degradation of fisheries and other aquatic resources.

Impairment in southern Alabama streams is thought to be due primarily to non-point sources, principally sedimentation. The obvious severity of the problem and the large quantities of fines suggest that fines play a principal role in driving aquatic communities.

The purpose of this chapter is twofold. First, a review of existing literature was conducted to provide background information on non-point source pollution and its effects on biota. Second, a case study was performed on depths of sediment fines in the Choctawhatchee–Pea watershed in southeast Alabama. Initial observations suggest fines would significantly impact steam ecology, but no previous studies attempted to quantify the impact of fines in this region.

6.2 BACKGROUND

6.2.1 Row Crops

Sediment, its related pollutants, and run-off from agriculture lands are the principal sources of non-point source pollutants (Duda and Johnson, 1985). High concentrations of pesticides and chemicals found in non-point source pollution pose major threats to the environmental integrity of aquatic ecosystems (Leonard et al., 1979; Trim, 1987; Huber et al., 2000). Non-point source agricultural pollution has become a major water quality concern not only in the United States, but also in Western Europe and parts of Africa (Mbagwu and Ita 1994; Hunt et al., 1999; Huber et al., 2000). Among all non-point sources of pollution affecting streams and rivers, agriculture in its several forms is by far the most important (U.S. Environmental Protection Agency (USEPA), 1990). Through the processes of runoff and leaching, pesticides and chemical fertilizers can cause surface water and groundwater contamination.

The movement of pesticide and chemical fertilizer residues in runoff from agricultural fields is a well documented source of surface water contamination (Yves-Caux et al., 1996; Troiano and Garretson, 1998). Commercial fertilizer residues containing nitrogen (N), phosphorus (P), potassium (K), secondary nutrients, and micronutrients are major sources of excess nutrients that degrade the integrity of aquatic systems (Spaling, 1995; Tian et al., 1996; Ma et al., 1999).

The runoff and leaching potential of pesticides and chemical fertilizers from agricultural land depend upon application rate, climate, land use, and geophysical characteristics of the land (Domagalski, 1996; USEPA, 1997; Huber et al., 2000). Rainfall events are the most important climatic factors that affect agricultural chemical concentrations in runoff (Dolmagalski, 1996). Rainfall is an effective pathway to move agricultural chemicals into soil matrices with high permeability, but it is an ineffective way to move chemicals into soils where runoff is produced due to low infiltration rates or sloping landscapes (Troiano and Garretson, 1998). Soil type and the physiochemical properties of the chemicals are the most important factors that affect chemical concentrations in groundwater due to leaching (Diaz-Diaz, 1998; Ma et al., 1999).

Other factors that affect the amounts of pesticides and chemical fertilizers found in non-point source pollution have been identified. Geological factors, such as the length and steepness of slope and soil composition affect the amounts of agricultural chemicals found in non-point source pollution. The proximity of water bodies to land where pesticides and chemical fertilizers have been depleted is also important in determining chemical yields (USEPA, 1997). The interdependence of these factors makes it difficult to ascertain their individual contributions (Hainly et al., 1994; Huber et al., 2000).

6.2.2 Livestock and Poultry

Animal agricultural production practices are increasingly scrutinized for their impacts on water quality throughout the United States (Martin, 1997). Impacts are due largely to the generation of excess sediment caused by livestock overuse of stream riparian areas (Waters, 1995). In addition to sedimentation, degrading effects of livestock include elevated stream water temperatures and loss of undercut banks used for cover and fish-rearing habitat. Most effects of riparian overuse include increased fine sediment generation from channel widening, decreased flow and shallow channels, sediment from slumping stream banks, and increased deposition on streambeds (Waters, 1995). Where riparian areas have been destroyed or degraded by overgrazing, damaging storm effects are much greater on stream banks and channels (Platts et al., 1985).

Run-off from areas of concentrated livestock use yields unnaturally high amounts of nutrients such as nitrogen and phosphorus, organic matter, chemicals, fecal bacteria, and other microorganisms that exceed the abilityof the local agricultural land base to use them for crop production (Oloya and Logan, 1980; Power and Schepers, 1989). Most nitrogen contamination to surface waters may occur as direct runoff in the form of nitrate (NO_3) and ammonia (NH_4). This can be enhanced with reduced soil infiltration, steep topography, increased nitrogen concentration at the soil surface from manure and fertilizer application, and a limited riparian buffer zone (Zebarth et al., 1999). Nitrate contamination of surface waters can cause accelerated growth of algae and aquatic plants, dissolved oxygen depletion, increased turbidity, and general water quality degradation.

Groundwater and surface water contamination from poultry wastes is also an issue in non-point source pollution. Poultry manures, litters, sludges, composts, and wastewaters originating from production operations are normally disposed of in large-scale land application programs and are rarely concentrated enough to be considered point sources of nitrogen or phosphorus (Sims and Wolf, 1994). Because poultry waste contains all essential plant nutrients (C, N, P, K, S, Ca, Mg, B, Cu, Fe, Mn, Mo, and Zn), it is commonly used as fertilizer. Excessive and poorly timed application of these fertilizers can result in nutrient losses by leaching, erosion, or runoff.

6.2.3 Silviculture

Land-disturbing activities may cause the introduction of large amounts of sediment into nearby streams and rivers. Erosion is particularly important on steep slopes devoid of vegetation as on agricultural fields, road cuts, and construction sites. A thick vegetative cover intercepts virtually all the kinetic energy of rainfall, therefore reducing soil erosion. The denser the vegetative cover, the lower the rate of erosion. This factor, which is controlled in turn by climate and land use, dominates the effects of all other controls. The southeast United States experiences heavy rainfalls, yet historically has one of the lowest geologic rates of erosion due to thick natural forest cover. The forests of the region were cleared for planting corn, tobacco, and cotton in the late eighteenth and early nineteenth centuries, resulting in higher erosion rates.

Forest roads are recognized as major sources of erosion from forest lands across the United States (Patrick, 1967; Elliot et al., 1994), historically accounting for as much as 90% of all sediment produced from forest land (Hoover, 1952; Anderson et al., 1967; Megahan and Kidd, 1972). Road side-slopes have been reported to account for 70 to 90% of the total soil loss from disturbed roadway areas (Swift, 1984). Roads accelerate erosion by increasing slope gradients and interrupting normal drainage patterns, which concentrates overland flow into ditches and channels. Erosion produced from forest road systems eventually reaches and degrades the quality of stream systems. Sediment from forest roads can shorten the lives of reservoirs, degrade drinking water, and clog fish spawning beds. Sediment losses from forest roads require special attention because sediment can be carried directly to waterways through ditches and crossings.

6.2.4 Sedimentation

It is generally agreed that agricultural activities are the main contributors to non-point source pollution in lakes and rivers. Sediment, by volume is the largest pollution source in the United States (Patrick et al., 1992; Stewart and Swinford, 1995). It is the predominant river non-point source pollutant. Sedimentation from agricultural fields occurs as a result of sheet, rill, and gully erosion. Eroded sediment is redeposited on the same field or transported from the field in runoff. Studies in North Carolina show erosion from agricultural land was the major contributor of sediment to streams and rivers (Lenat, 1984; Waters, 1995).

It was estimated that out of 77 million tons of eroded soil per year, 67% came from agricultural lands (Lenat, 1984). Sediment originating from surface soil is usually richer in nutrients due to past fertilizer applications, nutrient cycling, and other biological activities, and has greater pollution potential than subsurface soils (USEPA, 1997). Sediment from agricultural land contains higher proportions of finer and less dense particles than the original soil. This is due to the selective nature of soil erosion. Larger particles easily separate from the soil surface because they are less cohesive; however, they tend to settle out of suspension quickly because of their size. Smaller particles are not as easily detached and do not settle out as quickly. Selective erosion can cause an overall increase in pollutant delivery because small particles have much greater adsorption capacities than larger particles. As a result, eroded sediment generally has a higher concentration of nutrients and pesticides than the original soil (USEPA, 1997).

6.3 CASE STUDY

6.3.1 Introduction

The ultimate goal of this research is to describe the physical and chemical forces driving fish community structure and function in the Choctawhatchee–Pea watershed. Initial observations suggested that the amount of fines would significantly impact stream ecology. No previous studies attempted to quantify the impact of fines in this region. This preliminary study examines sediment fines and their effects on coastal plain stream fish communities.

6.3.2 Methods

6.3.2.1 Study Area

The Choctawhatchee–Pea watershed drains about 4,636 km^2 of southeastern Alabama (Figure 6.1) and was formed over 60 million years ago when the ocean receded from part of the state. That action isolated freshwater organisms into pockets and drains, which led to present-day levels of biological diversity (Hilton, 2000). The watershed experienced significant alteration in its natural landscape through land use practices. These alterations, combined with heavy rainfall and highly erodable soils, created a high potential for non-point source pollution. Agricultural activities, including chicken and hog farming, cultivation of sloping areas, and clear cutting of timber to stream edges, occurred in the basin. These activities contributed to point and non-point source pollution and streambank and surface erosion (U.S. Army Corps of Engineers, 1992; U.S. Department of Agriculture, 1993; Grace, 2000). In some counties in the watershed, unpaved roads provide more than half the sediment transported into streams (U.S. Department of Agriculture, 1993).

In recent years, non-point source pollution and landscape ecology research efforts focused attention on the need for improved understanding and management of the impacts of different land use activities, industrial processes, and human cultural practices on water quality along the river continuum from uplands to the sea (Omernik et al., 1981; Hunsaker and Levine, 1995; Johnson et al., 1995; Tufford et al., 1998).

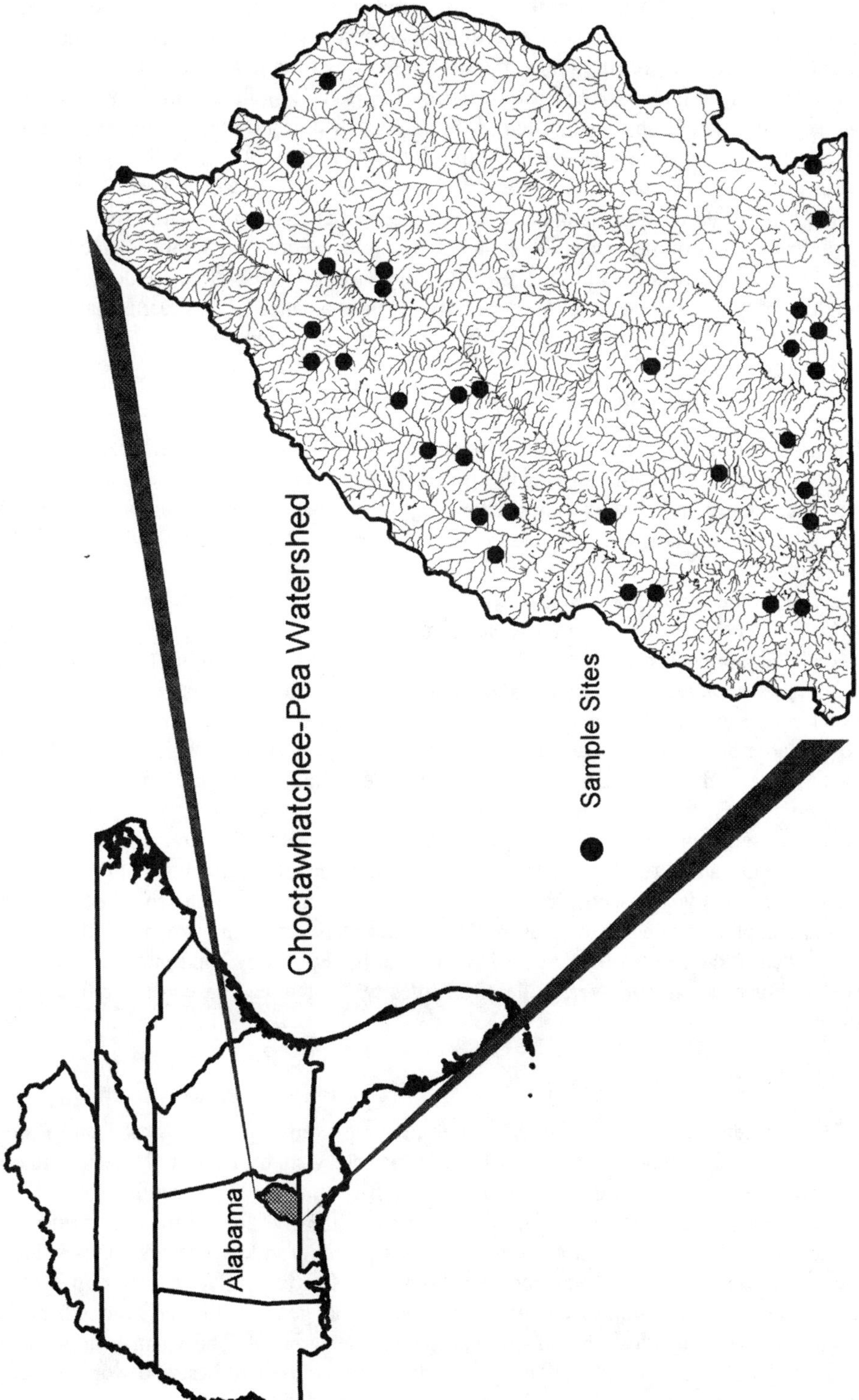

FIGURE 6.1 Map of the Choctawhatchee–Pea watershed, southern Alabama, indicating April 2001 sample locations.

6.3.2.2 Sample Design

U.S. Geological Survey 7.5-minute topographic maps were used to delineate the Choctawhatchee–Pea watershed. All possible access points (bridges and roads) along streams were identified and assigned site-specific codes. Stream order was then determined for the individual sites. Sites were grouped into three categories: Group 1 consisted of all stream sites determined to be first and second order. Group 2 consisted of third and fourth order streams, and the final group contained orders greater than fourth. Because this study was concerned with wadeable streams, all streams with orders higher than sixth (generally considered non-wadeable) were eliminated. All possible sampling locations were pooled based on their order groupings. The proportion of the entire sample that each group represented dictated the number of sites randomly selected from that group (i.e., if 90% of the possible sample sites fell within Group 1, 90% of the random draw would come from that group). This method allowed sampling to be concentrated on streams that comprised the majority of stream miles in the watershed without sacrificing randomness.

6.3.2.3. Field Sampling Protocols

Standard fish collecting gear was used, specifically a Smith-Root model 12-B® battery-powered backpack electrofisher. Fish sampling began upstream of the bridge access (beyond the bridge's influence on the stream) and continued upstream roughly 35 times mean stream width, 150 to 500 m. All representative habitats were sampled, and in larger segments, at least two riffle-pool/run series were collected.

Quantitative methods required placing captured fish in a live well until processing to minimize mortality. Fish collected during quantitative sampling were released immediately after they were identified, counted, and weighed. Every effort was made to minimize handling and holding times. Maximum and minimum lengths were recorded to the nearest millimeter along with a batch weight. Any fish that could not be identified with 100% confidence was preserved in 10% formalin and brought to the laboratory for verification with standard taxonomic keys (Menhinick, 1991; Etnier and Starnes, 1993; Jenkins and Burkhead, 1993; Mettee, et al., 1996) and confirmations by a professional taxonomist.

The depths of sediment fines were recorded at each stream reach. This consisted of measuring the depth of the water and the depths of the sand, silt, and other fine sediments that comprised the streambed. At each site, five transects proportional to the reach length were chosen. At each transect, water depth and depths of fines were measured at five intervals across the stream width. Depth of fines was recorded to the nearest 0.01 m by driving a calibrated rod by hand into the sediment as far as possible. For each site, an average depth of fines was calculated based on the five transect measurements.

6.3.2.4 Statistics

Relationships between sediment and fish communities were examined using SPSS® and Primer© statistical software (Clarke and Ainsworth, 1993). Data were visually inspected to determine the five "best" and five "worst" sites depending on the depth of fines and species diversity. The five sites with the deepest fines were grouped as the worst and the five lowest measured fines were grouped as the best. The same was done for the species diversity index. Species composition for the five best and worst sites was analyzed using the Bray-Curtis similarity matrix in a cluster analysis. Graphical representation of the similarity matrix was generated based on depth of fines and species data to show any clustering of sites based on these variables. The Mann-Whitney U-test was used to show significant differences in species diversity for the five best and worst sites.

6.4 RESULTS

Forty-five fish species representing 13 families were collected from the 34 sites. Collections were dominated by two minnow species, blacktail shiner (*Cyprinella venusta*) and silverjaw minnow (*Ericymba buccata*), three sunfish species, bluegill (*Lepomis macrochirus*), longear sunfish (*Lepomis megalotis*), and redspotted sunfish (*Lepomis miniatus*). The mean depth of fines was 0.37 meters (s.d. = 0.21), and fines were normally distributed ranging from zero to 0.85 meters. No significant difference in species diversity was found between the five best and five worst sites. Species composition in both groups did not show any tendencies to group in the cluster analysis (Figure 6.2) or clustering of sites based on depth of fines. No correlations were found between depth of fines and species diversity.

6.5 DISCUSSION

Receding waters from the coastal plains region in the middle to late Miocene left behind a thick layer of sediment creating a highly erodible substrate (Khudoley and Meyerhoff, 1971). The instability of this substrate caused extensive deposition from erosion and sedimentation in the southeastern coastal plains throughout the Tertiary that still occurs today (Swift, 1986). Decline in the native fish diversity in the southern United Sates can be attributed to habitat alteration and degradation (Angermeier, 1995; Warren et al., 1997). Clear cutting can affect the hydrological regimes and habitat complexities of stream systems (Poff et al., 1997). In the late 1800s, agricultural and silvicultural practices led to widespread alterations to the ecology of southern Alabama. The impacts on aquatic communities as a result of these ecological changes are unknown.

The buildup of fines in the Choctawhatchee–Pea system suggests that severe erosion and subsequent sediment deposition occurred, but we found no literature describing natural systems in

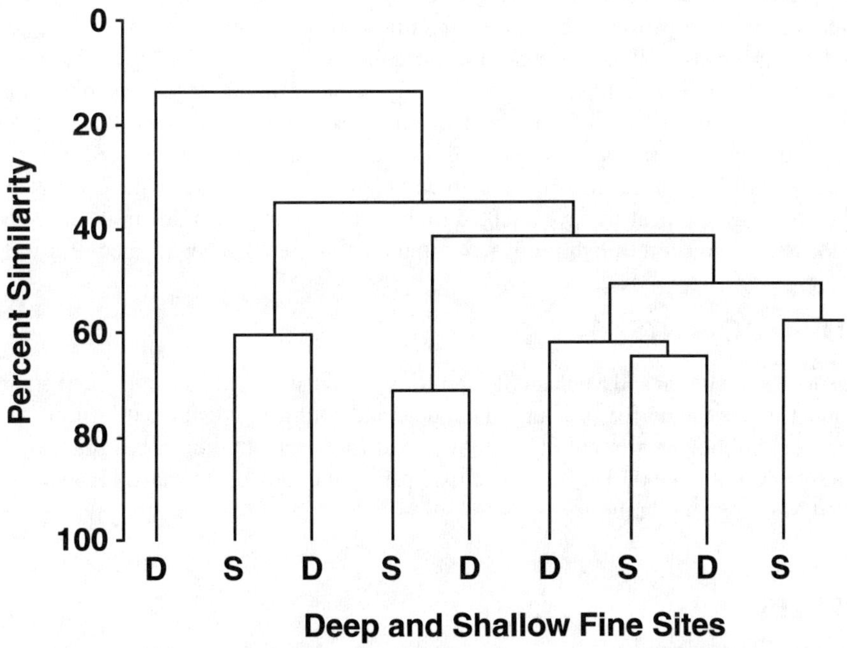

FIGURE 6.2 Dendogram depicting similarity of fish communities in the Choctawhatchee–Pea watershed at the five shallowest (S) and the five deepest (D) fines sites.

this area prior to 1800. The dose–response curve used to measure community response to pollution is based on a range of community health values, from reference conditions to heavily degraded conditions (Hughes, 1985; Hughes et al., 1994). Lack of historical information makes it difficult to describe a true reference condition, but it is anticipated that a substantial amount of time is required for a system to fully recover from disturbance. It is unlikely the Choctawhatchee–Pea watershed has fully recovered. It seems that sufficient time has passed to allow fish communities to respond and adjust to present conditions. We are forced to face the likelihood that the entire system is impaired; therefore reference condition expectations or "best obtainable conditions" in the watershed lack the degree of heterogeneity seen in other systems. Under the assumption that all streams in this system are degraded, detection levels were narrowed on the dose–response curve. Rather than using the entire curve to measure response, we focused on a small segment. The results of this study suggest that depth of fines is not the principal force driving fish communities in the watershed.

The shift in the dose–response curve may also be limited by other habitat variables. As fines increase, a significant shift in the curve may not be apparent due to other habitat factors buffering its response. Habitat factors, such as riparian width and overhanging vegetation, are known predictors of community health. Alabama waterways, for the most part, are well buffered with riparian vegetation, and extensive channelization has not occurred. However, Harding et al. (1998), found that fish communities in watersheds that currently contain dense forest cover and quality habitat may still be recovering from agricultural land use that occurred 40 years earlier. Fines may not become a driving force until after the "buffering" variables are impacted. In a system where fines are not a problem, the degradation of riparian buffers will elicit a given response. A system such as the Choctawhatchee–Pea watershed that has a problem with erosive sediments, may demonstrate a more severe response given the same degradation. We agree with McCormick et al. (2001) who believe that the generally impaired conditions of fish assemblages in southern Alabama coastal plains streams are reflections of the history of anthropogenic impacts that occurred throughout the region and the multiple stressors currently affecting fish communities.

Depth of fines was unable to explain the variability of species diversity in the watershed. The next step in this study will be multivariate analysis using many physiochemical and land use parameters in relation to the structures and functions of fish communities. Data from qualitative and quantitative habitat assessments obtained during fish sampling will be used to further explain the variability in fish community structures. Accumulation of sediments in this watershed continues, but fish communities appear to have adapted to these conditions. A multivariate analysis may help to describe the driving forces behind fish communities in the Choctawhatchee–Pea watershed.

ACKNOWLEDGMENTS

We thank Michael Mullen and Michael Stewart for their assistance in field collection and manuscript review and Thomas Simon for assisting in the field and confirming fish identifications. This project was supported by the Department of Biological and Environmental Sciences graduate program at Troy State University, the ALFA Insurance Company, and the Choctawhatchee, Pea, Yellow River Watershed Management Authority.

REFERENCES

Anderson, H.W., M.D. Hoover, and K.G. Reinhart. 1976. *Forest and Water. Effects of Forest Management on Floods*, Sedimentation, and Water Supply. PSW-18. USDA Forest Service, San Francisco, CA.

Angermeier, P.L. 1995. Ecological attributes of extinction-prone species: loss of freshwater fishes of Virginia, *Conservation Biology*, 9, 143–158.

Clark, K.A. and M. Ainsworth. 1993. A method of linking multivariate community structure to environmental variables, *Marine Ecology Progress Series*, 92, 205–219.

Cole, G.A. 1994. *Textbook of Limnology*, Fourth Edition. Waveland Press, Prospect Heights, IL.

Cook, M.R. and P. O'Neil. 2000. Implementation assessment for water resource *availability, protection, and utilization for the Choctawhatchee, Pea, and Yellow Rivers watersheds: surface water and biological resources. Geological Survey of Alabama*, Tuscaloosa, AL.

Diaz-Diaz, R., J. Garcia-Hernandez, and K. Loague. 1998. Leaching potential of four pesticides used for bananas in the Canary Islands, *Journal of Environmental Quality*, 27, 562–572.

Domagalski, J. 1996. Pesticides and pesticide degradation products in stormwater runoff: Sacramento River Basin, California, *Water Resources Bulletin*, 32, 953–964.

Duda, A.M. and R.J. Johnson. 1985. Commentary: cost effective targeting of agricultural nonpoint-source pollution controls, *Journal of Soil and Water Conservation*, 40, 108–110.

Elliot, W.J., R.B. Foltz, and M.D. Remboldt. 1994. *Predicting sedimentation from roads at stream crossings with the WEPP model.* Paper no. 9475 11. American Society of Agriculture Engineering (ASAE), St. Joseph, MI.

Grace, J.M. 2000. Forest roadside slopes and soil conservation techniques, *Journal of Soil and Water Conservation*, 55, 96–101.

Hainly, R.A. and J.M. Kahn, 1996. Factors affecting herbicide yields in the Chesapeake Bay watershed, June, 1994, *Water Resources Bulletin*, 32, 965–984.

Harding J.S., E.F. Benfield, P.V. Bolstad, G.S. Helfman, and E.B.D Jones III. 1998. Stream biodiversity: the ghost of land use past, *Proceedings of the National Academy of Sciences*, 95, 14843–14847.

Hayes, S., L. Newland, K. Morgan, and K. Dean. 1990. Septic tank and agricultural non-point source pollution within a rural watershed, *Toxicological and Environmental Chemistry*, 26, 137–315.

Hilton, J. 2000. Alabama's biological diversity. *Alabama Wildlife Magazine Archives,* 2000, 23–27.

Hoover, M.D. 1952. Water and timber management, *Journal of Soil and Water Conservation*, 7, 75–78.

Huber, A., M. Bach, and H.G. Frede. 2000. Pollution of surface water with pesticides in Germany: modeling non-point source inputs, *Agriculture, Ecosystems and Environment*, 80, 191–204.

Hughes, R.M. 1985. Use of watershed characteristics to select control streams for estimating effects on metal mining wastes on extensively disturbed streams, *Environmental Management*, 9, 253–262.

Hughes, R.M., S.A. Heiskary, W.J. Matthews, and C.O. Yoder. 1994. Use of ecoregions in biological monitoring, in S.L. Loeb and A. Spacie (Eds.). *Biological Monitoring of Aquatic Ecosystems*, Lewis Publishers, Chelsea, MI, 125–151.

Hunsaker, C.T. and D.A. Levine. 1995. Hierarchical approaches to the study of water quality in rivers, *Bioscience*, 45, 193–203.

Hunt, J.W., B.S. Anderson, B.M. Phillips, R.S. Tjeerdema, H.M. Puckett, and V. de Vlaming. 1999. Patterns of aquatic toxicity in an agriculturally dominated coastal watershed in California, *Agriculture, Ecosystems and Environment*, 75, 75–91.

Johnson, B.L., W.B. Richardson, and T.J. Naimo, 1995. Past, present, and future concepts in large river ecology, *Bioscience*, 45, 193–203.

Khudoley, K.M. and A.A. Meyerhoff. 1971. Paleogeography and geological history of *Greater Antilles*, Geological Society of America, Boulder, CO.

Lenat, D.R. 1984. Agriculture and stream water quality: a biological evaluation of erosion control practices, *Environmental Management*, 8, 333–334.

Leonard, R.A., G.W. Langdale, and W.G. Fleming. 1979. Herbicide runoff from upland Piedmont wastersheds: data and implications from modeling pesticide transport, *Journal of Environmental Quality*, 8, 223–229.

Ma, Q.L., A.E. Smith, and J.E. Hook. 1999. Water runoff and pesticide transport from a golf course fairway: observation vs. Opus model simulations, *Journal of Environmental Quality*, 28, 1463–1473.

Martin, J.H. Jr. 1997. The Clean Water Act and animal agriculture, *Journal of Environmental Quality*, 2, 1198–1203.

Mbagqu, I.G. and E.O. Ita. 1994. Pesticide use in the sub-humid zones of Nigeria: implications for conservation of aquatic resources, *Environmental Conservation,* 21, 214–219.

McCormick, F.H., R.M. Hughes, P.R. Kaufmann, D.V. Peck, J.L. Stoddard, and A.T. Herlihy. 2001. Development of an index of biotic integrity for the Mid-Atlantic Highlands region, *Transactions of the American Fisheries Society*, 130, 857–877.

Megahan, W.F. and W.J. Kidd, 1972. Effects of logging roads on sediment production rates in the Idaho Batholith, INT-123. USDA Forest Service, Intermountain Forest and Range Experiment Station, Ogden, UT.

Oloya, T.O. and T.J. Logan. 1980. Phosphate desorption from soils and sediments with varying levels of extractable phosphate, *Journal of Environmental Quality*, 9, 526–531.

Omernik, J.M., A.R. Abernathy, and L.M. Male, 1981. Stream nutrient levels and proximity of agricultural and forest lands to streams: some relationships, *Journal of Soil and Water Conservation*, 36, 227–337.

Paller, M.H. 2001. Comparison of fish and macroinvertebrate bioassessments from South Carolina coastal plain streams, *Aquatic Ecosystem Health and Management*, 4, 175–186.

Patrick, J.H. 1976. Soil erosion in the eastern forest, *Journal of Forestry*, 74, 671–677.

Patrick, R., F. Douglas, D.M. Palavage, and P.M. Stewart.1992. *Surface water quality: have the laws been successful?* Princeton University Press, Princeton, N.J.

Platts, W.S. and R.L. Nelson. 1985. Impacts of restoration grazing on stream banks in forested watersheds in Idaho, *North American Journal of Fisheries Management*, 5, 547–556.

Poff, N.L., J.D. Allen, M.B. Bain, J.R. Karr, K.L. Prestegaard, B.D. Richter, R.E. Sparks, and J.C. Stromberg. 1997. The natural flow regime: a paradigm for river conservation and restoration, *BioScience*, 47, 769–784.

Power, J.F. and J.S. Schepers, 1989. Nitrate concentration of groundwater in North America, Agriculture Ecosystems and Environment, 26, 165–188.

Sims, J.T. and D.C. Wolf. 1994. Poultry waste management: agricultural and environmental issues, in D.L. Sparks (Ed.). *Advances in Agronomy,* SPB Academic Press, Newark, DE. 1–83.

Spaling, H. 1995. Analyzing cumulative environmental effects of agricultural land drainage in southern Ontario, Canada, *Agriculture, Ecosystems and Environment*, 53, 279–292.

Stewart, P.M. and T.O. Swinford. 1995. Identification of sediment and nutrient sources impacting a critically endangered mussel species' habitat in a small agricultural stream, in J.R. Pratt, N. Bowers, and J.R. Stauffer (Eds), *Making Environmental Science. A Festschrift in Honor of John Cairns, Jr.*, Ecoprint, Portland, OR, 45–64.

Swift, L.W. Jr. 1984. Soil loss from roadbeds and cut and fill slopes in the southern Appalachian Mountains, *Southern Journal of Applied Forestry*, 8, 209–215.

Swift, G.C., C.R. Gilbert, S.A. Bortone, G.H. Burgess, and R.W. Yerger. 1986. Zoogeography of the freshwater fishes of the southeastern United States; Savannah River to Lake Ponchatrain, in C.H. Hocutt and E.O. Wiley (Eds.). *The Zoogeography of North American Freshwater Fishes,* John Wiley & Sons, New York.

Tian, X., G.J. Sabbagh, G.W. Cuperus, and M. Gregory. 1996. Evaluating potential environmental impact of insecticide application in a boll weevil eradication program, *Water Resources Bulletin*, 32, 1027–1037.

Trim, A.H. 1987. Acute toxicity of emulsifiable concentrations of three insecticides commonly found in non-point source runoff into estuarine waters and to the mummichog, *Fundulus heteroclitus, Bulletin of Environmental Contamination and Toxicology*, 38, 681–686.

Troiano, J. and C. Garreston. 1998. Movement of simazine in runoff water from citrus orchard row middles as affected by mechanical incorporation, *Journal of Environmental Quality*, 27, 488–494.

Tufford, D.L, H.N. McKellar, and J.R. Hussey, 1998. Insteam nonpoint source nutrient prediction with land-use proximity and seasonality, *Journal of Environmental Quality*, 27, 100–111.

U.S. Army Corps of Engineers. 1992. Reconnaissance Report: Choctawhatchee and Pea River Basins Study, Alabama and Florida CWIS 12814.

U.S. Department of Agriculture. 1993. Choctawhatchee and Pea River Basin Study, Alabama and Florida Reconnaissance Report. Water Resources Planning Staff Auburn, AL, Gainesville, FL.

U.S. Environmental Protection Agency. 1990. *The Quality of our Nation's Water: A Summary of the 1988 National Water Quality Inventory.* EPA 440/4–90–005. USEPA, Office of Water, Oceans, and Wetlands,Washington, D.C.

U.S. Environmental Protection Agency. 1997. *Guidance Specifying Management Measures for Sources of Non-point Sources of Pollution in Coastal Waters.* EPA-840-B-93–001c. USEPA, Office of Water, Washington, D.C.

Warren, M.L., P.L. Angermeier, B.M. Burr, and W.R. Haag. 1997. Decline of a diverse fish fauna: patterns of imperilment and protection in the southeastern United States, in G.W. Benz and D.E. Collins (Eds.). *Aquatic Fauna in Peril: The Southeastern Perspective*. Special Pub. 1, Southeast Aquatic Research Institute, Lenz Design and Communications, Decatur, GA. 105–164.

Waters, T.F. 1995. *Sediment in Streams: Sources, Biological Effects, and Control,* American Fisheries Society Monograph 7, Bethesda, MD.

Wilkinson, C.F. 1987. Soil conservationists and the uses of law, *Journal of Soil and Water Conservation*, 42, 304–311.

Yves-Caux, P., C. Bastien, and A. Crowe. 1996. Fate and impact of pesticides applied to potato cultures: the Nicolet River basin, *Ecotoxicology and Environmental Safety*, 33, 175–185.

Zebarth, B.J., J.W. Paul, and R. van Kleeck. 1999. The effect of nitrogen management in agricultural production on water and air quality: evaluation on a regional scale, *Agriculture, Ecosystems, and Environment*, 72, 35–52.

7 The Difficulty in Determining the Effects of Pesticides on Aquatic Communities

Scott A. Sobiech and Mary G. Henry

CONTENTS

All things are connected.

Chief Seattle

The dose makes the poison.

Paracelsus

7.1 INTRODUCTION

Pesticides can be found in nearly every home, business, farm, school, hospital, golf course, park, forest, stream, and lake in the United States (U.S. Fish and Wildlife Service, 1964; Odenkirchen and Eisler, 1988). Pesticides have been intentionally introduced into the environment, are intended to adversely affect "target organisms," are chemically dynamic, biologically mobile, and difficult to detect in natural systems (Ware, 1994). Many issues surrounding pesticides must be addressed if their role is to be better understood and adverse effects on non-target organisms prevented. Associated problems include organic carbon partitioning and degradation (Wilcock et al., 1993; Ankley et al., 1994), trophic transfer within the food web (Miller et al., 1966; DiPinto, 1996), differences in life stage effects (Stevens, 1992; Green et al., 1996), and extrapolating laboratory tests to field observations (Kersting and van der Brink, 1997).

0-8493-0905-0/03/$0.00+$1.50

Most pesticides pose some potential hazard to humans, animals, or the environment (U.S. Fish and Wildlife Service, 1964, 1989). They are designed to kill or otherwise adversely affect living organisms. Despite this drawback, pesticides are useful to society because of their ability to kill potential disease-causing organisms and control insects, weeds, and other pests. Many chemicals, such as cleaning products and antibacterial soaps used in and around the household are pesticides. Many of these products, such as kitchen and bath cleaning agents and antibacterial soaps, are washed down the drain and become part of the waste streams that reach wastewater treatment facilities that are not equipped to "treat" them. In addition, lawn and garden chemicals, swimming pool chemicals, and other pesticides used around the home can become part of the waste stream transported to water treatment facilities or may reach surface waters directly via surface runoff. Once in the environment, pesticides are often found in mixtures with other environmental pollutants (Wan et al., 1995). The potential antagonistic, synergistic, additive, and/or cumulative toxicological impacts from exposure to these chemical mixtures in the aquatic environment are unknown (Wilcock et al., 1993).

While the mere presence of pesticides detected in an aquatic environment may not be problematic, interpreting and evaluating what impact their presence may have on aquatic life is very complex (Stewart and Butcher, 1999; Lydy et al., Chapter 17). In addition to the problems of detection, the transformation, degradation, and metabolism of the parent pesticide through chemical, physical, and biological mechanisms can further complicate matters. These complications make the interpretation of potential affects of pesticides in the environment even more ambiguous.

Monitoring programs, such as the National Pesticide Synthesis Project, which is part of the National Water Quality Assessment Program (NAWQA) of the U.S. Geological Survey (USGS), have not been designed and implemented to be associated with specific pesticide use and application patterns. Virtually no monitoring program exists that can adequately monitor for adverse effects on non-target organisms. Often, limited funding for such monitoring programs dictates the number of pesticides that are included for chemical analyses. Therefore, observed environmental concentrations likely underestimate the peak concentrations and number of pesticidal products likely to be present in the environment.

These factors provide arguments for supporting increased efforts toward monitoring both abiotic and biotic components of aquatic ecosystems to better understand the presence, movement, and effects of pesticides in the aquatic environment. Aquatic assemblages will reflect the ultimate impacts due to pesticide exposure. Good species distribution, population, and assemblage baseline information combined with real use/real time pesticide detection, monitored over both the short and long terms will provide trend information that will aid in documenting biological changes due to pesticides in the environment (Wilhm and Dorris, 1968).

This chapter has three purposes: (1) to identify how pesticide use and occurrence in the aquatic environment have the potential to affect aquatic communities, (2) to present examples of documented impacts to non-target aquatic organisms reported by others that are representative of the types of unexpected effects from pesticides detected in aquatic systems, and (3) to make a case for monitoring so that the role of pesticides in the structure and function of aquatic communities can be better understood.

7.1.1 Pesticide Use

About 4.5 billion pounds of pesticides are used annually in the United States, and 4.63 billion pounds of chemicals (as active ingredients) were used as pesticides in 1997 (Aspelin and Grube, 1999). Pesticide use is increasing in the United States, with most uses on agricultural lands, forested lands, range lands, golf courses, lawns, and around homes. Agricultural use of pesticides accounts for 75% of the annual use in the United States. By volume, herbicides are the most widely-used pesticides, followed by insecticides (Aspelin and Grube, 1999). Many of these pesticides are broad-spectrum insecticides and herbicides that have the potential to impact non-target insects and plants.

The presence of pesticides and the potential for adverse impacts from the parent compound, the metabolized and degraded products, and inert ingredients in the environment are likely to increase. Many of these pesticides end up in aquatic systems via runoff into surface waters, drift, and through ground water (Larson et al., 1997).

7.1.2 Pesticide Registration and Regulation of Use

Pesticides have been intentionally introduced into the environment to harm, repel, control, or kill organisms viewed as pests. The general term *pesticide* encompasses more specific terms indicative of uses, such as insecticides (targeting insect pests), herbicides (plant pests), fungicides (fungi), rodenticides (rodents), disinfectants (bacteria), repellants, and other substances used to control pests. Pesticide use in the United States is regulated under the Federal Insecticide, Fungicide, and Rodenticide Act (FIFRA) (7 U.S.C.136 et seq.), which requires registration by the U.S. Environmental Protection Agency (USEPA) of all pesticides sold, distributed, or imported. To understand how pesticides might enter aquatic systems and affect biota, it is important to understand the registration process. To demonstrate that a pesticide does not pose an unreasonable risk to humans or the environment when used for its intended purpose, its manufacturer is required to provide information on the product and must conduct certain tests as benchmarks of toxicity including short-term acute lethality toxicity tests, partial life-cycle reproductive toxicity tests, and, if warranted, simulated and applied field studies.

The pesticide manufacturer must also supply information on the general chemistry of the product and its constituents along with limited studies on its fate, transport, and toxicological properties. This information is generally provided for the "active" ingredient (usually referred to as the technical grade active ingredient), rather than for the formulated product the end-user purchases. The active ingredient is the substance that delivers the efficacious or toxic effect to the target organism. The formulated product generally includes other ingredients identified on the label as "inert." The toxicological characteristics of the pesticide are based on exposure of standard laboratory test organisms to a single contaminant, the active ingredient for the pesticide in question, under controlled laboratory conditions. These limited laboratory studies are meant to oversimplify conditions of exposure and response, so that a clear picture of toxicity can be gained. These simplified results, devoid of many confounding variables, also provide a point of comparison between toxicities of various pesticides and among various species.

Few data regarding the toxicological impacts of pesticide use under natural or field conditions are available. Although USEPA has the ability to require field studies, they are not generally required for pesticide registration. They are thought to be of limited use because they are site-specific and are considered too expensive to perform on a routine basis. Standard laboratory test results are not necessarily indicative of real world exposures to the target and non-target organisms, nor are they meant to be, but real world exposure and effect information is needed if we are to understand the true risk that pesticide use poses to non-target organisms. Furthermore, information on the fate and effects of metabolites and degradates of an active ingredient is not always provided, nor is it required for registration. Thus, these limited studies further confound the laboratory-to-field extrapolation needed to predict the risk of pesticide use to non-target organisms under natural conditions.

While USEPA's pesticide registration requirements (40 CFR 158.490) address acute lethality and chronic exposure impacts of the active ingredients on reproduction, studies evaluating other sublethal endpoints from short term (acute) exposures or longer term (chronic) exposures, such as physiological, neurological, behavioral, or histological effects are not required for pesticide registration. While these sublethal effects may not result in adverse impacts to non-target organisms that may be manifested at the population or community level, all of these difficult-to-observe sublethal effects could significantly alter the structures and/or functions of aquatic communities (Waite and Carpenter, 2000).

FIFRA does not require that inert ingredients in pesticide formulations be disclosed to other federal, state, and local regulatory authorities, the general public, or the end users. An inert ingredient is any substance in the formulated product that is not intended to deliver the toxicological effect to the target organism. This does not mean that inert ingredients do not have toxicological properties or pose hazards to non-target organisms. This is problematic since little is known about the toxicological properties of most inert ingredients. In fact, an active ingredient in one pesticide formulation may serve as an inert ingredient in another pesticide formulation. The inert ingredient could be more toxic (to non-target organisms) than the active ingredient in the pesticide formulation.

Each year, USEPA continues to register new compounds for use as active ingredients and approves new uses for active ingredients. It approved 24 and 28 new active ingredients for use in 1996 and 1997, respectively (Aspelin and Grube, 1999). About 900 active ingredients in more than 20,000 formulations containing various inert ingredients and carriers are currently registered (Aspelin and Grube, 1999).

The 1996 amendments to the Safe Drinking Water Act (SDWA) (Public Law 104-182) and passage of the new Food Quality Protection Act (FQPA) (Public Law 104-170) required USEPA to develop a screening and testing program for endocrine-disrupting chemicals. Responding to these requirements, USEPA formed the Endocrine Disruptor Screening and Testing Advisory Committee (EDSTAC). Within two years of establishment, USEPA published the committee's final report in which it defined an endocrine disruptor as, "an exogenous chemical substance or mixture that alters the structure or function(s) of the endocrine system and causes adverse effects at the level of the organism, its progeny, populations, or subpopulations of organisms, based on scientific, data, weight-of-evidence, and the precautionary principle." (USEPA, 1998).

EDSTAC determined that the screening and testing program should address, in addition to the FQPA and SWDA requirements, the provisions of FIFRA, the Toxic Substances Control Act (15 U.S.C. 2601 et seq.), and the Federal Food, Drug, and Cosmetic Act (21 U.S.C. 30 et seq.). Therefore, EDSTAC recognized that the screening and testing program needed to address both human and ecological effects and recommended that the program "should include screening for adverse effects to wildlife, as well as humans, recognizing that wildlife are an inherently valuable element of ecosystems and their well-being can be an indication of the overall health of the environment in which humans live."

EDSTAC recommended that endocrine disrupting properties be evaluated for not only chemical substances, but also common mixtures of chemicals. Based on some of the ecological effects discussed below (Section 7.3) that have been observed and possibly attributed to pesticides, we believe that this program, once implemented, will aid in preventing, documenting, and discerning the likely effects of pesticides in many ecosystems including aquatic ecosystems and communities.

7.1.3 Regulation of Pesticides in the Aquatic Environment

Pesticides that do not cause unreasonable adverse effects to man or the environment when used in accordance with the approved label, taking into account the economic, social, and environmental costs and benefits of the use of any pesticide, are eligible for use, sale, distribution, and importation in the United States. FIFRA defines "environment" as, "... water, air, land, and all plants, and man and other animals living therein, and the interrelationships which exist among these." Therefore, in order to meet the eligibility standard for registration, it seems that pesticide use should not pose an unreasonable adverse impact on aquatic life.

The Federal Water Pollution Control Act (33 U.S.C. 1251–1376), commonly referred to as the Clean Water Act (CWA), sets forth a goal of restoring and maintaining the chemical, physical, and biological integrity of the waters of the United States. Specifically, Sections 303 and 304(a) of the CWA are directed toward achieving this goal related to the presence of chemical stressors in surface waters. Section 303(c) requires states to have water quality standards that serve the purpose and goal of the CWA. Section 304 of the CWA requires USEPA, based on the most current scientific information available, to derive numeric water quality criteria to protect aquatic life, including aquatic biological community diversity, productivity, and stability, from pollutants in any body of water. Similar to the ecological risk assessments used in pesticide registration, numeric water quality criteria are derived assuming exposure to and effects from a single contaminant, rather than exposure to the complex mixtures that are likely to be found in ambient aquatic environments.

Pesticides are often identified as one of the leading causes of impairment for streams included on the Section 303(d) lists of impaired waters. Models, such as the pesticide root zone model and exposure analysis model system (PRZMS, EXAMS) used by USEPA and pesticide manufacturers for risk assessment during pesticide registration, often predict a pesticide, *even when used according to the label*, will likely be found in surface runoff and eventually some surface water body. However, despite the predicted and observed pesticides in the various components of the hydrologic cycle, federal water quality criteria (pursuant to Section 304(a) of the CWA) intended for the protection of aquatic life have not been developed for most pesticides. USEPA's Office of Pesticide Programs and Office of Water do not require pesticide manufacturers to perform the toxicological studies necessary to derive aquatic life criteria as a requirement for pesticide registration. Of the 76 pesticides and seven degradation products detected in surface waters by USGS (1999) in the past decade, USEPA developed federal water quality criteria for only seven pesticides. In a *Federal Register* notice published on October 29, 1999, after decades of widespread use with essentially no water quality regulation, USEPA announced its intent to develop a water quality criterion for the herbicide atrazine. Many states may not have adopted these or alternative pesticide criteria as state water quality standards.

Despite growing evidence of distribution of pesticides in the aquatic environment, little or no state or federal regulation of pesticides entering the aquatic environment exists. When pesticide use results in predicted occurrence of that compound in the aquatic environment, the pesticide manufacturer should be required to generate the data necessary for USEPA to derive an aquatic life criterion. This will also assist those who perform monitoring in interpreting the results of assessment efforts. It seems that if a pesticide, when used according to the label, will result in non-compliance with a federal law intended to protect the environment, such as the CWA, the pesticide would fail to meet the eligibility requirement for registration: of posing no "unreasonable risk to man or the environment."

7.2 OCCURRENCE OF PESTICIDES IN AQUATIC ENVIRONMENTS

The combination of the widespread use of pesticides and the dynamics of their movements through the hydrologic cycle suggest that these products will occur in lakes, rivers, streams, and oceans. While much public, regulatory, and scientific concern has focused on the potential effects of pesticide use on human health and the environment, little effort has been expended to identify, quantify, and determine the fate and potential effects of residual pesticides in the environment on fish, wildlife, and invertebrates, especially in the aquatic environment. These concerns have stimulated increased efforts to document the occurrence and distribution of residual pesticides in the environment.

The occurrence of pesticidal products in the environment, particularly in aquatic ecosystems, was reviewed and the common occurrence of pesticides in water, sediment, and aquatic biota was discussed by Nowell et al. (1999). To better understand the presence of pesticides in surface waters, several studies over the last several decades tracked the distribution, trends, and factors affecting pesticides in surface waters (Larson et al., 1997). Investigators found that 76 of 118 pesticides or degradation pesticide products targeted for analyses in the reviewed studies were found in one or more surface water samples. Similarly, surface water sampling of 58 rivers and streams throughout the United States indicated that six or seven of the 46 pesticides or transformation products were present (Larson et al., 1999). Often, several pesticides were present in the same sample at concentrations that exceeded USEPA guideline values thought to be protective of aquatic life.

The USGS (1999) performed the most extensive evaluation of pesticides in surface and ground water by analyzing samples for 76 pesticides and 7 degradation products. Samples were collected from agricultural and urban areas under various climatic conditions and results suggest that pesticides may be a significant problem for aquatic life in urban and agricultural settings throughout the nation. Nearly every sample of water and fish taken from streams and major rivers, regardless of land use, contained at least one pesticide. Nearly every stream sample and about half of the groundwater samples collected from wells contained two or more pesticides. Five or more pesticides were found in more than half of the stream samples. The USGS also reported that two or more pesticides were detected in sediment and fish tissue samples. These results suggest that aquatic organisms are likely exposed to residual pesticides from more than water column exposure. The presence of pesticides in the sediment and food serve as additional potential routes of exposure for aquatic organisms.

Past studies confirmed that many pesticides and degradation products, despite being relatively short-lived in the environment, continue to be present in various environmental media including air, fog, rain, surface water, ground water, soil, sediment, and biota. This should be alarming to ecologists and regulators, especially since the impacts of these pesticides in these environmental media are unknown.

7.3 ECOLOGICAL EFFECTS OF PESTICIDES IN AQUATIC ENVIRONMENTS

One of the largest problems facing aquatic biologists who evaluate causes and effects of contaminants on aquatic communities is a consequence of pesticides. Pesticides are intentionally introduced into the environment to adversely affect living organisms at some level of biological organization, whether subcellular, cellular, or organismal. They have the potential to impact higher levels of biological organization, including the individual, population, or assemblage. Little baseline information is available about the distribution of individual species prior to the manufacture, production, and widespread use of pesticides. Even less baseline information on the biological aspects of species population distributions and community assemblages is available.

Lack of biological monitoring and assemblage information makes it difficult to discern the subtle community level impacts that likely occur due to pesticides. This lack of historic data complicates our ability to determine the effects of pesticides on aquatic populations and communities. Lacking this information, it is reasonable to assume that the use and subsequent occurrence of pesticides in the environment may alter the function and structure of aquatic communities.

Despite all the complicating factors dictating fate, persistence, and flow of pesticides through ecosystems and the resulting effects, impacts on certain biota have been correlated or associated with pesticide use. Diazinon, a widely used broad-spectrum insecticide, serves as one example. As of 1999, approximately 430 diazinon formulations were registered for use (USEPA, 1999). USEPA reported that outdoor homeowner use (39%), agricultural use (25%), professional lawn care use (19%), and indoor and outdoor pest control operator use (11%) constituted the primary distribution of diazinon in the United States. Scholz et al. (2000) demonstrated that chinook salmon *Oncorhynchus tshawytscha,* exhibited reduced responsiveness to predatory events after exposure to low levels of diazinon. They also reported reduced homing responses in salmon exposed to diazinon at levels typical of pulses observed in natural environments. Effects were observed at 0.3 to 1.0 mg/L, a concentration three to four orders of magnitude below the lethal concentration to 50% of a toxicity test population of Pacific salmon (Novartis, 1997).

Atlantic salmon *Salmo salar,* exposed to diazinon at environmentally relevant concentrations exhibited diminished olfactory responses and reduced responsiveness of male salmon to female pheromones (Moore and Waring, 1998). At a December 5, 2000 technical briefing for diazinon, USEPA stated that the use of diazinon "may affect" threatened and endangered listed salmonids.

Diazinon was to be phased out for indoor uses beginning March 2001 and for all lawn, garden, and turf uses by December 2003. However, some damage may already have been done, which raises the question whether adequate review actually precedes the registration of many pesticides.

Methoxychlor, a DDT substitute that has been banned, was one of many pesticides used historically in blueberry culture in Maine. It was found to disrupt normal reproduction and delay the onset of smolting in Atlantic salmon (Madsen et al., 1997).

Another example of suspected pesticide impacts on salmon was determined by the incidence of phenotypic sex reversal in wild, fall chinook salmon returning to spawn in the Hanford Reach of the Columbia River (Nagler et al., 2001). Their results suggest that a high percentage of phenotypic female chinook carry male-specific DNA and are thus, likely to be males that have undergone sex reversal. The authors acknowledge the possibility that temperature fluctuation potentially played a role in this phenomenon. They also suggest that environmental estrogens, i.e., pesticides with endrogenic properties (such as atrazine, lindane, carbofuran, methyl parathion, and dieldrin) detected in the Hanford Reach may also play a role. Laboratory studies on rainbow trout *Oncorhyncus mykiss,* confirmed the estrogenic qualities of these pesticides.

Regardless of the environmental media in which residual pesticides are present, they may ultimately end up in the aquatic environment because of the way the hydrologic cycle functions. While the mere presence of residual pesticides is not necessarily problematic, their potential impacts are unknown for many reasons. Often, the toxicological effects of pesticides to non-target organisms are unknown; however, while these impacts are not well understood, it is reasonable to suspect that aquatic communities are likely to be impacted.

7.4 INERT INGREDIENTS IN PESTICIDES

Inert ingredients in pesticides further complicate understanding biological responses associated with pesticidal products in aquatic ecosystems. Because inert ingredients represent proprietary information, their disclosure is not required. Thus, analytical monitoring programs are not likely to include monitoring for inert ingredients. Theoretically, effects to aquatic, non-target organisms from pesticide use may be attributable to the inert rather than the active ingredient. One example of inert ingredient toxicity is 4-nonylphenol found in Metacil 1.8D, an insecticide formulation historically used to control spruce budworm in Canada. The spatial and temporal use of Metacil 1.8D was correlated with declining abundance of Atlantic salmon (Fairchild et al., 1999) based on a negative relationship between chemical use of the inert ingredient and salmon catch in remote, forested areas of Maritime Canada.

7.5 MIXTURES

Potentially the most significant impact that the presence of pesticidal products may pose is due to the possible additive, synergistic, and/or cumulative impacts of mixtures. Realistic environmental exposures include exposure to a number of contaminants at one time. Thus, true environmental exposure consists of whatever mixture of contaminants is present in the environment. This is particularly the case in the aquatic environment where surface waters receive runoff from lawns, gardens, golf courses, agricultural fields, and orchards, and also receive point source discharges from private and municipal treatment facilities.

7.6 MONITORING

Environmental monitoring for pesticides is inadequately funded. While many monitoring programs suggest that pesticides are included in the list of analytes examined, the monitoring of pesticides is generally done only for conventional, more persistent pesticides (such as organochlorine compounds) that were once widely used. Many agriculture related organochlorine pesticides are still found in areas that have not been farmed in recent decades or are found far from where they are currently used. Unlike the conventional organochlorine pesticides such as DDT, toxaphene, and atrazine, which tend to have long half-lives and persist in the environment in various environmental media, newer pesticides often have short half-lives and are not easily detected. Monitoring and detection are complicated because many new pesticides require metabolism or degradation of the parent compound for the active ingredient to deliver the efficacious dose to the target organism. Most pesticide monitoring programs look for the active ingredient in the parent material, rather than the formulated product, metabolites, or degradates. As new products become available, the ability of monitoring programs to include all potential pesticides in use will become increasingly difficult. FIFRA Section 18, Emergency Use Exemptions, and Section 24(c), Special Local Needs Registrations, present further problems for comprehensive monitoring programs. Often the compounds are used and released into the environment without the knowledge of those performing the monitoring, and therefore, are not included in the list of analytes examined.

Aside from the analytical difficulties and costs of pesticide monitoring, the ability of existing monitoring programs to reflect actual concentrations is compounded because of the large number of pesticide products available. The applicator's choice, discretion, and lack of any required reporting of actual use and application of pesticides are problems. Evaluating the implications of pesticides in the aquatic environment by comparing analytical results of existing monitoring programs or studies provides little utility in evaluating biological signatures. While generalizations regarding use patterns may be drawn, the toxicological impact is a result of many other issues. The formulation, rate, timing, and number of applications and the location of the use, prevalent weather conditions, soil types, surrounding land use, and applicator's adherence to use limitations and restrictions affect the observed occurrence of the chemical and its distribution in the environment. Given these variables, trying to associate the occurrence and distribution of pesticides relative to use is nearly impossible. Biological monitoring of the status and trends of aquatic biota will provide an interpretation of the observed presence and distribution of the pesticides in the environment and if biologically monitored, the community will reflect the overall effects.

Most current pesticide monitoring in aquatic environments is not related to use patterns. The patterns are critical data elements for monitoring plans in order to integrate point with non-point source management, including development of total maximum daily loads (TMDLs) under the CWA. For example, most surface and ground water samples collected by USGS are targeted at periods of peak flow (hydrograph peaks). Additionally, urban areas and cropland have distinctive pesticide signatures. However, sampling location and collection of samples have not been targeted with specific pesticide use or site of application. Monitoring efforts primarily focus on land use and the assumed potential for pesticide use. Given the individual discretion of a landowner, the choice of pesticide could be quite wide ranging. Therefore, the presence of detectable concentrations of pesticides in surface waters warrants further investigation.

7.7 CONCLUSIONS

Kapustka et al. (1996) showed the disconnect between current pesticide registration data requirements and ecological impacts. They recommended (1) increased use of ecologically relevant tests; (2) anticipation of ecological consequences of exposure; and (3) monitoring to validate risk management decisions. We concur with these recommendations and further recommend that future efforts by local, state, and federal pesticide regulators should focus on (1) requiring longer term, sublethal exposure tests for aquatic and aquatic-dependent organisms; (2) field studies should be conducted to enhance understanding of effects on aquatic communities; and (3) increased efforts and funding should focus on new generation pesticide monitoring programs. We released Pandora from her box before we understood the implications of the release and without crafting a "recall plan." The most relevant question should be, if we cannot act where we know there is risk, how will we be able to act when we do not know the risk?

Comprehensive monitoring programs should increase the number of pesticides monitored, include monitoring for the transformed pesticidal products, and target monitoring to correlate with use patterns so that the "snapshot in time" that the monitoring reflects is indicative of the use of these products in the environment. For aquatic ecosystems, increased emphasis must be placed on longer term monitoring of aquatic system conditions and communities so that the implication of the presence of pesticidal mixtures in surface waters is better understood. Increasing efforts on monitoring programs will assist pesticide regulators in determining whether use limitations and restrictions are effective in keeping pesticidal products contained at a treatment site, per the registration and instructions on the label. If the label is the law, then the label should dictate the use and, thus, the spatial and temporal reality of exposure to target and non-target organisms.

REFERENCES

Ankley, G.T., D.J. Call, J.S. Cox, M.D. Kahl, R.A. Hoke, and P.A. Kosian. 1994. Organic carbon partitioning as a basis for predicting the toxicity of chlorpyrifos in sediment, *Environmental Toxicology and Chemistry,* 13, 621–626.

Aspelin, A.L. and A.H. Grube. 1999. Pesticide Industry Sales and Usage. 1996 and 1997 Market Estimates.U.S. Environmental Protection Agency, Biological and Economic Analysis Division. Office of Pesticide Programs. Office of Prevention, Pesticides, and Toxic Substances, Washington, D.C.

DiPinto, L.M. 1996. Trophic transfer of a sediment-associated organophosphate pesticide from meiobenthos to bottom feeding fish, Archives of Environmental Contamination and Toxicology, 30, 459–466.

Fairchild, W.L., E.O. Swansburg, J.T. Arsenault, and S.B. Brown. 1999. Does an association between pesticide use and subsequent declines in catch of Atlantic salmon (*Salmo salar*) represent a case of endocrine disruption? *Environmental Health Perspectives,* 107, 349–357.

Green, A.S., G.T. Chandler, and W.W. Piegorsch. 1996. Life-stage-specific toxicity of sediment associated chlorpyrifos to a marine, infaunal copepod, *Environmental Toxicology and Chemistry,* 15, 1182–1188.

Kapustka, L.A., B.A. Williams, and A. Fairbrother. 1996. Evaluating risk predictions at population and community levels in pesticide registration — hypotheses to be tested, *Environmental Toxicology and Chemistry,* 15, 427–431.

Kersting, K. and P.J. van der Brink. 1997. Effects of the insecticide Dursban® 4E (active ingredient: chlorpyrifos) in outdoor experimental ditches: responses of ecosystem metabolism, *Environmental Toxicology and Chemistry,* 16, 251–259.

Larson, S.J., P.D. Capel, and M.S. Majewski. 1997. *Pesticides in Surface Waters: Distributions, Trends, and Governing Factors.* Ann Arbor Press, Chelsea, MI.

Larson, S.J., R.J. Gilliom, and P.D. Capel. 1999. *Pesticides in Streams of the United States: Initial Results from the National Water Quality Assessment Program.* U.S. Geological Survey, Water-Resources Investigations Report 98–4222. Sacramento, CA.

Lydy, M.J., P.M. Stewart, and T.P. Simon. 2002. Relationship between fish assemblages and organochloride insecticides in sediment and fish tissues in south-central Kansas. Chapter 17, this volume.

Madsen, S.S., A.B. Mathiesen, and B. Korsgaard, 1997. Effects of 17ß-estradiol and 4-nonylphenol on smoltification and vitellogenesis in Atlantic salmon (*Salmo salar*), *Fish Physiology and Biochemistry,* 17, 303–312.

Moore, A. and C.P. Waring. 1998. Mechanistic effects of a triazine pesticide on reproductive endocrine function in mature male Atlantic salmon (*Salmo salar* L.) parr, *Pesticide Biochemistry and Physiology,* 62, 41–50.

Miller, C.W., B.M. Zuckerman, and A.J. Charig. 1966. Water translocation of Diaznon-C^{14} and Parathion-S^{35} off a model cranberry bog and subsequent occurrence in fish and mussels, *Transactions of the American Fisheries Society,* 95, 345–349.

Nagler, J.J., J. Bouma, G.H. Thorgaard, and D.D. Dauble. 2001. High Incidence of a Male-Specific Genetic Marker in Phenotypic Female Chinook Salmon from the Columbia River, *Environmental Health Perspectives,* 109, 67–69.

Nowell, L.H., P.D. Capel, P.D. Dileanis. 1999. *Pesticides in Stream Sediment and Aquatic Biota: Distribution, Trends, and Governing Factors*. Lewis Publishers. Boca Raton, FL.

Novartis Crop Protection. 1997. An Ecological Risk Assessment of Diazinon in the Sacramento and San Joaquin River Basins. Technical Report: 11/97. Novartis Crop Protection, Inc. Greensboro, NC.

Odenkirchen, E.W. and R. Eisler. 1988. *Chlorpyrifos Hazards to Fish, Wildlife, and Invertebrates: A Synoptic Review*. Biological Report 85 (1.13), U.S. Fish and Wildlife Service, Patuxent Wildlife Research Center, Laurel, MD.

Scholz, N.L., N.K. Truelove, B.L. French, B.A. Berejikian, T.P. Quinn, E. Casillas, and T.K. Collier. 2000. Diazinon disrupts antipredator and homing behaviors in chinook salmon (*Oncorhynchus tshawytscha*), *Canadian Journal of Fisheries and Aquatic Science,* 57, 1911–1918.

Stevens, M.M. 1992. Toxicity of organophosphorus insecticides to fourth-instar larvae of *Chironomus tepperi* Skuse (Diptera: Chironomidae), *Journal of the Australian Entomological Society,* 31, 335–337.

Stewart, P.M. and J.T. Butcher. 1999. The Effects of Cranberry Operations on Water Quality, Macroinvertebrate Communities, and Pesticide Concentrations of the St. Croix National Scenic Riverways. U.S. Geological Survey, Biological Resources Division, Porter, IN.

U.S. Environmental Protection Agency. 1998. Endocrine Disruptor Screening and Testing Advisory Committee Final Report. Washington, D.C. http://www.epa.gov/scipoly/oscpendo/history/finalrpt.htm.

U.S. Environmental Protection Agency. 1999. Reregistration Eligibility Decision Chapter for Diazinon. USEPA, Environmental Fate and Effects Division. Office of Pesticide Programs. Office of Prevention, Pesticides, and Toxic Substances, Washington, D.C.

U.S. Fish and Wildlife Service. 1964. The Effects of Pesticides on Fish and Wildlife. U.S. Department of the Interior, U.S. Fish and Wildlife Service, Circular 226, Washington, D.C.

U.S. Fish and Wildlife Service. 1989. *Summary of Factors Most Affected by the Pesticides Addressed in the Biological Opinion of June 14, 1989*. Dated September 14, 1989. U.S. Fish and Wildlife Service, Washington, D.C., Notice of availability was published in the *Federal Register* 55, 1168–1169.

U.S. Geological Survey (USGS). 1999. *The Quality of Our Nation's Waters — Nutrients and Pesticides*, U.S. Geological Survey Circular, 1225.

Waite, I.R. and K.D. Carpenter. 2000. Associations among fish assemblage structure and environmental variables in Willamette Basin streams, Oregon, *Transactions of the American Fisheries Society,* 129, 754–770.

Wan, M.T., S.Y. Szeto, and P. Price. 1995. Distribution and persistence of azinphos-methyl and parathion in chemigated cranberry bogs, *Journal of Environmental Quality,* 24, 589–596.

Ware, G.W. 1994. *The Pesticide Book*, 4th ed. Thomas Publications, Fresno, CA.

Wilcock, R.J., G.L. Northcott, and J.W. Nagels. 1993. Mass losses and changes in concentration of chlorpyrifos and *cis-* and *trans-*permethrin applied to surface of a stream, *Bulletin of Environmental Contamination and Toxicology,* 53, 337–343.

Wilhm, J.L. and T.C. Dorris. 1968. Biological parameters for water quality criteria, *BioScience,* 18, 477–481.

8 Ecological Effects of Metals on Benthic Invertebrates

Peter M. Kiffney and William H. Clements

CONTENTS

8.1 INTRODUCTION

The arthropods constitute a vast assemblage of animals. At least 75,000 species have been described — more than three times the number of all other animal species combined (Dorit et al., 1991). The arthropods successfully invaded a variety of habitats, such as wetlands, ponds, lakes, streams and rivers where they perform a variety of critical ecological functions, such as detritus processing (Wallace and Webster, 1996). They are relatively sessile and long-lived, and exhibit a range of sensitivities to various environmental stressors (Rosenberg and Resh, 1993). They also have a long history of serving as sentinels of water quality (Carpenter, 1924). Freshwater ecosystems are faced with a variety of insults including urbanization, agriculture, industrial effluent, and climate change. It is imperative that we understand how aquatic ecosystems are affected by these stresses and how they respond to management actions aimed at alleviating such insults. Because of their tremendous diversity, longevity, sensitivity, and critical roles in ecosystem function, benthic invertebrates provide excellent model systems for examining the effects of anthropogenic disturbances on aquatic ecosystems.

Multiple factors, such as food, temperature, predators, pathogens, and contaminants, interacting in complex ways, control benthic populations and communities. Large-scale extraction of metals presents environmental problems worldwide (Moore and Luoma, 1990); by the year 2000, mining had disturbed about 240,000 km^2 of the Earth's surface and milling and smelting released an estimated 70×10^3 metric tons of metals to the aquatic environment (Moore and Luoma, 1990). Such a large disturbance to terrestrial and aquatic ecosystems indicates that metals are important in structuring biological communities (Clements et al., 2000).

Most research on how contaminants interact with benthic invertebrate communities has focused on direct effects, such as changes in abundance of a particular species. In contrast, the indirect effects of metals (e.g., alterations in predator–prey interactions) are less understood. Regulation of

0-8493-0905-0/03/$0.00+$1.50

contaminants in the environment is based exclusively on direct effects (LC_{50} or the concentration at which 50% of the test species die), but contaminants may also have important indirect effects that may occur at concentrations lower than those allowed under current regulations (Kiffney, 1996).

Current regulations may not adequately capture all routes of metal exposure. Metals in the environment are regulated based on their direct effects (e.g., mortality) and uptake of water-borne metals is the only exposure pathway considered important to measure. Unfortunately, this has little ecological relevance for animals that often consume food items that contain high levels of metals (Kiffney and Clements, 1993; Farag et al., 1998); therefore, for some invertebrates, diet may be the main route of metal exposure (Timmermans et al., 1992, Wang and Fisher, 1999a).

In this chapter, we will first discuss characteristics of benthic invertebrates that make them such useful model systems to assess metal impacts on freshwater habitats. We will then summarize analytical approaches used in biological monitoring, with a focus on multimetric and multivariate techniques. Third, we review some of the direct and indirect effects that metals can have on benthic invertebrates. We argue that previous exposure to contaminants may influence how an individual may respond to novel stresses. Fourth, we present evidence suggesting that food may be as or more important than water as a route of metal uptake for aquatic animals. We focus our discussion on freshwater ecosystems, because we have the most experience and expertise with these habitats, and they have been the most intensely studied. We suggest, however, that concepts proposed in this chapter may be valid in other ecosystems. Based on our understanding of benthic invertebrate communities and metal pollution, we propose that regulators include ecological complexity, such as the role of food as an exposure pathway, into how they manage the release of contaminants into the environment. If ecological integrity is the regulatory goal, we should begin to incorporate ecological realism into the regulatory framework.

8.2 WHY MONITOR BENTHIC INVERTEBRATES?

Because organisms integrate conditions around them, they respond to physical, biological, and chemical changes in their environment (Karr and Chu, 1999). Therefore, biological endpoints provide more meaningful measures of ecological condition than simple chemical measures (e.g., dissolved oxygen). This attribute has been recognized for many years; thus, science has a long history of monitoring benthic invertebrate community structure to assess the effects of contaminants on aquatic ecosystems. Benthic invertebrates are routinely used by many local, state, and federal agencies in biological monitoring and assessment programs. We discuss some the attributes that make these invertebrates excellent model systems for examining environmental change.

Benthic invertebrates are ubiquitous, and therefore can be used to monitor the effects of metals and other contaminants in a wide range of systems and habitats within those systems. Since immature stages of invertebrates are sedentary and relatively long-lived, they integrate temporal heterogeneity of metal discharges, and spatial comparison (upstream reference vs. downstream polluted sites) of the effects of contaminants can be made (Rosenberg and Resh, 1993). Many taxa are closely associated with sediments (an important sink for metals) (Hare, 1992). Some taxa (e.g., Trichoptera: Hydropsychidae) can tolerate low to moderate metal concentrations, and metal concentrations in these and other invertebrates appear to be related to those in the environment (e.g., Kiffney and Clements, 1993). Invertebrates perform a number of important ecological functions (e.g., detritus processing), and changes in species composition or abundance can have important consequences for ecosystem function (Wallace and Webster, 1996).

For example, elimination of 90% of the insect biomass in a south Appalachian headwater stream through insecticide application reduced leaf litter breakdown and export of coarse particulate organic material compared to adjacent reference streams (Wallace et al., 1991). In a given habitat, the species present provide a spectrum of sensitivities to metals and other contaminants. Quantitative and qualitative collection of invertebrates is relatively easy, and they are also easy to manipulate for field and laboratory experiments. This characteristic is particularly appealing because it allows a researcher to compare results from field biomonitoring studies with controlled experiments, and this is a powerful method of establishing cause–effect relationships between metals and population and community changes (Kiffney and Clements, 1994). Because of these factors, benthic invertebrates provide excellent models systems for examining the effects of metals and other environmental stresses on ecological structure and function.

8.3 APPROACHES TO ASSESSING EFFECTS OF METALS ON INVERTEBRATES

A number of empirical approaches can determine the effects of metals on benthic invertebrate communities. These approaches include field surveys and laboratory and field experiments. In general, the cost, time, and logistical constraints of the experimental approach increase as ecological realism and statistical power increase.

Field biomonitoring is one of the oldest and most widely used approaches for assessing the effects of metals on invertebrate communities (Clements, 1992; Rosenberg and Resh, 1993). Descriptive field surveys attempt to explain patterns of invertebrate abundance by examining relationships among biotic and abiotic factors across a number of sites or times, some of which differ in specific environmental conditions (Cooper and Barmuta, 1993). For example, we have been monitoring the effects of heavy metal pollution on the upper Arkansas River basin (in Colorado) for 10 years (Clements, 1994). A series of fixed sample locations were located along a longitudinal, metal, and elevational gradient. Some sites were located upstream of metal inputs and served as reference sites. Downstream sites differed in the severity of metal loading with sites closest to the input having the highest water metal concentrations. This study also has an important temporal component, as we have been sampling twice a year and can thus determine when the most severe impacts occur. This sampling design also allows us to measure recovery of the system from a management action. Two water treatment plants were established in 1992 and 1994 to treat mine discharge into the Arkansas River; therefore, we can observe the recovery dynamics of the invertebrate community as metal loading declines.

Although this study design is temporally extensive, a number of statistical problems are associated with monitoring effects of metals on a single system like the Arkansas River. Treatments are not randomly allocated and samples are pseudoreplicated in space and time (Hurlbert, 1984). For example, the reference site on the Arkansas River is upstream of metal-polluted sites; therefore, one cannot establish a strong link with changes in community structure to metals because sites are not independent and other abiotic factors that influence benthic invertebrates also change from upstream to downstream. Specifically, any effects due to metals are confounded because habitat variables such as substrate composition, temperature, discharge, and water chemistry differ along a longitudinal and elevational gradient. Because of these design problems, extrapolation to systems other than the Arkansas River must be made with caution.

One way to address pseudoreplication in space is to sample more than one contaminated system with comparable habitat features. The use of replicated natural experiments to measure effects of anthropogenic disturbance on community structure is rare in stream ecology (Clements and Kiffney, 1995). Feldman and Conner (1992) used replicated natural streams to examine the effects of pH on benthic invertebrate communities. Clements and Kiffney (1995) sampled benthic invertebrates from six streams in the southern Rocky Mountains polluted by heavy metals, and found that changes in benthic community structure were relatively similar among streams. Because replication is included in this type of design, one may argue that researchers using this approach may make broader inferences about a population of streams.

Another approach to establishing a link between contaminants and benthic invertebrate communities is the use of experimental microcosms or mesocosms. The difference between microcosms (<10 m^3) and mesocosms (>10 m^3) is somewhat arbitrary and based on volume (Buikema and

Voshell, 1993). Clements et al. (1989a) developed an approach to test the effects of metals on natural assemblages of benthic invertebrates using stream microcosms. In this approach, benthic invertebrates colonize artificial substrata placed in a natural, reference stream for 30 to 40 days. A number of studies have shown that this is long enough for the benthic community to reach equilibrium (Clements et al., 1989a). More importantly, communities colonizing these trays were similar in abundance, species composition, and diversity to communities collected from the stream using conventional sampling approaches (Clements et al., 1989a; Kiffney and Clements, 1994). Following this colonization period, invertebrates were transferred to experimental channels, and acclimated for 1 to 2 days. The variable or variables (e.g., metal concentrations) of interest were then manipulated (Clements et al., 1988, 1989b; Kiffney and Clements, 1994; 1996; Kiffney, 1996).

An additional approach is to build artificial channels alongside a natural stream. Water containing algae, detritus, bacteria, and invertebrates is fed into channels via gravity from the natural stream. Additional benthic samples can be added to channels to ensure that they are fully colonized. After communities reach an equilibrium abundance and diversity, experimental manipulations, such as metal exposure, can be initiated (Richardson and Kiffney, 2000).

These experimental approaches have their limitations, including the problems of scale and artificiality (Schindler, 1977). Laboratory or field experimental channels isolate an organism from its physical, chemical, and biological environment. Water chemistry in experimental streams can differ from the chemistry of natural streams, and large-scale features such as groundwater flow and stream meandering cannot be replicated (Cooper and Barmuta, 1993). One way to address some of the weaknesses of using artificial streams in toxicity evaluations is to compare experimental results to natural systems. For example, in Virginia (Clements et al., 1988), Colorado (Kiffney and Clements, 1994; 1996), and British Columbia (Richardson and Kiffney, 2000) invertebrate community changes in experimental streams exposed to metals closely mimicked changes observed in field surveys of metal-polluted natural streams. Because experimental results agreed with field results, these studies provided strong evidence to support the hypothesis that changes in community structure were due to metal exposure.

Collection of pre-manipulation data is essential in field experiments because it allows the researcher to compare the dynamics of a response measure before introducing the treatment to the dynamics after treatment (Stewart-Oaten et al., 1986). This design monitors control and experimental sites before and after perturbations have occurred, and uses the difference between control and experimental areas as the response variable. The samples at each time are treated as replicates and mean differences between the two areas before and after the impact are compared by a t-test (Cooper and Barmuta, 1993). Other statistical procedures for this type of experimental design have been developed (see Rosenberg and Resh, 1993 and references therein). According to Cooper and Barmuta (1993) field experiments have a number of strengths: (1) calibration of monitoring programs; (2) greater control over relevant variables; and (3) determining the proximal mechanism or mechanisms.

We agree that manipulating isolated variables and studying the resultant effects on whole systems is a powerful way to assess the effects of contaminants on ecosystem structure and function. There are, however, a number of reasons why this approach is difficult to apply in practice. First are the ethical and regulatory issues pertaining to intentionally releasing toxic materials into the environment. Second, it is often difficult to replicate whole system experiments; therefore, psuedoreplication in space remains an issue. Third, whole system manipulations are often costly and studies can last several years. Because of these logistical constraints, we recommend combining field surveys with controlled experiments using microcosms or mesocosms. We need to extrapolate from small-scale experiments with care, but this approach is more cost-effective and provides a way to validate field results using controlled experiments, thereby linking patterns observed in the field with ecological processes.

8.4 MULTIMETRIC AND MULTIVARIATE APPROACHES TO MONITORING

There are many ways to analyze data collected from biological monitoring of benthic invertebrates, and each has its advantages and disadvantages. One of the current controversies in ecological monitoring is whether multimetric or multivariate analyses are the most appropriate ways to quantify benthic invertebrate communities. Although the multivariate method is complex, it allows for relationships to be quantified between environmental variables (e.g., metal levels) and invertebrate community structures. The multimetric method provides a simple graphical method of relating human activities to biological changes. This approach, however, is limited in that it only shows what sites are impaired compared to a reference condition, and does not indicate what environmental factors may drive the differences in scores.

Despite their limitations, which we will discuss later, biomonitoring approaches remain effective tools for evaluating ecological conditions and provide more meaningful information than simple chemical measures. In this section, we discuss some of the common approaches to analyzing data collected during biological monitoring. We also present ways of uniting the multimetric and multivariate approaches in analyzing benthic invertebrate data from biomonitoring studies.

Early on, scientists used various biological indices to classify the relative impairment of a site or system. Simple measures such as diversity and total abundance were some of the indices commonly used. These simple measures were eventually replaced with more complex indices to assess the effects of anthropogenic change on biological communities (Clements, 1997). For benthic macroinvertebrate indices, tolerance values are assigned to taxa and these values are integrated with measures of relative abundance in the field to calculate a pollution score. Hilsenhoff's biotic index is one of the most common, and it assigns scores to sampling locations based on the relative abundance of sensitive and tolerant groups (Clements, 1997). This index was developed to detect impairment due to organic enrichment and works reasonably well for this purpose, but it has been used for other types of pollution such as acidification and heavy metals with mixed results.

Indices have been criticized because (1) assignment of species to sensitive or tolerant classes is often subjective; and (2) responses to different classes of contaminants may be species-specific (Clements, 1997). One way to address these concerns is to derive tolerance values of invertebrates living in a defined system experimentally. For example, Clements et al. (1992) experimentally developed an index of metal impacts for benthic invertebrate communities in stream microcosms. Percent reduction of the 13 dominant taxa was determined by comparing abundance in copper-exposed streams to controls. Benthic samples were collected along a gradient of metal contamination from the Clinch River in Virginia, a system receiving Cu from a coal-fired power plant. Index values were calculated for field-collected samples and compared to values from the experiment. This index was an effective indicator of Cu exposure and was useful in delineating zones of impact and recovery in the Clinch River (Clements, 1992).

Other measures used to assess the effects of metals on benthic invertebrates include total species richness and number of species within certain taxonomic groups. One of the most commonly used richness measures is the EPT index (number of taxa in the Ephemeroptera, Plecoptera, and Trichoptera insect orders). This index is one of the most useful indicators of biological monitoring due to sample processing time and ease of interpretation (Clements, 1997). The assumption of this index is that mayflies, stoneflies, and caddisflies are most sensitive to environmental stress. This may be true for many classes of contaminants, but we have found that this index is not always a good indicator of heavy metal pollution because stoneflies and caddisflies can be highly tolerant to metals (Clements et al., 1992; Clements, 1994).

Mayfly abundance, especially Heptageniidae abundance, and species richness are biological indices we have found both in experiments and field surveys to be highly sensitive to detecting impacts of metal pollution in Rocky Mountain streams (Kiffney and Clements, 1994; 1996; Clements et al., 2000). Researchers in a wide variety of locations (see Section 8.5 and Chapter 19, this volume) have also found mayflies to be sensitive indicators of metal pollution.

No single measure of biological impairment will work equally for all contaminants in all aquatic habitats. Moreover, some individual metrics respond in unexpected directions (i.e., increased richness at polluted sites) (Clements, 1997). As a result, multimetric indices of ecological integrity have been proposed, especially in the United States. The most widely used multimetric index is Karr's index of biological integrity (IBI), which was originally designed to evaluate the effects of human impacts on fish populations (Karr, 1981). Fish IBI scores are used to classify sites based on three general categories: species richness and composition, trophic position, and fish abundance and condition. In general, reference sites are determined based on geographic and physical characteristics, and IBI scores from these reference sites are then compared to scores from test sites (i.e., those considered degraded).

An IBI for benthic invertebrate communities (BIBI) to distinguish degraded sites in the Tennessee Valley (Kerans and Kerr, 1994) and in the Pacific Northwest (Fore et al., 1996) was developed recently. Fore et al. (1996) found that they could detect changes in BIBI metric scores with intensity of road building, logging and other human activities. Proponents of this approach argue that simple graphical and statistical analyses that relate human activities to biological conditions are easier to explain and interpret than multivariate techniques (Fore et al., 1996). Some criticize the multimetric approach because not all information is used, metrics are often redundant in a combination index, and errors can be compounded (Reynoldson et al., 1997). It is also difficult to develop associations between BIBI scores and environmental correlates if only invertebrate data are collected.

Carlisle and Clements (1999) evaluated the sensitivity and variability of a suite of biological metrics for detecting the ecological effects of metals on stream invertebrate assemblages. Their results suggest that taxa richness was most sensitive to metals, had the lowest variability, and highest statistical power compared to functional group composition, abundance, and ratio metrics. Furthermore, they suggest that commonly used analyses and study designs are inadequate to detect changes in some abundance, feeding group, and ratio metrics given realistic sampling efforts.

A major rationale for using multimetric approaches is the expectation that at least some of the metrics will be sensitive to a diversity of human impacts. We know of little experimental verification of this assumption. Carlisle and Clements (1999) call for experimental testing of biological metrics under a variety of stressors to establish a direct link between metric scores and disturbance. Such an approach would greatly strengthen the applicability of multimetric indices and their use in the regulatory arena.

Multivariate approaches are used more commonly in Australia (Parsons and Norris, 1996), the U.K. (Wright et al., 1984; Moss et al., 1987), and Canada (Reynoldson et al., 1997) to detect human impacts on benthic invertebrate communities. In the multivariate approach, invertebrates and a wide range of environmental variables from a large number of unimpaired sites are collected. Unimpaired (reference) sites are then classified into groups based on uniformity via multivariate classification techniques (Norris, 1995). These groups are then used to describe the structure of the environmental data using multivariate techniques. A subset of environmental attributes that are little affected by most human activities (e.g., altitude) is chosen based on a multivariate selection procedure (typically discriminant function analysis) and this subset accounts for most of the variation between groups of sites. Environmental variables selected in the above procedure and invertebrates are then collected from test or impacted sites and are used to match test sites with reference sites.

After test and reference sites are matched, predictions are made as to the nature of the benthic communities at the test site. The taxa collected at the test site are then compared to predictions, and the severity of biological impact is based on how much the predicted community deviates from the reference site (see Norris, 1995 for more detail on the multivariate approach).

To determine how well the two methods correctly classified sites as unimpaired and impaired, Reynoldson et al. (1997) analyzed benthic invertebrate data from locations in the Fraser River basin in British Columbia, Canada. They concluded that the precision and accuracy of the multivariate assessment methods were consistently better than for the multimetric assessments. For example, the accuracy of the multimetric approach for correctly identifying a site as unimpaired was variable (ranged from 38 to 88%), whereas the multivariate assessments correctly identified all unimpaired sites (Reynoldson et al., 1997).

Multivariate approaches also have a number of disadvantages, such as difficulties in understanding the complexity of model construction (Reynoldson et al., 1997). After data are analyzed, it is often difficult to explain how the model works and what the data mean, especially to a non-technical audience. Fore et al. (1996) argue that even if multivariate analysis classified sites perfectly in relation to human disturbance, the inherent statistical complexity of this analysis distracts biologists from making clear, testable statements. They also suggest that multivariate analyses are most appropriate for exploratory analysis or constructing hypotheses about systems we know little about.

Although strict reliance on complex statistical algorithms may obscure important biological results, we feel that multivariate approaches are important tools for biological assessments of water quality. We also think that the multimetric approach has a number of important features, such as the incorporation of biological information into metric formation. We agree with the recommendations of Reynoldson et al. (1997) that multivariate and multimetric approaches are complementary and should be used together. For example, variables included in a multivariate analysis could include traditional metrics known to be sensitive to heavy metals (e.g., EPT, abundance of heptageniid mayflies, abundance of grazers). Alternatively, a multimetric index could be developed based on results of multivariate analyses. Canonical discriminant analyses, principal component analyses, and other multivariate procedures derive new axes based on linear combinations of multiple variables. Based on loading coefficients, the analysis also identifies variables that are most important for separation of groups (locations, sampling stations). Variables responsible for separation of reference and impacted stations could be included in a multimetric index.

In summary, there are disadvantages and advantages to multimetric and multivariate approaches in monitoring benthic invertebrate communities. One disadvantage shared by both is that they are not experimentally validated; therefore, they offer only possible hypotheses as to why benthic invertebrate communities may differ between impaired and reference sites. We suggest that both methods have their places in biological monitoring and we present ways they can be used together. The value of the multimetric approach, however, may be limiting unless other environmental data, such as substrate composition or water chemistry, are collected that may aid in developing relationships between metric scores and impairment. We recommend adding such measurements as a way to improve multimetric usefulness.

8.5 DIRECT EFFECTS OF METALS ON BENTHIC INVERTEBRATES

We will begin this section by discussing how metal exposure can directly impact benthic invertebrates across levels of biological organization from expression of genes in individual animals to biological processes that determine the structures of invertebrate populations and the communities to which they belong. At the end of the section, we propose a general model of how natural invertebrate communities will respond to metal contamination based on a number of studies performed in a wide range of geographic locations. These studies generally show that Ephemeroptera (mayflies) are the most metals-sensitive taxa and as a result may serve as useful biological indicators of metal contamination in streams.

Exposure to metals can directly affect benthic invertebrate communities at every level of biological organization. Studies demonstrated that invertebrate populations chronically exposed to heavy metals often exhibit increased tolerance relative to unexposed populations (e.g., Klerks and Levinton, 1992). The development of resistance may indicate selection pressure and provide evidence that a population is affected by metal exposure. A classic example showing the genetic basis for metal tolerance was demonstrated in Foundry Cove, New York (Klerks and Levinton, 1992). The oligochaete *Limnodrilus hoffmeisteri* populations from a site contaminated by high levels of cadmium (Cd) were more resistant to Cd-rich sediments than populations from a reference site. To determine whether this increased resistance was due to genetic adaptation or physiological acclimation, resistance of laboratory-reared offspring was investigated. Worms collected from Foundry Cove and their second-generation offspring exhibited significantly higher survival rates than worms from the control area. Evidence of enhanced metallothionein production was found in the tolerant populations, and it may have both physiological and genetic bases. These results show that the resistance of *L. hoffmeisteri* was largely genetic (Klerks and Levinton, 1992). Further laboratory experiments suggest that field-observed differences in resistance in *L. hoffmeisteri* could have been obtained in four to five generations (Klerks and Levinton, 1992).

Past exposure to metals may allow some individuals to be more tolerant to that stress than unexposed individuals, but this tolerance may have a cost (Wilson, 1988). Specifically, increased tolerance may be a tradeoff within individuals between the benefits of tolerance and the associated energetic costs, with consequences for populations and communities (Courtenay and Clements, 2000). For example, recent research suggests that benthic invertebrates from the metal-contaminated Arkansas River in Colorado were more sensitive to a novel stress (pH) compared to communities from a reference system. This study suggests that environmental history can be an important determinant of how stresses will impact invertebrate communities. We speculate that populations and communities from metal-stressed ecosystems may be more susceptible to future climate change, such as global warming, than unexposed communities.

Metal tolerance may be due to increased production of metal-binding proteins such as metallothionein (Roesijadi, 1992). Metallothioneins have been reported in several orders of aquatic invertebrates, including Diptera, Ephemeropta, and Plecoptera (Hare, 1992). Cadmium and copper (Cu) were bound to a metallothionein-type protein in nymphs of the mayfly *Baetis thermicus*, but zinc (Zn) was bound non-specifically to high molecular weight cellular constituents (Aoki et al., 1989). The metallothionein induced in *B. thermicus* could not be induced in two other intolerant *Baetis* sp. This suggests a protective role for the protein and provides a possible mechanism by which the distribution of closely related species can be determined by metal concentrations in the environment.

Metals can also be regulated and sequestered into less labile forms than a metal–protein association. High metal concentrations have been measured in intracellular granules in a wide variety of terrestrial and marine invertebrates and these granules have been implicated in the storage, sequestration, and excretion of trace metals (Hare, 1992). An invertebrate's ability to sequester metals in granules may determine, in part, its ability to survive in metal-contaminated environments. Granules may be excreted or stored; granules accumulated in the immature stages of aquatic invertebrates can be excreted during metamorphosis to the adult stage. Metal-rich granules have been associated with the exoskeletons of some benthic invertebrates, e.g., copper-containing granules in Trichoptera (Darlington and Gower, 1990). The mechanisms that organisms use to tolerate metals may have important implications for metal transfer to higher trophic levels. For example, metals in granules may be less bioavailable to predators than metals bound to metallothionein.

Exposure to sublethal concentrations of metals reduces growth, fecundity, and survival of aquatic organisms. This may arise from damage to metabolic processes or from energy investment made by organisms during detoxification (Vouri, 1994). It is possible that the behavior of an animal exposed to metals may be altered, which would affect the outcome of its interactions with other organisms (see Section 8.6). For example, metal stress is likely to influence the outcome of contests between the owner of a territory and an intruder (Vouri, 1994). Low Cd concentrations altered the competitive behavior of *Hydropsyche contubernalis* larvae after 24 hours, yet no visual signs of morphological damage were detected. The Cd-exposed intruder larvae spent significantly less time trying to enter the nets of resident larvae than did unexposed larvae (Vouri, 1994). This suggests that exposure to low metal concentrations reduces intraspecific competition in hydropsychids. The rapid behavioral change may make it possible for individuals to tolerate exposure to chemical stress by reducing time spent fighting (Vouri, 1994).

As metal concentrations increase, the quantities of metals entering invertebrates may exceed the processing capacity of their biochemical machinery, resulting in toxic effects (Hare, 1992). Studies using invertebrates to assess the impacts of metals typically show reduced abundance and number of species at downstream compared to upstream reference sites.

One of the first studies to show how benthic community structure was related to metal pollution was done in the 1920s by Carpenter (Carpenter, 1924). She reported reduced numbers of taxa and "impoverishment of the fauna" at sites downstream from lead (Pb) mining operations. Winner et al. (1980) found species numbers to be most sensitive to Cu exposure, whereas Clements (1992) found that total abundance was more sensitive than number of taxa. Macroinvertebrate abundance, however, was more variable than number of taxa, thus making it difficult to delineate zones of impact and recovery using this index.

In other cases, species diversity or total number of taxa was not a useful indicator of metal impacts (Clements, 1992). The relative insensitivity of diversity and richness to metals is a result of two factors. Species number may not change because tolerant taxa (e.g., hydropsychids) replace sensitive (e.g., mayflies) taxa. In addition, unlike organic enrichment, metal exposure may affect the abundances of all taxa equally. Some diversity measures, therefore, may not vary among metal-polluted and reference sites, as evenness is an important component of these measures.

Regardless of the importance of functional responses, relatively little effort has been devoted to assessing the effects of metals on ecosystem processes as they relate to invertebrate communities. Changes in ecosystem function include such processes as primary productivity, nutrient cycling, energy flow, and decomposition. Because structural measures are so intimately tied to changes in ecosystem function, the distinction between them is relatively arbitrary (Clements, 1997).

Some studies have shown that metals can affect primary production in aquatic ecosystems, which might have implications for secondary production. For example, McKnight (1981) observed reduced primary production for 10 days in a lake dosed with Cu. Similarly, Cd additions to experimental ponds depressed primary production, but rates returned to control levels after 10 days (Kettle and DeNoyelles, 1986). Cd additions to mesocosms at the Experimental Lakes Area in Canada produced no major effects on ecosystem processes (Schindler, 1987). In streams, Crossey and La Point (1988) observed significant increases in whole-system respiration and significant declines in the production/respiration ratios at sites contaminated with heavy metals. An increase in gross primary production was suggested, but rates were highly variable and changes could not be associated with metal impact.

Carlisle (2000) estimated energy flow in two food webs, one from a reference site and one from a metal-polluted site in the Rocky Mountains of Colorado. Although the trophic position (based on $\delta^{15}N$) of the stonefly predator *Megarcys signata* was similar in both streams, the energy reaching the upper trophic levels was predominantly from detrital sources in the metal-polluted site, which has been described as indicative of a stressed ecosystem (Odum, 1985). Metals, therefore, can initiate responses at every level of biological organization. For these responses to be useful in biological monitoring, however, they must possess a number of important attributes.

Useful indicators of metal pollution should be predictable and allow some degree of generalization among studies and locations (Clements, 1992). Most of the research examining effects of metals on stream invertebrates has shown that mayfly relative abundance declines, whereas chironomid relative abundance increases as metal concentrations increase. This response is supported by data collected from a variety of geographic locations (Winner et al., 1980 [Ohio]; Norris et al., 1982 [Australia]; Clements et al., 1988, 1989c [Virginia]; Leland et al., 1989 [California]; Short et al., 1990 [Kentucky]; Kiffney and Clements, 1994 [Colorado]; Poulton et al., 1995 [Montana]; Richardson and Kiffney, 2000 [British Columbia]). These studies show that abundance and species richness of mayflies may be a particularly useful indicator of metal pollution.

Clements et al. (2000), in a survey of 95 sites in the southern Rocky Mountain ecoregion in Colorado found one of the most consistent responses to heavy metal contamination was lower abundance of mayflies, especially heptageniid mayflies (Figure 8.1). The relatively high sensitivity of mayflies to heavy metals compared to other invertebrates included feeding habits (Smock, 1983; Leland et al., 1989), body size (Krantzberg and Stokes, 1989), presence of body coverings (Hodson et al., 1979), and presence of external platelike gills. Because both field biomonitoring and experimental studies have shown that mayflies are one of the most sensitive taxa to metals, we propose that mayfly abundance or species richness be used by state and federal agencies to assess the biological integrity of surface waters receiving trace metals.

In summary, the effects of contaminants may occur at all levels of biological organization. While biochemical and physiological alterations in an organism may occur rapidly and are often stressor-specific, the ecological relevance of these indicators is uncertain (Clements, 2000). There is a need for studies that address the mechanistic link between suborganismal effects of metals on individual species and subsequent changes to population and community structure. In particular, we suggest further research into the food web approach taken by Carlisle (2000), because it offers the possibility of linking population-level contaminant effects to ecosystem-level effects.

8.6 INDIRECT OR INTERACTIVE EFFECTS OF METALS

Metals can kill an animal directly, with death resulting from exposure to a toxic concentration, or metals may indirectly cause mortality by altering an animal's behavior, making it more susceptible to other stresses such as predation. We have a good understanding of the direct effects of metals on invertebrate communities. We know which groups are tolerant and sensitive, but we know little about how metals indirectly affect species diversity and abundance. Recent studies have addressed whether metals indirectly affect benthic communities (Kiffney, 1996; Clements, 1999; Lefcort et al., 1999). They suggest that low metal concentrations (i.e., below U.S. Environmental Protection Agency (USEPA) criteria) in water can impact the population structures of benthic invertebrates living in that aquatic environment.

Metals may indirectly affect stream invertebrate communities by altering the outcomes of biotic interactions. Stonefly predators are typically one of the most metal-tolerant groups, whereas some of their preferred prey (e.g., mayflies) are the most sensitive (Kiffney, 1996). It seems likely that predation rates may vary between metal-contaminated and pristine streams due to this difference in metal sensitivity. Clements et al. (1989b) observed that the vulnerability of net-spinning caddisflies (Hydropsychidae) to predation was greater in copper-dosed (6 μg/L Cu) microcosms than in controls. In a similar study, Kiffney (1996) found that the caddisfly *Hydropysche* sp. (Trichoptera: Hydropsychidae) and the stonefly *Pteronarcella badia* (Plecoptera: Pteronarcyidae) were more susceptible to predation at relatively low metal concentrations than in control streams.

Metal concentrations in both studies were one-half USEPA recommended chronic criterion levels for water of less than 50 mg/L $CaCO_3$ (USEPA, 1986). If biotic interactions are more important in determining population abundance for some taxa in environments contaminated by low metal concentrations, metals may indirectly affect the structures of benthic invertebrate communities. These studies were conducted in small microcosms under highly controlled conditions, which limits their applicability to natural ecosystems. Therefore, there is a need to perform similar studies under a more natural setting.

Clements (1999) addressed how metals mediated biotic interactions in laboratory microcosms and in the field. He examined biotic interactions in two invertebrate communities in stream microcosms. One community was sampled from the metal-contaminated Arkansas River (Colorado) and the other from a nearby reference stream (Poudre River). Stonefly predators from the Poudre River were then added to some of the microcosms while others served as controls. No additional metals were added to microcosms.

Predation rates on mayflies from the metal-polluted Arkansas River were higher compared to mayflies from the reference stream. An additional *in situ* cage experiment was performed in which communities from the two river systems were placed in mesh cages at a contaminated site on the Arkansas River. Stonefly predators from the Poudre River were added to half the cages

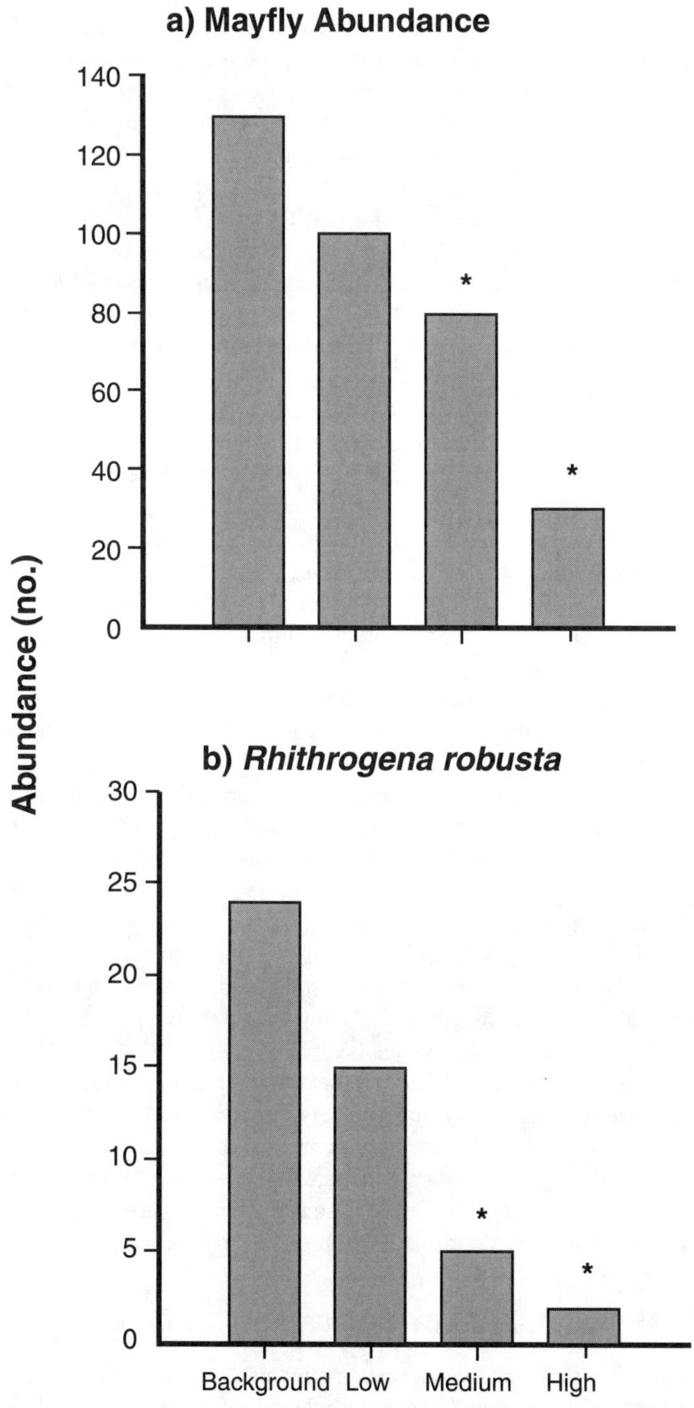

FIG 8.1 a) Mean mayfly abundance and b) *Rhithrogena robusta* at background, low-, medium-, and high-metal sites in Colorado streams. Asterisk indicates sites that were significantly different from background ($P < 0.05$) based on Dunnett's multiple comparison test.

while the other half served as controls. Results from the cage experiment were similar to those from the laboratory experiment; however, effects of predation were relatively weak because of high immigration.

The mechanism explaining increased susceptibility to predation is unknown, but mayflies rely primarily on tactile or chemosensory cues to detect presence of predators (Peckarsky, 1980). Because stoneflies in experimental streams were not exposed to metals and invertebrates collected from the Arkansas River were, Clements (1999) hypothesized that anti-predatory behavior was altered. Other studies have shown that anti-predatory behavior of invertebrates is affected by metal exposure. In contrast to snails (*Physella columbiana*) collected from reference lakes, snails collected from metal-polluted lakes downstream of a USEPA superfund site in Idaho failed to exhibit anti-predator behaviors (Lefcort et al., 2000). It was hypothesized that metals may have physiologically stressed the snails, because heavy metals such as Cd caused a partial shift from aerobic to anaerobic metabolic pathways (Lefcort et al., 2000). Metals may also indirectly regulate invertebrate communities by acting in concert with parasites. Parasitism gained attention recently as an important factor regulating populations of benthic invertebrates (Kohler and Wiley, 1997). For example, populations of the caddisfly *Glossosoma nigrior* in Michigan streams have experienced large population reductions due to an infection by the microsporidian *Courourdella* sp. Other species in stream communities also harbor parasites such as mermith nematodes, which parasitize *Baetis bicaudatus* (Vance and Peckarsky, 1997). As some invertebrates were more susceptible to predation when exposed to metals, we hypothesize that individuals infected by a parasite may be more susceptible to metal toxicity than uninfected conspecifics. Infected fish hosts are more sensitive to toxicants than uninfected conspecifics (Boyce and Yamada, 1977; Pascoe and Cram, 1977). Little research has been done to show how metals interact with parasites to affect benthic invertebrate populations.

A few studies support the hypothesis that infected individuals are more sensitive to toxicants than uninfected organisms. The amphipod *Gammarus pulex* is a widespread and abundant crustacean and is an important prey species for many fish. In England, infection of *G. pulex* by the acanthocephalan *Pomphohynchus laevis* is relatively common. Infected *G. pulex* were exposed to 2.1 μg/L Cd over an 80-day period (Brown and Pascoe, 1989). The LT_{50} (time required for 50% of test animals to die at each Cd concentration) was significantly less (17.5 days) for infected than uninfected conspecifics (42 days) (Brown and Pascoe, 1989). The parasite was unaffected by exposure to Cd. The mechanism underlying the greater sensitivity of infected amphipods was not apparent, but food consumption was 80% less when compared to uninfected conspecifics. The resulting metabolic stress may increase sensitivity to Cd and be a direct cause of toxicity. In a similar study, Guth et al. (1977) found that snails (*Lymnaea stagnalis*) infected with trematodes were more sensitive to zinc than uninfected snails.

Single-species toxicity experiments do not predict the indirect or interactive effects that metals can exert on benthic invertebrates. Consequently, it is important to perform multispecies tests at chronic metal concentrations to examine the significance of these complex effects on populations and communities. Laboratory experiments demonstrate that some invertebrates are more susceptible to various biotic stresses when exposed to metals. The next step is to test whether these interactions occur in the field and then determine their relative importance in regulating invertebrate populations.

8.7 THE ROLE OF FOOD AND WATER IN METAL UPTAKE

Benthic invertebrates living in metals-contaminated streams have two primary routes of exposure: direct exposure to metals in water and the accumulation of metals via food supply. Although invertebrates can accumulate metals from both food and water, few studies compare the relative importance of these exposure routes for freshwater systems. The available evidence suggests that food is more important than water for some freshwater invertebrates (e.g., Munger and Hare, 2000) and marine invertebrates (e.g., Wang and Fisher, 1999a). Despite the observation that food is a major route of metal uptake for some invertebrates and fish, agencies primarily regulate metals in the environment based on the dissolved phase (i.e., metals in water that passes through a 0.45 μm filter). This approach may not be protective of environmental quality, because it ignores food as an exposure pathway for metals.

Aquatic invertebrates are exposed to metals in both the dissolved and particulate phases. Metals dissolved in water may be accumulated by passive diffusion or facilitated transport. After ingestion of food, animals can accumulate particulate and sediment-bound metals. The relative contributions of food and water to metals accumulated by a benthic invertebrate depends on (1) the bioavailable concentrations of metal in each medium that are in contact with uptake sites on the invertebrate; (2) the rate at which water moves over body surfaces or food moves along the gut; (3) the relative surface areas of the absorptive regions; (4) the size, quantity, and quality of food particles and the manner in which metal is associated with food; (5) conditions such as pH and redox potential and the digestive enzymes present in the gut lumen; and (6) the ease with which a metal crosses the membranes involved in metal exchange (Hare, 1992; Wang and Fisher, 1999a).

Invertebrates have diverse feeding habits and are therefore exposed to contaminants via a number of food sources. A few studies have shown that food can be an important route of metal uptake for freshwater invertebrates. Table 8.1 compares the importance of exposure pathways for marine and freshwater invertebrates. Most of the Cd uptake in two aquatic invertebrate predators came from food (62%) and the majority of Zn uptake came from water (76%) (Timmermans et al., 1992). Munger and Hare (2000) found that food was the major route of Cd uptake in the predatory aquatic insect *Chaoborus punctipennis*. Food was also much more important than water as a Cd source of the predatory invertebrate *Sialis velata* (Roy and Hare, 1999). These studies indicate that food is an important route for metal uptake for some predatory freshwater invertebrates. It may also be an important route of exposure for functional feeding groups other than predators, because some groups (e.g., herbivores) feed on highly metal-rich materials, such as stream periphyton (Kiffney and Clements, 1993).

TABLE 8.1
Empirical Studies Determining the Relative Importance of Food and Water as the Dominant Routes of Metal Accumulation in Aquatic Invertebrates (Wang and Fisher, 1999a, b)

Metal	Animal	Food	Dominant Exposure Pathway	Reference
Ag	*Crossostrea virginica*	Phytoplankon, sediment	Water	Abbe and Sanders (1990)
	Palaemontes pugio	Detritus	Water	Connel et al. (1991)
As	*Eurytemora affinis*	Phytoplankton	Food	Sanders et al. (1989)
	Balanus improvisus	Phytoplankton	Food	Sanders et al. (1989)
Cd	*Chaoborus punctipennis*	Zooplankton	Food	Munger and Hare (1997; 2000)
	Limnesia maculata	Chironomidae	Food	Timmermans et al. (1992)
	Mystacides sp.	Chironomidae	Food	Timmermans et al. (1992)
	Sialis velata	Chironomidae	Food	Roy and Hare (1999)
Cr	*Pseudechinus novaezealandiae*	Phytoplankton	Food	Bremer et al. (1990)
Po	*Palaemon serratus*	Mussel	Food	Carvalho and Fowler (1976)
Se	*Meganyctiphanes norvegica*	Mussel	Food	Fowler and Benayoun (1976)
Zn	*Limnesia maculata*	Chironomidae	Water	Klaas et al. (1992)
	Mystacides sp.	Chironomidae	Water	Klaas et al. (1992)
	Orchestia gammarellus	Phytoplankton	Food	Weeks and Rainbow (1993)

Periphyton or biofilm generally has the highest metal concentrations compared to other compartments in metal-contaminated river ecosystems (Kiffney and Clements, 1993; Farag et al., 1998). Biofilm consists of attached algae, bacteria, and associated fine sediment and detrital material; it serves as a food source for grazers and collector-gatherers. Thus, biofilm may be an important link in the trophic transfer of metals in nature. Invertebrates (grazers and collectors) that feed on this material generally have more metals in their tissues than other functional feeding groups (e.g., predators) (Kiffney and Clements, 1993; Farag et al., 1998). Herbivores contained 22 μg/g of Cd compared to 6 μg/g for predatory invertebrates at the Coeur d'Alene basin in Idaho.

Invertebrates that feed on biofilm generally are members of the orders Ephemeroptera and Chironomidae. High numbers of these animals are frequently found drifting in the water columns of streams and they are main food items for fish predators (Radar, 1997). Therefore, grazers and collectors may serve as important links in the transfer of metals to higher trophic levels at a contaminated site. Woodward et al. (1994) found that diet-borne metals were more important than water-borne metals in reducing survival and growth of rainbow trout. Survival of fish was significantly reduced by exposure to a metal-contaminated diet, while survival of fish fed the reference diet was not affected by exposure to 0×, 1×, and 2× metal levels in water (1× = 1.1 μg Cd/L, 12 μg Cu/L, 3.2 μg Pb/L, and 50 μg Zn/L) (Woodward et al., 1994). Field studies have also shown that wild salmonids are physiologically impaired (e.g., higher metallothionein levels) in a metal-contaminated system compared to a reference site and this effect may be partially linked to feeding on contaminated invertebrates (Farag et al., 1995).

These studies support the hypothesis that diet can be more important than water in the transfer of metals in aquatic food webs. Nevertheless, metals are regulated in the environment based on dissolved fractions. One of the reasons for this is the difficulty of estimating the relative importance of food versus water because of the great number of metals and conditions that must be tested. However, despite logistical constraints, future research on the importance of metals in the food web is essential to developing biologically protective water and sediment quality criteria. We propose using the kinetic modeling approach to trace the transfer of metals in freshwater food webs.

Two approaches have been used to model metal accumulation in aquatic invertebrates (Landrum et al., 1992). The first assumes equilibrium partitioning of contaminants among various compartments (e.g., water, sediments, and animals). Because of equilibrium among compartments, the exposure pathway in this model is not important and metal concentrations in animals can be predicted based on measurement of metals in the aqueous phase (Wang and Fisher, 1999a; 1999b). Equilibrium partitioning between food (i.e., microorganisms) and water may occur relatively rapidly, especially if the living component has a short life span (Fisher et al., 1983). Equilibrium partitioning between animals and their food, however, may take much longer. Although this model appeals to regulators because of its simplicity, it may be less predictive of metal accumulations in natural systems (Wang and Fisher, 1999a).

A second approach is the bioenergetic-based kinetic model that treats contaminant accumulation as a first-order physiological process (Wang and Fisher, 1997). Metal uptake over time is described by the following equation:

$$\mathrm{d}C/\mathrm{d}t = (k_u - C_w) + (\mathrm{AE} * \mathrm{IR} * C_f) - (k_e + g) * C$$

where C is the metal concentration in the animals (μg/g), t is the time of exposure (days), k_u is the uptake rate constant from the dissolved phase (L/g/d), C_w is the metal concentration in the dissolved phase (μg/L), AE is the metal assimilation efficiency from ingested particles, IR is the ingestion rate of the animal (mg/g/d), C_f is the metal concentration in ingested particles (μg/mg), k_e is the efflux rate constant (per day), and g is the growth rate constant (per day).

These species-specific physiological parameters are measured experimentally and then incorporated into the model to predict metal concentrations in animals. The model can operate under steady- or nonsteady-state conditions. One critical parameter in understanding and modeling trophic transfer and accumulation of contaminants in aquatic systems is AE, as accumulation in aquatic invertebrates has been shown to be directly proportional to AE for metals. Not until recently have AEs been determined for aquatic invertebrates because food was thought unimportant for metal uptake, and most previous studies focused on uptake from the dissolved phase. In addition, standard techniques for measuring chemical assimilation have been lacking (Wang and Fisher, 1999b).

A promising new technique for evaluating the relative roles of food and water in metal uptake is the pulse-chase feeding method (Wang and Fisher, 1999b). Animals are fed radiolabeled materials for a short period (generally less than their gut passage time) to ensure that total ingestion can be accurately quantified. The animals are then fed nonradioactive food under identical conditions to purge their guts of undigested labeled materials. This approach assumes that animals under natural conditions are in constant contact with their food and that the assimilation process is not affected by chemical concentrations in the animals. To produce radioactive food for suspension feeders, phytoplankton or other suspended materials are labeled and resuspended in unlabeled water so that food is the only radioactive material presented and metal uptake from food can be isolated.

In the natural environment animals are exposed to a variety of conditions (e.g., food quality) that can affect their uptake of metals. In marine bivalves, food quality has its greatest effect on AEs. During the spring phytoplankton bloom in San Francisco Bay, a shift in seston composition to phytoplankton resulted in an increase in Cd, Zn, and Cr AEs in two clam species (Lee and Luoma, 1998). Assimilation of metals by bivalves is also associated with food quantity since AEs decrease as food levels increase. Food quantity and quality do not affect AEs in marine copepods (Wang and Fisher, 1999b).

In addition to these biological parameters, geochemical factors (e.g., metal speciation) can affect AEs. For example, the pH of the gut can influence metal desorption from ingested particles and therefore affect metal assimilation (Wang and Fisher, 1999b). These results indicate that the composition of food may be important in regulating the transfer of metals from diet to consumer organisms. Similarly for marine invertebrates, assimilation of metals by freshwater invertebrates may be related to AEs, particularly for metals accumulated primarily from ingested particles. The relative importance of metal uptake from aqueous and dietary sources is also dependent on AE. Differences in AEs may affect trophic transfer and biomagnification. Contaminants with low AEs in animals near the bottom of the food chain (e.g., zooplankton) are unlikely to be transferred up the food chain. In addition, the sensitivity of an animal to metals may be correlated with AEs; animals with high AEs for a metal may be more sensitive to that metal. Thus, we speculate that the sensitivity of mayflies to metals may be related to their consumption of metal-rich materials and high AEs compared to other taxa.

In summary, metal (e.g., Cd) uptake from food can be as important or more important than uptake from water for invertebrates. Models and an experimental approach have been used to determine the relative importance of food versus water as a route of metal transfer to marine invertebrates (e.g., Wang and Fisher, 1999a). We suggest that this approach be adopted by freshwater ecologists and ecotoxicologists interested in the transfer of metals in aquatic food webs. Understanding the relative importance of food and water for metal uptake is important to maintaining ecological integrity of natural systems, because water and sediment quality criteria for metals are based on the dissolved fractions. While regulators may favor this approach, it has little ecological relevance because the importance of metals in food is ignored.

8.8 CONCLUSIONS

Benthic invertebrate communities provide model systems for studying the impacts of contaminants on natural systems. Invertebrates have been used in biological monitoring for many years and, more recently, they have been used in laboratory and field toxicity studies. Approaches such as multimetric and multivariate methods have been developed to assess effects of metals and other anthropogenic stresses on benthic invertebrates. Each approach has its strengths and weaknesses and careful consideration should be given to each method and its specific application. One promising area of research is the exposure of natural communities of benthic invertebrates to contaminants under experimental conditions. Indicators developed from these experiments should then be tested in the field.

In addition, the role of food as an important route of metal uptake by benthic invertebrates must be incorporated into existing standards for environmental protection. Growing evidence suggests that food is the primary route of metal uptake for some metals (e.g., Cd) and is almost as important as water for other metals (e.g., Zn). Models and experimental methods used on marine invertebrates have been successful at delineating the importance of the two pathways for different metals and species. These approaches can be adopted to examine uptake of metals for freshwater benthic invertebrates

Earth is experiencing a variety of large-scale environmental changes, such as warming and increased exposure to ultraviolet radiation. It is quite likely that environmental history will affect how individuals, populations, and communities will respond to these large-scale stresses. We predict that benthic invertebrate communities from chronically stressed habitats will be more sensitive to this environmental change than unexposed communities.

ACKNOWLEDGMENTS

Thanks to J. Meador and N. Schultz for their comments on an earlier version of this manuscript.

REFERENCES

Abbe, G.R. and J.G. Sanders. 1990. Pathways of silver uptake and accumulation by the American oyster (*Crassostrea virginica*) in Chesapeake Bay, *Estuarine and Coastal Shelf Science*, 31, 113–123.

Aoki, Y., S. Hatakeyama, N. Kobayashi, Y. Sumi, T. Suzuki, and K. Suzuki. 1989. Comparison of cadmium-binding protein induction among mayfly larvae of heavy metal resistant (*Baetis thermicus*) and susceptible species (*B. yoshinensis* and *B. sahoensis*), *Comparative Biochemistry and Physiology*, 93C, 345–347.

Bremer, P.J., M.F. Barker, and M.W. Loutit. 1990. A comparison of direct absorption and phytoplankton ingestion in accumulation of chromium by sea urchin larvae, *Marine Environment Research*, 30, 233–241.

Boyce, N.P. and S.B. Yamada. 1977. Effects of a parasite, *Eubothrium salvelini* (Cestoda: Pseudophyllidea) on the resistance of juvenile sockeye salmon to zinc, *Journal of the Fisheries Research Board of Canada*, 34, 706–709.

Brown, A.F. and D. Pascoe. 1989. Parasitism and host sensitivity to cadmium: an acanthocephalan infection of the freshwater amphipod *Gammarus pulex*, *Journal of Applied Ecology*, 26, 473–487.

Buikema, A.L. and J.R. Voshell, Jr. 1993. Toxicity studies using freshwater benthic invertebrates, in D.M. Rosenberg and V.H. Resh (Eds.). *Freshwater Biomonitoring and Benthic Macroinvertebrates*, Chapman & Hall, New York, 344–398.

Carlisle, D.M. and W.H. Clements. 1999. Sensitivity and variability of metrics used in biological assessments of running waters, *Environmental Toxicology and Chemistry*, 18, 285–291.

Carlisle, D.M. 2000. Bioenergetic food webs as a means for linking toxicological effects across scales of ecological organization, *Journal of Aquatic Ecosystem Stress and Recovery*, 7, 155–165.

Carpenter, K.E. A study of the fauna of rivers polluted by lead mining in the Aberystwyth District of Cardiganshire, *Annals of Applied Biology*, 12, 1–23.

Carvalho, F.P. and S.W. Fowler. 1994. A double-tracer technique to determine the relative importance of water and food as sources of polonium-210 to marine prawns and fish, *Marine Ecology and Progress Series*, 103, 251–264.

Clements, W.H. 1992. Community responses of stream organisms to heavy metals: a review of observational and experimental approaches, in M.C. Newman and A.W. McIntosh (Eds.). *Ecotoxicology of Metals: Current Concepts and Applications*, Lewis Publishers, Boca Raton, FL, 363–391.

Clements, W.H. 1994. Benthic invertebrate community responses to heavy metals in the Upper Arkansas River Basin, Colorado. *Journal of the North American Benthological Society*, 13, 30–44.

Clements, W.H. 1997. Effects of contaminants at higher levels of biological organization in aquatic ecosystems, *Reviews in Toxicology*, 1, 107–146.

Clements, W.H. 1999. Metal tolerance and predator-prey interactions in benthic macroinvertebrate stream communities, *Ecological Application*, 9, 1073–1084.

Clements, W.H., D.S. Cherry, and J. Cairns, Jr. 1988. Impact of heavy metals on insect communities in streams: a comparison of observational and experimental results, *Canadian Journal of Fisheries and Aquatic Sciences*, 45, 2017–2025.

Clements, W.H., J.H. Van Hassel, D.S. Cherry, and J. Cairns, Jr. 1989a. Colonization, variability, and the use of substratum-filled trays for biomonitoring benthic communities, *Hydrobiologia*, 173, 45–53.

Clements, W.H., D.S. Cherry, and J. Cairns, Jr. 1989b. The influence of copper exposure on predator-prey interactions in aquatic insect communities, *Freshwater Biology*, 21, 483–488.

Clements, W.H., J.L. Farris, D.S. Cherry, and J. Cairns, Jr. 1989c. The influence of water quality on macroinvertebrate community responses to copper in outdoor experimental streams, *Aquatic Toxicology*, 14, 249–262.

Clements, W.H., D.S. Cherry, and J.H. Van Hassel. 1992. Assessment of the impact of heavy metals on benthic communities at the Clinch River (Virginia): evaluation of an index of community sensitivity, *Canadian Journal of Fisheries and Aquatic Sciences*, 49, 1686–1694.

Clements, W.H. and P.M. Kiffney. 1995. The influence of elevation on benthic community responses to heavy metals in Rocky Mountain streams, *Canadian Journal of Fisheries and Aquatic Sciences*, 52, 1966–1977.

Clements, W.H., D.M. Carlisle, J.M. Lazorchak, and P.C. Johnson. 2000. Heavy metals structure benthic communities in Colorado mountain streams, *Ecological Applications*, 10, 626–638.

Clements, W.H. 2000. Integrating effects across levels of biological organization: an overview, *Journal of Aquatic Ecosystem Stress and Recovery*, 7, 113–116.

Connell, D.B., J. Sanders, G. Riedel, and G.R. Abbe. 1991. Pathways of silver uptake and trophic transfer to estuarine organisms, *Environmental Science and Technology*, 25, 921–924.

Cooper, S.D. and L. Barmuta. 1993. Field experiments in biomonitoring, in D.M. Rosenberg and V.H. Resh (Eds.). *Freshwater Biomonitoring and Benthic Macroinvertebrates*, Chapman & Hall, New York, 399–441.

Courtenay, L.A. and W.H. Clements. 2000. Sensitivity to acidic pH in benthic invertebrate assemblages with different histories of exposure to metals, *Journal of the North American Benthological Society*, 19, 112–127.

Crossey, M.J. and T.W. LaPoint. 1988. A comparison of periphyton structural and functional responses to heavy metals, *Hydrobiologia*, 162, 109–121.

Darlington, S.T. and A.M. Gower. 1990 Location of copper in larvae of *Plectrocnemia conspersa* (Curtis) (Trichoptera) exposed to elevated metal concentrations in a mine drainage stream, *Hydrobiologia*, 196, 91–98.

Dorit, R.L., W.F. Walker, Jr., and R.D. Barnes. 1991. *Zoology*. Saunders College Publishing, Orlando, FL.

Farag, A.M., M.A. Stansbury, C. Hogstrand, E. MacConnell, and H.L. Bergman. 1995. The physiological impairment of free-ranging brown trout exposed to metals in the Clark Fork River, Montana, *Canadian Journal of Fisheries and Aquatic Sciences*, 52, 2038–2050.

Farag, A.M., D.F. Woodward, J.N. Goldstein, and J.S. Meyer. 1998. Concentrations of metals associated with mining waste in sediments, biofilm, benthic macroinvertebrates, and fish from the Coeur d'Alene River Basin, Idaho, *Archives of Environmental Contamination and Toxicology*, 34, 119–127.

Feldman, R.S. and E.F. Conner. 1992. The relationship between pH and community structure of invertebrates in streams of the Shenandoah National Park, Virginia, USA, *Freshwater Biology*, 27, 261–276.

Fisher, N.S., P. Bjerregaard, and S.W. Fowler. 1983. Interactions of marine plankton with transuranic elements. 1. Biokinetics of neptunium, plutonium, americium, and californium in phytoplankton, *Limnology and Oceanography*, 28, 432–447.

Fore, L.S. 2002. Biological assessment of mining disturbance on stream invertebrates in mineralized areas of Colorado, Chapter 19, this volume.

Fore, L.S., J.R. Karr, and R.W. Wisseman. 1996. Assessing invertebrate responses to human activities: evaluating alternative approaches, *Journal of the North American Benthological Society*, 15, 212–231.

Fowler, S.W. and G. Benayoun. 1976. Selenium kinetics in marine zooplankton, *Marine Science Communities*, 2, 43–62.

Guth, D.J., H.D. Blankespoor, and J. Cairns, Jr. 1977. Potentiation of zinc stress caused by parasitic infection of snails, *Hydrobiologia*, 55, 225–229.

Hare, L. 1992. Aquatic insects and trace metals: bioavailability, bioaccumulation, and toxicity, *Critical Reviews in Toxicology*, 22, 327–369.

Hodson, P.V., U. Borgmann, and H. Shear. 1979. Toxicity of copper to aquatic biota, in J. O. Nriagu (Eds.), *Copper in the Environment, II. Health Effects*, Wiley and Sons, New York, 308–372.

Hurlbert, S.H. 1984. Pseudoreplication and the design of ecological field experiments. *Ecological Monographs*, 54, 187–211.

Karr, J.R. 1981. Assessment of biotic integrity using fish communities, *Fisheries*, 6, 21–27.

Karr, J.R. and E.W. Chu. 1999. *Restoring Life in Running Waters. Better Biological Monitoring*. Island Press, Washington, D.C.

Kerans, B.L. and J.R. Karr. 1994. A benthic index of biotic integrity (B-IBI) for rivers of the Tennessee Valley, *Ecological Applications*, 4, 768–785.

Kettle, C.A. and F. DeNoyelles. 1986. Effects of cadmium stress on the plankton communities of experimental ponds, *Journal of Freshwater Ecology*, 3, 433–444.

Kiffney, P.M. 1996. Main and interactive effects of invertebrate density, predation, and metals on a Rocky Mountain stream macroinvertebrate community, *Canadian Journal of Fisheries and Aquatic Sciences*, 53, 1595–1601.

Kiffney, P.M. and W.H. Clements. 1993. Bioaccumulation of heavy metals by benthic invertebrates at the Arkansas River, Colorado, *Environmental Toxicology and Chemistry*, 12, 1507–1517.

Kiffney, P.M. and W.H. Clements. 1994. Effects of heavy metals on a Rocky Mountain stream macroinvertebrate assemblage in experimental microcosms, *Journal of the North American Benthological Society*, 13, 511–524.

Kiffney, P.M. and W.H. Clements. 1996. Effects of metals on stream assemblages from different altitudes, *Ecological Applications*, 6, 472–481.

Klerks, P.L. and J.S. Levinton. 1992. Evolution of resistance and changes in community-composition in metal-polluted environments: a case study on Foundry Cove, in R. Dallinger and P.S. Rainbow (Eds.). *Ecotoxicology of Metals in Invertebrates*, Lewis Publishers, Boca Raton, FL, 223–244.

Kohler, S.L. and M.J. Wiley. 1997. Pathogen outbreaks reveal large-scale effects of competition in stream communities, *Ecology*, 78, 2164–2176.

Krantzberg, G. and P.M. Stokes. 1989. Metal regulation, tolerance, and body burden in the larvae of the genus *Chironomus*, *Canadian Journal of Fisheries and Aquatic Sciences*, 46, 389–398.

Landrum, P.F., H. Lee, and M.J. Lydy. 1992. Toxicokinetics in aquatic systems: model comparisons and use in hazard assessment, *Environmental Toxicology and Chemistry*, 11, 1709–1725.

Lee, B. and S.N. Luoma. 1998. Influence of microalgal blooms on metal bioavailability: effects of microalgal biomass on absorption efficiency of Cd, Cr, and Zn by two bivalves from San Francisco Bay, *Limnology and Oceanography*, 43, 1455–1466.

Lefcort, H., S.M. Thomson, E.E. Cowles, H.L. Harowicz, B.M Livaudais, W.E. Roberts, and W.F. Ettinger. 1999. Ramifications of predator avoidance: predator and heavy-metal-mediated competition between tadpoles and snails, *Ecological Applications*, 9, 1477–1489.

Lefcort, H., E. Ammann, and S.M. Eiger. 2000. Antipredatory behavior as an index of heavy-metal pollution? A test using snails and caddisflies, *Archives of Environmental Contamination and Toxicology*, 38, 311–316.

Leland, H.V., S.V. Fend, T.L. Dudley, and J.L. Carter. 1989. Effects of copper on species composition of benthic insects in a Sierra Nevada, California, stream, *Freshwater Biology*, 21, 163–179.

McKnight, D. 1981. Chemical and biological processes controlling the response of a freshwater ecosystem to copper stress: a field study of the $CuSO_4$ treatment of Mill Pond Reservoir, Burlington, Mass., *Limnology and Oceanography*, 26, 518–531.

Moore, J.N. and S.N. Luoma. 1990. Hazardous wastes from large scale metal extraction: a case study, *Environmental Science and Technology*, 24, 1279–1285.

Moss, D., M.T. Furse, J.F. Wright, and P.D. Armitage. 1987. The prediction of macro-invertebrate fauna in unpolluted running water sites in Great Britain using environmental data, *Freshwater Biology*, 17, 41–52.

Munger, C. and L. Hare. 2000. Influence of ingestion rate and food types on cadmium accumulation by the aquatic insect *Chaoborus*, *Canadian Journal of Fisheries and Aquatic Sciences*, 57, 327–332.

Norris, R.H. 1995. Biological monitoring: the dilemma of data analysis, *Journal of the North American Benthological Society*, 14, 451–457.

Norris, R.H., P.S. Lake, and R. Swain. 1982. Ecological effects of mine effluents on the South Esk River, Northeastern Tasmania III. Benthic macroinvertebrates, *Australian Journal of Marine and Freshwater Research*, 33, 789–809.

Odum, E.P. 1985. Trends expected in stressed ecosystems, *Bioscience*, 35, 419–422.

Parsons, M. and R.H. Norris. 1996. The effect of habitat-specific sampling on biological assessment of water quality using a predictive model, *Freshwater Biology*, 36, 419–434.

Pascoe, D. and P. Cram. 1977. The effect of parasitism on the toxicity of cadmium to the three-spined stickleback, *Gasterosteus aculeatus* L., *Journal of Fish Biology*, 10, 467–472.

Poulton, B.C., D.P. Monda, D.F. Woodward, M.L. Wildhaber, and W.G. Brumbaugh. 1995. Relations between benthic community structure and metals concentrations in aquatic invertebrates: Clark Fork River, Montana, *Journal of Freshwater Ecology*, 10, 277–293.

Radar, R. 1997. A functional classification of the drift: traits that influence the invertebrate availability to salmonids, *Canadian Journal of Fisheries and Aquatic Sciences*, 1211–1234.

Reynoldson, T.B., R.H. Norris, V.H. Resh, K.E. Day, and D.M. Rosenberg. 1997. The reference condition: a comparison of multimetric and multivariate approaches to assess water-quality impairment using benthic macroinvertebrates, *Journal of the North American Benthological Society*, 16, 833–852.

Richardson, J.S. and P.M. Kiffney. 2000. Responses of a macroinvertebrate community from a pristine, southern British Columbia, Canada, stream to metals in experimental mesocosms, *Environmental Toxicology and Chemistry*, 19, 736–743.

Roesijadi, G. 1992. Metallothioneins in metal regulation and toxicity in aquatic animals, *Aquatic Toxicology*, 22, 81–114.

Rosenberg, D.M. and V.H. Resh. 1993. *Freshwater Biomonitoring and Benthic Macroinvertebrates*. Chapman & Hall, New York.

Roy, I. and L. Hare. 1999. Relative importance of water and food as cadmium sources to the predatory insect *Sialis velata* (Megaloptera), *Canadian Journal of Fisheries and Aquatic Sciences*, 56, 1143–1149.

Sanders, J.G., R.W. Osman, and G.F. Riedel. 1989. Pathways of arsenic and incorporation in estuarine phytoplankton and the filter-feeding invertebrates, *Eurytemora affinis*, *Balanus improvisus*, *Crassostrea virginica*, *Marine Biology*, 103, 319–325.

Schindler, D.W. 1977. Evolution of phosphorus limitation in lakes, *Science*, 195, 260–262.

Schindler, D.W. 1987. Determining ecosystem responses to anthropogenic stress, *Canadian Journal of Fisheries and Aquatic Sciences*, 44(suppl.), 6–25.

Short, T.M., J.A. Black, and W.J. Birge. 1990. Effects of acid-mine drainage on the chemical and biological character of an alkaline headwater stream, *Archives of Environmental Contamination and Toxicology*, 19, 241–248.

Smock, L.A. 1983. The influence of feeding habitat on whole-body metal concentrations in aquatic insects, *Freshwater Biology*, 13, 310–311.

Stewart-Oaten, A., W.W. Murdoch, and K.R. Parker. 1986. Environmental impacts assessment: "pseudoreplication" in time, *Ecology*, 67, 929–940.

Timmermans, K.R., E. Spijkerman, M. Tonkes, and H. Govers. 1992. Cadmium and zinc uptake by two species of aquatic invertebrate predators from dietary and aqueous sources. *Canadian Journal of Fisheries and Aquatic Sciences*, 49, 655–662.

U.S. Environmental Protection Agency. 1986. *Quality criteria for water 1986*. EPA-440/5–86–001. USEPA, Office of Water Regulations and Standards, Washington, D.C.

Vance, S.A. and B.L. Peckarsky. 1997. The effect of mermithid parasitism on predation of nymphal *Baetis bicaudatus* (Ephemeroptera) by invertebrates, *Oecologia*, 110, 147–152.

Vouri, K. 1994. Rapid behavioural and morphological responses of hydropsychid larvae (Trichoptera: Hydropsychidae) to sublethal cadmium exposure, *Environmental Pollution*, 84, 291–299.

Wallace, J. B., M.R. Whiles, S. Eggert, T.F. Cuffney, G.J. Lugthart, and K. Chung. 1991. A five-year study of export of fine organic particles from headwater streams: effects of season, extreme discharges and invertebrate manipulations, *Limnology and Oceanography*, 36, 670–682.

Wallace, J.B. and J.R. Webster. 1996. The role of macroinvertebrates in stream ecosystem function, *Annual Review of Entomology*, 41, 115–139.

Wang, W. and N.S. Fisher. 1999a. Delineating metal accumulation pathways for marine invertebrates, *The Science of the Total Environment*, 237, 459–472.

Wang, W. and N.S. Fisher. 1999b. Assimilation efficiencies of chemical contaminants in aquatic invertebrates: a synthesis, *Environmental Toxicology and Chemistry*, 18, 2034–2045.

Weeks, J.M. and P.S. Rainbow. 1993. The relative importance of food and seawater as sources of copper and zinc to talitrid amphipods (Crustacea; Amphipoda; Talitridae), *Journal of Applied Ecology*, 30, 1313–735.

Wilson, J.B. 1988. The cost of heavy metal tolerance, *Evolution*, 42, 408–413.

Winner, R.W., M.W. Boesel, and M.P. Farrel. 1980. Insect community structure as an index of heavy-metal pollution in lotic ecosystems, *Canadian Journal of Fisheries and Aquatic Sciences*, 37, 647–655.

Wright, J.F., D. Moss, P.D. Armitage, and M.T. Furse. 1984. A preliminary classification of running-water sites in Great Britain based on macroinvertebrate species and the prediction of community type using environmental data, *Freshwater Biology*, 14, 221–256.

Woodward, D.F.W.G. Brumbaugh, A.J. Delonay, E.E. Little, and C.E Smith. 1994. Effects on rainbow trout fry of a metals-contaminated diet of benthic invertebrates from the Clark Fork River, Montana, *Transactions of the American Fisheries Society*, 12, 51–62.

Section III

Method Advancement

9 A Method for Assessing Outfall Effects on Great River Fish Populations: The Traveling Zone Approach

Erich B. Emery and Jeffrey A. Thomas

CONTENTS

9.1 INTRODUCTION

The assessment of large and great river fish assemblages is complicated by sampling considerations (Simon and Sanders, 1999), metric development and testing (Simon, 1992; Simon and Emery, 1995; Simon and Stahl, 1998; Emery et al., 1999), and variation of individual IBI metrics (Gammon and Simon, 2000). Karr et al. (1985) indicated that large and great rivers of the United States do not receive adequate protection to conserve biological integrity. Great rivers are distinctive in that they are few in number, comprise the largest component of water resources (Vannote et al., 1980), and are disproportionately degraded (Karr et al., 1985; USEPA, 2000).

A pattern in community response to stress is used to determine biological integrity and ecological function (Karr and Dudley, 1981). It may be possible, by examining these discernable patterns in the response of a fish community, to determine certain biological response signatures

0-8493-0905-0/03/$0.00+$1.50

of the effluent (Yoder and Rankin, 1995a). Shifts in a fish community in a given area may be used to pinpoint specific stressors on a great resource such as the Ohio River.

9.2 BACKGROUND

9.2.1 Ohio River Valley Water Sanitation Commission (ORSANCO) Approach

Point source effects on biological communities of great rivers are relatively unexplored aspects of pollution control and prevention. Due to the large volume of water great rivers move, effluent mixing zones are typically extremely short (ORSANCO and Ohio River Users Group, 1999). Therefore, any immediate effect of the outfall on organisms residing in the river is generally slight and hard to quantify. It is necessary to examine biological community response in the smallest possible segments that will consistently show impairment trends.

Standard electrofishing zones for the Ohio River are 500 m in length (ORSANCO, 2000). The response of the fish community to an impairment, however, is often constrained to the first 100 to 200 m, allowing the community to adequately recover over the remaining 300 m of the zone. Therefore, by relying on a 500-m zone length, it is nearly impossible to describe the indicator response caused by the outfall or other pollution source. A method was needed to reveal changes in community structure and function too subtle to be captured by standard techniques.

9.2.2 Upstream versus Downstream Studies

Historically, the accepted method of determining the effect of an outfall on a stream was to compare the impacted area to an upstream, unimpaired "reference" condition. This method can work well and may be very effective in determining the extent of an impairment, but it also has several drawbacks that researchers must consider. The upstream site must reflect what the unimpaired study area conditions should be. Researchers should consider the importance of changes in microhabitat features (i.e., substrate type, depth, stream morphology) within the study area and the upstream reference area, carefully matching these conditions as closely as possible. We account for this variability by conducting a detailed examination of the microhabitats of the outfall zone and matching that data closely at an upstream location.

Another limitation of the upstream/downstream comparison is multiple impairments. It is often difficult, particularly in large and great rivers, to find an upstream reference site that matches the habitat of a study area and is not impacted by another outfall. It is common also for the study area to be impacted by multiple dischargers. Isolating the effect of one particular effluent in an area where several outfalls exist within a 500-m segment of a great river can be very difficult with a typical upstream/downstream study. However, it is possible to detect changes in the biological communities at the site of each impairment using the traveling zone (T-zone) method.

9.2.3 Ohio EPA Area Degradation Values

The Ohio Environmental Protection Agency (Yoder and Rankin, 1995b) developed a tool for interpreting biological response to impairments using multimetric data. This tool, the area of degradation value (ADV), allows examination of the recovery of a biological community by estimating the degree of departure from a biocriterion along a longitudinal continuum. The longitudinal resolution can be enhanced by spacing the sampling sites closer together, but this resolution can be limited by the required length of the sample site. Given the standard 500-m zone sampled by ORSANCO on the Ohio River (ORSANCO, 2000), determining the recovery of the biological community within that 500 m using the ADV is not possible. However, a similar result is accomplished by using the traveling zone method to assess the data on a finer scale.

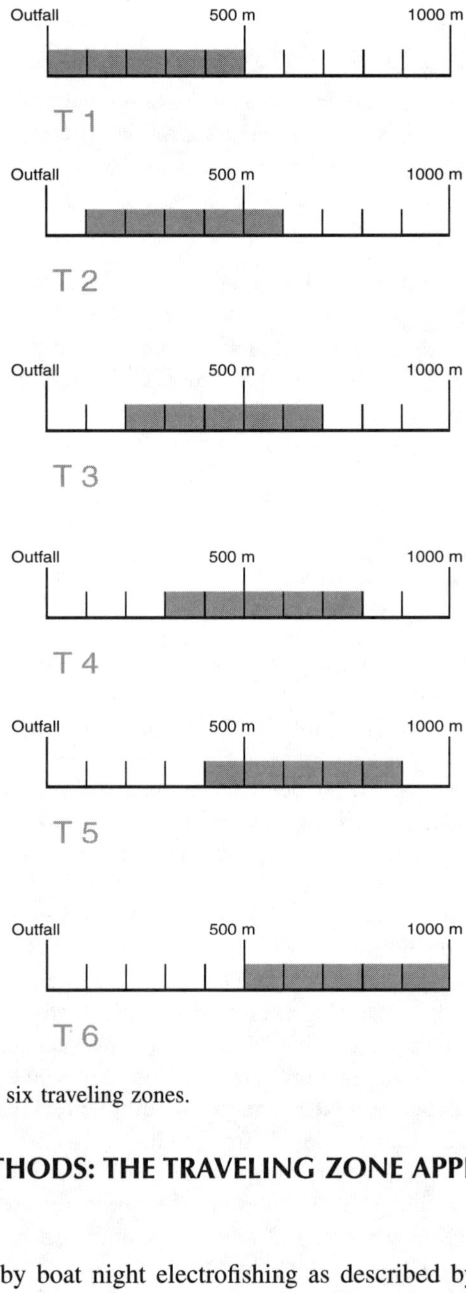

FIGURE 9.1 Diagram of the six traveling zones.

9.3 ASSESSMENT METHODS: THE TRAVELING ZONE APPROACH

9.3.1 Fish Methods

Sampling was conducted by boat night electrofishing as described by ORSANCO (2000). Two crews worked simultaneously to collect fish along 1000 m at each outfall. As one crew sampled the upper 500 m from the outfall downstream, the other crew started 500 m below the outfall and sampled the second half of the 1000-m zone. Each 500-m zone was sampled as a whole, with the fish from each 100-m segment recorded separately so that the data could be divided into ten 100-m segments and reconfigured for the T-zones (Figure 9.1).

9.3.2 Zone Design

The fish data were arranged so that the first T-zone (T1) consisted of the first five 100-m zones starting at the outfall. The second zone (T2) was the compilation of the data from the second to the sixth 100-meter zones, and so on downstream to T6, the last five 100-m zones.

9.3.3 Data Analysis

The six T-zones were created after the data were entered into a database and could be reconfigured. Thirteen fish metrics were calculated from data from these new 500-m zones (Emery et al., in preparation). The metrics were then graphed and appropriate statistical methods were applied to reveal trends observed from T1 to T6.

9.3.4 Water Chemistry

Chemical parameters were examined to determine the water quality of the effluent and track the downstream extent of the plume. Measurements of water temperature, pH, dissolved oxygen (DO), and conductivity were taken with a Hydrolab multiprobe water quality sampler at the outfall (0) and at each 100-m interval for the entire 1000 m. At point 0 and at every 100 m, bottom, middle, and surface measurements were taken at the shore, 15 m from shore, and 30 m from shore. This provided 54 data points per 500-m zone, which were used to reveal the water quality gradient along the zone.

9.4 RESULTS AND DISCUSSION

9.4.1 Defining Zones of Recovery

The traveling zone technique was successful in revealing gradients at the outfalls that were not stressed in the two normal concurrent 500-m zones. Differences in resolution between the T-zone method and the regular 500-m zones are shown in Figure 9.2. The percent of individuals as piscivores increased from the upper 500-m zone to the lower 500-m zone (Figure 9.2a). However, the T-zone approach (Figure 9.2b) better defined this increase. While looking at the two 500-m zones only, it can only be determined that the outfall no longer affects piscivores after 500 m. However, based on the T-zone approach, the effect may be diminished by T5, indicating that the effluent was diluted enough for the piscivore numbers to return to normal after 800 m.

This conclusion can be drawn because the last effluent effect on the percent of individuals as piscivores was seen at T4, which represents the compilation of data from the 500 m between 300 and 800 m. When evaluating the data from the last two 100-m zones, the percent of piscivores returned to expected conditions, suggesting an end of the effluent effect on the piscivore populations.

While Figure 9.2b shows an effluent that had an effect on a fish community for approximately 800 m, often the effect of an outfall on Ohio River fish does not extend that far downstream. Using 500-m zones, it is difficult to determine the distance an outfall effect may cover. Figure 9.3a shows an effluent that affected the centrarchid population within the first 500 m of the outfall. The effect appears diminished by the second 500-m zone, but it is impossible to determine precisely where the effect weakened. By examining the T-Zones at the same outfall (Figure 9.3b), it appears that the effect was only observed in the first 100 m of the outfall, since the number of centrarchids appears to have recovered by T2.

9.4.2 Gradient Patterns

The T-zone approach revealed gradients at most of the outfalls for each of the metrics examined (Figure 9.4). For example, the number of species steadily increased from the outfall to 1000 m downstream (Figure 9.4a). By looking only at T1 (upper 500 m) and T6 (lower 500 m), this gradient only appeared as a large jump. A similar trend was seen in the number of intolerant species as the effect began to level off after T4 (Figure 9.4b). The use of the T-zone method highlights the response to the stressor, provides a more robust sampling approach, and demonstrates responses that may have otherwise been overlooked.

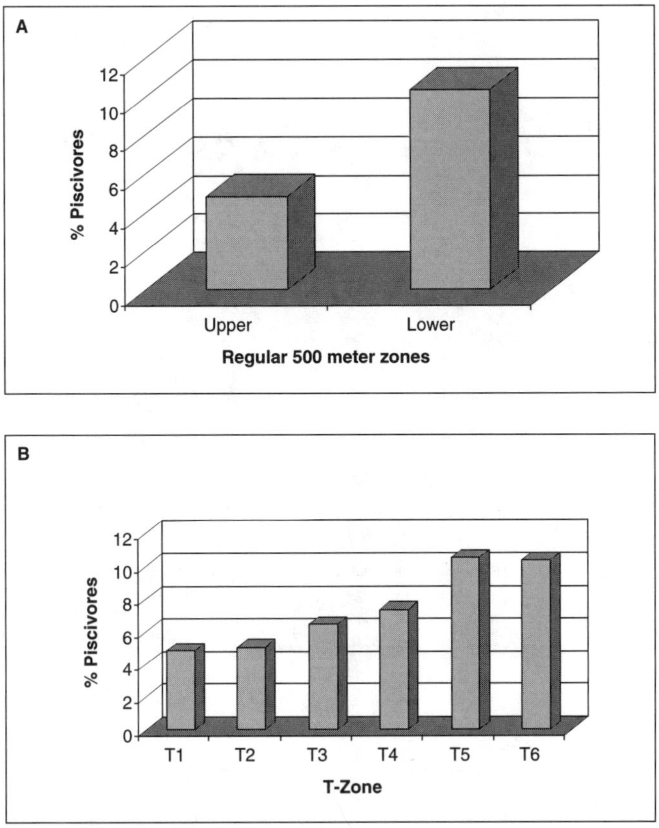

FIGURE 9.2 Resolutions of T-zones compared to those of two regular 500-m zones.

The T-zones showed the response gradient and validated the response of the multimetric index. By comparing the habitat and water chemistry parameters to the index score response for each T-zone, often a trend such as that illustrated in Figure 9.5 will brecome apparent. Figure 9.5 shows that as average surface temperature decreases from the outfall, a corresponding increase in multimetric index score was observed.

9.5 CONCLUSIONS

We developed a technique for evaluating fish community response, applicable for situations where the zone of impairment is too small to be adequately represented by a standard sized boat-electrofishing zone. By collecting data in 100-m increments along a continuous 1000 m, we are able to construct traveling zones or T-zones, each 500 m in length and incrementally 100 m further from the point of impact. This technique requires the sampling effort of two standard sized boat-electrofishing zones, but provides the equivalent of six standard sized boat-electrofishing zones. This overlapping technique provides 100-m resolution, increasing the ability to assess community response usually overlooked by standard 500-m zones.

ACKNOWLEDGMENTS

We would like to thank Matt Wooten, Jim Hawkes, and Robert Row for helping to collect the data used to analyze the effectiveness of the traveling zones. Thanks also to Frank McCormick for help in assessing the data and also for comments regarding this chapter.

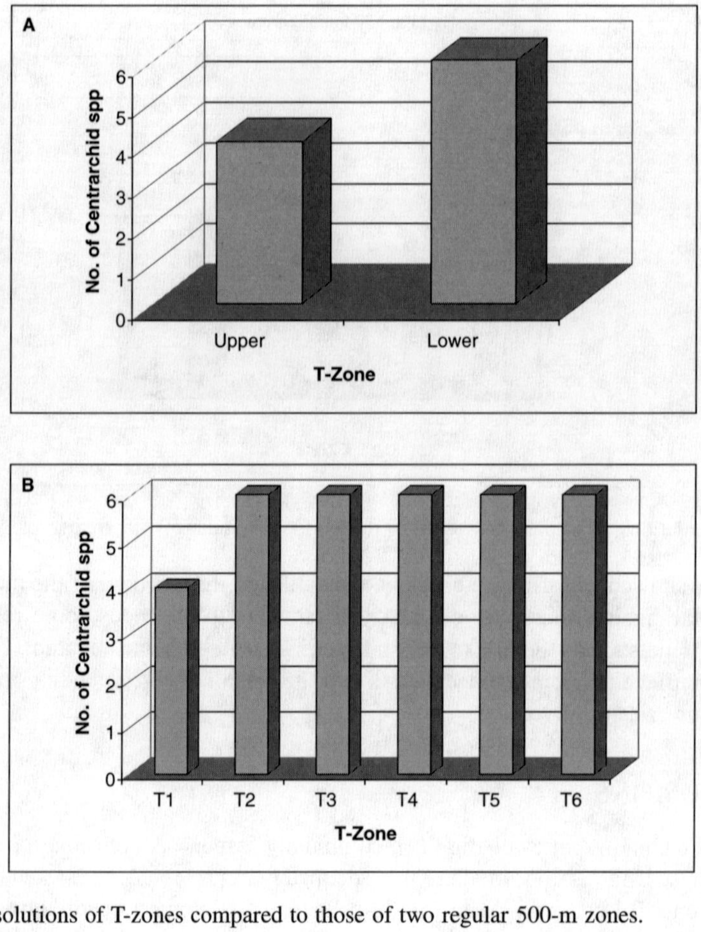

FIGURE 9.3 Resolutions of T-zones compared to those of two regular 500-m zones.

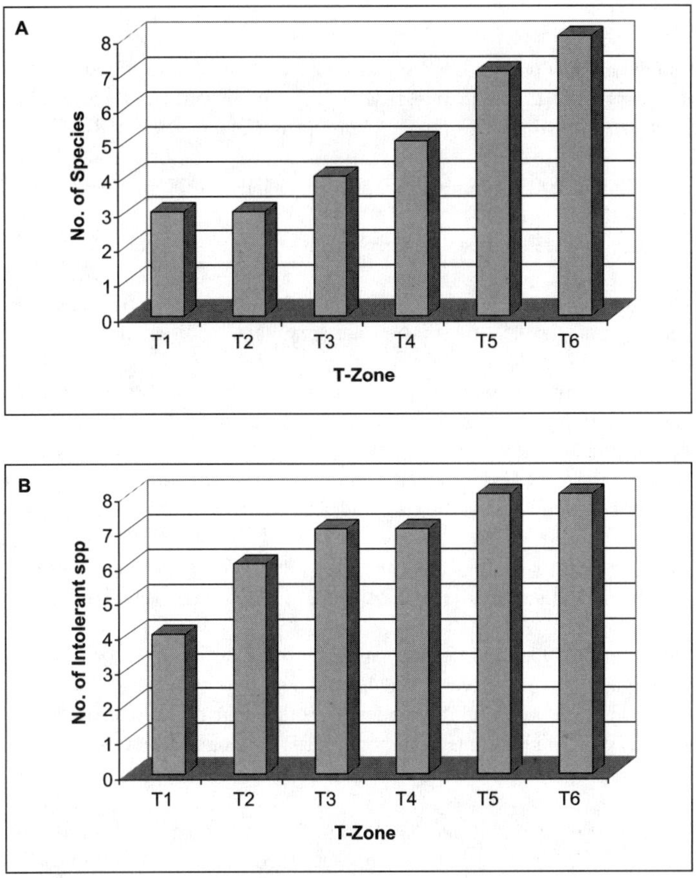

FIGURE 9.4 Examples of T-zone metric responses at two outfalls.

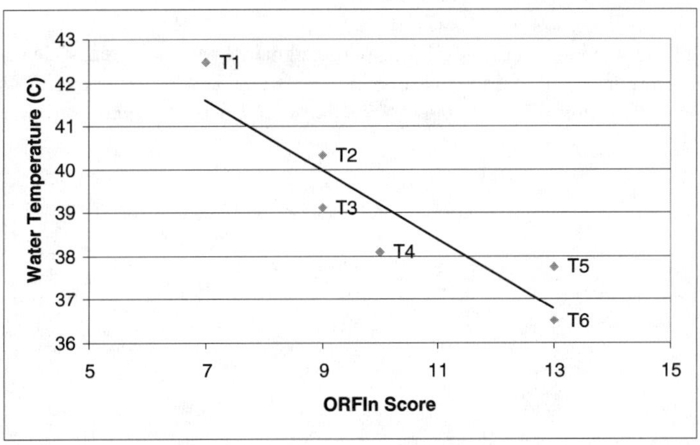

FIGURE 9.5 Surface water temperature versus ORFIn score at a thermal effluent.

REFERENCES

Emery, E.B., T.P. Simon, and R. Ovies. 1999. Influence of the family *Catostomidae* on the metrics developed for a great rivers index of biotic integrity, in T.P. Simon (Ed.). *Assessing the Sustainability and Biological Integrity of Water Resources Using Fish Communities*. CRC Press, Boca Raton, FL, 203–224.

Gammon, J.R. and T.P. Simon. 2000. Variation in a great river index of biotic integrity over a 20-year period, *Hydrobiologia,* 422/423, 291–304.

Karr, J.R. and D.R. Dudley. 1981. Ecological perspective on water quality goals, *Environmental Management*, 5, 55–68.

Karr, J.R., R.C. Heidinger, and E.H. Helmer. 1985. Sensitivity of the index of biotic integrity to changes in chlorine and ammonia levels from wastewater treatment facilities, *Journal of the Water Pollution Control Federation*, 57, 912–915.

Ohio River Valley Water Sanitation Commission. 2000. Quality Assurance Project Plan for the Collection of Fish Population Samples as Part of the Fish Community Biocriteria Development Program. ORSANCO, Cincinnati, OH.

ORSANCO and Ohio River Users Group. 1999. *Guidelines for Delineating Mixing* Zones for Ohio River Discharges. Part I: Calculation of Mixing and Review of State Policies. Prepared by Limno-Tech, Inc., Ann Arbor, MI.

Simon, T.P. 1992. Development of Biological Criteria for Large Rivers with an Emphasis on an Assessment of the White River Drainage, Indiana, EPA 905/R-92/006, U.S. Environmental Protection Agency, Region 5, Chicago, IL.

Simon, T.P. and E.B. Emery. 1995. Modification and assessment of an index of biotic integrity to quantify water resource quality in great rivers, *Regulated Rivers: Research and Management*, 11, 283–298.

Simon, T.P. and R.E. Sanders. 1999. Applying an index of biotic integrity based on great river fish communities: considerations in sampling and interpretation, in T.P. Simon (Ed.). *Assessing the Sustainability and Biological Integrity of Water Resources Using Fish Communities*. CRC Press, Boca Raton, FL, 475–506.

Simon, T.P. and J.R. Stahl. 1998. *Development of Index of Biotic Integrity Expectations for the Wabash River.* EPA 905/R-96/026. USEPA, Water Division, Watershed and Non-Point Source Branch, Chicago, IL

U.S. Environmental Protection Agency (USEPA). 2000. *Stressor Identification Guidance Document.* EPA 822/B-00/025. USEPA, Office of Water, Washington, D.C.

Vannote, R.L., G.W. Minshall, K.W. Cummins, J.R. Sedell, and C.E. Cushing. 1980. The river continuum concept, *Canadian Journal of Fisheries and Aquatic Sciences*, 37, 130–137.

Yoder, C.O. and E.T. Rankin. 1995a. Biological criteria program development and implementation in Ohio, in W.S. Davis and T.P. Simon (Eds.), *Biological Assessment and Criteria: Tools for Water Resource Planning and Decision Making*. Lewis, Boca Raton, FL, 263–286.

Yoder, C.O. and E.T. Rankin. 1995b. Biological response signatures and the area of degradation value: New tools for interpreting multimetric data, in W.S. Davis and T.P. Simon (Eds.), *Biological Assessment and Criteria: Tools for Water Resource Planning and Decision Making*. Lewis, Boca Raton, FL. 263–286.

10 Pioneer Species Metric Changes as a Result of Increased Anthropogenic Disturbance: Statewide Patterns and a Case Study of Four Ohio Streams

Edward T. Rankin and Thomas P. Simon

CONTENTS

0-8493-0905-0/03/$0.00+$1.50

10.1 INTRODUCTION

Smith (1971) coined the term "pioneer species" to represent the first fish species to recolonize an intermittent stream after water returned to small headwater streams in central Illinois. Smith described a list of species that were representative of adventitious taxa that were tolerant to extremes in environmental conditions including thermal, low dissolved oxygen, increased siltation and cementation of sediments, and generally included members of the herbivore or insectivore feeding guilds.

The Ohio Environmental Protection Agency, when modifying the index of biotic integrity (IBI) for headwater streams (≤20 square miles drainage area) in Ohio, used the percentage of pioneer species as a replacement metric for the percentage of carnivores (Ohio EPA, 1989). Carnivores are usually few in number in small headwater streams and generally are not consistent parts of the structural composition. The rationale Ohio used for this replacement is that streams that are either intermittent or go completely dry will have a group of fish species that are the first to reinvade after water has returned.

Simon and Lyons (1995) described the IBI as a family of multimetric indices adapted to a variety of regional settings. A number of other IBIs use the "percentage of pioneering species" to represent situations that reflected water quantity diversions or loss through agricultural withdrawal, wetland desiccation, and situations where a stream becomes subterranean (Simon, 1991; Goldstein et al., 1994; Simon, 1998; Simon and Stewart, 1998; Niemela et al., 1999; Simon et al., 1999).

Few studies other than Smith's (1971) have been conducted to determine whether the temporal trends observed in that small Illinois stream have application for other uses. The pioneer species metric has also been applied to situations where toxicity has caused the deaths of resident fishes and "pioneers" were the first to re-invade larger streams. In the broadest sense, pioneer species could be found in small to large watersheds where toxic disturbance is iterative and could cause loss of resident aquatic assemblages.

This chapter evaluates the pioneering species metric in relation to habitat and other stressors tracked in Ohio EPA's statewide database and in a number of site-specific situations that represent a range of stressor types including a manure spill, a mine spill with low pH and high metal levels, toxic runoff from a tire fire, and a long term exposure to ammonia and a mixture of industrial toxicants. The purpose of this chapter is to describe (1) the patterns of response of the pioneering species metric to habitat gradients as measured by the QHEI, (2) the return of pioneering fish species after the elimination or near-elimination of fish communities by various stressor types from several small Ohio streams, and (3) the implications for pioneering species metrics in IBI indices.

10.2 MATERIALS AND METHODS

10.2.1 Study Area

10.2.1.1 Statewide Data

The Ohio Environmental Protection Agency (OEPA) uses biological, chemical, and physical monitoring and assessment techniques in biosurveys to meet three major objectives: (1) determine the extent that designations assigned in the Ohio water quality standards (WQS) are attained or not attained; (2) determine whether use designations assigned to a given water body are appropriate and attainable; and (3) determine whether any changes in key ambient biological, chemical, or physical indicators have occurred over time, particularly before and after the implementation of point source pollution controls or best management practices. The statewide fish community database (OhioECOS) compiled from this process is comprised of data collected primarily with pulsed-DC electofishing methods across Ohio over the past 23 years following standard protocols set for Ohio EPA's stream sampling efforts (Ohio EPA, 1989).

The primary sources of data include the Ohio EPA Ecological Assessment Section, Ohio Department of Natural Resources (ODNR) in the Division of Natural Area and Preserves and the Division of Wildlife, the Ohio Department of Transportation (ODOT), universities in Ohio, consulting firms, and other entities. The entire pool of data consists of more than 17,000 samples. Subsets of this database include data on least impacted (reference) sites and physically modified reference sites that did not have point sources or acute impacts from livestock, spills, etc, but may have had elevated nutrients related to stream modification and encroachment or loss of riparian areas. Other subsets were created by querying the database by stream size (drainage area, square miles) and IBI score.

Habitat data using the qualitative habitat evaluation index (QHEI) are typically collected at each fish community sampling site. Methods and application of the QHEI are described in Ohio EPA (1989) and Rankin (1989, 1995). The QHEI is a visual index that estimates habitat quality within six major categories of macrohabitat: substrate condition, instream cover, channel condition, riparian width and bank erosion, pool and riffle quality and stream gradient adjusted by river size. Overall scores range from about 10 to 100. Individual habitat components have been associated with IBI scores at reference sites (natural and physically modified) and certain attributes are associated with higher and others with lower IBI scores (Rankin 1989, 1995).

10.2.1.2 Site-Specific Examples

Six streams whose fish assemblages were mostly or completely eliminated by spills, discharges, and other anthropogenic disturbances were selected. Recovery data and, in some cases, pre-impact data were examined to describe the effects of different stressors on pioneering species and patterns of recolonization (Figure 10.1). A brief description is provided for each of these streams and their stressors.

Lobdell Creek is a small wadeable stream in the Licking River watershed in Licking County. A large egg production operation is located in the upper reaches of the stream. A large spill of manure and other waste products from one of the large production barns on this property in May 1999 caused an extensive fish kill.

The upper portion of Lobdell Creek (upstream of RM 10.3) has been channelized, has poor habitat (QHEI range of 33 to 40), and is maintained in this condition under the Ohio drainage law as a petition ditch. The creek is designated a modified warmwater habitat (MWH). The habitat downstream of RM 10.3 is considered good (QHEI range of 63 to 75) and suitable to support the higher quality warmwater habitat (WWH) aquatic life use. In addition to the assemblage information that Ohio EPA collected in 1999 and 2000 after the manure spills, historic fish community data existed from an unpublished study of the Raccoon Creek watershed (that includes Lobdell Creek) made by Raymond Jezerinac between 1972 and 1974.

Hurford Run is a headwater stream in Canton, Stark County in northeast Ohio (Figure 10.1) with a drainage area of 8 sq. mi. Its watershed is in an urban setting with large industrial facilities and railroad yards adjacent to much of its length. It has a history of severe impairment from several point sources, including a steel operation and a refinery, and "legacy" impacts from contaminated sediments. The upper reaches have limited resource water (LRW) and modified warmwater habitat (MWH) aquatic life uses reflecting substantial habitat limitation, although the lower mile of the stream has better habitat and is designated a WWH stream. The site we examined is this WWH reach. While no "pre-discharge" data existed, the stream had a history of acute toxic impacts documented since the 1980s and some recovery in the late 1990s.

Sycamore Creek is a wadeable stream (drainage area = 67 sq. mi.) in the Sandusky River watershed that flows through Crawford and Wyandot counties in north-central Ohio. It has natural channel conditions and a well established riparian corridor with good habitat conditions (mean QHEI = 73.1). The stream is designated as a WWH aquatic life use (Ohio EPA, 2000). A large fire occurred at a tire recycling facility in August 1999 and subsequent rainstorms caused a spill of

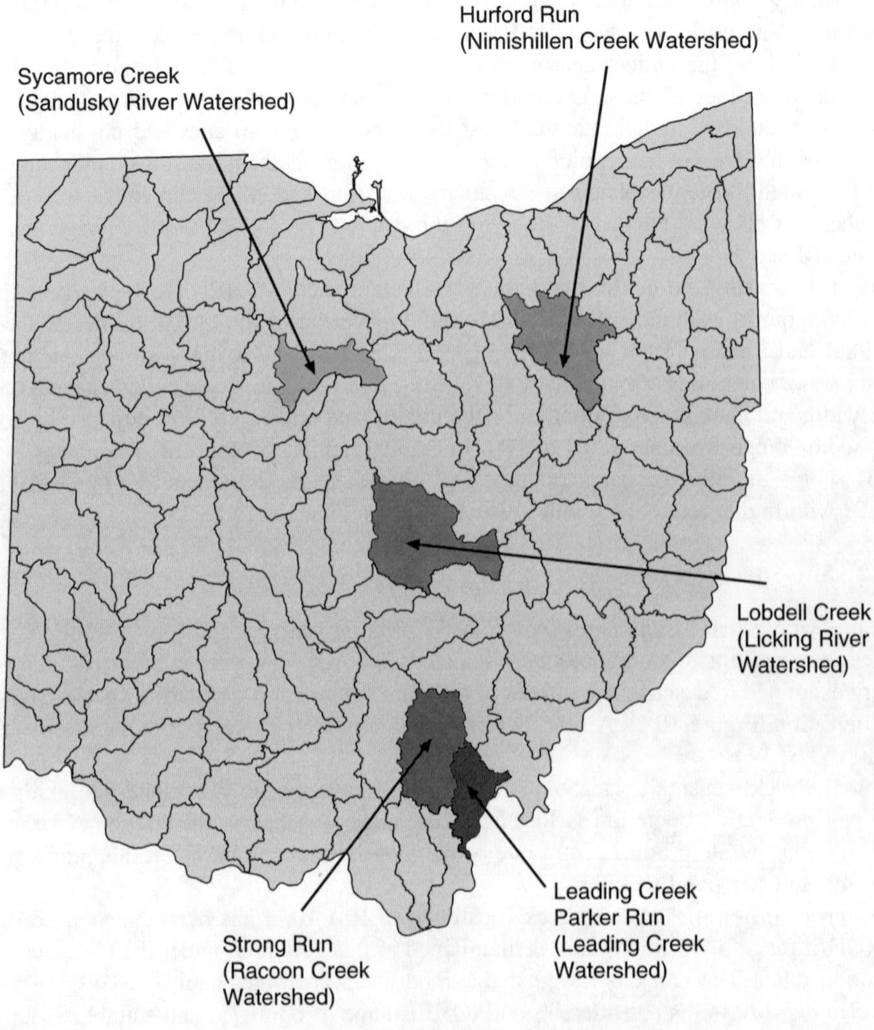

FIGURE 10.1 Study areas in Leading Creek and Raccoon Creek, tributaries of the Ohio River in Meigs County, southeastern Ohio. The creeks received acid mine effluent from the Meigs 31 Mine collapse in September 1993.

water contaminated with toxicants from melted tires. The spill eliminated most of the fish and macroinvertebrate communities for about 2.5 miles and had lesser impacts to the mouth of Sycamore Creek.

Meigs 31 Mine Spill — Leading Creek, Parker Run, and Strongs Run were among a number of streams affected by a large discharge of toxic mine water in southeast Ohio. The collapse and pumping of the Meigs 31 Mine in July 1993 was considered one of the largest environmental disturbances in southeastern Ohio. The mine pumped over 1.1 billion gal of low pH water, dissolved metals, and mine-associated contaminants into Parker Run, a tributary of Leading Creek, and into Strongs Run, Robinson Run, and Sugar Run, all tributaries of Raccoon Creek, to drain the largest long-wall coal mine in the United States. Pumping continued for 3 months (July through September 1993) and at one instant turned the Ohio River orange from the iron flocculent that moved through Leading Creek. A number of stations were selected in these affected streams in the Raccoon Creek and Leading Creek watersheds to monitor changes in the system based on fish and macroinvertebrate assemblage indicators (Ohio EPA, 1994). Patterns in these indicators (ecological endpoints) were used to determine when the streams eventually recovered to their previous (pre-discharge) conditions or warmwater habitat (WWH) designated use goals where pre-discharge data was unavailable (Ohio EPA, 1994; Simon et al., Chapter 4, this volume).

Leading Creek is a third order stream with a drainage area of 150 sq. mi. that begins in Athens County and flows southeast through Meigs County where it enters the Ohio River. It was a reference stream and state resource water for the Western Allegheny Plateau ecoregion. In addition, it harbored populations of sensitive species, such as the silver lamprey *Ichthyomyzon unicuspis*, a state threatened species. Nine stations were longitudinally distributed on Leading Creek including a single station upstream of the junction with Parker Run, the primary outlet for mine discharge. Two stations were established on Parker Run to monitor changes resulting from the mine effluents. Historical data from the 1980s existed for a site on Leading Creek at RM 10.3 that possessed good habitat conditions. In Parker Run, we examined a site (RM 1.5) where data were collected immediately before the discharge in 1993.

Raccoon Creek flows south from Hocking and Vinton Counties into Galia County where it enters the Ohio River. The mine pumped water into a number of small headwater streams including Sugar Run, Strongs Run, and Robinson Run. Pre-discharge data existed for a site at Strongs Run (RM 0.6).

10.2.2 Sample Methods

10.2.2.1 Statewide Database

Fish community data used in the statewide analysis were primarily collected with pulsed DC electrofishing methods following Ohio EPA protocols (Ohio EPA, 1989), using either a towboat or longline method and generally a T&J 1750 generator capable of supplying 125 or 250 volts. Catch-per-unit-effort was standardized by sampling distance (per 300 m) with time dependent on habitat complexity although minimum suggested sampling times are identified. Pioneering species (Table 10.1) were those identified in the state's IBI guidance (Ohio EPA, 1989), which were derived from the work of Smith (1971) in Illinois. Stream size for each site was measured as drainage area (sq. mi.) above that site.

TABLE 10.1
Pioneer Species Used in the Headwater IBI by Ohio EPA (1989) to Evaluate Intermittent Headwater Streams in Central Illinois

Common Name	Scientific Name
Creek chubsucker	*Erimyzon oblongus* (Mitchill)
Creek chub	*Semotilus atromaculatus* (Mitchill)
Silverjaw minnow	*Ericymba bucatta* Cope
Bluntnose minnow	*Pimephales notatus* Rafinesque
Fathead minnow	*Pimephales promelas* Rafinesque
Green sunfish	*Lepomis cyanellus* Rafinesque
Johnny darter	*Etheostoma nigrum* Rafinesque
Orangethroat darter	*Etheostoma spectabile* (Agassiz)

Source: Based Smith, P.W. 1971. Illinois Natural History Survey Biological Notes 76. With permission.

10.2.2.2 Site-Specific Examples

Most data used in the site-specific examples were collected via the methods outlined above, except for certain samples in the Meigs Mine stream system. Data include samples collected by the state and by consultants hired by the Southern Ohio Coal Company (SOCCO). As a result of a court settlement, all biologists responsible for sampling the aquatic biota of Leading Creek followed standard sampling procedures established by Ohio EPA (1989). Sampling was conducted two or three times per year by the consultants and typically once during the fall by the agencies. These procedures included distance-specified sampling for fish using tote barge electrofishing equipment. However, because conductivity in these streams varied considerably with discharge quality from the original spill, contributions from upstream unreclaimed areas, and inputs from an existing NPDES mine discharge, a large number of samples were collected using seining and with a more powerful Smith-Root 3.5 GPP electrofishing unit (3500 watts). The use of these alternate gear types likely resulted in more variability in abundance estimates for pioneering species. In addition, consultant-generated data consisted of "estimated" numbers of certain abundant species that may also add error to the estimates.

10.3 RESULTS AND DISCUSSION

The percent of individuals as pioneering species parameter has been used in a number of IBI derivations, but little assessment of this metric has been done beyond solid but anecdotal observations by biologists. It is useful to understand how this metric responds to various stressors in Midwest streams of various sizes since the original use focused on species that reinvade ephemeral streams at small drainage areas (Smith, 1971). The statewide analysis provides a robust examination of the trend in pioneering species response to stressors such as habitat degradation, while site-specific responses over time or space provide a more detailed assessment of pioneering species responses to various categories of multiple stressors (i.e., enrichment, mine effluent, toxic chemicals). This chapter does not examine the responses of specific species included in this metric although that would also be useful.

10.3.1 Statewide Database

The percent of individuals as pioneering species provides useful information on the responses of fish communities to ephemeral conditions in Ohio. The percent of the community as pioneering individuals is negatively correlated with the IBI in small Ohio streams (Figure 10.2; REF streams, drainage area = ≤30 sq. mi.). The relationship with IBI is strongest at the best sites and worst sites (as measured by the IBI); data at sites of intermediate quality show the greatest variability. Sites with low IBIs in this dataset have poorer habitat, are more likely to be nutrient enriched, and have poorer substrates than better sites. Increasing proportions of pioneering individuals are associated with other "negative" metrics such as percent of omnivores and, similar to the IBI, are negatively associated with positive metrics such as the number of headwater species, percent insectivores, and sensitive species (Table 10.2).

10.3.1.1 Pioneering Species and QHEI Habitat Metrics

The percent of individuals as pioneering species is negatively correlated with overall habitat quality in small Ohio streams. A box plot of pioneering species by ranges of the QHEI (Figure 10.3) show that sites with high QHEI scores (>80) have very diverse habitats and aquatic communities. The lack of scatter at the highest QHEI scores (>80) that are typically associated with the exceptional warmwater habitat (EWH) aquatic life use in Ohio illustrates the strong effect of habitat as a controlling factor for aquatic communities in small Ohio streams. In these high quality small streams, it is possible that the pioneering species do not compete as well as habitat specialist species

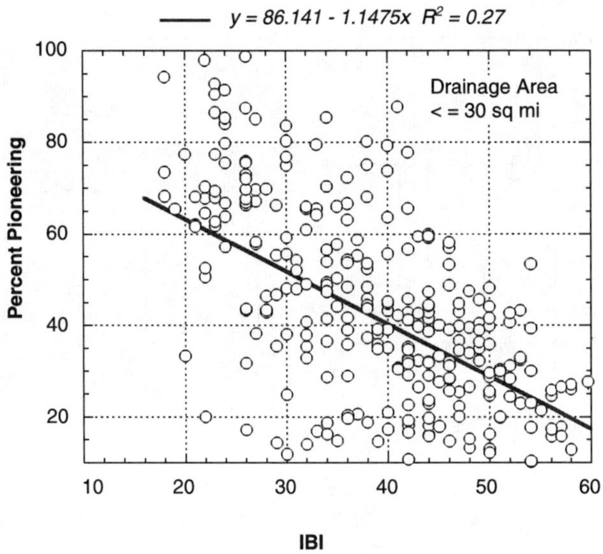

FIGURE 10.2 Correlation between the IBI and the percent of the catch as pioneering species at Ohio reference (least impacted and physically modified) sites in streams with drainage areas less than or equal to 30 sq. mi.

TABLE 10.2
Correlation of percent of individuals as pioneering species and several IBI metrics from 554 samples from reference and modified reference sites in Ohio in headwater streams (drainage area = <20 sq. mi.). Data collected between 1978 and 2000.

Metric	Percent of Individuals as Pioneering Species
Percent insectivores	–0.34
Percent omnivores	+0.52
Number of darter species	0.00
Total species	–0.10
Percent tolerants	+0.69
Number of sensitive species	–0.27
Number of headwater species	–0.32
Percent simple lithophil species	–0.48
Number of simple lithophil species	–0.27

or perhaps are eliminated by predators that are generally more common in streams with good habitats.

This relationship between pioneering species and habitat is reflected in most individual QHEI metrics, although variability is somewhat higher. Box and whisker plots of pioneering individuals with QHEI channel scores, substrate scores, cover scores, and riparian scores show that while pioneering species can reach high populations at upper ends of any individual metric (Figure 10.4), the sites where most of the high ranges of each metric co-occur (Figure 10.3, QHEI ≥80) represent sites where "generalized" niches are fewer and specialist species can compete well. Another consideration is that perhaps pioneering species are exposed to high rates of predation. Thus, while

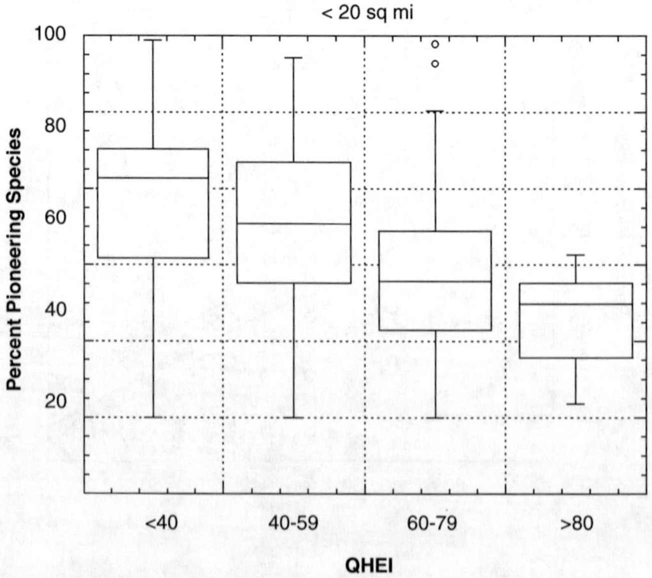

FIGURE 10.3 Box and whisker plot of the percent of the catch as pioneering species for ranges of the QHEI at Ohio reference sites in streams draining less than 30 sq. mi.

we can get samples with high proportions of pioneering species with high riparian scores alone, we rarely observe these high proportions where most of the combined habitat metrics indicate excellent habitat.

This pattern can also be observed by examining the proportion as pioneering species versus the number of warmwater habitat attributes (Figure 10.5). These attributes are often correlated with high IBI scores (Rankin, 1989). The more warmwater habitat features at a site, the higher the QHEI and the smaller the number of generalized habitat conditions. Sites with high numbers of these warmwater attributes rarely have more than 30 to 40% pioneering individuals.

10.3.1.2 Watershed Scale Effects on Pioneering Species

Habitat influences fish communities at multiple geographic scales. In certain large watersheds, such as the Maumee River watershed in Ohio, a large number of species have been extirpated from the drainage (Karr et al. 1985), and this constrains the pool of species for possible community recovery. Most extirpations are related to the substantial change in the habitat from the drainage of wetlands and loss of clear flowing rivers to increases in channelized agricultural ditches and turbid rivers.

We calculated the median watershed QHEIs for small streams in each of 93 watersheds in Ohio and correlated them with various upper percentiles of IBI scores as an estimate of the best attainable fish communities in these waters (Figure 10.6). More habitat degraded watersheds, as measured by the median QHEIs of all small stream QHEI scores in these watersheds, resulted in significantly lower "high-end" IBI percentiles (e.g., 90th and 95th percentiles). This suggests that widespread habitat loss affects the achievement of high IBI scores. For example, in a watershed where the median QHEI score is 60, the expected 95th percentile IBI is about 48, whereas in a watershed with a median QHEI of 30, the expected 95th percentile IBI is approximately 38 (Figure 10.6). In essence, the slope of this regression is the "slippery" one of cumulative habitat impacts. Although the analysis suffers from a non-random sample of habitat scores and uncertainty about the influence of other factors (e.g., differences in background chemistry among watersheds), it matches the strong anecdotal evidence of naturalists such as Milton Trautman who linked the decline of distribution in many fish species in Ohio to siltation and habitat loss (Trautman, 1981).

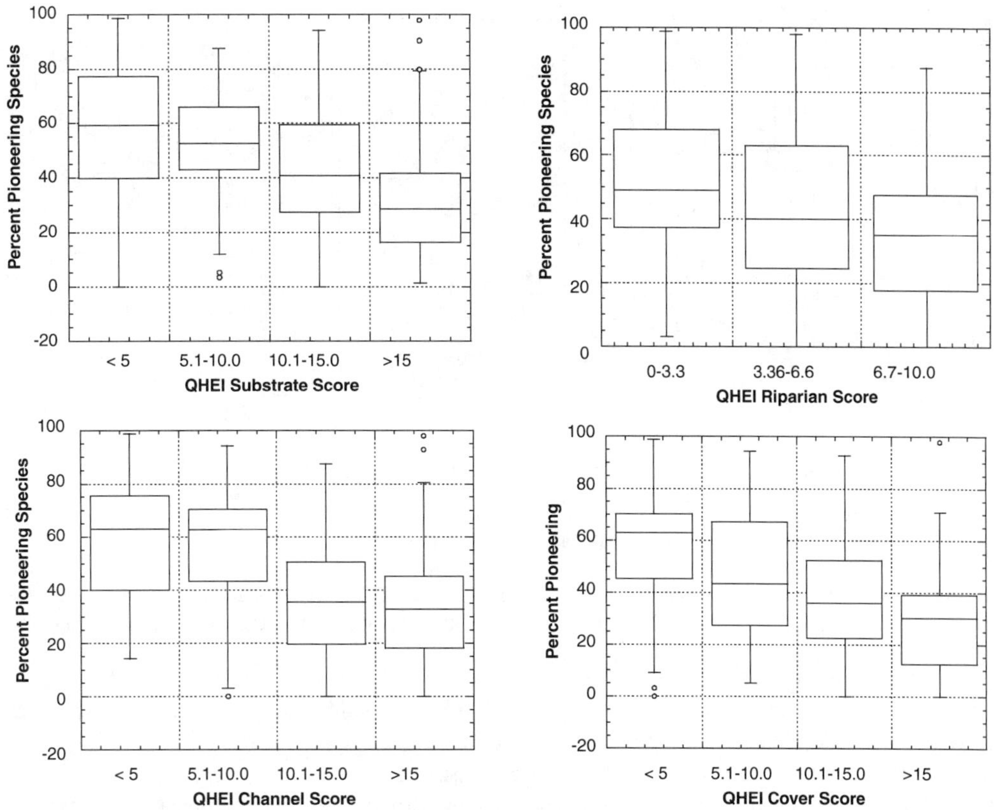

FIGURE 10.4 Box and whisker plots of the percent of the catch as pioneering species for ranges of the QHEI channel score, substrate score, cover score, and riparian score at Ohio reference sites in streams draining less than 30 sq. mi.

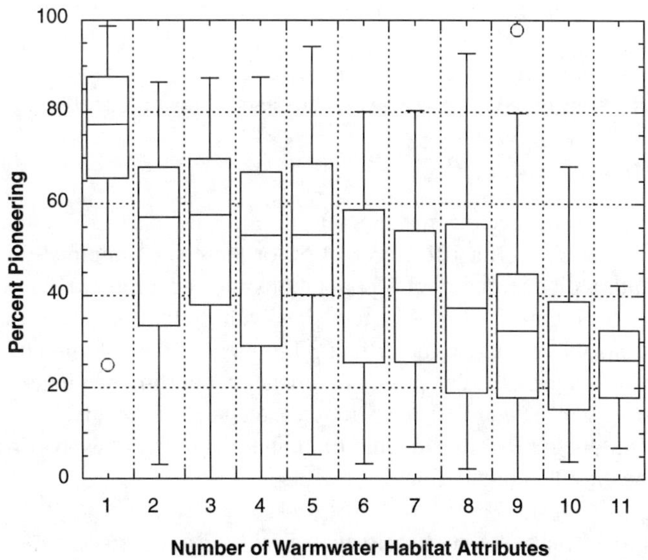

FIGURE 10.5 Box and whisker plots of the percent of the catch as pioneering species by number of warmwater (high quality) habitat attributes at Ohio reference sites in streams draining less than 30 sq. mi..

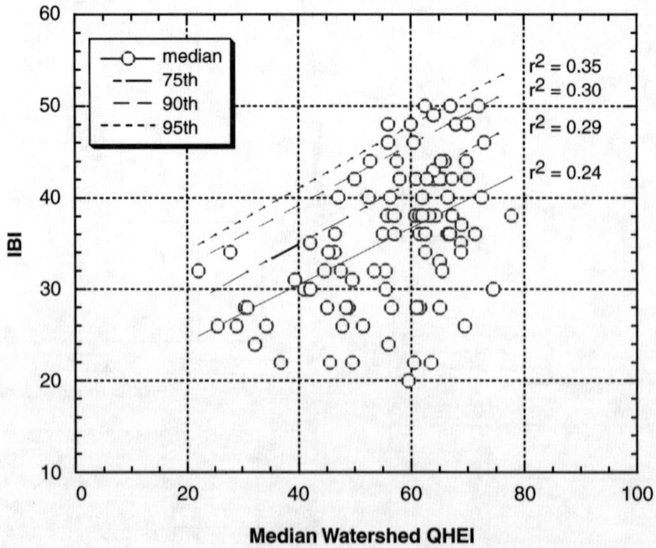

FIGURE 10.6 Correlation between median watershed habitat QHEI scores and median IBI scores and regression lines for 75th percentile, 90th percentile, and 95th percentile IBI scores in these same watersheds.

Watersheds with degraded habitats (low median QHEIs) were more often predominated by pioneering species and populations. Watersheds with higher quality habitats (higher median QHEIs) had fewer sites dominated by pioneering species and more with ranges of pioneering species between 30 and 40% (Figure 10.7). Although we suspected that the best individual small stream sites (QHEI scores >80) were always in watersheds with the highest median QHEIs, we were surprised that this was not always the case (Figure 10.8). Individual sites that score 80 or more were found in watersheds with median QHEIs between 45 and 70, although not where watershed medians were below 45. Such sites (individual QHEIs >80) rarely have high pioneering species populations, which suggests that even where overall watershed habitat quality is low, sites with very high habitat scores can have smaller populations of pioneering species. If some of the most sensitive insectivores (e.g., rosyface shiner *Notropis rubellus*, bigeye chub *Hybopsis amblops*) are eliminated, more tolerant (in a relative sense) insectivores and other non-pioneering species occur in higher proportions than pioneering species where habitat is still excellent.

10.3.2 Longitudinal Trends

Pioneering species generally decline with increasing drainage size in unimpacted streams. In a database of Ohio stream sites with IBI scores of 50 or more (i.e., minimally impacted or having good–excellent habitat), pioneering species populations showed a logarithmic decline as a proportion of the community (Figure 10.9a). Very few sites with drainages greater than 30 sq. mi. have pioneering populations that comprise more than 30 to 40% of the community. Sites that score in the poor to very poor range (IBI ≤26) possess a greater proportion of pioneering species even at larger stream and river sizes (Figure 10.9b), with the exception of great rivers (e.g., the Ohio River, >10,000 sq. mi.). Thus, degraded sites in Ohio often, but not always, respond with increases in the proportions of pioneering fish populations.

10.3.3 Site-Specific Responses of Pioneering Species Populations to Stressors

Although pioneering species often increase in relative abundance compared to more sensitive species at sites affected by various stressors, based on patterns observed in the statewide analysis,

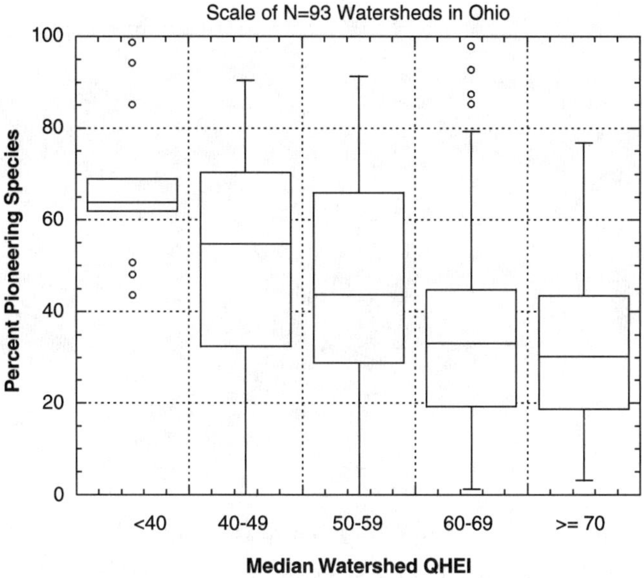

FIGURE 10.7 Box and whisker plot of the percent of the catch as pioneering species for ranges of median watershed QHEIs scores in these watersheds.

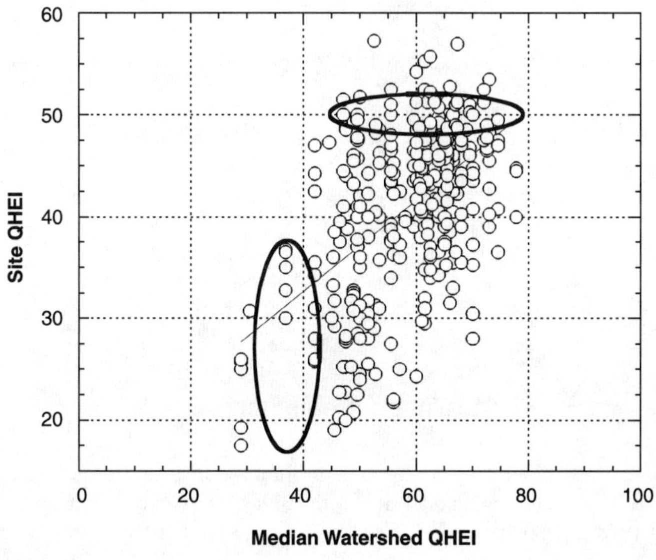

FIGURE 10.8 Correlation between median watershed habitat QHEI scores and sites specific habitat quality in small streams in these watersheds.

this is not a universal pattern. Poor and very poor sites sometimes have few pioneering species (see Figure 10.9b). Here we look at several site-specific case histories of spills and other impairments in which we observed recovery of communities over various time periods to examine trends in pioneering species populations.

10.3.3.1 Responses to Toxic Levels of Nutrient Additions: A Manure Spill

The first site-specific example is from a site in Lobdell Creek. The fish community in the upper reaches of this warmwater stream was essentially eliminated in 1999 due to a spill of manure from

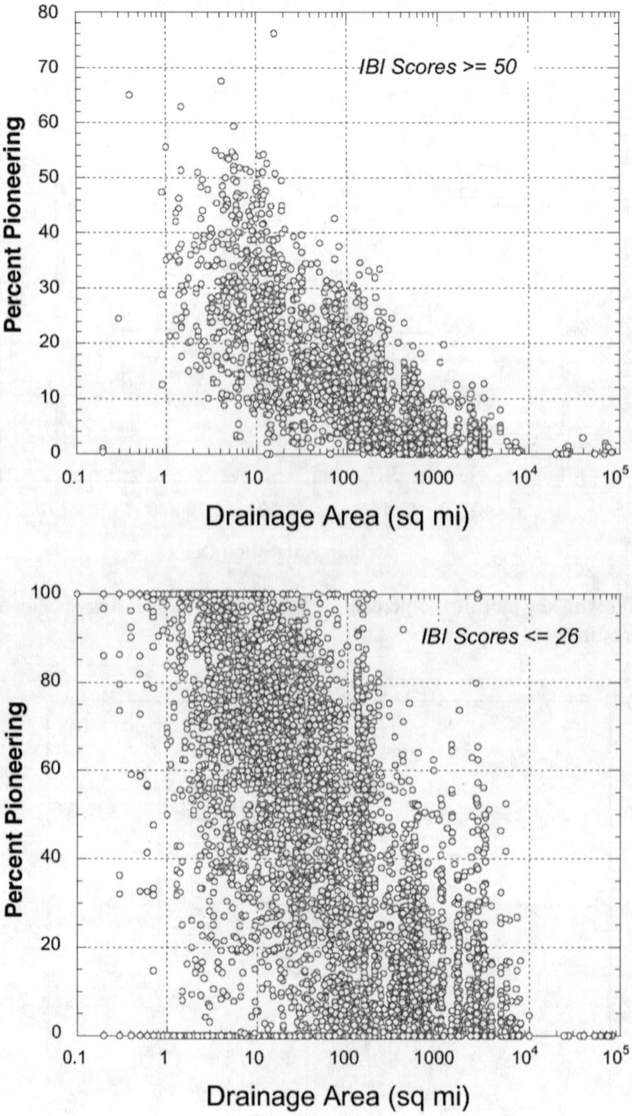

FIGURE 10.9 Percent of catch as pioneering species versus drainage area for all sites in the Ohio EPA fish database with IBI scores of 50 or more (top) and all sites with IBI scores of 26 or less (bottom). Data collected in Ohio streams between 1978 and 2000.

chickens in a large egg-laying facility. Some pre-discharge data was collected in 1998 from the upstream-most site (drainage area = 3.7 sq. mi.) and a single sample was collected downstream at the 9.8 sq. mi. drainage area site in 1972. The remaining data were collected after the spill. Figure 10.10 shows changes in the percent of individuals as pioneering species (top) and the IBIs (bottom) for three sites at various distances downstream from the source.

The IBI plot shows that all sites scored the minimum (12 points) immediately after the spill. The downstream sites recovered faster because of proximity to refugia, greater dilution, and increased distance from the spill location. The responses of pioneering species populations differed among sites. The small headwater site (drainage = 3.7 sq. mi.) was immediately re-invaded in 1999 by pioneering species (Figure 10.10, top). It contained only pioneering species in 1999 and the first

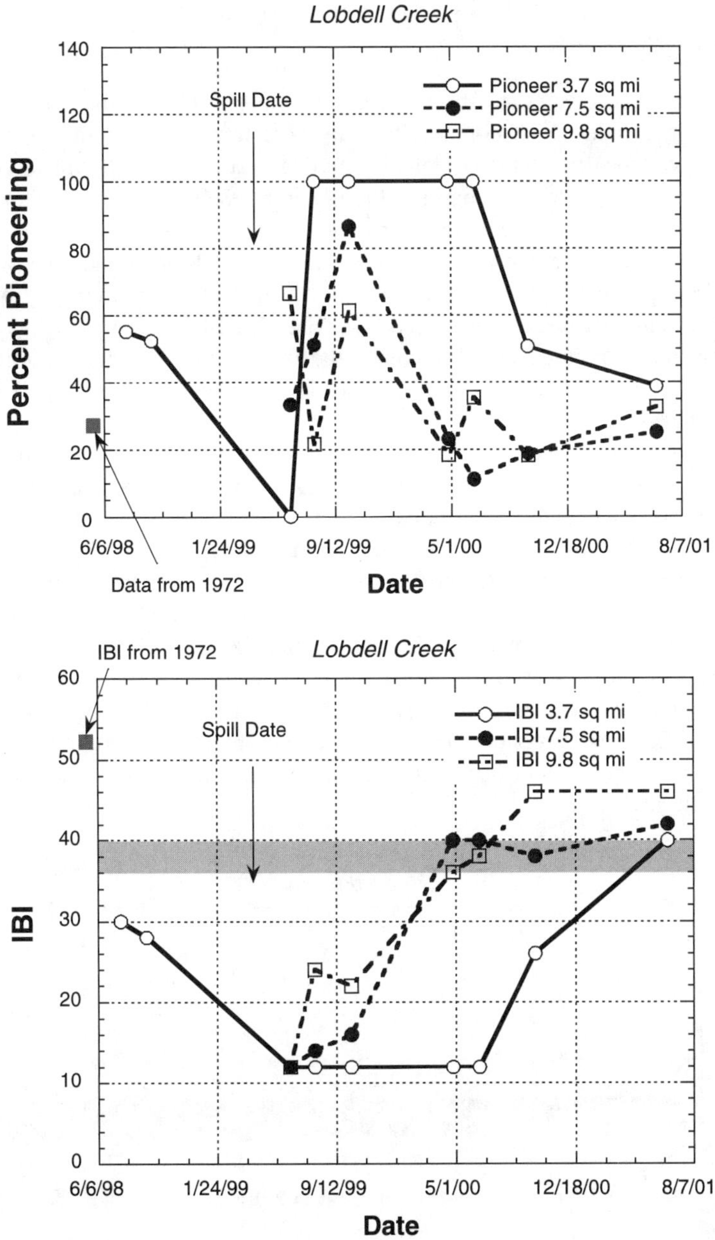

FIGURE 10.10 Percent of catch as pioneering species (top) and IBI scores (bottom) over time at three sampling locations in Lobdell Creek before and after a spill of chicken manure in 1999. Data from 1972 was collected by Raymond Jezerinac.

half of 2000, when non-pioneering species began to inhabit the site. The other sites showed different patterns with higher than background or predisturbance pioneering proportions after the spill and a more rapid return to background conditions for these species. The IBIs showed a similar pattern as the percent of pioneering species. Thus, for a manure/enrichment spill impact, the most headwater site showed the classic immediate response of pioneering species. The downstream headwater sites also followed this pattern, although they were less dominated by pioneering species immediately after the spill. This was likely due to nearby refugia and migration from downstream.

10.3.3.2 Hurford Run: A Case Study of an Industrial Source of Toxic Chemicals

Hurford Run is a small headwater stream in northeast Ohio near Canton that has been impaired by toxic chemicals for decades. Fish community data were first collected in the mid 1980s. The upstream reaches were nearly devoid of aquatic life due to a complex mix of stressors including toxic effluents of several industrial dischargers, thermal impacts, historic impacts, and contaminated sediments. Although habitat was degraded in reaches, it was considered adequate to support the MWH aquatic life use. The use was not attained in the mid 1980s and conditions were considered very poor. Fish communities in the upstream reaches were still considered very poor in 1999 in the vicinity of the industrial discharges (RM 1.8, IBI = 12; RM 2.1, IBI = 14), as were macroinvertebrates collected at the same sites. However, at the creek mouth, fish communities now achieve the WWH biocriteria for this ecoregion (Figure 10.11), although macroinvertebrates have only improved to the fair range.

The trend in pioneering species varied inversely with the IBI. When toxic impacts extended downstream to the mouth in the mid 1980s, the community was dominated by pioneering species. In 1998, with IBI scores achieving ecoregional expectations for headwater streams, the proportion of pioneering species was low. A common pattern of recovery from gradual abatement of point sources is a shrinking of the impairment "footprint" from downstream to upstream. Discharge quality is still poor enough to have severe effects on communities adjacent to the discharge, but impacts from toxicants are assimilated or volatize to reduce the impairment in downstream reaches. In the case of Hurford Run, it is likely that organic toxicants and ammonia contributed to the impairment in the mid 1980s and are still likely causes of impairment. The fish community structure pattern in Hurford Run has probably been constant elimination and migration of the aquatic community from a watershed pool of mostly tolerant and pioneering species from upstream and nearby small tributaries.

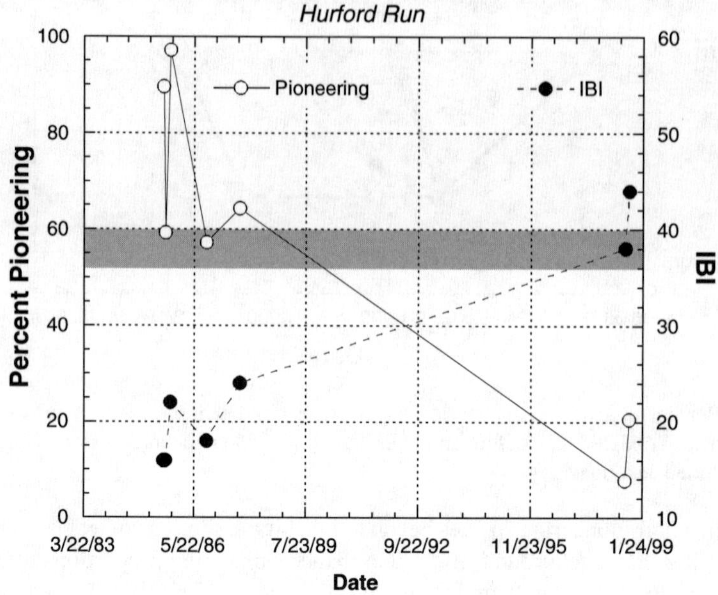

FIGURE 10.11 Percent of catch as pioneering species and IBI scores over time at the mouth of Hurford Run in Canton, Ohio.

10.3.3.3 Sycamore Creek: The Effects of a Non-Point Source Urban Spill

Sycamore Creek is a tributary of the Sandusky River in northwestern Ohio that suffered a spill of oil released from tires that melted in a large fire at Kirby's Tire Recycle, Inc. on August 21, 1999. Acutely toxic levels resulted from releases of anthracene, fluoranthene, naphthalene, copper and zinc, high biochemical oxygen demand, and chemical oxygen demand (Ohio EPA, 2000). Between August 25 and 27, dissolved oxygen in the lower 7 mi. of the stream was almost completely depleted. This wadeable stream had good habitat quality (mean QHEI = 73.1) for 6 mi. in the vicinity of the spill (RM mile 7.0). Figure 10.12 illustrates a longitudinal trend in pioneering species (top) and IBI (bottom) during four different sampling events in 1999 and 2000.

Pioneering species were low (<30%) upstream of the tire fire spill, as would be expected in a stream with good habitat that meets the WWH biocriteria for wadeable streams in that ecoregion. The tire fire eliminated most fish and those that survived or recolonized were not pioneering species. They were subsets of the range of insectivores and other species expected in that stream. No sudden increase in pioneering populations occurred in response to the fish kill. Sycamore Creek differed from Lobdell, Strongs Run, and Hurford Run because it was larger and had better quality habitat than those streams where we observed an early influx of pioneering species in lieu of other species.

10.3.3.4 Meigs Mine 31: A Large Spill of Contaminated Mine Water

In 1993, Ohio EPA permitted the discharge of over a billion gallons of mine-affected water to streams in the Leading Creek and Raccoon Creek watersheds. The aquatic communities were essentially eliminated in Parker Run, Leading Creek (downstream of Parker Run), and Strongs Run and communities were reduced in several others. Although the fish kills were catastrophic, the incident provided an opportunity to examine patterns of recolonization of fish and macroinvertebrate communities in a number of stream types, including the recolonization of pioneering species. This event also resulted in the derivation of explicit ecological recovery criteria based on Ohio's biocriteria, background community assemblage data, and a set of key aquatic species that needed to be restored to the impaired streams (Ohio EPA, 1994, Simon et al., 2002).

10.3.3.4.1 Parker Run

Parker Run received the greatest direct discharge of polluted mine water and continues to receive treated mine effluent from the operation of Meigs Mine 31. Fish communities were eliminated from reaches downstream of the discharge for about 2 mi. in June of 1993. This impacted the downstream confluence with Leading Creek for a distance of 15.6 mi. until it reached the Ohio River. In addition, Leading Creek upstream was impaired from where mine effluent pushed upstream into a low gradient pool. We will examine the impact trend in Parker Run at RM 1.6, which is a short distance downstream of the discharge point at a county road bridge. Data are also presented for Malloons Run, a similar sized, unimpacted headwater stream located nearby and also a tributary of Leading Creek.

The aquatic community was eliminated downstream of the discharge point and pioneering fish remained absent from the site for the rest of 1993. Refugia for Parker Run recolonization was limited with the poor quality of the larger Leading Creek acting as a block to migration. The site upstream of the discharge point in Parker Run and Little Parker Run was small and interstitial. It had fish communities and served as a minor source for immediate recolonization. In addition, the continuing discharge of mine effluent with high solids and conductivity and lower temperature may have been avoided by migrating fish.

Instead of the large, immediate influx of pioneering species observed at Lobdell Creek, Parker Run showed a slow gradual increase of pioneer populations (Figure 10.13). Many of the initial invaders were Ohio River species such as channel shiner *Notropis wickliffi* and gizzard shad *Dorosoma cepedianum*, which generally migrate from the Ohio River into streams in the fall. It is likely that the slow increase of pioneering species was a response to spawning movements and

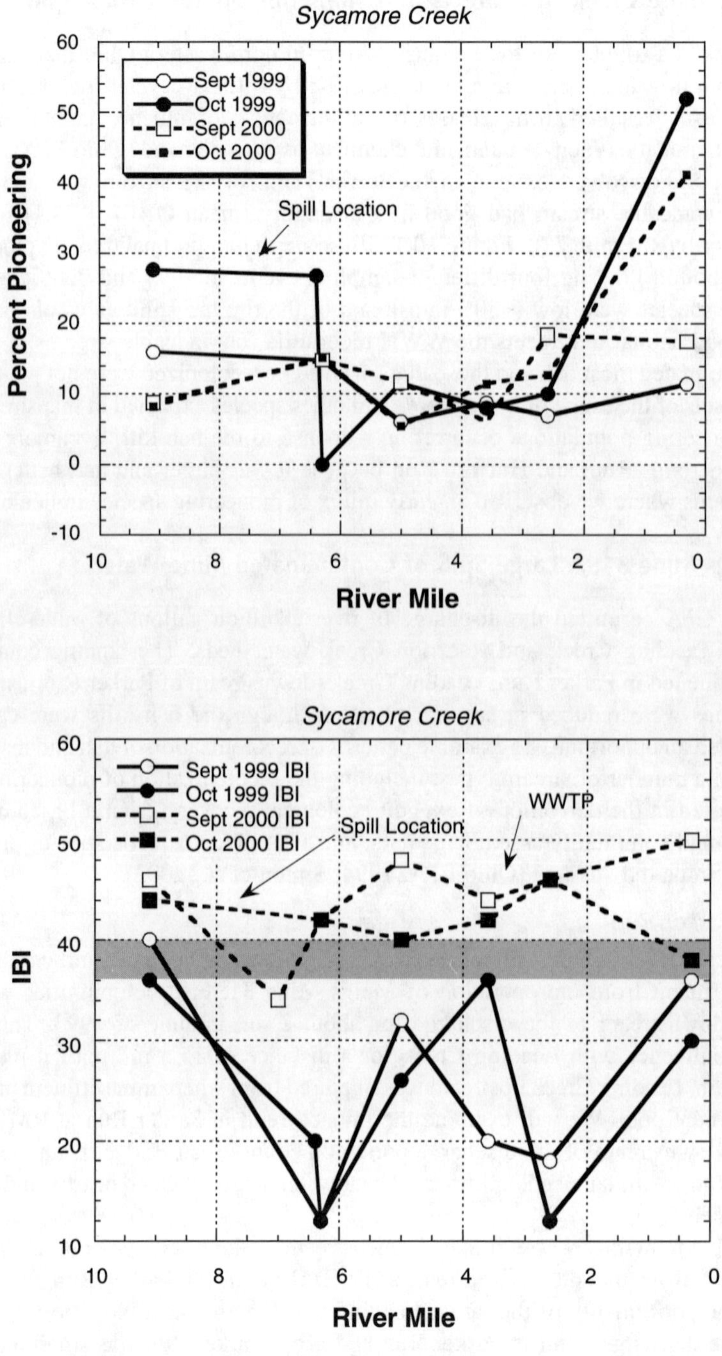

FIGURE 10.12 Longitudinal plot of percent of catch as pioneering species (top) and IBI scores (bottom) from upstream (left) to downstream (right) for four sampling passes over summers of 1999 and 2000 in Sycamore Creek near Sycamore, Ohio.

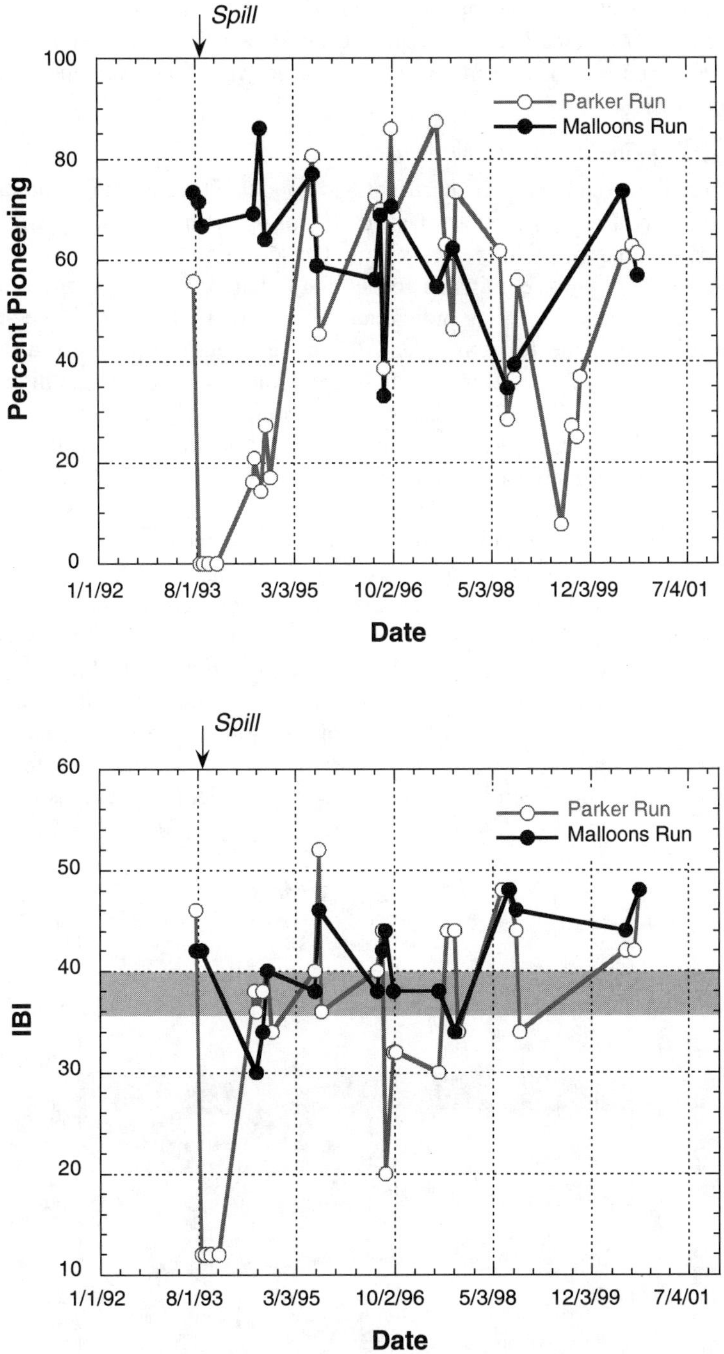

FIGURE 10.13 Percent of catch as pioneering species (top) and IBI scores (bottom) over time at a site in Parker Run at river mile 1.6 and at a nearby site on a stream not affected by the discharge (Malloons Run).

production in 1994 and later and perhaps was hindered by the existing effluent and periodic flood and drought events. The fluctuations in percent of pioneering fish in Parker Run were similar to those in the unimpacted Malloons Run, but showed wider variations, perhaps related to ongoing discharge (Figure 10.13).

10.3.3.4.2 Leading Creek at River Mile 10.3

This site was chosen for trend analysis because it had good habitat conditions and historic pre-spill fish collection data. With the exception of an early spring sample, Leading Creek at RM 10.3 contained 20 to 35% pioneering fish species prior to the spill (Figure 10.14, top). Immediately after the spill, the fish community was eliminated and similar to Parker Run, the proportion of the limited catch collected at this site as pioneering individuals (Figure 10.13, top) did not begin reaching pre-discharge proportions until the following year. A similar pattern was observed in the IBI (Figure 10.14, bottom). As with Parker Run, Leading Creek did not have a large, high quality stream that could quickly repopulate. The Ohio River did not contain a large source of the dominant fish species that could repopulate Leading Creek. In addition, mine effluent remained and increased in volume as the mine waters stored in the old mine were treated and discharged. This resulted in conductivities and total dissolved solids higher than pre-discharge conditions.

The lack of recovery is evident in the slow increase in biomass (Figure 10.15). Even in 1997, after 4 years of recovery, biomass was much less than pre-discharge. Rather than a situation where large populations in nearby unaffected waters can emigrate quickly, the pattern of recolonization was more likely a mix of emigration, growth, and reproduction slowed by the existing discharge and other impacts in the watershed (e.g., mining, agriculture, and livestock uses).

The natural community in Leading Creek, a WWH aquatic life use stream, was not dominated by pioneering species, but by insectivores. The pattern was slow recovery of pioneer species, instead of a large invasion of pioneering species. This pattern is likely a result of watershed size at the location we examined and the lack of nearby refugia for sufficient populations of species.

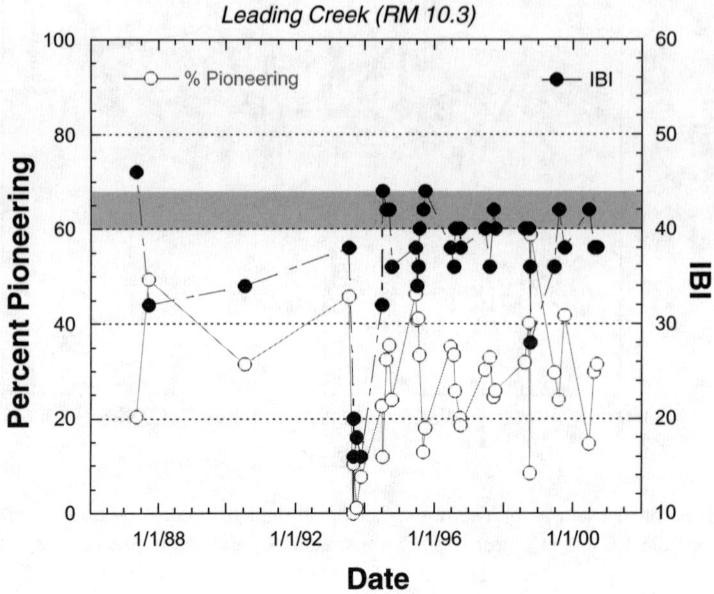

FIGURE 10.14 Percent of catch as pioneering species (top) and IBI scores (bottom) over time at a site in Leading Creek at river mile 10.3.

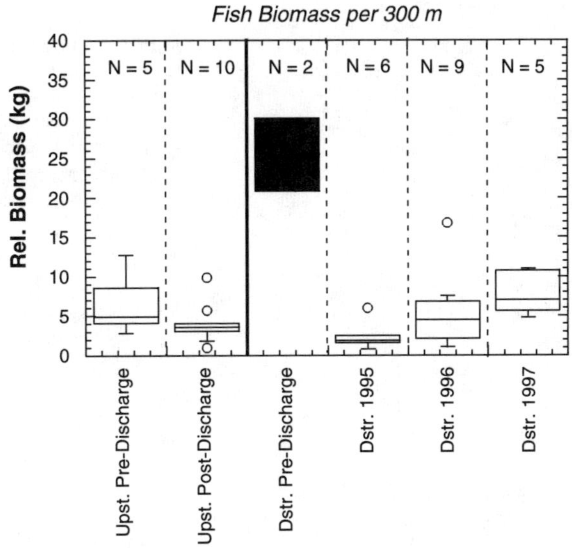

FIGURE 10.15 Pre-discharge to post-discharge changes in relative fish biomass collected in Leading Creek at a site upstream of Parker Run (discharge location) and at sites where biomass was collected downstream of Parker Run.

10.3.3.4.3 Strongs Run

Water was also pumped from the large Meigs Mine 31 complex into some streams that did not receive previous effluent. This water was discharged into Strongs Run, Sugar Run, and Robinson Run, which are Raccoon Creek watershed streams. Fish communities had not been completely eliminated in Sugar Run and Robinson Run. Although the discharge contained high levels of iron, the pH did not drop as it did in Strongs Run, where pH dropped to about 2 and caused a complete or near complete kill. The site near the mouth of this stream behaved differently from Leading Creek and Parker Run. The fish community was eliminated, but there was a dramatic immediate response of nearly 100% pioneering species followed by wide fluctuation around the pre-discharge conditions (Figure 10.16). The IBI took a bit longer to reach pre-discharge conditions and the recovery of more sensitive key species, particularly the southern redbelly dace *Phoxinus erythrogaster*, took several years. Thus, this small stream responded as predicted in response to a stressor that eliminated part of the community.

10.4 CONCLUSIONS

The responses of pioneering species to stressors generally followed the original description of Smith (1971): certain species are first to invade very small streams between cycles of flow and dewatering (ephemeral streams). The greatest response from pioneering species was observed in the smallest streams or where flow was lowest (e.g., Lobdell Creek, Hurford Run, Strongs Run). The pattern was affected by the presence or absence of refugia (i.e., Parker Run). Larger streams did not show strong proportional responses by pioneering species (e.g., Leading Creek, Sycamore Creek) although both streams had high quality habitats not generally associated with high proportions of pioneering species.

The pioneering species metric in the headwater IBI seemed to work well in the smaller streams we examined. The lack of response of pioneering species in larger streams suggests that this metric may be less useful in larger waters, at least as currently defined by its species composition. However, some proportion of larger streams sometimes show high proportions of pioneering fish species

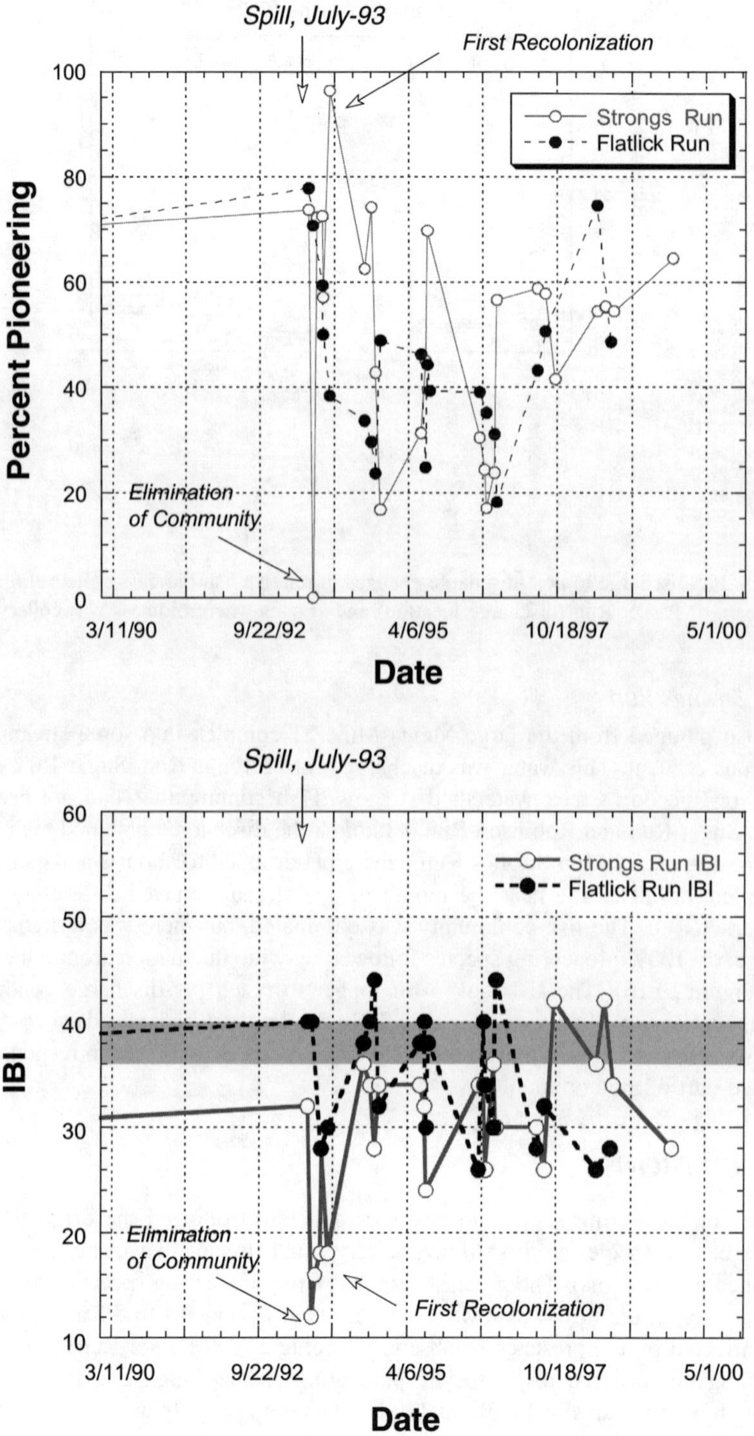

FIGURE 10.16 Percent of catch as pioneering species (top) and IBI scores (bottom) over time at a site in Strongs Run at river mile 0.6 and at a nearby site on a stream not affected by the discharge (Flatlick Run).

(Figure 10.9, bottom). Before the pioneering species is applied to larger rivers, a larger group of species should be examined to determine whether their life history characteristics could be "signature" for recovering biota. More work is also needed on the suite of pioneering species in small streams to examine whether the pattern would be more reliable with the inclusion of other species or even certain life stages of other species (e.g., young of white suckers in Midwest streams). Another alternative metric that others used successfully is presence of the largest or oldest expected year classes for each species (Oberdofff and Hughes, 1992; Hughes and Oberdorff, 1999). As with human life insurance tables, the presence of long-lived individuals suggests few environmental bottlenecks and consistency in water quantity and quality and habitat quality. In essence, such a metric would be the antithesis of the pioneering species metric as a signature for recent or repeated bottlenecks.

ACKNOWLEDGMENTS

The authors express appreciation to the Ecological Assessment Unit of the Ohio Environmental Protection Agency whose biologists collected and processed the majority of data used in these analyses and derived the IBIs formulated for Ohio waters. The biologists are Chris Yoder, Marc Smith, Roger Thoma, Dave Altfater, Chuck Boucher, Brian Alsdorf, Bob Miltner, Randy Sanders, Dennis Mishne, Mike Gray, Paul Albeit, and Ed Moore. This chapter is dedicated to the memory of the late Bernie Counts who is sorely missed at Ohio EPA. The opinions expressed do not necessarily represent those of the U.S. Fish and Wildlife Service. No official endorsement by that agency should be inferred.

REFERENCES

Goldstein, R.M., T.P. Simon, P.A. Bailey, M. Ell, K. Schmidt, and J.W. Emblom. 1994. Proposed metrics for the index of biotic integrity for the streams of the Red River of the North basin. *Proceedings of the North Dakota Water Quality Symposium.* March 30–31, 1994. Fargo, ND, 169–180.

Karr, J. R., L.A. Toth, and D.R. Dudley. 1985. Fish communities of midwestern rivers: a history of degradation, *BioScience*, 35, 90–95.

Niemela, S., E. Pearson, T.P. Simon, R.M. Goldstein, and P.A. Bailey. 1999. Development of an index of biotic integrity for the species depauperate Lake Agassiz Plain Ecoregion, North Dakota and Minnesota, in T.P. Simon (Ed.), *Assessing the Sustainability and Biological Integrity of Water Resources Using Fish Communities*, CRC Press, Boca Raton, FL, 339–366.

Oberdorff, T. and R.M. Hughes. 1992. Modification of an index of biotic integrity based on fish assemblages to characterize rivers of the Seine Basin, France, *Hydrobiologia*, 228, 117–130

Ohio Environmental Protection Agency. 1989. Biological Criteria for the Protection of Aquatic Life. Volume III. Standardized Biological Field Sampling and Laboratory Methods for Assessing Fish and Macroinvertebrate Communities. Ohio EPA, Division of Water Quality Planning and Assessment, Ecological Assessment Section, Columbus, OH.

Ohio Environmental Protection Agency. 1994. Ecological Recovery Endpoints for Streams Affected by the Meigs #31 Mine Discharges during July-September 1993. Ohio EPA, Division of Surface Water, Ecological Assessment Section, Columbus, OH.

Ohio Environmental Protection Agency. 2000. Biological and water quality study of Sycamore Creek and the Sandusky River, 1999. Ohio EPA, Division of Surface Water, Ecological Assessment Section, Columbus, OH.

Rankin, E.T. 1989. The qualitative habitat evaluation index (QHEI): rationale, methods, and application. Ohio EPA, Division of Water Quality Planning and Assessment, Columbus, OH.

Rankin, E.T. 1995. Habitat indices in water resource quality assessments, in W.S. Davis and T.P. Simon (Eds.), *Biological Assessment and Criteria: Tools for Water Resource Planning and Decision Making*. Lewis Publishers, Boca Raton. FL, 181–208.

Simon, T.P. 1991. Development of Ecoregion Expectations for the Index of Biotic Integrity. I. Central Corn Belt Plain. EPA 905–9–91–025. U.S. Environmental Protection Agency, Environmental Sciences Division, Chicago, IL.

Simon, T.P. 1998. Modification of an index of biotic integrity and development of reference condition expectations for dunal, palustrine wetland fish communities along the southern shore of Lake Michigan, *Aquatic Ecosystem Health and Management,* 1, 49–62.

Simon, T.P. and J. Lyons. 1995. Application of the index of biotic integrity to evaluate water resource integrity in freshwater ecosystems, in W.S. Davis and T.P. Simon (Eds.). *Biological Assessment and Criteria: Tools for Water Resource Planning and Decision Making*. Lewis Press, Boca Raton, FL, 245–262.

Simon, T.P. and P.M. Stewart. 1998. Application of an index of biotic integrity for dunal, palustrine wetlands: emphasis on assessment of nonpoint source landfill effects on the Grand Calumet Lagoons, *Aquatic Ecosystem Health and Management,* 1, 63–74.

Simon, T.P., R. Jankowski, and C. Morris. 1999. Modification of an index of biotic integrity for assessing vernal ponds and small palustrine wetlands using fish, crayfish, and amphibian assemblages along southern Lake Michigan, *Aquatic Ecosystem Health and Management,* 3, 407–418.

Simon, T.P., E.T. Rankin, R. Dufour, and S.A. Newhouse. 2002. Using biological criteria for establishing restoration and ecological recovery endpoints, Chapter 4, this volume.

Smith, P.W. 1971. Illinois streams: a classification based on their fishes and an analysis of factors responsible for the disappearance of native species. *Illinois Natural History Survey Biological Notes*, 76.

Trautman, M.B., 1981. *The Fishes of Ohio*. Ohio State University Press, Columbus, Ohio,

11 Correlation between Nutrient Stimulation and Presence of Omnivorous Fish along the Lake Erie Nearshore

Roger F. Thoma and Thomas P. Simon

CONTENTS

11.1 INTRODUCTION

The Great Lakes Water Quality Agreement of 1978 and subsequent revisions recognized the importance of reducing phosphorus loads to the Great Lakes as a mechanism of improving water quality. Bans on phosphorus use in detergents and the regulation of phosphorus loading from wastewater treatment facilities have substantially reduced its impact in the western basin of Lake Erie (Munawar et al., 1999). In the central and eastern basins, mean concentrations have returned to historic levels (DePinto et al., 1986; Hartig and Gannon, 1986; Markarewicz and Bertram, 1991; Charlton, 1999). Numerous species level indicators were studied during the recovery. They showed declines and improvements that assisted in formulating targets and determining the success of management objectives (Wood, 1973; Robbins et al., 1989; Schloesser et al., 1995; Haag et al., 1993; Krieger et al., 1996; Ludsin et al., 2001).

0-8493-0905-0/03/$0.00+$1.50

Despite the recognition that phosphorus may cause changes in ecosystem trophic dynamics that can lead to cascading trophic level effects (Karr et al., 1985; Nordin, 1985; Miltner and Rankin, 1998), few studies examined the relationship between phosphorus levels and possible effects on specific fish assemblage components. Studies have not examined patterns in omnivore fish species and no study cited the relationship between total IBI scores and phosphorus concentrations as a mechanism to establish management goals and criteria. This chapter examines relationships between the percent individuals as omnivore species and phosphorus concentration. In addition, patterns in omnivore species abundance from nearshore and bay habitats (open lake wetlands) and lacustuaries (drowned river mouths, and flooded estuary habitats) will be described to evaluate omnivore relationships to anthropogenic disturbance in Lake Erie.

11.1.1 Background

The U.S. Environmental Protection Agency (USEPA) and U.S. Department of Agriculture (USDA) are jointly developing nutrient criteria to reverse the increasing number of stream miles impaired by nutrient enrichment. Nutrient criteria are necessary because nitrogen and phosphorus can cause negative effects on aquatic life, human and domestic animals, and cause taste and odor problems (Welch et al., 1989; Dodds and Welch, 2000). Nutrient enrichment has caused alterations of aquatic communities, e.g., invertebrate assemblage structures and biomasses of fish, and is correlated with phosphorus concentrations (Nordin, 1985; Miltner and Rankin, 1998).

Agricultural runoff, sewage pollution, industrial contamination, and over-exploitation of fish stocks damaged the Lake Erie ecosystem (Charlton et al., 1999). Studies have documented the loss of dissolved oxygen (Burns, 1976; Charlton, 1994) that occurred in the central basin with the increase in biological oxygen demand in the 1970s and most recently as a result of zebra mussel respiration in the 1990s (Burns, 1976; Charlton, 1994).

Concerns about phosphorus loading to Lake Erie in the early 1970s prompted the Great Lakes Water Quality Agreement to focus on 50% targeted reductions in nutrient pollution. Typically, the total phosphorus loading had a strong west to east gradient because of substantial loads in the western basin. Phosphorus concentration goals were reached by the mid 1980s in the west basin and improvements in historic fish populations were noticed (Ryan et al., 1999). During nutrient load reductions, phosphorus decreased in the east and central basins, and with the introductions of zebra mussels, phosphorus was further decreased; it subsequently increased to previous mean concentrations (Charlton et al., 1999).

The largest change in phosphorus loads occurred in the western basin. Mean concentrations of 41 $\mu g\ P\ l^{-1}$ in the 1970s were reduced to 35.5 $\mu g\ P\ l^{-1}$ when nutrient controls went into effect. By the mid 1980s, decreases in phosphorus reached 20 $\mu g\ P\ l^{-1}$ or less (Lesht et al., 1991; Williams et al., 1998). Concentrations decreased by the mid 1990s and the mean concentrations of 12.5 to 25.4 $\mu g\ P\ l^{-1}$ that occurred in 1997 are typical.

Phosphorus concentrations in Lake Erie tributary streams remain high. Painter et al. (2000) reports highly elevated phosphorus concentrations in most of Ohio's tributary streams. Almost every stream from Maumee Bay east to the Cleveland metropolitan area displayed geometric mean phosphorus concentrations one to two times the established guidelines at most sites in their respective basins. The Grand and Ashtabula Rivers and Conneaut Creek displayed levels one time or less than guidelines at most sites. No phosphorus trend data for Ohio's tributary streams are currently reported in the literature.

11.2 METHODS

11.2.1 Study Area and Site Selection

The study area includes 90 sites collected between 1993 and 1996 at the nearshore and at the mouths of tributaries that flow into Lake Erie (Figure 11.1). The eastern basin is the deepest of the

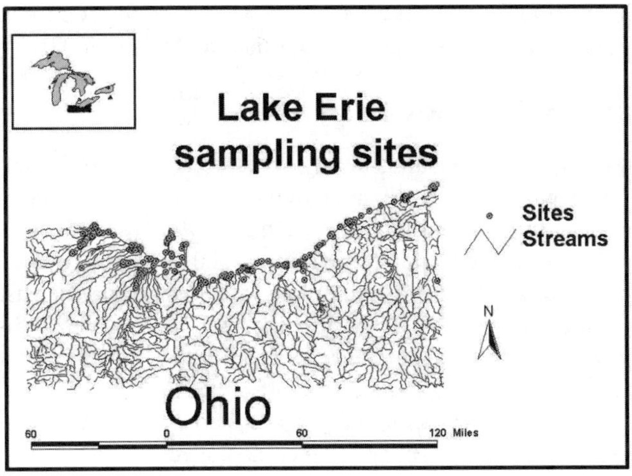

FIGURE 11.1 Locations of sample sites in nearshore and lacustuary habitats of Lake Erie study area collected between 1993 and 1996.

three basins of Lake Erie, and the western basin is the shallowest. Samples were taken from the central and western basins only. Mud bottoms predominate in the deeper offshore waters while much of the south shoreline consists of sandy beaches, precipitous shale cliffs, or shoreline protection structures. Shoreline modification dominates the south shore in Ohio.

Sites were selected based on quality of habitat types found in the Lake's nearshore areas and provided thorough coverage. A site was sampled every 5 mi. (or 8 km). Sites were located along harbor breakwaters, sand and gravel beaches, the shores of Lake Erie islands, bedrock cliffs, and modified shorelines. Open lake wetlands were sampled in Sandusky Bay, East Harbor State Park, and at Presque Isle, PA. Lacustuaries were sampled at 125 sites between 1982 and 1996. Sites were selected at the mouth, head, and midsection of each lacustuary.

11.2.2 Phosphorus Analysis

Grab water samples were collected for analysis of phosphorus at the same areas where fish sampling was performed. Three water samples were collected at each site during the spring, summer, and fall months. Samples were collected using a Kemmer water sampler (Welch, 1948) from mid-water depths or by buckets from bridges if located in the sampling zone. Samples were analyzed following USEPA (1986) methods (365.1, 365.4) as modified by Ohio EPA methods (260.1, 260.3) stipulated in the *Manual of Ohio EPA Surveillance Methods and Quality Assurance Practices* (Ohio EPA, 1989b).

Total Phosphorus — Samples were treated with a sulfuric acid and potassium sulfate reagent and digested for 1 hour at 175°C and 2.5 hours at 380°C in a block digester. The digestion reagent also contained mercuric oxide that served as a catalyst. Digested residue was then cooled, diluted to 10 mL and analyzed on an Astoric-Pacific rapid flow analyzer (RFA). The digestion converted all phosphorus to orthophosphate. On the RFA, the sample was diluted with a saline diluent, then treated with acid diluent to control the pH, followed by a molybdate tartrate reagent that reacted with orthophosphate in the sample. Ascorbic acid was then added to produce a blue color that was measured colorimetrically at 880 nm. The method quantification limit was 0.05 mg/L.

Low Level Phosphorus — When analyzing low level phosphorus, all phosphorus present in the sample is oxidized to orthophosphate by digestion with ammonium persulfate and sulfuric acid in an autoclave at 15 psi and 121°C for 30 minutes and then analyzed on the RFA. On the RFA, the sample was diluted with a saline diluent, followed by a molybdate tartrate reagent that reacted with the orthophosphate in the sample. Ascorbic acid was added to produce a blue color that was measured colorimetrically at 880 nm. Method quantification limit was 0.01 mg/L.

11.2.3 Fish Assemblage Collection

Fish assemblages were collected at all sites by downstream electrofishing during the day in lacustuaries and at night with the prevailing shoreline currents on open lake shorelines and bay wetlands with a 5.8-m modified V-hull jonboat (Thoma, 1999). DC current between 240 and 340 V providing 5 to 6 amps was applied to the water from a 7000-W generator and Smith Root pulsator set for 60 pulses sec.$^{-1}$. In low conductivity conditions, the voltage was increased between 360 and 500 and the frequency was increased to 120 pulses sec.$^{-1}$ in order to maintain the amperage. Anodes were two separately charged 1-m circumference electrospheres. Two 3-m articulated booms supported by distal floats were positioned about 2.1 m in front of the boat. The booms were positioned at 20-degree angles from the center lines to the port and starboard sides. This arrangement enabled the two electrospheres to be about 4.3-m apart when deployed.

The articulation of the booms allowed horizontal and vertical movements. This was especially important because the horizontal movement allowed the booms to flex if they encountered submerged obstacles, while the vertical movement allowed the electrospheres to ride with the wave action. Sampling was not conducted when wave size was 0.6 m or more. Swells or small waves less than 0.6 m caused stationary (non-articulated) booms to rise and pull the anodes from the water, thus interrupting current flow and allowing fish to escape. The placement of a flotation device at the distal end of the boom kept the electrosphere under the water surface at the proper depth of a few centimeters as the mechanism rode up and down with the waves.

The reach sampling distance was about 500 m, which allowed sampling of a complete habitat cycle (shore, littoral, and profundal) within 1 m of shore (Thoma, 1999). The amount of fishing time at each station was dependent on habitat complexity and ranged from 2000 to 5000 sec. The greater the number of fish captured and the greater the complexity of the shoreline, the more time spent in the reach. A crew of three individuals undertook all electrofishing efforts. One person was positioned on the bow and was the principal netter; a second person was mid-ship and served as an assistant to collect any fish the principal collector missed. The third person operated the boat, pulsator, and lights and collected any fish that surfaced at the stern. All fish were placed in a live well supplied with freshwater from a pump. Common carp were placed in their own live well to avoid excess oxygen consumption and the deaths of smaller fish.

All fish were identified to species and voucher specimens are housed in the Museum of Biodiversity at The Ohio State University in Columbus. At least 15 individuals (if available) of each species were weighed, examined for external anomalies, and returned to the water. Any remaining individuals not weighed were identified and counted. Anomalies included were deformities, eroded fins, lesions, and tumors (collectively called DELT) (Sanders et al., 1999).

Classification of fish species as omnivores followed Ohio EPA guidance (1989a). This guild collectively included fish that fed on a variety of materials including a minimum of 25% plants and 25% animal materials (Ohio EPA, 1989a).

11.2.4 Index of Biotic Integrity

The biological expectations for this study were based on reference conditions calibrated from two ecoregions in Lake Erie nearshore habitats (Thoma, 1999). Thoma developed reference conditions and an index of biotic integrity (IBI) based on original IBI rationale (Karr, 1981; Fausch et al., 1990; Karr and Chu, 1999). Fish assemblage data were grouped into twelve metrics in three categories including attributes based on species richness and composition, tolerance, trophic guild, abundance, reproductive guild, and individual health and condition (Table 11.1). Each metric received a numerical score of 5, 3, or 1, depending on whether the data was comparable to, deviated

TABLE 11.1
Metrics and Scoring Criteria for an IBI Calibrated for Nearshore and Tributary Lacustuaries of Lake Erie (Thoma, 1999)

	Scoring Criteria	
	Lake Erie	**Tributary**
Species Richness and Composition		
Total number of species	7–12	8–15
Number of sunfish species	3–4	4–6
Number phytophilic species	2–3	11–20%
Number of benthic species	3–5	3–5
Number of cyprinid species		3–4
Tolerance		
Number of intolerant species	3–4	3–5
Percent individuals as tolerant species	2.3–5%	11–22%
Trophic Guilds		
Percent individuals as omnivore species	17–38%	19–38%
Percent individuals as top carnivore species	14–28%	9–18%
Behavior Guilds		
Percent individuals as nonindigenous species	16–30%	9–14%
Percent individuals as lake-associated species	20–40%	
Abundance		
Relative number of individuals	650–1300	450–925
Individual Health and Condition		
Percent individuals with DELT anomalies	1–3%	0.1–2.5%

somewhat from, or deviated greatly from a reference or least impacted condition, and 0 for positive metrics having no representatives in the sample and negative metrics with values above the 95th percentile (Thoma, 1999). A sum of all metrics provided the final IBI score and allowed for placement of the site into an integrity class ranging from very poor to excellent.

Each metric individually provided information about a specific characteristic of the sampling site; together they characterized its underlying biological integrity (Karr et al., 1986). Our *a priori* hypothesis was that we would not anticipate the total IBI score to show statistically significant correlation with the concentration of phosphorus species, since the IBI incorporates multiple attributes that respond to a variety of disturbance gradients (Table 11.2). Spearman rank correlations between phosphorus concentration and metrics used in the Lake Erie index of biotic integrity were not significant.

11.2.5 Statistics

Relationships among total IBI score, the percent of individuals as omnivore species, and normalized phosphorus relationships were examined using Spearman's correlation procedures (SAS, 1996). Non-parametric correlations were used, since a few samples and some results were recorded as zeros. Fish assemblage data were examined for trends and patterns using box and whisker plots for select tributaries and open lake habitats. We used cluster analysis, Bray-Curtis hierarchical agglomerative clustering (Clarke and Warwick, 1994), to find any natural groupings in the similarity data among assemblages found at all sites and phosphorus concentrations. Finally, we used a simple linear regression analysis to evaluate the relationship of the percent of individuals as omnivore species, the concentrations of phosphorus, and the quality of habitat as measured by the qualitative habitat evaluation index (QHEI) (Rankin, 1995; SAS, 1996).

TABLE 11.2
Spearman's Correlations of Phosphorus Species with Omnivore and Total IBI Scores from Samples Collected along the Southern Shore of Lake Erie (n = 1149)

Metric	Total Phosphorus	Orthophosphate
Number of species	**0.411**	–0.183
Number of sunfish species	0.233	–0.295
Number of phytophilic species	–0.035	0.038
Number of benthic species	0.236	–0.143
Number of intolerant species	–0.118	–0.311
Percent individuals as tolerant species	0.197	–0.220
Percent individuals as omnivores	–0.118	–0.132
Percent individuals as top carnivores	0.153	–0.303
Percent individuals as non-indigenous species	**0.469**	**0.555**
Percent individuals as lake-associated species	0.080	0.151
Relative number of individuals	0.095	0.071
Percent individuals with DELT anomalies	**0.393**	0.068
IBI	**0.408**	–0.090

Note: bolded entries are significantly correlated ($P < 0.10$); bolded and underlined entries are significantly correlated ($P < 0.05$).

11.3 RESULTS AND DISCUSSION

11.3.1 Patterns of Phosphorus Concentrations along the Lake Erie Nearshore

Painter et al. (2000) documented moderate to high elevations of phosphorus concentrations in many of Ohio's Lake Erie tributaries. High levels of turbidity also existed in most lacustuaries sampled. The Ohio Lake Erie Commission (1998) documented highly elevated levels of suspended solids in Lake Erie tributaries. It appears that the high turbidity levels prevent or reduce the number and severity of algal blooms occurring in Ohio's Lake Erie lacustuaries. The consequent lack of algal blooms results in these areas not being recognized as eutrophic.

Studies addressed the relationships of phosphorus to observed patterns in the ambient environment. Assessed impacts of anthropogenic sources on biological integrity include studies on agricultural activities (Karr, 1981), wastewater treatment facilities (Karr et al., 1985), and non-point source effects of confined animal feedlots (Gammon and Simon, Chapter 20). To date, no studies have compared a specific chemical class (e.g., phosphorus) to the relationship of an individual metric. However, response patterns of select metrics within applications of the IBI have been evaluated by Emery et al. (1999), Eaton and Lydy (2000), Rankin and Simon (Chapter 10, this volume), and Lydy et al. (Chapter 17, this volume). They evaluated correlates among specific metrics or total IBI scores and select anthropogenic disturbance gradients. However, none of these studies evaluated the response of the percent of individuals as omnivore species metric to phosphorus concentration.

One of the major findings of Karr et al. (1985) was a significant difference between total IBI scores and chlorine and phosphorus concentrations at select sites downstream of wastewater treatment plants. They found a pattern of higher IBI scores at sites with lower concentrations of chlorine and slightly elevated concentrations of phosphorus. This pattern is interesting since stimulation of biological communities in warmwater streams causes slight increases in select species that bias the total IBI score.

Lyons et al. (1996) and Mundahl and Simon (1999) found that slight increases in phosphorus in coldwater streams caused a shift in the composition of the community membership that led to degradation of the typical coldwater fish assemblage. Lyons (1992) indicated that select species in warmwater fish assemblages may show artificial increases in abundance when nutrient stimulation was slightly to moderately elevated. Under gross stimulation they collapse into a few individuals of tolerant species.

11.3.2 Patterns of Omnivore Species Levels along the Lake Erie Nearshore

Examination of IBI and modified index of well-being (MIwb) values for the Lake Erie nearshore and lacustuary areas did not aid assessment of eutrophication since both biocriteria are compilation measures that represent numerous factors combined in a single number and may respond to factors other than enrichment. Our *a priori* hypothesis was that the omnivore metric (a single component of the IBI) would provide the best understood and most direct measure of trophic status (Table 11.3). Further analysis of biological community data, such at the obligate Great Lakes species metric (Moy et al., in review) or percent of individuals as detritivores (Goldstein and Simon, 1999), may reveal other measures of eutrophication in the Great Lakes and their river systems.

We evaluated patterns in omnivore species distributions and responses in most of the bays, nearshore habitats, and lacustuaries in Lake Erie (Figure 11.2). The diagram on the left depicts the trophic status of the Lake Erie nearshore, as measured by the abundance of omnivorous fishes in the lake by county. The most abundant populations of omnivores were found in western Lake Erie in Lucas County, which is associated with the Maumee River. The river is heavily loaded with nutrients from non-point and point pollution sources arising from the extensively farmed watershed and the urban center of Toledo (Painter et al., 2000), both of which contribute large quantities of untreated sewage to Maumee Bay. The other portion of the lake that showed elevated omnivore levels was in Lorain and Cuyahoga Counties. Both counties have large urban areas in their watersheds and that is reflected in their fish community compositions. An unknown portion of the elevated omnivore levels may be a response to chemical pollution from industrial activities and not entirely related to enrichment. This aspect of omnivore numbers needs further evaluation. An assessment of the trophic status of the open lake and nearshore habitats suggests that the Maumee Bay part of the western basin is decidedly eutrophic, while the rest of the lake should be considered mesotrophic.

Within the lacustuaries, conditions are different (Figure 11.2b). All but two tributaries have omnivore percentages greater than 20%, while four areas exhibit levels above 40%. We consider those four areas representative of areas that are highly eutrophic. These areas are the Maumee River/Ottawa River, Sandusky River/Little Muddy Creek, and the Cuyahoga and Chagrin Rivers.

The Cuyahoga River is eutrophic in response to habitat destruction and urban runoff, especially from combined sewer overflows (Thoma, 1999). The Chagrin River is eutrophic from urban runoff and, perhaps more importantly, from disturbance of bottom sediments by recreational boat traffic (Thoma, 1999). The Maumee and Sandusky areas are eutrophic in response to agricultural activities and sedimentary processes (Rasul et al., 1999). The Maumee area has added nutrient loads from municipal sewage.

Omnivore levels in the remaining lacustuaries, e.g., Portage River, Huron River, Old Woman Creek, Vermilion River, Black River, Rocky River, and Conneaut Creek ranged from 20 to 40%. These lacustuaries are considered mildly eutrophic. They suffer from a variety of environmental insults caused by agricultural runoff, habitat destruction (mostly from marinas), urban runoff (especially sewage overflows) and recreational boat traffic (Thoma, 1999). The Huron River is similar to the Chagrin in that boat traffic causes resuspension of nutrient-rich sediments into the

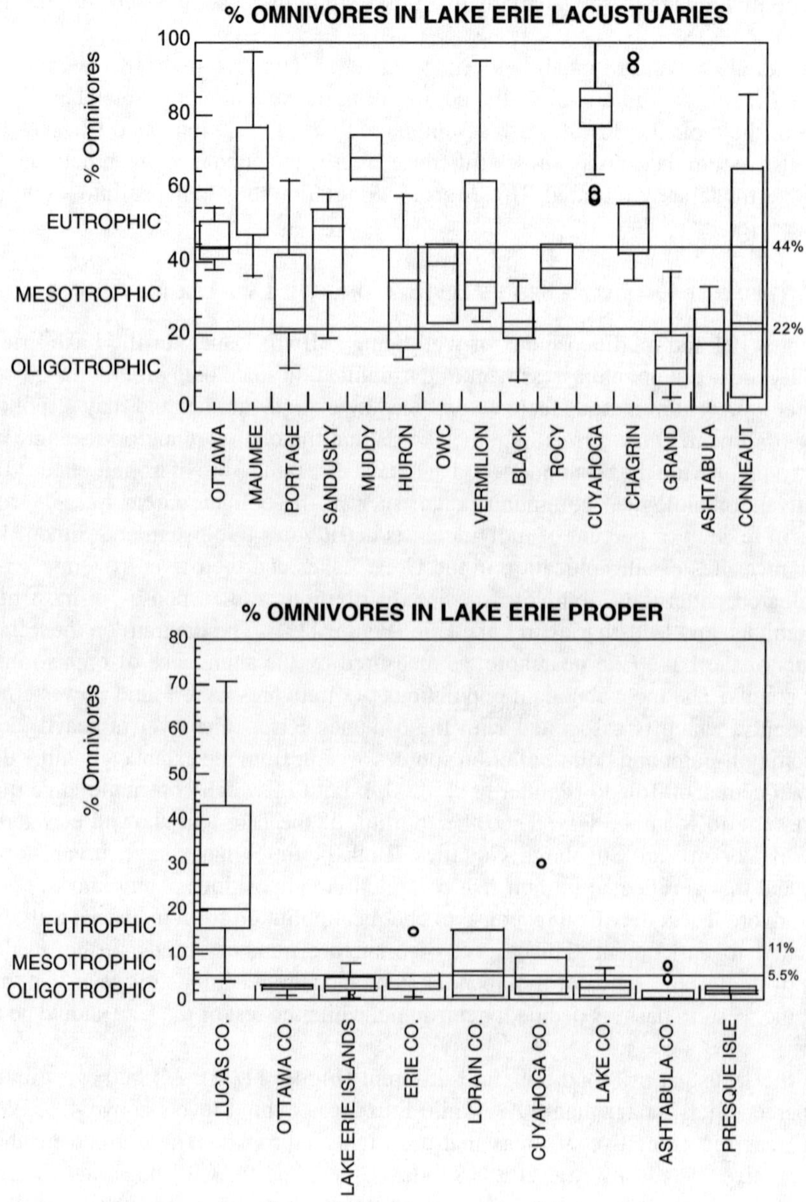

FIGURE 11.2 Percent individuals as omnivore species abundance relationships within select portions of Lake Erie habitats. Top: nearshore and bay. Bottom: lacustuaries and flooded estuary wetlands.

water column. Both rivers have biological communities in lotic stream reaches that display higher integrity and low omnivore numbers.

The Grand and Ashtabula Rivers are the only waterways that do not appear eutrophic. They may be the only remaining rivers along the Ohio shoreline that represent least-impacted conditions in Lake Erie. Both rivers have large undisturbed areas of watershed and minimal urbanization. In watershed use, Conneaut Creek is similar to the Grand and Ashtabula Rivers, but limited fish community samples taken primarily in a reach dominated by habitat destruction and coal handling facilities appeared to influence results. Analysis of further samples from the creek may reveal it is as good or better than the Grand and Ashtabula Rivers.

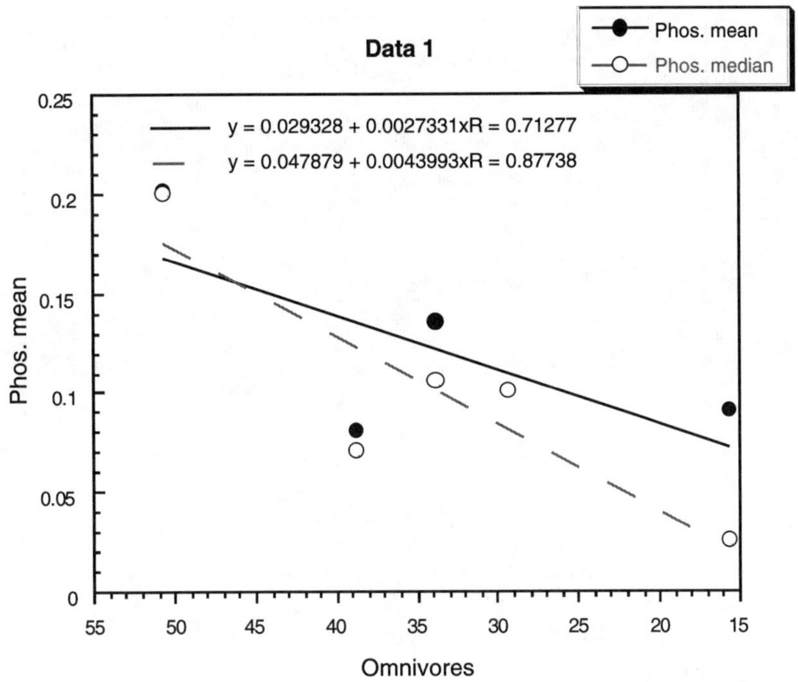

FIGURE 11.3 Linear regression relationship between percent individuals as omnivore species and phosphorus concentrations from Lake Erie lacustuaries.

11.3.3 Relationship between Omnivore Abundance and Concentrations of Phosphorus

We examined the relationship between mean phosphorus concentrations and the percent of individuals as omnivore species (Figure 11.3). A strong correlation exists between the mean (R = 0.713) and median (R = 0.877) concentrations of phosphorus and the percent of individuals as omnivores. Lower concentrations of phosphorus correlated with lower site omnivore species abundance. Higher concentrations of phosphorus correlated with higher omnivore species abundance. The number of sites for which omnivore fish abundance and phosphorus concentrations could be compared was limited by lack of available phosphorus data. The Lake Erie nearshore did not have enough phosphorous data to allow comparison of urban and rural site differences.

11.3.4 Relationship between Omnivore Abundance and Habitat Quality

An examination of 73 QHEI values correlated with the number of omnivorous fish individuals (Figure 11.4) showed a linear regression with a low relationship (R = 0.39). This is significantly less than the regression coefficient values obtained from correlations with phosphorus concentrations. The data for this analysis came from the lacustuaries of the Ashtabula River (15 records), Conneaut River (10 records), Cuyahoga River (15 records), Otter Creek (6 records), Vermilion River (12 records), and Lake Erie in Ashtabula County (15 records). These data provide a wide range of habitat quality and conditions that should, if habitat strongly influences omnivore abundance, provide enough data to effectively evaluate the relationship. The high quality sites (QHEI >90) were from Conneaut Creek at the most upstream end of the lacustuary bordering the lotic portions of the system.

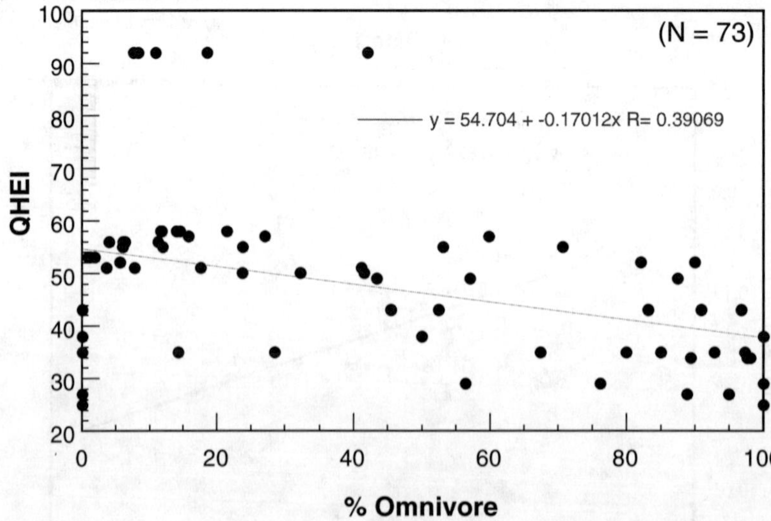

FIGURE 11.4 Linear regression relationship between percent individuals as omnivore species and habitat quality (QHEI) from Lake Erie and its lacustuaries.

11.4 CONCLUSIONS

This chapter examines the relationship of open lake and lacustuary site concentrations of phosphorus, the percent individuals as omnivore fish species, and patterns of omnivore species presence along the Ohio nearshore of Lake Erie. The trophic status of the nearshore, as measured by the abundance of omnivorous fishes showed that the most abundant populations of omnivores are found in western Lake Erie in Lucas County, the portion associated with the Maumee River (Figure 11.2). The other portion of the lake that showed elevated omnivore levels was the Lorain and Cuyahoga Counties area. An assessment of the trophic status of the lake nearshore habitats suggests that the Maumee Bay portion of the western basin is decidedly eutrophic, while the rest of the shore should be considered mesotrophic.

Conditions differ in the lacustuaries (Figure 11.2). All but two of the tributaries have omnivore percentages above 20%, while four areas exhibited levels above 40%. We consider the areas with percentages greater than 40% representative of areas that are highly eutrophic. These areas are the Maumee River/Ottawa River, Sandusky River/Little Muddy Creek, and the Cuyahoga and Chagrin Rivers. The Chagrin River is eutrophic from urban runoff and wastewater treatment effluent, and perhaps more important, from disturbance of bottom sediments by recreational boat traffic in the shallow lacustuary area. The Maumee and Sandusky areas have become eutrophic via agricultural activities and sedimentary processes (Rasul et al., 1999). The remaining lacustuaries, e.g., Portage River, Huron River, Old Woman Creek, Vermilion River, Black River, Rocky River, and Conneaut Creek contain 20 to 40% omnivores and are considered mildly eutrophic. The Grand and Ashtabula Rivers are the only streams that do not appear eutrophic. Conneaut Creek merits further investigation.

Total phosphorus concentration relationships to omnivore fish assemblage structure were evaluated (Figure 11.3). Results indicated significant correlation among the percent individuals as omnivores and mean and median concentrations of phosphorus. However the lack of least impacted sites in Lake Erie prohibited a further analysis of urban and rural site relationships. Analysis is also limited by the lack of an adequate phosphorus database. The authors are currently gathering data that will allow a more detailed analysis of the relationship between omnivorous fish and phosphorus concentrations and the association with habitat quality and toxic chemicals in sediments.

We have anecdotally observed that omnivorous fish increase in abundance in areas of low quality habitat and contaminated sediments. Lower quality habitat values can be, in part, the results of elevated phosphorus levels and their impacts on substrate, water column, and aquatic vegetation. A preliminary examination of habitat relationships to omnivores shows that habitat is not a controlling or predictive factor for omnivore abundance.

Intuitive conclusions concerning the interplay of phosphorus, habitat, and toxics must be avoided, as all three variables are frequently found at the same site in a complex union. Hypotheses concerning their relationship should be tested before acceptance. Some sample sites in lacustuaries with virtually no habitat quality (steel-lined ship channels over 7 m deep) displayed zero percent omnivores while others of similar habitat structure can contain 100% omnivores. Four lacustuaries in our study area (the Maumee, Black, Cuyahoga, and Ashtabula Rivers) have or recently had elevated sediment contamination levels. The Maumee, Black, and Cuyahoga Rivers all display elevated omnivore levels while the Ashtabula displays one of the lowest omnivore percentages.

ACKNOWLEDGMENTS

The authors would like to thank Robert Davic and Lauren Lambert for reviewing drafts of this chapter and encouraging the senior author to pursue further investigation and publication of the findings. The Lake Erie Protection Fund, Great Lakes Protection Fund, and Ohio EPA provided significant funds and resources. The opinions expressed do not necessarily represent those of the U.S. Fish and Wildlife Service. No official endorsement by that agency should be inferred.

REFERENCES

Bishop, C.A., B. Collins, P. Mineau, N.M. Burgess, W.F. Read, and C. Risley. 2000. Reproduction of cavity-nesting birds in pesticide-sprayed apple orchards in southern Ontario, Canada, 1988–1994, *Environmental Toxicology and Chemistry,* 19(3), 588–599.

Benke, A.C., G.E. Willke, F.K. Parrish, and D.L. Stites. 1981. *Effects of Urbanization on Stream Ecosystems,* ERCO7–81. Georgia Institute of Technology, Atlanta.

Burns, N.M. 1976. Oxygen depletion in the central and eastern basins of Lake Erie, 1970, *Journal of the Fisheries Research Board of Canada,* 33, 512–519.

Burton, G.A., Jr. 1991. Assessing the toxicity of freshwater sediments, *Environmental Toxicology and Chemistry,* 10, 1585–1627.

Charlton, M.N. 1994. The case for research of the effects of zebra mussels in Lake Erie: visualization of information from August and September 1993, *Journal of Biological Systems,* 2, 467–480.

Charlton, M.N., R. Le Sage, and J.E. Milne. 1999. Lake Erie in transition: the 1990s, in M. Munawar, T. Edsall, and I.F. Munawar (Eds.). *State of Lake Erie (SOLE) – Past, Present and Future,* Ecovision World Monograph Series, Backhuys Publishers, Leiden, The Netherlands, 97–123.

Clarke, K.R. and R.M. Warwick. 1994. *Change in Marine Communities: An Approach to Statistical Analysis and Interpretation,* Natural Environment Research Council, United Kingdom.

DePinto, J.V., T.C. Young, and L.M. McIlroy. 1986. Great Lakes water quality improvements, *Environmental Science and Technology* 20, 752–759.

Dodds, W.K. and E.B. Welch. 2000. Establishing nutrient criteria in streams, *Journal of the North American Benthological Society,* 19, 186–196.

Eaton, H.J. and M.J. Lydy. 2000. Assessment of water quality in Wichita, Kansas, using an Index of Biotic Integrity and analysis of bed sediment and fish tissue for organochlorine insecticides, *Archives of Environmental Contamination and Toxicology,* 39, 531–540.

Emery, E.B., T.P. Simon, and R. Ovies. 1999. Influence of the family *Catostomidae* on the metrics developed for a Great River Index of Biotic Integrity, in T.P. Simon (Ed.). *Assessing the Sustainability and Biological Integrity of Water Resources Using Fish Communities.* CRC Press, Boca Raton, FL, 203–224.

Fausch K.D., J. Lyons, J.R. Karr, and P.L. Angermeier. 1990. Fish communities as indicators of environmental degradation, *American Fisheries Society Symposium,* 8, 123–144.

Haag, W.R., D.J. Berg, D.W. Garton, and J.L. Farris. 1993. Reduced survival and fitness in native bivalves in response to fouling by the introduced zebra mussel (*Dreissena polymorpha*) in western Lake Erie, *Canadian Journal of Fisheries and Aquatic Sciences*, 50, 13–19.

Hartig, J.H. and J.E. Gannon. 1986. Opposing phosphorus and nitrogen trends in the Great Lakes, *Alternatives* 13, 19–23.

Karr, J.R. 1981. Assessment of biotic integrity using fish communities, *Fisheries*, 6, 21–27.

Karr, J. R. and E.W. Chu. 1999. *Restoring Life in Running Waters: Better Biological Monitoring*, Island Press, Covelo, CA.

Karr, J.R., R.C. Heidinger, and E.H. Helmer. 1985. Effects of chlorine and ammonia from wastewater treatment facilities on biotic integrity, *Journal of the Water Pollution Control Federation,* 57, 912–915.

Karr, J.R., K.D. Fausch, P.L. Angermeier, P.R. Yant, and I.J. Schlosser. 1986. *Assessing the Biological Integrity in Running Waters: A Method and its Rationale*, Illinois Natural History Survey Special Publication 5, 28.

Krieger, K.A., D.W. Schloesser, D.W. Manny, C.E. Trisler, S.E. Hedy, J.J.H. Ciborowski, and K.M. Muth. 1996. Recovery of burrowing mayflies (Ephemeroptera: Ephemeridae: *Hexagenia*) in western Lake Erie, *Journal of Great Lakes Research,* 22, 254–263.

Lesht, B.M., T.D. Fontaine III, and D. Dolan. 1991. Great Lakes total phosphorus model: post audit and regionalised sensitivity analysis, *Journal of Great Lakes Research,* 17, 3–17.

Lydy, M.J., A.J. Strong, and T.P. Simon. 2000. Development of an Index of Biotic Integrity for the Little Arkansas River Basin, Kansas, *Archives of Environmental Contamination and Toxicology,* 39, 523–530.

Lydy, M.J., P.M. Stewart, and T.P. Simon. 2002. Relationship between fish assemblages and organochloride insecticides in sediment and fish tissue in South-central Kansas, Chapter 17, this volume.

Lyons, J. 1992. *Using the Index of Biotic Integrity (IBI) to Measure Environmental Quality in Warmwater Streams of Wisconsin*, General Technical Report, NC-149, U.S. Department of Agriculture, Forest Service, North Central Forest Experiment Station, St. Paul, MN.

Ludsin, S.A., M.W. Kershner, K.A. Blocksom, R.L. Knight, and R.A. Stein. 2001. Life after death in Lake Erie: nutrient controls drive fish species richness, rehabilitation, *Ecological Applications,* 11, 731–746.

Markarewicz, J.C. and P. Bertram. 1991. Evidence for the restoration of the Lake Erie ecosystem, *Bioscience* 41, 216–223.

Miltner, R.J. and E.T. Rankin. 1998. Primary nutrients and the biotic integrity of rivers and streams, *Freshwater Biology,* 40, 145–158.

Moy, P.B., T.P. Simon, and C.L. Morris. In review. Modification of an index of biotic integrity for evaluating harbors and embayments of southern Lake Michigan, *Journal of Great Lakes Research.*

Munawar, M., T. Edsall, and I.F. Munawar (Eds.). 1999. *State of Lake Erie (SOLE) – Past, Present and Future,* Ecovision World Monograph Series, Backhuys Publishers, Leiden, The Netherlands.

Mundahl, N.D. and T.P. Simon. 1999. Development and application of an index of biotic integrity for coldwater streams of the Upper Midwestern United States. in T.P. Simon (Ed.). *Assessing the Sustainability and Biological Integrity of Water Resources Using Fish Communities.* CRC Press, Boca Raton, FL, 383–415.

Nordin, R.N. 1985. Water quality criteria for nutrients and algae (technical appendix). Water Quality Unit, Resources Quality Section, Water Management Branch, British Columbia Ministry of the Environment, Victoria, B.C. (Available from http://www.env.gov.bc.ca).

Ohio Environmental Protection Agency. 1989a. *Biological Criteria for the Protection of Aquatic Life: Volume III. Standardized Biological Field Sampling and Laboratory Methods for Assessing Fish and Macroinvertebrate Communities.* Ohio EPA, Division of Water Quality Planning and Assessment, Columbus, Ohio.

Ohio Environmental Protection Agency. 1989b. *Ohio EPA Manual of Surveillance Methods and Quality Assurance Practices, updated edition.* Ohio EPA, Division of Environmental Services, Columbus, Ohio.

Ohio Lake Erie Commission, 1998. *State of Ohio 1998 State of the Lake Report, Lake Erie Quality Index.*

Painter, S., D.N. Myers, and J. Letterhos. 2000. Characterization of Data and Data Collection Programs for Assessing Pollutants of Concern to Lake Erie. Lake Erie Lakewide Management Plan (LaMP), Technical Report Series. Columbus, OH, 61 pp.

Rankin, E.T. 1995. The use of habitat assessment in water resource management programs, in W.S. Davis and T.P. Simon (Eds.). *Biological Assessment and Criteria: Tools for Water Resource Planning and Decision Making*. Lewis Publishers, Boca Raton, FL. 181–208.

Rankin, E.T. and T.P. Simon. 2002. Pioneer species metric changes as a result of increased anthropogenic disturbance: statewide patterns and a case study of four Ohio streams, Chapter 10, this volume.

Rasul, N., J.P. Coakley, and R. Pippert. 1999. Sedimentary environment of western Lake Erie: geologic setting, sediment distribution and anthropogenic effects, in M. Munawar, T. Edsall, and I.F. Munawar (Eds.). *State of Lake Erie (SOLE) – Past Present and Future,* Ecovision World Monograph Series, Backhuys Publishers, Leiden, The Netherlands, 57–74.

Robbins, J.A., T. Keilty, D.S. White, and D.N. Edgington. 1989. Relationships among tubificid abundances, sediment composition, and accumulation rates in Lake Erie, *Canadian Journal of Fisheries and Aquatic Sciences*, 46, 223–231.

Ryan, P.A., L.D. Witzel, J. Paine, M. Freeman, M. Hardy, S. Scholten, L. Sztramko, and R. MacGregor. 1999. Recent trends in fish populations in eastern Lake Erie in relation to changing lake trophic state and food web, in M. Munawar, T. Edsall, and I.F. Munawar (Eds.). *State of Lake Erie (SOLE) – Past Present and Future,* Ecovision World Monograph Series, Backhuys Publishers, Leiden, The Netherlands, 241–289.

Sanders, R.E., R.J. Miltner, C.O. Yoder, and E.T. Rankin. 1999. The use of external deformities, erosion, lesion, and tumors (DELT anomalies) in fish assemblages for characterizing aquatic resources: a case study of seven Ohio streams, in Simon T.P. (Ed.). *Assessing the Sustainability and Biological Integrity of Water Resources Using Fish Communities*. CRC Press, Boca Raton, FL, 225–248.

SAS Institute Inc. 1996. *SAS User's Guide: Statistics.* SAS Institute, Cary, N.C.

Schloesser, D.W., T.B. Reynoldsen, and B.A. Manny. 1995. Oligochaete fauna of western Lake Erie, 1961 and 1982: signs of sediment quality recovery, *Journal of Great Lakes Research,* 21, 294–306.

Simon, T.P. and J. Lyons. 1995. Application of the index of biotic integrity to evaluate water resource integrity in freshwater ecosystems, in W.S. Davis and T.P. Simon (Eds.). *Biological Assessment and Criteria: Tools for Water Resource Planning and Decision Making*, Lewis Publishers, Boca Raton, FL, 245–262

Strong, A.J., S.A. Wilkinson, and M.J. Lydy. 1998. Fish communities in the Little Arkansas River Basin, Kansas 1884–1996, *Transactions of the Kansas Academy of Science,* 101, 17–24.

Thoma, R.F. 1999. Biological monitoring and an index of biotic integrity for Lake Erie's nearshore waters, in T.P. Simon (Ed.). *Assessing the Sustainability and Biological Integrity of Water Resources Using Fish Communities*, CRC Press, Boca Raton, FL, 417–461.

U.S. Environmental Protection Agency. 1986. Method #365.1, 365.4. *Test Methods for Evaluating Solid Waste: Physical/Chemical Methods*. EPA-SW846, Office of Solid Waste and Emergency Response, Washington, D.C.

Welch, E.B., R.R. Horner, and C.R. Patmont. 1989. Predictions of nuisance periphytic biomass: a management approach, *Water Research*, 23, 401–405.

Williams, D.J., K.W. Kuntz, S. L'Italien, and V. Richardson. 1998. Lake Erie Surveillance Program: Spatial and Temporal Trends of Select Parameters, with Emphasis on 1994–1995 Results. Environment Canada Report Number EHD/ECB-OR/98–05/1.

Wood, K.G. 1973. Decline of *Hexagenia* (Ephemeroptera) nymphs in western Lake Erie, in W.L. Peters and J.G. Peters (Eds.). *Proceedings of the First International Conference on Ephemeroptera, August 17–20, 1970*, Florida Agricultural and Mechanical University, 26–32.

12 Regional Ecological Normalization Using Linear Models: A Meta-Method for Scaling Stream Assessment Indicators

Michael J. Wiley, Paul W. Seelbach, Kevin Wehrly, and Joan S. Martin

CONTENTS

0-8493-0905-0/03/$0.00+$1.50

12.1 INTRODUCTION

Multiple state and national agencies and various regional groups often collect ecological data from the same river systems using disparate methodologies chosen to maximize individual program goals. Differences in sampling design, sampling methodology, and analytical procedures produce a paucity of integrated regional analyses, despite abundant regional data (National Research Council, 1995; Ward, 1996). For this reason recent technical and academic discussions have focused on standardization of methods (Barbour et al., 1999) and evaluation of competing assessment methodologies (Reynoldson et al., 1997; Hawkins et al., 2000), with the goal of reaching a common methodological basis for integrated large-scale stream assessments.

Stream resource inventories enumerate and classify ecological resources across a region of interest. They are often undertaken to aid proactive development of regional management and conservation strategies (Biggs et al., 1990, New Zealand; Claessen et al., 1994, The Netherlands; Seelbach and Wiley, 1997, Michigan; Higgins et al., 1998, Great Lakes Basin; Sowa et al., 1999, Missouri). Environmental impact assessments retrospectively evaluate changes in stream resource condition related to human activities (e.g., Gilliom et al., 1995, NAWQA; Yoder and Rankin, 1995, Ohio EPA; U.S. Environmental Protection Agency (USEPA), 1997, EMAP). Both resource inventory and environmental assessment programs are typically interested in estimating the current ecological status of stream ecosystems. Status in this context can be defined as the observed condition relative to some potential condition, i.e., *reference condition* (Karr et al., 1986; Gallant et al., 1989; Claessen et al., 1994; Hakanson, 1996; Reynoldson et al., 1997; Barbour et al., 1999).

Determining the appropriate reference condition for a given indicator metric at a specific site is a fundamental methodological challenge common to all ecological assessment work (Reynoldson et al., 1997; Seelbach et al., 2001). Erroneous specification of the reference condition leads to a biased assessment. Likewise, failure to account for variance in indicator measurements due to (1) natural geographic and temporal processes and (2) site- and gear-related biases in sampling methodology, will be translated into systematic errors in assessments of status. The high natural variability characteristic of stream ecosystems (Bryce and Clarke, 1996; Wiley et al., 1997) is likely to exacerbate these effects. Without a clear understanding of the variances around estimates of site condition and reference condition, we cannot distinguish between deviations from expectation due to (1) natural variation, (2) those due to measurement error, (3) those due to erroneous specification of reference condition, and (4) those that should be attributed to human impacts.

A second major challenge facing the integration of regional assessment data is the large diversity of indicator variables (e.g., fish, invertebrates, periphyton, habitat, and water chemistry) employed by various organizations. The multimetric approach embraces the incorporation of multiple indicators as conceptually desirable (Karr et al., 1986). However, the scoring schemes required to integrate component metrics make the overall multimetric values difficult to compare quantitatively to other types of measurements. Fundamentally different kinds of measurement units, often reported using arbitrary scalings for convenience, make combining data from different organizations and jurisdictions problematic.

How can we integrate the assessment products of different organizations and agencies, each with its own unique rationale, survey method, and model for generating reference expectations? Are overlapping surveys a redundant and inefficient use of public resources? Or can we improve both precision and accuracy in ecological assessments by combining disparate datasets and the views they offer? In Michigan the ecological character of streams and rivers is strongly shaped by their specific landscape positions and the resulting hydrogeomorphic contexts (Seelbach and Wiley, 1997; Wiley et al., 1997; Wehrly et al., 1998; Wehrly, 1999; Wiley and Seelbach, 2001; Baker et al., 2001; Baker et al., 2001a, 2001b; Zorn et al., 2002). Might a landscape perspective be brought to bear on the problem of resource assessment? If so, how can we introduce landscape context into assessment protocols and metrics that are already well established and useful in their institutional contexts?

This chapter presents a general method for standardizing and correcting geographical and other systematic biases in assessment data by explicitly modeling reference conditions using standard linear statistical techniques. We then employ these landscape-sensitive models to normalize biological, chemical, and physical indicator data. Finally, we illustrate (at both statewide and river basin levels) how systematic biases in Michigan assessment data can be removed, and how very different metrics and datasets can be combined to provide more comprehensive regional assessment of the status of streams and rivers.

Specifically, in this chapter we will (1) present a generalized approach for estimating site-specific reference conditions and normalizing indicator scores using landscape-sensitive models, (2) present two case studies that illustrate the use of regional normalization models, and (3) discuss relationships between the ecological normalization procedure described and other common assessment approaches now in use. In the first case study, we evaluate and combine statewide fish community datasets of two agencies to provide a single comprehensive statewide assessment. In the second case study, macroinvertebrate, chemistry, and physical habitat assessment data are combined to assess ecological status in a rapidly developing watershed.

12.2 REGIONAL ECOLOGICAL NORMALIZATION: AN EXPLICIT MODEL-BASED APPROACH TO EVALUATING INDICATOR METRIC DATA

Since genuine historical observations are seldom available, reference conditions for a site must usually be estimated using a *model* (i.e., an approximation) of some type. Approaches to modeling reference condition to date can be identified as involving (1) explicit mathematical modeling of reference conditions (reviewed in Seelbach et al., 2001), (2) implicit modeling by means of classification (e.g., Karr et al., 1986 and related IBI-type methods), or (3) a combination of both (e.g., Wright et al., 1988 and related RIVPACS-type methods; see Figure 12.1).

Stream classification in this context is used as a basis for extrapolating sampled attributes from a set of streams judged to be "least disturbed" (reference streams), to a stream of interest (Davis and Henderson, 1978; Zonneveld, 1994). Classification categories can be based on extensive collections of site-specific data (bottom-up classification; e.g., HGM, Hauer and Smith, 1998; RIVPACS, Wright et al.,1988) or on a top-down regionalization approach based on geographic affinities (e.g., ecoregional and watershed stratification, Gallant et al., 1989; Seelbach et al., 2001). Regardless of the modeling method, the accuracy and precision of the reference condition model is fundamental to the accuracy and precision of an assessment.

We use the term *regional normalization* to refer to a generalized procedure in which indicator data are centered and scaled using explicit site-specific models constructed from regional datasets (Figure 12.2). *Normalization* is generally used in statistics and engineering to refer to a mathematical transformation that alters or cancels the units of a parameter in a way that facilitates interpretation. In this case, we begin by centering a *raw indicator data score* (I_{raw}; for example, the number of native fish species at a site) as a deviation from the score we would normally expect (I_{exp}):

$$\textit{deviation score}\ (I_{dev}) = \text{observed score} - \text{expected score} = I_{raw} - I_{exp}$$

Our expectation (I_{exp}) is an explicit estimation of the reference condition at a particular site, and is the product of a normalizing model that predicts scores for specific metrics from local habitat and catchment characteristics. In the examples that follow these normalizing models are linear statistical models but in fact could be any type of model that provides reasonably accurate and precise estimation. Ideally the deviation score tells us how similar a test site score is to the regional average for sites having similar local habitat and catchment characteristics. The deviation score could also be expressed as a percentage or ratio, rather than as a difference (e.g., this site has 50%

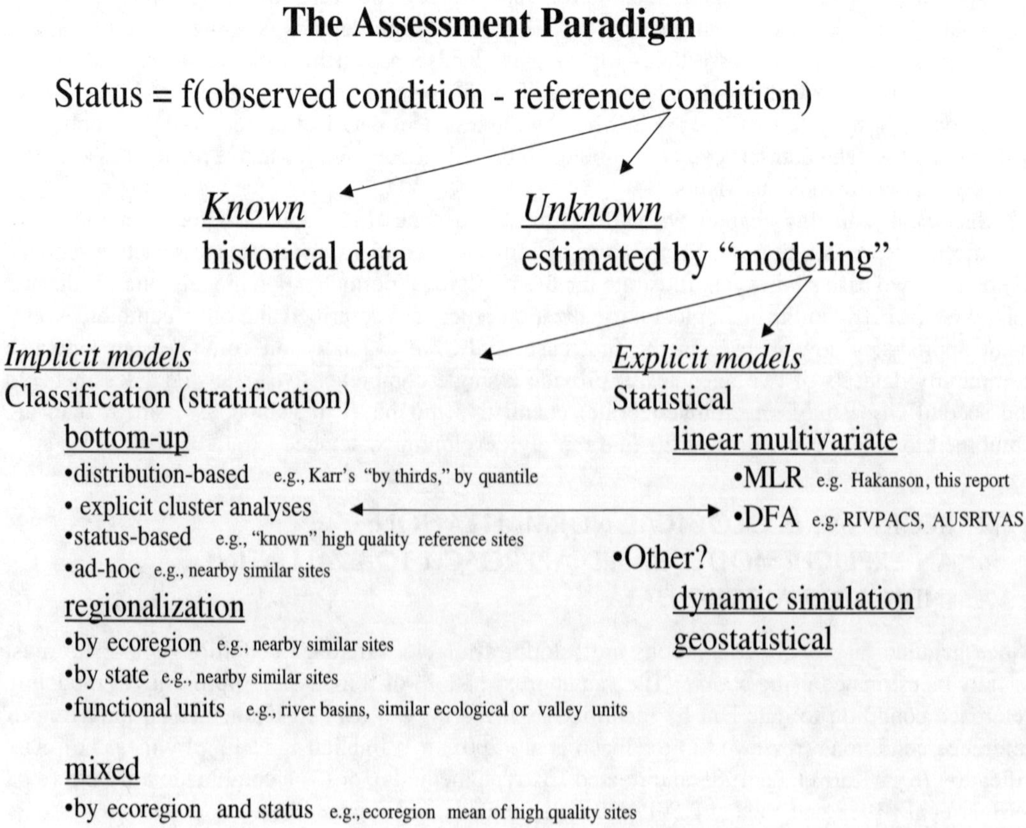

FIGURE 12.1 Basic conceptual model of approaches to reference modeling in ecological assessment.

of the native fish species that the normalizing model predicts it should have). Deviation scores are then scaled to create a *normalized score.* This is accomplished by dividing the deviation score by a measure of the natural variation around the predicted score:

$$I_{norm} = (I_{raw} - I_{exp}) \sqrt{Var\ (I_{exp})}$$

The idea is to express the deviation of a particular metric in terms of its predicted regional variability. Operationally we propose to scale metric deviation scores in units of estimated *standard deviations (SDs) of the reference*. This type of scaling is useful because (1) the larger the unexplainable variance a metric exhibits, the less any given absolute deviation from an expected value tells us about relative status, and (2) this scaling allows us to directly compare normalized scores computed from very different metrics (for example, native fish taxa, macroinvertebrate families, water temperature, and conductivity).

A direct analogy from multiple linear regression (MLR) analysis is the use of standardized regression coefficients that allow comparisons of the relative magnitude of effects of independent predictor variables having different units (Sokal and Rohlf, 1995). The normalized scores can also be used to construct *ad hoc* summary metrics (i.e., multimetrics), and compare trends within and between classes of indicators (typical goals of assessment programs; USEPA, 1997). An overview of the procedure we are proposing appears below.

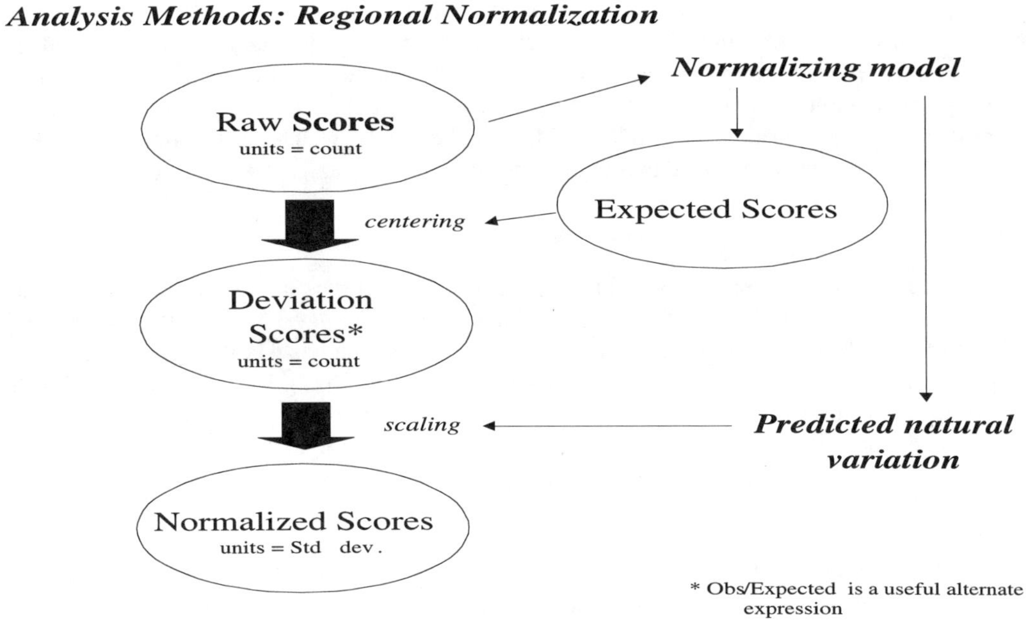

FIGURE 12.2 Schematic overview of the regional normalization procedure.

12.2.1 Modeling Expected Scores

The clear success of the multivariate approach pioneered in Europe, England, and Australia (Wright et al., 1988, 1989; Norris and Georges, 1993; Turak et al., 1999; Simpson and Norris, in press) demonstrates that linear statistical modeling is well suited to predicting stream community characteristics. MLR modeling of hydrologic, chemical, fish community, and other ecological parameters in streams and rivers has a particularly well documented history (Bowlby and Roff, 1986; Fausch et al., 1988; Osborne and Wiley, 1988; Kleiman, 1995; Wiley et al., 1997, 1998; Wehrly et al., 1998; review by Seelbach et al., 2001; Wiley and Seelbach, in press). Modeling expected values for use in a regional normalization does not necessarily require a particular modeling approach. It needs only a reasonably accurate quantitative estimate of the "expected value" and an associated procedure for estimating the variance of that estimate.

Our experience has been that relatively simple MLR models of indicator metrics of interest are adequate for producing the required estimates. Predictor (independent) variables in the regressions can consist of landscape-level variables derived from map analysis alone (Seelbach et al., 2001), in which case predictions can be made prior to site visitation based on GIS analysis of regional and catchment properties. The regression models can also employ site-specific data (e.g., substrate distributions, bank condition, water quality measures; E.A. Baker et al., 2001). The overall accuracy of the normalizing model can be easily examined and communicated using standard statistical methods. For reasons that should become clear below, one desirable characteristic of the normalizing model that is crucial to this method is incorporation of effects related to human impacts. Thus, the dataset used to generate the normalizing model, in contrast to typical IBI and RIVPACS reference sets, should include representation of sites displaying the full range of impacts and qualities that exist throughout the region.

Moving from MLR model construction to prediction of expected values (I_{exp}) requires an adjustment of the MLR equation if predictor variables include parameters that the analyst clearly wants held to some "least impacted" standard in the assessment. For example, if in an MLR analysis of regional fish data, *percent urban land use in the catchment* has a statistically significant negative effect on species richness, then the analyst is faced with a choice. He or she must decide whether to predict the expected species richness given observed urbanization or to predict the expectation in the absence of urbanization. The second choice implies a more rigorous standard and the coefficient for *percent urban land use* in the normalizing model should be set to zero or to the minimum value in the regional dataset. The normalizing model then can be seen as a special case (adjustment) of the equation derived from MLR modeling, in which certain variables may be controlled to generate site-specific expectations under minimally impacted conditions.

12.2.2 Normalizing Scores by Their Variances

To produce normalized scores, deviation scores are scaled in terms of standard deviations. This scaling can be accompliished in at least three distinctly different ways. Scores can be normalized by the variance around the prediction. Small negative deviations can yield large negative normalized scores if the normalizing model is very accurate (i.e., the analyst has great confidence in what the expected value should be). Conversely, large deviations will yield small normalized deviations if the normalizing model explained little of the parameter variance. This property is attractive in regulatory settings since it scales expectations to our demonstrable knowledge. Difficulties with this approach, however, are considerable:

1. The variance is a variable that changes with the specific values of the predictor variables. This leads to normalization by a site-specific standard that is conceptually difficult to interpret.
2. The validity of the estimated variance around the prediction can be heavily influenced by the multivariate normality of the dataset employed.
3. The variances around the normalized model predictions are fairly difficult to calculate and are not routinely available through most statistical software packages.

A second approach is to use a single estimate related to the goodness-of-fit of the normalizing model to the regional calibrating dataset. Scaling in this way avoids most of the difficulties above but still retains a normalization that depends on both the natural variability of the modeled indicator and the ability of the analyst to model (account for) it. The standard error around the regression is a reasonable candidate.

The third approach is to scale the deviations by an estimate of the natural variation in the indicator. This variation can be estimated by computing the variance of the predictions made by the "adjusted" MLR model (normalizing model). Normalized scores are less sensitive to larger deviations from expected values in indicator variables that vary widely across the region, and more sensitive to smaller deviations from expectation in variables that are spatially more consistent. It is certainly possible that this type of variation in the indicator might be related to the other two types (variance around the predictions and residual variance about the regression). It also seems possible that for some indicators these types of variation might not correspond (cases where model fit is poor because of lack of linearity or latent variables).

We usually opt for the third approach for practical reasons. Scaling by the predicted indicator variance gave us more consistently sized standard errors across both biological and physical indicator metrics. This consistency helps in graphic comparisons of different indicator metrics and the construction of multimetric summaries.

12.2.3 Constructing Summary (Multimetric) Indicators

Since normalized indicator scores are in standard deviation units, they can be mathematically combined into summarizing scores. Following the general guidelines used for combining habitat suitability indices (Terrell et al., 1982), indices that are not statistically independent can simply be averaged. Individual metrics can be differentially weighted if desired. Alternately, if metrics are statistically and conceptually independent, it may be desirable to compute geometric means. In this case, an offset greater than the largest negative score should be used before the values are multiplied, for example, the geometric mean of I_1 and I_2 = $((I_1 + \text{offset}) * (I_2 + \text{offset})^{.5} - \text{offset}$, where the offset is a positive integer large enough to ensure all scores are positive before the multiplication.

12.2.4 Comparing Normalized Data to Other Assessment Output

Indicator values of any kind can be regionally normalized. IBI scores, for example, can be regionally normalized by simply subtracting the mean of regional reference site values and dividing by the standard deviation of that mean. The "region" can be any meaningful classification of sites that reduced intragroup variance. Alternately, an MLR model could be fit relating IBI values to catchment and local habitat variables and used to construct a normalizing model. Note that it is unlikely that the different methods would yield identical normalized values, since the first uses the mean of existing regional values as its model expectation, while the second could presumably correct for various components of natural intersite variation. Like the methods of Wright et al. (1988), normalizing models of individual metric components can also be combined to produce expected IBI scores. RIVPACS-type analyses (Wright et al., 1988,1989; Norris and Georges, 1993; Simpson and Norris, in press) produce predicted species lists that can be combined into community metrics and compared to observed values (i.e., deviation scores). These constructed metrics could also be normalized using variance estimates from the modeling dataset or based on variance estimates for discriminant function analysis.

A common problem facing large-scale assessment programs that cross jurisdictional or administrative boundaries is the lack of standardization in sampling protocols, which leads to significant biases between datasets collected by different regions/agencies (the first example application described below is a good case in point). Likewise, the often incompatible sets of indicator metrics employed by various organizations make combining data from different organizations problematic. This is clearly one of the major obstacles facing a national assessment of stream quality based on integrating state and regional surveys. Because normalized indicator metrics all have the same units, it is possible to pool normalized scores from differing datasets and/or different indicator metrics as long as regionally appropriate normalizing models are used. In this way, otherwise incompatible datasets and metrics can be normalized individually, and the normalized scores combined to form a larger regional analysis.

12.3 EXAMPLE APPLICATIONS

12.3.1 Case Study 1: Modeling Reference Conditions for Michigan Stream Fish Communities

Two state agencies in Michigan conduct inventories and assessments of stream fish communities. Although both agencies sample across the state, they have different objectives, sampling strategies, and methodologies. The Michigan Department of Environmental Quality (MDEQ) uses a multimetric IBI-based (Procedure 51 or P51; 1997) bioassessment protocol in wadable streams using primarily single-pass sampling with back-pack DC electrofishing gear and occasionally tow-barge electrofishing equipment in mid-sized rivers. The goal of this sampling is to produce a representative species list and reasonable estimates of relative abundance for use in regulatory assessment and federal reporting.

The Michigan Department of Natural Resources (MDNR), Fisheries Division separately compiles statewide data on fish communities, using more rigorous sampling protocols. Its Michigan River Inventory (MRI) Project database (Seelbach and Wiley, 1997) includes full community population estimates for several hundred sites based on 3-pass, depletion electrofishing using tow-barge mounted gear, or rotenone sampling. The objective of MRI sampling was to compile a complete species list with population estimates (numbers and weights). The MRI database was designed for ecological research focused on management of fish communities and game fishes.

The two programs also employ very different methods for specifying reference condition. MDEQ uses a regionalization approach to develop generalized site reference conditions for scoring IBI-type metrics, while MRI has employed landscape-based modeling to develop site-specific reference predictions (Wiley et al., 1997; Zorn et al., 2002; Wehrly et al., 1998, Seelbach et al., 2002). The objective of the analysis reported here is to explore the feasibility of combining these two datasets to facilitate interagency analyses and collaboration in fisheries-related assessment activities.

12.3.1.1 Introduction and Methods

To test our ability to combine MRI and MDEQ fish assemblage data, we used MRI data from 215 sites located throughout the Lower Peninsula and MDEQ data from 174 sites. The MDEQ sites were selected at random from an existing larger dataset. For each dataset, we calculated the ten metrics used in MDEQ's modification of Karr's IBI that represent measures of richness, composition, tolerance, and trophic structure or habitat requirements (MDEQ, 1997).

Landscape variables describing catchment and reach characteristics associated with each of the MRI sites had been previously determined or modeled using existing databases developed as part of the Michigan River Inventory Project, following methods described by Seelbach and Wiley (1997). These geographic data layers are maintained and accessed using a geographic information system (GIS) housed at the University of Michigan's School of Natural Resources and Environment. A parallel dataset on landscape and reach characteristics was developed for the MDEQ sites following the same methodology used to characterize MRI sites. Individual upstream catchment boundaries were delineated for each site based on subwatershed divides mapped by MDNR from United States Geological Survey (USGS) 1:24,000-scale topographic maps.

Watershed boundaries were then locally modified for the MDEQ sample sites following a 3 arc-second digital elevation model (at a scale of 1:250,000). Linear (stream network) and circular (sample point) buffers were generated using Arc/Info software (ESRI, Inc.); land use, surficial geology, and soils were summarized for each catchment and buffer (see Seelbach and Wiley, 1997 for map metadata). Groundwater flux data were summarized for each catchment and buffer based upon a spatial model (map layer) predicting maximum potential groundwater velocity based on Darcy's Law (Wiley et al., 1997; M.E. Baker et al., 2001b). Measurements of channel morphology (width, mean depth, and length sampled) were included in the MDEQ dataset.

12.3.1.2 Results and Evaluation

12.3.1.2.1 Compatibility of the MDEQ and MRI Datasets

Initial comparisons of the sample of MDEQ data ($n = 174$) with the MRI dataset ($n = 215$) were based on patterns of increasing taxa richness with increasing site catchment area — a common correlation observed in Michigan and elsewhere (Angermeier and Schlosser, 1989; Wiley et al., 1997; Zorn et al., 1997). Procedure 51 species richness was lower than MRI sample richness for streams of equivalent size, and this discrepancy increased with catchment area (i.e., increased in larger rivers; Figure 12.3). Statistical comparisons of slopes and intercepts confirmed that the two datasets had fundamentally different relationships with stream size, indicating that simple pooling of the datasets for analysis would be inappropriate (ANCOVA $p < 0.05$).

The observed pattern of bias (undersampling) in the MDEQ data was very consistent with what we would expect from a less efficient sampling method. The nature of the sample run (single pass) and the typical gear employed (DC backpack) reduced sampling efficiency compared with the more intensive MRI sampling methods. Because of electrode positions and current capacity, backpack units have smaller field sizes even when current flow through the field is equivalent. Consequently, smaller fish are more likely to detect field edges and escape, and/or operator motion is more likely to drive smaller fish away (because the effective field edge is closer to the operators). Similarly, larger fish are more prone to escape because they can burst-swim out of the effective field more

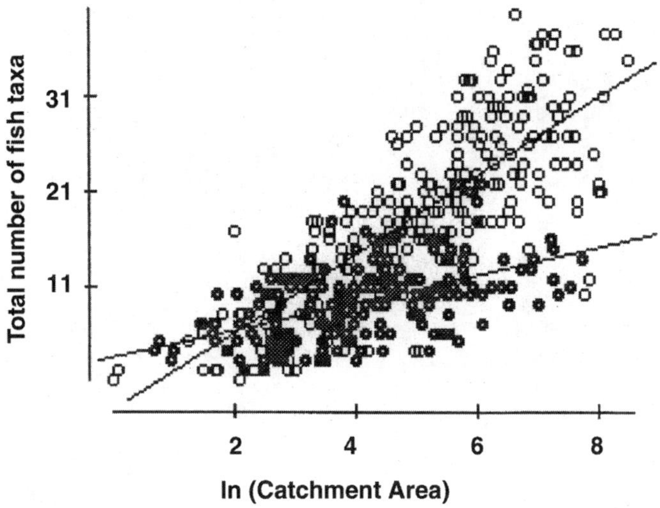

FIGURE 12.3 Comparison of MDEQ (sample; dark circles) and MRI (light circles) datasets. Axes are total number of species plotted against $\log_{10}$ of site catchment drainage area.

easily. Electrofishing efficiencies decline rapidly in larger rivers (Bayley and Dowling, 1993). In the MRI dataset, larger river sites were sampled with rotenone, which is relatively efficient even in larger rivers (Seelbach et al., 1994). Hence the difference in diversity between the two datasets became greater as the size of the river increased.

12.3.1.2.2 Landscape-Based Modeling of Reference Conditions

Although the two datasets could not be jointly analyzed, regional normalization allowed the results of separate assessment analyses to be meaningfully pooled and compared. Based on exploratory analysis, we chose a consistent normalizing model (EQS 1–3) for each indicator and fit the models independently to each dataset. For the purposes of illustration, landscape-sensitive MLR models were generated for *total number of fish species* (TOTspp) and for *number of intolerant fish species* (INTOLspp), two of the primary metrics used in the P51 score (MDEQ Procedure 51 and most other IBI-type assessment indices). These two variables explained 58% and 65% of the variation in P51 scores for the MDEQ and MRI datasets, respectively. We also built a MLR model for the Procedure 51 score. The general models for each of these three variables were as follows:

$$\ln(\text{TOTspp}) = \alpha + \beta_1 \ln X_1 + \beta_2 X_2 + \beta_3 \ln X_3 + \beta_4 \ln X_4 + \beta_5 \ln X_5 + \beta_6 \ln X_6 \tag{12.1}$$

where X_1 = catchment area, X_2 = stream width, X_3 = stream gradient, X_4 = percent urban land use in catchment, X_5 = ground water index (Wiley et al., 1997; M.E. Baker et al., 2001b) value for 4-km site buffer, and X_6 = percent open water cover for a 4-km site buffer.

$$\text{INTOLspp} = \alpha + \beta_1 \ln X_1 + \beta_2 X_2 + \beta_3 \ln X_3 + \beta_4 \ln X_4 + \beta_5 \ln X_5 + \beta_6 \ln X_6 + \beta_7 \ln X_7 \tag{12.2}$$

where X_1 = catchment area, X_2 = percent urban land use, X_3 = percent agricultural land use, X_4 = percent open water cover for a 4-km site buffer, X_5 = ground water index value for a 4-km site buffer, X_6 = stream width, and X_7 = binary flag (1 if more than 90% of species are coldwater taxa).

$$\text{P51} = \alpha + \beta_1 \ln X_1 + \beta_2 X_2 + \beta_3 \ln X_3 \tag{12.3}$$

where X_1 = catchment area, X_2 = stream width, and X_3 = percent urban land use.

TABLE 12.1
Regression Model Diagnostics from MLR Normalizing Models for Various Dataset Partitions[a]

		MDEQ Procedure 51		Intolerant Fish Taxa		Total Fish Taxa	
Dataset	***df***	**R^2**	**F-Statistic**	**R^2**	**F-Statistic**	**R^2**	**F-Statistic**
DEQ North	*11*	32.1	3.2	47.2	5.9	45.9	5.7
DEQ South	*155*	25.5	18.7	27.8	11.2	30.4	12.4
MRI North	*35*	59.3	19.4	48.1	7.3	75.8	26.5
MRI South	*170*	50.4	59.5	51.7	31.9	58.3	41.3
DEQ all	*184*	26.9	22.2	28.5	13.7	36.0	18.8
MRI all	*217*	51.9	77.3	45.9	31.4	72.0	96.6

[a] The first four rows were used in the normalization analyses. All F-Statistics were significant at $p < 0.05$ except the P51 regression with the DEQ North data (F = 3.2).

Exploratory analysis also indicated significant differences in model parameter estimates for the TOTspp and INTOLspp datasets and between sites in northern (northern lakes and forests (NLF) and north central hardwood forests (NCHF)), and southern (Huron–Erie lake plains (HELP), Southern Michigan–Northern Indiana till plains (SMNITP), and eastern corn belt plains (ECBP)) ecoregions of the state. Therefore, models were fit separately for northern DEQ, southern DEQ, northern MRI and southern MRI sites. All MLR models generated were significant (F statistics ranged from 3.2 to 59.5; p-values, from 0.05 to 0.00001; adjusted R-squared values, 0.26 to 0.76). Fits to MDEQ data were consistently poorer than to MRI data across both northern and southern ecoregions (Table 12.1).

Regional normalizing models for each indicator metric (TOTspp, INTOLspp, and P51) were constructed from the MLR models. In the normalizing models, urban and agricultural land use values were set to zero. Deviation scores were calculated as observed minus expected scores, with the order reversed (expected minus observed) if > 90% of the species were cold water taxa so that the normalized score would appropriately decline when taxa increased at cold groundwater-dominated sites. Deviation scores were scaled in terms of the standard deviation of the values predicted by the normalizing model (by dataset). Once normalized, indicator metrics from both datasets and ecoregions were recombined for analysis with respect to catchment land use and ecoregion.

Normalization generally improved the sensitivity of indicator metrics to landscape composition (Table 12.2). The sensitivities of scores from the combined datasets for raw P51 scores and for regionally normalized INTOLspp and P51 scores to key landscape attributes are illustrated in Figure 12.4. Raw P51 scores declined strongly with *increasing percent urban landcover*, but showed a small positive response to *percent agricultural landcover*. Despite a scoring protocol based on stream width, P51 values appeared strongly biased with respect to stream size (Figure 12.4c). Normalized P51 scores retained sensitivity to urban land use, while the excessive correlation with stream size was corrected. Prior to normalization, P51 scores calculated from the MDEQ dataset were significantly lower than those calculated from the MRI data (mean difference was –3.6). After normalization, however, the mean of the P51 scores did not differ significantly between datasets (t-test; $p > 0.05$). Normalized P51 scores showed little relationship to *percent agricultural land use* ($r = 0.03$).

Raw INTOLspp (not shown) scores were negatively correlated ($r = -0.21$) with *percent urban land use* and positively with *stream size* ($r = 0.61$), but not with *percent agriculture land use* ($r = -0.05$). Regionally normalized scores for intolerant species (Figures 12.4g and h) were negatively and significantly correlated with both urban and agricultural land uses. A strong negative correlation with stream size was also found, suggesting a widespread loss of intolerant species in

TABLE 12.2
Pearson Correlation Coefficients between Catchment Variables and Fish Community Assessment Indicators

Indicator Metric	% Urban Land Cover	% Ag. Land Cover	ln (Catchment Area)
P51	–0.27	+0.05	+0.65
Normalized P51	–0.26	+0.03	+0.10
Intolerant fish taxa	–0.21	–0.05	+0.61
Normalized intolerant fish taxa	–0.12	–0.33	–0.31
Total fish taxa	–0.13	+0.22	+0.71
Normalized total fish taxa	–0.23	+0.08	+0.04

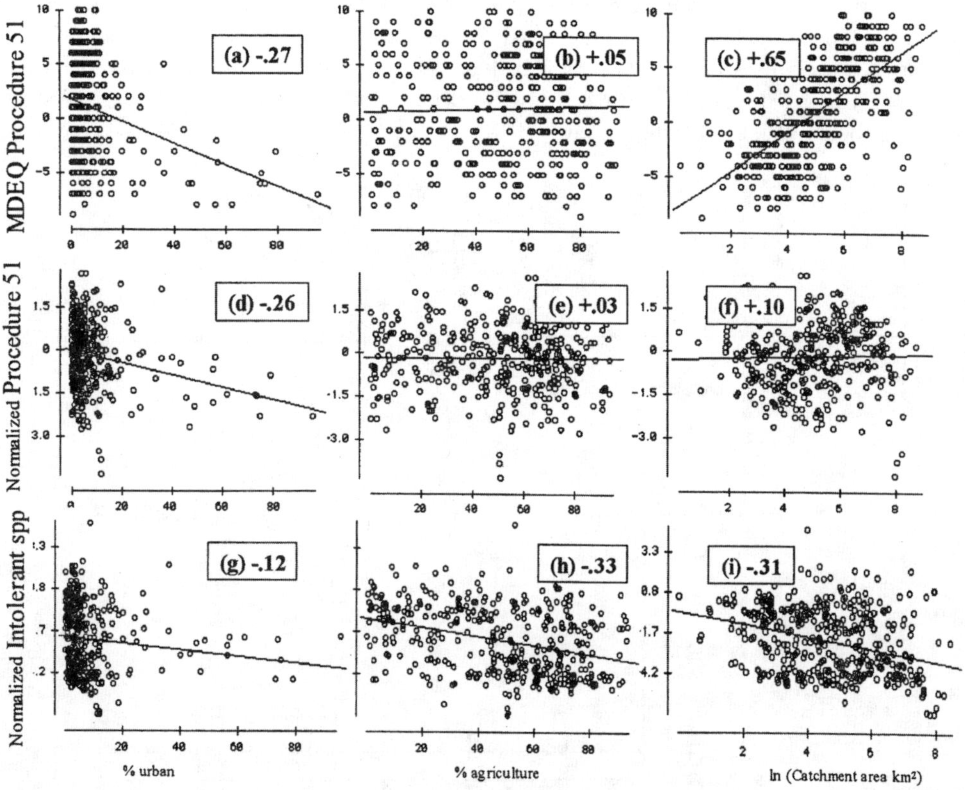

FIGURE 12.4 Sensitivity of raw (a-c) and normalized (d-i) scores to landscape variables. Numbers in boxes are simple correlations coefficients. The X-axis for a, d, and g is the proportion of the upstream catchment in urban land cover. The X-axis for b, e, and h is the proportion of the upstream catchment in agricultural land use. The X-axis for c, f, and i is the natural log of the upstream catchment area.

larger river systems. As in the P51 score, raw intolerant species scores differed significantly between datasets, but normalized intolerant species scores did not. Normalization scores for total species richness (not shown) also improved sensitivity to urban land use (correlations with *percent urban land use* were –0.13 and –0.23 for raw and normalized scores, respectively), but neither raw nor normalized total species scores were significantly related to agricultural land use. A strong size bias in the raw total species index was also corrected (r = 0.71 and 0.04 for raw and normalized counts, respectively).

12.3.1.2.3 Overall Status of Michigan Streams Based on Pooled Datasets

Centering and scaling indicator scores using landscape-based models provided several benefits. Regional normalization allowed us to compare and combine MDEQ and MRI assessment data and compute comparable biological metrics based on both datasets. It also corrected the inherent catchment size bias of the P51 indicator and improved sensitivity to landscape alteration in total species and intolerant species metrics.

Pooling normalized scores from the MDEQ and the MRI databases also provided a powerful regional sample for evaluating the overall status of Michigan rivers. We created a combined fish community metric by averaging all three normalized scores for the P51, number of intolerant fish taxa, and number of total fish species metrics derived from both surveys. Overall, if we used a –1 standard deviation criterion to identify serious degradation, the pooled analysis suggested that almost half the sites examined statewide were seriously degraded. Geographically, however, we noted considerable variation in biological integrity.

In a comparison by ecoregion (Figure 12.5a), the HELP sites (n = 75) were clearly the most degraded. About 75% of the sites sampled in that ecoregion had negative overall scores; the median score was –1.18 (SD). About 67% of these HELP sites had scores more than 1 SD below modeled expectations and could be considered seriously degraded. The distribution of scores for the SMNIPT sites (n = 252) was similar, but they had a less deviant median score of –0.68, with 61% of the sites falling below the 1 SD criterion level.

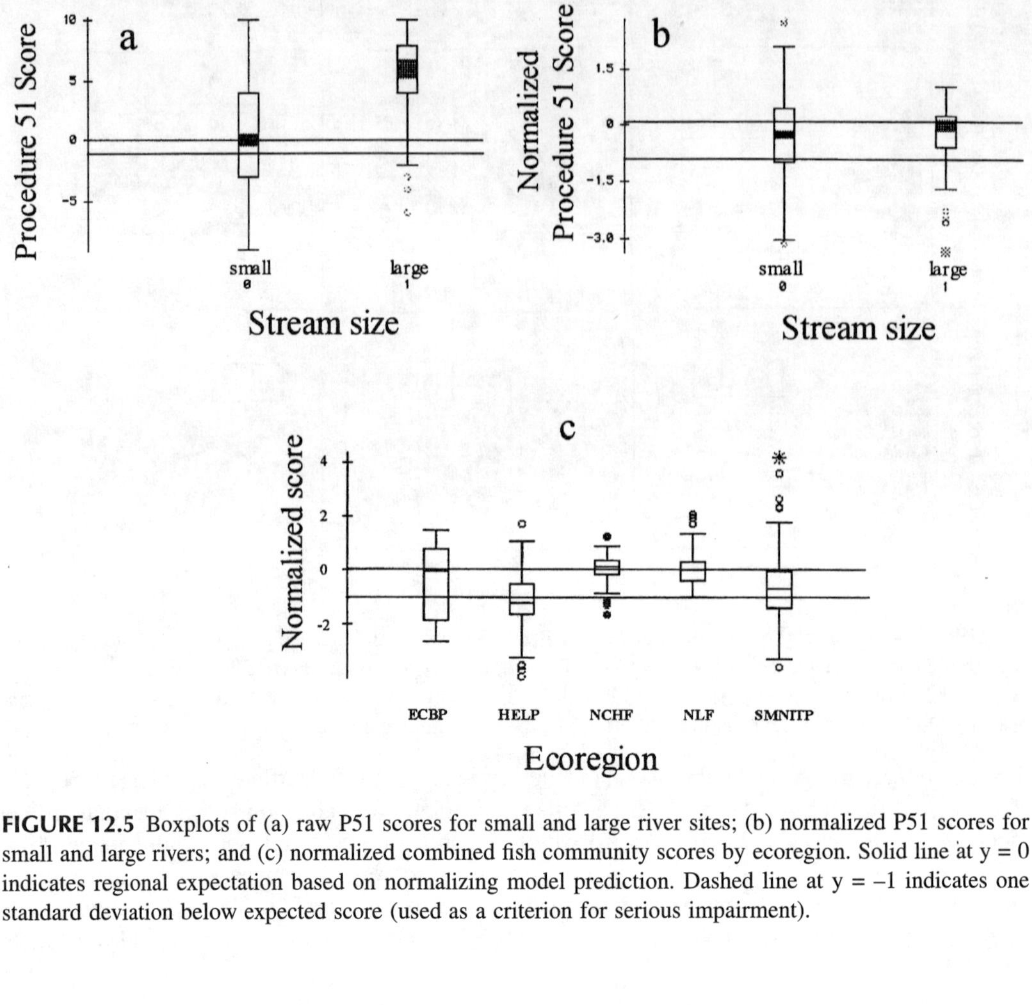

FIGURE 12.5 Boxplots of (a) raw P51 scores for small and large river sites; (b) normalized P51 scores for small and large rivers; and (c) normalized combined fish community scores by ecoregion. Solid line at y = 0 indicates regional expectation based on normalizing model prediction. Dashed line at y = –1 indicates one standard deviation below expected score (used as a criterion for serious impairment).

In contrast, sites in the NLF ecoregion had a median score slightly better than that predicted by the regional model (0.016; n = 55) and only 7% of the sites were more than 1 SD below expectation. NCHF sites had a median value very close to expectation (–0.0070; n = 30) and no sites fell below the 1 SD criterion level. The ECBP ecoregion was poorly represented in our combined dataset (n = 17). ECBP sites we sampled had a median score of –0.016 and 44% had scores indicating serious degradation.

Compared to the raw Procedure 51 index scores, regionally normalized fish metrics gave a strikingly different picture of the status of Michigan streams. Raw Procedure 51 scores suggested that most smaller streams in this state (arbitrarily defined as having catchment areas <1000 km^2) were in poor condition while most of the larger sites were in good condition (Figure 12.5b). This result was driven largely by strong correlations between raw taxa count data and stream size (catchment area) described above.

The normalized scores statistically corrected for this trend and in contrast, indicated little difference between smaller and larger river sites (Figure 12.5c). Indeed, larger river sites appeared to be in slightly worse condition overall because of the declining intolerant species scores described above. About 27% of the smaller sites had P51 values over 1 SD below expectation, as did 17% of the larger streams. However, 86% of the larger stream sites failed to meet expectations in terms of the number of intolerant fish taxa compared with 61% failing the 1 SD criterion in smaller streams. Using the overall metric described above, 66% of the larger river sample sites and 46% of the smaller river sample sites appeared degraded.

12.3.2 Case Study 2: Macroinvertebrate and Habitat Monitoring Data from the Huron River

12.3.2.1 Introduction and Methods

In southeast Michigan, the Huron River Watershed Councils' Stream Adopter Program (HRWC-SAP) is an example of a smaller regional government group involved in stream assessment. Its volunteers have been sampling invertebrates, chemistry, and stream habitat characteristics for the purposes of stewardship and advocacy since 1993. They have developed their own sampling design and methods that are distinct from those employed by MDEQ and MRI. This case study demonstrates the application of regional normalization to macroinvertebrate, chemical, and physical habitat data with the goal of building multimetric evaluations of stream condition by combining normalized versions of several different biological and physical indicator metrics (Wiley and Martin, 2000). Much of the Huron River basin is being rapidly urbanized and impacts of development on stream quality were the primary interests of the watershed council program.

HRWCSAP collection protocols were similar to those employed by many state and federal rapid bioassessment programs (Martin, 1998). Macroinvertebrate data were collected twice a year from 41 tributary creeks distributed throughout the 2352 km^2 basin of the Huron River. In most years, additional winter sampling was carried out with collection efforts focused exclusively on winter stonefly fauna. Physical data describing channel characteristics (average depth, substrate, bank condition, etc.) were also collected by volunteers on different dates and less frequently. Complete sets of biological and physical data were available for 36 of the 41 sites. For this analysis, status assessment was based on collections averaged from 1996 through 1999. Normalizing models were based on average site values for 1997 and 1998. Biological metrics used in the analysis included (1) total number of insect families, (2) number of EPT (Ephemeroptera, Plecoptera, and Trichoptera) families, (3) number of sensitive families based on the classification of Hilsenhoff (1988), and (4) a winter stonefly diversity index. Physical data were summarized using a habitat metric developed by MDEQ and an "excessive" conductivity score (>800 μs). Map-derived variables included in normalizing models were catchment area, reach slope, and other landscape descriptors.

12.3.2.2 Results and Evaluation

Statistically significant normalizing models were constructed for each of the biological and habitat indicator metrics (Table 12.3). MLR models included four to eight independent variables, and all were statistically significant. Models explained 43 to 54% of the variation in their respective indicator values. Coefficients for excess conductivity and bad odors were set to zero in all models where they occurred when the normalizing models were used to generate expected indicator values. Likewise, if the bank condition score (range of 1 to 10) was less than 7 (7 = fair to good), it was reset to a value of 7 when used to generate expected indicator values. An overall biological metric was computed as the average of the four biological scores, based on the assumption that the scores were not methodologically independent (sampling biases were correlated). An overall physical habitat score was computed as the geometric mean (computed with offsets to retain the negative values) of the normalized habitat and conductivity scores based on the assumption that these parameters were independent. We also computed an overall condition index as the geometric mean of the overall physical and overall biological scores.

Based on normalized data, we found considerable variation in the current status of sites and sub-basins within the Huron River system. Urbanized streams in the Ann Arbor–Ypsilanti area had the lowest scores for many indicators. All indicators were negatively correlated with both *percent of the catchment in urban/suburban land cover* and *percent of catchment in impervious ground surface*. This was true of both raw and normalized scores. Normalization did not improve apparent sensitivity of metrics to urban or agricultural land use. In a few instances, raw scores were more negatively correlated than normalized scores. Normalization did, however, allow us to compute combined (multimetric) scores summarizing both groups of physical and biological indicators. Summary physical and biological scores correlated with each other ($r = 0.56$, $p < 0.05$) as we might expect, although physical scores tended to be more negative than matched biological scores (Figure 12.6a and b). Both summary metrics were negatively correlated with *percent urban land cover* and *percent impervious surface.*

TABLE 12.3
Summary of Normalizing Models Used in the Huron River Watershed Assessment[a]

	EPT families	Insect Families	Sensitive Insect Families	DEQ Habitat Quality Index
Transformation	ln(x + 1)	ln(x + 1)	ln(x + 1)	ln(x)
R^2	0.62	0.48	0.44	0.61
F-Statistic	15.0	30.2	34.3	11.8
Standard errror	0.320	0.287	0.353	0.123
Sites (n)	35	35	35	35
Intercept	No	No	No	Yes
ln catchment area	X	X	X	X
Excess Conductivity	X	X	X	
ln slope	X	X	X	
Bad odor	X	X	X	
ln macrophyte cover	X	X	X	
% fines	X			X
% fines 2	X			
Bank stability		X		X
ln % fines			X	
Embeddedness				X

[a] Variables included in each model shown by X.

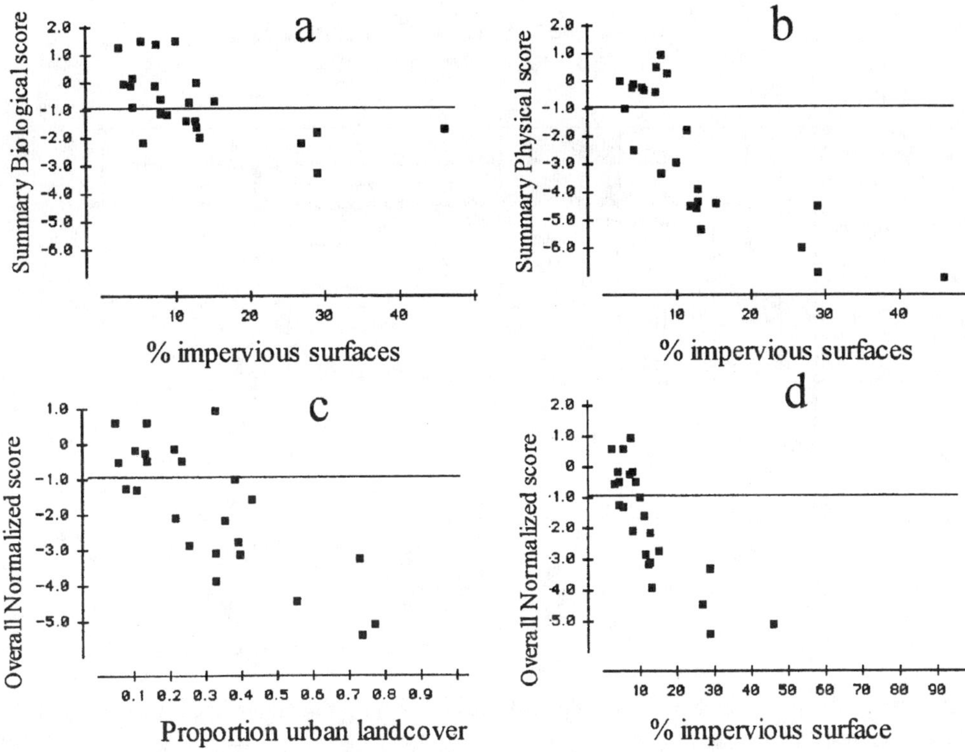

FIGURE 12.6 Responses of normalized summary metrics to landscape development in the Huron River (Michigan) watershed. The *Summary Biological Score* combines indices for total macroinvertebrate taxa, EPT taxa, intolerant taxa, and winter stoneflies. The *Summary Physical Score* includes indices for physical habitat in the channel and water conductivity. The *Overall Normalized Score* combines all biological and physical data. The dashed line at y = –1 indicates minus one SD criterion for serious impairment.

The overall normalized multimetric score, combining four biological and two physical metrics, was also negatively correlated with *percent urbanization* (Figure 12.6c). Likewise, this index of general ecological status declined rapidly with imperviousness; marked degradation occurred at imperviousness levels above approximately 10% (Figure 12.9d). Normalized scores for all indicator metrics were pooled for the entire river and compared directly to provide an overview of the status of the entire basin (Figure 12.7). Explicit modeling of the expected values allowed deviation scoring and regional normalization of the disparate indicators. Normalization in turn facilitates a truly multivariate ecological assessment. The HRWC staff set normalized scores below –1.0 SD as a programmatic threshold for concern. Based on that, 46% of sites in the analysis had biological scores indicating ecological degradation, 58% had physical and overall scores below –1 SD, and 33% had overall normalized scores >–2.0 SD below expectation, indicating a substantial basis for concern throughout the Huron River system. With respect to rapid urbanization, regional normalization provided a useful format for systematically communicating the widespread evidence of declining chemical, physical channel, and biological condition of the Huron's tributary streams.

12.4 DISCUSSION

In both example applications, regional normalization of indicator metrics facilitated a more accurate and integrated multimetric analysis of stream ecosystem condition and responses to differing land cover/use patterns. In the case of the MDEQ fish data, systematic stream size biases in the IBI-type scoring procedure were corrected and sensitivity of the metric to urban land use was enhanced.

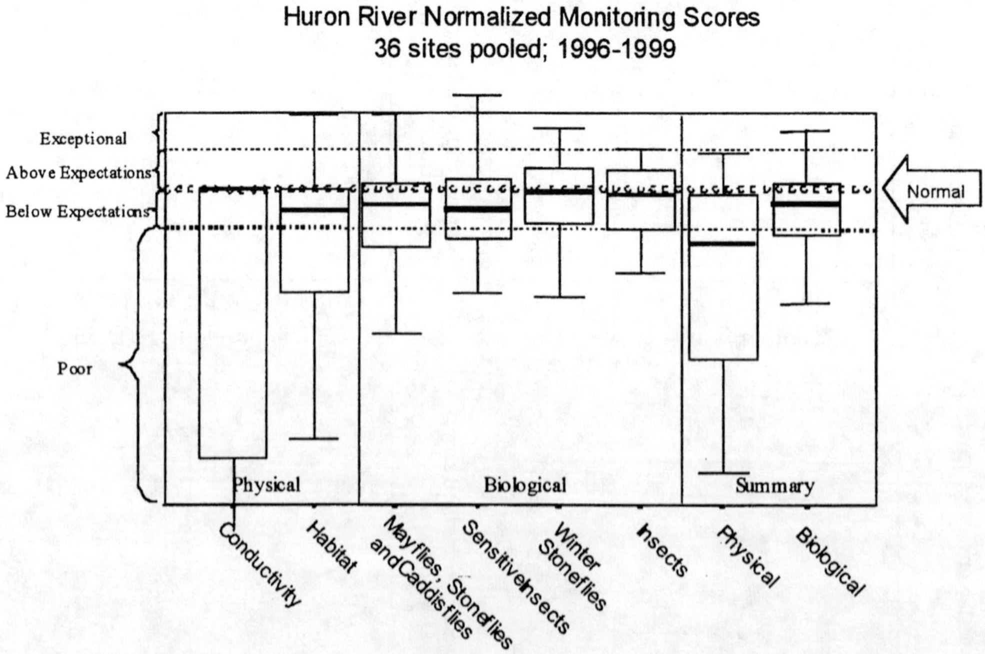

FIGURE 12.7 Overview of ecological status for the Huron River. A suite of normalized and summary physical and biological metrics for all sample sites combined is shown as a series of boxplots.

These benefits derive from two basic aspects of the normalization procedure: the empirical incorporation of measured regional covariates into the reference model and unit standardization in terms of regional variance.

The normalizing model plays a pivotal role in both cases. It provides an explicit and testable hypothesis for reference condition at a particular site, and can be used to generate a conservative estimate of pre-impact regional variation. Most widely used indicator metrics and methods to date, while implicitly relying on some kind of reference model, tend to obscure its derivation and underlying assumptions (if only by leaving them unstated) and provide no vehicle for model evaluation or communication. In a regulatory setting, this may be counter-productive because even the possibility of scoring biases devalues the indicator as a neutral and quantitative measure of ecological integrity.

Likewise, to the extent that the learning dataset (Seelbach et al., 2002) used to generate scoring procedures for indicator metrics is derived from small collections of reference sites identified *a priori*, implicit reference models are unlikely to incorporate the wide degree of natural variation that exists among the larger regional set of sites analyzed. The result will be assessments in which status is systematically biased downward because sites wrongly appear deviant from the reference condition. The bias against small streams that we found in the raw P51 scores is a good example of this effect. All apparent deviation, because of the context of the analysis (assessment of human impact), is attributed to human impact but in reality can be due to errors or inadequacies in the specification of the reference model.

The importance of accounting for sources of natural variation in site characteristics while specifying reference conditions has been evident for some time. Variation in taxa richness with catchment area is a good example of natural variation in community structure that is almost universally observed and usually is built into the scoring procedure for IBI-type metrics (Angermeier and Schlosser, 1989; Karr and Chu, 1999).

Other natural factors also affect species richness, and if not explicitly incorporated into scoring procedures, can result in systematic biases in indicator performance. For example, thermal conditions so strongly shape fish community composition (Wehrly et al., in press) that assessment methods that use taxa counts as a metric require major adjustments when applied across cold water and warm water systems (Lyons et al., 1996; Mundahl and Simon, 1999). Similarly, immigration from nearby stream confluences can dramatically affect species richness. Species richness near confluences more strongly correlate with confluent catchment area than with site upstream catchment area (Osborne and Wiley, 1992). Failure to account for confluence effects results in unrealistic scoring standards and low IBI scores (Osborne et al., 1992). The solution to these and other similar effects is the explicit statistical control of all major covariates for the indicator variable.

12.4.1 Relationship of Regional Ecological Normalization to Other Assessment Procedures

Regional ecological normalization of indicator data is a meta-method in the sense that it does not address the primary issues in assessment of indicator choice and sampling methodology. It can be applied to any indicator methodology where the indicator metric exhibits significant spatial variation. Regional normalization is not a replacement for any specific indicator methodology but is an adjustment that corrects geographic and statistical biases and standardizes output. Indicator species, various biotic indices, physical and chemical measures, and multimetric indices like the IBI can all be regionally normalized. The main requirement is a spatially robust dataset from which regional models can be constructed.

Raw IBI scores (and other population and community multimetrics) that use procedural scoring criteria are potentially biased by regional variations in ecological factors not explicitly addressed in their scoring procedures. In the IBI, the reference model is implicit in the scoring criteria. Therefore any significant geographic variance in the variables used to estimate component metrics (not related to human perturbations) theoretically requires compensatory changes in scoring criteria. In a practical sense, this means that the scoring criteria must be regionally adjusted to compensate for regional variations as envisioned by Karr et al. (1987).

This is the reason that ecological classifications (e.g., ecoregions; Gallant et al., 1989) and other spatial stratification approaches are closely associated with multimetric assessments. The strategy has been to adjust metrics using regional strata whenever necessary. Presumably, finer spatial stratification could achieve more accurate implicit reference criteria, but at the cost of reduced methodological standardization and with a requirement for more reference sites on which to base the scoring criteria. The major difficulty with this approach is that regionalization (use of strata-wide means) does a poor job of describing landscape processes and structures, which are organized as fine-grained mosaics.

Local variations in hydrology and water temperature are good examples of ecological factors that strongly affect fish community structure (Wiley et al., 1987; Zorn et al., 2002; Werhly et al., in press) but cannot be meaningfully summarized across an entire ecoregion or jurisdictional unit. A second serious problem arising from increased spatial stratification is that it is often difficult or impossible to locate a sufficient number of unimpacted reference sites from which strata-specific scoring criteria must be developed. The alternative approach of statistically modeling site-specific reference conditions from regional datasets addresses both of these issues (Seelbach et al., 2002).

The fact that we can easily build regional normalizing models for IBI-type metrics is perhaps the strongest argument that explainable (and therefore unnecessary) spatial variance in IBI scores exists and can be removed. In the examples above, we built acceptable statistical models for several IBI-type indices from multiple datasets. We recently completed an assessment survey of streams across the entire northern lakes and forest ecoregion using the EMAP sampling design (Baker et al., 2001). Despite the relatively pristine condition of these watersheds, data-based normalization models were again successfully generated. Normalized IBI scores led to significantly more realistic interpretations of ecological status.

Regional ecological normalization and the analyses developed by Wright, Norris, and others (e.g., RIVPACS, AUSRIVAS) share a number of important similarities. Both approaches use explicit linear statistical models to make predictions about the expected indicator values employing site- and catchment- specific information as model inputs. RIVPACS uses discriminant function analysis, while we generally employ MLR analysis, but one could use almost any kind of accurate model. Since both approaches make explicit predictions about expectations at a focal site, both evaluate the site in terms of a deviation score. Deviation scores are expressed in RIVPACS analyses typically as a ratio (observed/expected) whereas deviation in the form of a difference (observed minis expected) is more useful in a normalizing calculation. Obviously, both ways of expressing deviation from reference are related and are useful in specific contexts.

The approaches differ in scope and strategy. RIVPACS and related procedures are specific indicator methods for employing macroinvertebrate data in stream assessments. They include specific protocols for dealing with and modeling large multispecies communities. These are specific indicator methods in contrast to the regional normalization described here, which we characterize as a meta-method. The RIVPACS approach provides what we described above as a "centering" of the indicator data relative to an explicitly modeled reference community. It does not, however, provide a scaling of that result in a way that incorporates natural variability or model precision.

Another clear difference is the use of classification in model building. Classification is used in RIVPACS initially to specify *a priori* the set of sites to be modeled as the reference condition and by means of cluster analysis to reduce variation in community composition to a smaller set of guilds for discriminant functions analysis (Turak et al., 1999). We consider RIVPACS to be more specific than regional normalization in that it specifies in detail the type of model to be employed. Strategically, this and related approaches refine model expectations for the focal site using classification to simplify and reduce variation (e.g., building the reference model based on best sites only, substituting guilds for more numerous species). In contrast, regional ecological normalization refines modeled expectation by explicitly incorporating measured covariates and then controlling for them statistically.

RIVPACS output in any of its flexible forms could presumably be regionally normalized. To the extent that a RIVPACS analysis does an adequate job of incorporating key regional habitat characteristics, biases should be minimized by the method itself. The only benefit of regional normalization is scaling the RIVPACS-generated scores to make them comparable to other types of metrics and data. The process of building a normalizing model, however, is a very clear way to test and validate the RIVPACS-type approach. If biases were found, regional normalization could be used to provide corrections.

12.4.2 Scaling by Variance (Uncertainty) and Its Relationship to Risk Assessment

The final step in the ecological normalization procedure we propose is the scaling of the indicator deviation score (i.e., observed minus expected indicator value) by the inverse of a variance measure for the indicator. Operationally this scaling provides unit conformity, which allows easy comparison of different types of variables. It also makes convenient the assembly of normalized scores into multimetric summaries as we illustrated above. This scaling by variance has another important conceptual implication. It incorporates into the normalized score a formal evaluation of the uncertainty associated with the deviation score. Variance in this context can be thought of as a measure of the uncertainty about the true expected value of the reference condition and therefore the deviation score.

Deviation values in indices with large expected variances are scaled down in magnitude by variance normalization. Conversely, small deviation scores can be inflated by scaling if expected variances are low. The result is that the normalized score incorporates an explicit evaluation of uncertainty and casts the deviation scoring into a risk assessment framework. Larger scaled deviations imply higher probabilitites (risk) that observed index values are abnormal. Smaller scaled deviations imply that an observed deviation from the reference condition, even if large, is unlikely to be distinguishable from background variation and/or modeling error.

Two important issues related to the interpretation of normalized scores become clearer when viewed in a risk assessment framework (USEPA, 1998). First, the criterion we choose to interpret normalized scores as impaired or acceptable is clearly arbitrary. To the extent that the normalized scores represent a continuous probability of difference from expectation, there is no absolutely correct interpretation. As in all probability-based analyses, the analyst must arbitrarily set meaningful critical levels in order to evaluate a hypothesis of difference. In the analyses reported above, we used a ± 1 SD criterion to evaluate departures from the normalizing model. This seemed a conservative and interpretable but also obviously arbitrary criterion. If underlying index values and modeling errors are relatively normally distributed, interpretation might be possible in terms of confidence limits or probabilities. For example, a criterion of ± 2 SD might be associated with 95% confidence intervals.

A second issue is the importance and meaning of the way we parameterize the variance term used in normalization. In the examples above, we used the variance of the predicted regional reference conditions to normalize deviation scores. As suggested earlier, the standard error of regression is also an attractive source for the variance estimate since it indicates the confidence we can place in the normalizing model predictions. If we want to interpret the normalized scores in a risk assessment framework, then perhaps the best solution would be to incorporate both types of variances in the scaling term.

Though seldom measured, aquatic ecosystems experience large, regionally coherent and incoherent temporal variations as part of their natural dynamics (Magnuson et al., 1990; Urquhart et al., 1993; Kohler and Wiley, 1997; Wiley et al., 1997). In the spatially intensive, temporally sparse sampling characteristic of stream inventory and assessment studies, it seems clear that temporal variance can be erroneously masked and then modeled as spatial variance (Wiley et al., 1997). Even if modeling of observed data were error-free, we would expect significant natural temporal variance in community parameters associated with internal and exogenous large-scale dynamics. As a result we should expect some uncertainty (variance) associated with the deviation score reflecting temporal dynamics that a regional statistical model of the expected reference condition simply cannot predict.

Likewise, if it were possible that an indicator variable had no intrinsic variation across a region, we might still have significant uncertainty about the true value of the deviation score if our model for that indicator is imperfect. Although it should not discourage us from modeling, the fits of the linear models we employed in normalizations described above (see Table 12.1) are imperfect and make it clear that substantial predictive error is possible. It seems appropriate that deviation scores for indices that are more difficult to model accurately should be discounted in the normalization scaling compared to indices that can be more successfully modeled. For example, we found the number of intolerant fish taxa harder to model accurately (Table 12.1) than the total number of fish taxa. If we scaled the deviation score by the inverse of the model standard error of regression the intolerant taxa metric would be appropriately devalued relative to the total taxa metric in the calculation of a summary index. Together, the regional variance around the metric value and the variance around the prediction of that value represent both major sources of uncertainty in the deviation score. Using a function of their sum in variance scaling may be the optimal solution.

12.5 CONCLUSIONS

Since its introduction in the Clean Water Act of 1972, the concept of biological integrity has played a key conceptual role in the development of assessment and inventory methodologies. While no completely satisfactory definition of biological integrity has been proposed, a widely used definition includes: " …the ability to support and maintain a … biological community … *comparable to natural lakes and streams of the region*" (emphasis ours; Frey, 1977; Karr and Dudley, 1981; USEPA, 1991; National Research Council, 1995). Regional ecological normalization can help guarantee the ecological comparability implied in that definition. It is a relatively simple quantitative analysis applicable *post hoc* to almost any existing indicator methodology. Using explicit, regionally-derived, site-specific reference models and scaling results by explicit estimates of uncertainty, regional ecological normalization can help clarify assessment assumptions, integrate multiple types of data and indicator metrics, and communicate relative risks of impairment.

We demonstrated the practicality and utility of this analytical method by presenting two case studies from our stream assessment work in Michigan. In an analysis of state-wide fish community data collected by different agencies, we illustrated how normalization allowed us to combine statistically distinct datasets for a region-wide assessment. Regional ecological normalization improved sensitivity of several indicator metrics and removed much of the inappropriate size bias in the MDEQ data. In an analysis of data from a watershed-based citizen monitoring program, we illustrated the flexible ways in which normalized metric scores can be combined to provide comprehensive muti-metric assessments. Both physical and biological metrics were combined to examine strong negative effects of increasing urban landcover and imperviousness.

Regional ecological normalization of indicator metrics uses statistical modeling to estimate site-specific reference conditions. It is in this respect related methodologically to the analyses developed by Wright, Norris, and others (e.g., RIVPACS, AUSRIVAS analyses). It differs from these approaches primarily in (1) the generality of its application, (2) its focus on the prediction of indicator metrics instead of community composition, (3) the explicit identification and evaluation of the reference model, and (4) the use of variance scaling to incorporate uncertainty into the assessment procedure.

ACKNOWLEDGMENTS

We gratefully acknowledge Theresa Dakin and Paul Rentschler for the hours of database organization and analysis they provided. Thanks also to the staff and volunteers of the Huron River Watershed Council's Adopt-a-Stream Program and to the Michigan Department of Environmental Quality for their support and willingness to share their data.

REFERENCES

Angermeier, P.L. and I.J. Schlosser. 1989. Species-area relationships for stream fishes, *Ecology*, 70, 1450–1462.

Baker, E.A., K.E. Wehrly, P.W. Seelbach, M.J. Wiley, L. Wang, and T.P. Simon. 2001. *Regional normalization models for assessing ecological integrity in streams of the Northern Lakes and Forest Ecoregion.* U.S. Environmental Protection Agency, R-EMAP, Final Report for Grant R-82620701–2, Duluth, MN.

Baker, M.E., M.J. Wiley, and P.W. Seelbach. 2001a. GIS-based hydrologic models of riparian areas: implications for stream water quality, *Journal of the American Water Resources Association*, 37(6), 1–14.

Baker, M.E., M.J. Wiley, and P.W. Seelbach. 2001b. Spatially-Explicit Models of Groundwater Loading in Glaciated Landscapes: Considerations and Development in Lower Michigan. Michigan Department of Natural Resources, Fisheries Research Report, Ann Arbor, MI.

Barbour, M.T., J. Gerritsen, B.D. Snyder, and J.B. Stribling. 1999. *Rapid Bioassessment Protocols for Use in Streams and Wadeable Rivers: Periphyton, Benthic Macroinvertebrates, and Fish, 2nd Edition*. EPA 841-B-99–002, U.S. Environmental Protection Agency, Office of Watershed, Oceans, and Wetlands, Washington, D.C.

Bayley P.B. and D.C. Dowling. 1993. The effect of habitat in biasing fish abundance and species richness estimates when using various sampling methods in streams, *Polskie Archiwum Hydrobiologii,* 40, 5–14.

Biggs, B.J.F., M.J. Duncan, I.G. Jowett, J.M. Quinn, C.W. Hickey, R.J. Davies-Colley, and M.E. Close. 1990. Ecological characterization, classification, and modelling of New Zealand rivers: an introduction and synthesis, *New Zealand Journal of Marine and Freshwater Research,* 24, 277–304.

Bowlby, J.N. and J.C. Roff. 1986. Trout biomass and habitat relationships in southern Ontario streams, *Transactions of the American Fisheries Society,* 115, 503-514.

Bryce, S.A. and S.E. Clark. 1996. Landscape-level ecological regions: linking state-level ecoregion frameworks with stream habitat classifications, *Environmental Management,* 20, 297–311.

Claessen, F.A. M., F. Klijn, J.P.M. Witte, and J.G. Nienhuis. 1994. Ecosystem classification and hydro-ecological modeling for national water management, in F. Klijn (Ed.). *Ecosystem Classification for Environmental Management*. Kluwer Academic Publishers, Dordrecht, The Netherlands, 199–222.

Davis, L.S. and J.A. Henderson. 1978. Many uses and many users: some desirable characteristics of a common land and water classification system, in *Classification, Inventory, and Analysis of Fish and Wildlife Habitat.* FWS/OBS-78/76, U.S. Fish and Wildlife Service, Washington, D.C., 13–34.

Fausch, K.D., C.L. Hawkes, and M.G. Parsons. 1988. *Models that predict standing crop of stream fish from habitat variables: 1950-85*. General Technical Report PNW-GTR 213, U.S. Department of Agriculture, Forest Service, Pacific Northwest Field Station, Portland, OR.

Frey, D.G. 1977. Biological integrity of water – an historical approach, in R.K. Ballantine and L.J. Guarraia (Eds.). *The Integrity of Water, Proceedings of a Symposium, March 10–12, 1975.* U.S. Environmental Protection Agency, Washington, D.C., 127–140.

Gallant, A.L., T.R. Whittier, D.P. Larsen, J.M. Omernik, and R.M. Hughes. 1989. *Regionalization as a Tool for Managing Environmental Resources.* EPA/600/3–89/060.U.S. Environmental Protection Agency, Corvallis, OR.

Gilliom, R.J., W.M. Alley, and M.E. Gurtz. 1995. *Design of the National Water-Quality Assessment Program: Occurrence and Distribution of Water-Quality Conditions*. U.S. Geological Survey, Circular 1112.

Hakanson, L. 1996. Predicting important lake habitat variables from maps using modern modelling tools. *Canadian Journal of Fisheries and Aquatic Sciences* 53(1), 364–382.

Hawkins, C.P., R.H. Norris, J.N. Hogue, and J.W. Feminella. 2000. Development and evaluation of predictive models for measuring the biological integrity of streams, *Ecological Applications,* 10, 1456–1477.

Hauer, F.R. and R.D. Smith. 1998. The hydrogeomorphic approach to functional assessment of riparian wetlands: evaluating impacts and mitigation on river floodplains in the U.S.A., *Freshwater Biology,* 40, 517–530.

Higgins, J., M. Lammert, M. Bryer, M. DePhilip, and D. Grossman. 1998. *Freshwater Conservation in the Great Lakes Basin: Development and Application of an Aquatic Community Framework.* The Nature Conservancy, Great Lakes Program Office, Final Project Report, Chicago, IL.

Hilsenhoff, W.L. 1988. Rapid field assessment of organic pollution with a family level biotic index, *Journal of the North American Benthological Society,* 7, 165–68.

Karr, J.R. and E.W. Chu. 1999. *Restoring Life in Running Waters: Better Biological Monitoring*. Island Press, Covelo, CA.

Karr, J.R., and D.R. Dudley. 1981. Ecological perspectives on water quality goals, *Environmental Management,* 5, 55–68.

Karr, J. R., K.D. Fausch, P.L. Angermeier, P.R. Yant, and I.J. Schlosser. 1986. *Assessing Biological Integrity in Running Waters: A Method and Its Rationale*. Illinois Natural History Survey, Special Publication 5, Champaign, IL.

Kleiman, R. 1995. The effects of season, land use/cover, and hydrology on stream water chemistry in Michigan's lower peninsula. M.S. thesis, University of Michigan, Ann Arbor, MI.

Kohler, S.L. and M.J. Wiley. 1997. Pathogen outbreaks reveal large-scale effects of competition in stream communities, *Ecology* 87(7), 2164–2176

Lyons, J., L. Wang, and T.D. Simonson. 1996. Development and validation of an index of biotic integrity for coldwater streams in Wisconsin, *North American Journal of Fisheries Management,* 16, 241–256.

Magnuson, J.J., B.J. Benson and T.K. Kratz. 1990. Temporal coherence in the limnoogy of a suite of lakes in Wisconsin, USA, *Freshwater Biology,* 23, 145–159.

Martin, J. S. 1998. Final 319 Report to MDEQ. Huron River Watershed Council, Report, Ann Arbor, MI.

Michigan Department of Environmental Quality. 1997. *A Strategic Environmental Quality Monitoring Program for Michigan's Surface Waters*. MI/DEQ/SWQ-96/152, Michigan Department of Environmental Quality, Lansing. MI.

Mundahl, N.D. and T.P. Simon. 1999. Development and application of an Index of Biotic Integrity for coldwater streams of the upper Midwestern United States, in T.P. Simon (Ed.). *Assessing the Sustainability and Biological Integrity of Water Resources Using Fish Communities*. CRC Press, Boca Raton, FL, 383–416

National Research Council. 1995. *Review of EPA's Environmental Monitoring and Assessment Program: Overall Evaluation*. National Academy Press, Washington, D.C.

Norris, R.H. and A. Georges. 1993. Analysis and interpretation of benthic macroinvertebrate surveys, in D.M. Rosenberg and V.H. Resh (Eds). *Freshwater Biomonitoring and Benthic Macroinverterates*. Chapman & Hall, New York, 234–286

Osborne, L.L. and M.J. Wiley. 1988. Empirical relationships between land use/cover and stream water quality in an agricultural watershed, *Environmental Management,* 26, 9–27.

Osborne, L.L. and M.J. Wiley. 1992. The influence of tributary spatial position on the structure of warmwater fish communities, *Canadian Journal of Fisheries and Aquatic Sciences,* 49, 671–681.

Osborne, L.L., S.L. Kohler, P.B. Bayley, D. Day, W. Bertrand, M.J. Wiley, and R. Sauer. 1992. Influence of stream spatial location in a drainage network on the Index of Biotic Integrity, *Transaction of the American Fisheries Society,* 121, 635–643.

Reynoldson, T.B., R.H. Norris, V.H. Resh, K.E. Day, and D.M. Rosenberg. 1997. The reference condition: a comparison of multimetric and multivariate approaches to assess water-quality impairment using benthic macroinvertebrates, *Journal of the North American Benthological Society,* 16, 833–852.

Seelbach, P.W., R.N. Lockwood, and J.R. Ryckman. 1994. Efficiency of Sampling River Fishes with Rotenone. Michigan Department of Natural Resources, Fisheries Research Report 2009, Ann Arbor, MI.

Seelbach, P.W. and M.J. Wiley. 1997. Overview of the Michigan Rivers Inventory Project. Michigan Department of Natural Resources, Fisheries Technical Report 97–3, Ann Arbor, MI.

Seelbach, P.W., M.J. Wiley, P.A. Soranno, and M.T. Bremigan. 2002. Aquatic conservation planning: using landscape maps to predict ecological reference conditions for specific waters, in K. Gutzwiller (Ed.), *Concepts and Applications of Landscape Ecology in Biological Conservation.* Springer-Verlag, New York, Chapter 25.

Simpson, J. and R.H. Norris. In press. Biological assessment of water quality: development of AUSRIVAS models and outputs, in J.F. Wright, D.W. Sutcliffe, and M.T. Furse (Eds). *RIVPACS and Similar Techniques for Assessing the Biological Quality of Freshwaters.* Freshwater Biological Association and Environmental Agency, U.K.

Sokal, R.R. and F.J. Rohlf. 1995. *Biometry,* 3rd ed. W.H. Freeman, New York.

Sowa, S.P. 1999. *Implementing the Aquatic Component of Gap Analysis in Riverine Environments: A Training Workbook.* Missouri Resource Assessment Partnership, Columbia, MO.

Turak E., L.K. Flack, R.H. Norris, J. Simpson, and N. Waddell. 1999. Assessment of river condition at a large spatial scale using predictive models, *Freshwater Biology,* 41, 283–298.

Urquhart, N.S., W.S. Overton, and D.S. Birkes. 1993. Comparing sampling designs for monitoring ecological status and trends: impact of temporal patterns, in V. Barnett and K. Tuchman. (Eds.). *Statistics for the Environment,* John Wiley & Sons, London, 71–85.

U.S. Environmental Protection Agency. 1991. EMAP Surface Waters Monitoring and Research Strategy. U.S. Environmental Protection Agency, USEPA, Office of Research and Development, Pacific Ecology Division, Corvallis, OR.

U.S. Environmental Protection Agency 1998. Guidelines for Ecological Risk Assessment, EPA/630/R-95/002F. USEPA. Office of Research and Development, Risk Assessment Forum, Washington, D.C.

U.S. Environmental Protection Agency. 1997. Environmental Monitoring and Assessment Program (EMAP) Research Strategy. EPA/620/R-98/001. USEPA, Office of Research and Development, Washington, D.C.

Ward, R.C. 1996. Water quality monitoring: where's the beef? *American Water Resources Association,* 32, 673–680.

Wehrly, K.E. 1999. The influence of thermal regime on the distribution and abundance of stream fishes in Michigan, Ph.D. dissertation, University of Michigan, Ann Arbor, MI.

Wehrly, K.E., M.J. Wiley, and P.W. Seelbach. 1998. Landscape-based models that predict July thermal characteristics of lower Michigan rivers. Michigan Department of Natural Resources, Fisheries Research Report 2037, Ann Arbor, MI.

Wehrly, K.E., M.J. Wiley, and P.W. Seelbach. 2002. Classifying regional variation in thermal regime using stream fish community patterns, *Transactions of the American Fisheries Society.* In press.

Wiley, M. and J. Martin. 2000. Current conditions, recent changes, and major threats to the Huron River: a report based on eight years of citizen monitoring. Huron River Watershed Council, Ann Arbor, MI. http://rivers.snre.umich.edu/mri/huron/hrwcrpt/index.htm.

Wiley, M.J. and P.W. Seelbach. In press. Hydrology of rivers in Michigan's Lower Peninsula. Michigan Department of Natural Resources, Fisheries Research Report 2039, Ann Arbor, MI.

Wiley, M.J., S.L. Kohler, and P.W. Seelbach. 1997. Reconciling landscape and site based views of aquatic stream communities, *Freshwater Biology,* 37, 133–148.

Wiley, M.J., P.W. Seelbach, and S.P. Bowler. 1998. Ecological Targets for Rehabilitation of the Rouge River. Final Report RPO-PI-SR21.00, Rouge Project Office, Wayne County, MI.

Wright, J.F., P.D. Armitage, M.T. Furse, and D. Moss. 1988. A new approach to the biological surveillance of river quality using macroinvertebrates, *Verhandlungen der Internationale Vereinigung fur Theoretische und Angewande Limnolgie,* 23, 1548–1552.

Wright, J.F., P.D. Armitage, M.T. Furse, and D. Moss. 1989. Prediction of invertebrate communities using stream measurements, *Regulated Rivers: Research and Management,* 4, 147–155.

Yoder, C.O. and E.T. Rankin. 1995. Biological criteria program development and implementation in Ohio, in W.S Davis and T.P. Simon (Eds.). *Biological Assessment and Criteria: Tools for Water Resource Planning and Decision Making,* Lewis Publishers, Boca Raton, FL, 109–144.

Zonneveld, I.S. 1994. Basic principles of classification, in F. Klijn (Ed.). *Ecosystem Classification for Environmental Management*, Kluwer Academic Publishers, Dordrecht, The Netherlands, 23–47.

Zorn, T.G., P.W. Seelbach, and M.J. Wiley. 2002. Distributions of stream fishes and their relationship to stream size and hydrology in Michigan's Lower Peninsula, *Transactions of the American Fisheries Society,* 131, 70–85.

Section IV

Land Use Modification Patterns

13 Fish and Benthic Macroinvertebrate Assemblages as Indicators of Stream Degradation in Urbanizing Watersheds

Lizhu Wang and John Lyons

CONTENTS

0-8493-0905-0/03/$0.00+$1.50

13.1 INTRODUCTION

Watershed urbanization is widespread and increasing throughout the United States. Conversion of rural lands to urban lands is driven by a combination of population growth, net movements of people from rural areas to urban zones, and increased rates of urban land use per capita. The population of the United States reached 273 million in 1999 and is predicted to be 403 million by 2050, a 48% increase (all census data in this article are from www.census.gov/population/estimate/nation). In 1930, about 56% of the population (123 million people) lived in urban areas. By 1990 this percentage increased to 75% (248 million people) and is expected to rise above 80% by 2025. The amount of urban land in the United States has more than tripled from 8,065 sq. mi. in 1950 to 27,838 sq. mi. in 1990. Most of this increase was caused by "urban sprawl" as people moved from higher density downtown zones to newly built lower density suburbs encircling the central city. In 1950, the population density per square mile was 6,121 in urbanized areas, and declined to 3,411 people per sq. mi. by 1990 as cities spread out.

The trend of increasing urban land use in the United States does not bode well for stream ecosystems. Urban development in a watershed almost inevitably harms the streams that drain that watershed (Booth and Jackson, 1997; Wang et al., 2000, 2001). Urbanization modifies watershed land cover in a way that dramatically affects stream hydrology, habitat, water quality, and biota. Major impacts on stream biota can occur at surprisingly low levels of urban development, in watersheds that remain largely rural in character. In the early stages of watershed urbanization, biological changes in a stream may be easier to detect than the physical and chemical changes that cause them.

In this chapter, we review and summarize the literature on the major impacts of urbanization on stream ecosystems, with emphasis on the effects of urban sprawl on fish and macroinvertebrate assemblages. Our goal is to identify appropriate and effective biological indicators of the early stages of urbanization and to describe particular biological "signatures" — specific structural, compositional, or functional characteristics of assemblages that are diagnostic of urbanization (Yoder and Rankin, 1995). Such signatures may provide insight into the specific physical and chemical stresses caused by urban development and suggest possible mitigation strategies.

13.2 PHYSICAL AND CHEMICAL EFFECTS OF URBANIZATION

Urban and suburban land uses in a watershed invariably have dramatic effects on streams (Figure 13.1). At relatively low levels of urbanization, typical of newly developing suburbs, the primary impacts are on hydrology (Leopold, 1968; Holis, 1975; Simmons and Reynolds, 1982; Booth, 1991; Schueler, 1994; Booth and Jackson, 1997). In humid temperate regions, an undeveloped watershed usually has the capacity to assimilate much of the rain that falls on it during normal storm events. As such a watershed urbanizes, it is increasingly covered by buildings, roads, parking lots, and other impervious surfaces that do not allow precipitation to infiltrate the soil and enter the groundwater. Even where the soil in the watershed remains uncovered, compaction from construction or other human activities often greatly reduces its ability to absorb water. Instead, precipitation drains across the land surface of the watershed and directly into the stream. Runoff from impervious surfaces and compacted soils is rapidly conveyed to stream channels through a network of gutters, storm sewers, and other drainage networks.

The runoff generated from urban and suburban areas has a profound influence on stream flow regime and channel morphology. In an undeveloped watershed, stream flows are mainly generated by subsurface and groundwater flows, and surface runoff is minimal except during major storm events. In urban watersheds, the greatly increased surface runoff changes both the patterns and magnitudes of stream flows. Typically, the magnitude and frequency of flooding are much greater than would have occurred before urbanization (Booth and Jackson, 1997). Peak flows may be two to five times greater than under non-urban conditions, and bankfull or greater flows may occur ten

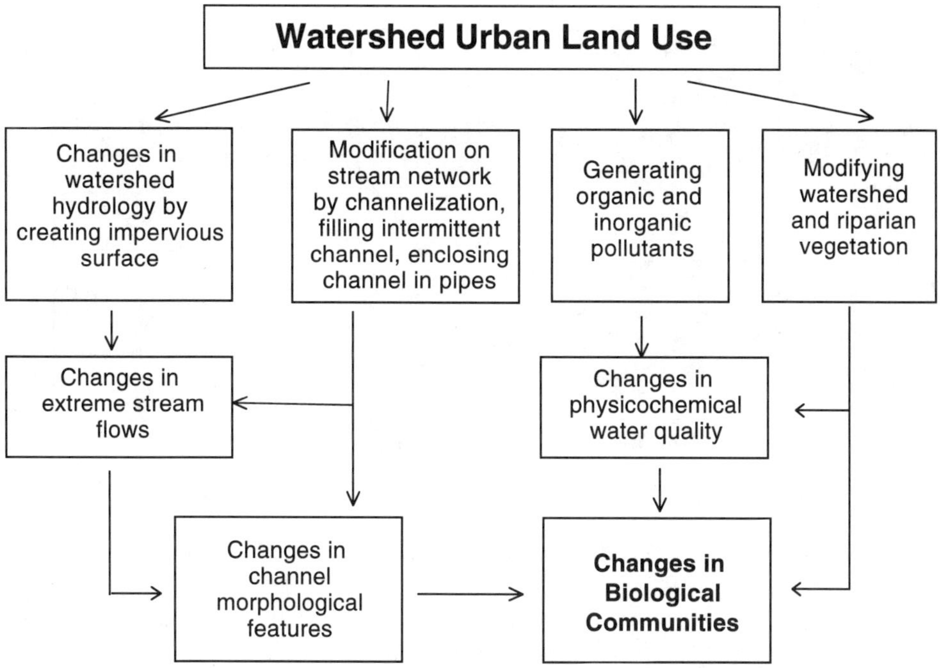

FIGURE 13.1 Watershed urban land use causes a series of modifications in physicochemical water quality, hydrological regime, and channel morphological features, all of which are reflected in the characteristics of biological communities.

times more often (Hollis, 1975; Krug and Goddard, 1986; Barker et al., 1991, cited in Booth and Jackson, 1997). More frequent and severe flooding can directly harm stream biota in a variety of ways, including loss of habitat, disruption of reproduction, reductions in feeding and growth, and displacement and mortality of aquatic organisms, leading to major shifts in biotic communities.

Changes in flood patterns modify stream channel morphology. In undisturbed watersheds, channel morphology is largely determined by bankfull flows (Leopold et al., 1964). When watershed urbanization increases peak discharges and the frequency of high flows, the equilibrium between stream flows and channel resistance is broken. This process results in either quasi-equilibrium channel expansion, where cross-sectional area increases in near-proportion to the discharge increase through channel widening into the floodplain or down-cutting into the streambed (Schueler, 1994), or catastrophic channel incision, where the channel down-cuts far out of proportion to the discharge increase (Booth, 1990). In either case, the results are unstable stream banks and streambeds, heavy erosion, and greatly modified pool–riffle structure, all of which cause a decline in habitat quality for aquatic and riparian organisms (Hammer, 1972; Booth, 1990; Booth and Jackson, 1997; May et al., 1997). Habitat losses are exacerbated by the efforts to increase drainage that often accompany watershed urbanization including stream channel ditching and straightening and the installation of culverts and concrete channels.

Increased runoff also reduces the amount of water available to filter into the soil, recharge the groundwater, and maintain the water table (Booth and Jackson, 1997; Finkenbine et al., 2000). This in turn leads to declines in stream baseflows during periods without precipitation, sometimes to the extent that perennial streams become intermittent. For example, in an analysis of long-term gauging date from six streams on the south shore of Long Island, New York, Simmons and Reynolds (1982) found that urbanization over the previous three decades had significantly reduced stream base flow. In an urban area with both sanitary and storm sewers, base flow had been reduced by 80% whereas in an unsewered urban area the reduction was 16%. Flows in two streams that remained rural did not change.

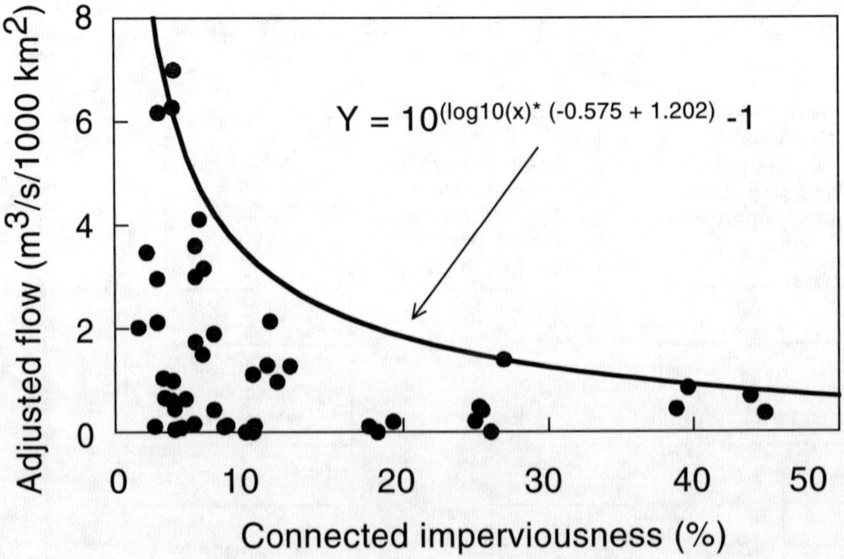

FIGURE 13.2 As the percentage of connected imperviousness increases in a watershed, stream base flow yield (base flow per unit area of the watershed) decreases sharply. (*Source:* Wang, L. et al., 2001. *Environmental Management,* 28, 255–266. With permission.)

In predominantly forested watersheds of the Piedmont region of Maryland, Klein (1979) reported that base flow per unit watershed area was negatively linearly correlated with watershed percent impervious surface area. More recently, Wang et al. (2001) examined 47 southeastern Wisconsin streams in a predominantly agricultural region and found that base flow per unit area was negatively and non-linearly related with urbanization (Figure 13.2). At very low urbanization levels, base flows were highly variable, but at least some were high. As urban development increased, maximum values of base flow declined sharply, and base flows were consistently low at modest to high urban levels. Reduced ground water inputs and lowered flows can have profound impacts on stream biota through reductions in habitat volume and quality, increases in water temperature variability and extremes, and decreased water quality. Pumping of groundwater to supply the urban population can further reduce stream flows and compound the impacts on biota.

Urban development in a watershed also alters stream water quality (USEPA, 1983, 1994, 2000). The early obvious deterioration of water quality caused by urbanization was associated with point source pollution from industrial and commercial operations and from domestic sewage (Novotny and Olem, 1994). Since the U.S. Clean Water Act and associated state legislation successfully reduced the effects of point source pollution, the magnitude of urban pollution from diffuse surface runoff has become more apparent. Runoff from impervious surfaces carries sediment and a variety of pollutants to streams (Benke et al., 1981; Masterson and Bannerman, 1994; USEPA, 1994; Crunkilton et al., 1996; Wernick et al., 1998). Toxic metals, oxygen-demanding organic materials, nutrients, pesticides, petroleum products, and road salts often build up on roadways and in parking lots during dry periods and then are delivered in high concentrations to streams during storms. Reduced baseflows from lowered water tables result in lower volumes of water available to dilute pollutants delivered by runoff and increases in summer water temperature that may enhance the toxicity of the pollutants.

There are numerous sources and types of diffuse urban pollution. Novotny and Olem (1994) summarized nine major sources from urban development and runoff: (1) pollution deposited in precipitation; (2) erosion of pervious lands; (3) accumulation of dry atmospheric deposits and build-up of street refuse, including litter, street dust and dirt, organic residues from vegetation and animal population, and traffic emission; (4) solids accumulation and growth in sewers; (5) leaching of pollutants from septic and other sources; (6) application, storage, and wash-off of deicing and other chemicals; (7) application of pesticides and fertilizers onto grassed urban lands; (8) discharge of pollutants, such as car oils, detergents, and other household and commercial solvents and chemicals into the drainage systems; and (9) cross-connection of sewage and industrial wastes from sanitary sewers, failing septic tanks, and other sources into storm sewers.

Urban runoff contains numerous types of pollutants. Based on the U.S. National Urban Runoff Project (NURP), 14 potentially hazardous inorganic constituents were detected (USEPA, 1983). About 91% of the samples contained copper, lead, and zinc. Other frequently detected inorganic pollutants included arsenic, chromium, cadmium, nickel, and cyanide. The metallic toxicity of urban runoff was also found to vary with land use — runoff from high traffic areas was more toxic than runoff from residential areas (Bannerman, 1991, cited in Novotny and Olem, 1994). Diverse organic pollutants have also been found in urban runoff. In the NURP study, the most commonly found organic pollutant was the plasticizer bis (2-ethylhexyl) phthalate (22% of samples), followed by the pesticide α-hexachlorocyclohexane (20% of samples). Eleven other organic pollutants were reported with detection frequencies between 10 and 20% (USEPA, 1983). The annual loading of organic pollutants, such as petroleum hydrocarbons and polycyclic aromatic hydrocarbons, was also found to vary with land uses. Loading was highest from highways and industrial lands and lower from commercial and residential lands (Hoffman, 1985).

Riparian urban development has particularly acute impacts on stream water and habitat quality (Booth and Jackson, 1997; Wang et al., 2001). Removal of riparian vegetation destabilizes stream banks and increases erosion, directly degrading habitat and water quality. It also destroys a natural filter that can remove pollutants from runoff. Even if a buffer of riparian vegetation is retained along a stream during development, storm sewers and drains can allow runoff to bypass the vegetation. If streambank vegetation is lost to development, so too is the shade that moderates water temperatures and the supplies of leaf litter and woody debris that provide essential food and shelter for aquatic organisms (Finkenbine et al., 2000).

13.3 IMPERVIOUSNESS AS A MEASURE OF URBANIZATION

The urban landscape is a complex mosaic of different lands that serve a wide range of human uses such as housing, commerce, industry, transportation, communication and utilities, government, and recreation. A major challenge in evaluating the effects of urbanization on stream ecosystems is identifying a simple yet effective indicator of the intensity of urban land use. The ideal urban indicator could be used to display land use patterns, analyze relations between land uses and stream conditions, and enhance communications among scientists, land developers, planners, and water resource managers (Booth and Jackson, 1997).

A variety of urban indicators have been cited in the literature. The easiest to understand and apply is the total surface coverage of all urban land uses within a watershed. Several studies have found significant negative relations between stream ecosystem attributes and the percentage of the watershed in urban lands (Steedman, 1988; Limburg and Schmidt, 1990; Lenat and Crawford, 1994; Weaver and Garmen, 1994; Wang et al., 1997; Klauda et al., 1998). However, this measure does not distinguish the relative effects of different urban land use categories with clearly different impacts on streams. For instance, urban recreational lands such as parks almost certainly cause less harm to streams than heavily developed industrial or commercial lands (Wang et al., 2001). Thus total urban land use is only useful for examining very general and broad-scale effects of urbanization. Other commonly used measures of urbanization, such as human population density or house density, have similar limitations. Although negative relations have been reported between population density (Jones and Clark, 1987; Schueler, 1997) and house density (Benke et al., 1981) and stream biota, neither measure is a precise indicator of relative urban influences. Indeed, high

TABLE 13.1
Coefficients of Determination (r^2) for Regressions between Urban Land Use Variables (Percent of Watershed Area) and Fish Variables and Bank Erosion (Wang et al., 2001)

Land Cover Variables (%)	Species Number (Ln)	IBI Score (Ln)	Shannon Index (Ln)	Individuals as Tolerant Fish (%)	Fish Number/ 100 m² (Ln)	Bank Erosion
Connected imperviousness	0.55*(–)	0.32*(–)	0.50*(–)	0.19*(+)	0.39*(–)	0.27*(+)
Highway, street, parking	0.48*(–)	0.23*(–)	0.48*(–)	0.17*(+)	0.28*(–)	0.25*(+)
Commercial land	0.41*(–)	0.33*(–)	0.30*(–)	0.21*(+)	0.31*(–)	0.21*(+)
Urban land	0.35*(–)	0.22*(–)	0.33*(–)	0.13 (+)	0.28*(–)	0.21*(+)
Residential land	0.23*(–)	0.16*(–)	0.20*(–)	—	0.22*(–)	0.20*(+)
Government land	0.30*(–)	0.31*(–)	0.20*(–)	0.11 (+)	0.29*(–)	—
Industry	0.08 (–)	—	0.14 (–)	—	—	—
Other transportation	0.18*(–)	0.15*(–)	—	—	0.15*(–)	—
Recreational land	0.14 (–)	—	0.22*(–)	0.10 (+)	0.10 (–)	—

Note: Coefficients listed only for regression slopes that were significant at $p < 0.05$; * indicates significant at $p < 0.01$; (+) indicates the relationship is positive and (–) indicates it is negative. Ln = Natural log transformed variables) for 47 Wisconsin warmwater streams.

impact urban land uses such as industrial or commercial areas often have relatively low population and house densities.

The consensus from recent literature is that the amount of impervious surface in a watershed is the best indicator of urbanization effects on streams (Klein, 1979; Schueler and Galli, 1992; Luchetti and Fuersteburg, 1993; Scheuler, 1994; Arnold and Gibbons, 1996; Booth and Jackson, 1997; May et al., 1997; Walsh, 2000; Wang et al., 2000, 2001). For example, Wang et al. (2001) compared relations of nine variables associated with urbanization and six variables describing stream habitat and fish assemblages (Table 13.1) and reported that impervious surface area was the best variable for describing impacts of urbanization. Not only does imperviousness accurately reflect the relative intensities and impacts of various types of urban land uses (Figure 13.3), but it also directly relates to one of the main mechanisms by which urbanization affects streams, i.e., surface runoff. Even more importantly, imperviousness is one of the few urban land use variables that can be explicitly quantified, managed, and controlled at each stage of land development (Schueler, 1994; Arnold and Gibbons, 1996). Although total impervious surface is strongly related to stream physical, chemical, and biological characteristics, the amount of connected (with a direct drainage connection to a stream) and effective (includes nominally pervious surfaces that have been sufficiently compacted to function as essentially impervious) impervious surface is a better indicator (Booth and Jackson, 1997; Walsh, 2000; Wang et al., 2000, 2001).

13.4 BIOLOGICAL INDICATORS OF URBAN DEGRADATION

Although watershed urbanization exerts many physical and chemical effects on streams, these effects are often difficult to detect, particularly during early phases of urban development. For example, water quality impacts associated with increased surface runoff are often episodic, associated with rain storms and snow melts, and can only be assessed through continuous monitoring, which is expensive and time consuming. Even with continuous monitoring, runoff volumes and characteristics tend to be highly variable, and long time data series are required to evaluate trends (Benke et al., 1981; Novotny and Chester, 1981; Hoffman, 1985; Novotny and Olem, 1994). Even when changes in the loading or concentration of a pollutant are detected, the effects of the pollutant

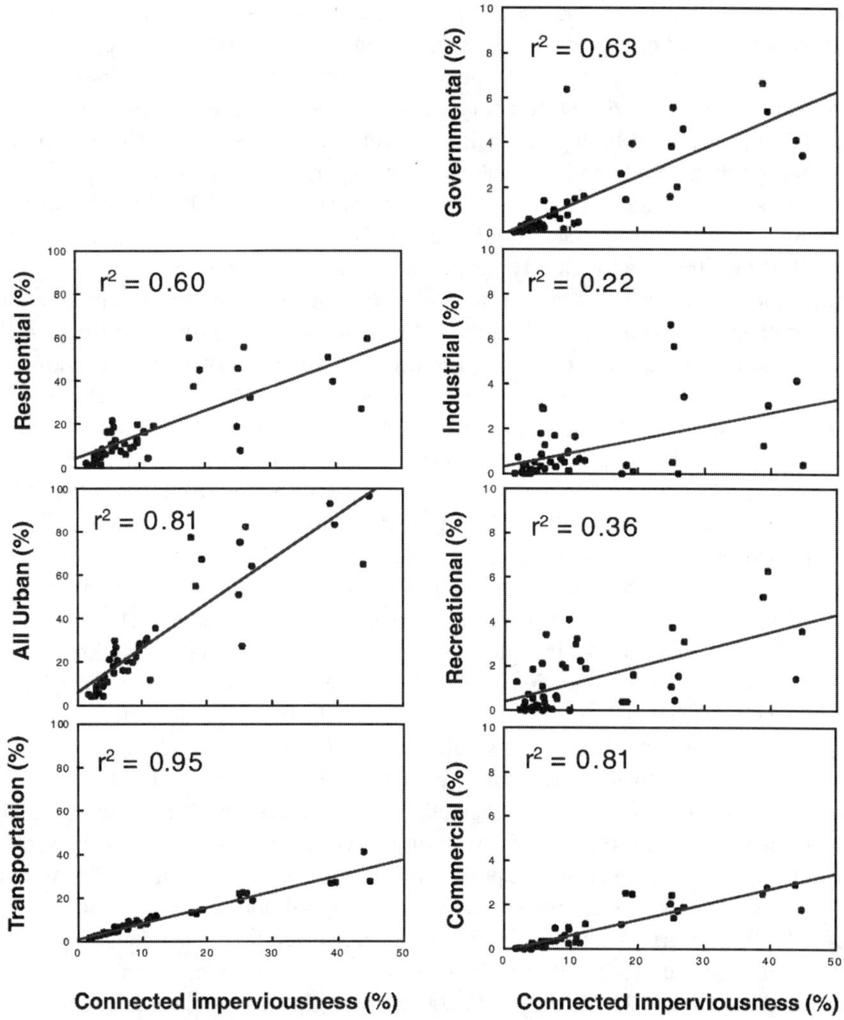

FIGURE 13.3 Watershed-connected imperviousness represents all the urban land use components because of its significant correlation with other urban characteristics. (*Source:* Wang, L. et al., 2001. *Environmental Management,* 28, 255–266. With permission.)

on the stream ecosystem are often unclear. Toxicological studies of runoff can provide insight into biological effects (e.g., Bannerman, 1991; Crunkilton et al., 1996), but results are often highly site-specific.

A cost effective and ecologically relevant way to assess urban impacts on streams is to look directly at responses of biological assemblages such as fish or aquatic insects (Karr, 1981, 1987, 1993; Karr and Chu, 1999). Because fish and insect assemblages consist of a variety of species with different life histories, sensitivity to degradation, and function in the ecosystem, they respond to a range of urbanization effects. Biotic assemblages represent the endpoint of the combined influences of hydrology, channel morphology, and water quantity and quality, and one or a few appropriate samples of the assemblage can provide unique insight into the condition of the stream and the causes of degradation. Assemblages have proven to be accurate and easily measured indicators of the overall quality or health of stream ecosystems (Karr and Chu, 1999).

Numerous assemblage-level indicators are available to assess urban impacts in streams. The most commonly used and effective indicators for urbanizing watersheds can be grouped into three categories: tolerance measures, species richness or diversity, and multimetric biotic indices (Fausch et al., 1990; Karr and Chu, 1999). Different types of indicators tend to be most sensitive to different environmental impacts. Most biological indicators were developed empirically, based on observed responses to a gradient of degradation rather than experimental tests of sensitivity to specific stressors. This is appropriate because watershed urbanization invariably has multiple interactive impacts on stream ecosystems that are often diffuse and cumulative. The best indicators are sensitive to most or all of the stressors that are typical of watershed urbanization.

Tolerance measures are based on the documented relative sensitivity of particular species (or genera or families) to urbanization. Broadly defined, tolerance measures can also include total abundance or biomass of all species. The presence or abundance of taxa of a known degree of sensitivity at a site is thus a measure of the presence or degree of urbanization effects on a stream. For example, Hilsenhoff (1987) developed an aquatic macroinvertebrate index to assess organic pollution, a common consequence of urbanization. Each taxon was assigned a rating based on its sensitivity to organic pollution. The index was the weighted average (based on relative abundance) of the ratings of all of the taxa collected in a semiquantitative sample. If most of the taxa and individuals present were sensitive to organic pollution, then the index score would be low. If the assemblage were dominated by taxa tolerant of pollution, the index score would be high. Tolerance measures have a long history in urban watersheds, dating back to the early 1900s and initial efforts to document effects of sewage and industrial pollution on rivers and lakes (Kolkwitz and Marson, 1908; Wright and Tidd, 1933; Chandler, 1970; Davis, 1995). By their nature, tolerance measures tend to be relatively narrow in their sensitivity (Fausch et al., 1990), but they nonetheless have been found to be useful in recent urbanization studies (e.g., Stepenuck et al., 2002).

Species richness and diversity measures are based on the premise that both the number of species (or higher taxa) present and the evenness of the distribution of individuals among species are related to the amount and type of urbanization present in the watershed. Usually richness and diversity are assumed to be inversely proportional to urbanization, although this may not hold for coldwater streams where fish assemblages are dominated by salmonids (Lyons et al., 1996; Mundahl and Simon, 1999). In warmwater streams, strong negative relations have been reported between watershed urbanization and both species richness and diversity (e.g., Klein, 1979; Linke et al., 1999; Gregg and Stednick, 2000; Wang et al., 2000, 2001). Species richness and diversity measures were first applied to stream assessment in the 1960s (Wilhm and Dorris, 1966) and remain popular, both as stand-alone indicators and as components of multimetric indices (Washington, 1984; Fausch et al., 1990; Norris and Georges, 1993; Davise, 1995; Simon and Lyons, 1995; Karr and Chu, 1999).

Multimetric indices combine a variety of different measures, including tolerance and species richness/diversity, into a single index that reflects structural, compositional, and functional attributes of assemblages (Barbour et al., 1995; Karr and Chu, 1999). By incorporating several different measures, multimetric indices are sensitive to a wide range of urbanization impacts. Most multimetric indices are also calibrated such that they take into account and correct for natural variation in assemblages owing to biogeography or ecological conditions in the stream. For example, it is well established that in the absence of human perturbations larger streams tend to have more fish species than smaller streams and certain river basins have richer fish faunas than other basins (Fausch et al., 1984). Similarly, coldwater streams tend to have fewer fish species than similarly sized and located warmwater streams (Lyons et al., 1996). Well designed multimetric indices take these natural differences into account and have metrics and standards that are specific to the region and type of stream they are designed for.

Currently the most widely used family of multimetric indices is the index of biotic integrity (IBI), first developed by Karr (1981) for stream fishes in the central United States and now applied to both fish and aquatic macroinvertebrates in streams, rivers, and lakes throughout the world (Simon and Lyons, 1995; Hughes and Oberdorff, 1999). Variants of the IBI are often known by other names, such as the invertebrate community index (Ohio EPA, 1988; DeShon, 1995), but all versions and variants share several features. Each metric represents a different attribute of the structure, composition, or function of the assemblage. Natural environmental or biogeographic factors that influence these attributes are taken into account in the application of the index. Metrics are chosen and calibrated based largely on empirical data; standards for each metric are derived from representative relatively unimpacted sites. Each metric is sensitive to one or more types of environmental degradation. The overall index score is the sum of the scores for individual metrics. Over the past 15 years, IBIs have been broadly applied to document urban impacts on streams (e.g., Steedman, 1988; Karr and Chu, 1999; Schleiger, 2000; Wang et al., 1997, 2000, 2001).

Regardless of the biological indicator used, impacts of urbanization on assemblages must be distinguished from impacts from other sources, such as forestry practices or agriculture, and from natural variation owing to climate fluctuations. Several different sampling designs have been applied to identify urban impacts. The most common approach has been to sample a large number of watersheds across a gradient of low to high urbanization and then use correlation or regression methods to identify a relation between degree of urbanization and characteristics of the stream biota (e.g., Klein, 1979; Jones and Clark, 1987; Steedman, 1988; Weaver and Garman, 1994; Maxted and Shaver, 1996; May et al., 1997; Wang et al., 1997, 2001; Klauda et al., 1998; Schleiger, 2000).

Consistent with this design, stream sampling sites and watersheds must be chosen to have similar physical, chemical, and biological conditions aside from the differences in urban land use. However, identifying a sufficient number of inherently similar streams is often difficult. This design is relatively straightforward to apply and interpret, but can only identify correlations between urbanization and biological characteristics, not cause and effect. A variant on this approach is to compare among sites that can be categorized into discrete groups based on relative level of urbanization, such as high, medium, and low, and then to use categorical statistics, such as analysis of variance, to identify patterns related to degree of urbanization (e.g., Garie and McIntosh, 1986; Scott et al., 1986; Matthews and Gelwick, 1990; Victor and Fufeyin, 1993; Lenat and Crawford, 1994; Couch, 1995, in Schueler, 1997).

An alternative approach, less commonly applied but arguably more powerful, is to sample stream sites before and after watershed urbanization occurs (Marsh and Minckley, 1982; Weaver and Garman, 1994; Wichert, 1994, 1995; Baker and Sharp, 1998; Fitzgerald et al., 1998). This historical comparison has the potential to detect more subtle effects of urbanization, particularly in settings where watersheds had already been impacted by agriculture prior to urbanization (Wang et al., 2000). The relative scarcity of appropriate historical data from urbanizing areas has limited the application of this approach.

The best approach to documenting urban impacts on stream ecosystems is to compare biological data collected from multiple streams before and after urban development and compare them with data collected from similar reference streams where no urbanization has taken place (Moscrip and Montgomery, 1997; Wang et al., 2000). This approach has rarely been used because of the lack of appropriate historical data and the difficulty in finding similar streams that differ only in the amount of urbanization in their watersheds. This sampling design is considered the most appropriate and powerful for ecological field evaluation when sample sizes are limited and serial autocorrelation among samples is likely (Stewart-Oaten and Murdoch, 1986). It is particularly effective in evaluating the effectiveness of urban watershed best management practices (BMPs) for improving stream conditions (Jones et al., 1996; Maxted and Shaver, 1996).

13.4.1 Responses of Macroinvertebrate Assemblages to Urbanization

13.4.1.1 Tolerance Measures

Total abundance of benthic macroinvertebrates is generally not strongly related to degree of watershed urbanization, but the absolute or relative abundance of various macroinvertebrate taxa often is. Jones and Clark (1987) studied 22 stream sites in northern Virginia and found that the amount of watershed development had little influence on total insect density, although abundance of specific taxa did shift. The percent abundance of the order Diptera, a relatively tolerant group, was 12 to 36% at rural sites, 14 to 66% at moderately urbanized sites, and 33 to 99% at heavily urbanized sites. Karr and Chu (1999) reported no systematic relationship between total benthic invertebrate abundance and percent watershed impervious area for lowland streams in Puget Sound, Washington.

In North Carolina, the relative abundance of Ephemeroptera, Plecoptera, and Trichoptera (EPT), three of the most sensitive aquatic insect orders, was an order of magnitude lower in an urban stream than a forested or agricultural stream (Lenat and Crawford, 1994). They also reported that a modified Hilsenhoff biotic index (HBI) was significantly higher in the urban stream, indicating more degradation. Kemp and Spotila (1997) reported that Amphipoda, Ephemeroptera, Chironomidae, Plecoptera, and Trichoptera were more abundant at a rural Pennsylvania stream and Isopoda and Oligochaeta dominated at an urbanized stream. Stepenuck et al. (2002) studied benthic macroinvertebrates at 43 of the 47 warmwater streams in southeastern Wisconsin studied by Wang et al. (2000, 2001) for fish. They found that watershed percent connected imperviousness was significantly negatively correlated with EPT relative abundance and positively correlated with the HBI. However, these relationships were not linear. Some streams had good values (i.e., high EPT, low HBI) whereas others had poor values where watershed imperviousness was less than 12%. When watershed imperviousness exceeded 12%, all streams had poor values.

The compositions of different macroinvertebrate feeding groups were used with some success to detect the effects of urbanization on stream food webs. In their study comparing forested, agricultural, and urban watersheds in North Carolina, Lenat and Crawford (1994) reported that shredders (taxa feeding on coarse particulate materials) were found only in the forested stream (4%). Both the agricultural and urban streams had relatively high proportions of scrapers (taxa feeding on attached algae, 16 to 21%). Filter-feeders were important in both the agricultural and forested sites (46 to 47%), but accounted for only 10% of the macroinvertebrates at the urban site. The urban site was dominated by collector-gatherers.

In comparing macroinvertebrates collected from nine reaches in the Laurel Creek watershed in southern Ontario, Canada, Winter and Duthie (1998) found that the proportion of deposit feeders in urban reaches was much higher than in the rural reaches. Stepenuck et al. (2002) studied 43 southeastern Wisconsin warmwater streams and found complex relations between macroinvertebrate feeding groups and urbanization. For watersheds with connected imperviousness <12%, the percentages of various feeding groups were highly variable. Scrapers varied from 0 to 83%, collectors from 40 to 100%, and gatherers from 2 to 92%. However, above 12% imperviousness, the percentage of scrapers was always low and the percentages of collectors and gatherers were always high. Karr and Chu (1999) reported no consistent relation between urbanization and macroinvertebrate feeding groups in studies from Washington state and argued against their use as urban indicators.

13.4.1.2 Richness and Diversity

A decline in taxa richness is generally one of the most common indicators of watershed urbanization for stream benthic macroinvertebrates. Klein (1979) sampled 15 small watersheds in Maryland and found a sharp decline in the number of macroinvertebrate taxa with increasing watershed urban land use. For forested or agricultural watersheds (total imperviousness <1%), the taxa richness ranged from 9 to 19, with a mean of 15. However, for watersheds with 17 to 56%

total imperviousness, taxa richness decreased to 2 to 8 with a mean of 5. In a study of eight sampling sites along Shabakunk Creek, New Jersey, Garie and McIntosh (1986) found a dramatic decline in macroinvertebrate taxa richness, from 13 in relatively undeveloped upstream areas to 4 in heavily developed downstream areas.

Lenat and Crawford (1994) also reported that a North Carolina urban stream had much lower total and EPT taxa richness than forest and agricultural streams. In the Atlanta area, Georgia, Benke et al. (1981) studied 21 streams that had 0 to 98% urban land use and 0 to 941 houses per square mile. They found a strong negative relationship between house density and taxa richness, but no apparent relation with diversity because of an uneven distribution of abundance for small-body sized individuals. In southeastern Wisconsin streams, Stepenuck et al. (2002) also reported negative but nonlinear relations between percent watershed connected imperviousness and taxa richness and diversity. Streams with low imperviousness had variable richness and diversity, whereas streams with high imperviousness always had low richness and diversity.

13.4.1.3 Multimetric Indices

Multimetric macroinvertebrate indices have been less widely developed and applied than multimetric fish indices, and only a few studies examined the relations between urbanization and multimetric macroinvertebrate indices. Shaver et al. (1994) reported preliminary results from a study on 19 stream sites in Delaware and found a negative nonlinear relation between percent watershed total imperviousness and macroinvertebrate biological index score expressed as percent of reference stream scores. Streams with low watershed imperviousness had variable biological scores, whereas streams with high imperviousness had consistently low scores. Similarly, Horner et al. (1996) and May et al. (1997) reported studies from Washington state and described a negative nonlinear relationship between percent watershed total imperviousness and benthic IBI values. Only reaches with total impervious area less than 3.9% exhibited excellent scores. All good benthic IBI values were associated with watersheds having less than 11% imperviousness. For watersheds with more than 11% imperviousness, all benthic IBI values were fair to poor and generally benthic IBI scores were negatively correlated with imperviousness. Eight sites with particularly wide riparian buffers had better benthic IBI scores than would have been expected based on the amount of impervious surface in their watersheds.

13.4.2 Responses of Fish Assemblages to Urbanization

13.4.2.1 Tolerance Measures

A variety of studies related fish abundance to amount of watershed urbanization. In comparing fish data collected in 1958 and 1990 from Tuckahoe Creek, Virginia, Weaver and Garman (1994) reported that total fish abundance and abundance within each of the major trophic guilds (carnivore, herbivore, omnivore) were substantially lower in 1990 than in 1958. Across their six study sites, declines ranged from more than 95 to 58% with an average of 80%. During the same period, urban land use in the watershed increased from 7 to 20%. Wang et al. (2000) reported that the mean catches of fish for 47 warmwater streams in Wisconsin decreased from 75 to 44 per 100 m^2 from mid 1970s to 1997. They attributed the change in abundance to the increase in urban land use; about 40% of the watersheds had less than 10% urban land in 1970 but only 20% of the watersheds had less than 10% urban land in 1990.

A common response to urbanization is the loss of sensitive fishes and an increase in the relative proportion of tolerant species. In a study of four similar watersheds with different levels of urbanization in the Maryland Piedmont, Schueler (1994) reported that two sensitive fish species were lost as imperviousness increased from 10 to 12% and four more were lost when imperviousness increased to 25%. In comparing three similar North Carolina streams with different land uses, Lenat and Crawford (1994) reported that intolerant fishes were absent from the urban stream but were found in the forested and agricultural streams. Onorato et al. (1998) also reported that several sensitive fish species became less abundant in areas impacted by urban sources of pollution in the upper Cahaba River in Alabama.

Wang et al. (2000) compared fish data collected in mid 1970s and 1997 from 47 southeastern Wisconsin test streams with different watershed urbanization levels. To examine the relation between urban land use change and fish community change, they divided the streams into four groups based on 1970s data. Group A consisted of watersheds with more than 10% connected imperviousness. Group B consisted of watersheds with less than 10% connected imperviousness and 10 or fewer fish species. Group C contained watersheds with less than 10% connected imperviousness and more than 10 fish species. The reference group consisted of four streams with minimal urban development. During both the 1970s and 1997, the four reference streams had significantly higher percentages of intolerant species and individuals than any one of the other three groups. Intolerant fish species were not found in Group A streams and they were significantly fewer in Group B than in Group C streams. The reference and Group A streams had no significant changes in percentages of tolerant fish between the 1970s and 1997. Group B showed a significant decrease, and Group C showed a significant increase in tolerant fish over the study period.

Anadromous fishes may be particularly sensitive to watershed urbanization. Limburg and Schmidt (1990) studied 16 tributaries of the Hudson River in New York state, which represented an urban–rural gradient of watershed land uses, including tributaries in the New York City metropolitan area. They reported a strong relationship between densities of fish eggs/larvae and an index of urbanization. The densities of fish eggs and larvae declined dramatically, from greater than $30/m^3$ to less than $1.6/m^3$ for anadromous fishes and from more than $2.5/m^3$ to less than $1.6/m^3$ for residential fishes, when watershed urban land use increased from 0 to 10%. Densities for both anadromous and residential fishes were consistently low, less than $1.6/m^3$, when watershed urbanization exceeded 10%. Moscrip and Montgomery (1997), studying streams in Washington state, found that the abundance of migratory salmon (*Oncorhynchus* species) significantly decreased from the early 1970s to the early 1990s when watershed urban land use grew from 20 to 50% in Flett Creek and from near zero to 45% in Swamp Creek. In contrast, May Creek, a nearby stream where watershed urbanization only increased from near zero to 12%, no clear trends in salmon abundance were detected during the same period. Scott et al. (1986) compared the fish assemblage in Kelsey Creek, Bellevue, Washington, with that of a rural reference stream and reported that urbanization of the Kelsey Creek watershed was associated with a shift in species relative abundances. Urban degradation had a greater impact on migratory coho salmon (*Oncorhynchus kisutch*) and resident nonsalmonid species than on cutthroat trout (*Oncorhynchus clarki*). Age 0 and 1 cutthroat trout comprised the majority of fish found in Kelsey Creek, whereas the control stream supported a diverse assemblage of coho salmon of various ages, other salmon and trout species, and numerous nonsalmonids. When data from other streams were included, the percentage of cutthroat trout in Lake Washington tributary streams was positively related to the proportion of watershed impervious area. Studies from other streams in this region of Washington also reported that urbanization altered the ratio of juvenile coho salmon to juvenile cutthroat trout (Horner et al., 1996, May et al., 1997). Coho salmon tended to dominate in undeveloped watersheds, whereas cutthroat were relatively more common in urbanized streams.

When watersheds become urbanized, native fishes may be replaced by introduced fishes, which are often more tolerant of pollution and habitat degradation (Schueler, 1997). Marsh and Minckley (1982) compared fish collections from the Salt River and associated canals in the Phoenix metropolitan area in Arizona between 1890 and 1981. During this period, the river system changed from perennial to intermittent due to urban water demands and stream channelization. As a result, a historic assemblage of 15 native fish species declined to four and 29 introduced species had been recorded since 1926. Couch et al. (1995, in Schueler, 1997) compared eight streams that had 70 to 90% urban land use with one forest-dominated reference stream near Atlanta, Georgia. They reported that, on average, nonnative fish comprised 25% of the fish species in the urban streams but only 6% in the reference stream.

13.4.2.2 Richness and Diversity

Generally, increasing watershed urbanization is associated with declining fish species richness and diversity. Klein (1979) examined 27 small watersheds and 78 published fish collections in the Piedmont province of Maryland and documented a strong negative relation between watershed total impervious area and fish species diversity. In comparing four similar Maryland streams with different amounts of urban land use, Schueler (1994) reported that the number of fish species decreased from 12 to 2 when watershed impervious area increased from 10 to 55%. Lenat and Crawford (1994) reported that the number of stream fish species was substantially greater in forested (19 species) and agricultural watersheds (19 species) than in an urban watershed (9 species) in North Carolina.

Weaver and Garman (1994) found that fish diversity decreased substantially for all six of their sampling sites in Tuckahoe Creek watershed, Virginia, between 1958 and 1990, a period when urbanization of the watershed increased dramatically. In a comparison of eight urban streams with one forested reference stream in Georgia, Couch et al. (1995, in Schueler, 1997) determined that the mean number of fish species in urban streams was lower than in the reference stream (12 vs.17). Wang et al. (2000, 2001) documented a nonlinear negative relation between watershed connected imperviousness and fish diversity and richness in 47 southeastern Wisconsin streams. Sites below about 8% imperviousness had a wide range of diversity and richness values, but at least some were high. Maximum diversity and richness values dropped sharply from 8 to 12% imperviousness, and maximum values were consistently low above 12% imperviousness.

13.4.2.3 Multimetric Indices

Urbanization is negatively related to fish IBI scores. Steedman (1988) examined 209 stream sites in 10 watersheds near Toronto, Ontario, and reported that fish IBIs calculated at both basin and site scales were inversely related to percent urban land use at the scale of both the watersheds ($r^2 = 0.67$) and sites within watersheds ($r^2 = 0.27$ to 0.82). Wang et al. (1997) analyzed fish and land use data from 134 sites on 103 streams located throughout Wisconsin and found a negative relationship between fish IBI score and percent watershed urban land use. For watersheds with less than 20% urban land use, some streams had high IBI scores and others had low scores. However, streams in watersheds with more than 20% urban lands had consistently low IBI scores.

DeVivo et al. (1997) studied 23 second to fourth order streams with different levels of watershed urban land use in Georgia. They found that fish IBI score was inversely related to watershed population density. Once watershed population density exceeded four per acre, the stream fish IBI score was consistently poor. Also in Georgia, Schleiger (2000) sampled 507 sites on 340 streams and found a negative correlation between IBI scores and water quality measures associated with low-density urban lands. Klauda et al. (1998) found a strong negative correlation between fish IBI score and urban land use for 61 subwatersheds with urbanization levels ranging from 0 to 95% in the Patapsco River basin, Maryland. More recently, Wang et al. (2001) studied 47 small Wisconsin warmwater streams with different levels of watershed urban land use. Using nonlinear quantile regression, they found that the maximum possible IBI score declined exponentially with increasing amount of connected impervious surfaces in the watershed. They noted a sharp decline in maximum possible IBI scores between 8 and 12% watershed imperviousness.

13.5 BIOLOGICAL SIGNATURES OF URBANIZATION

Results from these studies indicate that common responses of macroinvertebrate and fish assemblages can serve as "signatures" (Yoder and Rankin, 1995) of the impacts of watershed urbanization. For macroinvertebrates, the decline in abundance or complete loss of sensitive taxa, particularly among the orders Ephemeroptera, Plecoptera, and Trichoptera, is a typical consequence of urban development. However, total macroinvertebrate density usually does not change in particular patterns because tolerant taxa become more abundant and offset the loss of sensitive individuals. As a result, the Hilsenhoff biotic index — a measure of the average tolerance of an assemblage to organic pollution — usually increases (i.e., the assemblage becomes more tolerant). Macroinvertebrate taxa richness and diversity almost always decline as a watershed is developed, as do biotic integrity index scores. Some investigators reported effects of urbanization on the relative abundance of various feeding groups, but this response is not consistent among studies.

Stream fish assemblages change in a variety of ways as a consequence of watershed urbanization. Usually, overall fish density decreases. The abundance of anadromous species (if present) and species sensitive to habitat degradation and water pollution decline the most. As a result, the proportion of exotic species and species tolerant of wide ranges of habitat and water quality conditions may increase even though the densities of these species may remain fairly constant or even decline. Species richness and diversity drop as watershed development proceeds, although this response is most pronounced in the richer assemblages of warmwater streams and may not be apparent at low levels of urbanization for the more depauperate salmonid-dominated assemblages of coldwater streams (Lyons et al., 1996). For both warmwater and coldwater streams, biotic integrity index scores decrease sharply with increasing urbanization.

The changes in macroinvertebrate and fish assemblages caused by watershed urbanization are not unique and are typical of the general response of stream biota to other sources of degradation, such as industrial point-source pollution or watershed agriculture (Karr and Chu, 1999). What is unique, however, is the low level of watershed urbanization that triggers the biological response and the strength and magnitude of this response. For example, Wang et al. (1997) found that agriculture was not associated with a clear change in Wisconsin stream fish assemblages until it covered more than 40 to 50% of the watershed. Biotic integrity was typically not greatly reduced until agricultural coverage exceeded 60 to 70%. Conversely, fish assemblages were affected at 10% urban coverage and biotic integrity was consistently poor above 20% urban coverage. Stream fish assemblages already greatly degraded by watershed agriculture became significantly worse as a small portion of that agriculture was replaced by urbanization (Wang et al., 2000).

13.6 LEVELS OF IMPERVIOUSNESS THAT CAUSE BIOLOGICAL DEGRADATION

Over the last 25 years, increasing numbers of studies considered the relation between watershed imperviousness and measures of the quality of fish or macroinvertebrate assemblages in streams of the United States. A pattern emerges when these studies are considered collectively. Streams in rural watersheds with little or no impervious surface may have highly variable biological assemblages, but the conditions of at least some of these are rated fair to good. Generally, effects of urbanization on stream biota are manifest at surprisingly low levels of imperviousness. Initial thresholds of impact are remarkably consistent across the country, varying from 4 to 8% effective imperviousness and 8 to 12% total imperviousness. Beyond these thresholds, biological conditions decline sharply with small increments in impervious surface. Above a relatively low level of imperviousness, usually 12 to 15% effective and 14 to 20% total, biological conditions are consistently poor.

13.6.1 Findings from Different Regions

13.6.1.1 Mid-Atlantic States

Maryland studies indicate that 8 to 12% total impervious area is a lower threshold level of urbanization associated with a sharp decline in the quality of stream biota. Klein (1979) studied macroinvertebrates in 13 streams in the Maryland Piedmont. Five watersheds had total imperviousness between 0 and 1% and 9 to 19 macroinvertebrate taxa. The remaining eight watersheds had imperviousness between 17 and 56% and 2 to 8 taxa. No data were collected for watersheds with imperviousness between 2 and 17%.

Klein (1979) also measured stream base flow for 23 watersheds that ranged in total imperviousness from 0 to 56%, and found that base flow per unit watershed area did not decrease obviously until imperviousness exceeded 25%. He also summarized 78 published fish collections from the same area and reported a clear decreasing trend in fish species diversity as total imperviousness increased. Fish diversity declined sharply from 12 to 25% total imperviousness, although a fair amount of variation was found among the sampling sites. In summarizing all the different measures, Klein concluded that stream quality impairment was first evidenced when total watershed imperviousness reached 12% but that impairment did not become severe until it reached 30%.

Schueler and Galli (1992) examined 23 headwater stream sites in the Anacostia watershed and found that all stream sites with less than 10% total imperviousness had good to fair macroinvertebrate diversity. Nearly all sites with 12% or more impervious area had poor diversity. Klauda et al. (1998) surveyed 61 stream sites in the Patapsco basin with watershed urbanization that ranged from 0 to 95%. They reported a strong negative relationship between percent urban land use and fish IBI score. Although they did not determine the amount of watershed impervious area, their data indicated a sharp decline in IBI score between 15 and 35% urban land use, which corresponds to about 8 to 15% total imperviousness.

Studies from other Mid-Atlantic states also indicated that 8 to 12% total imperviousness was a level above which major declines in stream biota became evident. Jones and Clark (1987) monitored 22 stream sites in northern Virginia and concluded that macroinvertebrate diversity declined markedly after watershed population density exceeded four or more individuals per acre. Schueler (1994) translated this population density into 10 to 20% imperviousness. Shaver et al. (1994) analyzed data from 19 stream sites in Delaware and suggested that macroinvertebrate biological score dropped dramatically when imperviousness increased above 8 to 15%. Beyond this imperviousness level biological quality was remarkably low. Similarly, Schueler (1994) compared a number of studies from Maryland, Virginia, Delaware, and New Jersey. These studies were mainly unpublished and covered a range of methods and stream indicators. Schueler concluded that stream degradation occurred at 10 to 20% watershed imperviousness.

13.6.1.2 Pacific Northwest States

Thresholds of impact were observed beginning at 8 to 10% effective imperviousness and 5 to 20% total imperviousness. Booth and Jackson (1994) analyzed long term stream flow patterns and stream channel morphology data collected from numerous streams in King County, Washington, in the Seattle metropolitan area. They found that 10% effective imperviousness was the level that distinguished stable and unstable stream channels. In the same study, they classified stream physical habitats as excellent, fair, or poor along 140 km of stream channels in two drainage basins. Marked degradation of stream physical habitat occurred consistently above 8 to 10% imperviousness. Finkenbine et al. (2000) studied streams near Vancouver, British Columbia, and found evidence of declines in stream habitat quality for fish at 20% total imperviousness. In the Puget Sound lowlands in Washington state, Horner et al. (1996) and May et al. (1997) determined that benthic invertebrate IBI scores declined sharply from 5 to 12% total imperviousness and were consistently low above

12%. They also found that the ratio of coho salmon to cutthroat trout numbers dropped dramatically between 5% and 14% total imperviousness.

13.6.1.3 Midwestern States

In Wisconsin and Minnesota, biological effects of urbanization were first apparent at 8 to 10% connected imperviousness. Wang et al. (2000) compared fish data collected in the mid 1970s and 1997 relative to urban land use data from 1970 and 1990 for 47 warmwater streams in southeastern Wisconsin. Fish species richness and IBI scores were highly variable for watersheds with less than 8% connected imperviousness, but were consistently low for watersheds with more than 12% imperviousness. Maximum species richness values and IBI scores declined sharply from 8 to 12% imperviousness. Stream sites that had less than 8% imperviousness and fewer than 10 fish species in the 1970s suffered the greatest relative increases in imperviousness and declines in species numbers over the study period. Wang et al. (2001) examined the same 47 Wisconsin streams and used 90% quantile nonlinear regressions to estimate the maximum possible values of fish species richness, IBI scores, and base flow per unit watershed area that could be achieved for a given level of watershed connected imperviousness. This analysis quantified the sharp drop in stream quality at 8 to 12% imperviousness (Figures 13.2 and 13.4).

Stepenuck et al. (2002) evaluated benthic macroinvertebrate assemblages in 43 of the 47 streams studied by Wang et al. (2000, 2001). They found a major decrease in the Shannon diversity index, percent EPT individuals, taxa richness, various feeding group variables, and an increase in the HBI between 8 and 12% connected imperviousness. The preliminary results from a study examining 37 current or former trout streams across an urbanization gradient in Wisconsin and Minnesota

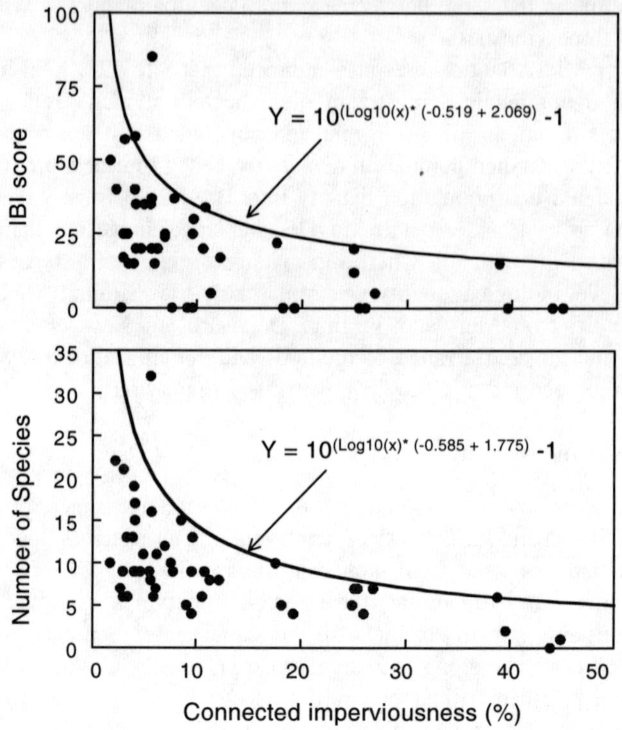

FIGURE 13.4 Relations between the percentage of connected imperviousness in the watershed and the number of fish species and the IBI score for sites on 47 Wisconsin warmwater streams. Fish species richness and IBI values decline sharply at imperviousness levels between 8 and 12%. (*Source:* Wang, L. et al., 2001. *Environmental Management,* 28, 255–266. With permission.)

indicated that both abundance of coldwater fishes and IBI score were significantly negatively correlated with percent watershed-connected imperviousness (Wang et al., unpublished data). Declines were evident at 5% connected imperviousness, and fish abundance and biotic integrity were consistently low above 8%. This result is consistent with that from Washington state (May et al., 1997) where the ratio of coho salmon to cutthroat trout numbers declined sharply between 5 and 14% total imperviousness.

13.6.1.4 Other Regions

Although studies of urbanization impacts have been conducted elsewhere in the United States and southern Canada, these studies have not expressed urbanization in terms of impervious surface. However, in a few cases, it was possible to convert their findings into levels of imperviousness, and the results were consistent with findings from the Mid-Atlantic, Pacific Northwest, and Midwestern states. For example, DeVivo et al. (1997) surveyed 21 watersheds in the Upper Chattahoochee Basin near Atlanta, Georgia, and documented that fish biotic integrity decreased after watershed population density exceeded four or more individuals per acre, which was equivalent to 10 to 20% total imperviousness (Schueler, 1994).

13.6.2 Modifying Factors

13.6.2.1 Vegetative Buffer

A wide, continuous, and mature riparian forest buffer appears to increase the amount of watershed urbanization that a stream can withstand before it becomes degraded (Steedman, 1988). May et al. (1997) developed a model relating the percentage of a naturally vegetated riparian corridor more than 30 m wide to the percentage of watershed total impervious area to predict macroinvertebrate biotic integrity for Puget Sound lowland area streams in Washington state. Based on this model, if the entire stream length had a riparian buffer, it would take about 55% total imperviousness to degrade biotic integrity from good to fair. However, if only half the stream length had a buffer, it would only take 12% imperviousness for the same change in biotic integrity. Similarly, with a complete buffer 82% total imperviousness would be required to drop biotic integrity from fair to poor, but it would take only 40% imperviousness if only half the stream length was buffered. Because of the nature of the data used to develop the model, the authors cautioned that it was difficult to use the model to make an exact judgment as to how much riparian corridor was appropriate for each specific development scenario.

13.6.2.2 Urbanization Location

The spatial distribution of urban land use in a watershed seems to also influence the level of urbanization necessary to degrade the stream ecosystem. Wang et al. (2001) examined the relative influence of the amount of connected imperviousness within a 50-m buffer, a 50- to 100-m buffer, beyond a 100-m buffer, and within a 1-mi. radius, a 1- to 2-mi. radius, and beyond a 2-mi. radius from the sampling sites for 47 warmwater streams in southwestern Wisconsin. Wang concluded that fish species richness was more strongly influenced by the amount of imperviousness within the 50-m buffer, the 50- to 100-m buffer, and within the 1-mi. radius than by imperviousness further from the stream (Figure 13.5).

13.6.2.3 Best Management Practices (BMPs)

Urban BMPs are designed and implemented to reduce the negative hydrological, physicochemical, and biological effects of urbanization. Presumably, therefore, installation of BMPs should permit a greater amount of urbanization before a particular impact on a stream ecosystem is observed.

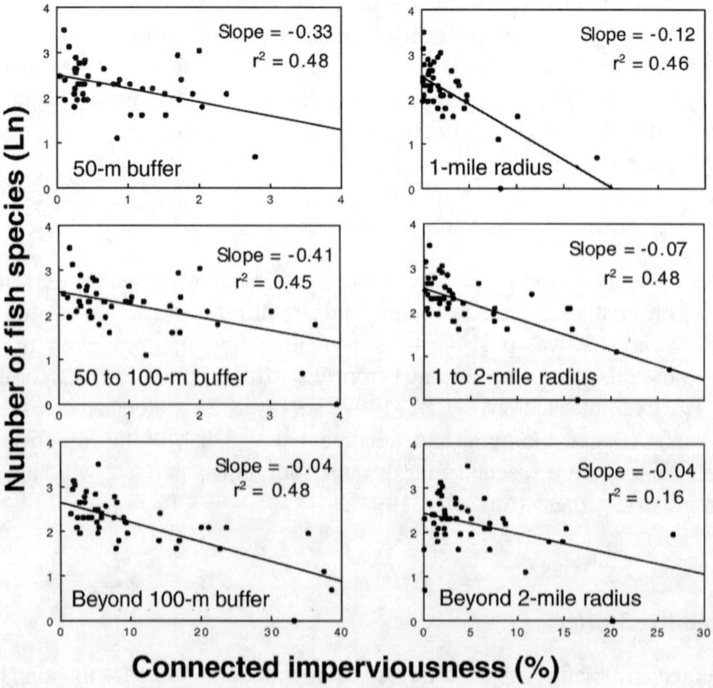

FIGURE 13.5 Relations between the number of fish species and connected impervious land uses within different buffer and radius scales expressed as percentages of watershed area. The regression slopes for the 50- and 100-m buffer plots are significantly higher than that for the >100-m buffer, indicating that urbanization closer to the stream is more influential than urbanization far from the stream. The X-axes for the 50- and 50- to 100-m buffers are scaled from 0 to 4% and that for the >100-m buffer is scaled from 0 to 40%. The regression slope for the 1-mi. radius plot is significantly greater than those for the 1- to 2-mi. and >2-mi. radius plots, again indicating that urbanization nearer a sampling site is more important than distant urbanization. (*Source:* Wang, L. et al., 2001. *Environmental Management,* 28, 255–266. With permission.)

However, studies to date have shown little evidence of major biological benefits from urban BMPs. Maxted and Shaver (1996) collected macroinvertebrates from eight streams in Delaware that had at least 20% watershed imperviousness and implemented stormwater management facilities at least 2 years before data collection. These streams were compared with 38 streams that had no urban BMPs. The overall macroinvertebrate community measured as a community index and a sensitive species index was not statistically different between the BMP and non-BMP sites. The BMPs evaluated did not prevent the loss of sensitive macroinvertebrate species after the watershed exceeded 20% imperviousness. The authors cautioned that their results were not conclusive regarding the value of stormwater controls for protecting stream biota because of the small size and complexity of their data set.

Jones et al. (1996) used macroinvertebrates and fish to assess the ability of urban BMPs to mitigate storm-water impacts for a suburban watershed in Virginia. They found that appropriately designed and properly sited BMPs could provide some relief from stormwater impacts on stream biota. However, the macroinvertebrate and fish assemblages differed greatly from those in undeveloped watersheds and reflected a fundamental alteration in stream biotic diversity, structure, and function.

Finally, Finkenbine et al. (2000), examined the value of constructing stormwater detention basins to rehabilitate urbanized watersheds around Vancouver, British Columbia. The detention basins caused few if any improvements in hydrology or habitat for fish if the stream had already been altered by watershed urbanization. They concluded that establishment of vegetated riparian buffer zones and additions of large woody debris were the best ways to stabilize stream channels and improve fish habitats.

13.7 CONCLUSIONS

Watershed urbanization is a major and increasing source of environmental degradation for streams in the United States. Urbanization affects stream hydrology, habitat and water quality, and biota, and impacts are apparent at surprisingly low levels of development, when much of the watershed remains rural and the land-use is best characterized as suburban. The best single measure of urbanization in a watershed is the amount of impervious surfaces, that is, surfaces such as roads, parking lots, and buildings that prevent precipitation from soaking into the ground and create surface runoff. The amount of impervious surface directly relates to the main mechanisms by which urban land use degrades streams — greater flooding, reduced baseflow, destabilized and degraded habitat, and increased water pollution — and can be directly quantified and managed as a watershed develops.

Macroinvertebrate and fish assemblages appear particularly sensitive to watershed urbanization. Various measures of the quality of these assemblages, encompassing tolerance, species richness and diversity, and multimetric indices, have been related to watershed urbanization through broad spatial comparisons among streams along a broad urban gradient and historical comparisons of the same streams as they urbanize over time. The most common and consistent responses of macroinvertebrate assemblages to urbanization include a decline in sensitive taxa such as EPT, an increase in the relative abundance of taxa tolerant of organic pollution and habitat degradation, a decrease in species richness and diversity, and a drop in overall biotic integrity. The relative abundance of various feeding groups may shift, but responses varied among studies. For fish assemblages, typical responses were a drop in fish density, a loss of species sensitive to habitat and water quality degradation, and decline in species richness, diversity, and overall biotic integrity. Anadromous species may be particularly sensitive to urbanization, and exotic species may replace natives in some urbanizing situations.

The levels of imperviousness at which biological communities first show the effects of watershed urbanization are remarkably consistent across the United States. Impacts first appear at 4 to 8% effective and 8 to 12% total imperviousness in the watershed, and beyond these thresholds minor increases in imperviousness cause major declines in quality of the biota. Above 12 to 15% effective and 14 to 20% total imperviousness macroinvertebrate and fish assemblages are almost always in poor condition. If streams have well vegetated riparian buffers or the impervious surface is located far from the stream, then good quality assemblages may persist at somewhat higher levels of imperviousness. However, current stormwater BMPs appear to provide little obvious benefit to stream biota in urbanizing watersheds.

ACKNOWLEDGMENTS

We thank Paul Kanehl and numerous temporary field technicians for field data collection and laboratory data processing for the studies conducted in Wisconsin. Roger Bannerman, Edward Emmons, and others at the Wisconsin Department of Natural Resources (WDNR) provided support for Wisconsin's studies and insightful discussions. Support in preparation of this chapter came from WDNR Bureau of Watershed Management and the Federal Aid in Sport Fishery Restoration Program, Project F-95-P, Study SSIF, administered through the Fish and Habitat Research Section of the WDNR Bureau of Integrated Science Services.

REFERENCES

Arnold, C.L., Jr. and C.J. Gibbons. 1996. Impervious surface coverage: the emergence of a key environmental indicator, *Journal of the American Planning Association*, 62, 243–258.

Baker, S.C. and H.F. Sharp Jr. 1998. Evaluation of the recovery of a polluted urban stream using the Ephemeroptera-Plecoptera-Trichoptera index, *Journal of Freshwater Ecology*, 13, 229–234.

Bannerman, R. 1991. *Pollutants in Wisconsin Stormwater*, unpublished report, Wisconsin Department of Natural Resources, Madison, Wisconsin.

Barbour, M.T., J.M. Stribling, and J.R. Karr. 1995. Multimetric approach for establishing biocriteria and measuring biological condition, in W.S. Davis and T.P. Simon (Eds.), *Biological Assessment and Criteria: Tools for Water Resource Planning and Decision Making*, Lewis Publishers, Boca Raton, FL. 63–77.

Barker, B.L., R.D. Nelson, and M.S. Wigmosta. 1991. Performance of detention ponds designed according to current standards, *Puget Sound Research 91*, Puget Sound Water Quality Authority.

Benke, A.C., G.E. Willeke, F.K. Parrish, and D.L. Stites. 1981. *Effects of urbanization on stream ecosystem.* Completion Report OWRT Project No. A-055-GA. U.S. Department of Interior, Office of Water Research and Technology, Ft. Collins, CO.

Booth, D.B. 1990. Stream-channel incision following drainage-basin urbanization, *Water Resources Bulletin*, 26, 407–417.

Booth, D.B. 1991. Urbanization and the natural drainage system-impacts, solutions, and prognosis, *The Northwest Environmental Journal*, 7, 93–118.

Booth, D.B. and C.R. Jackson. 1994. Urbanization of aquatic systems — degradation thresholds and the limits of mitigation. American Water Resources Association, Summer Symposium on Effects of Human-Induced Changes in Hydrologic Systems, Jackson Hole, Wyoming, 425–434.

Booth, D.B. and C.R. Jackson. 1997. Urbanization of aquatic systems: degradation threshold, stormwater detection, and the limits of mitigation, *Journal of the American Water Resources Association*, 33, 1077–1090.

Chandler, J.R. 1970. A biological approach to water quality management, *Water Pollution Control*, 69, 415–421.

Couch, C. 1995. Fish dynamic in urban streams near Atlanta, Georgia, in T. Schueler (Ed.). 1997. *Watershed Protection Techniques*, 2, 511–514.

Crunkilton, R., J. Kleist, J. Ramcheck, W. DeVita, and D. Villeneueve. 1996. Assessment of the response of aquatic organisms to long-term *in situ* exposure of urban runoff, in L.A. Roesner (Ed.), *Effects of Watershed Development on Aquatic Ecosystems. Proceedings of an Engineering Foundation Conference*, American Society of Civil Engineers, New York.

Davis, W.S. 1995. Biological assessment and criteria: building on the past, in W.S. Davis and T.P. Simon (Eds.). *Biological Assessment and Criteria: Tools for Water Resource Planning and Decision Making*, Lewis Publishers, Boca Raton, FL, 15–29.

DeShon, J.E. 1995. Development and application of the invertebrate community index (ICI), in Davis, W.S. and T.P. Simon (Eds.), *Biological Assessment and Criteria: Tools for Water Resource Planning and Decision Making*, Lewis Publishers, Boca Raton, FL, 217–243.

DeVivo, J.C., C.A. Couch, and B.J. Freeman. 1997. Use of preliminary index of biotic integrity in urban streams around Atlanta, Georgia. *1997 Georgia Water Resources Conference*, 40–43.

Fausch, K.D., J.R. Karr, and P.R. Yant. 1984. Regional application of an index of biotic integrity based on stream fish communities, *Transactions of the American Fisheries Society*, 113, 39–55.

Fausch, K.D., J. Lyons, J.R. Karr, and P.L. Angermeier. 1990. Fish communities as indicators of environmental degradation, *American Fisheries Society Symposium*, 8, 123–144.

Finkenbine, J. K., J. W. Atwater, and D.S. Mavinic. 2000. Stream health after urbanization, *Journal of the American Water Resources Association*, 36, 1149–1160.

Fitzgerald, D.G., E. Kott, R.P. Lanno, and D.G. Dixon. 1998. A quarter century of change in the fish communities of three small streams modified by anthropogenic activities, *Journal of Aquatic Ecosystem Stress and Recovery*, 6, 111–127.

Garie, H.L. and A. McIntosh. 1986. Distribution of benthic macroinvertebrates in a stream exposed to urban runoff, *Water Resources Bulletin*, 22, 447–455.

Gregg, D.C. and J.D. Stednick. 2000. Variability in measures of macroinvertebrate community structure by stream reach and stream class, *Journal of American Water Resources Association*, 36, 95–103.

Hammer, T.R. 1972. Stream channel enlargement due to urbanization, *Water Resources Research*, 8, 1530–15640.

Hilsenhoff, W.L. 1987. An improved biotic index of organic stream pollution, *The Great Lakes Entomologist*, 20, 31–39.

Hoffman, E.J. 1985. Urban runoff pollutant inputs to Narragansett Bay: comparison to point sources. In *Perspectives on Nonpoint Pollution, Proceedings, National Conference*, EPA 440/5–85–001, U.S. Environmental Protection Agency, Washington, D.C., 159–164.

Hollis, G.E. 1975. The effects of urbanization on floods of different recurrence interval, *Water Resources Research*, 66, 84–88.

Horner, R.R., D.B. Booth, A.A. Azous, and C.W. May. 1996. Watershed determinants of ecosystem functioning, in L.A. Roesner (Ed.). *Effects of Watershed Development and Management on Aquatic Ecosystems,* Proceedings of the ASCE Conference, Snowbird, UT.

Hughes, R.M. and T. Oberdorff. 1999. Applications of IBI concepts and metrics to waters outside the United States, in T.P. Simon (Ed.). *Assessing the Sustainability and Biological Integrity of Water Resources Using Fish Communities*, CRC Press, Boca Raton, FL, 79–93.

Jones, R.C. and C.C. Clark. 1987. Impact of watershed urbanization on stream insect communities, *Water Resource Bulletin*, 23, 1047–1055.

Jones, R.C., A. Via-Norton, and D.R. Morgan. 1996. Bioassessment of BMP effectiveness in mitigating stormwater impacts on aquatic biota, in L.A. Roesner (Ed.). *Effects of Watershed Development and Management on Aquatic Ecosystems*, Proceedings of an Engineering Foundation Conference, New York.

Karr, J.R. 1981. Assessment of biotic integrity using fish communities, *Fisheries*, 6, 21–27.

Karr, J.R. 1987. Biological monitoring and environmental assessment: a conceptual framework, *Environmental Management*, 11, 249–256.

Karr, J.R. 1993. Measuring biological integrity: lessions from streams, in S. Woodley, J. Kay, and G. Francis (Eds). *Ecological Integrity and the Management of Ecosystems*, St. Lucie Press, Delray Beach, FL.

Karr, J.R. and E.W. Chu. 1999. *Restoring Life in Running Water, Better Biological Monitoring*. Island Press, Washington, D.C.

Kemp, S.J. and J.R. Spotila. 1997. Effects of urbanization on brown trout *Salmo trutta*, other fishes and macroinvertebrates in Valley Creek, Valley Forge, Pennsylvania, *American Midland Naturalist*, 138, 55–68.

Klauda, R., P. Kazyak, S. Stranko, M. Southerland, N. Roth, and J. Chaillou. 1998. Maryland biological stream survey: a state agency program to assess the impact of anthropogenic stress on stream habitat quality and biota, *Environmental Monitoring and Assessment*, 51, 299–316.

Klein, R.D. 1979. Urbanization and stream quality impairment, *Water Resources Bulletin*, 15, 948–963.

Kolkwitz, R. and M. Marsson. 1908 (translated in 1967). Ecology of plant saprobia, in L.E. Keup, W.M. Ingram, and K.M. Mackenthum (Eds.). *Biology of Water Pollution.* Federal Water Pollution Control Administration, Washinton, D.C., 47–52.

Krug, W.R. and G.L. Goddard. 1986. Effects of urbanization on streamflow, sediment loads, and channel morphology in Pheasant Branch Basin near Middleton, Wisconsin. *U.S. Geological Survey Water-Resources Investigations Report*, 85–4068.

Lenat, D.R. and J.K. Crawford. 1994. Effects of land use on water quality and aquatic biota of three North Carolina piedmont streams, *Hydrobiologia*, 294, 185–199.

Leopold, L.B. 1968. The Hydrologic Effects of Urban Land Use: Hydrology for Urban Land Planning — A Guidebook of the Hydrologic Effects of Urban Land Use. U.S.Geological Survey Circular 554.

Leopold, L.B., M.G. Wolman, and J.P. Miller. 1964. *Fluvial Processes in Geomorphology.* W. H. Freeman, San Francisco, CA.

Limburg, K.E. and R.E. Schmidt. 1990. Patterns of fish spawning in Hudson River tributaries: response to an urban gradient? *Ecology*, 71, 1238–1245.

Linke, S., R.C. Bailey, and J. Schwindt. 1999. Temporal variability of stream bioassessments using benthic macroinvertebrates, *Freshwater Biology*, 42, 575–584.

Luchetti, G. and R. Fuersteburg. 1993. Relative fish use in urban and non-urban streams, *Proceedings, Conference on Wild Salmon*, Vancouver, British Columbia, Canada.

Lyons, J., L. Wang, and T.D. Simonson. 1996. Development and validation of an index of biotic integrity for coldwater streams in Wisconsin, *North American Journal of Fisheries Management*, 16, 241–256.

Marsh, P.C. and W.L. Minckley. 1982. Fishes of the Phoenix metropolitan area in central Arizona, *North American Journal of Fisheries Management*, 4, 395–402.

Masterson, J.P. and T.T. Bannerman. 1994. Impacts of stormwater runoff on urban streams in Milwaukee County, Wisconsin, Wisconsin Department of Natural Resources Report to U.S. Environmental Protection Agency, Grant Number CP995252–01.

Matthews, W.J. and F.P. Gelwick. 1990. Fishes of Crutcho Creek and the North Canadian River in central Oklahoma: effects of urbanization, *The Southwestern Naturalist*, 35, 403–410.

Maxted, J. and E. Shaver. 1996. The use of retention basins to mitigate stormwater impacts on aquatic life, in L.A. Roesner (Ed.). *Effects of Watershed Development and Management on Aquatic Ecosystems, Proceedings of an Engineering Foundation Conference*, New York.

May, C.W., R.R. Horner, J. R. Karr, B.W. Mar, and E.B. Welch. 1997. Effects of urbanization on small streams in the Puget Sound Lowland Ecoregion, *Watershed Protection Techniques,* 2, 485–494.

Moscrip, A.L. and D.R. Montgomery. 1997. Urbanization, flood frequency, and salmon abundance in Puget Lowland streams, *Journal of the American Water Resources Association*, 33, 1289–1297.

Mundahl, N.D. and T.P. Simon. 1999. Development and application of an index of biotic integrity for coldwater streams of upper Midwestern United States, in T.P. Simon (Ed.). *Assessing the Sustainability and Biological Integrity of Water Resources Using Fish Communities*, CRC Press, Boca Raton, FL, 383–411.

Norris, R.H. and A. Georges. 1993. Analysis and interpretation of benthic macroinvertebrate surveys, in D.M. Rosenberg and V.H. Resh (Eds.). *Freshwater Biomonitoring and Benthic Macroinvertebrate.* Chapman & Hall, New York, 234–285

Novotny, V. and G. Chesters. 1981. *Handbook of Nonpoint Pollution: Sources and Management*, Van Nostrand Reinhold, New York.

Novotny, V. and H. Olem. 1994. Water Quality Prevention, Identification, and Management of Diffuse Pollution. Van Nostrand Reinhold, New York.

Ohio EPA. 1988. *Biological Criteria for the Protection of Aquatic Life*, Vol. 1–3. Ecological Assessment Section, Division of Water Quality Monitoring and Assessment, Ohio Environmental Protection Agency, Columbus, OH.

Onorato, D., K.R. Marion, and R.A. Angus. 1998. Longitudinal variations in the ichthyofaunal assemblages of the upper Cahaba River: possible effects of urbanization in a watershed, *Journal of Freshwater Ecology*, 13, 139–154.

Schleiger, S.L. 2000. Use of an index of biotic integrity to detect effects of land uses on stream fish communities in west-central Georgia, *Transactions of the American Fisheries Society*, 129, 1118–1133.

Schueler, T. 1994. The importance of imperviousness, *Watershed Protection Techniques,* 1, 100–111.

Schueler, T. 1997. Fish dynamics in urban streams near Atlanta, Georgia, *Watershed Protection Techniques,* 2, 511–514.

Schueler, T. and J. Galli. 1992. Environmental impacts of stormwater ponds, in *Watershed Restoration Source Book.* Anacostia Restoration Team, Metropolitan Washington Council of Governments, Washington, D.C.

Scott, J.B., C.R. Steward, and Q.J. Stober. 1986. Effects of urbanization on fish population dynamics in Kelsey Creek, Washington, *Transactions of American Fisheries Society*, 115, 555–567.

Shaver, E., J. Maxted, G. Curtis, and D. Carter. 1994. Watershed protection using an integrated approach, *Proceedings from Stormwater NPDES Related Monitoring Needs. Mount Crested Butte, Colorado, August 7–12, 1994.* American Society of Civil Engineers, New York.

Simon, T.P. and J. Lyons. 1995. Application of the index of biotic integrity to evaluate water resource integrity in freshwater ecosystems, in W.S. Davis and T.P. Simon (Eds.), *Biological Assessment and Criteria: Tools for Water Resource Planning and Decision Making*, Lewis Publishers, Boca Raton, FL, 245–262.

Simmons, D.L. and R.J. Reynolds. 1982. Effects of urbanization on base flow of selected south-shore streams, Long Island, New York, *Water Resource Bulletin*, 18, 797–805.

Steedman, R.J. 1988. Modification and assessment of an index of biotic integrity to quantify stream quality in southern Ontario, *Canadian Journal of Fisheries and Aquatic Sciences*, 45, 492–501.

Stepenuck, K.F., R.L. Crunkilton, and L. Wang. 2002. Impacts of urban land use on macroinvertebrate communities in southeastern Wisconsin streams. *Journal of American Water Resources Association*, in press.

Stewart-Oaten, A. and W.W. Murdoch. 1986. Experimental impact assessment: "pseudo-replication" in times? *Ecology*, 67, 929–940.

U.S. Environmental Protection Agency. 1983. *Results of the Nationwide Urban Runoff Program. Vol. I. Final Report*, USEPA, Water Planning Division, Washington, D.C..

U.S. Environmental Protection Agency. 1994. *National Water Quality Inventory, 1992 Report to Congress.* EPA 841-R-94–001. USEPA, Office of Watersheds, Oceans and Wetlands, Washington, D.C.

U.S. Environmental Protection Agency. 2000. *Liquid Assets 2000: American Water Resources at a Turning Point.* EPA-840-B-00–001. USEPA, Office of Watersheds, Oceans and Wetlands, Washington, D.C.

Victor, R. and P. Fufeyin. 1993. Fish communities of a stretch of river affected by urban disturbance in Nigeria, *Tropical Zoology*, 6, 1–10.

Walsh, C.J. 2000. Urban impacts on the ecology of receiving waters: a framework for assessment, conservation and restoration, *Hydrobiologia*, 431, 107–114.

Wang, L., J. Lyons, P. Kanehl, and R. Gatti. 1997. Influences of watershed land use on habitat quality and biotic integrity in Wisconsin streams, *Fisheries*, 22, 6–12.

Wang, L., J. Lyons, P. Kanehl, R. Bannerman, and E. Emmons. 2000. Watershed urbanization and changes in fish communities in southeastern Wisconsin streams, *Journal of the American Water Resources Association*, 36, 1173–1189.

Wang, L., J. Lyons, P. Kanehl, and R. Bannerman. 2001. Impacts of urbanization on stream habitat and fish across multiple spatial scales, *Environmental Management*, 28, 255–266.

Washington, H.G. 1984. Diversity, biotic and similarity indices, a review with special relavance to aquatic ecosystems, *Water Research*, 18, 653–694.

Weaver, L.A. and G.C. Garman. 1994. Urbanization of a watershed and historical changes in a stream fish assemblage, *Transactions of the American Fisheries Society*, 123, 162–172.

Wernick, B.G., K.E. Cook, and H. Schreier. 1998. Land use and streamwater nitrate-N dynamics in an urban-rural fringe watershed, *Journal of the American Water Resources Association*, 34, 639–650.

Wichert, G.A. 1994. Fish as indicators of ecological sustainability: historical sequences in Toronto area streams, *Water Pollution Research Journal of Canada*, 19, 599–617.

Wichert, G.A. 1995. Effects of improved sewage effluent management and urbanization on fish associations of Toronto streams, *North American Journal of Fisheries Management*, 15, 440–456.

Wilhm, J.L. and T.C. Dorris. 1966. Species diversity of benthic macroinvertebrates in a stream receiving domestic and oil refinery effluents, *American Midland Naturalist*, 76, 427–449.

Winter, J.G. and H.C. Duthie, 1998. Effects of urbanization on water quality, periphyton and invertebrate communities in a southern Ontario stream, *Canadan Water Resources Journal*, 23, 245–257.

Wright, S. and W.M. Tidd. 1933. Summary of limnological investigation in western Lake Erie in 1929 and 1930, *Transactions of the American Fisheries Society*, 63, 271–285.

Yoder, C.O. and E.T. Rankin. 1995. Biological response signatures and the area of degradation value: new tools for interpreting multimetric data, in W.S. Davis and T.P. Simon (Eds.). *Biological Assessment and Criteria: Tools for Water Resource Planning and Decision Making*, Lewis Publishers, Boca Raton, FL, 263–286.

14 Conservatism of Confined Disposal Facilities Based on the Biological Stability and Integrity of Plant Communities: A Case Study in the Laurentian Great Lakes Basin

Gerould S. Wilhelm, Thomas P. Simon, and Paul M. Stewart

CONTENTS

0-8493-0905-0/03/$0.00+$1.50

14.1 INTRODUCTION

In order to restore degraded habitats along the Great Lakes, human manipulation and management are necessary to remove and isolate sediments considered toxic or hazardous to fish, wildlife, and humans (Great Lakes Commission, 1999; Stewart et al., Chapter 5, this volume). Human modifications of these near-shore environments include dredging and disposal, beach nourishment, and other industrial activities that disturb native plant communities.

Limited attempts have been made to construct a multimetric index that evaluates the conditions of riverine and palustrine plant assemblages, with the exception of Stewart et al. (1999) and Simon et al. (2001). The plant index of biotic integrity (PIBI) is a multimetric measure that evaluates species composition, habitat guild structure, relative abundance, and plant health of individuals for palustrine and riverine areas along southern Lake Michigan (Simon et al., 2001). Swink and Wilhelm (1994) developed an index of floristic quality that incorporates tolerance and fidelity to native plant community membership as important components.

This chapter is a series of case studies based on six confined disposal facilities (CDFs) on the shores of the Great Lakes that document the recovery success of dredged sediment receiving areas. These CDFs receive a variety of dredged sediments from near-shore areas that include toxic and nutrient-enriched sediments. U.S. Army Corps of Engineers regulations require that contaminated sediments be confined in such a manner as to be restricted from contact with ambient waters. No restoration efforts have been made at these sites. Instead, a default plant assemblage trajectory has been allowed to progress randomly.

14.1.1 Potential Impacts Associated with Confined Disposal Facilities

CDFs are near-shore areas along the Great Lakes where dredged materials from rivers, harbors, and other in-channel construction sites are deposited. Allen and Hardy (1980) reviewed the literature on the effects of navigation on fish and wildlife. Confined disposal facilities are usually sited in heavily industrialized areas, often becoming the only "green space," and thus attract an array of animal assemblages. An example of this is the Cleveland CDF where local birders have verified that many migrant birds use this site as a temporary stopover. No attempt has been made to document the diversity of these areas compared to pre-industrial insect and bird diversity. No evidence suggests that these areas succeed to a diverse native plant community. Observations show that CDFs possess depauperate plant communities of weedy and non-native species (G. Wilhelm, personal observation).

14.1.2 Assumptions Underlying the Restoration Ecology of Disturbed Systems

The traditional hypothesis for vegetation development on disturbed landscapes is that if left alone over a long enough time, the plant community that develops will succeed on a trajectory that results in a climax community. Casual observations of CDFs do not support this traditional view. It is not known whether this is due to lack of validity of the hypothesis or related to contamination and the nature of dredged material. Rankin and Simon (Chapter 10) evaluated the effects of recovery after a complete kill of the aquatic assemblage of Leading Creek from a mine spill. The biological integrity of fish assemblages after complete loss has not recovered after 8 years. The original diversity has not returned to pre-spill conditions despite the removal of further acid mine drainage effects and the threshold response seems to be insurmountable.

In other disturbed sites where native plant species have been introduced and established and appropriate management protocols implemented, relatively diverse plant communities have developed (Swink and Wilhelm, 1994; Packard and Mutel, 1997). It is possible that CDFs will respond in a similar manner once appropriate management practices are established and implemented.

It is not known whether contaminant contributions to CDFs prohibit plant succession. In areas where significant contamination exists, plant communities of high natural quality still persist. For example, plant diversity in areas surrounded by high ambient contaminant concentrations, such as the Indiana Dunes area and the Clark and Pine Nature Preserve are considered among the highest natural plant assemblages in Northwest Indiana (G. Wilhelm, personal observation). We only have limited understanding of the fate and effects of both autochthonous and allochthonous contaminants on native plant communities (Stewart et al., 1999; Stewart et al., Chapter 5, this volume).

14.2 METHODS

14.2.1 Study Area

The Laurentian Great Lakes include six large waterbodies with connecting channels and wetlands that flow easterly into the Atlantic Ocean. The six CDFs evaluated represent a variety of case studies along the near-shore areas of Lake Michigan, Lake Huron, and Lake Erie (Figure 14.1). These CDFs are located near or receive contaminated dredge spoils from Great Lakes areas of concern (International Joint Commission 1983). These areas are considered to contain severe environmental contamination that limits the ability of local areas to meet designated uses (Hartig and Zarull, 1992).

Vegetation was evaluated at two sites on Lake Michigan. The Alsip CDF (ALS) near Chicago, Illinois, was inventoried in 1976 and the Milwaukee CDF (MIL), in Wisconsin, was sampled in 2000. A single CDF site was sampled on Lake Huron at the Saginaw River CDF (SAG) near Bay City, Michigan in 2000. Three sites were investigated on Lake Erie. Point Mouille (PTM) south of Detroit, Michigan, and the Cleveland CDF (CLE) in Ohio, were sampled in 2000. The Times Beach site (BUF) at Buffalo, New York, was examined in 1985 and again in 1996. Kawkawlin Prairie (KP) (Lake Huron) and Grand Beach Prairie (GRB) are natural areas that were sampled to evaluate reference condition expectations during 2000.

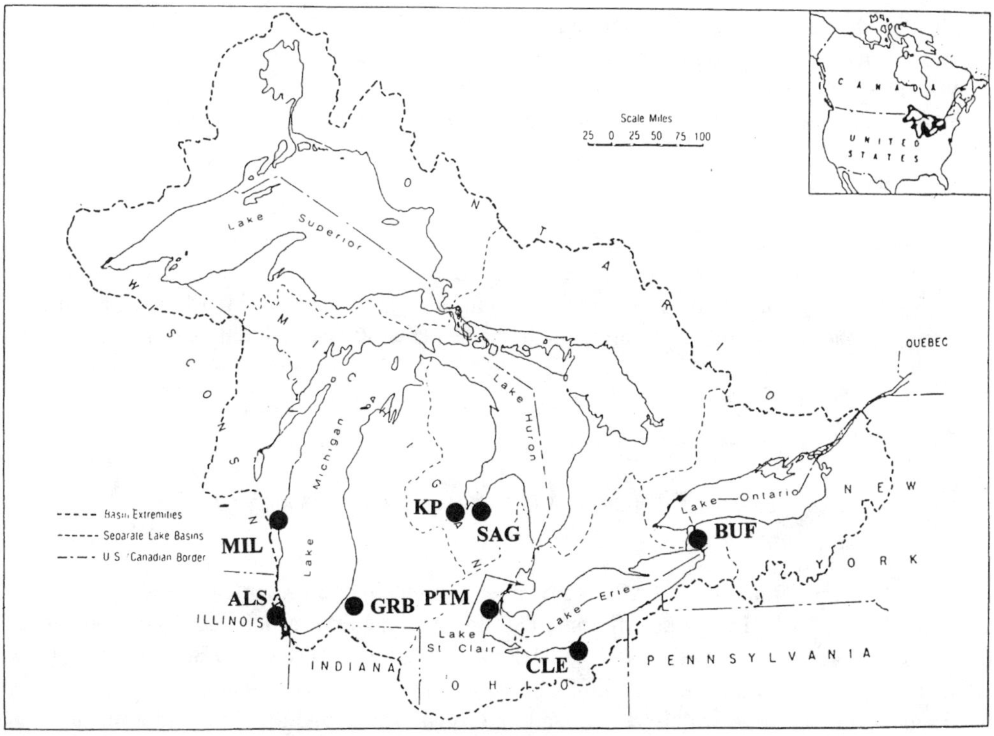

FIGURE 14.1 Map of Greats Lakes region showing confined disposal facilities (CDFs) and reference remnant prairie site locations. MIL = Milwaukee, Wisconsin, ALS = Alsip, Illinois, GRB = Grand Beach Prairie, Michigan, KP = Kawkawlin Prairie, Michigan, SAG = Saginaw River, Michigan, PTM = Point Mouille, Michigan, CLE = Cleveland, Ohio, and BUF = Times Beach, Buffalo, New York.

14.2.2 Sampling Methods

14.2.2.1 Plant Sampling and Enumeration

A floristic inventory of vascular vegetation in CDFs consisted of inspections of representative areas of all plant community types. A general description of the areas consisted of a total inventory of all identifiable vascular plant species. At the Times Beach and Milwaukee CDFs, representative transect surveys were performed to describe the relative abundance, diversity, and quality of 10-m diameter areas. At each site, species were assigned estimated percent cover values ranging as low as 1% for trace occurrences and as high as 100%.

14.2.2.2 Coefficient of Conservatism

The conservatism (C) value is a coefficient ranging from 0 to 10 assigned by Swink and Wilhelm (1994) to native plant species of the Chicago region. Non-native species receive default values of zero. The C-values represent the confidence that any particular species is likely to be found in a stable remnant plant community (conservative species). Coefficients ranging from 0 to 3 provide low confidence that a species was found in a natural plant community with full system integrity. Species with C-values from 7 to 10 show high affinity to remnant natural areas. Species with values between 4 and 6 are likely to come from moderately degraded remnant areas.

A mean C-value lower than 3.0 suggests a high degree of disturbance and represents species not indicative of a natural system with high biological integrity. Areas with mean C-values between 3 and 4 are usually fairly degraded but contain a few conservative species; these values are usually more than what is achievable in routine wetland restorations. Mean C-values above 4 suggest the presence of fair remnant biodiversity with sufficient integrity that restoration attempts are unlikely to replicate. Most seral or successional areas, even those that are quite aged, are unable to achieve species matrices with enough conservative diversity to yield mean C-values greater than 4 (G. Wilhelm, personal observation). High quality remnant systems have mean C-values from 5 to 6 and comprise less than 0.1% of the land area in Illinois (Swink and Wilhelm, 1994).

14.2.2.3 Floristic Quality Index

The floristic quality index (FQI) is a univariate biocriterion that includes aspects of fidelity to natural areas (C-value) and number of species. To derive the FQI, the C-values are summed and divided by the total number of plant taxa providing a mean C-value, which is multiplied by the square root of the total number of native plant species (Swink and Wilhelm, 1994; Herman et al., 1996, 1997).

14.3 RESULTS AND DISCUSSION

14.3.1 Patterns in Mean Coefficient of Conservatism Values and Floristic Quality Index Scores — Case Studies

All dominant species found in the CDFs were either non-native or represented species with C-values less than 4 (Table 14.1). In contrast, floristic data for Kawkawlin Prairie, a degraded remnant lake plain prairie located near the Saginaw River CDF about 2 km west of Saginaw Bay showed relatively high mean C (mean C-value = 3.3) and FQI values (22; Table 14.2). The other natural area examined, the Grand Beach Prairie, a widely recognized high-quality lake plain prairie, had a mean C-value of 5.7 and an FQI of 48. Both of the metrics scored much higher for natural sites than any of the CDF sites.

TABLE 14.1
List of Plant Species Found, C-Values, and Presence or Absence of Plant Species at Six Confined Disposal Facility Sites and Two Reference Sites in the Great Lakes. C-Values are Those Given for Michigan (Hermann et al., 1996).

		Site[a]								
Species	C-value	MIL	SAG	PTM	CLE	ALS	BUF85	BUF96	KP	GRB
Abutilon theophrasti	0	X								
Acer negundo	0	X							X	X
Acer rubrum	1									
Acer saccharinum	2								X	
Achillea millefolium	1						X		X	X
Acnida altissima	0	X				X				
Agalinis purpurea	7									X
Agrimonia parviflora	4								X	
Agropyron repens	0	X			X	X				
Agrostis alba	0						X	X		
Agrostis hyemalis	4									X
Ailanthus altissima	0				X	X				
Alisma plantago-aquatica	1								X	
Alliaria petiolata	0	X								
Amaranthus blitoides	0		X							
Amaranthus hypochondriacus		X								
Amaranthus powellii	0		X							
Amaranthus retroflexus	0	X	X							
Amaranthus tuberculatus	6	X								
Ambrosia artemisiifolia elatior		X		X						
Ambrosia trifida	0	X	X			X				
Andropogon gerardii	5								X	
Andropogon virginicus	4								X	
Anthemis cotula	0		X							
Apocynum sibiricum	3					X			X	
Arctium minus	0		X	X		X	X		X	
Arenaria serpyllifolia	0				X					

TABLE 14.1 (CONTINUED)
List of Plant Species Found, C-Values, and Presence or Absence of Plant Species at Six Confined Disposal Facility Sites and Two Reference Sites in the Great Lakes. C-Values are Those Given for Michigan (Hermann et al., 1996).

		Site[a]								
Species	C-value	MIL	SAG	PTM	CLE	ALS	BUF85	BUF96	KP	GRB
Aristida oligantha	0					X				
Aronia prunifolia	5									X
Artemisia annua	0	X	X							
Artemisia biennis	0	X								
Artemisia vulgaris	0	X			X					
Asclepias incarnata	4	X					X		X	
Asclepias syriaca	0	X			X				X	
Aster brachyactis	0	X								
Aster dumosus	7								X	X
Aster ericoides	3								X	
Aster lanceolatus	2							X		
Aster lateriflorus	2						X			
Aster novae-angliae	4	X				X	X	X		
Aster pilosus	0		X			X	X			X
Atriplex patula	0	X	X		X	X				
Barbarea vulgaris	0							X		
Bidens cernua	3		X				X			
Bidens comosa	5		X							
Bidens connata	5					X				
Bidens frondosa	1	X			X	X				X
Bidens polylepis	0					X				
Bidens vulgata	1		X							
Brasenia schreberi	6									X
Brassica kaber	0				X	X				
Brassica nigra	0		X	X			X		X	
Bromus commutatus	0				X					
Bromus japonicus	0					X				

Bromus tectorum	0				X					
Capsella bursa-pastoris	0	X								
Carduus nutans	0	X			X					
Calamagrostis canadensis	3					X	X	X		
Carex brevior	3							X		
Carex cristatella	3						X	X		
Carex granularis	2								X	
Carex longii	6									X
Carex lurida	3							X		
Carex pellita	2									X
Carex stipata	1							X	X	
Centaurea maculosa	5		X						X	
Cephalanthus occidentalis	7								X	
Chenopodium album	0	X	X		X	X				
Chenopodium glaucum	0	X	X	X	X					
Cichorium intybus	0	X	X		X					
Cirsium arvense	0	X			X	X	X	X		
Cirsium vulgare	0	X	X		X	X	X	X		
Citrullus lanatus	0		X							
Comandra umbellata	5								X	
Conium maculatum	0				X					
Convolvulus arvensis	0	X								
Convolvulus sepium	2	X			X	X	X	X		
Conyza canadensis	0		X		X					
Cornus foemina	1								X	
Cornus stolonifera	2				X		X	X	X	
Cucumis melo	0			X						
Cynodon dactylon	0	X								
Cynoglossum officinale	0			X						
Cyperus erythrorhizos	2	X	X	X		X				
Cyperus odoratus	3		X							
Daucus carota	0		X			X	X			X
Desmodium canadense	3							X		
Dianthus sylvestris	0				X					

TABLE 14.1 (CONTINUED)
List of Plant Species Found, C-Values, and Presence or Absence of Plant Species at Six Confined Disposal Facility Sites and Two Reference Sites in the Great Lakes. C-Values are Those Given for Michigan (Hermann et al., 1996).

		Site[a]								
Species	C-value	MIL	SAG	PTM	CLE	ALS	BUF85	BUF96	KP	GRB
Digitaria sanguinalis	0		X							
Drosera intermedia	8									X
Echinochloa crusgalli	0	X	X	X	X	X				X
Eleocharis engelmannii	8								X	
Eleocharis erythropoda	4					X			X	
Eleocharis melanocarpa	9								X	
Eleocharis obtusa	3									X
Eleocharis olivacea	7									X
Eleocharis smallii	5									X
Elymus canadensis	4					X				
Epilobium ciliatum	3	X								
Epilobium hirsutum	0						X			
Equisetum hyemale	2								X	
Eragrostis hypnoides	8		X							
Eragrostis pectinacea	0		X							
Erechtites hieracifolia	2									X
Erigeron annuus	0	X			X					
Eupatorium altissimum	0				X					
Eupatorium maculatum	4					X				
Eupatorium perfoliatum	4					X			X	
Eupatorium serotinum	0			X	X				X	
Euthamia graminifolia	3					X		X		
Euthamia remota	10									X
Festuca elatior	0					X				
Fimbristylis autumnalis	6						X		X	
Fragaria virgininiana	2								X	X
Fraxinus pens subintegerrima	2	X					X			

Galium aparine	0				X					
Gaura biennis pitcheri	2				X					
Geum canadense	1						X	X		
Glyceria striata	4						X			
Gnaphalium obtusifolium	2								X	
Helenium autumnale	5								X	
Heleochloa schoenoides	0	X								
Helianthus annuus	0	X	X							
Helianthus petiolaris	0					X				
Hibiscus trionum	0	X								
Hordeum jubatum	0	X			X	X				
Hypericum boreale	5									X
Hypericum majus	4									X
Hypericum mutilum	5									X
Hypericum perforatum	0			X						
Impatiens capensis	3	X			X		X	X		
Iris virginica shrevei	5	X								
Juncus alpinus	5									X
Juncus biflorus	8									X
Juncus dudleyi	1							X		X
Juncus effusus	3							X	X	X
Juncus scirpoides	9									X
Juncus torreyi	4									X
Kochia scoparia	0		X			X				
Lactuca serriola	0	X			X	X				
Leersia oryzoides	3		X		X	X	X	X		X
Lemna minuscula	5	X								
Lepidium densiflorum	0		X							
Lepidium virginicum	0				X					
Leptochloa fascicularis	0	X	X	X						
Liatris spicata	8									X
Lilium michiganense	5								X	
Linaria vulgaris	0	X			X					
Lindernia anagallidea	8									X

TABLE 14.1 (CONTINUED)
List of Plant Species Found, C-Values, and Presence or Absence of Plant Species at Six Confined Disposal Facility Sites and Two Reference Sites in the Great Lakes. C-Values are Those Given for Michigan (Hermann et al., 1996).

Species	C-value	Site[a]								
		MIL	SAG	PTM	CLE	ALS	BUF85	BUF96	KP	GRB
Lolium perenne	0					X				
Lotus corniculata	0				X					
Ludwigia alternifolia	8									X
Lychnis alba	0		X				X			
Lycopersicum esculentum	0	X	X							
Lycopodiella inundata	7							X	X	
Lycopus americanus	2		X						X	
Lycopus europaeus	0				X					
Lysimachia ciliata	4						X			
Lysimachia quadriflora	10							X		
Lysimachia terrestris	6									X
Lythrum salicaria	0	X					X	X	X	
Malva neglecta	0		X							
Medicago lupulina	0				X					
Melilotus alba	0		X			X	X			X
Melilotus officinalis	0				X					
Mollugo verticillata	0	X								
Monarda fistulosa	2								X	
Morus alba	0			X		X				
Muhlenbergia mexicana	3							X		
Myosoton aquaticum	0				X					
Nepeta cataria	0		X	X		X	X		X	
Nyssa sylvatica	9									X
Oenothera biennis	0	X	X		X	X		X		
Osmunda regalis	5									X
Panicum capillare	1	X	X	X		X				
Panicum dichrotomiflorum	0	X	X	X	X					
Panicum lindheimeri	8									X

Panicum miliaceum	0		X							
Panicum rigidulum	7									X
Panicum spretum	9									X
Panicum virgatum	4								X	X
Parthenocissus inserta	4				X					
Pastinaca sativa	0	X								
Phalaris arundinacea	0	X			X	X	X	X		
Phragmites australis	1	X	X	X		X	X	X		
Physalis subglabrata	0					X				
Phytolacca americana	2				X					
Plantago lanceolata	0				X				X	
Plantago major	0	X	X							
Poa annua	0			X						
Poa compressa	0				X		X			
Poa palustris	3							X		
Poa pratensis	0		X			X		X	X	X
Polygala cruciata	9									X
Polygonum aviculare	0	X	X	X	X	X				
Polygonum cuspidatum	0			X						
Polygonum hydropiperoides	2	X							X	
Polygonum lapathifolium	0	X	X	X	X	X				
Polygonum pensylvanicum	0	X	X	X	X	X			X	
Polygonum persicaria	0	X		X		X				
Polygonum punctatum	5						X			
Polygonum scandens	2		X			X				
Pontederia cordata	8									X
Populus deltoides	2	X	X	X	X	X				
Portulaca oleracea	0		X							
Potamogeton bicupulatus	10								X	
Potentilla norvegica	0	X				X				
Potentilla recta	0								X	
Potentilla simplex	2									X
Proserpinaca palustris	6								X	
Prunella vulgaris	0								X	
Puccinellia distans	0	X								
Pycnanthemum virginianum	5							X		

TABLE 14.1 (CONTINUED)
List of Plant Species Found, C-Values, and Presence or Absence of Plant Species at Six Confined Disposal Facility Sites and Two Reference Sites in the Great Lakes. C-Values are Those Given for Michigan (Hermann et al., 1996).

Species	C-value	Site[a]								
		MIL	SAG	PTM	CLE	ALS	BUF85	BUF96	KP	GRB
Quercus palustris	8									X
Ranunculus sceleratus	6	X	X		X					
Rhamnus cathartica	0					X		X		
Rhamnus frangula	0				X			X	X	
Rhexia virginica	9									X
Rhus typhina						X		X		
Rhynchospora capitellata	6								X	
Rhynchospora macrostachya	9								X	
Rorippa palustris	1		X							
Rotala ramosior	8									X
Rubus flagellaris	1									X
Rubus hispidus	4									X
Rubus strigosus	2		X							
Rudbeckia hirta	1								X	
Rumex altissimus	2	X			X					
Rumex crispus	0		X	X		X	X	X		
Rumex maritimus	5		X							
Rumex mexicanus	1	X								
Rumex obtusifolius	0						X	X		
Sagittaria graminea	10									X
Sagittaria latifolia	1		X							
Salix amygdaloides	3		X							
Salix discolor	1						X			X
Salix eriocephala	2		X		X		X			
Salix exigua	1			X	X	X		X	X	X
Salix fragilis	0							X		
Salix interior	1		X				X			
Salix pentandra	0		X							
Sambucus canadensis	3				X					

Saponaria officinalis	0								X	
Schoenoplectus purchianus	8								X	
Schoenoplectus tabernaemontani	44					X			X	
Scirpus atrovirens	3						X	X		
Scirpus cyperinus	5									X
Scirpus fluviatilis	4					X				
Scirpus pendulus	3								X	
Scirpus validus creber	5	X	X			X				
Scleria reticularis	10									X
Scutellaria lateriflora	5				X					
Setaria faberi	0		X				X			
Setaria glauca	0			X						
Setaria viridis	0			X			X			
Silene pratensis	0		X							
Sisymbrium altissimum	0			X						
Smililacina stellata	5								X	
Solanum dulcamara	0	X	X		X	X	X	X		
Solanum ptycanthum	1	X	X							
Solanum tuberosum	0		X							
Solidago altissima	1	X				X	X	X		
Solidago canadensis	1	X	X						X	
Solidago gigantea	4	X								
Solidago graminifolia nuttallii	3			X						
Solidago nemoralis	2								X	
Solidago riddellii	6								X	
Solidago rugosa	3									X
Sonchus asper	0					X				
Sonchus uliginosus	0	X				X				
Sorghastrum nutans	6								X	
Spergularia media	0	X								
Spiraea alba	4									X
Spiraea tomentosa	5									X
Spiranthes cernua	4									X
Stachys hyssopifolia	10									X
Stellaria media	0		X							
Taraxacum officinale	0						X	X		

TABLE 14.1 (CONTINUED)
List of Plant Species Found, C-Values, and Presence or Absence of Plant Species at Six Confined Disposal Facility Sites and Two Reference Sites in the Great Lakes. C-Values are Those Given for Michigan (Hermann et al., 1996).

		Site[a]								
Species	C-value	MIL	SAG	PTM	CLE	ALS	BUF85	BUF96	KP	GRB
Tragopogon dubius	0				X					
Triadenum fraseri	6									X
Trifolium repens	0	X			X					
Toxicodendron radicans	2							X		
Typha angustifolia	1					X	X			X
Typha x glauca	1	X	X				X	X		
Typha latifolia	1						X			
Ulmus americana	1				X					
Urtica dioica	1	X		X					X	
Utricularia radiata	10									X
Verbascum blattaria	0				X					
Verbascum thapsus	0	X			X					
Verbena bracteata	0				X					
Verbena hastata	4	X				X				
Viburnum opulus	0						X	X		
Viola lanceolata	8									X
Vitis riparia	3					X		X	X	X
Xanthium strumarium	0	X	X	X	X					
Zea mays	0			X						
Number of native species	41	32	8	30	42	32	19	42	72	
Number of adventive species	48	41	5	50	30	12	17	10	2	
Total number of species	89	73	13	80	72	44	36	52	74	

Note: MIL = Milwaukee Harbor CDF, SAG = Saginaw River CDF, PTM = Point Mouillee CDF, CLE = Cleveland Dike 9, ALS = Alsip CDF, BUF85 = Times Beach CDF (1985), BUF96 = Times Beach (1996), KP = Kawkawlin Prairie (reference site), and GRB = Grand Prairie (reference site).

TABLE 14.2
Comparisons of Floristic Quality Parameters for CDF Inventories

	Floristic Quality Parameters					
Site	NT	NT w/a	MC	MC w/a	FQI	FQI w/a
Milwaukee CDF	41	89	1.6	0.7	10	7
Saginaw CDF	32	73	2.1	0.9	12	8
Point Mouille CDF	8	13	1.3	0.8	4	3
Cleveland CDF	30	80	1.4	0.5	8	5
Alsip CDF	42	72	1.6	0.9	10	8
Times Beach 1985	32	44	2.5	1.8	14	12
Times Beach 1996	19	36	2.0	1.1	9	6
Mean ± SD of CDFs	29 ± 11	58 ± 26	1.8 ± 0.4	1.1 ± 0.5	10 ± 3	7 ± 3
Kawkawlin Prairie	42	52	3.3	2.7	22	19
Grand Beach Prairie	72	74	5.7	5.5	48	47

Note: NT = native taxa, MC = mean C, FQI = floristic quality index, w/a = with adventives.

A significant correlation exists between the number of native taxa with and without adventives, and the FQI with and without adventives as shown by Spearman rank order correlations (Table 14.3). For example, the mean C-value and FQI are highly correlated ($R = 0.950$, $p < 0.001$), as are FQI and the number of native taxa ($R = 0.814$, $p = 0.008$). The number of native species with adventives and mean C-values with adventives were not correlated ($p = 0.417$). This shows that attributes of the FQI including the mean C-values are highly correlated and the addition of adventive species in most cases did not change the relationships. In addition, when the two remnant prairies were removed from the Spearman rank correlations, many of the relationships continued to exist but at lower correlations.

14.3.2 Dominant Species

In older areas of the Milwaukee CDF, on sediment that may have exceeded 10 years of age, the dominant vegetation was *Phalaris arundinacea* (reed canary grass, C = 0), *Salix interior* (sandbar willow, C = 1), and *Urtica procera* (tall Nettle, C = 2). These invasive wind-pollinated species represent a mean C-value of 1.0 (Table 14.1). In areas that were scarcely one year old, the vegetation was sparse with only scattered growth of annuals, rosettes, and first-year perennials. The dominant species included, in order of relative importance, *Polygonum lapathifolium* (heartsease, C = 0), *Arctium minus* (common burdock, C = 0), *Chenopodium album* (lamb's quarters, C = 0), *Echinochloa crusgalli* (barnyard grass, C = 0), *Helianthus annuus* (annual sunflower, C = 0), *Impatiens capensis* (spotted touch-me-not, C = 3), *Polygonum pensylvanicum* (pinkweed, C = 0), *P. persicaria* (lady's thumb, C = 0), *Abutilon theophrasti* (velvetleaf, C = 0), and *Acnida altissima* (water hemp, C = 0), which represented a mean C-value of 0.3.

Some of the sites had halophytic species, such as *Aster brachyactis* (rayless aster), *Atriplex patula* (common orach), *Chenopodium glaucum* (oak-leaved goosefoot), *Leptochloa fascicularis* (salt meadow grass), *Hordeum jubatum* (squirrel tail grass), *Puccinellia distans* (alkali grass), and *Spergularia media* (salt spurrey). Such species were evident in and around sites where snow plowed from salted streets may have been stored or up slope from them. On a few of the sites, *Lythrum salicaria* (purple loosestrife) was recorded, which also suggests that chlorides and other industrial salts are in the ambient hydrology.

Additional species that routinely thrive on industrial waste sites include *Artemisia vulgaris* (mugwort, C = 0), *Heleochloa schoenoides* (false timothy, C = 0), *Helianthus annuus* (annual

TABLE 14.3
Spearman Rank Order Correlations (p Levels) for Six CDF and Two Reference Sites Sampled in the Laurentian Greak Lakes

Variable	All Sites	CDF Sites
NT and NT w/a	0.487 (0.183)	0.613 (0.144)
NT and Mean C	0.654 (0.056)	0.327 (0.474)
NT and Mean C w/a	0.532 (0.141)	0.045 (0.923)
NT and FQI	0.814 (0.008)	0.655 (0.111)
NT and FQI w/a	0.852 (0.004)	0.718 (0.069)
NT w/a and Mean C	0.050 (0.898)	0.018 (0.969)
NT w/a and Mean C w/a	–0.310 (0.417)	–0.559 (0.192)
NT w/a and FQI	0.218 (0.574)	0.270 (0.558)
NT w/a and FQI w/a	0.176 (0.651)	0.198 (0.670)
Mean C and Mean C w/a	0.903 (0.001)	0.791 (0.034)
Mean C and FQI	0.950 (<0.001)	0.891 (0.007)
Mean C and FQI w/a	0.929 (<0.001)	0.845 (0.017)
Mean C w/a and FQI	0.811 (0.008)	0.591 (0.162)
Mean C w/a and FQI w/a	0.832 (0.005)	0.636 (0.124)
FQI and FQI w/a	0.987 (<0.001)	0.973 (<0.001)

Note: Statistically significant correlations are bolded. NT = native taxa, MC = mean C, FQI = floristic quality index, w/a = with adventives.

TABLE 14.4
Comparisons of Floristic Quality Parameters between the Milwaukee CDF (n = 29) and the Times Beach CDF (n = 19), 1985 and 1996

	Floristic Quality Parameters at the Quadrat Level					
Site	NT	NT w/a	MC	MC w/a	FQI	FQI w/a
Milwaukee CDF	6.1 ± 3.6	15.0 ± 7.8	1.3 ± 1.1	0.6 ± 0.6	2.7 ±.2.3	1.8 ± 1.8
Times Beach 1985	4.3 ± 3.7	5.9 ± 4.1	1.4 ± 1.2	1.1 ± 0.9	3.5 ± 3.0	3.1 ± 2.7
Times Beach 1996	3.3 ± 1.9	5.4 ± 2.9	1.2 ± 0.6	0.7 ± 0.3	2.3 ± 1.4	1.8 ± 1.1

Note: NT = native taxa, MC = mean C, FQI = floristic quality index, w/a = with adventives.

sunflower, C = 0), *Phragmites australis* (common reed, C = 1), and *Typha* X *glauca* (hybrid cattail, C = 1). The older sites were largely characterized either by dense colonies of *Phalaris arundinacea* (reed canary grass, C = 0) with only scattered presence of other ruderal species, while *Salix interior* (sandbar willow, C = 1) dominated three sites.

14.3.3 Temporal and Spatial Patterns

Plant assemblages that develop on CDFs and other industrial or fallow agricultural soils lack a diversity of conservative native species. The species tend to be either non-native weeds or native species that do not require a diverse suite of native insects for pollination, propagation, and dispersal. Many of the ruderal species are wind-pollinated or are easily pollinated by a few ubiquitous insects and are not highly dependent on complex mycorrhizal and rhizosphere development (Miller and Jastrow, 1992a, b).

Seral soils, such as those of deposited dredge materials from CDFs, are inherently bereft of system diversity and are frequently isolated from refugia of remnant diversity. The biological effect of soil contamination with metals or organics on vegetation is difficult to determine (Stewart et al., 1999). It is probable that organic and inorganic contaminants have lethal, sublethal, and chronic effects on biological organisms and systems. These effects, however, are difficult to isolate from noisy systems, which are unable to succeed to a high-quality, biodiverse, stable system. Wiley et al. (Chapter 12) evaluated relationships among multimetrics and found that the greatest "noise" was found in highly disturbed systems. For example, the Times Beach CDF changed relatively little (declined slightly) between 1985 and 1996 (Table 14.2). Such systems tend to result in sterile monocultures of coarse weeds such as *Phalaris arundinacea* (reed canary grass, C = 0), *Phragmites australis* (common reed, C = 1), *Lythrum salicaria* (purple loosestrife, C = 0), or *Typha* (cattail, C = 1) species. Under most circumstances, these weakened systems will "boom and bust" in successional patterns that involve only non-native or weedy native species.

14.3.4 Case Studies

14.3.4.1 Milwaukee CDF

The survey of the Milwaukee CDF included inventories at 26 sediment core sample areas. A general description of the core areas consisted of a total inventory of all identifiable plant species within a 10-m radius of each sediment core sample point that included the estimated percent cover. All dominant species noted at Milwaukee were either non-native or represented species with C-values below 4. Overall, the mean C-value of the 89 species was 0.7, while the 41 native species had a mean C-value of 1.6 (Table 14.2).

14.3.4.2 Saginaw CDF

An inventory of the Saginaw CDF at Bay City, Michigan, produced an aggregate mean C-value of 0.9 (Table 14.2). Annual, biennial, or weakly perennial weeds, many of Eurasian origin, dominated the most recent lifts of dredge material. This assembly, however, quickly changed to one of three essential systems. In areas that had regular overflow of sluice or sediment, but not so deep as to bury the rhizomes, the tendency was to shift to monoculture of *Phalaris arundinacea* (reed canary grass, C = 0). In older areas under the regular influence of chlorides or other industrial salts, the tendency appeared to be a change to *Phragmites australis* (giant reed, C = 1). In the older and higher elevations, the dominant species were clones of *Salix exigua* (sandbar willow, C = 1), with an undergrowth of *Urtica dioica* (tall nettle, C = 1). All these dominant species are wind pollinated.

14.3.4.3 Point Mouille CDF

One of the Pointe Mouille cells (Cell 5), had a mean C-value of 0.8 (Table 14.1). This area was dominated by annual weeds, such as *Echinochloa crusgalli* (barnyard grass, C = 0), and *Polygonum* species in the section *Persicaria* (all with C-values = 0). In the wetter areas to the south, the presence of small annuals and halophytic species suggested high chloride concentrations. Within a few years, this flat will probably be dominated by *Phragmites australis* (giant reed, C = 1), which already lines the dike and all the older ambient areas.

14.3.4.4 Other CDFs in the Lower Great Lake Region

The CDFs at Dike 9 in Cleveland, Ohio, Times Beach in Buffalo, New York, and Alsip, near Chicago, Illinois, all had low mean C-values (Table 14.1). The Alsip CDF was less than 10 years old when inventoried. The Cleveland CDF has areas of various ages and wetness levels, but its plants are similar to those of other CDFs. The situation at Times Beach is somewhat different. Since the 1970s, it has been going through a form of old field succession with several different assemblages. It was only partially filled, with a topographic gradient comprised of high ground and coarse particles and progressively finer particles down gradient. At its lowest elevations, it has several emergent zones and deeper water.

14.3.4.5 Comparison with Remnant Areas

As a comparison, floristic data are provided for remnant lake plain prairies (Table 14.2). The Kawkawlin Prairie and Grand Beach Prairie, had relatively high native mean C-values compared to those found at CDF sites. Even though the Kawkawlin Prairie was quite degraded from chronic fire suppression and grazing, its floristic parameters are notably higher than those associated with CDFs. The Grand Beach Prairie is one of the better lake plain remnants in Michigan. Surveys of the Grand Beach Prairie in July 1991 found 74 native species, with average C-values of 5.7 that were higher than values at all other sites.

14.4 CONCLUSIONS

Vegetation management at CDFs is limited and defaults to weed species that can invade from adjacent industrial and agricultural areas. Plant assemblage stability on CDFs will continue to remain problematic unless we achieve a general understanding of the basic macroecological successional patterns and constraints, after which the more subtle effects of contaminants can be studied.

A directed approach to CDF restoration and land management would enlist the diversity of native species that would place the system on a more predictable trajectory that indicates it is recruiting diversity and macrostability. Biologically available contaminants tend to concentrate in progressively fewer species in extremely simple rhizospheres, which can magnify their impact in unpredictable arrays, depending on the "boom or bust" cycle. We hypothesize that greater biological diversity would reduce the cumulative impacts of contaminants and surpluses of nutrients. The more complex and stable the enzyme systems in the rhizosphere, the greater is the likelihood that organic contaminants will be metabolized or transformed into less toxic or innocuous components. Although our data are limited, no relationship appears to exist between conservatism and system age in CDFs, which suggests the continued lack of a trajectory toward a natural system with high biological integrity.

ACKNOWLEDGMENTS

We would like to thank John Simmers and Richard Price, U.S. Army Corps of Engineers, WES, Vicksburg, MS, for support and collaboration. The opinions expressed in this chapter do not necessarily reflect those of the U.S. Army Corps of Engineers or the U.S. Fish and Wildlife Service. No official endorsement should be inferred.

REFERENCES

Allen, K.O. and J.W. Hardy. 1980. *Impacts of Navigational Dredging on Fish and Wildlife: A Literature Review.* FWS/OBS-80/07. U.S. Fish and Wildlife Services, Department of the Interior, Office of Biological Services, Washington, D.C.

Great Lakes Commission. 1999. *Dredging and the Great Lakes.* Great Lakes Commission, Ann Arbor, MI.

Hartig, J.H. and M.A. Zarull (Eds.). 1992. *Under RAPs: Towards Grassroots Ecological Democracy in the Great Lakes.* University of Michigan Press, Ann Arbor, MI.

Herman, K., M. Penskar, A. Resnicek, G. Wilhelm, L. Masters, and W. Brodowicz. 1996. The Michigan floristic quality assessment system with wetland categories, Michigan Natural Features Inventory, Lansing.

Herman, K.D., L.A. Masters, M.R. Penskar, A.A. Resnicek, G.S. Wilhelm, and W.W. Brodowicz. 1997. Floristic quality assessment: development and application in the State of Michigan (USA), *Natural Areas Journal*, 17, 265–276.

International Joint Commission. 1983. Report on Great Lakes Water Quality – Appendix A – Areas of Concern in the Great lakes Basin: 1983 update of class "A" areas. Great Lakes Water Quality Board, Windsor, Ontario. 113 pp.

Miller, R.M. and J.D. Jastrow. 1992a. The application of VA mycorrhizae to ecosystem restoration and reclamation, in M.A. Allen, (Ed.). *Mycorrhizal Functioning*, Chapman & Hall, New York, 438–467.

Miller, R.M. and J.D. Jastrow. 1992b. The role of mycorrhizal fungi in soil conservation, in G. Bethlensalvay and R. Linderman, (Eds.). *Mycorrhizae in Sustainable Agriculture,* American Society of Agronomy, Crop Science Society of America, and Soil Science Society of America, Madison, WI, 29–44.

Packard, S. and C.F. Mutel. 1997. *The Tall Grass Restoration Handbook for Prairies, Savannas, and Woodlands,* Island Press, Washington, D.C.

Rankin, E.T. and T.P. Simon. 2002. Pioneer species metric changes as a result of increased anthropogenc disturbance: statewide patterns and a case study of four Ohio streams, Chapter 10, this volume.

Simon, T.P., P.M. Stewart, and P.E. Rothrock. 2001. Development of multimetric indices of biotic integrity for riverine and palustrine wetland plant communities along southern Lake Michigan, *Aquatic Ecosystem Health and Management,* 4, 293–309.

Stewart, P.M., E.P. Garza, and J.T. Butcher. 2002. Potential effects of contaminated dredge spoils on plant communities: a literature review, Chapter 5, this volume.

Stewart, P.M., R. Scribaillo, and T.P. Simon. 1999. The use of aquatic macrophytes in monitoring and assessment of biological integrity, *Environmental Science Forum,* 96, 275–302.

Swink, F. and G.S. Wilhelm. 1994. *Plants of the Chicago Region*. Indiana Academy of Sciences, Indianapolis, IN.

Wilhelm, G.S. and L.A. Masters. 1999. *Floristic Quality Assessment and Computer Applications*. Conservation Research Institute, Elmhurst, IL.

Wiley, M.J., P.W. Seelbach, K. Wehrly, and J. Martin. 2002. Regional ecological normalization using linear models: a meta-method for scaling stream assessment indicators, Chapter 12, this volume.

15 Macroinvertebrate Assemblages Associated with Patterns in Land Use and Water Quality

Daren M. Carlisle, Paul M. Stewart, and Jason T. Butcher

CONTENTS

15.1 INTRODUCTION

15.1.1 Dose-Response Relationship

The dose-response relationship is the most fundamental and pervasive concept in toxicology (Eaton and Klaassen, 1996). The shape of the dose-response relationship reveals mechanisms of toxicity and thresholds. Standardized evaluations of dose response are used to predict biologically safe toxicant concentrations (e.g., LD_{50}). The dose response also provides an intuitive means of comparing the toxicity of a wide variety of xenobiotics. Traditional aquatic toxicology evaluates dose responses for individuals and populations (Rand et al., 1995). Scientific progress in ecotoxicology will require the extension of dose responses to communities and ecosystems (Cairns, 1992). Toward that end, quantification of ecological response signatures for specific anthropogenic stressors is crucial.

The goal of ecotoxicology is to understand the dose-response relationship between anthropogenic stressors and ecosystems (Figure 15.1). This process is complicated for at least three reasons. First, a multitude of chemical, physical, and biological modifiers influence the bioavailability of contaminants to organisms (Sprague, 1995). As a result, accurate estimates of dose (exposure) are difficult. Second, experimental contaminant exposures of whole ecosystems have statistical, practical, and ethical constraints (Cairns et al., 1996; Schlinder, 1998). Consequently, experimental studies are often limited to "lower" levels of biological organization (e.g., populations, assemblages), then

0-8493-0905-0/03/$0.00+$1.50

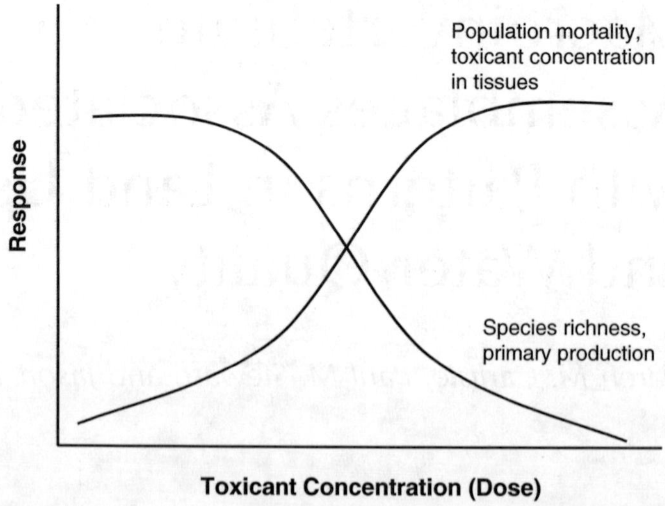

FIGURE 15.1 Example of how the dose-response paradigm can be applied to populations, communities, and ecosystems. The response in traditional dose-response models is usually population mortality. Responses at community and ecosystem levels may be positive or negative or linear or nonlinear, depending on the ecological attribute represented.

extrapolated to the ecosystem and higher levels. The limitations of this approach are well documented (Kimball and Levin, 1985; Cairns and Pratt, 1989; Gonzalez and Frost, 1994). Finally, quantifying ecological responses is complicated because they (and usually the stressors) are often multivariate and nonlinear. Many attributes of communities and ecosystems (e.g., taxa abundance, process rates) co-vary along stress gradients (Schaeffer et al., 1988). The challenge is to balance the simplification of complex responses with preservation of ecological realism.

Two general approaches are used to reduce the complexity (e.g., dimensionality) of ecological responses. The first is to create aggregated multimetric indices to represent a wide range of ecological attributes. Examples include the index of biotic integrity (IBI) (Karr, 1981) for fish communities, similar methods developed for macroinvertebrate (Barbour et al., 1999), and aquatic plant assemblages (Simon et al., 2001). These indices are intended to represent species diversity, community composition, pollution tolerance, and trophic structure (Barbour et al., 1999). Through calculating five to ten individual metrics and combining them into a single index, the complex response is summarized into a single, composite variable (or dimension). Examples of this approach are found throughout this volume. Most state agencies in the United States use some form of multimetric procedure for biological assessment (Resh et al., 1995).

An alternative approach aggregates population responses with multivariate statistics. Multivariate approaches primarily use species occurrences or abundances rather than aggregated metrics of community attributes (Landis et al., 1994). Ordination is a multivariate method that identifies gradients along which species abundances or occurrences co-vary (ter Braak, 1995). The ecological foundation of ordination is the theory of gradient analysis (ter Braak and Prentice, 1988), which predicts that community composition changes along environmental gradients because species possess unique ranges of tolerance to environmental conditions (e.g., light, temperature). Ordination has a long history in ecological research (Gauch, 1982) and environmental assessments (Johnson et al., 1993; Landis et al., 1994), and is appropriate for nonlinear relationships. Since ordination reduces many population responses to just a few, the ordination gradient (axis) could be used as a community response variable. Inasmuch as anthropogenic stressors impose a gradient in space or time on environmental conditions, ordination may be an informative way to characterize ecological responses to those stressors.

15.1.2 Non-point Sources

Pollution from non-point sources, such as urban and agricultural runoff, is the leading cause of degradation of surface waters in the United States (USEPA, 1998; Morris, et al., Chapter 6). Throughout the U.S., forests and grasslands have been cleared for agricultural production and rural watersheds have been developed into suburban communities. Most non-point source pollution involves multiple stressors (Andoh, 1994), including sediments, nutrients, metals, hydrocarbons, and pesticides. Urbanization of the landscape also alters hydrologic regimes and stream habitats (Poff, et al., 1997). The complexity of urban, agricultural, and natural systems assures that the effects of non-point source runoff on ecosystems is often idiosyncratic and difficult to predict (Pitt, 1995).

15.1.3 Study Objectives

Most national parks were designated to preserve significant natural resources. Park borders often reflect political rather than ecological boundaries. Consequently, catchments of many streams are only partially within park boundaries, and are therefore subject to land use changes and potential contamination from non-point sources outside the park. The National Park Service has initiated a program to monitor natural resources, particularly those at risk from land use changes surrounding the parks (Silsbee and Peterson, 1991). This effort requires the identification of response signatures indicative of the ecological effects of human activities. The goal of this chapter is to identify a biological response signature (e.g., indicator assemblages) for tributary streams in Cuyahoga Valley National Park. More than 20 first to fourth order tributary streams enter the Cuyahoga River within park boundaries. Many of these catchments are outside park boundaries and under suburban development. The purpose of this research is to provide park managers with a monitoring tool for identifying the extent and degree of aquatic resource degradation due to land use changes in tributary catchments.

15.2 METHODS

15.2.1 Environmental Variables

Water chemistry data from routine water quality monitoring were combined with additional sampling of 21 streams in 1994 and 1995 (Table 15.1). Water chemistry was monitored using standard methods (APHA, 1998) at most sites from April to October. Annual means were used for this analysis from sites where multiple observations were available. Stream habitat was assessed using the methods of Yoder and Rankin (1995), with some additional variables measured. Quantitative habitat variables used in this analysis included stream width and depth and percent riffle/pool (Stewart et al., 1998). A geographic information system analysis was used to calculate drainage area, drainage density, and land use classifications for each catchment.

15.2.2 Benthic Macroinvertebrate Assemblages

Benthic macroinvertebrates were collected in the summers of 1994 and 1995. Three surber samples and multiple d-nets were collected in each stream (Stewart et al., 1998). With the exception of the use of surber samplers, all macroinvertebrate collecting and sample processing followed the methods of Ohio EPA (1989). One goal of this sampling was a biological inventory of tributary streams, so multiple methods were employed to obtain a representative sample of the taxa present. A reference collection was created from these collections and archived for the park. Taxonomic data were used to compute ten individual metrics that were summed to produce a multimetric index score for each stream for each year (Ohio EPA, 1987). Because abundance estimates from quantitative samples were highly variable and misrepresented large and mobile taxa, we used the presence/absence of each taxon (across all sampling methods) in multivariate ordination. Rare taxa (found at $<10\%$ of

TABLE 15.1
Environmental Conditions (Range of Means) Measured in 21 Tributary Streams in Cuyahoga Valley National Park, 1994 and 1995

Variable	Code	1994	1995
Alkalinity (mg/L)	ALK	85–216	73–286
Ammonia (mg/L)[a]	—	0.000–0.120	
Conductivity (μmhos/cm)	CON	438–1442	348–1220
Total Cu (mg/L)[a]	—	5–40	
Dissolved oxygen (mg/L)	DO	9.2–11.9	8.3–11.3
Total hardness (mg/L)[a]	HARD	208–446	
Total Fe (mg/L)[a]	FE	140–1270	
Total phosphorus (mg/L)	TP	0.00–0.01	0.00–0.06
Turbidity (FTU)	TRB	1–166	3–146
pH (standard units)	—	7.2–7.8	7.1–7.8
Total Zn (mg/L)[a]	ZN	10–36	
Stream width (m)[a]	—	1–19	
Stream depth (cm)[a]	—	3–25	
Percent riffle[a]	RIFF	40–79	
Drainage area (km^2)	DRNG	1–147	
Drainage density (m/ha)	—	16–34	
Percent of watershed with impervious surface	URB	0–16	
Percent of watershed in agriculture	AGR	6–29	

[a] Variable measured only in 1994.

— Variable excluded from analyses due to high correlation with other independent variables.

sites) were excluded because they may have been present accidentally and could have obscured the relationship between taxa distributions and environmental gradients.

15.2.3 Statistical Methods

Statistical analyses consisted of two steps. First, assemblage response signatures were created with multivariate ordination of the taxonomic data (Norris and Georges, 1993). Environmental data were not used to characterize assemblage responses because community composition is usually a better indicator of environmental conditions than a given set of chemical or physical variables (ter Braak, 1995). In fact, the basis of biological assessment is that species integrate temporally and spatially stochastic environmental conditions (Rosenberg and Resh, 1993). The final step was to determine whether the multivariate response signatures were related to natural or anthropogenic variation in environmental data.

Detrended correspondence analysis (DCA) was used for the ordination of taxonomic data. The ecological basis of DCA is that species tend to be most abundant where environmental conditions are optimal. Because species possess unique environmental requirements, species replacement occurs in space and time as environmental conditions change (ter Braak and Prentice, 1988). Detrended correspondence analysis estimates major gradients of community change by iteratively deriving latent variables (axes) that maximize the dispersion of taxa distributions. Each latent variable or axis is mathematically constrained to be uncorrelated with all previous ones, and therefore represents additional (and progressively less important) gradients that summarize variation in community composition.

Latent variables are constructed without reference to measured environmental variables, but can be compared *post hoc* to actual environmental data. Although DCA is based on unimodal species responses to the environment, it performs satisfactorily under any monotonic response (ter Braak, 1996). There were no firm *a priori* grounds to assume a linear or nonlinear relationship between species abundance and non-point source pollution. The perceived nature of taxa-specific responses to environmental degradation is influenced by how completely the contamination gradient (e.g., analogous to exposure in the dose-response model) is represented. For example, short sections of unimodal responses may be perceived as linear.

The strength of the association between ordination axes and anthropogenic stressors was used to evaluate how well the underlying biological response signatures were identified. Inferences from correlations of community composition and environmental variables are fraught with statistical challenges. Because biological communities are influenced by many independent variables that often interact, the number of possible models that explain variation in community composition is large. Consequently, selecting the best model is the most critical process in making valid inferences. Most model selection methods are based on hypothesis testing (e.g., stepwise regression), despite the lack of supporting statistical theory for doing so (Burnham and Anderson, 1998). For example, there are no formal criteria for using P-values from multiple, often nested, models with differing statistical power for selecting the "best" model. Further, models selected on the basis of hypothesis testing are often overfitted and therefore of dubious value in making strong inferences (Wilkinson, 1998).

We used an information-theoretic approach (Burnham and Anderson, 1998) for selecting models that best describe the relationship between ordination axes (biological response signatures) and environmental variables. Information-theoretic methods emphasize prior knowledge of study systems sufficient to formulate alternative hypotheses in the form of explicit models. Given a set of *a priori* candidate models, information-theoretic approaches provide a quantitative assessment of the "weight of evidence" in the data for the most parsimonious model (Burnham and Anderson, 1998). Alternative models are compared on the basis of Akaike's information criterion (AIC), which is computed as:

$$\text{AIC} = -2 \log (\text{L}(\hat{\theta}|y)) + 2K$$

where $\log (\text{L}(\hat{\theta}|y))$ is the value of the log-likelihood function at its maximum and K is the total number of estimable parameters in the model. A heuristic interpretation of AIC is that the likelihood term in the equation is a measure of lack of model fit, and the second term ($2K$) is a "penalty" for increasing the number of model parameters. This heuristic view shows that AIC incorporates the bias versus variance tradeoff, which is necessary to identify the most parsimonious model.

Pearson correlation analyses were performed on log-transformed environmental data and used to select a subset of the environmental variables that were not strongly (e.g., $|r| < 0.6$) correlated. From this subset of variables, candidate multiple regression models that represented alternative hypotheses about the factors that control stream community structure (Table 15.2) were assembled. The residual sum of squares for each multiple regression model was obtained from SYSTAT (1998), and used to calculate a special case of Akaike's information criterion (AIC_c) for small sample sizes (Burnham and Anderson, 1998). The AIC_c value of each model was standardized to the model with the smallest AIC_c value (ΔAIC_c), which allowed a ranking of alternative models. In general, alternative models with ΔAIC_c values within two or three (hence two to three times the value of the model with the smallest AIC_c) are equally supported by the data (Burnham and Anderson, 1998). Models with ΔAIC values >10 are considered not supported by the data. Finally, the Akaike weights (Burnham and Anderson 1998) were calculated for each model, which are interpreted as the estimated probability that each model is the best given the data at hand and the set of candidate models under consideration.

TABLE 15.2
Candidate *a Priori* Models for the Relationship between Ordination Axes and Environmental Variables Measured in 21 Tributaries in Cuyahoga Valley National Park

Model Description and Structure	Parameters
Habitat	
Y = RIFF + DRNG[a]	4
Habitat with interactions	
Y = RIFF + DRNG + RIFF*DRNG	5
"Natural" chemistry	
Y = ALK + HARD[b] + CON + FE[b]	6(4)
Natural chemistry, with interactions	
Y = ALK + HARD + CON + FE + FE*HARD	7
Natural chemistry and habitat	
Y = ALK + HARD[b] + CON + FE[b] + RIFF + DRNG	8(6)
Natural chemistry and habitat, with interactions	
Y = ALK + HARD[b] + CON + FE[b] + RIFF + DRNG + RIFF*DRNG + FE*HARD[b]	10(7)
Land use	
Y = URB + AGR	4
Chemical stressors	
Y = DO + TP + TRB + ZN[b]	6(5)
Land use and chemical stressors	
Y = URB + AGR + DO + TP + TRB + ZN[b]	8(7)
Land use and habitat	
Y = URB + AGR + RIFF + DRNG	6
Land use, chemical stressors, and habitat	
Y = URB + AGR + DO + TP + TRB + ZN[b] + RIFF + DRNG	10(9)
Global model	
Y = ALK + HARD[b] + CON + FE[b] + RIFF + DRNG + RIFF*DRNG + FE*HARD[b] + URB + AGR + DO + TP + TRB + ZN[b]	16(12)

Note: Each model was applied to the first two axes of detrended correspondence analysis, with site score as the dependent variable. Variable codes are as in Table 15.1. Number of parameters = intercept + regression coefficients + estimate of σ^2.

[a] Highly correlated with stream width and depth measured at sampling point.

15.3 RESULTS AND DISCUSSION

One hundred fifty-one taxa were collected in 1994. The first two DCA axes (eigenvalues: $\lambda_1 = 0.3$, $\lambda_2 = 0.2$) explained 19% of the total variation in community structure. Although the explained variation seems low, additional facts suggest otherwise. Explaining 100% of the variation (even with inordinately complex models) is impossible in ecological data because of natural stochasticity and inevitable sampling error. It is notable, therefore, that nearly one fifth of the variation in >150 taxa was explained in only two dimensions. Ordinations that explain a relatively small percentage of total community variation may still be ecologically informative (Gauch, 1982). Two hundred taxa were collected in 1995. The first two DCA axes ($\lambda_1 = 0.2$, $\lambda_2 = 0.1$) explained 22% of the total variation in community structure.

Axes from DCAs were interpreted with additional graphical and statistical analyses. The presence/absence of all taxa compared to site scores on DCA axes was examined visually and then tested for statistical significance with logistic regression (SYSTAT, 1998). The taxa whose occurrences were significantly associated with Axis 1 are presented (Tables 15.3 and 15.4). In general,

TABLE 15.3
Taxa whose Probability of Occurrence was Significantly Correlated with Site Scores from the First Axis of Detrended Correspondence Analysis in 1994

Taxon	Coefficient (se)	LR	χ^2
Centroptilum sp.	0.9 (0.5)	4.16	0.041
Cambarus sp.	3.1 (1.6)	12.69	0.000
Leuctra sp.	1.3 (0.7)	14.26	0.007
Aquarius remigis	0.1 (0.5)	4.56	0.032
Brillia sp.	1.2 (4.7)	4.67	0.031
Atherix lantha	1.1 (0.6)	4.64	0.031
Oligochaeta	–2.3 (1.1)	12.06	0.001
Baetis flavistriga	–2.6 (1.2)	11.29	0.001
B. intercalaris	–1.8 (0.9)	9.40	0.001
Boyeria vinosa	–1.2 (0.6)	6.62	0.020
Cheumatopsyche sp.	–4.7 (2.8)	15.63	0.001
Hydropsyche depravata	–0.8 (0.5)	3.84	0.050
Hydropsyche sp.	–1.3 (0.6)	7.04	0.008
Dicranota sp.	–1.1 (0.6)	4.30	0.038
Pilaria sp.	–2.2 (1.2)	6.58	0.010
Ceratopogonidae	–3.9 (2.5)	12.61	0.000
Nilotanypus sp.	–1.2 (0.6)	5.14	0.023
Eukiefferiella sp.	–2.9 (1.8)	7.67	0.006
Microtendipes sp.	–0.9 (0.5)	3.80	0.051
Dicranota sp.	–1.1 (0.6)	4.30	0.038

Note: LR = likelihood ratio statistic. χ^2 = probability for the likelihood ratio statistic which tests whether the coefficient for the axis predictor variable is significantly different from zero in logistic regression.

DCA axes were representative of a different suite of taxa each year. Only five taxa were related to Axis 1 in both years, but the direction of their relationships was opposite from year to year. Annual variation in the composition of ordination axes could have resulted from phenological processes or environmental conditions. Sampling occurred earlier in the summer in 1995 than in 1994 (June vs. August), and therefore may have yielded slightly different assemblages. In addition, stream flow was higher in 1995 than in 1994, which could also have caused annual differences in assemblage composition.

The first ordination axis was associated with anthropogenic stressors for both years, and therefore represented a community response signature. Data from 1994 supported models with land use and habitat variables (Table 15.5). The best model contained land use variables, particularly the proportion of the catchments with impervious surfaces (log-linear regression: $F = 11.234$, $P = 0.001$, adjusted $R^2 = 0.52$). The relationship between land use and assemblage composition was nonlinear; hence small differences in urbanization among catchments were associated with substantial variation in stream community composition (Figure 15.2). Habitat characteristics (drainage area and percent riffle) were also important predictors of community composition. The presence of habitat in the second, third, and fourth best models suggested that much of the community response was due to differences in habitat among streams. Because the collective likelihood for the top four models was 0.97, we are confident that a combination of land use and habitat were the best predictors of community composition in 1994. We quantified the relative importance of these variables by summing the Akaike weights (Burnham and Anderson, 1998) of models

TABLE 15.4
Taxa whose Probability of Occurrence was Significantly Correlated with Site Scores from the First Axis of Detrended Correspondence Analysis in 1995

Taxon	Coeff. (se)	LR	χ^2
Caecidotea sp.	4.7 (2.3)	9.90	0.002
Crangonyx sp.	9.0 (4.6)	14.88	0.000
Baetis flavistraga	6.2 (2.8)	10.67	0.001
Baetis intercalaris	2.6 (1.5)	4.39	0.036
Callibaetis sp.	5.8 (3.0)	9.48	0.002
Enallagma sp.	3.1 (1.7)	5.46	0.020
Trichocorixa sp.	6.4 (3.3)	10.30	0.001
Hydropsyche morossa	2.9 (1.4)	6.27	0.012
H. sparna	6.3 (2.9)	13.49	0.000
H. depravata	5.1 (2.8)	7.53	0.006
Dixsella sp.	3.3 (1.6)	7.43	0.006
Ablabesmyia sp.	19.5 (11.2)	20.84	0.000
Natarsia sp.	5.2 (2.4)	11.63	0.001
Cricotopus sp.	3.2 (1.6)	6.70	0.010
Heterotrissocladius sp.	5.9 (2.8)	12.21	0.000
Paracladopelma sp.	3.0 (1.5)	5.75	0.016
Menetus sp.	3.4 (1.8)	6.23	0.013
Pisidium sp.	12.2 (6.6)	18.72	0.000
Cambarus sp.	–4.3 (1.9)	14.88	0.000
Nixe sp.	–3.7 (1.7)	8.33	0.004
Calopterygidae	–4.1 (2.2)	6.97	0.008
Leuctra sp.	–2.9 (1.6)	5.00	0.025
Amphinemura sp.	–4.0 (1.1)	9.04	0.003
Pycnopsyche sp.	–7.6 (3.5)	15.28	0.000
Ormosia sp.	–6.3 (3.1)	12.09	0.001
Anopheles sp.	–3.4 (1.8)	6.17	0.013
Diamesa sp.	–5.5 (2.7)	11.01	0.001
Hemerodromia sp.	–4.8 (2.2)	10.89	0.031

Note: LR = likelihood ratio statistic. χ^2 = probability for the likelihood ratio statistic which tests whether the coefficient for the axis predictor variable is significantly different from zero in logistic regression.

containing land use and habitat. Land use variables ($\Sigma w_i = 0.68$) were more important than habitat ($\Sigma w_i = 0.44$) in predicting community structure, given the set of models we considered.

Data from 1995 supported models with land use and chemical stressors (Table 15.6). The most plausible model contained both land use and chemical stressors (linear regression: F = 7.369, P = 0.002, adj. $R^2 = 0.64$). The second best model included chemical stressors only, but was half as plausible as the best model. Because the collective likelihood of the top two models was >0.90, we are confident that land use and chemical stressors were the best predictors of community composition in 1995. The sum of Akaike weights showed that chemical stressors ($\Sigma w_{i\,=}\, 0.91$) were better predictors of community structure than land use ($\Sigma w_{i\,=}\, 0.62$) in the context of the set of models we considered.

Natural chemistry was a significant predictor of some community variation. Natural chemistry (e.g., alkalinity, hardness, Fe, and specific conductance) was the predictor variable in the best

TABLE 15.5
Model Selection Results (ΔAICc) and Akaike Weights (w_i) for Candidate Models of the Relationship between Ordination and Multi-Metric Axes and Environmental Variables Measured in 21 Tributaries in Cuyahoga Valley National Park in 1994

	DCA Axis 1		DCA Axis 2		ICI Axis	
Model	ΔAICc	w_i	ΔAICc	w_i	ΔAICc	w_i
1	3.22	0.11	***0.94***	***0.31***	***4.15***	***0.09***
2	***2.25***	***0.18***	***3.09***	***0.11***	6.15	0.03
3	8.49	0.01	7.02	0.02	***0.00***	***0.72***
4	13.60	<0.01	5.47	0.03	5.41	0.05
5	7.64	0.01	17.79	<0.01	10.89	<0.01
6	20.33	<0.01	12.14	<0.01	19.72	<0.01
7	***0.00***	***0.54***	***0.00***	***0.50***	***5.08***	***0.06***
8	8.08	0.01	9.09	0.01	5.49	0.05
9	9.53	<0.01	15.92	<0.01	16.75	<0.01
10	***2.71***	***0.14***	6.17	0.02	11.57	<0.01
11	25.37	<0.01	30.10	<0.01	32.39	<0.01
12	225.66	<0.01	255.93	<0.01	267.74	<0.01

Note: The top three models best supported by the data are in italicized bold type. Akaike weights are interpreted as the relative likelihood that model i is the best based on the data and the set of models under consideration. Model numbers are as described in Table 15.2. ICI = invertebrate community index (Ohio EPA, 1989).

[a]Model 4 was not tested with 1995 data because total hardness and Fe were not measured.

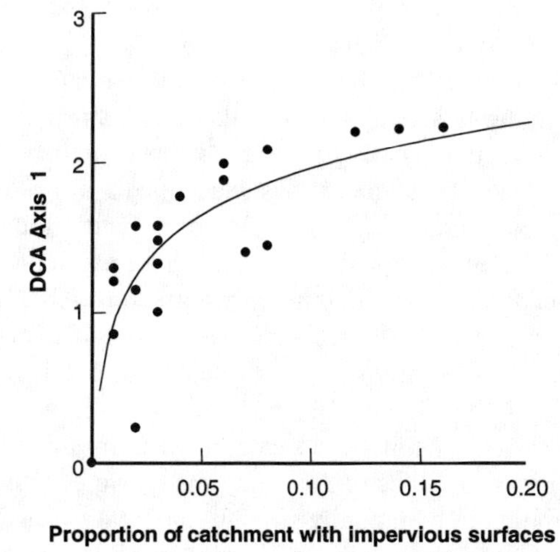

FIGURE 15.2 Stream macroinvertebrate community composition (as summarized by detrended correspondence analysis) as a function of catchment land use in 20 tributaries of the Cuyahoga River, Cuyahoga Valley National Park, 1994.

TABLE 15.6
Model Selection Results (ΔAICc) and Akaike Weights (w_i) for Candidate Models of the Relationship between Ordination and Multi-Metric Axes and Environmental Variables Measured in 21 Tributaries in Cuyahoga Valley National Park in 1995

	DCA Axis 1		DCA Axis 2		ICI Axis	
Model	ΔAICc	w_i	ΔAICc	w_i	ΔAICc	w_i
1	14.84	<0.01	34.08	<0.01	7.78	0.01
2	18.33	<0.01	31.58	<0.01	***4.01***	***0.09***
3	***5.14***	***0.05***	***0.00***	***0.85***	***0.00***	***0.69***
4[a]						
5	6.26	0.03	***5.11***	***0.07***	7.92	0.01
6	11.43	<0.01	***4.68***	***0.08***	7.32	0.02
7	7.76	0.01	36.90	<0.01	7.39	0.02
8	***1.42***	***0.30***	17.32	<0.01	***2.95***	***0.16***
9	***0.00***	***0.61***	26.48	<0.01	11.95	<0.01
10	14.39	<0.01	36.42	<0.01	14.30	<0.01
11	20.83	<0.01	44.11	<0.01	35.20	<0.01
12	47.71	<0.01	47.73	<0.01	51.73	<0.01

Note: The top three models best supported by the data are in italicized bold type. Akaike weights are interpreted as the relative likelihood that model i is the best based on the data and the set of models under consideration. Model numbers are as described in Table 15.2. ICI = invertebrate community index (Ohio EPA, 1989).

models for the second DCA axis of 1995 data. Alkalinity and natural metals are strong predictors of among-stream variations in productivity and community structure (Koetsier et al., 1996). Although some "natural" constituents can be influenced by anthropogenic activities (e.g., acid deposition and alkalinity), basin-scale differences in geology are the most likely controlling variables of cations and Fe in CUVA watersheds.

Results from this study support the general theory that stream biotic community structure responds to landscapes at multiple spatial scales (Allan and Johnson, 1997). The environmental variables controlling stream community structure can be conceptualized in a spatial hierarchy that progresses from habitat to reach to segment to subcatchment to basin (Hawkins et al., 1992). Interactions of processes across hierarchical levels are poorly understood and temporally dynamic (Poff, 1997; Wiley et al., 1997). Furthermore, few studies have explicitly tested the relative importance of local versus landscape controls on community structure (Allan and Johnson, 1997).

Allan et al. (1997) found that reach-level factors were of secondary importance to catchment land use in predicting benthic community structure. In contrast, benthic communities in Michigan streams (Richards et al., 1997) and Lake Michigan wetlands (Stewart et al., 2000) were more strongly influenced by local scale habitat than catchment processes. Annual differences in the relative importance of predictor variables in this study may have been caused by annual variation in weather. Most of the variation in community structure in 1994 was associated with processes at the subcatchment (land use) scale, and secondarily to the reach (percent riffle) scale.

Chemical contaminants were the most important predictors of community variation in 1995, followed by land use. The importance of habitat in explaining variation in community structure in 1994 may have been due to extremely low stream flow conditions during that year (P.M. Stewart, personal observation). Conversely, when stream flow was near normal in 1995, habitat was less important in explaining community variation than were chemical contaminants. Although these

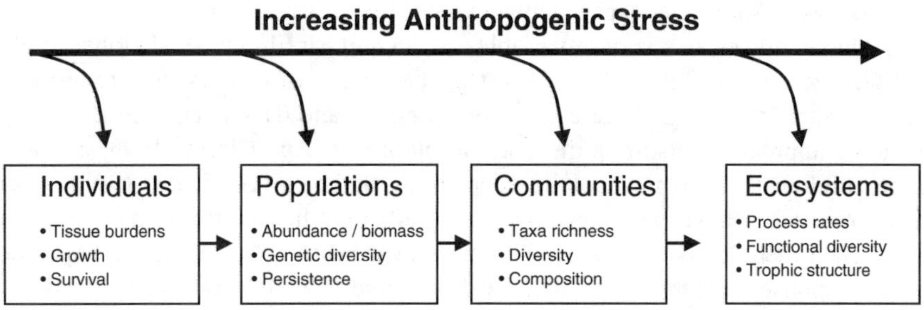

FIGURE 15.3 Hypothetical relationship between human disturbance level and expected biological responses along the ecological hierarchy.

possible mechanisms are speculative, our results suggest that ecological response signatures are temporally dynamic (Wiley et al., 1997), partially due to periodic stress imposed by variation in habitat availability.

Although non-point source pollution was not rigorously quantified in this study, land use may be a reasonable estimate of pollution potential. Non-point source pollution is difficult to quantify because the chemical stressors occur mostly during rare storm events and as complex mixtures. Hawkins et al. (2000) used the proportion of catchment logged within 30 years as an independent measure of human disturbance in California mountains. Land use predicted stream chemical conditions in the midwestern U.S. and Asia (Hirose and Kuramoto, 1981, Johnson et al., 1997). However, the proportion of a catchment that contains impervious surfaces may co-vary with the number and size of point source discharges. This would limit our ability to conclude that non-point sources are the leading stressor in this study.

In contrast to multivariate response signatures, multimetric ICI scores were not associated with anthropogenic stressors (Tables 15.5 and 15.6). Although several of the constituent metrics were correlated with habitat and contaminants (Stewart et al., 1998), the combined index appeared to be more sensitive to natural differences among streams than to anthropogenic influences. The best predictive model contained natural chemistry variables for both years. However, none of the multiple regression models containing ICI as the dependent variable were statistically significant ($P < 0.05$). Theory of the effects of contaminants across ecological organizational levels predicts that individuals may respond at the lowest stress levels through changes in tissue burdens, growth, and survival (Figure 15.3). Ecosystem level effects are presumably least sensitive (Clements, 1997), and occur only after stress-induced changes at lower levels of ecological organization.

Most multimetric approaches measure community level attributes. In contrast, multivariate approaches generally operate on the abundance and distribution of populations. The model in Figure 15.3 predicts that alterations in population abundance and distribution are precursors to community level responses. It is possible; therefore, that multivariate approaches are able to detect more subtle ecological effects than multimetric approaches, and the anthropogenic stress gradient in this study was not large enough to elicit a multimetric response. Brinkhurst (1993) suggested that simplified multimetric indices might be capable of detecting only gross disturbance levels, a supposition supported by our study.

This chapter is not intended to advocate multivariate approaches over multimetrics. Discussion of the relative merits and problems of each approach is extensive (e.g., Suter, 1993; Karr and Chu, 1999) and beyond the scope of this chapter.

Multimetrics have broad support in the academic and regulatory worlds as tools for biological assessment (Davis and Simon, 1995; Yoder and Rankin, 1995). Implicit in the multimetric approach is the assumption that ecological attributes are accurately portrayed in each component metric and reasonably aggregated into the overall index. Wallace et al. (1996) provided the only experimental evidence that variation in cumulative metrics is associated with variation in community and ecosystem processes. Additional experimental evidence is clearly needed to validate the ecological realism of additive indices as response variables (Resh and McElravy, 1993; Johnson et al., 1993, Norris and Georges, 1993; Suter, 1993). The statistical rigor of metrics in long-term monitoring also requires further study (e.g., Fore et al., 1994; Carlisle and Clements, 1999).

Multivariate approaches also require more development (e.g., Clarke, 1993) and applicability to regulatory biologists who may not be adept with statistics (Gerritsen, 1995). User-friendly multivariate approaches have been developed, so perhaps much of the remaining apprehension about multivariate methods is due to unwillingness to learn (Norris, 1995). Perhaps practitioners of biological response signatures would do well to remember that multimetric and multivariate approaches are mere tools and effective use of tools requires learning. Each tool has its own strengths and weaknesses and reliance on a single approach seems unwise. Metaphorically, if one's repertoire is limited to a hammer, then every problem appears to be a nail (Smilaeur, 1992). Finally, recent developments in geographic information systems, multivariate statistics, and model selection procedures provide a fertile area for the development of innovative tools to examine the complex relationships between ecosystems and human imprints on the landscape (Johnson and Gage, 1997).

15.4 SUMMARY AND CONCLUSIONS

Multivariate ordination was used to describe invertebrate assemblage responses to potential Allan, J.D., D.L. Erickson, and J. Fray. 1997. The influence of catchment land use on stream integrity across multiple spatial scales, Freshwater Biology, 37, 149–162.

Allan, J. D. and L.B. Johnson. 1997. Catchment-scale analysis of aquatic non-point source pollution. Detrended correspondence analysis was performed on the distribution of 200 taxa over 2 years among 21 tributary streams of the Cuyahoga River, Cuyahoga Valley National Park. Two dimensions of community structure collectively summarized one fifth of the overall community variation in each year's data. To determine whether these dimensions or axes were ecological responses to anthropogenic stressors (as in traditional dose-response models), we used them as dependent variables in a model selection procedure designed to identify which natural and anthropogenic variables explained the most variation in community structure.

Ordination axis interpretations and the dominant explanatory variables were different each year. Land use (e.g., percent imperviousness) and habitat were the most important explanatory variables the first year, while chemistry and land use were most important the following year. Different collection periods and stream flow conditions between years are probable causes of annual variation in our results. Taxa most strongly associated with the ordination axes are the best candidates for indicator assemblages, but the underlying relationships between community composition and anthropogenic stressors were obscured by environmental and sampling inconsistencies. We suggest that multivariate ordination and model selection procedures are powerful tools for isolating ecological response signatures from natural variability, and advocate more integration of multivariate and multimetric approaches.

ACKNOWLEDGMENTS

We thank Meg Plona for project development and location assistance and Anthony Gareau for GIS analyses, both of the National Park Service's Cuyahoga Valley National Park. Rich Durfee identified macroinvertebrates and created the reference collection. Tom Swinford, Patrick Hudson, and Robert Hesselberg assisted with the project. The thoughtful comments of Nicole MacRury, Gary Vequist, and Steve Cinnamon improved earlier versions of the manuscript. This project was partially supported by funding from the Water Resources Division of the National Park Service. This chapter is contribution 1156 of the USGS Great Lakes Science Center.

REFERENCES

Allan, J.D., D.L. Erickson, and J. Fray. 1997. The influence of catchment land use on stream integrity across multiple spatial scales, *Freshwater Biology,* 37, 149–162.

Allan, J.D. and L.B. Johnson. 1997. Catchment-scale analysis of aquatic ecosystems, *Freshwater Biology,* 37, 107–111.

Andoh, R.Y.G. 1994. Urban runoff: nature, characteristics, and control, *Journal of the Institution of Water and Environmental Management,* 8, 371–378.

American Public Health Association. 1998. *Standard Methods for the Examination of Water and Wastewater,* 20th ed., APHA, Washington, D.C.

Barbour, M.T., J. Gerritsen, B.D. Snyder, and J.B. Stribling. 1999. *Rapid Bioassessment Protocols for Use in Streams and Wadeable Rivers: Periphyton, Benthic Macroinvertebrates, and Fish.* 2nd ed., EPA 841-B-99–002. U.S. Environmental Protection Agency, Office of Water, Washington, D.C.

Brinkhurst, R.O. 1993. Future directions in freshwater biomonitoring using benthic macroinvertebrates, in D.M. Rosenberg and V.H. Resh (Eds.). *Freshwater Biomonitoring and Benthic Macroinvertebrates.* Chapman & Hall, New York, 442–460.

Burnham, K.P. and D.R. Anderson. 1998. *Model Selection and Inference: A Practical Information-Theoretic Approach.* Springer-Verlag, New York.

Cairns, J., Jr. 1992. Paradigms flossed: the coming of age of environmental toxicology, *Environmental Toxicology and Chemistry,* 11, 285–287.

Cairns, J., Jr., J.R. Bidwell, and M.E. Arnegard. 1996. Toxicity testing with communities: microcosms, mesocosms, and whole-system manipulations, *Reviews of Environmental Contamination and Toxicology,* 147, 45–69.

Cairns, J., Jr. and J.R. Pratt. 1989. The scientific basis of bioassays, *Hydrobiologia,* 188/189, 5–20.

Carlisle, D.M. and W.H. Clements. 1999. Sensitivity and variability of metrics used in biological assessments of running waters, *Environmental Toxicology and Chemistry,* 18, 285–291.

Clarke, K.R. 1993. Non-parametric multivariate analyses of changes in community structure, *Australian Journal of Ecology,* 18, 117–143.

Clements, W.H. 1997. Effects of contaminants at higher levels of biological organization in aquatic ecosystems, *Reviews in Toxicology,* 1, 107–146.

Davis, W.S. and T.P. Simon. 1995. *Biological Assessment and Criteria: Tools for Water Resource Planning and Decision Making.* Lewis Publishers, Boca Raton, FL.

Eaton, D.L. and C.D. Klaassen. 1996. Principles of toxicology, in C.D. Klaasen (Ed.). *Casarett and Doull's Toxicology: The Basic Science of Poisons.* McGraw-Hill, New York, 13–34.

Fore, L.S., Karr, J.R., and L.L. Conquest. 1994. Statistical properties of an index of biological integrity used to evaluate water resources, *Canadian Journal of Fisheries and Aquatic Sciences,* 51, 1077–1087.

Gauch, H.G., Jr. 1982. *Multivariate Analysis in Community Ecology.* University of Cambridge Press, New York.

Gerritsen, J. 1995. Additive biological indices for resource management, *Journal of the North American Benthological Society,* 14, 451–457

Gonzalez, M.J. and T.M. Frost. 1994. Comparisons of laboratory bioassays and a whole-lake experiment: rotifer responses to experimental acidification, *Ecological Applications,* 4, 69–80.

Hawkins, C.P., J.L. Kerschner, P.A. Bisson, M.D. Bryant, L.M. Decker, S.V. Gregory, D.A. McCoullough, C.K. Overton, G.H. Reeves, R.J. Steedman, and M.K. Young. 1992. A hierarchical approach to classifying stream habitat features, *Fisheries,* 18, 3–11.

Hawkins, C.P., R.H. Norris, J.N. Hogue, and J.W. Feminella. 2000. Development and evaluation of predictive models for measuring biological integrity of streams, *Ecological Applications,* 10, 1456–1477.

Hirose, T. and N. Kuramoto. 1981. Stream water quality as influenced by land use patterns in the Kakioka Basin, Japan, *Journal of Environmental Quality,* 10, 184–188.

Johnson, L.B. and S.H. Gage. 1997. Landscape approaches to the analysis of aquatic ecosystems, *Freshwater Biology,* 37, 113–132.

Johnson, L.B., C. Richards, G.E. Host, and J.W. Arthur. 1997. Landscape influences on water chemistry in Midwestern stream ecosystems, *Freshwater Biology,* 37, 193–208.

Johnson, R.K., T. Widerholm, and D.M. Rosenberg. 1993. Freshwater biomonitoring using individual organisms, populations, and species assemblages of benthic macroinvertebrates, in D.M. Rosenberg and V.H. Resh (Eds). *Freshwater Biomonitoring and Benthic Macroinvertebrates.* Chapman & Hall, New York, 40–158

Karr, J.R. 1981. Assessment of biotic integrity using fish communities, *Fisheries*, 6, 21–27.

Karr, J.R. and E.W. Chu. 1999. *Restoring Life in Running Waters: Better Biological Monitoring*. Island Press, Washington, D.C.

Kimball, K.D. and S.A. Levin. 1985. Limitations of laboratory bioassays: the need for ecosystem-level testing, *BioScience*, 35, 165–171.

Koetsier, P., G.W. Minshall, and C.T. Robinson. 1996. Benthos and macroinvertebrate drift in six streams differing in alkalinity, *Hydrobiologia*, 317, 41–49.

Landis, W.G., G.B. Matthews, and R.A. Sergeant. 1994. Application of multivariate techniques to endpoint determination, selection and evaluation in ecological risk assessment, *Environmental Toxicology and Chemistry,* 13, 1917–1927.

Morris, C.C., J.A. Sawyer, H.H. Bennett, and C.D. Robinson. 2002. Effects of sediment quantity on the health of aquatic ecosystems: a case study on depth of fines in Coastal Plain streams in Alabama, Chapter 6, this volume.

Norris, R.H. 1995. Biological monitoring: the dilemma of data analysis, *Journal of the North American Benthological Society,* 14, 440–450.

Norris, R.H. and A. Georges. 1993. Analysis and interpretation of benthic macroinvertebrate surveys, in D.M. Rosenberg and V.H. Resh (Eds.). *Freshwater Biomonitoring and Benthic Macroinvertebrates.* Chapman & Hall, New York, 24–286.

Ohio Environmental Protection Agency. 1987. *Biological Criteria for the Protection of Aquatic Life. Volume I: The Role of Biological Data in Water Quality Assessment.* Division of Water Quality Monitoring and Assessment, Columbus, OH.

Ohio Environmental Protection Agency. 1989. *Biological Criteria for the Protection of Aquatic Life. Volume III: Standardized Biological Field Sampling and Laboratory Methods for Assessing Fish and Macroinvertebrate Communities.* Division of Water Quality Monitoring and Assessment, Columbus, OH.

Pitt, R.E. 1995. Effects of urban runoff on aquatic biota, in D.J. Hoffman, B.A. Rattner, G.A. Burton, Jr., and J.Cairns, Jr. (Eds). *Handbook of Ecotoxcology.* Lewis Publishers, Boca Raton, FL, 609–630.

Poff, N.L. 1997. Landscape filters and species traits: towards mechanistic understanding and prediction in stream ecology, *Journal of the North American Benthological Society,* 16, 391–409.

Poff, N.L., J.D. Allan, M.B. Bain, J.R. Karr, K.L. Prestegaard, B.D. Ritcher, R.E. Sparks, and J.C. Stromberg. 1997. The natural flow regime, *BioScience*, 47, 769–784.

Rand, G.M., P.G. Wells, and L.S. McCarty. 1995. Introduction to aquatic toxicology, in G.M. Rand (Ed). *Fundamentals of Aquatic Toxicology: Effects, Environmental Fate, and Risk Assessment,* 2nd ed., Taylor & Francis, Washington, D.C., 3–67.

Resh, V.H. and E.P. McElravy. 1993. Contemporary quantitative approaches to biomonitoring using benthic macroinvertebrates, in D.M. Rosenberg and V.H. Resh (Eds). *Freshwater Biomonitoring and Benthic Macroinvertebrates.* Chapman & Hall, New York, 159–194.

Resh, V.H., R.H. Norris, and M.T. Barbour. 1995. Design and implementation of rapid assessment approaches for water resource monitoring using benthic macroinvertebrates, *Australian Journal of Ecology,* 20, 108–121.

Richards, C., L.B. Johnson, and G.E. Host. 1996. Landscape-scale influences in stream habitats and biota, *Canadian Journal of Fisheries and Aquatic Sciences*, 53 (Suppl.1), 295–311.

Rosenberg, D.M. and V.H. Resh. 1993. *Freshwater Biomonitoring and Benthic Macroinvertebrates.* Chapman & Hall, New York.

Schaeffer, D.J., E.E. Herricks, and H.W. Kerster. 1988. Ecosystem health: I. Measuring ecosystem health, *Environmental Management*, 12, 445–455.

Schindler, D.W. 1998. Replication versus realism: the need for ecosystem-scale experiments, *Ecosystems*, 1, 323–334.

Silsbee, D.G. and D.L. Peterson. 1991. Designing and Implementing Comprehensive Long-Term Inventory and Monitoring Programs for National Park System Lands. Natural Resources Report NPS/NRUW/NRR-91/04. U.S. Department of the Interior, National Park Service, Washington, D.C.

Smilaeur, P. 1992. *Canodraw.* Microcomputer Power, Ithaca, NY.

Simon, T.P., P.M. Stewart, and P.E. Rothrock. 2001. Development of multimetric indices of biotic integrity for riverine and palustrine wetland plant communities along southern Lake Michigan, *Aquatic Ecosystem Health and Management*, 4(3), 293–309.

Sprague, J.B. 1995. Appendix C: factors that modify toxicity, in G.M. Rand (Ed). *Fundamentals of Aquatic Toxicology: Effects, Environmental Fate, and Risk Assessment*, 2nd ed., Taylor & Francis, Washington, D.C, 1012–1051.

Stewart, P.M., J.T. Butcher, and T.O. Swinford. 2000. Land use, habitat, and water quality effects on macroinvertebrate communities in three watersheds of a Lake Michigan associated marsh system, *Aquatic Ecosystem Health and Management*, 3, 179–189.

Stewart, P.M., P. Hudson, J.T. Butcher, and R. Hesselberg. 1998. Benthic Macroinvertebrate and Polycyclic Aromatic Hydrocarbon Inventory in Tributaries to the Cuyahoga River at the Cuyahoga Valley National Recreation Area. U.S. Geological Survey, Great Lakes Science Center, Lake Michigan Ecological Research Station, Porter, IN.

Suter, G.W., II. 1993. A critique of ecosystem health concepts and indexes, *Environmental Toxicology and Chemistry*, 12, 1533–1539.

SYSTAT. 1998. *Version 8.0*. SPSS, Inc., Chicago, IL.

ter Braak, C.J.F. 1995. Ordination, in R.H.G. Jongman, C.J.F. ter Braak, and O.F.R. Van Tongeren (Eds). *Data Analysis in Community and Landscape Ecology*. Cambridge University Press, New York, 91–169.

ter Braak, C.J.F. 1996. *Unimodal models to relate species to environment*. DLO Agricultural Mathematics Group. Wageningen, The Netherlands.

ter Braak, C.J.F. and I.C. Prentice. 1988. A theory of gradient analysis, *Advances in Ecological Research*, 18, 271–317.

U.S. Environmental Protection Agency. 1998. National Water Quality Inventory: 1998 Report to Congress. EPA 841-R-00–001. USEPA, Office of Water. Washington, D.C.

Yoder, C.O. and E.T. Rankin. 1995. Biological criteria program development and implementation in Ohio, in W.S. Davis and T.P. Simon (Eds). *Biological Assessment and Criteria: Tools for Water Resource Planning and Decision Making*. Lewis Publishers, Boca Raton, FL, 109–144.

Wallace, J.B., J.W. Grubaugh, and M.R. Whiles. 1996. Biotic indices and stream ecosystem processes: results from an experimental study, *Ecological Applications*, 6, 140–151.

Wiley, M.J., S.L. Kohler, and P.W. Seelbach. 1997. Reconciling landscape and local views of aquatic communities: lessons from Michigan trout streams, *Freshwater Biology*, 37, 133–148.

Wilkinson, L. 1998. Linear models, in L. Wilkinson (Ed). *SYSTAT 8: Statistics*. SPSS, Inc., Chicago, IL, 335–368.

16 Effects of Metals on Freshwater Macroinvertebrates: A Review and Case Study of the Correspondence of Multimetric Index, Toxicity Testing, and Copper Concentrations in Sediment and Water

Christopher A. Mebane

CONTENTS

0-8493-0905-0/03/$0.00+$1.50

16.1 INTRODUCTION

The use of multimetric indices (MMIS) as interpretive tools in biological surveys is widespread, particularly in the United States. However, two considerations related to their use are whether pollutant-specific indexes are needed and the need for independent validations of index responses. In this chapter, the performance of the benthic macroinvertebrate index of biotic integrity (B-IBI) is tested with a known, specific pollutant (copper). The B-IBI is a nonspecific index that is promoted as a broadly effective tool (Karr and Chu, 1999). Patterns of effects of elevated metals on macroinvertebrate assemblages are described, and a case study relating multimetric assemblage indexes to sediment toxicity and concentrations of copper in water and sediment are described. The chaper concludes with recommendations on the use of multimetric indexes in relation to other assessment tools.

Differing views have been offered on the need for specific indices and the need for index validation. On the first point, one criticism of the MMI as a useful interpretive tool has been that commonly used components (metrics) may respond differently to different stressors and thus confound interpretation. For example, Suter (1993) criticized the designation of tolerant species in the index of biological integrity (IBI) because a species such as green sunfish may thrive in waters degraded by organic enrichment and be highly sensitive to other stressors (selenium) that could produce "nonsense results" (see Simon, 1999). Macroinvertebrate indices have been highly successful for estimating effects of certain classes of pollutants (e.g., organic enrichment, Hilsenhoff, 1987) but these measures may be limited because responses of benthic macroinvertebrates may be chemical-specific.

Clements and Kiffney (1996) questioned the general applicability of benthic community measures to assess water quality because of this specificity. For example, the EPT index indicates taxa richness of the Ephemeroptera, Plecoptera, and Trichoptera orders and is highly sensitive to organic enrichment. However, the EPT index was not sensitive for assessing effects of heavy metals due to the tolerance of many Plecoptera and Trichoptera that replace metals-sensitive Ephemeroptera in metal-impacted streams (Clements, 1994; Clements and Kiffney, 1994). Clements and Kiffney (1996) suggest that despite the need for general indicators of stress on ecoregional scales, it may be necessary to develop more chemical-specific measures of effect (e.g., Clements et al., 1992). In contrast, Karr and Chu (1999) suggest that one index, the benthic macroinvertebrate index of biotic integrity (B-IBI) is generally applicable for running waters, and developing a MMI on a case-by-case basis may not provide much more information despite the effort.

Validation of biosurvey results generally involves co-occurrence studies with statistical or qualitative comparisons of instream index responses, toxicity testing or environmental correlates (de Flaming and Norberg-King, 1999; Chapman, 1995; USEPA, 2000). These study results are not expressed commonly as multimetric indexes, but as other community endpoints such as species richness and density. Coherence of results supports a weight-of-evidence conclusion that the index or results are probably "valid" responses.

For example, laboratory toxicity tests may predict field observations and correlations between chemical or habitat features may indicate such associations are unlikely to occur by chance. However, lack of coherence between effects may simply reflect limitations of the tools, rather than demonstrate that index results of apparent effects to natural systems are wrong. A multimetric index of fish or invertebrate assemblage integrates conditions over time, whereas samples collected for laboratory testing could miss episodic stresses. The toxicity test could simply be insensitive to the stressor — standard toxicity test species are selected for properties unrelated to their importance in natural systems, such ease of culture, robustness to laboratory stresses, and availability in sufficient numbers throughout the year (Chapman, 1995). Damage to aquatic ecosystems may be subtle, and missed by standard bioassessment endpoints or indices. Nevertheless, coherence between multimetric index results and other lines of evidence would support the realism and applicability of index results.

Stressor-specific response patterns in macroinvertebrate assemblages have been reported for organic enrichment (Hilsenhoff, 1987), fine-grained sediment (Relyea et al., 1999; Mebane et al., in press), and certain metals. Metals have been shown to have predictable and distinctive effects on macroinvertebrate assemblages, particularly copper and zinc. Multimetric indexes can reflect changes in fish assemblages associated with metals enrichment (Rankin and Yoder, 1999; Mebane, in press), but a review of the effects of metals on fish is beyond the scope of this chapter. Sorenson (1991) provides an excellent review of the effects of metals on fish. One other remarkable study on the shifts in fish assemblages due to sublethal copper concentrations bears mention. Geckler et al. (1976) described the effects on fish assemblages observed in long term experimental dosing of a natural stream with copper. These results have not been widely available, which is unfortunate since this study of community responses to a known stressor is very relevant to biocriteria development. Further, controlled dosing studies at the scale conducted are unlikely to be repeated in the U.S. under modern environmental laws.

Macroinvertebrates may bioaccumulate metals with no apparent assemblage shifts or other obvious adverse effects, but with significant adverse effects to higher trophic levels. While mercury and selenium are probably best known, other metals such as copper and zinc have also been implicated as adversely affecting fish through bioaccumulation in macroinvertebrates. The effects of metals transferred via macroinvertebrates to higher trophic levels are also beyond the scope of this chapter. See Farag et al. (1994, 1999), Canton and VanDeveer (1997), Jarvinen and Ankley (1999) or articles in Beyer et al. (1996) for recent research and reviews of this aspect of metals and aquatic macroinvertebrates.

Macroinvertebrate assemblages are responsive to several metals including copper, zinc, and lead, which are perhaps the most common metals reported at elevated concentrations and producing adverse effects. Other cationic metals such as cadmium, nickel, silver and cobalt may adversely affect macroinvertebrates, but their effects on macroinvertebrate assemblages are less frequently described in the literature. Elevated aluminum concentrations accompany acidification of most lakes and streams, with resulting shifts in benthic macroinvertebrate assemblages and loss of fish populations (Bergman and Mattice, 1990; Stephenson et al., 1994). This chapter emphasizes the effects of copper on macroinvertebrate assemblages in lotic circumneutral waters, i.e., those without low pH reductions to acid-mine drainage or acid deposition.

16.1.1 Review of Macroinvertebrate Assemblages as Indicators of Metal Pollution

Benthic macroinvertebrates are essential components for energy cycling in aquatic ecosystems and serve as the primary food sources for invertivore fish such as salmonids and sculpins. Field surveys of benthic macroinvertebrate communities are often used for ecological assessments of sediment and water quality monitoring. They have several features that make them significant for aquatic ecological assessments. First, indigenous benthic macroinvertebrates are ecologically important as an intermediate trophic level between microorganisms and fish. They are abundant in most streams; they have limited migration patterns or are sessile, which makes them suitable for site-specific impacts. Their life spans of several months to a few years allow them to be used as continuous indicators of sediment and water quality by integrating spatial and temporal variation, rather than a snapshot of conditions at one space in time (MacDonald et al., 1991; Rosenberg and Resh, 1993).

Macroinvertebrate community structure analyses have proven to be reliable and sensitive indicators of metal pollution in the water column. Shifts in benthic community structure commonly associated with adverse effects of metals include declines in the abundance of mayflies, reduced numbers of different mayfly species, reduced overall numbers of species, and increased dominance by midges, true flies, and worms. Declines in mayfly abundance and loss of mayfly taxa have been reported consistently as sensitive and reliable indicators of metals pollution, especially for copper and zinc (Winner et al., 1980; Kiffney and Clements, Chapter 8, this volume).

16.1.2 Field Studies

Sprague et al. (1965) conducted a field survey of macroinvertebrate patterns in relation to copper–zinc wastewater from a mine draining into the Mirimachi River in New Brunswick, Canada. Metal-polluted sections of the river showed reduced species richness, a decline in representation of mayflies, and a proportional increase in chironomids and caddisflies. Subsequent correlative field studies showed similar results. In all cases reviewed, metals more consistently affected the Ephemeroptera than other orders. Members of the Tricoptera, Plecoptera, Coleoptera, and Diptera orders showed variable patterns; at least some members were often tolerant of elevated metals concentrations (Chadwick et al., 1986; Clements et al., 1988; Clements, 1994; Beltman et al., 1999; Clements et al., 2000; Mebane, 2001).

16.1.3 Single Invertebrate Species Testing with Waterborne Metals

Interpretation of field biomonitoring can be confounded by patterns of multiple natural and anthropogenic variables, stream continuum differences, and upstream–downstream pseudoreplication. Toxicity testing, in contrast, can provide a causal link between a contaminant and a biological response because other factors are controlled. Because most aquatic insects have aquatic larval stages and flying, terrestrial adult life stages, they do not make very convenient test species to develop standard methods for and to routinely culture in ecotoxicological laboratories. Instead, more conveniently cultured organisms such as daphnids are routinely used (Lewis et al., 1994).

A review of the U.S. Environmental Protection Agency (USEPA) water quality criteria documents shows that metals toxicity testing with freshwater macroinvertebrates is less frequently reported than for fish, except for species such as daphnids that are routinely cultured as laboratory test organisms (USEPA, 1996, 2001). Daphnids are frequently used and are ranked among the most sensitive freshwater species tested.

Standard methods have also been developed for the amphipod *Hyalella azteca*, which can be quite sensitive to metals (USEPA, 2001), and for the *Chironomus* midge (ASTM, 1991). Convenience notwithstanding, the latter genus is of questionable relevance as a standard test species to use for metals. Although *Chironomus tentans* have been sensitive to copper in laboratory tests (Gauss et al., 1985), field surveys show *Chironomus* may thrive in copper-contaminated ambient waters and may regulate (i.e., excrete amounts in excess of their dietary needs) copper and zinc body burdens (Cain et al., 1992).

Nebeker et al. (1984) exposed *Clistornia*, a caddisfly typical of mountain streams in the Pacific Northwest, to copper in soft water in life cycle tests. Copper concentrations of ≈17 μg/l prevented completion of the life cycle, and the no-observed-effect level for copper was 8 μg/l. These results indicate that at least some caddisflies are as sensitive as mayflies to copper contamination, despite most field and experimental stream studies findings that some caddisflies can be relatively tolerant of elevated metals.

Clements and Kiffney (1994) compared toxicity to *Ceriodaphnia* in zinc-contaminated water samples collected from sites where they also collected benthic macroinvertebrates. *Ceriodaphnia* toxicity reliably predicted shifts in the *in situ* macroinvertebrate assemblage. However, *in situ* differences were observed at sites that were not toxic to *Ceriodaphnia.* This suggests that *Ceriodaphnia* may not be as sensitive as resident species to metals or that *in situ* benthic macroinvertebrates are better integrators of contaminant effects than the discrete water samples collected for *Ceriodaphnia* testing. The latter may often be a factor in toxicity testing of ambient waters, considering the low likelihood of collecting water grab samples at the most severe ambient conditions. Even when resident stream invertebrates that are sensitive to metals (i.e., mayflies) are used in toxicity testing the measured toxicity endpoints may not match field patterns. Endpoints that can be measured and interpreted in static single species tests are usually limited to growth and lethality. In streams, other responses may be more sensitive such as attempted avoidance of contaminated conditions by drift.

In ecotoxicology literature, single species toxicity tests have long been criticized because they are overly simplistic, are ecologically unrealistic, cannot predict effects at higher levels of organization, and cannot predict indirect effects (Buikema and Voshell, 1993; Clements and Kiffney, 1996). de Vlaming and Norberg-King (1999) argue that more ecologically realistic alternatives such as mesocosm testing may not show greater sensitivity, predictive power, or greater interpretability than less expensive single species tests.

16.1.4 Experimental Stream Studies with Macroinvertebrate Assemblages

Winner et al. (1980) may have reported the first experimental stream study with metals and macroinvertebrate assemblages. Subsequently, W.H. Clements and colleagues extensively used the approach to mimic field conditions under controlled conditions to predict effects of perturbations on stream macroinvertebrate communities.

Winner et al. (1980) compared water chemistries and macroinvertebrate community structures in two streams. One stream received copper, chromium, and zinc effluent from a metal plating facility, while the other was experimentally dosed with copper for 30 months to a rather constant intermediate to low (maximum of 120 μg/l) level. In the most heavily stressed sections of both streams, macroinvertebrates (other than tubificid worms and chironomids) were virtually eliminated from rock rubble riffle habitats. Midge larvae still comprised 75 to 86% of the insect communities at the least polluted stations. The correlation coefficient for percentage of chironomids in relation to copper concentrations was +0.93 ($P < 0.01$), showing that the percent chironomids in benthic samples was highly correlated with copper concentration in impacted streams. Winner et al. concluded that the macroinvertebrate community gave a "predictable, graded response to heavy metals."

Clements et al. (1988, 1992) reported that exposure to copper significantly reduced both the total number of individuals and number of taxa during each season, with the greatest effects observed in summer. The relative abundance of Ephemeropterans (mayflies) decreased in treated streams during each season. The responses of other aquatic insects, including Dipterans (true flies) and Plecopterans (stoneflies), varied between seasons, but these groups were generally less sensitive to copper exposure. The relative abundance of the dominant chironomids, Orthocladiini, increased in copper-treated streams in winter and spring. Clements also noted that the combination of measuring numbers of taxa, overall abundance, abundance and diversity of the sensitive mayflies, and the relative abundance of insensitive Orthoclad chironomids were reliable indications of heavy metals contamination.

Other metrics such as species diversity and functional groups varied more with season than with treatment. Clements et al. (1992) compared benthic communities at copper–zinc stressed stations with copper dosed outdoor experimental streams. Sensitivities of 13 dominant taxa were measured in outdoor experimental streams by exposing organisms to copper (25 μg/l) for 10 days. Sensitivities, defined as proportional reductions in abundance, in treated streams relative to controls ranged from 1.00 for Ephemeroptera that were completely eliminated to –0.14 for taxa that increased in treated streams (Orthocladiini chironomids). Clements et al. (1988) observed similar shifts in communities in the experimentally copper dosed streams and in the copper–zinc stressed streams.

Richardson and Kiffney (2000) experimentally dosed small, pristine forest streams with a mixture of metals typical of small urban streams. They found great variation in the sensitivities of various community measures to contaminant exposure. The Chironomidae, the most abundant group, showed no significant effects of metals even at concentrations well in excess of established water quality criteria. In contrast, several mayfly genera showed dramatic declines in abundance and increases in emigration rates as metal concentrations increased. Responses were highly taxon-specific, and even among species shown to be sensitive, the degree of decline and the dose at which that decline was noticeable differed among taxa.

Richardson and Kiffney (2000) found great variation between experimental units although they were set up as similarly as possible and suggested that it may be even more difficult to compare field sites for the effects of contaminants given all the other differences inherent in field studies. Nevertheless, their results confirm field observations that metals in solution have significant impacts on benthic communities.

16.1.5 Macroinvertebrates and Metals-Contaminated Sediment

In addition to considering the relationships between metals concentrations in water and macroinvertebrates, sediment sorbed metals may be important in assessments for several reasons. Metals concentrations in sediments may be five or more orders of magnitude higher than in surface waters, causing sediments to act as sinks or sources of metals in surface water. Organisms living in or near sediments may be adversely affected by contaminants in sediment, even though the overlying water quality criteria are not exceeded, and benthic macroinvertebrates are integrally linked to sediments (Horowitz, 1991; Burton, 1992a).

Relating concentrations of metals in sediments with effects on macroinvertebrate assemblages can be complex due to differences such as the chemical bioavailabilities of metals in sediments and the different life stages of species exposed to metals in sediments and in water. Further, in running waters, contaminated sediments and areas for macroinvertebrate sampling may be spatially separated. Macroinvertebrate assemblages are usually sampled in riffles, areas of relatively fast water. In contrast, contaminants in sediments are disproportionably concentrated in the fine grained sediments that accumulate in stream margins, pools, and other relatively slow moving water areas.

16.1.6 Bioavailability of Metals in Sediments

Because of their ubiquitous distribution and persistent nature, metals such as copper, lead, and zinc are often found at elevated concentrations (above crustal background concentrations) in aquatic sediments. The techniques of sampling and analyzing metals in sediment have customarily been similar to sampling metals in soils, with concentrations reported on a dry weight normalized basis (mg metal/kg dry sediment) (Horowitz,1991). However, considerable uncertainty and controversy surround metals concentrations that pose significant ecological risks because of the effects of co-varying chemical mixtures, differences between laboratory and field exposures, and chemical binding of metals, sulfides, and organic carbon that affect bioavailability of the metals (Ankley et al., 1996a; McDonald et al., 2000).

As an example of the ranges of metals effects and sediment quality guidelines in the literature, sediment-sorbed copper concentrations associated with biological effects in freshwater studies are listed in Table 16.1. These effect concentrations were developed from several approaches: (1) spiked sediment bioassays in which a known quantity of the test chemical mixed into the sediment is allowed to equilibrate and is tested for toxicity; (2) co-occurrence analyses of benthic community structures with mixtures of chemical concentrations; (3) the apparent effects threshold (AET) defined as the maximum concentration of a chemical that did not reduce the survival of the particular indicator (e.g., amphipod survival, benthic communities); and (4) the sediment quality triad, an effects-based approach that incorporates measures of sediment chemistry, sediment toxicity, and benthic community structure and has been used to develop numeric criteria.

MacDonald et al. (2000) derived so-called consensus-based sediment quality guidelines (SQGs) by taking the geometric means of other dry weight normalized metals' SQGs. These threshold effects concentrations, below which harmful effects on sediment dwelling organisms are not expected, and probable effects concentrations, above which harmful effects are likely, are also shown. An additional approach is based upon a model of equilibrium partitioning between sediment sorbed metals and interstitial pore waters. The concentrations of dissolved metals in the interstitial pore waters are predicted to be controlled by the formation of metal–sulfide or metal–organic carbon complexes. If the molar concentrations of extractable sulfides, commonly called acid-volatile sulfides, exceed the molar concentrations of simultaneously extracted metals, the metals in the

TABLE 16.1
Selected Concentrations of Copper in Sediments Associated with Effects (mg Copper/kg Sediment Dry Weight)

Copper	Biological effects
20	Highly toxic to amphipod *Hyalella azteca*, Waukegan Harbor, IL (a)
26	LOEC (reduced growth) in 48-hour spiked sediment bioassays with the freshwater clam *Corbicula fluminea* larvae (b)
31	Consensus-based threshold effects concentrations (below which harmful effects are unlikely to be observed) (c)
36	LC50 for 14-day *Ceriodaphnia dubia* spiked sediment bioassays (d)
37	Benthic macroinvertebrate AET (depression of mayflies and amphipods), Keweenaw Waterway, MI (e)
40	LC50 for 14-day *Daphnia magna* spiked sediment bioassays (d)
71	Threshold observed effects concentration, Panther Creek, ID (geometric mean of NOEC and LOEC) (f)
86	LOEC (reduced survival) in 10-day spiked sediment bioassays with the clam *Corbicula fluminea* juveniles (b)
149	Consensus-based probable effects concentrations (above which harmful effects are probable) (c)
162	LC50 for 14-day *Pimepales promelas* (fathead minnow) spiked sediment bioassays (d)
170	LC50 for 48- hr. *D. magna* spiked sediment bioassays (d)
207	NOEC (survival) in 10-day spiked sediment bioassays with amphipod *H. azteca* (b)
216	Reduced growth for 14-day *Chironomus tentans* spiked sediment bioassays (d)
262	LC50 for 10-day *Hyalella* spiked sediment bioassays (d)
287	Sediment quality triad, Clark Fork River, MT (g)
325	Amphipod *Hyalella azteca* chronic (reduced growth) AET, Clark Fork River, MT (g)
347	LOEC (reduced survival) in 10-day spiked sediment bioassays with amphipod *H. azteca* (b)
386	Lowest concentration in sediments toxic to the amphipod *H. azteca*, Panther Creek, ID in 10-day tests. All sediments with higher concentrations also showed toxicity (f)
480	*Daphnia magna* 48-hour AET, Keweenaw Waterway, MI (e)
681	LC50 48-hour *Daphnia magna* spiked sediment bioassays, Soap Creek Pond, OR (h)
857	LC50 10-day*Chironomus tentans* spiked sediment bioassays, Soap Creek Pond, OR (h)
878	Nontoxic to rainbow trout sac fry in 21-day exposure (g)
890	Acutely toxic in 10-day tests to the mayflies *Hexagenia* and *Hyalella*, Steilacoom Lake, WA (i)
918	LC50 for 10-day *Hyalella* spiked sediment bioassays, Silver Creek, WA (j)
1026	LC50 for 14-day *Chironomus tentans* spiked sediment bioassays (d)
1078	LC50 for 10-day *Hyalella* spiked sediment bioassays, Soap Creek Pond, OR (h)
2296	LC50 for 10-day *C. tentans* spiked sediment bioassays, Tualatin River, OR (h)

Note: NOEC = no observed effects concentration; LOEC = lowest observed effects concentration. (a) Ingersoll and Nelson, 1990. (b) Lynde et al., 1993. (c) MacDonald et al., 2000. (d) Suedel et al., 1994. (e) Malueng et al., 1984. (f) This study. (g) Kemble et al., 1994. (h) Cairns et al., 1984. (i) Bennett and Cubbage, 1992. (j) Vandersypen, 1993.

sediments are predicted to be non-toxic (Ankley et al., 1996b). These studies are not included in Table 16.1 (or in the consensus guidelines) because that approach requires simultaneous measurements of an extractable fraction of iron sulfides and metals.

The concentrations listed in Table 16.1 of adverse effects to benthic invertebrates associated with sediment in copper range over two orders of magnitude at first glance seem to indicate that dry weight normalized copper concentrations are not particularly useful benchmarks for predicting risk to benthic macroinvertebrates. However, the low values reported are from co-occurrence studies, where it is likely that other contaminants contributed to the apparent effects. At the high concentrations listed, it is possible that bioavailability is limited due to sulfide or organic carbon complexes (Ankley et al., 1996a), or that relatively insensitive endpoints were used.

16.2 CASE STUDY METHODS

16.2.1 Study Area

Aquatic ecosystems that receive metals pollution from discrete sources may provide an ironic benefit to those interested in biocriteria development: having a measurable gradient of stress against which biological response patterns may be objectively evaluated. Such is the case for the Panther Creek watershed in central Idaho. The following section describes the evaluation of the response pattern of a macroinvertebrate multimetric index to a gradient of copper concentrations in water, and comparing the macroinvertebrate MMI patterns to amphipod toxicity measured in the laboratory with sediment collected near the macroinvertebrate sampling sites.

The Blackbird Mine, on a high divide in the Salmon River Mountains at the edge of the River of No Return Wilderness in central Idaho, is about 36 km southwest of the town of Salmon. The mine was operated as a large combined open-pit mine that produced cobalt, copper ore, and waste from about 1950 to 1982. Runoff from the mine ran into two streams, Big Deer Creek to the north and Blackbird Creek to the south (Figure 16.1). Both streams flow east into Panther Creek at 20 km and 40 km upstream, respectively, from its mouth. Panther Creek is a third to fifth Strahler-order stream in the study area; overall it is about 71 km long with an average gradient of 1.9% and a drainage area of 1380 km^2.

About 98% of the watershed is located within the Salmon National Forest. This forested, lightly roaded, watershed has few significant anthropogenic disturbances other than mine drainage containing elevated copper and cobalt concentrations (Mebane, 1994; Beltman et al., 1999). The Panther Creek watershed underwent extensive characterization of chemical conditions from 1992 to 1998 as part of natural resource damage assessment and site restoration studies of discharges from a hard rock mine. Recently, sources of metals pollution from the mine have been greatly reduced with the goal of fully restoring aquatic life communities in downstream waters (Renner, 1998). With few confounding variables, it is possible to compare the responses of macroinvertebrate metrics to chemical concentrations.

The mine runoff produced elevated cobalt and copper concentrations in downstream streams (Mebane, 1994; Beltman et al., 1999). Copper is well known to be toxic to aquatic life, but much less has been reported about cobalt (reviewed in Mebane, 1994). In paired acute tests with rainbow trout, copper was about 100 times more toxic than cobalt at 96 h (Marr et al., 1998). Most reports also suggest that cobalt is less toxic to invertebrates than copper (e.g., Diamond et al., 1992; USEPA, 1996; but see Biesinger and Christensen, 1972). Because of copper's apparent higher toxicity and because the metals tend to co-vary in the study area, the effects cannot be easily distinguished. For this reason the study focused on copper. The possibility cannot be excluded that effects associated with copper are due to cobalt or the mixture of the two metals.

16.2.2 Reach Selection and Collection Methods

Water in the Panther Creek watershed has been extensively sampled and analyzed as part of natural resource damage assessment and remedial investigation of Blackbird Mine pollution. Water chemistry samples were collected from macroinvertebrate-sampling reaches and filtered with 0.45 µm filters. Copper detection limits were lower than the maximum (acute) criterion for protection of aquatic life (≤6 µg/l).

The water chemistry data were compared to macroinvertebrate samples at 31 locations sampled by Idaho Department of Environmental Quality and by Salmon National Forest personnel. Both sampling programs used similar field methods and specified similar taxonomic efforts, so data from both collections should have been comparable. Both programs used 0.1-m^2 Hess samplers with 500-µm mesh and collected three replicates from riffle habitats during summer–fall stable flow regimes. Elevations of sample locations ranged from 2200 to 1000m. Ephemeroptera, Plecoptera, and Trichoptera were usually identified to species; other insects (except Chironomidae) were

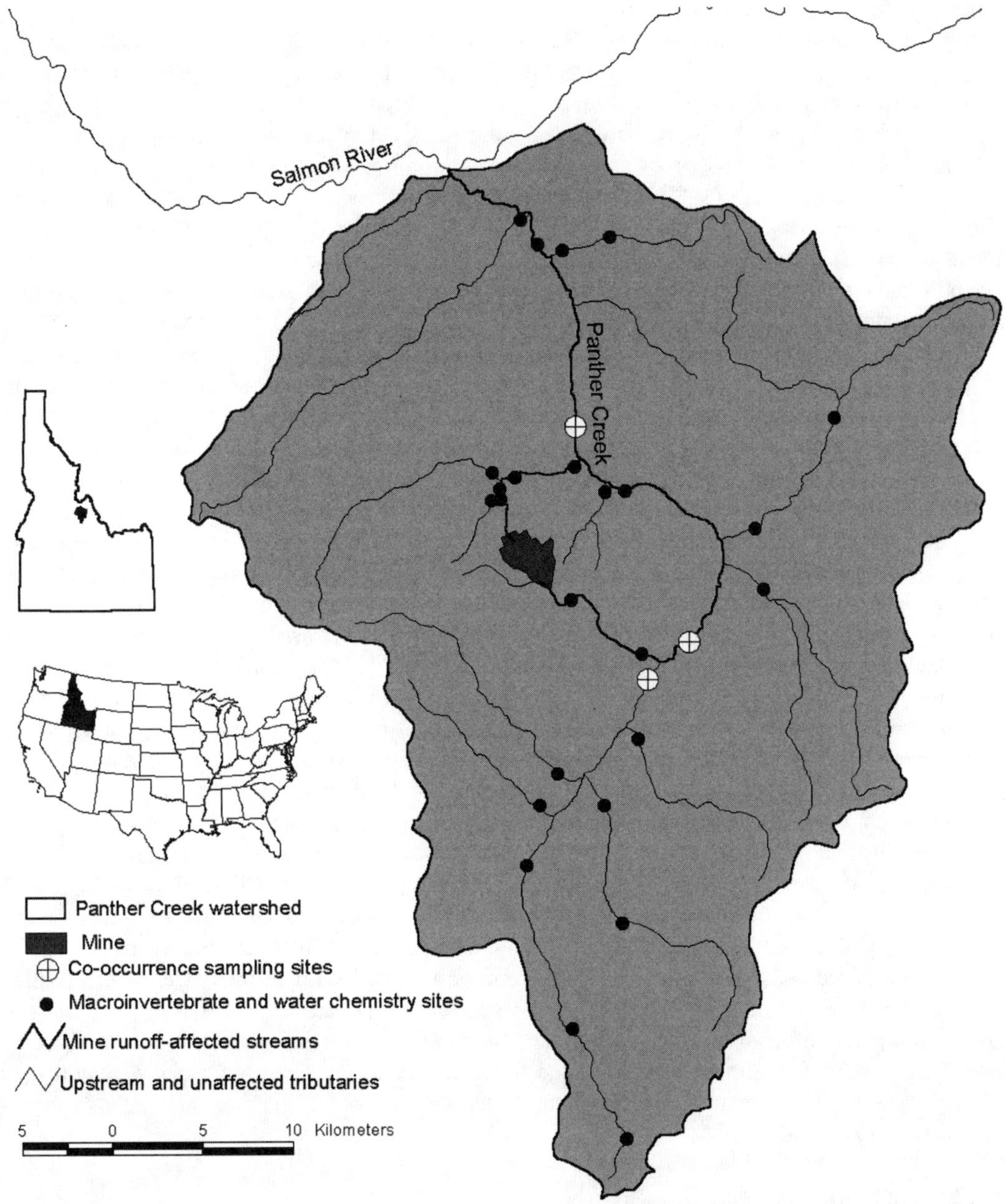

FIGURE 16.1 Sample sites in the Panther Creek watershed in central Idaho, U.S. Runoff from the Blackbird Mine drains into two tributaries of Panther Creek. The remainder of the watershed has few anthropogenic disturbances; 98% of the land use in the watershed is in national forest lands.

identified to genus. Molluscs, copepods, and most other invertebrates were identified to order. Contract laboratories made all taxonomic identifications.

Sediment sample sites were selected to include reaches upstream of mine runoff and conditions downstream of both mine-influenced tributaries, Blackbird Creek and Big Deer Creek (Figure 16.1). The specific locations within these reaches corresponded with macroinvertebrate trend stations established earlier by the U.S. Forest Service. Sediment collection methods were invented for this study with the goal of collecting sediments from lotic habitats. Method references for sample collection for sediment toxicity testing focused on soft-bottomed, low energy systems, such as lakes, estuaries, and slow moving rivers (Håkason, 1984; ASTM, 1990; MacKnight, 1991; Burton, 1992b). These methods focused on soft-bottomed waterways because significantly higher metal concentrations tend to appear on clay- and silt-sized fractions of sediments rather than on larger sand and gravel fractions. This variability in grain sizes could hide a metals dispersion pattern arising from anthropogenic sources and confound environmental investigations.

However, Moore et al. (1989) concluded that these grain size controls may be inappropriate for high gradient, coarse-grained rivers that drain mining areas because the relationship of increasing metal concentration to decreasing particle size is subdued. Larger particles stay in place longer, often in shallow oxygenated areas of a stream and therefore, in these systems, may have more time to accumulate oxide coatings and associated trace metals than smaller particles (Moore et al., 1989).

A more fundamental objection to the usual sediment sampling methods for biological testing is that soft-bottomed areas are unrepresentative of the stone-bottomed substrate that constitutes the predominant habitat in the high energy streams of the study area. One objective was to compare invertebrate patterns observed instream to invertebrate responses in laboratory settings when exposed to sediments from the site. The instream invertebrates were collected from stone-bottomed riffle areas, not from soft-bottomed depositional areas.

Common stream invertebrates differ in their abilities to thrive in different substrates. Chironomid midges and oligochaete worms do well in silts and mud, but the larger mayflies, stoneflies, and caddisflies prefer mixtures of coarse sands and gravels. Stream salmonids build their nests in coarse-grained sandy, gravelly sediments and juveniles are closely associated with coarse-grained substrates in cobble–rubble areas (Bjornn and Reiser, 1991). For these reasons, the sampling targeted depositional microhabitats located in the riffle-run macroinvertebrate sampling transects. Pockets of fine-grained sediments were sampled from depositional microhabitats located in the riffle-run macroinvertebrate sampling locations. These pockets included eddies behind boulders mid-channel and fine sediments embedded in the cobble interstices. Midstream samplings at the centers of the three sediment sampling locations (Figure 16.1) were unsuccessful because of fast water, scouring and large substrate so sediments were collected only from slow water areas near the stream margins.

In the slow water along the stream margins, sediment was scooped from the top 2 cm with stainless steel spoons. In fast water locations, the current would wash the fines from the spoons, so samples in those locations were collected with kitchen basters. Albeit difficult, by using a baster, sand and silt-sized sediments were extracted from the mid-riffle cobble interstices where benthic macroinvertebrates would likely occur. Sediments for each replicate were stirred in stainless steel bowls until color and texture were uniform, and then transferred to glass sample jars with Teflon® lids. A total of 2.5 L sediment were collected from each replicate; three replicates were collected from each location. Sediments were analyzed for particle size, organic carbon, and metals concentrations (ASTM, 1990).

16.2.3 Sediment Toxicity Testing

Adverse effects from sediment sorbed metals contamination are limited by the bioavailability of the metals to aquatic organisms. The extent and magnitude of contamination are not necessarily biologically significant if bioavailability is limited. Although sediments may contain relatively high concentrations of toxic heavy metals, this presence may not necessarily cause adverse effects to sediment-dwelling organisms. Bioavailability of contaminants is difficult to predict from chemical concentrations. The factors that determine the sorptive behavior of metals are complex (Ankley et al., 1996). The only means of measuring bioavailability is measuring a biological response via laboratory bioaccumulation or toxicity testing (Chapman et al., 1992). For this study, bioavailability was measured by exposing the amphipod, *Hyalella azteca*, a freshwater crustacean, to sediments collected from Panther Creek in a 10-day toxicity test.

Hyalella (Amphipoda), is a small freshwater crustacean routinely used to screen the toxicity of contaminated sediments. *Hyalella* is not indigenous to the study area, although it is common throughout lakes and low gradient streams in North America. Several characteristics make it desirable as a test organism including a short life cycle (3 to 4 weeks), widespread and abundant distribution, and a wide tolerance of sediment grain size. *Hyalella* is an epibenthic detritivore and will burrow in the top 1 cm of a sediment surface in search of food (ASTM, 1991).

The endpoints measured in the 10-day tests were survival and growth (measured as dry weight). In addition to the upstream reference site, clean silica aquarium sand was used as negative control sediment to test *Hyalella* performance. Five laboratory replicates were tested with each of the three field replicates that were collected at each location.

16.2.4 B-IBI (Multimetric Index) Calibration

One study question was to test the performance of an "off-the-shelf" general-purpose multimetric index across a gradient of disturbance. Recent approaches to developing a multimetric index usually involve evaluating candidate metrics, calibration, and aggregating the metrics into an index (Barbour et al., 1999, Fore, Chapter 19, this volume). Since different pollutants (e.g., sediments, nutrients, and metals) can have different biological signals, pollutant-specific indexes or effects measures have been suggested (Kiffney and Clements, 1996; USEPA, 2000). Similarly, multimetric indexes for specific locales based on ecoregions or watersheds have been developed (Robinson and Minshall, 1998; Jessup and Gerritsen, 2001). In contrast, Karr and Chu (1999) suggested that one index, the benthic index of biological integrity (B-IBI) is generally applicable for running waters, and developing a MMI on a case-by-case basis may not provide much more information for the effort.

This part of the case study examines the B-IBI performance with copper. The B-IBI is one of several multimetric indices used in North America to infer water quality conditions based on benthic macroinvertebrate assemblages. All multimetric indices are conceptually similar since they use several metrics (variables) that evaluate structural elements and functional processes of macroinvertebrate assemblages. They normalize the metrics to a unitless score based on regional reference conditions, and the metrics added together comprise an overall index score. The indices are typically appropriate over wide geographic areas with minor modifications to calibrate the metrics for different areas of the country or for different stressors (Barbour et al., 1999).

This study focused on the B-IBI because it has been used to determine impacts from common pollutants in the Pacific Northwest and elsewhere. Examples include grazing impacts in streams in the high desert ecoregion of eastern Oregon (Fore and Karr, 1995), forest roads (Fore et al., 1996), and urbanization of watersheds (Fore, 1999). No specific testing or use of the B-IBI with metals has been reported. Other nonpollutant specific MMI indexes with similar metrics and index structure as those used in the B-IBI would presumably perform similarly to the B-IBI (e.g., Barbour et al., 1999; Jessup and Gerritsen, 2001).

While the metrics in the B-IBI have been widely used, their scoring ranges must still be calibrated based on regional reference expectations. Sixteen reference sites were selected based upon their locations and chemical data. Sites never exceeding Idaho acute water quality standards for dissolved copper ($\leq$6 µg/l) were considered reference sites; sites from locations where exceedances were observed were considered metals impaired sites. Metric values and scores were assigned using a taxa attribute list for Idaho based upon ratings compiled by Wisseman (1996) and Merritt and Cummins (1996). Chironomidae, Oligiochaeta, Nematoda, and Tubificidae taxa were added as generally tolerant, even though there are exceptions to this generalization (e.g., Chapman and Brinkhurst, 1984; Clements et al., 1992).

A score of 1.0 was assigned to the 90th percentile metric scores for the reference sites, 0.8 for the median reference scores, and 0.25 for the 10th percentile of reference scores. To calculate intermediate metric scores, fitted simple curves were then to these points and used the equations

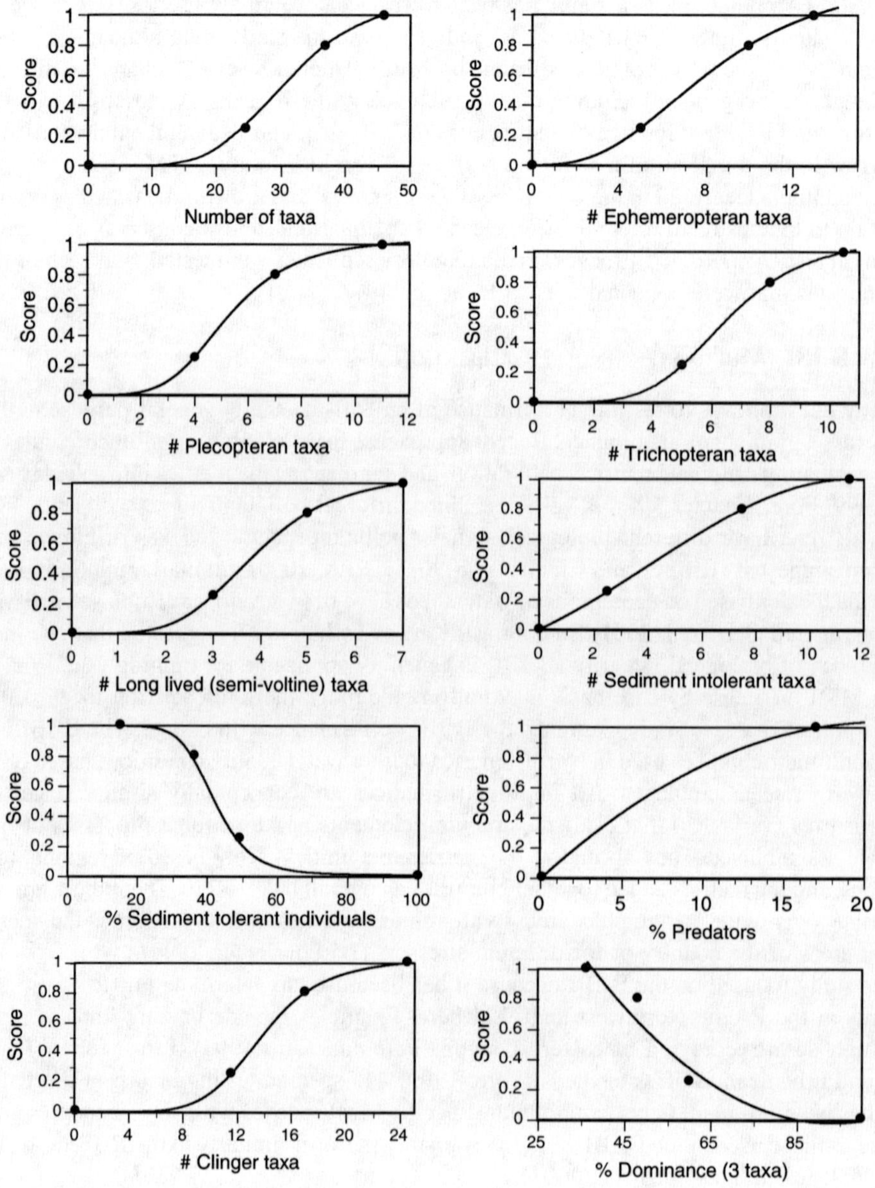

FIGURE 16.2 Multimetric index scoring curves for Salmon River tributaries, Idaho. The 90th percentile of each metric value from the reference sites was assigned a score of 1.0; the median reference value, 0.75; the 10th percentile reference value, 0.25; the minimum possible metric value was assigned a zero score.

for the curves to generate scores (Figure 16.2). The resulting scores were multiplied by five to make the range of B-IBI scores similar to previously published versions (0 to 50).

16.2.5 Statistics

Summary statistics, hypothesis testing, and correlations were used, respectively, to summarize data, determine the probability that apparent differences were simply due to chance, and quantify relationships of variables. Water chemistry data and macroinvertebrates often were collected in different years. In these cases, mean water concentrations for the stable flow seasons were used as the best estimates of the chemical concentrations experienced by the macroinvertebrates.

The Mann-Whitney nonparametric U-test was used to determine differences between metric values at reference and mining-influenced sites. For the amphipod survival data, Tukey's "honest significant difference" procedure was used to test the hypothesis that survival was not significantly different when the organisms were exposed to sediments from different locations in Panther Creek. Since survival data are proportionally rather than normally distributed, they were square root arcsine-transformed. A nonparametric analog to the Tukey test based on the Kruskall-Wallis test statistic was also used (Zar, 1984).

16.3 RESULTS AND DISCUSSION

16.3.1 Macroinvertebrate Multimetric Index (MMI) Testing

When grouping metrics from the MMI by sites with low or high ambient copper concentrations, all but one of the metrics were statistically different and varied in the expected direction (Figure 16.3). The exception was the percent predators metric, which was no different at high or low copper sites. The rationale for this metric is that the percentage of animals that are obligate predators provides a measure of trophic complexity. Less disturbed sites would be expected to support a greater diversity of prey items and a variety of habitats in which to find them (Fore, Chapter 19, this volume).

Other researchers reported variable results between this metric and metals. Clements et al. (2000) found a distinct gradient of decreasing percentages of predators with increasing metals concentrations in a regional study of Colorado streams. In contrast, Fore (Chapter 19) reported significant *positive* correlation between percent predators and metal concentrations from the Eagle River in Colorado. Carlisle and Clements (1999) found that functional group metrics were insensitive to metals pollution, highly variable, or both.

The previous comparison lumped sites into low and high copper concentration groups. In Figure 16.4, the MMI and selected individual macroinvertebrate samples are compared to associated copper concentrations. The B-IBI performed strongly, with consistently lower scores at sites that exceeded the maximum aquatic life criteria for copper. Similar comparisons using a different seven-metric MMI developed for the ecoregions of Idaho did not perform as well as the B-IBI (Mebane, 2001). Total density of benthic macroinvertebrates showed an erratic response to copper concentrations. Some samples from moderately affected copper sites were as abundant as at reference sites. Abundances at some sites with moderate copper contamination had abrupt seasonal declines in abundance. Perhaps hatches from areas with low taxa richness could remove much of the invertebrate assemblage.

The density, relative density, or taxa richness of Ephemeroptera, Plecoptera, and Trichoptera (mayflies, stoneflies, and caddisflies) is an important component in nearly all MMIs developed for temperate running waters (Barbour et al., 1999). In general, their representation in the invertebrate assemblage is expected to decline in response to degraded conditions. In stone-bottomed streams in Idaho with minimal anthropogenic disturbances, these three orders are usually well represented and often dominate the assemblage (Andrews et al., 1979; Robinson and Minshall, 1995; Jessup and Gerritsen, 2001). This was generally the case with the Panther Creek data.

Although some Trichopterans thrive in metals-enriched waters, both the Trichoptera taxa and the combined EPT metrics still corresponded fairly well to copper concentrations. Densities of EPT individuals and mayflies did not necessarily correspond to percentages of EPT individuals and mayflies (Figure 16.4). Some researchers advocate using relative density (e.g., % EPT) *in lieu* of actual reported densities. Reasons for not using macroinvertebrate densities as a metric are that densities (1) may be naturally highly variable and (2) may be particularly difficult to compare across studies because of differences in sampling and taxonomic target counts (Barbour et al., 1999; Fore, Chapter 19, this volume). However, in the Panther Creek data, the plot of mayfly density and copper shows that mayflies have been virtually extirpated from copper-affected sites (Figure 16.4),

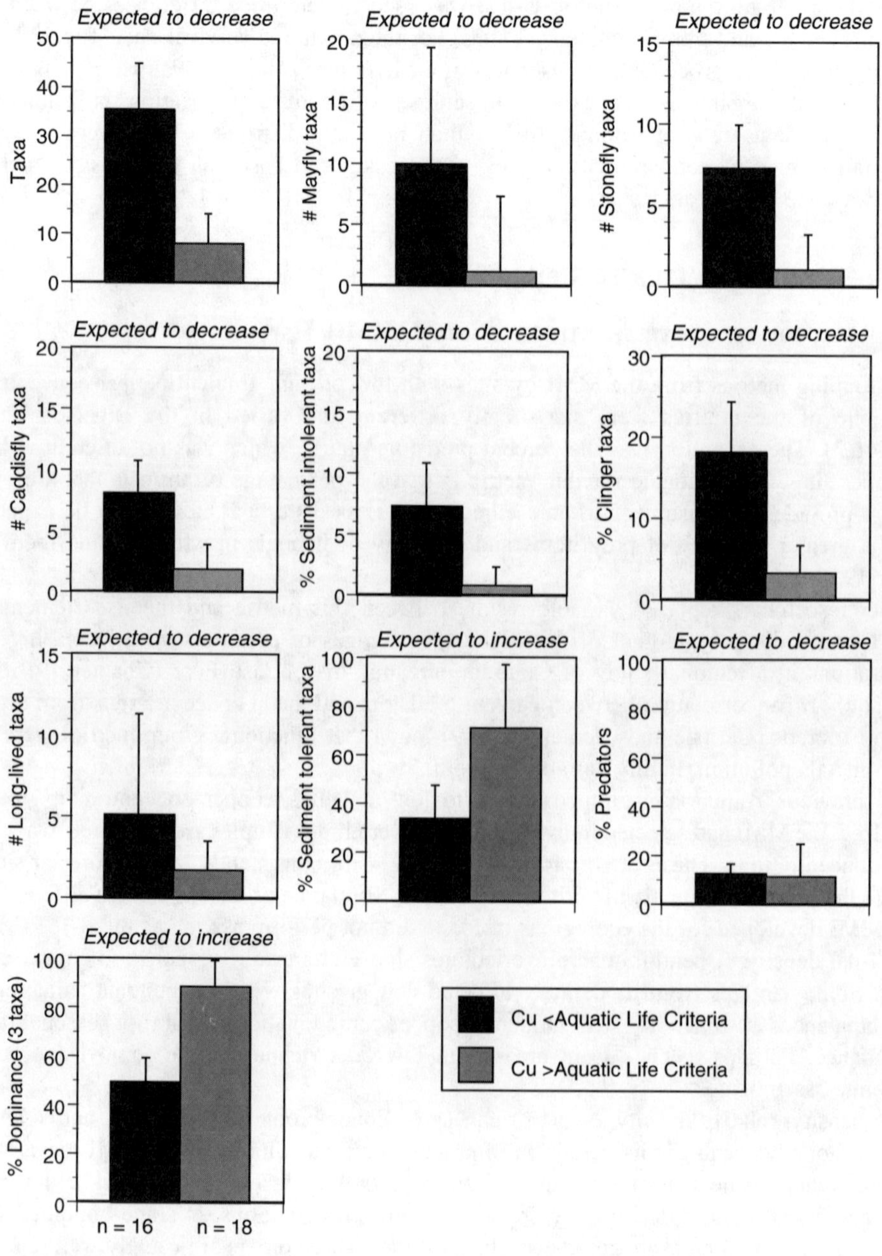

FIGURE 16.3 Metric values used for the multimetric index from reference sites and sites exceeding copper aquatic life criteria. Expected responses of each metric to degraded conditions are shown. Values for all groups were statistically different at $P < 0.05$ except for percent predators using the Mann-Whitney U-test.

but the plot of percent mayflies and copper does not reflect the true severity of effects. Most sites with high copper levels have low percentages of mayflies, but some sites (those with very low overall densities of macroinvertebrates) are composed of up to 50% mayflies. Carlisle and Clements (1999) also found mayfly abundance was a sensitive metric in experimental streams exposed to a mixture of metals. This is not necessarily a counter argument to the criticisms of using actual densities in MMIs. Rather, for all their utility, MMIs and statistical summaries do not supplant the value of inspecting data for patterns.

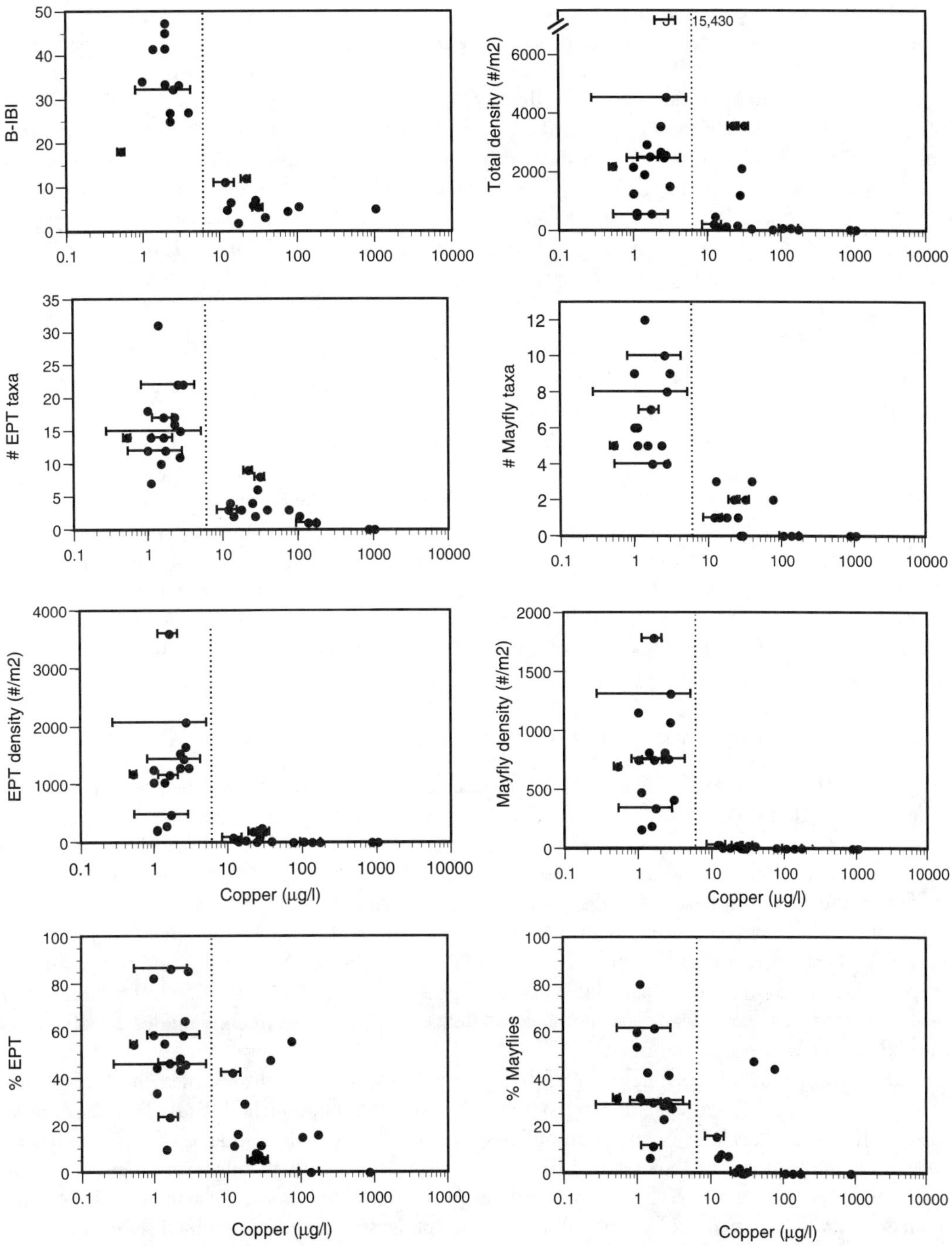

FIGURE 16.4 Relationships of multimetric index (B-IBI) scores, dissolved copper concentrations, and macroinvertebrate metrics in the Panther Creek watershed. Vertical dashed line shows approximate acute copper water quality criteria (≈6 µg/l). Mean copper concentrations are shown for locations that were not sampled for metals at the same time as the macroinvertebrate collections. These locations are indicated by values with error bars. Error bars show ± one standard deviation of the mean.

The effects of copper apparently overwhelmed other factors that may have affected the macroinvertebrate assemblage. Stream size or elevation may affect taxa richness or the sensitivity of macroinvertebrates to metals. The macroinvertebrate sites in the comparison were from a range of stream sizes (2nd to 5th orders) and elevations (1000 to 2200 m). Taxa richness generally increases downstream from headwaters to mid-order streams (Vannote et al., 1980) and macroinvertebrates may be more sensitive to metals at either headwater streams or at higher elevations than midsize streams at lower elevations (Kiffney and Clements, 1994, 1996). Reference sites had higher taxa richness than copper-influenced sites even though the copper-influenced sites would otherwise be expected to have higher richness since they tended to be at lower elevations and were larger waters than the reference sites (Figure 16.3).

16.3.2 Sediment Toxicity

Sediment samples from Panther Creek above Blackbird Creek, below Blackbird Creek, and below Big Deer Creek, were tested for acute toxicity to the freshwater crustacean *Hyalella azteca*. The samples from the stations below Blackbird Creek and Big Deer Creek had survival rates of 25 and 24%, respectively. Survival from the reference sediments collected from upstream of Blackbird Creek was 63% (Table 16.2). The stations below both creeks had significantly reduced survival ($P < 0.01$) from the upstream reference station. Survival rates between the two stations influenced by the contaminated tributaries were not statistically different. Since the downstream stations showed reduced survival rates, statistical tests of the survivors for differences in growth were not made. However, the only copper-contaminated sample that had similar survival rates as the reference samples (58 versus 63%) had lower growth than the reference at the end of the 10-day test (average weight of 0.08 versus 0.20 mg).

16.3.3 Invertebrate Toxicity Related to Sediment Contamination

High levels of metals were present in Panther Creek sediments below the tributaries draining the Blackbird mining district, and these sediments were acutely toxic to the epibenthic amphipod, *Hyalella*. *Hyalella* toxicity was correlated with metal concentrations in the streambed sediments and organic carbon is significantly associated with metals bioavailability (Table 16.3). Of the variables measured in these tests, the strength of association with toxicity was Co>Cu>As. Normalizing metals concentrations to the amount of organic carbon in the sediment increased the strength of the correlations for each of these metals. Both linear (Pearson's r) and nonparametric (Spearman's rho) values were calculated and were similar. Copper and cobalt concentrations in the sediment samples were nearly perfectly correlated ($r = 0.99$), so effects cannot be statistically attributed to one metal or the other.

Coarse-grained, mid-channel sediments had much lower metals levels than the fine-grained sediments collected from near the stream banks (C. Mebane, unpublished data). These sediments also had very low levels of organic carbon, and were well oxygenated because of their more open interstitial pore spaces and location in faster water. Despite lower bulk metals levels, these sediments were acutely toxic to the *Hyalella*, with no survival in any replicates. This increased mortality cannot be entirely explained by a possible *Hyalella* intolerance to coarse-grained sediments, since some of the reference sediments also had high proportions of coarse-grained sediments. These results suggest that in high gradient, coarse-grained systems that are contaminated with metals, targeting the fine-grained sediments for toxicity testing may underestimate the exposure of aquatic life to metals contamination.

Various physical and chemical factors affect sediment–trace element chemistry and bioavailability, including grain size, organometallic bonding, complexing with iron and manganese oxides, and sulfides (Horowitz, 1991). Iron normalization affects correlations between sediment lead and arsenic concentrations and bioaccumulation (Horowitz, 1991; Cain et al., 1992). Normalizing

TABLE 16.2
Metals Concentrations in Sediments from Reference and Mining-Influenced Sites and Corresponding Survival Rates and Weights of *Hyalella azteca* after Exposure to Sediments for 10 Days

Panther Creek Location	Arsenic (mg/kg)	Cobalt (mg/kg)	Copper (mg/kg)	Iron (%)	TOC (%)	Mean Survival (± S.D.)	Mean Dry Wt. (mg)
Upstream of Blackbird Creek Mean	9.5	4.2	8.3	1.21	0.88	5.8 ± 2.2	0.17 ± 0.08
	6.8	4.5	8.3	1.40	0.87	6.4 ± 2.9	0.21 ± 0.07
	10.7	3.8	12.9	1.36	3.78	6.8 ± 1.8	0.18 ± 0.03
	9.0	4.2	9.8	1.3	1.8	6.3 ± 2.2	0.19 ± 0.02
Downstream of Blackbird Creek Mean	766	436	2280	7.45	2.70	3.4 ± 0..9	0.17 ± 0.07
	867	507	2700	8.56	2.71	4.0 ± 2.5	0.18 ± 0.07
	888	554	2890	8.41	2.85	0.0 ± 0.0	0.0 ± 0.0
	840	499	2623	8.1	2.8	2.5 ± 2.3	0.12 ± 0.10
Downstream of Big Deer Creek Mean	14	76	386	0.93	0.12	0.0 ± 0.0	0.0 ± 0.0
	27	91	750	1.09	0.28	1.4 ± 1.3	0.07 ± 0.07
	31	74	889	1.20	0.51	5.8 ± 0.8	0.08 ± 0.04
	24	80	675	1.1	0.3	2.4 ± 2.7	0.05 ± 0.04
Negative Control	—	—	—	—	—	9.2 ± 1.3	0.15 ± 0.04

Note: Concentrations in dry weight; 10.0 represents 100% survival (10 amphipods per beaker). Sample mean is the mean of five replicates. Location mean is mean of the three samples. TOC = total organic carbon.

TABLE 16.3
Correlation between Amphipod Toxicity and Metal Concentrations in Bed Sediments Collected from Panther Creek

	Copper	Cobalt	Arsenic	Molar Sum (Cu + Co)	Molar Sum (Cu + Co + As)
Bulk metal (dry weight)	–0.47	–0.48	–0.37	–0.47	–0.45
Normalized to organic carbon	–0.72[a]	–0.81[b]	–0.57	–0.83[b]	–0.78[a]

Note: Pearson correlation coefficients, r, and significance levels are shown. Significance levels : [a]$p < 0.05$; [b]$p < 0.01$.

sediment metal concentrations to iron concentrations in the sediments did not significantly change the correlations between sediment metal concentrations and survival. However, the factors affecting bioavailabilities of metals are complex and poorly understood.

Some studies support porewater metal concentrations and other sediment sorbed contaminants as major controlling factors for copper toxicity. Porewater concentrations appear to be determined by binding phases including sulfides and organic carbon, which regulate partitioning of copper between sediments and porewater (Ankley et al., 1996b; Kemble et al., 1994). However, these authors reported that while porewater concentrations of copper and other metals predicted toxicity, porewater concentrations could not be predicted by normalizing to conventionals such as acid volatile sulfides. Vandersypen (1993) reported that porewater copper concentrations did not correlate well with *Hyalella* toxicity and suggested that binding with organics also reduced copper toxicity. Deaver et al. (1993) reported that *Hyalella* response to copper exposure in sediments was related to organic carbon content of the sediments. Increased organic matter content of sediment corresponded with decreased toxicity. A higher organic carbon content in the sediment (4%) was amended with five times more copper sulfate than a lower organic carbon sediment (0.9%) to achieve the same level of toxicity. Similarly, higher organic carbon content appears to have been related to reduced metals toxicity in the Panther Creek samples tested.

Hyalella mortality following laboratory exposure to field-collected sediments suggests that copper and cobalt in Panther Creek sediments are readily bioavailable and are toxic to macroinvertebrates. These contaminated sediments likely contributed to adverse effects on the Panther Creek aquatic ecosystem, even if the overlying water did not.

16.3.4 Comparison of MMI, Copper Concentrations, and Amphipod Toxicity Testing

Laboratory sediment toxicity testing with *Hyalella* corresponded well with the MMI and the selected benthic macroinvertebrate metrics that also responded to waterborne copper concentrations (Figure 16.5). While both toxicity results and the macroinvertebrate metrics and index corresponded with the presence of elevated metals, the magnitude of response was greater with the MMI and some of the macroinvertebrate metrics. For example the B-IBI score at the reference site was 5.8 times higher than the higher of the metals-influenced sites; the survival rates at the reference site averaged 2.5 times higher than the higher of the metals-influenced sites (Figure 16.5). Sediment toxicity tests conducted with clean overlying water showed that metals in the sediments could be harmful to invertebrates. However, the field collected benthic macroinvertebrates could have been exposed to both contaminated sediments among the riffles and waterborne copper.

Since the water and sediment copper concentrations were correlated ($r = 0.88$, $P < 0.01$), definitive conclusions that metals in water or sediment are mostly responsible for the macroinvertebrate effects shown by the MMI are not possible. For bioassessment purposes, this distinction may not always be practical. Integrating the data from the different exposure and effects characterization may be more informative.

The sediment quality triad (SQT) is a descriptive framework that can be adapted to graphically illustrate the degree of degradation and the magnitude of stressors and effects. The SQT was developed as an effects-based approach to integrate the toxicity of marine sediments, benthic community structure, and chemical analyses of the sediment (Chapman et al., 1992). The concentrations and effects were plotted on triaxial or "sunray" plots scaled against criteria or reference values.

Other investigators modified the approach by adding axes for bioaccumulation or toxicity testing with more species. Although these plots included up to five axes, the adaptions are still called sediment quality triads (e.g., Canfield et al., 1994; Borgmann et al., 2001). The sediment quality triad approach to integrating assessments can be used where three or more stressor or response variables are of interest.

In Figure 16.6, the water and sediment chemistry (stressor variables) and the MMI and toxicity testing (response variables) are plotted as a four-axis water quality triad. The stressor variables are scaled according to published quality criteria values and the response variables are scaled according to their optimal (best possible) values. Following the SQT, optimal values occur at the graph origin so that big plot areas indicate degraded conditions. The graphs show that in Panther Creek all four axes indicate near optimal scores for the station upstream of the mine effluent; however there was some unexplained *Hyalella* mortality at reference stations. For the two stations downstream of contaminated tributaries, all four axes of the graphs indicate severely elevated exposures and

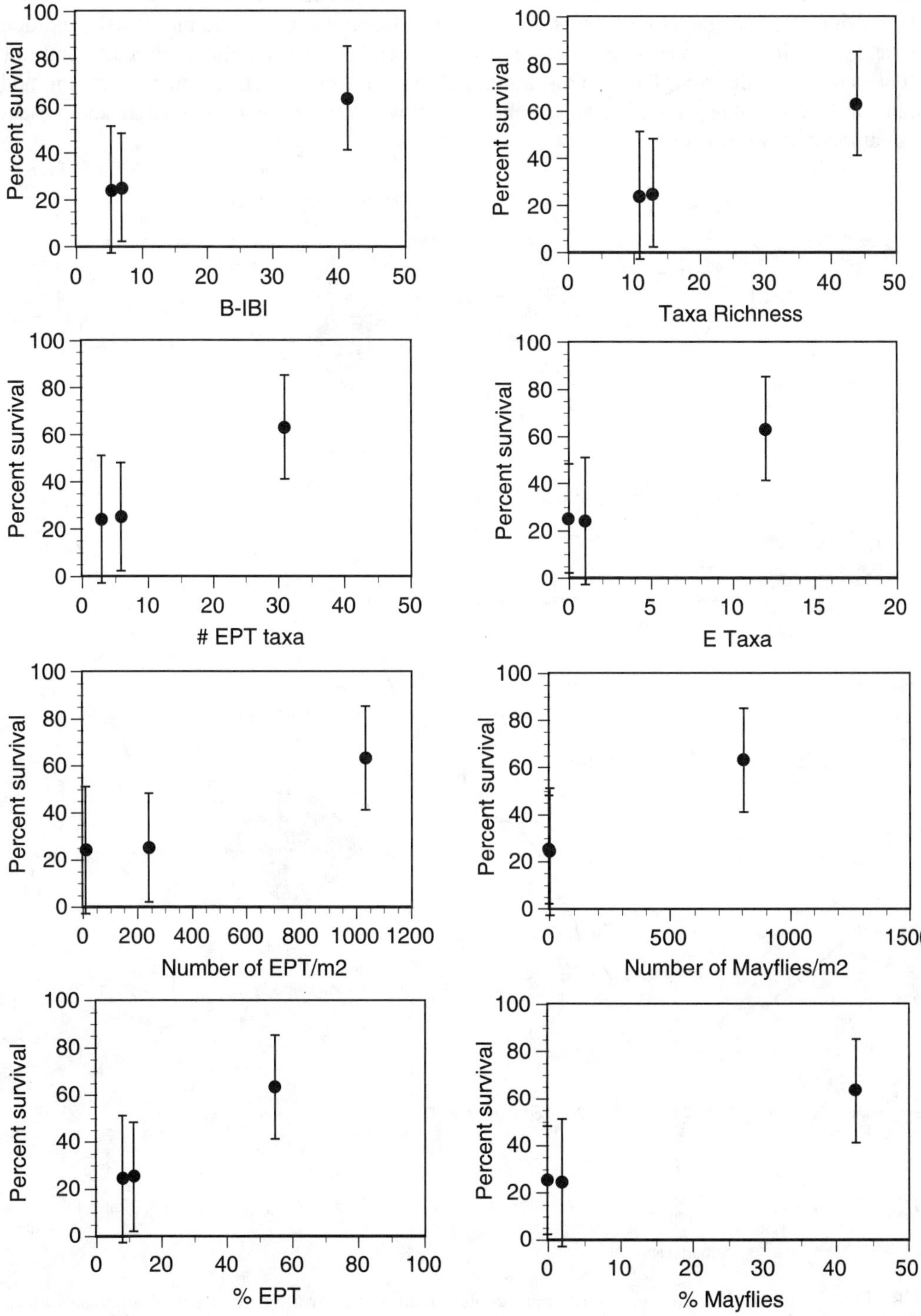

FIGURE 16.5 Correspondence between macroinvertebrate multimetric indices and selected metrics with *Hyalella* 10-day survival laboratory toxicity tests with sediment collected near macroinvertebrate sampling riffles.

degraded conditions. The station downstream of Big Deer Creek had more moderate sediment contamination (5 times guidelines versus 17 times guidelines). However, the biological condition axes were very similar (severely degraded) at both sites with high invertebrate mortality and low MMI scores. While these SQT-type plots would not supplant more formal statistical testing and defining of impairment benchmarks, they can be a useful way to graphically summarize and present a large amount of information.

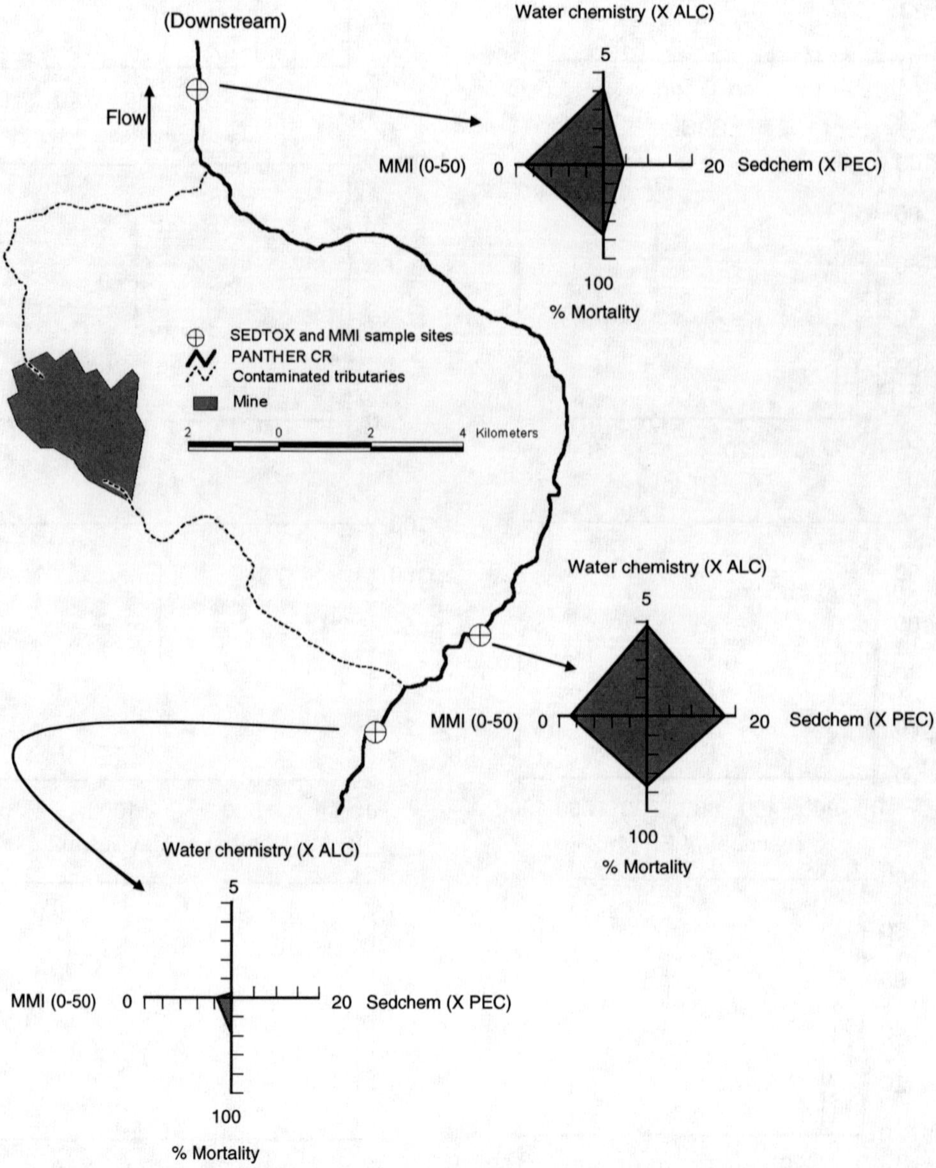

FIGURE 16.6 Water quality triad plots of copper contamination and invertebrate effects. Larger plot areas indicate more severe contamination or effects. Axes are scaled as follows from optimal to severe conditions: water chemistry, 0 to 5 times the acute copper aquatic life criteria (about 6 μg/l); sediment chemistry, 0 to 20 times sediment probable effect concentration (149 mg/kg; Table 16.1); *Hyalella* mortality, 0 to 100%; multimetric index, 50 to 0.

16.4 CONCLUSIONS

Two considerations with the use of multimetric indexes are whether pollutant specific indexes and independent validations of index responses are needed. The B-IBI, a nonpollutant-specific macroinvertebrate MMI, was tested with a known pollutant (copper). Copper and other metals have been shown to affect invertebrate assemblages in stream field studies, water and sediment toxicity testing with single invertebrate species, and in experimental streams with macroinvertebrate assemblages. Therefore, since invertebrate assemblage shifts are known to occur with metals contamination, if it is effective, the general application MMI should show a strong response between copper-affected and unaffected sites. Otherwise, the MMI should not be used.

The case study showed strong relationships between multimetric index results, laboratory toxicity testing, and metals concentrations in water and sediment. The results of the macroinvertebrate MMI changes in benthic assemblage structure, *Hyalella* laboratory toxicity tests of field-collected sediment, and copper concentrations in water and sediment all converged to show harmful effects of metals released from mine runoff. The macroinvertebrate MMI showed a larger difference between reference and test sites than did the sediment toxicity testing.

The dissolved metals in the water column or the sediment sorbed metals could have independently affected the macroinvertebrate assemblage. The toxicity to sediments observed in the laboratory was independent of the ambient water column metals since the tests were conducted with overlying clean water from the laboratory. The close correspondence between the riffle collected macroinvertebrate assemblage and dissolved copper concentrations suggests that water column copper could have independently affected the macroinvertebrates. The combined exposure routes may have produced more severe effects than either route alone would.

The study results show that the MMI was responsive to copper concentrations, with no high scores occurring at sites with water concentrations exceeding the acute aquatic life criteria for copper. An integrated assessment comparing the MMI, toxicity testing, water quality criteria, and sediment quality guidelines showed good agreement among the four lines of evidence.

Study goals should dictate whether a pollutant-specific or general application multimetric index would be preferable. If the study goals were to detect subtle changes in a spatially or pollutant-focused setting, then a spatially or stressor-specific multimetric index or other assessment endpoint would be preferable. Fore (Chapter 19, this volume) demonstrates the power of a spatially and pollutant-focused multimetric index for monitoring the recovery of the Eagle River in Colorado. In contrast, goals of recent studies in Idaho have been broad regional surveys of stream biological and habitat conditions to help set restoration or nondegradation priorities. For broad application, general application multimetric indexes such as the B-IBI tested here or the Jessup and Gerritsen (2001) stream macroinvertebrate index would be suitable.

ACKNOWLEDGMENTS

Bruce Smith provided macroinvertebrate data collected by the Salmon National Forest Ecosystem Baseline Monitoring Program. Steve Robinson directed additional macroinvertebrate collections as part of the Idaho Beneficial Use Reconnaissance Program. Kate Forster and Deb Carter of the Salmon National Forest assisted in sediment sampling. Alyce Fritz, Jay Field, Mary Baker Matta, and Nancy Beckvar of the NOAA Hazardous Materials Response and Assessment Division provided support and advice on the toxicity testing study design. Toxicity testing was conducted at EVS Consultants' laboratories, North Vancouver, British Columbia, Canada. Completion of the study was also made possible by the Lemhi County Sheriff's Department and the staff of the Steele Memorial Hospital (Salmon, Idaho) who retrieved and repaired the author after a mishap during field sampling.

REFERENCES

American Society for Testing and Materials. 1990. Standard guide for collection, storage, characterization, and manipulation of sediments for toxicological testing. Method E1391–90, in *Annual Book of ASTM Standards*, Vol. 11.04. American Society for Testing and Materials, Philadelphia.

American Society for Testing and Materials. 1991. Standard guide for conducting sediment toxicity tests with freshwater invertebrates. Method E1383–90, in *Annual Book of ASTM Standards*, Vol. 11.04. American Society for Testing and Materials, Philadelphia.

Andrews, D.A. and G.W. Minshall. 1979. Distribution of benthic invertebrates in the Lost Streams of Idaho, *American Midlands Naturalist,* 102, 140–149.

Ankley, G.T., D.M. Di Toro, D.J. Hansen, and W.J. Berry. 1996a. Assessing the ecological risk of metals in sediments, *Environmental Toxicology and Chemistry,* 15, 2053–2055.

Ankley, G.T., D.M. Di Toro, D.J. Hansen, and W.J. Berry. 1996b. Technical basis and proposal for deriving sediment quality criteria for metals, *Environmental Toxicology and Chemistry,* 15, 2056–2066.

Barbour, M.T., J. Gerritsen, B.D. Snyder, and J.B. Stribling. 1999. *Rapid Bioassessment Protocols for Use in Streams and Rivers: Periphyton, Benthic Macroinvertebrates, and Fish.* 2nd ed. EPA 841-B-99–002. Environmental Protection Agency, Washington, D.C. (http://www.epa.gov/owow/monitoring/bioassess.html)

Beltman, D., W.H. Clements, J. Lipton, and D. Cacela. 1999. Benthic invertebrate metals exposure, accumulation, and community-level impacts downstream from a hard-rock mine site, *Environmental Toxicology and Chemistry,* 18, 299–307.

Bennett, J. and J. Cubbage. 1992. Copper in Sediments from Steilacoom Lake, Pierce County, Washington, Environmental Investigations and Laboratory Services, Washington Department of Ecology, Olympia, WA.

Bergman, H.L. and J.S. Mattice. 1990. Lake acidification and fisheries project: brook trout (*Salvelinus fontinalis*) early life stages, *Canadian Journal of Fisheries and Aquatic Sciences,* 47, 1578–1579.

Beyer, W.N., G.H. Heinz, and A.W. Redmon-Norwood (Eds.) 1996. *Environmental Contaminants in Wildlife: Interpreting Tissue Concentrations*, Lewis Publishers, Boca Raton, Florida

Biesinger, K.E. and G.M. Christensen. 1972. Effects of various metals on survival, growth, reproduction, and metabolism of *Daphnia magna, Journal of the Fisheries Research Board of Canada,* 29,1691–1700.

Bjornn, T.C. and D.W. Reiser. 1991. Habitat requirements of salmonids in streams. *American Fisheries Society Special Publication,* 19, 83–138.

Borgmann, U., W.P. Norwood, T.B. Reynoldson, F. Rosa. 2001. Identifying cause in sediment assessments: bioavailability and the sediment quality triad, *Canadian Journal of Fisheries and Aquatic Sciences,* 58,950–960

Buikema, A.L., Jr. and J.R. Voshell, Jr. 1993. Toxicity studies using freshwater benthic macroinvertebrates, in Rosenberg, D.M. and V.H. Resh (Eds.). *Freshwater Biomonitoring and Benthic Invertebrates,* Chapman & Hall, New York.

Burton, G.A. (Ed.). 1992a. *Sediment Toxicity Assessment,* Lewis Publishers, Chelsea, MI.

Burton, G.A. 1992b. Sediment collection and processing: Factors affecting realism, in Burton, G.A. (Ed.). *Sediment Toxicity Assessment,* Lewis Publishers, Chelsea, MI.

Cain, D.J., S.N. Luoma, J.L. Carter, and S.V. Fend. 1992. Aquatic insects as bioindicators of trace metal contamination cobble-bottom rivers and streams, *Canadian Journal of Fisheries and Aquatic Sciences,* 49, 2141–2154.

Cairns, M.A.A.V. Nebeker, J.N. Gakstatter, and W.L. Griffis. 1984. Toxicity of copper-spiked sediments to freshwater invertebrates, *Environmental Toxicology and Chemistry,* 3, 435–445.

Canfield, T.J., N.E. Kemble, W.G. Brumbaugh, F.J. Dwyer, C.G. Ingersoll, and J. Fairchild. 1994. Use of benthic invertebrate community structure and the sediment quality triad to evaluate metal-contaminated sediment in the upper Clark Fork River, Montana, *Environmental Toxicology and Chemistry,* 13, 1999–2012.

Canton, S.P. and W.D. Van Derveer. 1997. Selenium toxicity to aquatic life: an argument for sediment-based water quality criteria, *Environmental Toxicology and Chemistry,* 16, 1255–1259.

Carlisle, D.M. and W.H. Clements. 1999. Sensitivity and variability of metrics used in biological assessments of running waters, *Environmental Toxicology and Chemistry* 18, 285–291.

Chadwick, J.W., S.P. Canton, and R.L. Dent. 1986. Recovery of benthic invertebrate communities in Silver Bow Creek, Montana, following improved metal mine wastewater treatment, *Water, Air, and Soil Pollution*, 28, 427–438,

Chapman, P.M. 1995. Extrapolating laboratory toxicity test results to the field, *Environmental Toxicology and Chemistry*, 14, 927–930.

Chapman, P.M. and R.O. Brinkhurst. 1984. Lethal and sublethal tolerances of aquatic oligochaetes with reference to their use as a biotic index of pollution, *Hydrobiologia*, 115, 139–144.

Chapman, P.M., E.A. Power, and G.A. Burton, Jr. 1992. Integrative assessments in aquatic ecosystems, in Burton, G.A. (Ed.). *Sediment Toxicity Assessment*. Lewis Publishers, Chelsea, MI, 313–340.

Clements, W.H. 1994. Benthic invertebrate community responses to heavy metals in the upper Arkansas River basin, Colorado, *Journal of the North American Benthological Society*, 13, 30–44.

Clements, W.H. and P.M. Kiffney. 1994. Integrated laboratory and field approach for assessing heavy metals at the Arkansas River, Colorado, *Environmental Toxicology and Chemistry*, 13, 387–404.

Clements, W.H. and P.M. Kiffney. 1996. Validation of whole effluent toxicity tests: integrated studies using field assessments, microcosms, and mesocosms, in D.L. Grothe, K.L. Dickson, and D.K. Reed-Judkins (Eds.). *Whole Effluent Toxicity Testing: An Evaluation of Methods and Prediction of Receiving System Impacts*, SETAC Press, Pensacola, FL. 229–244.

Clements, W.H., D.M. Carlisle, J.M. Lazorchak, and P.C. Johnson. 2000. Heavy metals structure benthic communities in Colorado mountain streams, *Ecological Applications*, 10, 626–638.

Clements, W.H., D.S. Cherry, and J. Cairns. 1988. Structural alterations in aquatic insect communities exposed to copper in laboratory streams, *Environmental Toxicology and Chemistry*, 7, 715–722.

Clements, W.H., D.S. Cherry, and J.H. Van Hassel. 1992. Assessment of the impact of heavy metals on benthic communities at the Clinch River (Virginia): evaluation of an index of community sensitivity, *Canadian Journal of Fisheries and Aquatic Sciences*, 49, 1686–1694

de Vlaming, V. and T.J. Norberg-King. 1999. A review of single species toxicity tests: are the tests reliable predictors of aquatic ecosystem community response? EPA 600/R/97/114, U.S. Environmental Protection Agency, Office of Research and Development, Duluth, MN.

Deaver, E., B.C. Suedel, and J.H. Rogers, Jr. 1993. Copper sulfate as a reference toxicant for use in sediment toxicity tests, Abstracts, 14th annual meeting, 14–18 November 1993. Society of Environmental Toxicology and Chemistry, Pensacola, FL, 313.

Diamond, J.M., E.L. Winchester, D.G. Mackler, W.J. Rasnake, J.K. Fanelli and D. Gruber. 1992. Toxicity of cobalt to freshwater indicator species as a function of water hardness, *Aquatic Toxicology*, 22, 163–180.

Farag, A.M., C.J. Boese, D.F. Woodward, and H.L. Bergman. 1994. Physiological changes and tissue metal accumulations of rainbow trout exposed to foodborne and waterborne metals, *Environmental Toxicology and Chemistry*, 13, 2021–2030

Farag, A.M., D.F. Woodward, W. Brumbaugh, J.N.Goldstein, E. McConnell, C. Hogstrand, and F.T. Barrows. 1999. Dietary effects of metals-contaminated invertebrates from the Coeur d'Alene River, Idaho on cutthroat trout, *Transactions of the American Fisheries Society*, 128, 578–592

Fore, L.S. 1999. *Measuring the Effects of Urbanization on Bellevue Streams*. Statistical Design. Seattle, WA.

Fore, L.S. 2002. Biological assessment of mining disturbances on stream invertebrates in mineralized areas of Colorado, Chapter 19, this volume.

Fore, L.S. and J.R. Karr. 1995. *Benthic Invertebrate Response to Human Activities East of the Cascades*, Institute of Environmental Studies, University of Washington, Seattle, WA.

Fore, L.S., J.R. Karr, and R.W. Wisseman. 1996. Assessing invertebrate response to human activities: evaluating alternative approaches, *Journal of the North American Benthological Society*, 15, 212–231.

Gauss, J.D., P.E. Woods, R.W. Winner, and J.H. Skillings. 1985. Acute toxicity of copper to three life stages of *Chironomus tentans* as affected by waterhardness-alkalinity, *Environmental Pollution* (Series A), 37, 149–157.

Geckler, J.R., W.B. Horning, T.M. Nieheisel, Q.H. Pickering, E.L Robinson, and C.E. Stephan. 1976. Validity of laboratory tests for predicting copper toxicity in streams, EPA 600/3–76–116, U.S. Environmental Protection Agency, Ecological Research Service, Cincinnati, OH.

Håkason, L. 1984. Sediment sampling in different aquatic environments: statistical aspects, *Water Resources Research*, 20, 41–46.

Hilsenhoff, W.L. 1987. An improved biotic index of organic stream pollution, *Great Lakes Entomological Society*, 20, 31–39.

Horowitz, A.J. 1991. *A primer on sediment-trace element chemistry, second edition*, Lewis Publishers, Chelsea, MI.

Ingersoll, C.G. and M.K. Nelson. 1990. Testing sediment toxicity with *Hyalella azteca* (Amphipoda) and *Chironomus riparius* (Diptera) *ASTM STP 13th Symposium on Aquatic Toxicology and Risk Assessment.* American Society for Testing and Materials, Philadelphia, PA, 93–109.

Jarvinen, A.W. and G.T. Ankley. 1999. *Linkage of Effects To Tissue Residues: Development of a Comprehensive Database for Aquatic Organisms Exposed to Inorganic and Organic Chemicals,* SETAC Press, Pensacola, FL.

Jessup, B. and J. Gerritsen. 2001. Stream macroinvertebrate index, in Grafe, C.S (Ed.) *Idaho Small Streams Ecological Assessment Framework.* Idaho Department of Environmental Quality, Boise, ID (available at www2.state.id.us/deq), ch. 3.

Karr, J.R. and E.W. Chu. 1999. *Restoring Life in Running Waters: Better Biological Monitoring*, Island Press, Washington, D.C.

Kemble, N.E., W.G. Brumbaugh, E.L. Brunson, F.J. Dwyer, C.G. Ingersoll, D.P. Monda, and D.F. Woodward. 1994. Toxicity of metal-contaminated sediments from the Clark Fork River, Montana, to aquatic invertebrates and fish in laboratory exposures, *Environmental Toxicology and Chemistry,* 13, 1985–1998.

Kerans, B.L. and J.R. Karr. 1994. A benthic index of biotic integrity (B-IBI) for rivers of the Tennessee Valley, *Ecological Applications,* 4, 768–785.

Kiffney, P.M. and W.H. Clements. 1994. Structural responses of benthic macroinvertebrate communities from different stream orders to zinc, *Environmental Toxicology and Chemistry,* 13, 389–395.

Kiffney, P.M. and W.H. Clements. 1996. Effects of metals of stream macroinvertebrate assemblages from different altitudes, *Ecological Applications,* 6, 472–481.

Kiffney, P.M. and W.H. Clements. 2002. Ecological effects of metals on benthic macroinvertebrates, Chapter 8, this volume.

Lewis, P.A., D.J. Klemm, J.M. Lazorchak, T.J. Norberg-King, W.H. Peltier, and M.A. Heber (Eds.). 1994. *Methods for Estimating the Chronic Toxicity of Effluents and Receiving Waters to Freshwater and Marine Organisms,* EPA 440/600/4–91/002, U.S. Environmental Protection Agency, Washington, D.C.

Lynde, S.R., D.S. Cherry, D.S. Buikema, and A.L. Lauth. 1993. Comparison on invertebrate impairment and microbial colonization on leaves in river sediment exposed to copper, *Abstracts, 14th Annual Meeting, 14–18 November 1993. Pensacola: Society of Environmental Toxicology and Chemistry,* 310.

MacDonald, D.D., C.G. Ingersoll, and T.A. Berger. 2000. Development and evaluation of consensus-based sediment quality guidelines for freshwater ecosystems, *Archives of Environmental Contamination and Toxicology,* 39, 20–31.

MacDonald, L.H., A.W. Smart, and R.C. Wissmar. 1991. *Monitoring Guidelines to Evaluate Effects of Forestry Activities on Streams in the Pacific Northwest and Alaska,* EPA 910/9–91–001, U.S. Environmental Protection Agency, Seattle, WA.

MacKnight, S.D. 1991. Selection of bottom sediment sampling stations, in Mudroch A. and S.D. MacKnight (Eds.). *Handbook of Techniques for Aquatic Sediments Sampling*, CRC Press, Boca Raton, FL.

Malueng, K.W., G.S. Schuytema, D.F. Krawczyk, and J.H. Gakstatter. 1984. Laboratory sediment toxicity tests, sediment chemistry and distribution of benthic macroinvertebrates in sediments from Keweenaw Waterway, Michigan, *Environmental Toxicology and Chemistry,* 3, 233–242.

Marr, J.C.A., J.A. Hansen, J.S. Meyer, C. Caecela, T. Podrabshy, J. Lipton, and H. Bergman, H.L. 1998. Toxicity of cobalt and copper to rainbow trout: application of a mechanistic model from predicting survival, *Aquatic Toxicology,* 43, 225–238.

Mebane, C.A. 1994. *Blackbird Mine Prelininary Natural Resource Survery,* U.S. National Oceanic and Atmospheric Administration, Hazardous Materials Assessment and Response Division, Seattle, WA. (also available from the author at cmebane@deq.state.id.us).

Mebane, C.A., T.R. Maret, and R.M. Hughes. In press. An index of biological integrity (IBI) for Pacific Northwest rivers. *Transactions of the American Fisheries Society.*

Mebane, C.A. 2001. Testing bioassessment metrics: macroinvertebrate, sculpin, and salmonid responses to stream habitat, sediment, and metals, *Environmental Assessment and Monitoring*, 67, 293–322

Merritt, R.W. and K.W. Cummins. 1996. *An Introduction to the Aquatic Insects of North America,* 3rd ed., Kendall Hunt, Dubuque, IA.

Moore, J.N., E.J. Brook, and C. Johns. 1989. Grain size partitioning of metals in contaminated, coarse-grained river floodplain sediment: Clark Fork River, Montana, USA, *Environmental Geology and Water Sciences,* 14, 107–115.

Nebeker, A.V., C. Savonen, R.J. Baker, and J.K. McCrady. 1984. Effects of copper, nickel, and zinc on the life cycle of the caddisfly *Clistoronia magnifica* (Limnephilidae), *Environmental Toxicology and Chemistry,* 3, 645–649.

Rankin, E.T. and C.O. Yoder. 1999. Methods for deriving maximum species richness lines and other threshold relationships in biological field data, in Simon, T.P. (Ed). *Assessing the Sustainability and Biological Integrity of Water Resources Using Fish Communities.* CRC Press, Boca Raton, FL, 611–624.

Relyea, C.D., G.W. Minshall, and S. Rushforth. 1999. Development of the Fine Sediment Bioassessment Index for use in Monitoring Northwestern United States Streams, report to the National Council of the Paper Industry for Air and Stream Improvement, Inc., Corvallis, OR, Idaho State University and Brigham Young University.

Renner, R.A. 1998. Calculating the cost of natural resource damage, *Environmental Science and Technology,* 32(3), 86A-90A.

Richardson, J.S. and P.M. Kiffney. 2000. Responses of a macroinvertebrate community from a pristine, southern British Columbia, Canada, stream to metals in experimental mesocosms, *Environmental Toxicology and Chemistry,* 19, 736–743.

Robinson, C.T. and G.W. Minshall. 1995. Biological Metrics for Regional Biomonitoring and Assessment of Small Streams in Idaho, final report to the Idaho Division of Environmental Quality, Stream Ecology Center, Department of Biological Sciences, Idaho State University, Pocatello, ID.

Robinson, C.T. and G.W. Minshall. 1998. Regional assessment of wadable streams in Idaho, USA, *Great Basin Naturalist,* 58, 54–65.

Rosenberg, D.M. and V.H. Resh. 1993. Introduction to freshwater biomonitoring and benthic macroinvertebrates, in Rosenberg, D.M. and V.H. Resh (Eds.). *Freshwater Biomonitoring and Benthic Invertebrates.* Chapman & Hall, New York, 1–9.

Simon, T.P. 1999. Introduction: biological integrity and use of ecological health concepts for application to water resource characterization, in T.P. Simon (Ed.). *Assessing the Sustainability and Biological Integrity of Water Resources Using Fish Communities,* CRC Press, Boca Raton, FL. 3–16.

Sorensen, E.M.B. 1991. *Metal Poisoning in Fish,* CRC Press, Boca Raton, FL.

Sprague, J.B., P.F. Elson, and R.L. Saunders. 1965. Sublethal copper-zinc pollution in a salmon river: a field and laboratory study, *International Journal of Air and Water Pollution,* 9, 531–543.

Stephenson, M.E.G. Mierle, R.A. Reid, and G.L. Mackie. 1994. Effects of experimental and cultural lake acidification on littoral benthic macroinvertebrate assemblages, *Canadian Journal of Fisheries and Aquatic Sciences,* 1147–1161

Suedel, B.C., E. Deaver, and J.H. Rodgers, Jr. 1994. Experimental factors that may affect toxicity of aqueous and sediment-bound copper to freshwater organisms, University of Mississippi, Department of Biology, Biological Field Station. Hattiesburg, MS.

Suter, G.W. II. 1993. A critique of ecosystem health concepts and indexes, *Environmental Toxicology and Chemistry,* 12, 1533–1539.

U.S. Environmental Protection Agency. 1996. *1995 Updates: Water Quality Criteria Documents for the Protection of Aquatic Life in Ambient Water,* USEPA 820-B-96–001. U.S. Environmental Protection Agency, Washington, D.C.

U.S. Environmental Protection Agency. 2001. *2001 Update of Ambient Water Quality Criteria for Cadmium,* EPA 822-R-01–001. USEPA, Office of Research and Development, Washington, D.C.

Vandersypen, J.P. 1993. The acid volatile sulfide approach as an indicator of the bioavailability of copper and cadmium in the sediment of Silver Creek, Washington, M.S. thesis, Western Washington University, Bellingham, WA.

Vannote, R.L., G.W. Minshall, K.W. Cummins, J. R. Sedell, and C.E. Cushing. 1980. The river continuum concept, *Canadian Journal of Fisheries and Aquatic Sciences,* 37, 130–137.

Winner, R.W., M.W. Boesel, and M.P. Farrell. 1980. Insect community structure as an index of heavy-metal pollution in lotic ecosystems, *Canadian Journal of Fisheries and Aquatic Science,* 37, 647–655.

Wisseman, R.W., 1996. *Benthic Invertebrate Biomonitoring and Bioassessment in Western Montane Streams.* Aquatic Biology Associates, Corvallis, OR 97330.

Zar, J.R. 1984. *Biostatistical Analysis,* 2nd ed. Prentice-Hall, Englewood Cliffs, NJ.

17 Relationships between Fish Assemblages and Organochloride Insecticides in Sediment and Fish Tissue in South Central Kansas

Michael J. Lydy, Paul M. Stewart, and Thomas P. Simon

CONTENTS

17.1 INTRODUCTION

Regardless of varying bans on their use, organochlorine (OC) insecticides can still be detected in surface and ground water throughout the United States (Schmitt et al., 1990). Organochlorine insecticides have a number of common characteristics that make them problematic. All are lipophilic in nature, and thus concentrate in organisms. The toxicity of OCs varies; however, all act as neurotoxicants and produce both acute and chronic effects in field and laboratory tests (Hoffman et al., 1995; Sobiech and Henry, Chapter 7, this volume). OC insecticides exert deleterious effects on reproductive viability in fish (Nebeker et al., 1974; Leatherland and Sonstegard, 1978) and birds (Kubiak et al., 1989; Bishop et al., 2000), and bioaccumulate and biomagnify within aquatic and terrestrial food chains (Kawano et al., 1988).

0-8493-0905-0/03/$0.00+$1.50

Even though OC pesticides have been found to affect non-target organisms including fish (Saiki and Schmitt, 1986; USGS, 1997; Pereira et al., 1996), few studies have examined the relationship of OC pesticide contamination to possible effects on fish assemblages. Therefore, a study was initiated by Eaton and Lydy (2000) on the Arkansas River in Wichita, Kansas. They examined bed sediment and fish tissue OC pesticide concentrations and compared them to fish assemblage data collected from sites with different land uses, using an index of biotic integrity (IBI). They found significantly greater total organochlorine concentrations in fish tissue and sediments in urban areas than agricultural areas and significant correlations among total organochlorine concentrations in fish tissues and sediments. No relationship was found, however, between total IBI scores and fish and sediment organochlorine concentrations.

This chapter further examines the data collected by Eaton and Lydy (2000) and evaluates fish tissue and sediment OC pesticide concentrations in relationship to individual metrics and the structure and function of the fish assemblage.

17.2 METHODS

17.2.1 Study Area and Site Selection

The study area included the Arkansas River, Little Arkansas River, and Gypsum, Jester, Cowskin, and Chisholm Creeks in South Central Kansas (Figure 17.1). Twenty sample sites, located in Sedgwick and Sumner Counties near Wichita were selected to compare organochlorine insecticide concentrations in areas of urban and agricultural land uses. Sample reaches were representative of each stream, incorporating all available habitats, including at least one riffle, run, and pool.

17.2.2 Sediment and Fish Tissue Collection

Fish and sediment were collected for analysis of OCs concurrently with fish sampling. Three sediment samples were collected at each site, one at each end of the sample reach and one in the middle. Grab sediment samples were collected using 100-ml glass jars, not exceeding 6 cm in depth because the upper 2 cm represented the biologically active portion of the sediment (Burton, 1991). When possible, five carp (*Cyprinus carpio*) were collected for tissue analysis at each site. Sixty-three carp were collected, 39 at nine urban sites and 24 at six agricultural sites.

17.2.3 Organochlorine Analysis

Sediment and fish samples were analyzed for the following organochlorine pesticides: alpha, beta, delta, and gamma BHC (1,2,3,4,5,6-hexachlorocyclohexane), endrin aldehyde, endrin ketone, heptachlor, heptachlor epoxide, 4,4′-DDD, 4,4′-DDE, 4,4′-DDT, and the alpha and gamma isomers of chlordane. Chemical analyses were performed using a Hewlett Packard Series 6890 gas chromatograph system (HP6890GC) equipped with an electron capture detector (ECD) following modified protocols from USEPA Method 8080A (1990).

Quality control/quality assurance included dual-column confirmation (DB608™ and DB5™), an extraction blank, and either a matrix spike (fish samples) or blank spike (sediment samples) for each extraction batch. Furthermore, 200 μl of the surrogate recovery standard dibromooctoflourobiphenyl (DBOFB) was added to all samples prior to extraction. The surrogate recovery data for sediments and fish tissue were within acceptable levels (40 to 120%), while the matrix spike (fish) and blank spike (sediment) values ranged from 80 to 105% and 93 to 104%, respectively.

Prior to analysis, the three sediment samples collected from each site were homogenized. Overlying water was decanted and discarded as were foreign objects such as sticks, leaves, and rocks. Percent moisture was determined for the composite sample. The extraction method for the sediment was a modification of USEPA Method 3550 (sonication extraction) for low concentrations

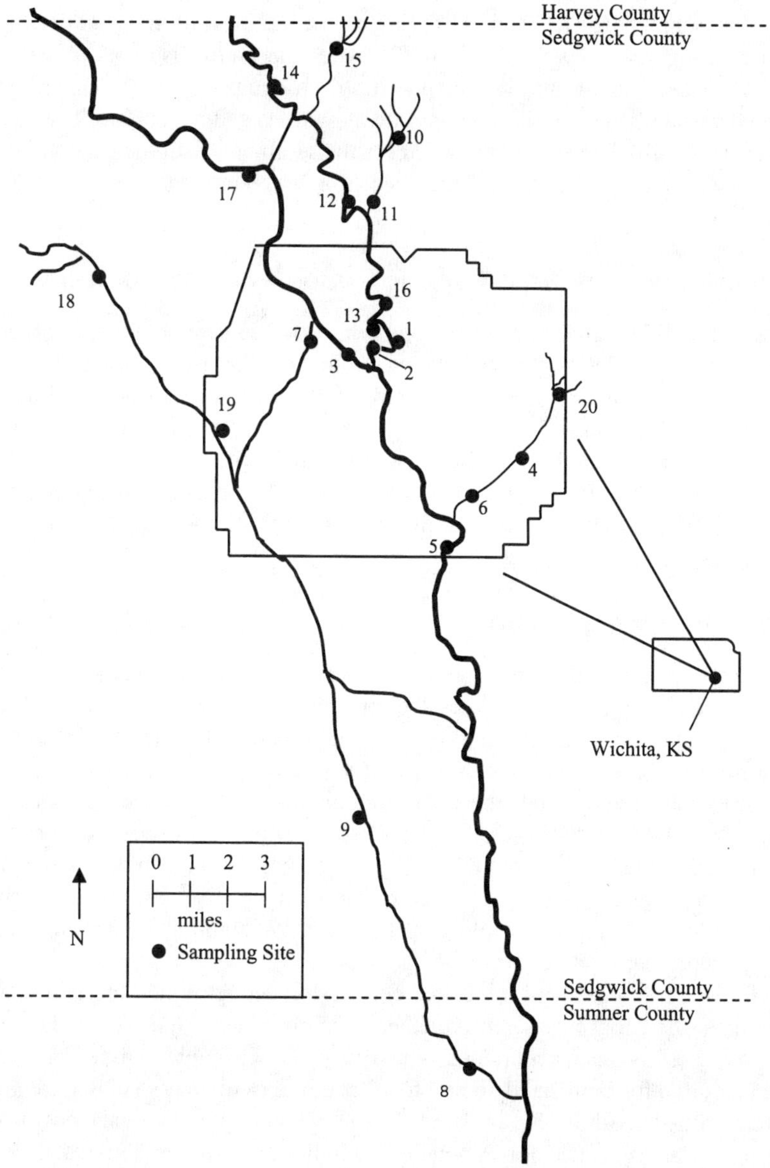

FIGURE 17.1 Locations of the 20 sample sites on the Arkansas and Little Arkansas Rivers and their tributaries in the Wichita study area.

of organics and pesticides (1986). Further details concerning the sediment extraction procedures can be found in Eaton and Lydy (2000).

This study used fish fillets (skin removed) for OC analysis, since federal standards are not available for whole fish tissue concentrations (Nowell and Resek, 1994). Five fish were analyzed for OCs at 11 of the 15 sites where fish were collected for tissue analysis. At Sites 7 and 12, only one fish was analyzed, Three fish were analyzed at Sites 15 and 19. Site-specific mean OC concentrations are the average values from all fish collected at a site, while land use-specific mean concentrations are the average values from all fish collected from all sites within that land use designation (Eaton and Lydy, 2000).

The OC analysis of fish tissues is a modification of the procedure used by the Kansas Department of Health and Environment (KDHE) and U.S. Environmental Protection Agency (USEPA) for their fish tissue monitoring programs (USEPA, 1980). After the entire fish was partially thawed and filleted, the fillet tissue was ground in a blender until no chunks were present. About 30 g (±1.0 g) of the ground tissue was mixed thoroughly with anhydrous sodium sulfate. The tissue extraction procedure was essentially the same as that used for the sediment, with the exception that no sulfur clean-up was necessary. Furthermore, lipid digestion was performed on the fish tissue using concentrated H_2SO_4.

17.2.4 Normalization for Total Organic Carbon and Lipid Content

Percent lipids in fish tissue and total organic carbon (TOC) in sediment were determined. Lipids were extracted from a 5 g (±0.5 g) tissue sample from each fish using 10 ml of 50:50 methylene chloride:acetone (v/v). Each sample was sonicated for 30 sec, filtered through a Whatman 41 filter, thoroughly rinsed with solvent, dried at 50°C overnight, and the resulting lipid weighed. TOC analysis was performed at the National Oceanographic and Atmospheric Administration (NOAA) Great Lakes Environmental Research Laboratory in Ann Arbor, MI, using protocols detailed in Harkey et al. (1994). The method detection limit (MDL) for all compounds analyzed in sediment and fish tissue was 1.00 μg/kg.

17.2.5 Fish Assemblage Collection

Fish assemblages were collected at all 20 sites by upstream wading or boating. The reach sampling distance was chosen to be about 15 times the stream width (100 to 300 m), which allowed a complete habitat cycle (run, riffle, pool) to be sampled. In small headwater streams (<10 m wide), a Smith-Root Model 15C backpack electroshocker was used. Fish sampling in the larger streams (>10m wide) was performed using one or a combination of the following methods: backpack electroshocker, Smith-Root tote-barge electroshocker, or Smith-Root boat electroshocker. We collected fish using pulsed direct current (DC) with 2- to 4-amp output. In addition, each reach was sampled with a 4.5-m long, 6.35-mm mesh common minnow seine. Sampling continued until no new species were found by repeat sampling of the reach. We combined our data from seining and electrofishing to represent relative abundance at each site.

Upon collection, fish were held in aerated coolers to keep specimen mortality at a minimum. All fish were identified to species and voucher specimens were housed in the Environmental Toxicology Core Facility at Wichita State University. Thirty individuals of each species were weighed and measured for both standard and total length. External anomalies were noted and then the fish were returned to the river or stream. Any remaining individuals not processed were identified, counted, and checked for anomalies including deformities, eroded fins, lesions, and tumors (collectively called DELT) (Sanders et al., 1999).

17.2.6 Index of Biotic Integrity

The biological expectations for this study were based on reference conditions from the Central Great Plains Ecoregion (Table 17.1; Lydy et al., 2000). Fish assemblage data were grouped into twelve metrics in six categories, including attributes based on species richness and composition, tolerance, trophic guild, abundance, reproductive guild, and individual health and condition (Table 17.1). Each metric received a numerical score of 5, 3, or 1 depending on whether the data were comparable to, deviated somewhat from, or deviated greatly from a reference or least impacted condition (Lydy et al., 2000). A sum of all metrics provided the final IBI score and allowed for placement of the site into an integrity class ranging from very poor to excellent. Each metric individually provided information about a specific characteristic of the sampling site; together they characterized the underlying biological integrity of the site (Karr et al., 1986). The total number

TABLE 17.1
Metrics and Scoring Criteria for an Index of Biotic Integrity Calibrated for the Central Great Plains Ecoregion

	Scoring Criteria		
	5	3	1
Species Richness and Composition			
Total number of species	Varies with drainage basin		
Number of minnow species	≥5	3–4	≤2
Number of centrarchid species	≥3	2	≤1
Number of benthic invertivore species	≥3	2	≤1
Tolerance			
Number of sensitive species	≥4	2–3	≤1
Percent individuals as green sunfish	<15%	15–30%	>30%
Trophic Guilds			
Percent individuals as detritivores	<15%	15–30%	>30%
Percent individuals as invertivores	>40%	20–40%	<20%
Percent individuals as carnivores	>10%	5–10%	<5%
Abundance			
Relative number of individuals	Varies with drainage basin		
Reproductive Guild			
Percent individuals as simple lithophils	>15%	8–15%	<8%
Individual Health and Condition			
Percent individuals with DELT	<0.1%	0.1–1.3%	>1.3%

Sources: Strong, A.J., S.A. Wilkinson, and M.J. Lydy. 1998. *Transactions of the Kansas Academy of Science*, 101, 17–24 and Lydy, M.J., A.J. Strong, and T.P. Simon. 2000. *Archives of Environmental Contamination and Toxicology*, 39, 523–530. With permission.

of species and relative number of individual metrics were found to correlate with the size of the drainage basin (Lydy et al., 2000); therefore, these scores were normalized by drawing maximum species richness (MSR) curves (Strong et al., 1998; Lydy et al., 2000).

17.2.7 Statistics

Relationships among IBI metrics, species diversity indices, and normalized fish tissue and sediment OC concentrations were examined using Spearman's correlation procedures (SAS, 1996). Nonparametric correlations were used, since only 15 to 20 samples were involved and many results were recorded as zeros. Fish assemblage data were examined for trends and patterns using cluster analysis, Bray-Curtis hierarchical agglomerative clustering (Clarke and Warwick, 1994), to find any natural groupings in the similarity data among assemblages found at all sites. The entire species assemblage was analyzed for similarity and this similarity matrix was compared to the normalized sediment and fish tissue OC concentration using Spearman's correlation to determine relationships using the Biota-Environmental (BIOENV) matching subroutine in Primer 5© for Windows (Plymouth Marine Laboratories, U.K.).

17.3 RESULTS AND DISCUSSION

17.3.1 Background and Interpretation

An index of biotic integrity (IBI) is an ecological approach to biomonitoring incorporating multiple parameters of a fish community in a composite index that reflects anthropogenic disturbances in aquatic systems (Karr et al., 1986, Fausch et al., 1990). While the IBI approach is widely accepted, multimetric approaches in general are criticized because of a reported lack of environmental realism (Suter, 1992). Responses to these criticisms were made by Simon and Lyons (1995) and Karr and Chu (1999) and are major driving forces behind this chapter. Still, arguments remain and the strength of IBI interpretation would be increased if real-world relationships were found.

Studies have addressed this problem and assessed the impacts of anthropogenic sources on biological integrity including predominant land use practices in relation to acidity of streams (Hall et al., 1996), contaminated sediments in Chesapeake Bay (Hartwell et al., 1997, Hartwell, 1999), agricultural activities (Karr, 1981), wastewater treatment facilities (Karr et al., 1985), and nonpoint source effects of a landfill (Simon and Stewart, 1998). However, no studies to date have compared a specific chemical class (OC pesticides) to the structure and function of a fish assemblage, with the exception of Eaton and Lydy (2000) who evaluated correlates among organochlorine concentrations and total IBI scores.

One of the major findings of Eaton and Lydy (2000) was that no significant differences were found among total IBI scores at urban versus agricultural sites, but a pattern with higher IBI scores was noted at the agricultural sites. In general, both urban and agricultural land uses have been shown in the literature to impact biological integrity in aquatic systems. For example, Benke et al. (1981) found a negative relationship between benthic species richness and degree of watershed urbanization, while Steedman (1988) found a strong negative relationship between the amount of urban land and IBI scores in streams in Toronto, Ontario.

On the other hand, a study of Michigan streams showed that forested land use had a positive relationship with IBI scores, while predominantly agricultural areas had a negative relationship with IBI scores (Roth et al., 1996). The lack of a significant relationship in the Eaton and Lydy study may be due to the negative effects of both land uses on biotic integrity or the fact that most land uses occur on a larger watershed scale, and their impacts on fish communities in streams often cannot be fully understood by looking only at the surrounding land use.

Eaton and Lydy (2000) also found that total IBI scores were not statistically correlated with OC concentrations in fish or sediment. This finding was not surprising, since degradation observed at a site seldom results from a single chemical contaminant or class of chemicals (Thurmann et al., 1990). Due to their lipophilic nature, OCs tend to accumulate in sediments and lipids, and not in the water column. Carp, a bottom-feeding fish that interacts directly with sediments, was the only species analyzed for OCs.

The IBI assesses the diversity and health of the entire fish community. A better relationship may have been found between IBI scores and OC concentrations if fish from many different habitats (e.g., benthic, pelagic, epipelagic) had been analyzed. Hartwell et al. (1997) found similar results for the Chesapeake Bay when they compared an IBI to a toxicological risk ranking system Hartwell (1997) developed to quantify the degree of risk due to the presence of toxic contaminants. The model reduced an array of ambient toxicity data into a site-specific metric appropriate for comparison with other metrics, such as the IBI or other community diversity indices. Hartwell found that sediment toxicity data were not correlated with IBI scores for the entire fish community, and were strongly correlated only with the IBI scores of the bottom-dwelling fish community.

We further evaluated the Eaton and Lydy data set and found that several individual OC pesticides in sediments were correlated with individual metrics (Table 17.2). Significant correlations were found between the percent individuals as carnivores and gamma BHC, alpha chlordane, gamma chlordane, 4, 4′-DDE, and 4, 4′-DDT. The number of minnow species correlated with 4, 4′-DDD. Total and individual OC pesticides, however were not significantly correlated with the overall IBI.

Individual organochlorine pesticides in fish tissues had fewer significant correlations with metrics than those found in sediments (Table 17.3). The relative number of individuals was significantly correlated with 4,4′-DDT and heptachlorepoxide. Overall, the results do not show causation, and suggest a relationship between sediment concentrations of pesticides and the number of minnow and predator species.

TABLE 17.2
Spearman's Correlations of Sediment Organochlorine Concentrations with Metrics and IBI from Samples Collected near Wichita, Kansas

Metric	alfbhc	alfchl	gamchl	dde	ddd	ddt
No. species	**0.411**	–0.183	–0.209	–0.099	–0.353	0.132
No. minnow species	0.233	–0.295	–0.299	**–0.388**	***–0.496***	–0.276
No. centrarchid species	–0.035	0.038	0.000	0.110	–0.051	–0.097
No. benthic invert. species	0.236	–0.143	–0.164	–0.287	–0.240	–0.100
No. sensitive species	–0.118	–0.311	–0.311	**–0.400**	–0.244	–0.229
Percent ind. as green sunfish	0.197	–0.220	–0.248	–0.096	–0.134	–0.085
Percent ind. as detritivores	–0.118	–0.132	–0.153	–0.151	0.216	–0.169
Percent ind. as invertivores	0.153	–0.303	0.321	**–0.396**	–0.268	–0.252
Percent ind. as carnivores	***0.469***	***0.555***	***0.473***	***0.648***	0.373	***0.604***
Relative no. individuals	0.095	0.071	0.096	0.122	–0.166	0.109
Percent ind. as simple lithophils	0.080	0151	0.172	0.178	–0.240	0.051
Percent ind. with DELT	**0.393**	0.068	0.100	0.009	–0.315	–0.074
IBI	**0.408**	–0.090	–0.117	–0.124	–0.338	–0.104

Note: N = 20. Bolded entries are significantly correlated (P < 0.10). Bolded and underlined entries are significantly correlated (P < 0.05). Column headings: alfbhc = alpha BHC (1,2,3,4,5,6-hexachlorocyclohexane), alfchl = alpha chlordane, gamchl = gamma chlordane, dde = 4,4′ DDE, ddd = 4,4′ DDD, ddt = 4,4′ DDT. Other contaminants did not produce enough positive responses to be included in analysis.

TABLE 17.3
Spearman's Correlations of Fish Tissue Organochlorine Concentrations with Metrics and IBI from Samples Collected near Wichita, Kansas

Metric	gambhc	alfchl	gamchl	dde	ddt	hepepo	endket
No. species	–0.268	–0.114	–0.033	0.253	0.270	0.327	0.106
No. minnow species	0.013	–0.169	–0.255	0.008	0.230	0.271	0.096
No. centrarchid species	–0.344	–0.186	–0.138	–0.164	–0.351	–0.378	–0.166
No. benthic invert. species	–0.291	–0.155	–0.189	0.156	0.350	0.389	0.159
No. sensitive species	**–0.475**	–0.254	–0.173	–0.33	–0.047	–0.189	0.111
Percent ind. as green sunfish	0.166	0.018	–0.170	0.406	0.247	0.133	0.195
Percent ind. as detritivores	–0.301	–0.302	–0.291	–0.339	0.044	0.214	0.314
Percent ind. as invertivores	–0.236	0.065	0.033	0.221	0.270	0.327	**0.479**
Percent ind. as carnivores	–0.289	0.209	0.235	–0.359	–0.247	–0.133	–0.195
Relative no. individuals	–0.175	0.098	0.156	**0.444**	***0.559***	***0.576***	0.359
Percent ind. as simple lithophils	0.091	0.372	0.223	0.284	–0.085	–0.240	–0.352
Percent ind. with DELT	**0.459**	0.120	0.081	0.434	0.143	0.134	–0.078
IBI	–0.175	–0.076	–0.106	0.220	0.207	0.311	0.200

Note: N = 15. Bolded entries are significantly correlated (P < 0.10). Bolded and underlined entries are significantly correlated (P < 0.05). gambhc = gamma BHC (1,2,3,4,5,6-hexachlorocyclohexane), alfchl = alpha chlordane, gamchl = gamma chlordane, dde = 4,4′ DDE, ddt = 4,4′ DDT, hepepo = heptachlor epoxide, endket = endrin ketone. Other contaminants did not produce enough positive responses to be included in analysis.

17.3.2 Multivariate Evaluation of Fish Assemblage Structure and OC Pesticide Contaminants

Bray-Curtis similarity of the fish assemblage at all sites showed three main clusters (Figure 17.2). Cluster 1 was comprised of sites 7, 19, 18, 8, and 20. This cluster had the lowest mean IBI score (25), and included agricultural and urban sites. The common characteristic of these sites was significant habitat modification, coupled with the direct impact of urban and/or agricultural land use. Habitat modifications included channelization for floodway development and removal of in-stream and riparian habitats. In addition to habitat modification, these sites were unique in that they were the only ones directly impacted by new residential development directly adjacent to the streams and/or local livestock operations.

The second cluster (sites 6, 9, 15, 10, and 11) was part of a larger group and included small tributary sites located in row crop agricultural areas. These sites were not channelized or dammed, and most had fairly intact riparian zones. The only major impacts on these sites other than row crop agriculture were fluctuations in water levels in the summer months. These sites had a much higher aggregate IBI score (34) in comparison to cluster 1. The third cluster was comprised of two related groups (1, 2, 13, 16 and 12, 17, 3, 5, 4, and 14). These sites were all main stem Arkansas and Little Arkansas sites, mostly from urban settings, that were partially channelized and dredged. The first group was located exclusively on the Little Arkansas River, while the second group contained a combination of Little and Big Arkansas River sites. Mean IBI scores were 31 and 39, respectively. The higher IBI scores associated with the second group probably resulted because these sites still retained fair amounts of in-stream habitat and had less direct urban impacts.

The bioenvironmental analysis (BIOENV), which was an examination of the Bray-Curtis similarity matrix of the entire fish assemblage, at all 20 sites and the sediment OC data showed that only 4,4′-DDD was significantly correlated with the fish similarity matrix ($r = 0.487$, $p = 0.032$, $n = 20$). This correlation was fairly good, considering the small sample size, and suggests some degree of relationship may exist between fish assemblages and sediment OCs. There was low correlation (ns) between the fish similarity matrix and OCs in fish tissues.

Overall, we were unable to distinguish between IBI scores from agricultural versus urban sites. Both disturbance types cause a decline in IBI with such response signatures as a reduction in native species, an increase in exotics, and declines in other metrics. The fact that in some cases the urban and rural sites clustered together lend support to the argument that both types of impairment push the fish assemblage down a similar response indicator path. The watershed had no least impacted sites to compare with the urban and rural sites.

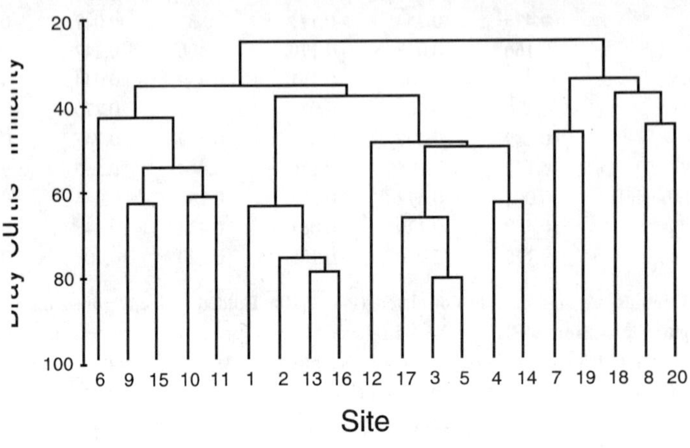

FIGURE 17.2 Cluster analysis of fish assemblage similarity matrix for 20 sampling sites.

17.4 CONCLUSIONS

This chapter further examined the data collected by Eaton and Lydy (2000) and evaluated fish and sediment OC pesticide concentrations in relationship to individual IBI metrics and the structure and function of the fish assemblage. Results indicated some correlations among individual IBI metrics (e.g., percent individuals as carnivores and number of minnow species) and individual OC pesticides in fish and sediment, but the total number of significant correlations was small and the significance could be due simply to experiment-wide type I error. The bioenvironmental analysis (BIOENV) showed that 4,4′DDD was significantly correlated with the fish similarity matrix, suggesting some degree of relationship between fish assemblages and sediment OCs, but 4,4′DDD was the only chemical that showed a significant relationship.

Finally, the Bray-Curtis similarity results indicated three major clusters. The importance of OC pesticides in defining those clusters is not clear. It appears that a number of factors are controlling the fish community including anthropogenic factors coupled with land use activites. Even though we did not find a number of meaningful relationships among fish and sediment OC pesticide concentrations, individual IBI metrics, and the structure and function of the fish assemblage, this is still important because our *a priori* hypotheses showed significant relationships.

The number of species, number of minnow species, number of sensitive species, percent individuals as invertivores and carnivores, and those with deformities, eroded fins, lesions, and tumors (DELT anomalies) all showed significant correlations with one or more sediment pesticides. Concentrations of fish tissue OC concentrations were significantly correlated with number of sensitive species, percent individuals as invertivores, relative number of individuals, and percent individuals with DELT anomalies. Future projects should include a full habitat assessment coupled with consideration of spatial distribution patterns of land use and a more in-depth OC assessment over multiple feeding guilds.

ACKNOWLEDGMENTS

The opinions expressed do not necessarily represent those of the U.S. Fish and Wildlife Service. No official endorsement by that agency should be inferred.

REFERENCES

Bishop, C.A., B. Collins, P. Mineau, N.M. Burgess, W.F. Read, and C. Risley. 2000. Reproduction of cavity-nesting birds in pesticide-sprayed apple orchards in southern Ontario, Canada, 1988–1994, *Environmental Toxicology and Chemistry*, 19(3), 588–599.

Benke, A.C., G.E. Willke, F.K. Parrish, and D.L. Stites. 1981. *Effects of urbanization on stream ecosystems*, ERCO7–81. Georgia Institute of Technology, Atlanta.

Burton, G.A., Jr. 1991. Assessing the toxicity of freshwater sediments, *Environmental Toxicology and Chemistry*, 10, 1585–1627.

Clarke, K.R. and R.M. Warwick. 1994. *Change in Marine Communities: An Approach to Statistical Analysis and Interpretation*, Natural Environment Research Council, U.K.

Eaton, H.J. and M.J. Lydy. 2000. Assessment of water quality in Wichita, Kansas, using an Index of Biotic Integrity and analysis of bed sediment and fish tissue for organochlorine insecticides, *Archives of Environmental Contamination and Toxicology*, 39, 531–540.

Fausch K.D., J. Lyons, J.R. Karr, and P.L. Angermeier. 1990. Fish communities as indicators of environmental degradation, *American Fisheries Society Symposium*, 8, 123–144.

Hall, L.W., Jr., M.C. Scott, W.D. Killen, and R.D. Anderson. 1996. The effects of land-use characteristics and acid sensitivity on the ecological status of Maryland coastal plain streams, *Environmental Toxicology and Chemistry*, 15, 384–394.

Harkey G.A., M.J. Lydy, J. Kukkonen, and P.F. Landrum. 1994. Feeding selectivity and assimilation of PAH and PCB in *Diporeia* spp., *Environmental Toxicology and Chemistry*, 13, 1445–1455.

Hartwell, S.I. 1997. Demonstration of a toxicological risk ranking method to correlate measures of ambient toxicity and fish community diversity, *Environmental Toxicology and Chemistry*, 16, 361–371.

Hartwell, S.I., C.E. Dawson, E.Q. Durell, R.W. Alden, P.C. Adolphson, D.A. Wright, G.M. Coelho, J.A. Magee, S. Ailstock, and M. Norman. 1997. Correlation of measures of ambient toxicity and fish community diversity in Chesapeake Bay, USA, tributaries urbanizing watersheds, *Environmental Toxicology and Chemistry*, 16(12), 2556–2567.

Hartwell, S.I. 1999. Empirical assessment of an ambient toxicity risk ranking model's ability to differentiate clean and contaminated sites, *Environmental Toxicology and Chemistry*, 18(6), 1298–1303.

Hoffman, D.J., B.A. Rattner, G.A. Burton, Jr., and J. Cairns, Jr. 1995. *Handbook of Ecotoxicology*. Lewis Publishers, Boca Raton, FL, 424–468.

Karr, J.R. 1981. Assessment of biotic integrity using fish communities. *Fisheries*, 6, 21–27.

Karr, J.R. and E.W. Chu. 1999. *Restoring Life in Running Waters: Better Biological Monitoring*, Island Press, Covelo, CA.

Karr, J.R., R.C. Heidinger, and E.H. Helmer. 1985. Effects of chlorine and ammonia from wastewater treatment facilities on biotic integrity, *Journal of the Water Pollution Control Federation*, 57, 912–915.

Karr, J.R., K.D. Fausch, P.L. Angermeier, P.R. Yant, and I.J. Schlosser I.J. 1986. *Assessing the Biological Integrity in Running Waters: A Method and its Rationale*, Illinois Natural History Survey Special Publication 5, Springfield.

Kawano, M., T. Inoue, T. Wada, H. Hidaka, and R. Tatsukawa. 1988. Bioconcentration and residue patterns of chlordane compounds in marine animals: invertebrates, fish, mammals and seabirds, *Environmental Science and Technology*, 22, 792–801.

Kubiak, T.J., H.J. Harris, L.M. Smith, T.R. Schwartz, D.I. Stalling, J.A. Trick, L. Sileo, D.E. Docherty, and T.C. Erdman. 1989. Microcontaminants and reproductive impairment of the Foster's tern on Green Bay, Lake Michigan,1983, *Archives of Environmental Contamination and Toxicology*, 18, 706–715.

Leatherland, J.F. and R.A. Sonstegard. 1978. Lowering of serum thyroxine and triiodothyronine levels in yearling coho salmon (*Oncorhynchus kisutch*) by dietary mirex and PCBs, *Journal of Fisheries Research Board of Canada*, 4, 9–15.

Lydy, M.J., A.J. Strong, and T.P. Simon. 2000. Development of an Index of Biotic Integrity for the Little Arkansas River Basin, Kansas, *Archives of Environmental Contamination and Toxicology*, 39, 523–530.

Nebeker, A.V., F.A. Puglisi, and D.L. Dofoe. 1974. Effect of polychlorinated biphenyl compounds on survival and reproduction of the fathead minnows and flagfish, *Transactions of the American Fisheries Society*, 103, 562–568.

Nowell, L.H. and E.A. Resek. 1994. Summary of National Standards and Guidelines for Pesticides in Water, Bed Sediment, and Aquatic Organisms and Their Application to Water-Quality Assessment, U.S. Geological Survey Open-File Report 94, Washington, D.C.

Pereira, W.E., J.L. Domagalski,, F.D. Hostettler, L.R. Brown, and J.B. Rapp. 1996. Occurrence and accumulation of pesticides and organic contaminants in river sediment, water and clam tissues from the San Joaquin River and tributaries, California, *Environmental Toxicology and Chemistry*, 15, 172–180.

Roth, N.R., J.D. David, and D.L. Erickson. 1996. Landscape influences on stream biotic integrity assessed at multiple spatial scales, *Landscape Ecology*, 11, 141–156.

Saiki, M.K. and C.J. Schmitt. 1986. Organochlorine chemical residues in bluegills and common carp from the irrigated San Joaquin Valley Floor, California, *Archives of Environmental Contamination and Toxicology*, 15, 357–366.

Sanders, R.E., R.J. Miltner, C.O. Yoder, and E.T. Rankin. 1999. The use of external deformities, erosion, lesion, and tumors (DELT anomalies) in fish assemblages for characterizing aquatic resources: a case study of seven Ohio streams, in T.P. Simon (Ed.). *Assessing the Sustainability and Biological Integrity of Water Resources Using Fish Communities*. CRC Press, Boca Raton, FL, 225–248.

SAS Institute, Inc. 1996. *SAS User's Guide: Statistics*. SAS Institute, Cary, NC.

Schmitt, C.J., J.L Zajicek, and P.H. Peterman. 1990. National contaminant biomonitoring program: residues of organochlorine chemicals in U.S. freshwater fish, 1976–1984, *Archives of Environmental Contamination and Toxicology*, 14, 225–231.

Simon, T.P. and J. Lyons. 1995. Application of the index of biotic integrity to evaluate water resource integrity in freshwater ecosystems, in W.S. Davis and T.P. Simon (Eds.). *Biological Assessment and Criteria: Tools for Water Resource Planning and Decision Making*, Lewis Publishers, Boca Raton, FL, 245–262.

Simon, T.P. and P.M. Stewart. 1998. Application of an Index of Biotic Integrity for dunal, palustrine wetlands: emphasis on assessment of nonpoint landfill effects on the Grand Calumet Lagoons, *Aquatic Ecosystem Health and Management,* 1, 63–74.

Strong, A.J., S.A. Wilkinson, and M.J. Lydy. 1998. Fish communities in the Little Arkansas River Basin, Kansas 1884–1996, *Transactions of the Kansas Academy of Science,* 101, 17–24.

Suter, G.W., II. 1992. A critique of ecosystem health concepts and indexes, *Environmental Toxicology and Chemistry,* 12, 1533–1539.

Steedman, R.J. 1988. Modification and assessment of an index of biotic integrity to quantify stream quality in southern Ontario, *Canadian Journal of Fisheries and Aquatic Sciences,* 45, 492–501.

Thurmann, E.M., M.T. Meyer, M.L. Pomes, C.A. Perry, and A.P. Schwab. 1990. Enzyme-linked immunoassay compared with gas chromatography/mass spectrometry for the determination of triazine herbicides in water, *Analytical Chemistry,* 62, 2043–2048.

U.S. Environmental Protection Agency. 1980. Modification of Mills, Onley, Gaither method for the determination of multiple organochlorine pesticides and metabolites in human or animal adipose tissue. In *Manual of Analytical Methods for the Analysis of Pesticides in Human and Environmental Samples.* Section 5, A, (1), (a), EPA-600/8–801–038, Washington, D.C.

United States Environmental Protection Agency. 1986. *Method #3550, Sonication Extraction. Test Methods for Evaluating Solid Waste: Physical/Chemical Methods.* EPA-SW846, Office of Solid Waste and Emergency Response, Washington, D.C.

United States Environmental Protection Agency. 1990. *Method #8080A, Organochlorine Pesticides and Polychlorinated Biphenyls by Gas Chromatography. Test Methods for Evaluating Solid Waste, Physical/Chemical Methods.* 3rd ed., Proposed Updates I and II, EPA-SW846 Office of Solid Waste and Emergency Response, Washington, D.C.

United States Geological Survey. 1997. Occurrence of selected organochlorine compounds in fish tissue from eastern Iowa streams, 1995. Fact Sheet FS-027–97, Iowa City, IA.

18 Effects of Silviculture on Indices of Biotic Integrity for Benthic Macroinvertebrate and Fish Assemblages in Northeastern Minnesota's Northern Lakes and Forest Ecoregion (USA)

Thomas P. Simon and Joseph A. Exl

CONTENTS

18.1 INTRODUCTION

Few boreal waters are managed in a sustainable manner since cumulative effects are generally not considered (Schindler 1998). Fisheries and water quality have declined in most large water bodies of the southern boreal forest; however, some effects are results of overexploitation, alteration of

0-8493-0905-0/03/$0.00+$1.50

flow patterns, introductions of non-native species, and discharge of euthrophic nutrients and persistent contaminants (Bodaly et al., 1984a, b, 1984, 1993).

Relatively few aquatic and terrestrial environments are not impacted by human activities to some degree. Thus, the ability to determine biological integrity becomes more difficult (Davis and Simon, 1995; Hughes, 1995). Few pristine or presettlement areas remain, making the ability to determine reference conditions for a region paramount to the success of determining biological integrity for a class of water bodies. Karr and Dudley (1981) defined biological integrity as "the capability of supporting and maintaining a balanced, integrated, adaptive community of organisms having species composition, diversity, and functional organization comparable to that of natural habitats of the region."

Northeastern Minnesota's Northern Lakes and Forest (NLF) Ecoregion has a minimally impacted landscape. Most of the interior is sparsely populated or uninhabited; most development is located only near the Lake Superior shoreline (Minnesota Planning Land Management Information Center). The NLF Ecoregion is known for its scenic beauty and outdoor recreational opportunities within the many state and national forests. Beyond the sparse population, the main disturbances in this area are atmospheric deposition, silviculture, mineral exploration, and beavers.

This chapter will test whether multimetric indices are capable of determining differences in stream conditions between natural disturbances caused by beavers from human disturbance caused by silviculture. We wanted to determine whether multimetric indices can differentiate between human and natural disturbances within watersheds.

18.2 BACKGROUND

Human activity within a watershed can alter seasonal and long-term variation of lake and stream hydrology and chemistry (Schindler, 1998). Improper management of watersheds and airsheds also causes degradation of aquatic ecosystems. Frelich and Reich (1998) tested three disturbance severity models for boreal forests, (1) continuous, where changes in disturbance severity cause proportional and continuous changes in stand composition, (2) discontinuous, where a threshold disturbance exists and coexistence of two alternative compositional states is possible, and (3) the cusp, where thresholds exist and coexistence of two alternative compositional states is possible at the same disturbance severity. They found that most forest management models assume forest succession is a static process and that forests repeat the same regeneration and successional stages after each disturbance. These changes in the forest also affect the resident biological assemblages (Exl and Simon, in press).

Clear-cut logging, climatic warming, acid precipitation, and stratospheric ozone depletion are among the more important indirect stressors. Clear-cut logging tends to increase flow rates, sedimentation, and chemical inputs with most effects disappearing within 2 to 3 years (Nicolson, 1975). Clear cutting in Minnesota has caused higher water levels in wetlands (Verry, 1986). Additionally, ecosystems influenced by humans usually exhibit reduction in resilience to natural disturbances such as droughts, fire, pests, and disease (De Leo and Levin, 1997). Schindler (1998) indicated that most boreal lakes contain 2 to 5 times background mercury levels as a result of atmospheric deposition.

Beavers are considered natural disturbances in northern watersheds since they modify biological, physical, and chemical characteristics. Beaver dam construction effects hydrology by blocking the stream flow to create ponds. This inundation of water behind the dams destroys certain habitats and significantly modifies the riparian zone.

Boreal freshwaters contain very few species, making them very susceptible to perturbations (Schindler, 1990). Unproductive lakes result in relatively few species and slow growth rates. Mature fish found in these systems devote most of their energy to reproduction. Benthic invertebrate communities are similarly poor in species richness. Small lakes may often contain no more than a few hundred species (Schindler et al., 1989). Some species may have no functional replacement with similar food and habitat requirements (Schindler, 1998). Loss of species within a functional group may reduce long term-stability and produce noticeable changes in short-term system dynamics (Levin, 1997).

18.3 MATERIALS AND METHODS

18.3.1 Study Area Description

The study area is in the "arrowhead" region of Minnesota (Figure 18.1); it is bordered by Canada to the north and Lake Superior to the south. The NLF Ecoregion occupies an area of 180,500 km^2 in Northern Minnesota, Wisconsin, and Michigan. The landscape has been modified by glaciation events that scoured and deposited large amounts of glacial till, creating till plains, terminal moraines, outwash plains, and kettle basins (Omernik et al., 2000). These glacial features and the bedrock ridges that remain typify the study area. The land cover is dominated by large expanses of wetlands, lakes and conifer/deciduous forests on nutrient poor soils (Table 18.1).

Lowland streams in the NLF include palustrine and riverine wetland systems. Trees, shrubs, and persistent emergent vegetation dominate the palustrine systems. The riverine systems include all wetlands and deep-water habitats within the stream channel. Unlike the palustrine system, trees, shrubs, and persistent emergent vegetation do not dominate riverine systems. Beaver activity constantly modifies these systems through construction and modification of existing dams and creating various sized ponds, depending on topography. Beaver ponds serve as barriers to fish

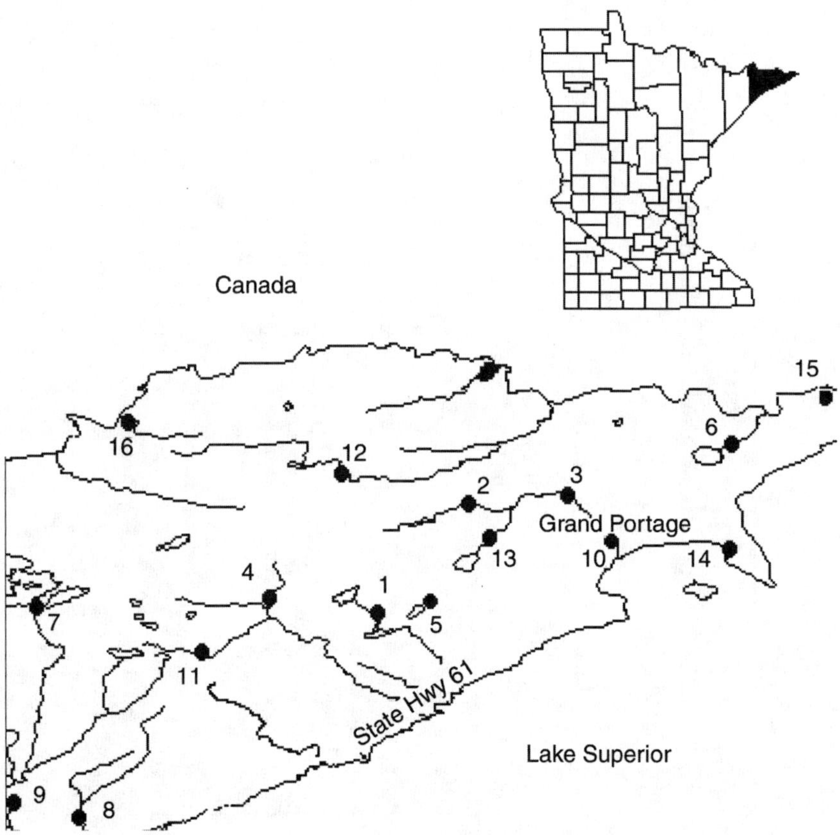

FIGURE 18.1 Map of the "arrowhead" region of northeastern Minnesota and the locations of sites used to determine biological integrity.

TABLE 18.1
Index Scores, Land Use, Stream Modification Type, and Biotic Integrity Ratings for Sixteen Watersheds Sampled in Streams from Northeastern Minnesota

Site	Name	Land Use	Flood Plain Quality	Stream Modification	QHEI	BCI	Biotic Integrity Rating	IBI	Biotic Integrity Rating
1	Center Lake Tributary	Timber	Wetland	None	53	12	Poor	13	Very Poor
2	Mount Maude Tributary	Timber	Wetland	Beaver Dam	58	10	Poor	0	Very Poor
3	Grand Portage Creek Tributary	Natural	Forest	None	75	36	Good	0	Very Poor
4	Hollow Rock Creek Tributary	Natural	Wetland	None	60	14	Poor	36	Fair
5	Cuffs Lake Outlet	Natural	Forest	None	64	34	Good	0	Very Poor
6	Teal Lake Stream	Bog	Wetland	None	63	12	Poor	54	Good
7	Swamp Lake Tributary	Timber	Forest, Wetland	None	55	12	Poor	49	Fair
8	Sleepy Hollow Creek	Timber	Forest	Beaver Dam	70	14	Poor	0	Very Poor
9	Reservation River Tributary	Natural	Forest, Wetland	Beaver Dam	71	32	Good	50	Good
10	Grand Portage Creek Tributary	Natural	Forest	None	73	46	Exceptional	0	Very Poor
11	Hollow Rock Tributary	Natural	Forest, Wetland	None	42	12	Poor	0	Very Poor
12	Chevans Outlet	Timber	Forest, Wetland	Beaver Dam	72	24	Fair	50	Good
13	Dutchman Impoundment	Timber	Forest, Wetland	Impounded	64	14	Poor	21	Very Poor
14	Little Creek		Forest, Wetland	Beaver Dam	72	20	Poor	13	Very Poor
15	Pigeon Point Tributary	Backfilled	Wetland	None	48	10	Poor	16	Very Poor
16	Monument Creek	Natural	Forest, Wetland	Beaver Dam	64	12	Poor	0	Very Poor

movement, cause an increase in temperature and a decrease in dissolved oxygen levels, and can increase sunlight penetration and algal production as canopy cover is lost.

A total of sixteen watersheds were sampled in the study area. Timber harvests occurred in six of the watersheds. Center Lake Tributary (Site 1), Mount Maude Tributary (Site 2), Swamp Lake Tributary (Site 7), Sleepy Hollow Creek (Site 8), Chevans Outlet (Site 12) and Dutchman Impoundment (Site 13). Grand Portage Creek Tributary (Site 3), Hollow Rock Creek Tributary (Site 4), Cuffs Lake Outlet (Site 5), Reservation River Tributary (Site 9), Grand Portage Creek Tributary (Site 10), Hollow Rock Tributary (Site 11), and Monument Creek (Site 16) were considered undisturbed watersheds. Teal Lake Stream (Site 6) could be considered undisturbed, but was located in a bog. A collapsing bank from an adjacent parking lot influenced Pigeon Point Tributary (Site 15). Land use information was unavailable for Little Creek (Site 14) due to a lack of longitude/latitude coordinates. Sites were further influenced by wetlands, forests, or both (Table 18.1). Watersheds 2, 8, 9, 12, 14, and 16 had single or multiple beaver dam constructions within the sampling reach.

18.3.2 Design Approach and Collection

We evaluated stream condition in least impacted, naturally disturbed, and anthropogenically modified watersheds with varying levels of human disturbance using a paired watershed approach. Variability within biological indicators was determined at nonhuman impact sites and least impacted sites.

Macroinvertebrate communities were collected using a multihabitat composite sample at each site (Chirhart, 1998; Butcher et al., in review). Sampling occurred in an upstream manner, covering a distance determined by multiplying the average stream width by 35. A total of 20 sampling efforts occurred at each site using a D-frame net with 600-micron mesh. Each sampling effort consisted of two D-frame sweeps in a common habitat. The 20 sampling efforts were divided into broad habitat types: (1) hard bottom, (2) vegetation, (3) undercut banks, (4) snags, (5) leaf packs, and (6) soft bottom. Samples were collected in proportion to habitat type availability based on preliminary observations of the stream segment, with the exception of soft bottom habitats. For example, a segment consisting of 50% vegetation would incorporate 10 sampling units from this habitat type. No more than two efforts were given to soft bottom habitats because of their nonproductivity. Sites with approximately 50% soft bottom received one effort. Those sites comprising more than 50% soft bottom habitat received two efforts.

After sampling was completed at a site, the samples were composited and large pieces of debris were removed. Each sample was transferred to a sieve to remove fine particulate organic matter. The remaining material was then transferred to a sample container and filled to the top with ethyl alcohol. Site labels were placed on the inside and outside of each container. In the laboratory, a 300-organism subsample was obtained from the debris. All organisms were identified to the lowest practical level for analysis.

Fish sampling covered the same stream reaches where the macroinvertebrate samples were collected (Niemela and Feist, 2000; Wang et al., 1998). Fish were obtained in an upstream manner with a Wisconsin model battery-operated backpack electrofisher. All fish were placed in a live well until the reach was finished or overcrowding occurred. Fish were identified, batch weighed (to the nearest tenth of a gram) and a minimum and maximum size range (nearest millimeter) was determined for each species. Fish were then returned to the stream except for voucher specimens, which were placed into appropriately labeled sample containers.

After macroinvertebrate and fish sampling, a qualitative habitat analysis was done along each stream reach. Information was obtained on its riparian zone, instream habitat availability, and channel morphology. In addition, basic water chemistry was also obtained via a multiparameter water quality meter.

18.3.3 Indicators

Two indices were tested: the benthic community index (BCI) (Butcher et al., in review) and the index of biotic integrity (IBI) (Niemela and Feist, 2000). Both use the structure and function of stream macroinvertebrate and fish assemblages, and sum structural (e.g., number of Diptera or Cyprinids), functional (e.g., number of scrapers or simple lithophils), and conditional metrics (e.g., HBI or percent DELT anomalies) to determine a final score. Each metric corresponds to a biological attribute that responds in a predictable manner to varying levels of human perturbation.

The IBI was calibrated on a watershed scale, while the BCI used an ecoregion framework. Niemela and Feist (2000) calibrated the fish IBI based on data from the St. Croix River basin to develop expectations for the NLF. Butcher et al. (in review) used benthic macroinvertebrate data from the entire NLF in Minnesota, Michigan, and Wisconsin. This application of the IBI and BCI went beyond the original authors' intended use of the indices. Butcher et al. (in review) suggested the BCI was not adapted for use in riverine wetland habitats. In addition, our application of the IBI for fish assemblages was outside the area where it was calibrated. Both the IBI and BCI represented the best indices available for the NLF.

The qualitative habitat evaluation index (QHEI) was developed by the Ohio Environmental Protection Agency to evaluate physical habitat in lotic systems (OEPA, 1989). The QHEI is based on the sum of metric scores for sediment composition, instream cover, riparian quality, pool and riffle quality, and erosion to obtain a habitat score. In addition to habitat and biotic measures, water chemistry data were collected at each site. Measurements included, pH, dissolved oxygen (mg/L), conductivity (mS/cm), salinity (%), turbidity (NTU), and water temperature (°C). This information was used to correlate with patterns in IBI, BCI, and QHEI findings.

18.3.4 Statistics

We used nonparametric Spearman rank order correlation analysis to explore interrelationships of individual water quality parameters, land use, index scores, and habitat information (Sokal and Rohlf, 1995). Interquartile box and whisker plots were generated to compare individual patterns. Analyses were preformed using Statistica software (StatSoft, Inc., 1995).

18.4 RESULTS

18.4.1 Patterns in Indicator Responses

Water chemistry varied with both land use and stream modification (Figures 18.2 and 18.3, and Table 18.2). In watersheds where logging occurred, decreases in median values were observed for salinity, conductivity, and pH, while increased stream temperatures compared to those for least impacted watersheds. Turbidity was 60% lower and dissolved oxygen was 11.4% higher in logged watersheds. Watersheds with beaver activity showed reductions in median pH, turbidity, and dissolved oxygen and increased in stream temperature when compared to nonbeaver sites. No strong correlation was observed between water quality parameters and land use ($n = 16$, $p > 0.05$), with the exception of stream temperature, which positively correlated with stream modification ($n = 16$, $rs = 0.4995$, $p < 0.05$).

Habitat measures varied with land use and degree of stream modification (Figures 18.4 and 18.5, and Table 18.2). In watersheds where logging occurred, median stream width and depth increased while QHEI scores decreased, when compared to natural watersheds. Beaver activity caused an increase in stream width and depth. However, QHEI scores increased with beaver activity. Land use was negatively correlated with stream depth ($n = 15$, $rs = -0.5341$, $p < 0.05$). Stream modification positively correlated with stream width ($n = 16$, $rs = 0.5204$, $p < 0.04$). BCI scores were more responsive to watershed and instream modification than the IBI (Tables 18.1, 18.3, and 18.4).

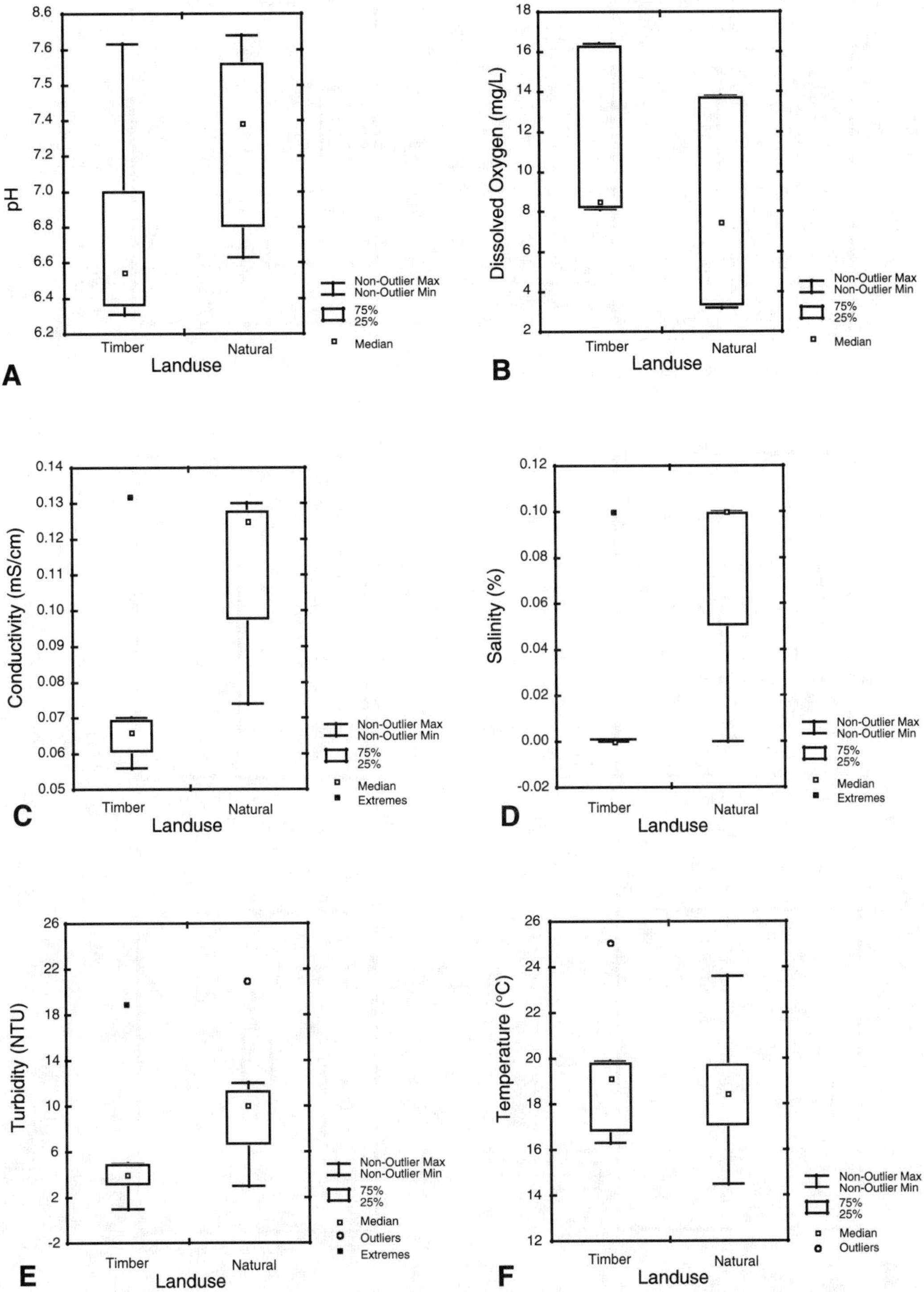

FIGURE 18.2 Inter-quartile box and whisker graphs showing the relationships of individual water chemistry parameters as influenced by land use within the watershed. A = pH, B = dissolved oxygen, C = conductivity, D = salinity, E = turbidity, and F = temperature.

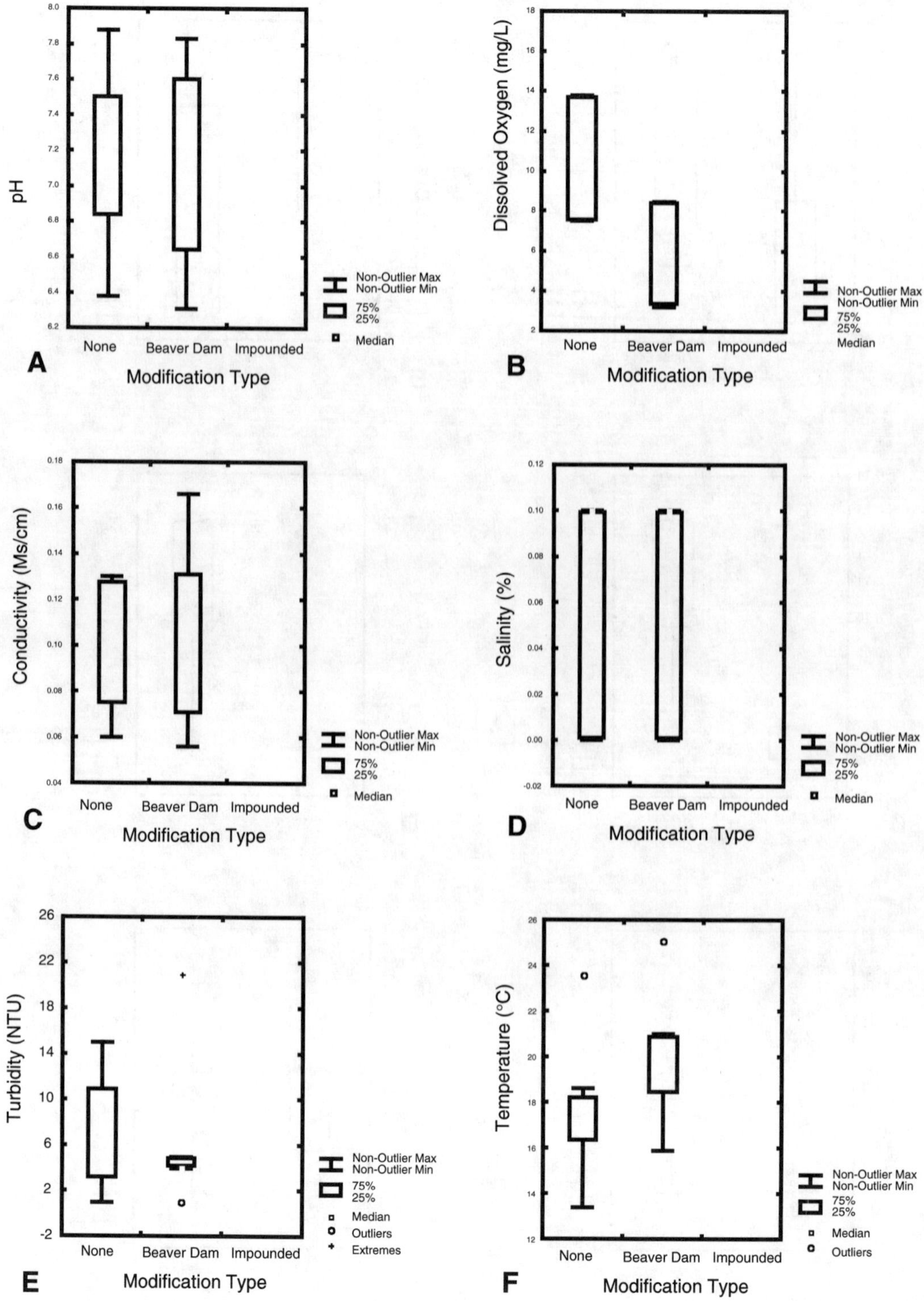

FIGURE 18.3 Inter-quartile box and whisker graphs showing the relationships of individual water chemistry parameters as influenced by three stream modification types. A = pH, B = dissolved oxygen, C = conductivity, D = salinity, E = turbidity, and F = temperature.

TABLE 18.2
Physical and Water Chemistry Values for Each of the Sixteen Watershed Sampled in Streams for Northeastern Minnesota

Site	Name	Stream Width (m)	Stream Depth (m)	Flow (ft/sec)	pH	Dissolved Oxygen (mg/L)	Conductivity (Ms/cm)	Salinity (%)	Turbidity (NTU)	Temperature (°C)
1	Center Lake Tributary	1.1	1	0.01	6.38	NA	0.068	0	3	16.3
2	Mount Maude Tributary	18.4	1	0.01	7.83	NA	0.07	0	4	25.1
3	Grand Portage Creek Tributary	1.2	0.1	0.125	7.52	NA	0.128	0.1	11	16.7
4	Hollow Rock Creek Tributary	2	0.7	0.001	6.83	NA	0.13	0.1	12	17.4
5	Cuffs Lake Outlet	0.6	0.2	0.03	7.88	NA	0.074	0	3	18.3
6	Teal Lake Stream	3.3	1	0.01	7.25	NA	0.128	0.1	9	23.6
7	Swamp Lake Tributary	3.2	1	0.02	7.01	NA	0.06	0	1	16.8
8	Sleepy Hollow Creek	1.2	0.4	0.001	6.72	8.12	0.132	0.1	5	19.9
9	Reservation River Tributary	6.2	1	0.02	7.61	NA	0.128	0.1	4	21
10	Grand Portage Creek Tributary	1	0.3	1	7.85	13.78	0.114	0.1	11	18.6
11	Hollow Rock Tributary	0.4	0.15	Intermittent	6.76	7.51	0.08	0	9	14.5
12	Chevans Outlet	15.3	1	0.25	6.31	8.48	0.056	0	4	18.4
13	Dutchman Impoundment	2.8	0.8	1.1	6.35	16.42	0.064	0	19	19.9
14	Little Creek	1.2	0.2	0.001	6.83	NA	0.166	0.1	1	15.9
15	Pigeon Point Tributary	0.3	0.2	0.01	7.08	8.3	0.124	0.1	15	13.4
16	Monument Creek	1.9	0.6	0.001	6.63	3.2	0.122	0.1	21	18.6

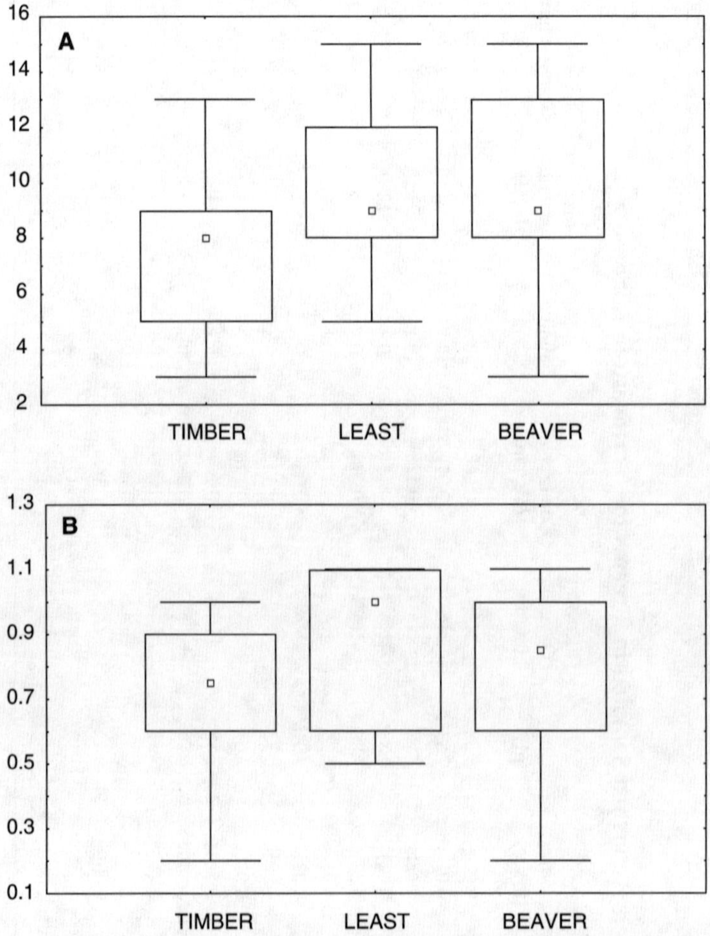

FIGURE 18.4 Inter-quartile box and whisker graphs showing the relationships of least impacted, timbered, and beaver-influenced watersheds for select BCI metrics. A = number of species, B = Shannon-Weiner diversity index.

18.4.2 Multimetric Indices Response

18.4.2.1 Benthic Macroinvertebrate Assemblage

Species Richness and Composition — Species richness is perhaps one of the single most important metrics that explains differences between disturbed and least impacted sites (DeShon, 1995). Typically, the higher the number of species, the higher the biological integrity. The number of species, Shannon-Weiner diversity index, number of Ephemeroptera, percent individuals as Trichoptera, and percent individuals as crustacea and mollusca were not significantly different among timbered, least impacted, and beaver-influenced watersheds (Figures 18.4 and 18.5).

Tolerance and Sensitive Species — As disturbance gradients increase, the number or percentage of tolerant species increases and the number of sensitive species declines. The number of Diptera species metric was statistically significant between the beaver-influenced and timbered watersheds (ANOVA, df = 3, F = 31.333, p = 0.031), but not between least impacted and either the timbered (p = 0.637) or beaver-influenced (p = 0.674) watersheds. The percent individuals as Trichoptera was significantly different (ANOVA, df = 3, F = 50.29167, p = 0.10) between least impacted and beaver-influenced watersheds.

TABLE 18.3
Benthic Community Index (BCI) Scores and Individual Metric Values for Sixteen Watersheds from Northeastern Minnesota Streams

Site	# of Ephemeroptera	# of Diptera	Richness	Shannon-Weiner diversity	% Trichoptera	% Crustacea and Mollusca	# of Filterers	# of Scrapers	# of EPT	HBI	BCI	Narrative
1	0(1)	1(1)	5(1)	0.7(1)	7(3)	63(1)	1(1)	1(1)	1(1)	7.28(1)	12	Poor
2	3(1)	1(1)	9(1)	0.2(1)	1(1)	69(1)	1(1)	2(1)	4(1)	7.85(1)	10	Poor
3	3(3)	4(1)	11(3)	1(5)	61(5)	7(3)	6(3)	5(3)	15(5)	2.51(5)	36	Good
4	1(1)	3(1)	9(3)	0.5(1)	1(1)	90(1)	1(1)	3(3)	3(1)	7.55(1)	14	Poor
5	2(3)	3(1)	9(3)	1.1(5)	29(5)	9(3)	5(3)	6(5)	11(3)	5.08(3)	34	Good
6	0(1)	0(1)	7(1)	0.9(3)	0(1)	82(1)	1(1)	2(1)	0(1)	7.27(1)	12	Poor
7	1(1)	0(1)	8(1)	0.9(3)	2(1)	67(1)	1(1)	2(1)	3(1)	7.39(1)	12	Poor
8	0(1)	1(1)	3(1)	0.6(1)	6(3)	75(1)	1(1)	2(3)	1(1)	6.16(1)	14	Poor
9	4(1)	6(3)	15(5)	1.1(3)	48(5)	19(1)	9(5)	5(3)	14(3)	4.85(3)	32	Good
10	5(5)	3(1)	12(5)	1(5)	73(5)	2(5)	7(5)	6(5)	16(5)	2.81(5)	46	Exceptional
11	0(1)	2(1)	5(1)	0.6(1)	1(1)	76(1)	1(1)	2(3)	1(1)	7.73(1)	12	Poor
12	2(1)	6(3)	13(5)	1(3)	8(3)	47(1)	6(3)	2(1)	9(3)	6.6(1)	24	Fair
13	0(1)	3(1)	9(1)	0.8(3)	2(1)	72(1)	4(3)	1(1)	5(1)	7.75(1)	14	Poor
14	1(1)	3(1)	9(3)	1(5)	5(1)	8(3)	1(1)	2(3)	2(1)	6.41(1)	20	Poor
15	0(1)	2(1)	5(1)	0.5(1)	2(1)	64(1)	0(1)	0(1)	1(1)	6.77(1)	10	Poor
16	0(1)	2(1)	8(1)	0.7(1)	2(1)	86(1)	2(1)	4(3)	4(1)	6.99(1)	12	Poor

Note: Metric values are shown with metric scores in parentheses.

TABLE 18.4
Index of Biotic Integrity (IBI) Scores and Individual Metric Values for Sixteen Watersheds from Northeastern Minnesota Streams

Site	Total # of Spp.	# of Headwater Spp.	# of Minnow Species	% Tolerant Spp.	% Dominant 2 Spp.	# of Invertivore Spp.	% Simple Lithophils	# of Fish/100m	% DELT	IBI	Biotic Integrity Rating
1	1(0)	0(0)	0(0)	100(0)	100(0)	1(2)	0(0)	7(0)	0(10)	13	Very Poor
2										0	Very Poor
3										0	Very Poor
4	3(0)	1(5)	1(2)	22(10)	88(0)	2(5)	0(0)	3(0)	0(10)	36	Fair
5										0	Very Poor
6	7(5)	1(5)	3(5)	68(7)	41(10)	4(7)	0(0)	8(0)	0(10)	54	Good
7	4(2)	1(5)	1(2)	6(10)	96(0)	3(5)	0(0)	76(10)	0(10)	49	Fair
8										0	Very Poor
9	6(5)	1(5)	2(5)	73(5)	93(0)	3(5)	0(0)	13(10)	0(10)	50	Good
10										0	Very Poor
11										0	
12	3(0)	2(5)	2(5)	10(10)	90(0)	2(5)	0(0)	56(10)	0(10)	50	Good
13	1(0)	0(0)	0(0)	100(0)	100(0)	0(0)	0(0)	867(10)	0(10)	21	Very Poor
14	1(0)	0(0)	0(0)	100(0)	100(0)	1(2)	0(0)	4(0)	0(10)	13	Very Poor
15	3(0)	0(0)	0(0)	100(0)	87(2)	1(2)	0(0)	10(0)	0(10)	16	Very Poor
16										0	Very Poor

Note: Metric values are shown with metric scores in parentheses.

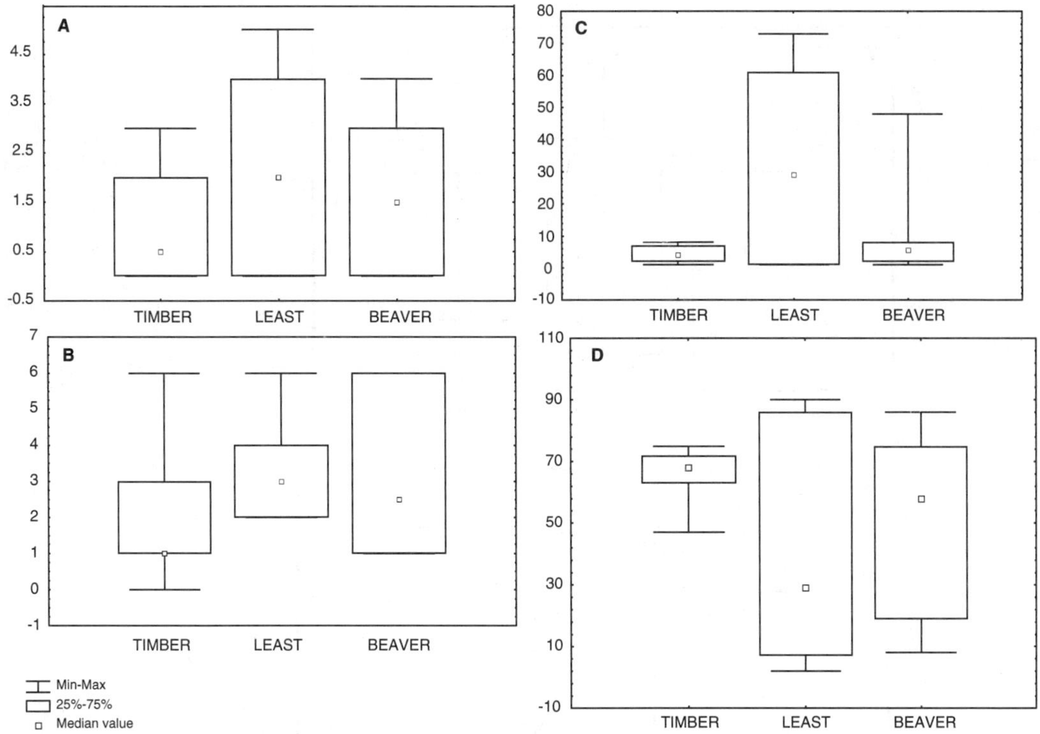

FIGURE 18.5 Inter-quartile box and whisker graphs showing the relationships of least impacted, timbered, and beaver-influenced watersheds for select BCI metrics. A = number of Ephemeroptera species, B = number of Diptera species, C = percent individuals as Trichoptera species, and D = percent individuals as crustacea and mollusca species.

Trophic Function — The number of filtering species and number of scraping species increased with increasing biological integrity (Butcher et al., in press). None of the function metrics were significantly different in the least impacted, timbered, and beaver-influenced watersheds (Figure 18.6). Little overlap was noted among the results of the Hilsenhoff biotic index and the number of scraping species between least impacted and affected watersheds. No significant difference was observed in the number of filtering species metric between naturally influenced and timbered sites and least impacted sites.

18.4.2.2 Fish Assemblage

Species Richness and Composition — The *a priori* hypothesis for stream fish communities was that increasing numbers of species occurred in warmwater streams with increasing biological integrity (Simon and Lyons, 1995), while Mundahl and Simon (1999) found that declining numbers of species should be apparent for coldwater streams and increasing biological integrity. The "arrowhead" region consisted primarily of coolwater and coldwater streams. No statistically significant difference was observed between least impacted and timbered or beaver-influenced watersheds for the total number of species, number of headwater species, or number of minnow species metrics.

Tolerance and Sensitivity — Beaver-influenced watersheds have higher proportions of pools and elevated temperatures as a result of less canopy cover. Tolerant species increase with declining biological integrity, as does the percent individuals as dominant species. The percent individuals as tolerant species and percent individuals as dominant species were significantly different in least impacted and both timbered and beaver-influenced watersheds. Percent individuals as tolerant species had mean values that were significantly different between timbered (mean = 70%; ANOVA,

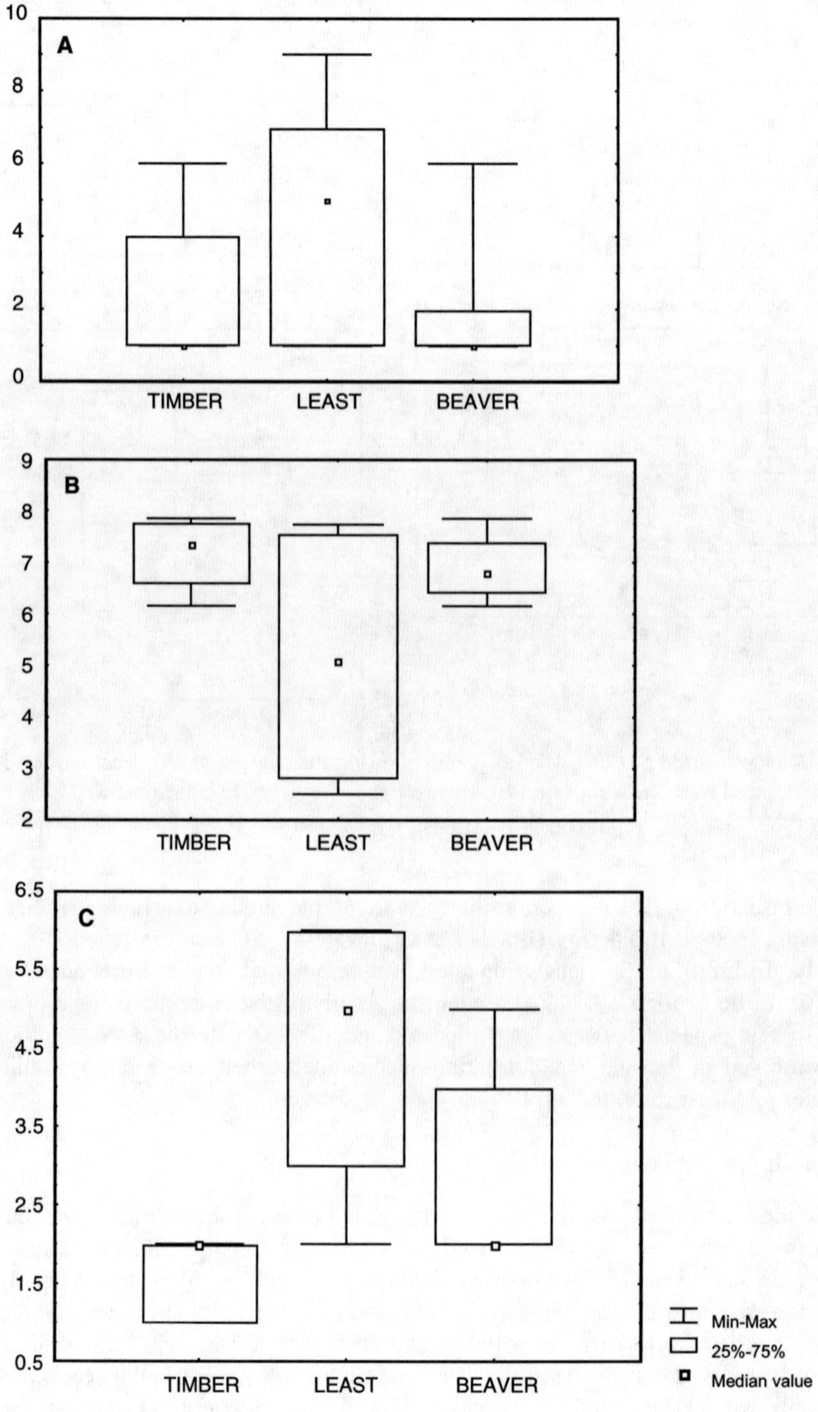

FIGURE 18.6 Inter-quartile box and whisker graphs showing the relationships of least impacted, timbered, and beaver-influenced watersheds for select BCI metrics. A = number of filtering species, B = Hilsenhoff biotic index, and C = number of scraping species.

df = 3, F = 21.667, $p = 0.032$) and beaver-influenced (mean = 61%; ANOVA, df = 3, F = 32.554, $p = 0.048$) watersheds and least impacted (mean = 33.7%) watersheds.

Functional Feeding and Reproductive Guilds — Trophic and reproductive guilds are functional attributes of fish assemblages that become useful metrics since they are able to separate assemblages based on balanced feeding relationships and determine whether sediment quality has blocked pore spaces, prohibiting spawning of sensitive species. The number of invertivore species and percent individuals as simple lithophils were not significantly different between least impacted and either the timbered or beaver-influenced watersheds. No simple lithophils were collected in any of the watersheds, and fewer than three invertivore species were collected.

Individual Condition and Relative Abundance — Deformities, eroded fins, lesions, and tumors (DELT anomalies) are found at heavily disturbed sites (Sanders et al., 1999). These conditions persist at the lowest extremes of biological integrity. In addition, the relative number of individuals increases with increasing biological integrity, except when exaggerated as a result of nutrient input (Thoma and Simon, Chapter 11, this volume). No DELT anomalies were found associated with any of the least impacted, timbered, or beaver-influenced watersheds. Typically, DELT anomalies are associated with chemical contamination. Relative abundance of fish was expected to be at least 25 individuals in a representative reach. No fish were collected from six watersheds and fewer than 25 individuals were collected from seven watersheds. The low numbers of individuals probably resulted because many of these sites are upstream of torrential rapids, waterfalls, and steep gradients. If fish were ever able to colonize these sites, recolonization would have been difficult after timbering. The relative abundance of fish was highest at timbered sites (mean = 46.3) and beaver-influenced (mean = 34.5) watersheds. Few fish were collected from least impacted watersheds, which is what is generally expected from coolwater and coldwater headwater streams.

18.4.3 Patterns in Watershed Multimetric Index Scores

The median IBI values for natural watersheds was 0, while the BCI median score was 23 (Figure 18.7). An IBI score of 0 was considered very poor, which normally indicates that a site is heavily impacted (Niemela and Feist, 2000). In this case, scores were likely a function of streams that had intermittent flows during some point of the year or a result of stream size. A BCI score of 23 received a fair designation (Butcher et al., in review) and 62.5% of sites with zero scores were in least impacted watersheds. Watersheds with logging had median IBI scores of 12 or very poor. The BCI evaluated these sites similarly with median score of 13 or poor. BCI scores were higher in stream segments with beavers than watersheds without beavers. The IBI, however, showed lower scores for stream segments housing beavers. The BCI showed a highly significant positive correlation to the QHEI ($n = 16$, $rs = 0.8452$, $p < 0.00004$). The IBI was positively correlated with both stream depth ($n = 16$, $rs = 0.6425$, $p < 0.008$) and width ($n = 16$, $rs = 0.5382$, $p < 0.04$).

18.5 DISCUSSION

Boreal ecosystems display considerable long-term variations, even in the absence of anthropogenic disturbances. Natural phenomena such as fire, windthrow, and insect outbreaks have all been shown to effect boreal terrestrial ecosystems. Most tree stands in boreal forests originated from high severity disturbances, such as fire and logging (Frelich and Reich, 1998). Lakes and streams in these watersheds are affected by these disturbances and by climate and weather (Wright, 1976; Schindler et al., 1980, 1990; Bayley et al., 1992a; Schindler, 1998b). Following fire or other disturbances, increases in chemical and hydrological outputs throughout the disturbance area can last from a few to several years, depending on severity, climate, and weather (Schindler et al., 1980, 1996). Base cations (Ca, Mg, Na, and K) increase and even larger increases occur in strong acid anions (SO_4, NO_3, Cl) after fire produces a net acidifying effect on streams (Bayley et al., 1992b).

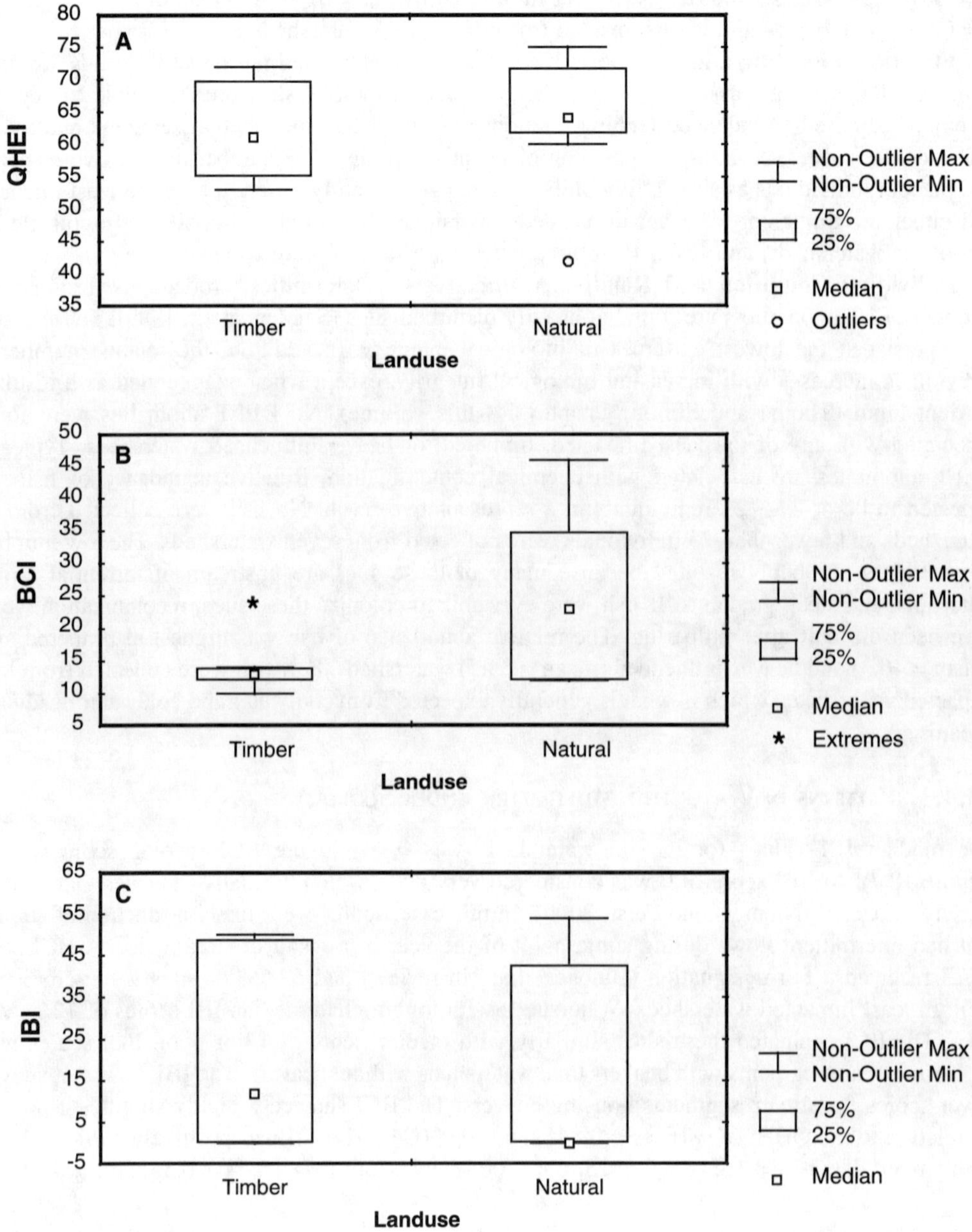

FIGURE 18.7 Inter-quartile box and whisker graphs showing the relationships of index scores for three indicators (QHEI, BCI, and IBI) based on land use within the watershed in natural and timbered areas. A = QHEI, B = BCI, C = IBI.

Wetlands capture much of the sulfate and nitrate entering a watershed from natural and anthropogenic sources (Bayley et al., 1986). Logging can exert effects similar to a severe fire.

Silviculture manipulates forest vegetation for a variety of objectives such as sustaining wildlife habitat, maintaining hydrologic processes, restoring ecosystems, conserving biodiversity, and producing wood products (Graham and Jain, 1998). However, logging does not always mimic natural disturbances because of differences in severity (Ahlegren, 1970; Noble et al., 1977; Abrams and Dickman, 1982; Carleton and MacLellan, 1994).

The BCI distinguished human disturbances from natural ones in the 16 watersheds. Our *a priori* hypothesis was that all 16 watersheds would be considered to have reference condition qualities for the NLF Ecoregion. However, 12 of the 16 (75%) sites scored poor or fair. Scores ranged from 10 to 24 (Tables 18.1 and 18.3). Further investigation of aerial photographs and land use census data revealed logging and mining activities had occurred in the study area. Poor or fair scores could be explained by logging and mining activities within the watershed at 50% of the sites. Site 15 appeared to be influenced by a bank collapse of a parking lot on a high gradient into the stream. Poor scores at two other sites could be attributed to natural disturbances including sampling in an intermittent stream and a bog.

Exl and Simon (in press) hypothesized that distribution of BCI scores at the nonhuman impact sites would depict a bell-shaped curve rather than a skewed distribution toward good and excellent conditions. They found that CI scored naturally impacted (beaver) sites higher than anthropogenic disturbed (silviculture) sites (Tables 18.1 and 18.3). Naturally impacted sites had greater diversity of habitat because of beaver modifications. The BCI was highly correlated with the QHEI, and select habitat variables may be limiting factors for benthic macroinvertebrate assemblages. Higher BCI scores occurred at sites with beaver dams. Such sites generally had greater QHEI scores compared to sites without dams (Figure 18.8 and Table 18.1). Beaver dams created more habitat by providing marginal wetland vegetation, woody debris, and deep pools within the stream channel.

The IBI generally classified sites in poorer condition than the BCI. The version of the IBI was developed for the St. Croix River Basin, approximately 100 miles south of our study area in Minnesota. Differences in aquatic community expectations may be variable even within the same ecoregion. Nine of 16 sites (56.3%) typically, would not have been scored had they been located in the St. Croix River Basin, following the methods used by Niemela and Feist (2000). No fish or fewer than 25 fish were collected at these sites (Table 18.4). Streams without fish present may have been the results of local extirpation or low water levels. Differences in stream condition were seen between BCI and IBI scores (Table 18.1), with the exception of the Reservation River tributary, which was the only site where the two indices evaluated stream conditions similarly, and the IBI scored naturally impacted (beaver) sites lower than those influenced by logging in the watershed (Figures 18.7 and 18.8).

The QHEI closely matched trends in the BCI (Figure 18.9). Although no habitat expectations were designated for this ecoregion, the scores indicate good habitat for aquatic life. Barbour et al. (1999) showed that habitat values generally followed a direct correlation with the indicator multimetric indices. The QHEI scored sites impacted by logging lower than natural landscapes (Figure 18.7). Streams with resident beaver populations had higher scores due to the increased instream habitat stability offered by ponds.

Index scores varied in response to complex interaction of land use, stream modification, and flood plains. BCI scores ranged from 12 to 46 at least impacted sites, but many of the lower scores could also be attributed to wetland floodplain influence (Figure 18.10). Wetland flood plains had substantially lower QHEI scores than forested flood plains. Low BCI scores could be a response to reduced habitat availability in these watersheds. Goldstein et al. (in press) showed that relationships in fish communities in the NLF were explained by embeddedness, the percent of large and small substrate particles, the mean channel width, the mean distance between geomorphological units, the amount of woody debris, the amount of forested land use, and the amount of stream shading.

The IBI scored forested flood plains lower than wetlands. Some of these sites were high gradient first order streams that may have precluded fish. The IBI was expected to perform with less precision than the BCI since it was calibrated at a watershed scale. Since so many of our sites did not contain fish, the fish community may not be a good indicator for headwater wetland streams. The use of fish needs to be more finely calibrated for drainage area effects. We must find a way to determine whether a fish assemblage was expected in these watersheds or whether local extirpation occurred as a result of silviculture practices.

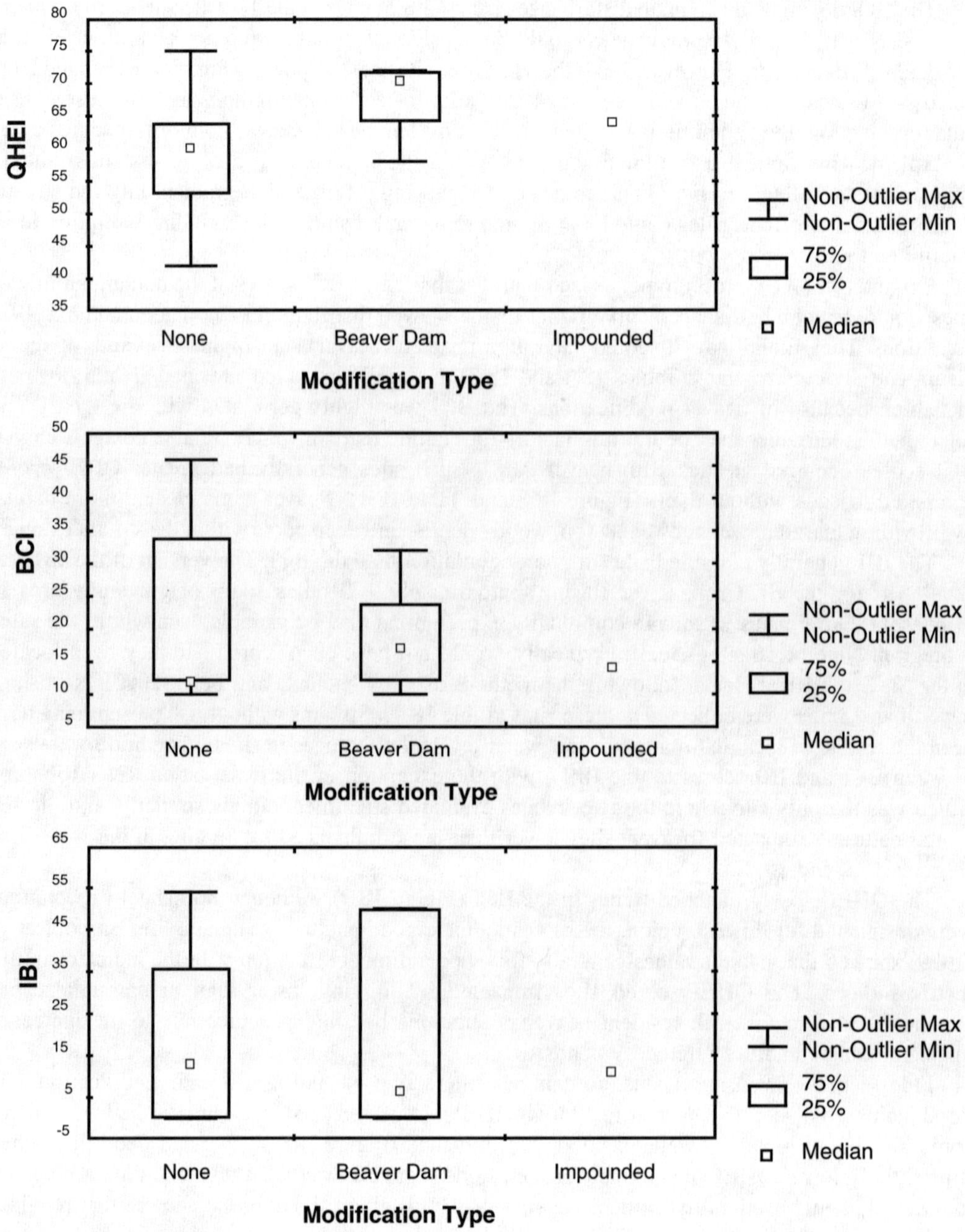

FIGURE 18.8 Inter-quartile box and whisker graphs showing the relationships of index scores for three indicators (QHEI, BCI, and IBI) as influenced by three stream modification types. A = QHEI, B = BCI, C = IBI.

18.6 CONCLUSIONS

Relatively few aquatic and terrestrial environments have not been impacted by human activities to some degree. This makes the ability to determine biological integrity more difficult. Northeastern Minnesota's Northern Lakes and Forest Ecoregion has a minimally impacted landscape. Water chemistry and habitat varied with both land use and stream modification. In watersheds where

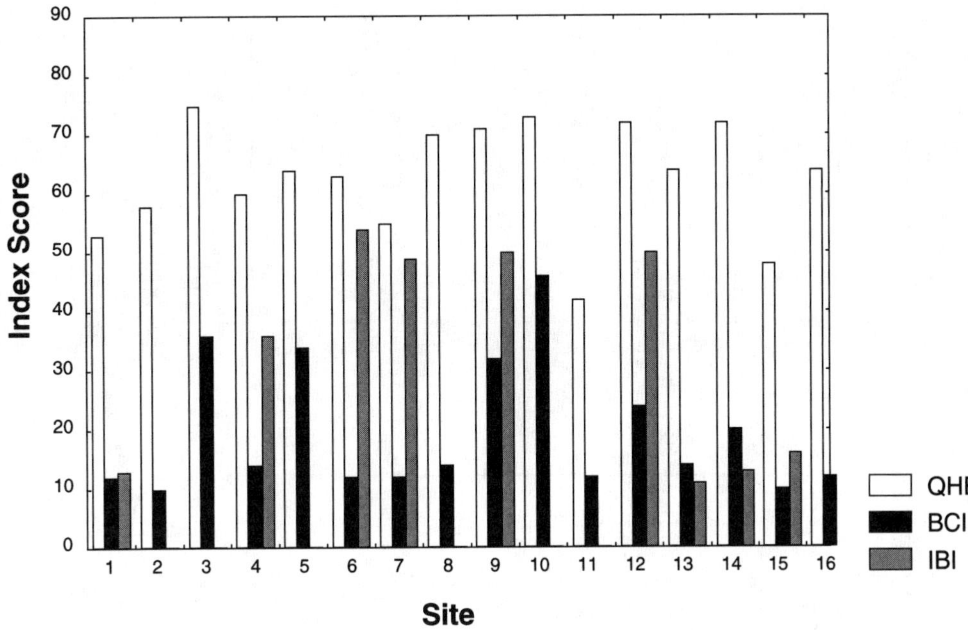

FIGURE 18.9 Patterns in index scores for three indicators (QHEI, BCI, and IBI) for 16 watersheds in Northeastern Minnesota streams.

logging occurred, decreases in median values were observed for salinity, conductivity, pH, and QHEI while median stream width, depth and stream temperatures increased, compared to least impacted watersheds. Turbidity was 60% lower and dissolved oxygen was 11.4% higher in logged watersheds. Watersheds with beaver activity showed reductions in median pH, turbidity and dissolved oxygen and increases in QHEI scores, stream width, depth, and temperature, when compared to nonbeaver sites. The IBI median value for natural watersheds was 0, while the BCI median score was 23. Least impacted watersheds represented 62.5% of the sites with scores of zero. Watersheds with logging had median IBI scores of 12 or very poor. The BCI evaluated these sites similarly with median scores of 13 or poor. Benthic community index scores were higher in stream segments with beavers. The IBI, however, showed lower scores for stream segments with beavers. The BCI showed a highly significant positive correlation to the QHEI.

ACKNOWLEDGMENTS

Special thanks to Ron Carlson, Arthur Lubin, Kevin Pierard, and Linda Holst of the U.S. Environmental Protection Agency, for their assistance on this project. This project was funded through the U.S. Environmental Protection Agency's Regional Environmental Monitoring and Assessment Program (REMAP) and Region 5, Assistance Identification 826206–01–0, Request 9728GH0023. Field assistance was provided by partners on this project including Paul Kanehl, Wisconsin DNR, Bureau of Research; Scott Niemela, Mike Feist, Joel Chirhart, and Konrad Schmidt, Minnesota Pollution Control Agency; Ed Baker, Michigan DNR; Margaret Watkins and John Johnson, Grand Portage Chippewa Nation; and Jason Butcher, U.S. Geological Survey. The opinions expressed herein do not necessarily reflect those of the U.S. Fish and Wildlife Service or U.S. Environmental Protection Agency, although portions of this study may have been funded wholly or in part by those agencies.

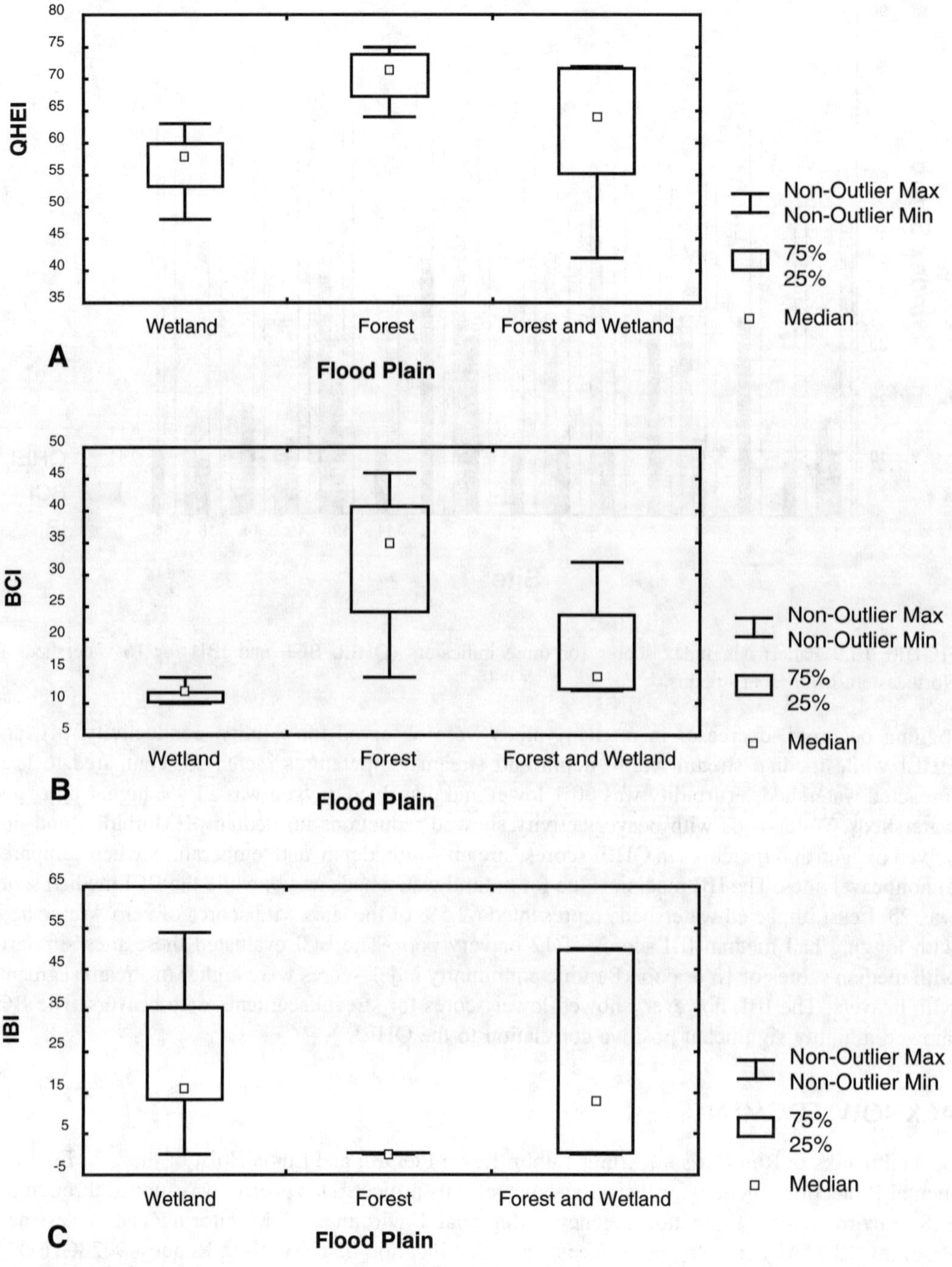

FIGURE 18.10 Inter-quartile box and whisker graphs showing the relationships of index scores for three indicators (QHEI, BCI, and IBI) and three stream types. A = QHEI, B = BCI, C = IBI.

REFERENCES

Abrams, M.D. and D.E. Dickman. 1982. Early revegetation of clearcut and burned jackpine sites in northern lower Michigan, *Canadian Journal of Botany,* 60, 946–954.

Ahlgren, C.E. 1970. Some Effects of Prescribed Burning on Jack Pine Reproduction in Northeastern Minnesota. University of Minnesota Agricultural Experiment Station. Miscellaneous Report 94, Forestry Series 5–1970.

Bayley, S.E., D.W. Schindler, K.G. Beaty, B.R. Parker, and M.P. Stainton. 1992a. Effects of multiple fires on nutrient yields from streams draining boreal fores and fen watersheds: nitrogen and phosphorus, *Canadian Journal of Fisheries and Aquatic Sciences,* 49(3), 584–596.

Bayley, S.E., R.S. Behr, and C.A. Kelly. 1986. Retention and release of S from a freshwater wetland, *Water, Air, and Soil Pollution*, 31, 101–114.

Barbour, M.T., J. Gerritsen, B.D. Snyder, and J.B. Stribling. 1999. *Rapid Bioassessment Protocols for Use in Streams and Wadeable Rivers: Periphyton, Benthic Macroinvertebrates and Fish,* 2nd ed. EPA 841-B-99–002. U.S. Environmental Protection Agency; Office of Water; Washington, D.C.

Bodaly, R.A., R.E. Hecky, and R.J.P. Fudge. 1984a. Increases in fish mercury levels in lakes flooded by the Churchill River diversion, northern Manitoba, *Canadian Journal of Fisheries and Aquatic Sciences,* 41, 682–691.

Bodaly, R.A., T.W.D. Johnson, R.J.P. Fudge, and J.W. Clayton. 1984b. Collapse of the lake whitefish (*Coregonus clupeaformis*) fishery in southern Indian Lake, Manitoba, following lake impoundment and river diversion, *Canadian Journal of Fisheries and Aquatic Sciences*, 41, 692–700.

Bodaly, R.A., J.W.M. Rudd, R.J.P. Fudge, and C.A. Kelly. 1993. Mercury concentrations in fish related to size of remote Canadian Shield lakes, *Canadian Journal of Fisheries and Aquatic Sciences*, 50, 980–987.

Bodaly, R.A., V.L. St. Louis, M.J. Paterson, R.J.P. Fudge, B.D. Hall, D.M. Rosenberg, and J.W.M. Rudd. 1997. Bioaccumulation of mercury in the aquatic food chain in newly flooded areas, in H. Segel and A. Sigel (Eds.). *Mercury and Its Effects on Environment and Biology,* Marcel Dekker, New York, 259–287.

Butcher, Jason T., P.M. Stewart, and T.P. Simon. In review. Development of a benthic community index for the Northern Lakes and Forest Ecoregion, *Ecological Indicators.*

Butcher, Jason T., P.M. Stewart, and T.P. Simon. In review. Classification effects on a benthic community index for the Northern Lakes and Forest Ecoregion, *Ecological Indicators.*

Carleton, T.J. and P. MacLellan. 1994. Woody Responses to fire versus clear-cutting logging: a comparative survey in the central Canadian boreal forest, *Ecoscience,* 1, 141–152.

Chirhart, Joel. 1998. Invertebrate sampling procedures for Northern Lakes and Forest streams, in T.P. Simon and P.M. Stewart (Eds.). *Standard Operating Procedures for Development of Watershed Indicators in REMAP: Northern Lakes and Forest Streams.* U.S. Environmental Protection Agency, Chicago, IL.

Davis, W.S. and T.P. Simon. 1995. Introduction, in W.S. Davis and T.P. Simon (Eds.), *Biological Assessment and Criteria: Tools for Water Resource Planning and Decision Making.* Lewis Publishers, Boca Raton, FL, 3–6.

De Leo, G.A. and S. Levin. 1997. The multifaceted aspects of ecosystem integrity, *Conservation Ecology,* 1(1), 3.

Exl, J.A. and T.P. Simon. In press. Biological integrity of several streams located in northeastern Minnesota's Northern Lakes and Forest Ecoregion (USA) with emphasis on differentiating natural from anthropogenic disturbances, in A.H. Ansari (Ed.). *Dimensions in Pollution*, Bophal, India.

Frelich, L.E. and P.B. Reich. 1998. Disturbance severity and threshold responses in the boreal forest, *Conservation Ecology*, 2(2), 7.

Goldstein, R.M., L. Wang, T.P. Simon, and P.M. Stewart. In press. Development of a stream habitat index for the Northern Lakes and Forest Ecoregion, *North American Journal of Fisheries Management.*

Graham, R.T. and T.B. Jain. 1998. Silviculture's role in managing boreal forests, *Conservation Ecology,* 2(2), 8.

Hughes, R.M. 1995. Defining acceptable biological status by comparing with reference conditions, in W.S. Davis and T.P. Simon (Eds.). *Biological Assessment and Criteria: Tools for Water Resource Planning and Decision Making.* Lewis Publishers, Boca Raton, FL, 31–48.

Karr, J.R. and D.R. Dudley. 1981. Ecological Perspective on Water Quality Goals, *Environmental Management,* 5, 55–68.

Levin, S.A. 1997. Biodiversity: Interfacing Populations and Ecosystems, in T. Abe, S.A. Levin, and M. Higashi (Eds.), Springer-Verlag, New York, 277–288

Mundahl, N.D. and T.P. Simon. 1999. Development and application of an index of biotic integrity for coldwater streams of the upper midwestern United States, in T.P. Simon (Ed.), *Assessing the Sustainability and Biological Integrity of Water Resources Using Fish Communities*. CRC Press, Boca Raton, FL, 383–416.

Nicolson, J.A. 1975. Water quality and clearcutting in a boreal forest ecosystem, NRCC Number 15195, 11–14 August, Winnipeg, Manitoba, Canada, *Proceedings of Canadian Hydrology Symposium*, 75, 734–738.

Niemela, S.A. and M. Feist. 2000. *Index of Biotic Integrity (IBI) Guidance for Coolwater Rivers and Streams of the St. Croix River Basin in Minnesota*. Minnesota Pollution Control Agency Biological Monitoring Program.

Noble, M.G., L.K. DeBoer, K.L. Johnson, B.A. Coffin, L.G. Fellows, and N.A. Cristensen. 1977. Quantitative relationships among some *Pinus banksiana–Picea mariana* forests subjected to wildfire and post-logging treatments, *Canadian Journal of Forest Research*, 7, 368–377.

Ohio Environmental Protection Agency. 1989. *Biological Criteria for the Protection of Aquatic Life*. Vol. III. *Standardized Biological Field Sampling and Laboratory Methods for Assessing Fish and Macroinvertebrate Communities*. Ohio EPA, Division of Water Quality Monitoring and Assessment, Columbus, OH.

Ohio Environmental Protection Agency. 1987. *Biological Criteria for the Protection of Aquatic Life*. Vol. I. *The Role of Biological Data in Water Quality Assessment*. Ohio EPA, Division of Water Quality Monitoring and Assessment, Columbus, OH.

Omernik, J.M., S.S. Chapman, R.A. Lillie, and R.T. Dumke. 2000. Ecoregions of Wisconsin, *Transactions of the Wisconsin Academy of Arts, Sciences, and Letters*, 88, 77–103.

Schindler, D.W. 1998a. Sustaining aquatic ecosystems in boreal regions, *Conservation Ecology*, 2(2), 18.

Schindler, D.W. 1998b. A dim future for boreal waters and landscapes: cumulative effects of climatic warming, stratospheric ozone depletion, acid precipitation, and other human activities, *BioScience*, 48, 157–164.

Schindler, D.W., S.E. Bayley, B.R. Parker, K.G. Beaty, D.R. Cruikshank, E.J. Fee, E.U. Schindler, and M.P. Stainton. 1996. The effects of climatic warming on the properties of boreal lakes and streams at the experimental lakes area, Northwestern Ontario, *Limnology and Oceanography*, 41, 1004–1017.

Schindler, D.W. 1990. Experimental perturbations of whole lakes as tests of hypothesis concerning ecosystem structure and function, Proceedings of the 1987 Crafoord Symposium, *Oikos*, 57, 25–41.

Schindler, D.W., K.G. Beaty, E.J. Fee, D.R. Cruikshank, E.D. DeBruyn, D.L. Findlay, G.A. Linsey, J.A. Shearer, M.P. Stainton, and M.A. Turner. 1990. Effects of climatic warming on lakes of the central boreal forest, *Science*, 250, 967–970.

Schindler, D.W., S.E.M. Kasian, and R.H. Hesslein. 1989. Losses of biota from american aquatic communities due to acid rain. Proceeding of the World Wildlife Conference, Estes Park, Colorado, 14–18 September 1987, *Environmental Monitoring and Assessment*, 12, 269–285.

Schindler, D.W., R.W. Newbury, K.G. Beaty, J. Prokopowich, T. Ruszcynski, and J.A. Dalton. Effects of a windstorm and forest fire on chemical losses from forested watersheds and on the quality of receiving streams, *Canadian Journal of Fisheries and Aquatic Sciences*, 37, 328–334.

Simon, T.P. and J. Lyons. 1995. Applicationof the index of biotic integrity to evaluate water resource integrity in freshwater ecosystems, in W.S. Davis and T.P. Simon (Eds.). *Biological Assessment and Criteria: Tools for Water Resource Planning and Decision Making*. Lewis Publishers, Boca Raton, FL, 245–262

StatSoft, Inc. (1995). *STATISTICA for Windows*. Tulsa, OK: StatSoft, Inc., http://www.statsoft.com.

Verry, E.S. 1986. Forest harvesting and water: the lake states experience, *Water Resources Bulletin*, 22, 1039–1047.

Wang, L., J. Lyons, P. Kanehl, and T.P. Simon. 1998. Fish sampling procedures for Northern Lakes and Forest Streams. in T.P. Simon and P.M. Stewart (Eds.). *Standard Operating Procedures for Development of Watershed Indicators in REMAP: Northern Lakes and Forest Streams*. U.S. Environmental Protection Agency, Chicago, IL.

Wright, R.F. 1976. The impact of forest fire on the nutrient influxes to small lakes in northeastern Minnesota, *Ecology*, 57, 649–663.

19 Biological Assessment of Mining Disturbance on Stream Invertebrates in Mineralized Areas of Colorado

Leska S. Fore

CONTENTS

0-8493-0905-0/03/$0.00+$1.50

19.1 INTRODUCTION

Scientific methods for biological assessment of surface waters have been developed and applied broadly in the United States, primarily within the regulatory context of the Clean Water Act (Ransel, 1995; Davis et al., 1996). The widespread contamination of headwater streams in the western United States by historic and contemporary mining operations represents an opportunity to extend biological assessment to a new regulatory context under the Comprehensive Environmental Response, Compensation, and Liability Act (CERCLA) which designates Superfund sites for remediation and monitoring.

The Belden mine site on the Eagle River was mined from the 1870s through the 1980s. The state of Colorado filed a lawsuit against the mine owners in 1984 and remediation began soon afterward. Chemical monitoring began in the 1980s; fish and invertebrate monitoring began in 1990. Large improvements in metal levels were observed in the early 1990s and steady improvement has been noted since then. All heavy metals except zinc are currently below chemical standards defined by the U.S. Environmental Protection Agency (USEPA). Improvements in the biological condition of the river have also been observed; brown trout (*Salmo trutta*) have increased in abundance and sensitive mayfly taxa, e.g., heptageniids, are now present. Although brown trout do not represent a return to natural conditions (they were introduced from Europe as sport fish), their renewed presence indicates that the Eagle River can support some fish species.

Although USEPA requires states to develop biological criteria for surface waters, many including Colorado have yet to do so and there is no current biological standard to apply to the Eagle River (Karr, 1991). The purpose of this study was to use a regional data set and a more intensive data set for the Eagle River to define a biological endpoint for remediation of the Belden mine site. Biological criteria will be used to judge whether the invertebrate community has recovered sufficiently to consider remediation successful and complete. This chapter describes the results of metric testing, the methods used to develop a multimetric index, and guidelines for biological assessment derived from statistical power analysis of the multimetric index.

19.2 METHODS

19.2.1 Sampling Design

19.2.1.1 Eagle River

Invertebrates were collected using Hess samplers from eight sites along the Eagle River from 1992 through 1999 (Figure 19.1). Five replicate samples were collected at each site during the spring, summer and fall. Two sites were located upstream of the influence of the mine, three sites were in the mine area, and three sites were located further downstream (Table 19.1). The downstream sites were influenced by metals from the mine and by urban development. In addition, a railroad line and highway run the length of this section of the river. Because the valley is narrow, the railroad and highway were built within or very near the riparian area of the river. Of the 124 invertebrate site visits, 57 had information for both invertebrates and total recoverable metals (unfiltered samples).

19.2.1.2 Regional Environmental Monitoring and Assessment Program

Between 1994 and 1995, stream sites in the Southern Rocky Mountain ecoregion were sampled as part of USEPA's Regional Environmental Monitoring and Assessment Program (REMAP). REMAP sites were selected according to a probability sampling design that randomly selects stream reaches from a specified geographic area (Paulsen et al., 1998; Clements et al., 2000). Because sites were selected randomly, they are representative of conditions in the Southern Rocky Mountain ecoregion

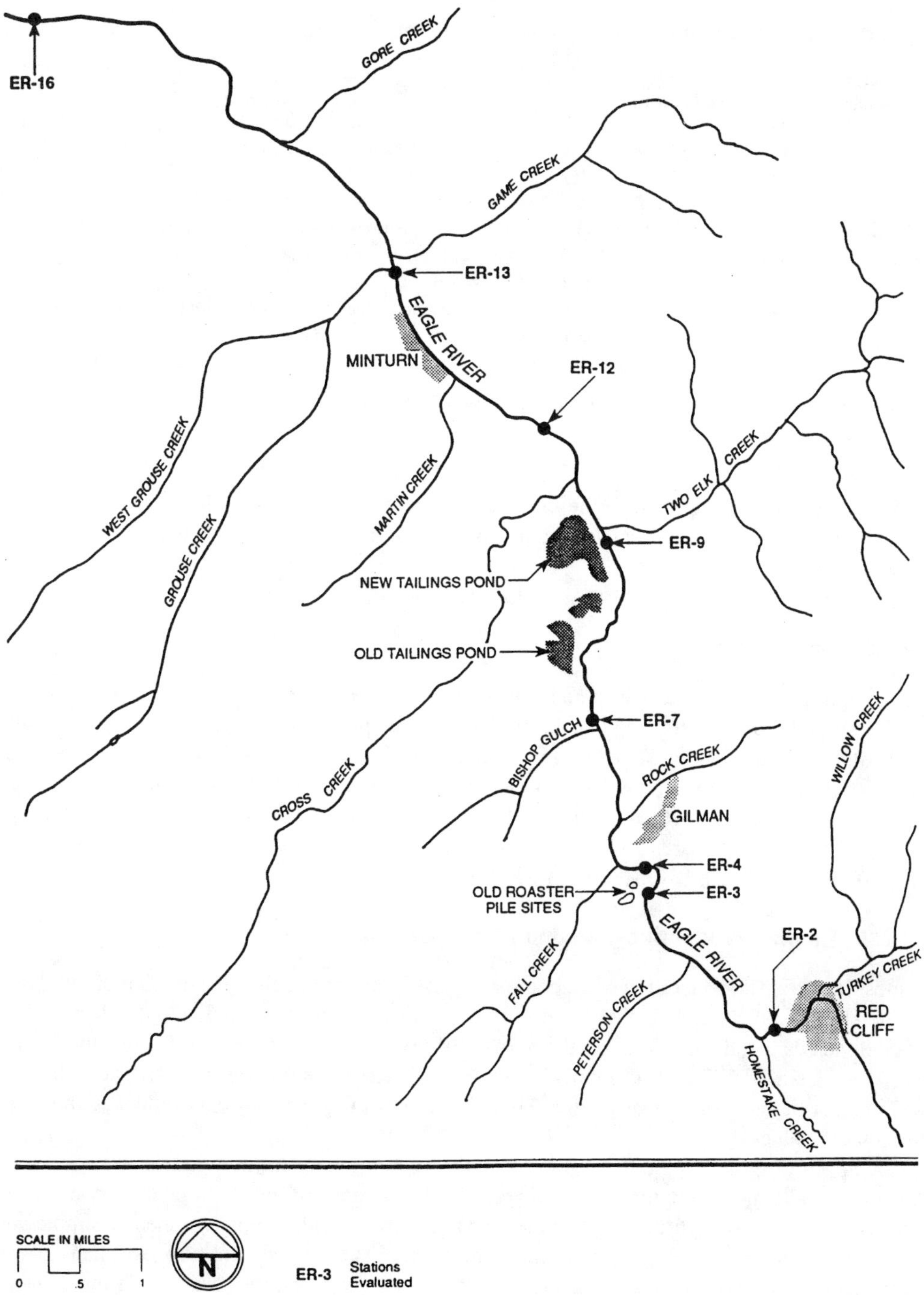

FIGURE 19.1 Invertebrate sampling locations along the Eagle River, CO.

TABLE 19.1
Site Codes for Invertebrate and Chemistry Sampling Stations

Site Code: Invertebrates	Chem.	River Mile	Site Description
ER2	E1R	2.0	DS of Red Cliff; US of Homestake Creek and mining. Reference site.
ER3	E3	3.5	US from mine. Reference site.
ER4	E5	4.0	DS from New House Tunnel and former roaster piles. Mine site.
ER7	E11	5.0	DS from Eagle Mine; US from old tailing pile. Mine site.
ER9	E13	6.5	US from Two Elk Creek; DS from old tailing pile; just DS from mine WWTP. Mine site.
ER12	E15	7.5	DS from Cross Creek and mined area; US from Minturn. Downstream site.
ER13	E22	9.5	DS from Game Creek. Downstream site.
ER16	E28	17.5	DS from Avon. Urban influence from Avon, Eagle and Vail. Downstream site.

Note: Approximate river mile measured from the headwater of Homestake Creek. Description notes whether site was a reference, mine, or downstream site. DS = downstream. US = upstream.

and their results can be applied to the entire region with a known measure of uncertainty. Over two years, 95 sites on 73 streams were sampled for fish, invertebrates and metals.

19.2.2 Comparison of Sampling and Laboratory Protocols

19.2.2.1 Field Sampling

Field sampling on the Eagle River and the REMAP study were comparable (Table 19.2). The REMAP study used kicknet samples and the Eagle River protocol sampled a smaller area with a Hess sampler. Both methods sampled the substrates within riffle areas. The sampling period for REMAP (August through September) fell between the summer (August) and fall (October) sampling periods on the Eagle River. I selected fall samples as more comparable to the REMAP data because summer temperatures and maturation of invertebrates typically occur later at higher elevations such as the Eagle River (7500 to 9500 m).

19.2.2.2 Laboratory Identification and Metric Calculation

The most important consideration when comparing laboratory protocols is the number of individuals identified per sample. Taxa richness is correlated with the number of individuals identified, the more individuals identified, the larger the number of taxa found in the sample (Larsen and Herlihy, 1998; Doberstein et al., 2000; Li et al., 2001). Many metrics included in multimetric indices are based on taxa richness of certain groups. Thus, laboratory effort can strongly influence the ranges of the metric values.

For the REMAP and Eagle River protocols, the target numbers of individuals for each sample were similar, 300 and 250 individuals, respectively. The largest difference in laboratory effort was related to chironomid identification. For the REMAP data, all chironomids in the subsample were identified; in contrast, the Eagle River chironomids were subsampled and no more than 25 were identified from each sample. Level of taxonomic effort for chironomids was much more intense for the REMAP study, which increased measures of total taxa richness for REMAP sites over those from the Eagle River sites.

19.2.3 Calculation of Metal Concentrations

Metal concentrations in stream samples were summarized following the approach used by Clements and Carlisle (1998) and Clements et al. (2000). The chemical criteria for aluminum, cadmium,

TABLE 19.2
Comparison of Field and Lab Protocols for the REMAP Study and Eagle River Hess Sampling

Protocol Element	REMAP	Eagle River
	Field Methods	
Sampler type	Modified kicknet	Hess
Mesh size	595 μm	500 μm
Area sampled	9×0.5 m^2	5×0.1 m^2
Composite	Yes	No
Replicate samples	No	Yes
Microhabitat	Riffles (pools excluded)	Riffles
Channel location	Left, right or middle	Left, right or middle
Time sampled	9×20 sec	5×90 sec
Years of data	1994–1995	1992–1998
Seasons sampled	Summer (Aug.–Sept.)	Spring, 1995–1999 (March, April) Summer 1995–1998 (August) Fall 1992 (Nov.), 1993–1998 (Oct.)
	Laboratory Methods	
Target number of individuals	300	250 (does not include chironomids and oligochaetes)
Subsampled	Yes, thirds; then gridded tray with 40 squares	Yes, 2×2 gridded tray
Taxonomic level: insects	Approximately species	Approximately species
Chironomidae	Genus/species	Genus[a]
Annelida (Oligochaeta)	Approximately family; 2 taxa	Species, 7 taxa
Turbellaria	Class (Turbellaria)	Genus, 6 taxa
Crustacea and Mollusca	Yes	None collected
Counts recorded as	Number of individuals	Density (number/m^2)
Large and rare search	No	Yes, included in data

[a] Chironomids were subsampled separately; no more than 25 identified.

copper, iron, lead, manganese and zinc were used and adjusted for hardness where necessary (USEPA, 1986). I divided the observed value for each metal by its USEPA criterion value and summed values for all metals to obtain a single measure of metal concentration in terms of cumulative criterion units (CCUs).

19.2.4 Description of Candidate Metrics

Metrics are attributes of the biological assemblage associated with degradation caused by human disturbance. The first step in constructing a multimetric index is to identify an appropriate set of candidate metrics and test them for correlation with independent measures of human disturbance. Metrics that are not redundant in terms of the taxa counted or the biological concept measured are selected from those significantly associated with disturbance. Candidate metrics were identified and developed from studies that used the same REMAP data set or were developed for indices from the Pacific Northwest, Idaho, or Japan (Clements and Carlisle, 1998; Karr, 1998; Clements et al., 2000; Jessup and Gerritsen, 2000; Fore et al., 2001; Karr and Rossano, 2001; Morley and Karr, 2002). In total, for the Eagle River data set, 16 metrics were tested for correlation with CCUs. For the REMAP data set, 11 metrics were tested for correlation with CCUs. Some metrics could not be calculated for the REMAP data set because natural history information was unavailable or density of individuals was not recorded.

19.2.4.1 Taxa Richness Metrics

Taxa richness metrics were calculated as the total number of taxa (i.e., families, genera, or species) within a specific group. Total taxa richness includes all the different invertebrates collected from a stream site (e.g., mayflies, caddisflies, stoneflies, true flies, midges, clams, snails, and worms). Taxa richness declines as flow regimes are altered, habitat is lost, chemicals are introduced, energy cycles are disrupted, and alien taxa invade (Karr et al., 2000).

Mayfly taxa richness is expected to decline in response to most types of human influence, although it can increase with nutrient enrichment (Miltner and Rankin, 1998). Many mayflies graze on algae and are particularly sensitive to metals and other toxic pollutions that interfere with their food sources (Kuwabara, 1985; Genter and Lehman, 2000).

Stonefly taxa richness is expected to decline as human disturbance increases. Many stoneflies are predators that stalk their prey and hide around and between rocks. Hiding places between rocks are lost as sediment washes into a stream. Many stoneflies are shredders and depend on leaf litter falling from riparian vegetation for food.

Caddisfly taxa feed in a number of ways including spinning nets to trap food, scraping food on top of exposed rocks, and collecting detritus. Many caddisflies build gravel or wood cases to protect them from predators; others are predators. Even though caddisflies are diverse in habits, taxa richness declines steadily as humans eliminate the variety and complexity of their stream habitats.

Metal-intolerant taxa richness is expected to decline as metal concentration increases. Taxa included in this metric were mayflies of the species *Drunella doddsi* and the genera *Cinygmula, Epeorus, Paraleptophlebia,* and *Rhithrogena*; stoneflies in the genera *Skwala, Suwallia,* and *Sweltsa*; caddisflies in the genus *Rhyacophila*; and true flies in the genus *Pericoma* (W. Clements, Colorado State University, personal communication, February 2000). These taxa were selected based on changes in their abundance in response to mining disturbance (Clements et al., 2000).

Taxa defined as clingers have physical adaptations such as ventral suckers, dorso-ventral flattening, or well-developed tarsal claws, or they construct retreats that they attach to substrates (Merritt and Cummins, 1996). These animals are able to cling to smooth substrates in fast water and typically occupy the open areas between rocks and cobbles along the bottoms of streams. Thus, they are particularly sensitive to fine sediments that fill these spaces and eliminate the variety and complexity of these small habitats. Clingers may use these areas to forage, escape from predators, or lay their eggs. Sediment also prevents clingers from moving down deeper into the stream bed or hyporheos of the channel. Whether an insect is described as a clinger does not depend on life stage or instar as it does for some functional feeding group assignments.

Long-lived (semi-voltine) invertebrates included stoneflies (species in the families Pteronarcyidae and Perlidae), caddisflies (species in Arctopsychinae and Brachycentridae), and beetles (species in Dytiscidae and Elmidae). These species require more than one year to complete their life cycles and are thus exposed to all the human activities that influence the stream throughout one or more years. Long-lived taxa richness is expected to decline with disturbance particularly when an increase in flood events occurs.

Predator taxa represent the top of the food web and depend on reliable sources of other invertebrates for food. The percentage obligate predators provides a measure of the trophic complexity supported by a site. This metric is expected to decline with human disturbance.

19.2.4.2 Percentage or Relative Abundance Metrics

Percent relative abundance was calculated as the number of individuals in a specific group divided by the total number of individuals in the sample. Heptageniid mayflies are particularly sensitive to heavy metals, possibly because the metals damage their gills (Clements et al., 2000). The relative abundance of this group is expected to decline with heavy metals.

Taxa identified as scrapers vary according to different authors. Invertebrates can also switch from one feeding strategy to another as they mature. This metric was calculated using designations made by Merritt and Cummins (1996) and modifications of their designations made by R. Wisseman for western taxa (R. Wisseman, Aquatic Biology Associates, Inc., Corvallis, OR, personal communication). Both versions were tested because the discrepancies involved several mayfly genera (e.g., *Ameletus, Drunella, Epeorus,* and *Rhithrogena*) known to be sensitive to metals.

19.2.5 Index Development

I selected metrics for a multimetric index based on their association with CCUs for both the REMAP and the Eagle River data sets. Correlation was tested with Spearman's r, a nonparametric test based on ranks. The data from the REMAP data set was used to define scoring criteria for the metrics because these sites represented a random sample of regional conditions and provided a good estimate of the possible range of metric values within the Southern Rocky Mountain ecoregion.

Metrics were combined into a multimetric index by assigning scores to ranges of values for each metric (Karr, 1981; Karr and Chu, 1999; Barbour et al., 1999). High values of taxa richness were assigned scores of 5 to indicate biological conditions similar to those found at undisturbed sites. Moderate values were assigned values of 3 to indicate moderate disturbance. Low values were scored as 1 to indicate that the biological condition diverged significantly from what would be expected in the absence of human disturbance. The benthic index of biotic integrity (B-IBI) represented the sum of these metric scores.

19.2.6 Data Analysis

Analysis of the data was structured to answer the following four questions.

1. What attributes of the invertebrate assemblage were significantly correlated with metal waste from past mining activities? Potential metrics were tested against a gradient of CCUs for Eagle River and REMAP sites. Mining sites typically had higher CCUs than reference sites along the Eagle River, indicating that CCUs represented a valid measure of mining disturbance.
2. Was biological condition at the Eagle River mining sites degraded compared to reference sites upstream? Do sites downstream of the mine show an improvement in biological condition? Grouping site visits by year and season, index values along a longitudinal gradient were compared from two upstream reference sites, past three sites in the mined area, and downstream to three sites that experienced multiple human influences.
3. Has biological condition improved over time? Index values were plotted for the three mine sites against years to evaluate changes over time.
4. Can we use the multimetric index to evaluate the success of remediation efforts at the Eagle River mine sites? Statistical power analysis was used to determine the amount of change in index values required to indicate a statistically significant difference. I calculated the number of distinct categories of biological condition that B-IBI could distinguish for the sample design used to monitor Eagle River sites. The minimum detectable difference was calculated (MDD; Zar, 1984) for the index based on a two-sample t-test with five replicates. Alpha of 0.05 and power (1 – beta) of 0.80 (Peterman, 1990; Carlisle and Clements, 1999 were used). The sample variance was estimated using the residual, or error, variance from ANOVA.

$$\text{MDD} = \sqrt{\frac{2s_p^2}{n}}(t_{\alpha,\,\nu} + t_{\beta(1),\,\nu})$$

where s^2 = estimated variance for replicate Hess samples, n = the number of replicate Hess samples, 5 in our case, t = Student's t for alpha = 0.05 (2-sided), beta = 0.20 (1-sided), and ν 2 $(n - 1) = 8$ for 5 replicate samples.

The use of normality-based statistical models is appropriate for multimetric indices, and no transformation of B-IBI was necessary before computing statistical power. Multimetric indices are mathematically similar to averages, which are typically normally distributed, and satisfy the assumptions of parametric tests such as ANOVA (Fore et al., 1994).

19.3 RESULTS

19.3.1 Water Chemistry

In order to compare the intensities of metal concentration in the Eagle River with those from other regional stream sites, quartiles for CCUs were estimated from the REMAP data. Based on a comparison with REMAP samples, 8 Eagle River samples fell in the lowest quartile, 5 in the second quartile, 8 in the third, and 14 in the upper quartile indicating the highest metal concentration. Thus, 13 sites were below and 22 above the 50th percentile. This indicates that metal concentrations at Eagle River sites were relatively high compared to a random selection of sites in the ecoregion. The range of CCU values observed along the Eagle River represented a broad range of concentrations and therefore provided a good test for correlation of biological metrics and metal concentration.

CCU measurement was dominated by zinc for Eagle River sites. Zinc contributed 48% of the units to the total value. Following zinc were lead (18%), cadmium, copper, iron (~10% each), and manganese (4%). For REMAP samples collected throughout the ecoregion, the largest contributor to CCU was copper (47%), followed by aluminum (28%), zinc (11%), cadmium, iron, lead, and manganese (< 10% each).

19.3.2 Invertebrate Sampling

For the Eagle River, 264 out of 616 Hess samples contained fewer than 250 individuals (the target number). Samples with fewer than 100 individuals rarely had more than 25 total taxa, while larger samples had up to 35 to 40 (Figure 19.2). Taxa richness of mayflies was also limited for sample sizes that had less than 100 individuals identified. For metrics calculated as percentages, sample size did not lower metric values, but some metrics were more variable for lower sample sizes, such as percent heptageniid mayflies.

Samples with fewer than 100 individuals per sample were excluded from metric testing because taxa richness was constrained for small sample sizes. About 9% of the Hess samples were excluded (54 of 616). Many invertebrate samples in the REMAP data also failed to reach the specified target of 300; 68 of 96 samples had less than 300 individuals identified. Sixteen REMAP site visits with fewer than 100 individuals were excluded from metric testing. Elimination of these sites probably made the criteria for metric selection a bit more strict because these sites included the seven with the highest levels of CCU and consequently the poorest biological condition.

19.3.3 Correlation of Biological Metrics with Metal Concentration

For the Eagle River, biological metrics showed stronger correlations with CCU for spring and fall samples than for summer samples (Table 19.3 and Figures 19.3 through 19.6). This was due to the range of possible CCU values in the summer that was approximately a third of the range observed

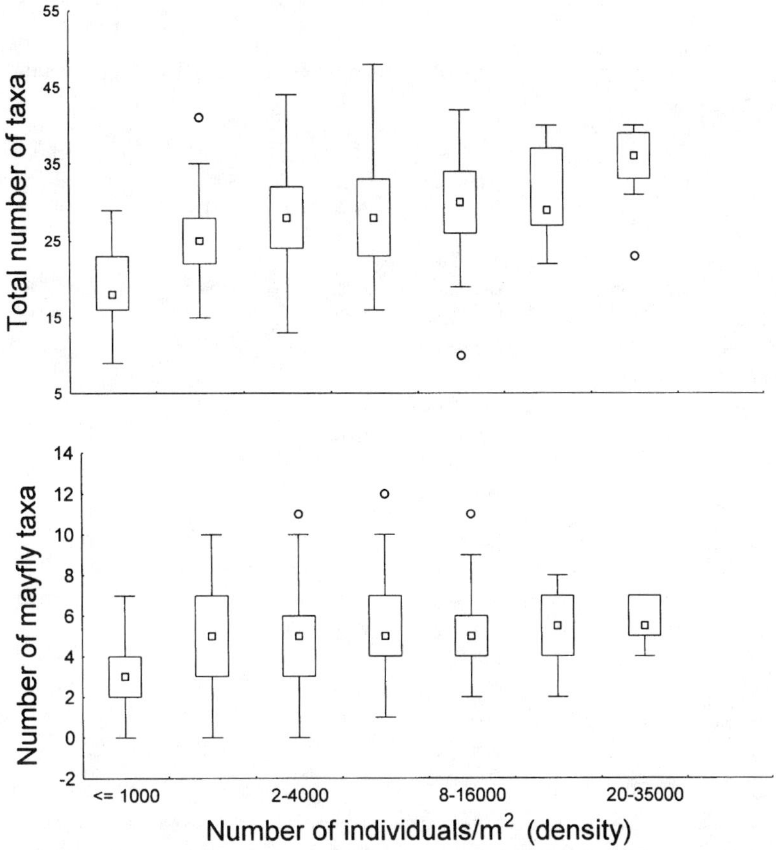

FIGURE 19.2 For Eagle River Hess samples with 100 or less individuals (density < 1000/m^2) the total number of taxa and the number of mayfly taxa was less than the number observed for samples with more individuals collected. I eliminated samples with less than 100 individuals from metric testing because the maximum value for taxa richness measures was restricted by the sample size.

for spring and half of the range for fall. In other words, metal concentrations were lower in summer; consequently, summer samples did not provide enough difference in CCU values to test metrics. Therefore, the discussion below refers to correlations for spring and fall samples.

Of the 16 candidate metrics tested for the Eagle River, 10 were significantly correlated with CCU for both spring and fall samples in the direction predicted. Total number of taxa (with and without chironomids); number of mayfly, caddisfly, metal-intolerant, clinger, and intolerant taxa; percentages of heptageniid mayflies and scrapers; and density of individuals all declined significantly as CCU increased.

Two candidate metrics, long-lived taxa richness and percent predators, were significantly correlated with CCU for the Eagle River, but in the *opposite* direction predicted from other studies. Both increased with CCU primarily due to the large number of beetles at sites with higher CCUs.

For REMAP sites, 5 of the 11 metrics tested declined as metal concentration increased (Figure 19.7). These included total number of taxa (excluding chironomids), number of mayfly, stonefly, and metal-intolerant taxa, and the percentage of heptageniid mayflies.

When scrapers were designated according to Merritt and Cummins (1996), percentage scrapers increased with CCUs for Eagle River for spring samples when metal concentrations were highest. In contrast, when scrapers were designated according to modifications by R. Wisseman, percentage scrapers *declined* significantly for fall and spring samples. Neither version of this metric was significant for the REMAP data.

TABLE 19.3
Spearman's Correlation for Candidate Metrics and Total CCU for Eagle River and REMAP Samples

Metric Name	Predicted Response	Eagle Spring N = 22	Eagle Summer 16	Eagle Fall 19	REMAP Fall 80
	B-IBI (Colorado)				
Total taxa richness without chironomids	Decrease	–0.52**	–0.54*	–0.63***	–0.25*
Mayfly taxa richness	Decrease	–0.64***		–0.71***	–0.42***
Stonefly taxa richness	Decrease				–0.34***
Caddisfly taxa richness	Decrease	–0.46*	–0.73***	–0.5*	
Metal intolerant taxa richness	Decrease	–0.55***		–0.62***	–0.49***
Clinger taxa richness	Decrease	–0.42*	–0.49*	–0.46*	
% Heptageniid mayflies	Decrease	–0.67***		–0.63***	–0.51***
	Other Metrics				
Total taxa richness (with chironomids)	Decrease	–0.48*	–0.49*	–0.56**	
% Mayflies (excluding heptageniids)	Decrease				
% Predator	Decrease	_0.58***_		_0.5*_	
% Scraper (M & C)	Decrease	_0.53**_			
% Scraper (RW)	Decrease	–0.71***		–0.65***	NC
Intolerant taxa richness (RW)	Decrease	–0.38*	–0.49*	–0.45*	NC
Long-lived taxa richness	Decrease	_0.41*_		_0.58***_	NC
% Tolerant (RW)	Increase		_–0.49*_		NC
Density per m^2	Decrease	–0.57***		–0.4*	NC

Note: Sample season and sample size for each comparison are given. For Eagle River, metric values were averaged for replicate Hess samples. Tests for significance were one-sided; $p < 0.05$ = *; $p < 0.01$ = **; $p < 0.005$ = ***. Two versions of percent scrapers were tested based on assignments by Merritt and Cummins (M & C; 1996) and R. Wisseman (RW). Underlined correlations indicate significance in the opposite direction predicted. NC = metric was not calculated.

19.3.4 Development of a Benthic Index of Biotic Integrity for Colorado

Seven metrics were selected and included in a benthic index of biotic integrity (B-IBI). Total taxa richness (excluding chironomids), number of mayfly, stonefly, caddisfly, metal intolerant, and clinger taxa, and percent heptageniid mayflies declined as CCUs increased. Total taxa richness (including chironomids) was not significant for REMAP sites but was for the Eagle River, probably because very few chironomids were identified for Eagle River sites. Total taxa richness was significant for the REMAP data when chironomids were excluded.

Stonefly taxa richness was only significantly correlated with CCU for the REMAP data set. Eagle River sites had consistently high numbers of stonefly taxa; 14 of 19 sites had more than 7 taxa. In contrast, only a third of the REMAP sites had values equal to or more than 7. This metric was retained because its strong association with mining disturbance in the larger region suggested that other aspects of mining disturbance not found in the Eagle River, such as habitat loss, may be influencing stoneflies.

Caddisfly and clinger taxa richnesses were not significantly correlated with metal concentrations for the REMAP data. The broad range of values for these metrics at sites with lower CCUs indicated that other disturbances may have limited the number of taxa in these groups. However, the consistently low values for sites with high CCU values supported the idea that metal concentration limited the number of taxa in these groups.

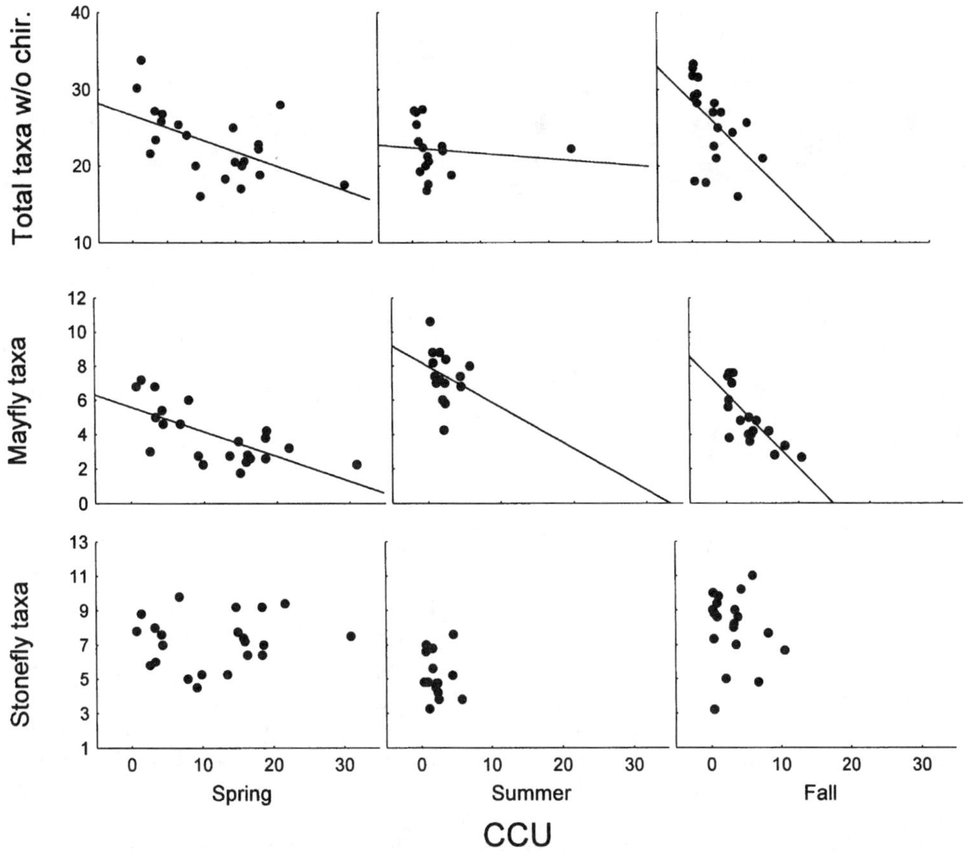

FIGURE 19.3 Metric values plotted against CCUs by season. Metric values are averages of replicate Hess samples. Only Hess samples with more than 100 individuals identified are included. Strongest correlations were found for spring and fall samples because the range of metal concentration was less for summer samples. See Table 19.3 for correlation statistics. Total taxa richness (excluding chironomids) and the number of mayfly taxa decreased as metal concentration (CCU) increased. Number of stonefly taxa was not significantly correlated with CCU.

Although intolerant taxa richness was correlated with CCUs it was somewhat redundant with metal intolerant taxa richness. Long-lived and predator taxa richness significantly correlated with disturbance, but in the opposite direction predicted. These metrics include many beetles and the observed reversal may be related to the different roles they play in high altitude streams or may indicate a specific response to mining. Additional testing is required to show that this response was consistent for mining disturbances before it could be included in an index.

Scoring criteria for metrics were based on the range of metric values for REMAP site visits (Table 19.4; Figure 19.8). All metrics but one showed a similar range of values for both data sets, which further supports the previous conclusion that the field and laboratory protocols were comparable for the data sets. The range of metric values should be similar because the range of CCU values was similar for both data sets. Scoring criteria for clinger taxa richness differed because REMAP samples included more taxa of chironomids (some of which were clingers).

19.3.5 Statistical Precision of the Index

The minimum detectable difference (MDD) for a two-sample *t*-test was calculated based on five replicate Hess samples. Power analysis determined that the MDD for the index was 8.2 based on

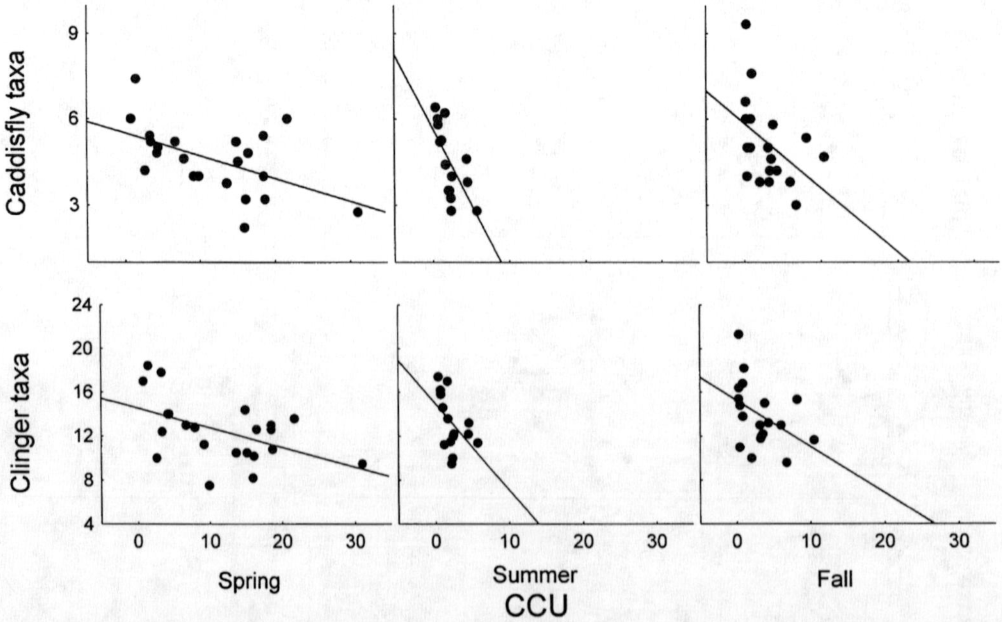

FIGURE 19.4 Metric values plotted against CCUs by season. Metric values are averages of replicate Hess samples. Only Hess samples with more than 100 individuals identified are included. Strongest correlations were found for spring and fall samples because the range of metal concentration was less for summer samples. See Table 19.3 for correlation statistics. The number of caddisfly and clinger taxa decreased for all seasons as metal concentration increased for Eagle River samples.

an estimated variance of 16.5 for all samples. Sample variance was also estimated separately for spring (s^2 = 15.3), summer (19.2) and fall (14.8) samples. Variance for summer samples was slightly higher but all three estimates were close.

Values for B-IBI ranged from 7 (for all seven metrics with scores of 1) to 35 (for all seven metrics with scores of 5). I divided the potential range of the index (35 – 7 = 28) by the MDD to get 3.4 distinct categories of biological conditions. For the three ranges of values, 7 through 16 were defined as poor, 17 through 25 as fair, and greater than 25 as good. It was also tested whether index values calculated for samples with larger numbers of invertebrates would increase the precision of the index. Other studies showed that a larger number of individuals per sample is associated with lower variability of metrics and indices (Larsen and Herlihy, 1998; Doberstein et al., 2000). To test this idea, a subset of site visits from the Eagle River data set were selected for testing. Fall samples from 1996 and 1997 from all eight monitoring sites were used. Two new composite samples from the original five replicates were created by combining the three samples with the fewest individuals and the two samples with the most individuals. Thus, each of the new samples included more individual invertebrates than did the original replicate samples.

Estimates of variability for the index were lower for combined samples than for the original five replicate Hess samples, 11.25 compared to 16.5. For a sampling protocol based on five replicate samples, each with a similar, larger sample size, the number of categories of biological condition that could be detected using B-IBI increased from 3.4 to 4.1, a 20% increase in precision.

19.3.6 Correlation of B-IBI with Eagle River Mining Disturbance

B-IBI values declined significantly as metal concentration increased (Figure 19.9) for sites on the Eagle River. Although CCUs were typically lowest for reference sites, they varied for mine sites and sites downstream. B-IBI values were lowest for mining sites; upstream reference sites were

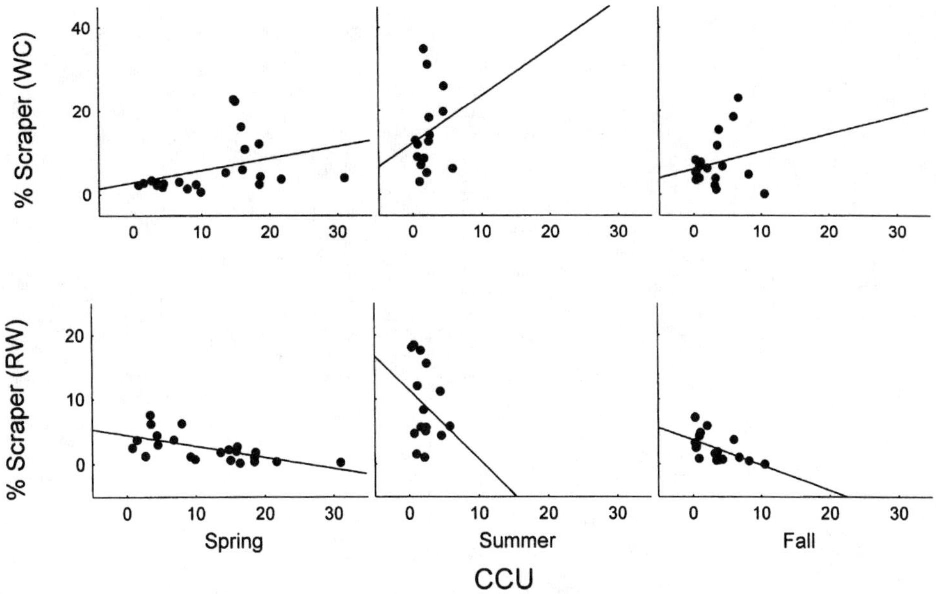

FIGURE 19.5 Metric values plotted against CCUs by season. Metric values are averages of replicate Hess samples. Only Hess samples with more than 100 individuals identified are included. Strongest correlations were found for spring and fall samples because the range of metal concentration was less for summer samples. See Table 19.3 for correlation statistics. Percent scrapers increased slightly with CCU (the opposite of the predicted response) for assignments based on Merritt and Cummins (1996). Assignments made by R. Wisseman declined with metals.

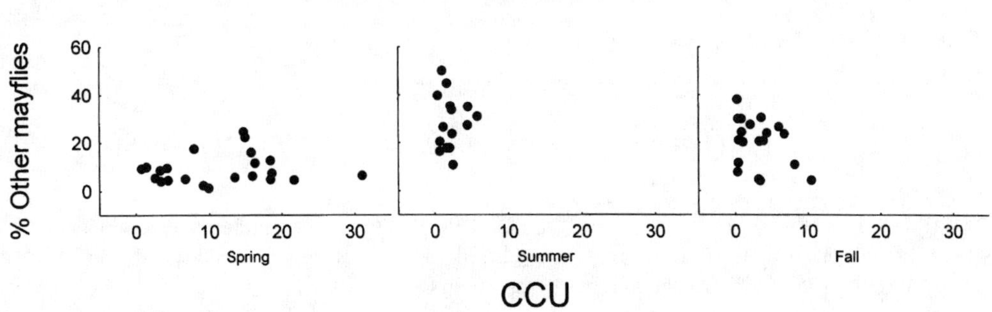

FIGURE 19.6 Metric values plotted against CCUs by season. Metric values are averages of replicate Hess samples. Only Hess samples with more than 100 individuals identified are included. Strongest correlations were found for spring and fall samples because the range of metal concentration was less for summer samples. See Table 19.3 for correlation statistics. Percent heptageniid mayflies declined as CCU increased. Percent other mayflies was not correlated with CCU.

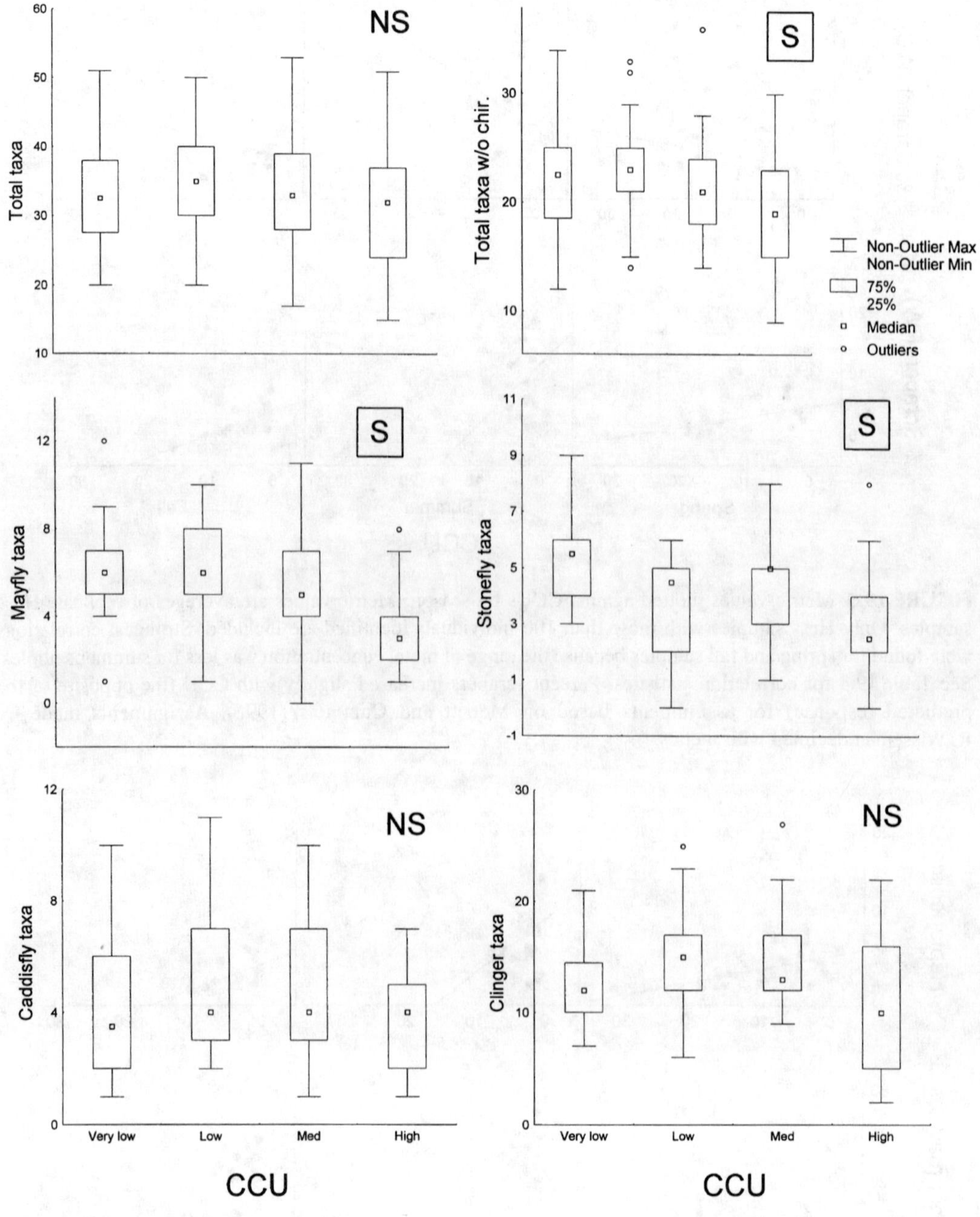

FIGURE 19.7 Eleven metrics plotted against CCUs for REMAP sites. S = a significant correlation with CCU. NS = not significant (n = 80). Box categories were based on quartiles for CCU values and defined as very low (<0.75), low (0.75 to 1.3), medium (1.3 to 3), and high (>3).

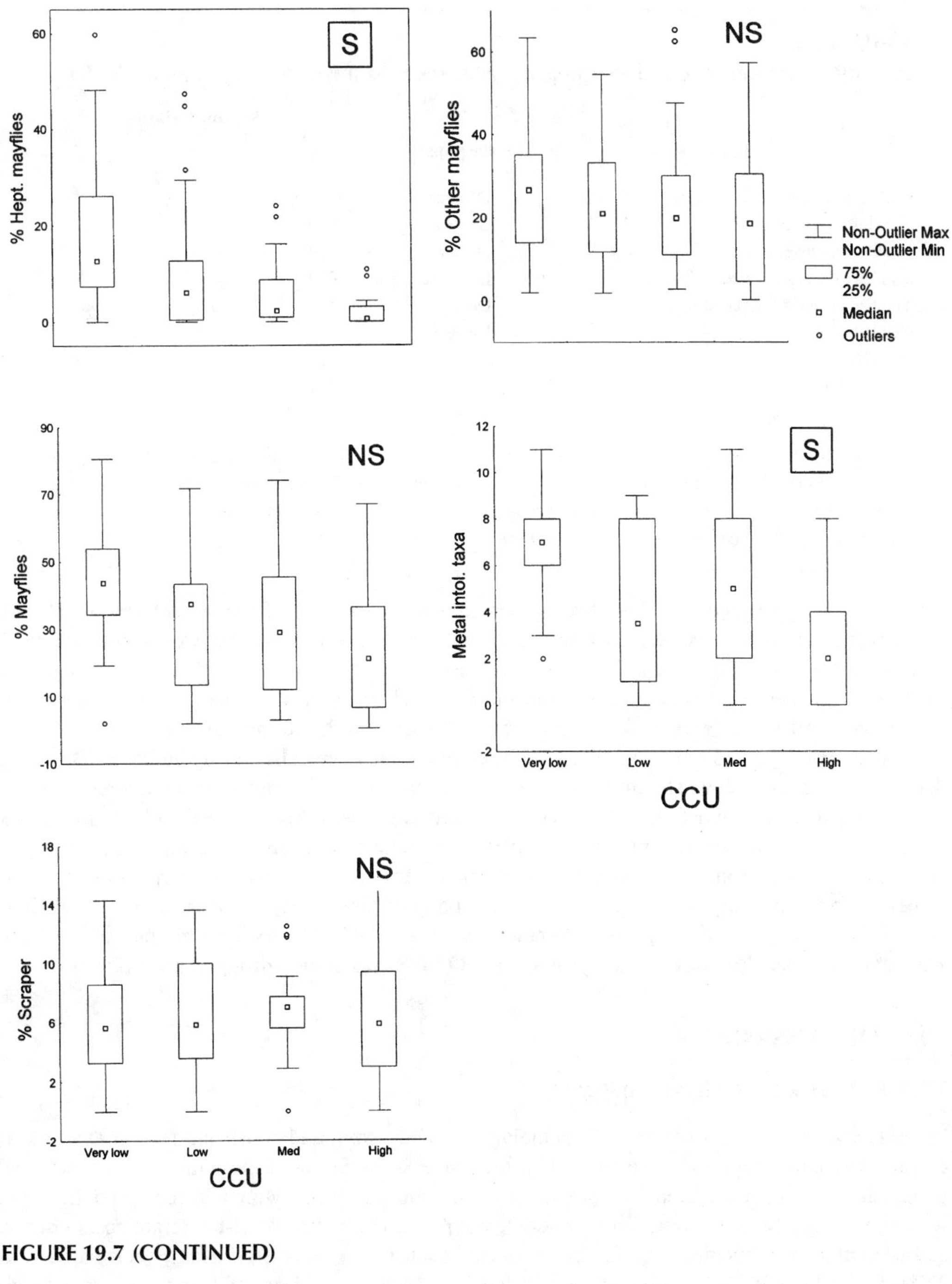

FIGURE 19.7 (CONTINUED)

TABLE 19.4
Scoring Assignments Used to Integrate Metrics into a Multimetric Index (B-IBI)

Metric	Response	Scoring Criteria 1	3	5
Total number of taxa *excluding chironomids*	Decrease	[0, 20]	(20, 28]	>28
Number of mayfly taxa	Decrease	[0, 4]	(4, 8]	>8
Number of stonefly taxa	Decrease	[0, 4]	(4, 7)	> = 7
Number of caddisfly taxa	Decrease	[0, 4]	(4, 7]	>7
Number of metal intolerant taxa	Decrease	[0, 3]	(3, 6]	>6
Number of clinger taxa	Decrease			
REMAP		[0, 11]	(11, 19]	>19
Eagle River		[0, 11]	(11, 14)	> = 14
% Heptageniid mayflies	Decrease	[0, 3.2]	(3.2, 18]	>18

Note: Different scoring assignments are given for Eagle River for clinger taxa because of the many fewer chironomids identified for Eagle River samples. Square brackets indicate closed intervals. The value next to the bracket is included in the range. Round brackets indicate open intervals. The value is not included in the range. For example, for a sample with a total of 20 taxa, the metric score would be 1.

almost always higher and sites downstream from the mine were often higher as well (Figure 19.10). For spring and fall samples, reference sites had higher index values than the two most downstream mine sites for almost all years. One site, ER4, had index values that reflected its location as a transitional site between reference condition and mine influence. In some years, ER4's index values were more similar to reference sites and other years more similar to the mine sites.

Fall samples showed a similar pattern with reference sites, typically having higher B-IBI values than mine sites. Sites downstream of the mined area showed some improvement and recovery for some years in spring and fall, but the final site downstream of Avon tended to have low index values, possibly indicating the influences of urbanization. Patterns were similar for summer samples but much less clear. Summer index values were more variable than values for spring or fall. B-IBI values improved slightly at two of the three mining sites (Figure 19.11). At sites ER4 and ER9, index values for spring samples were somewhat higher in recent years. Fall samples at the same sites increased steadily until the last year sampled (1998) when they dropped dramatically.

19.4 DISCUSSION

19.4.1 Patterns in Metric Response

Several measures of the invertebrate assemblage showed strong and consistent responses to metal contamination on a regional scale and within the Eagle River. Seven of these metrics were selected to include in a B-IBI calibrated for Colorado and the Southern Rocky Mountain ecoregion. Included were total taxa richness (without chironomids), mayfly taxa richness, metal-intolerant taxa richness, and percentage heptageniid mayflies in the index because they were significantly correlated with CCUs for both the REMAP and Eagle River data sets. Including chironomids in total taxa richness was not significant for REMAP sites, probably because an intense laboratory effort was used in the REMAP project to identify chironomids to genus and often species. Chironomids, unlike most other benthic insects, prefer soft sediment, which is typically associated in mountain streams with human activities that cause erosion and sedimentation.

Stonefly richness strongly correlated with CCUs for the REMAP data but not for the Eagle River. Stoneflies are among the most sensitive groups and several of the metal-intolerant taxa are stoneflies. This metric was included in the index because of its strong correlation with metals at the regional scale and because the number of stonefly taxa found at Eagle River sites was consistently high (over seven at most sites). This suggests that some activity associated with mining at other sites was not present on the Eagle River.

Caddisfly and clinger taxa richness were included as metrics because they were strongly correlated with metal concentration in the Eagle River. The primary human disturbance at Eagle River was related to mining and both metrics were also higher for reference sites than for mining sites. For the regional REMAP data, caddisfly and clinger taxa richness were not significantly correlated with CCU; however, both were limited by high metal concentrations. Lack of statistical significance was due to a mix of high and low numbers of taxa in sites with low metal concentrations. Of the 16 sites with the highest metal concentrations (CCUs more than 6 times the EPA limit), none had more than 4 caddisfly taxa. In contrast, sites with CCU values below the EPA limit had up to 9 different taxa.

Density was not included as a metric because it is very difficult to compare across studies. Density estimates are extremely variable and depend on time spent sampling, intensity of sampling, type of sampler, and crew experience. Density can also vary according to seasonal and climatic changes and other sources of natural variability. Nonetheless, very low density is a strong indicator of metal contamination.

The relationship between percentage scrapers and metal concentration depended on the assignments used to designate invertebrates as scrapers. This metric was excluded because of its inconsistent response to metals and its dependence on the invertebrates that are included. For designations based on Merrit and Cummins (1996), percentage scrapers were not significantly correlated with metal concentration, but tended to show an increase; for designations made by Wisseman, percentage scrapers *decreased* significantly for Eagle River sites.

Two metrics included in B-IBI were specifically related to metal contamination: metal intolerant taxa richness and percent heptageniid mayflies. The other five metrics were also associated with metal contamination but are known to be sensitive to other types of human disturbance as well. An index based on the sensitivity of invertebrates to metals alone could be developed, but might fail to capture other disturbances associated with mining such as loss of riparian cover, flow alteration, or loss of microhabitat in the cobble due to sediment from erosion (Karr et al., 2000).

19.4.2 Measuring Human Influence

The most difficult aspect of testing biological metrics, particularly on a large regional scale, is defining a realistic measure of human disturbance. Livestock grazing, agriculture, urbanization, and mining occur at different spatial scales and at varying intensities across the landscape. No perfect solution or variable exists for quantifying human disturbance and every measure is a compromise. The REMAP and the Eagle River sampling designs focused on a single type of human disturbance (i.e., mining) and a single aspect of mining (i.e., heavy metal contamination). The emphasis on mining limits our approach to metric testing for this region, and the exclusive focus on metals narrows it further.

Mining involves more than metal contamination. Development and construction associated with mining increase erosion and sediment in stream channels, alter hydrological flow patterns through mining tunnels and adits, damage riparian vegetation, and restructure the food web when primary producers such as algae are lost due to metal contamination. Metric selection for B-IBI reflects the general and specific nature of human disturbance. Metal-intolerant taxa richness and percent heptageniid mayfly metrics were selected specifically for their sensitivity to metal contamination. Other metrics address more general aspects of disturbance associated with mining and other human activities. Clinger taxa richness declines with sedimentation, stonefly taxa richness declines with loss of riparian vegetation, and an overall decline in total taxa richness follows the general loss of habitat complexity.

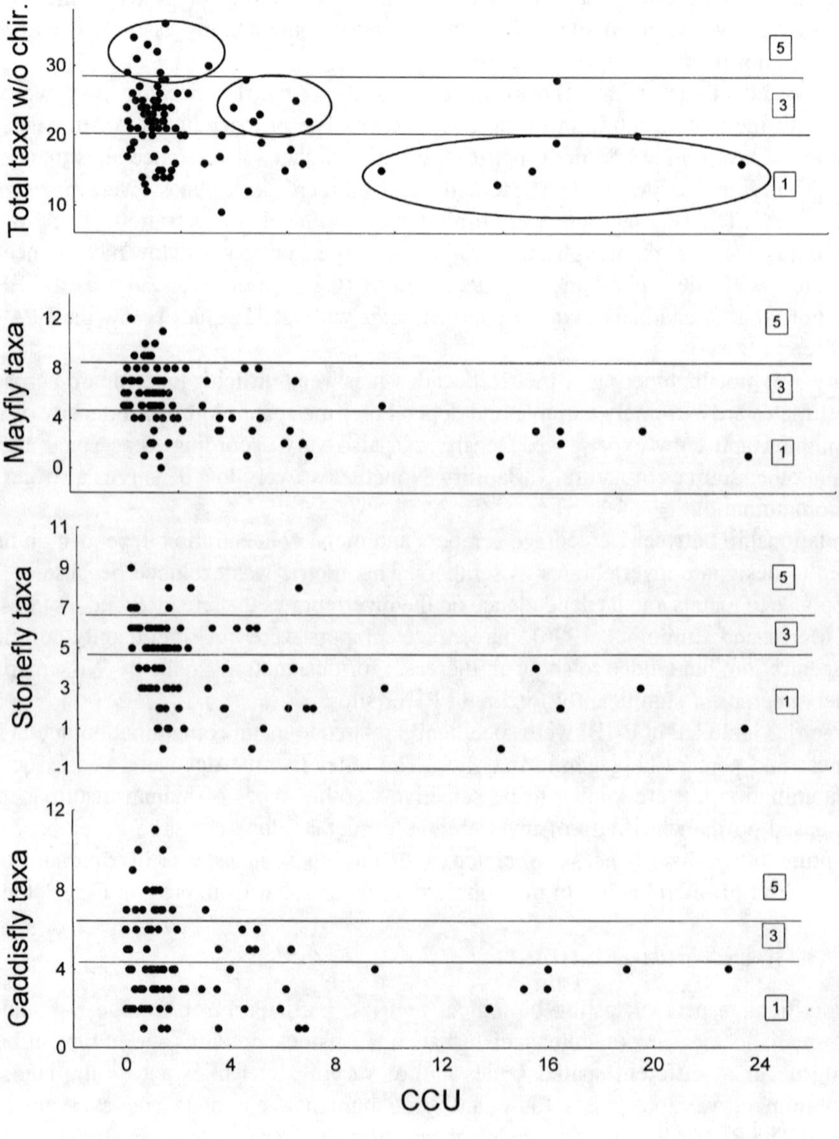

FIGURE 19.8 B-IBI metrics plotted against CCUs for REMAP samples (n = 89). Scoring criteria for B-IBI are indicated by lines and scores (1, 3, or 5) on the right of each graph. Circles in the top panel indicate groups of sites used to set scoring criteria.

In the Southern Rocky Mountain ecoregion, heavy metal contamination is widespread, so metal concentration was an appropriate measure of human influence. Other activities known to degrade streams were not quantified for this study. For example, caddisfly and clinger taxa richness metrics were not significantly correlated with metal concentration for the REMAP data set; although they were much lower at sites with high metal concentrations. Many sites with low metal concentrations also had a mix of high and low values for these metrics. Lack of statistical significance was probably due to other types of disturbance, e.g., grazing, farming, or urbanization, that degraded these sites in the absence of metals. Measures of land cover and land use derived from satellite data would probably help explain poor biological conditions at sites without heavy metals.

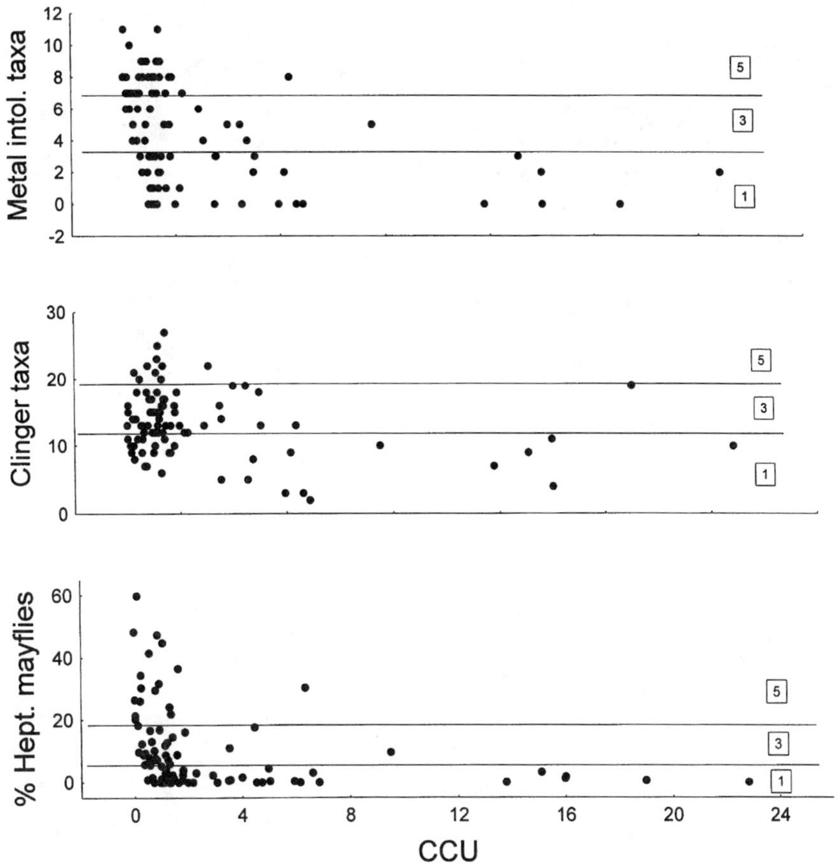

FIGURE 19.8 (CONTINUED)

For reference sites and mine sites in the Eagle River, confounding human influences were less of a concern because the primary difference between the two groups of sites was the presence of mining activities. However, further downstream urbanization and development also influenced the invertebrate assemblage.

19.4.3 Pseudoreplication

Power analysis assumes independence of sample sites and replicates (Peterman, 1990; Hurlbert, 1984). Replicates within a river are problematic because some of the same water flows through all the sites (Heffner et al., 1996). A strict definition of independence specifies, for example, that multiple fish within a single tank are not true, independent replicates because all are affected by the same water in the tank. In contrast, fish in separate tanks are true replicates. Interpretation of independence can be taken to extremes, for example, requiring that the fish in separate tanks also be located in different rooms or cities. The definition of independence for replicate samples is not an absolute and depends somewhat on the situation.

When conducting statistical tests, such as hypothesis testing or power analysis, the primary concern when sampling units are not independent is that another shared factor not considered in the sampling design is actually causing the observed differences rather than the independent factors tested by the experiment (Dunham and Vinyard, 1997). For example, some variable such as the

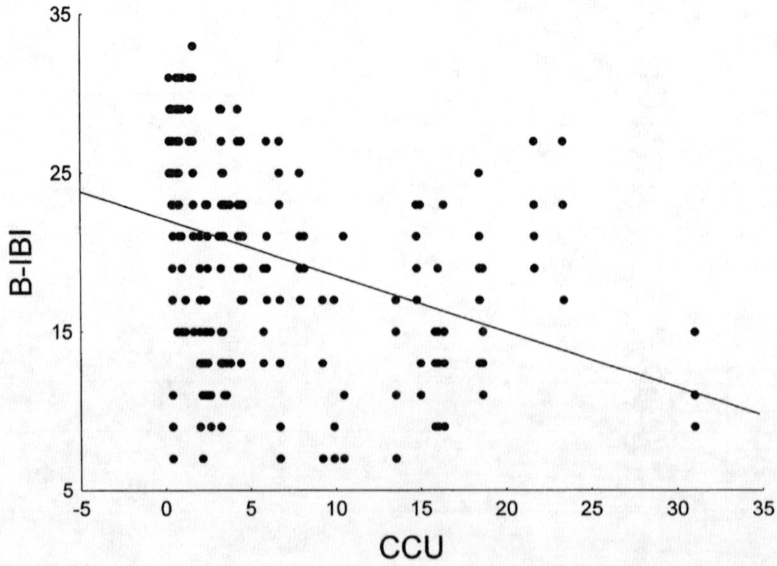

FIGURE 19.9 B-IBI was significantly correlated with total CCU for Eagle River Hess samples collected in the fall (Spearman's $r = -0.47$, $p << 0.001$). B-IBI is shown for each replicate sample with more than 100 individuals identified ($n = 236$). The least-squares regression line is drawn.

presence of a poisonous plant rather than metal from a mine may be influencing the invertebrate community of the Eagle River. By sampling across an independent population of sites, such as different drainages in the ecoregion, the influence of site-specific differences is randomized and, hopefully, eliminated so that the influence of the target variable, e.g., metal concentration, can be fairly tested.

What is statistically preferable is not always logistically possible (Stewart-Oaten et al., 1992). In such cases, one relies on logical argument to connect the evidence to the conclusions (Beyers, 1998). For the Eagle River, it is reasonable to assume that higher B-IBI values at reference sites and lower values at mine sites are results of mining activities. Analysis and conclusions based on the larger scale REMAP data support this conclusion. Alternative explanations may be considered, but a similar level of evidence would have to be supplied for them to be credible.

A second concern is that similar sites will lead to underestimates of variability. We expect smaller differences between sites within a river because they have been influenced by the same processes and events, such as geological formation or soil type. For this reason, variability may be underestimated and power estimates higher than for an analysis based on truly independent samples (Dunham and Vinyard, 1997). Because same-day replicate samples from the Eagle River were used to estimate statistical power, pseudoreplication should be considered before applying these results outside the Eagle River watershed.

19.4.4 Defining Thresholds for Impairment

Statistics can be used to define the number of categories of biological condition that B-IBI can detect, but stakeholders must decide what level of biological condition represents an acceptable standard and what represents impairment. The statistical precision of B-IBI was estimated using power analysis. One way to understand this approach is to think of B-IBI as a yardstick ranging from 7 (very poor biological condition) to 35 (excellent condition). Power analysis defines how many tick marks are on the yardstick. For sampling protocols used on the Eagle River, the tick marks were 8.2 units apart. In other words, B-IBI values for sites must be at least 8.2 units apart to be significantly different. One way to apply this result to the Eagle River would be to define

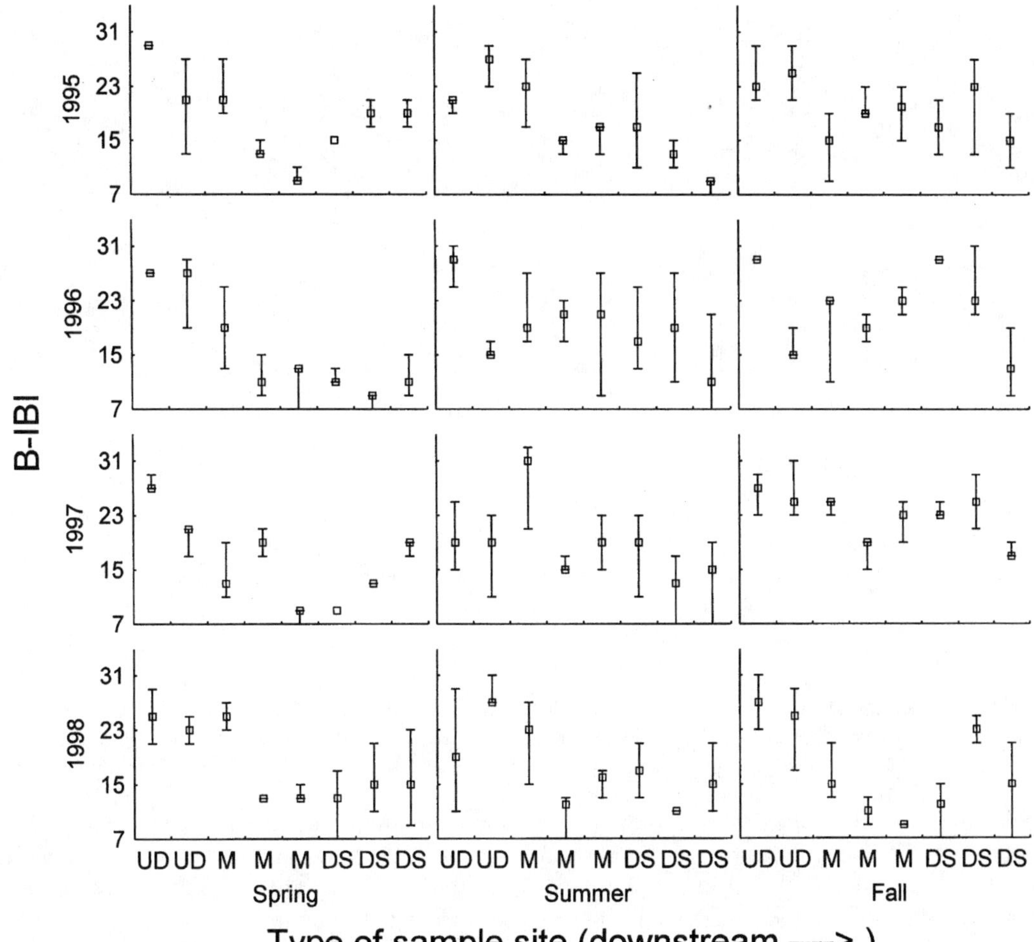

FIGURE 19.10 Ranges for B-IBI vales are shown for Eagle River sites by year and season. B-IBI values were higher for undisturbed sites (UD) located upstream than for mine sites (M) for most years. Differences in B-IBI values for undisturbed and mine sites were most extreme during spring and fall and less consistent for summer samples. Downstream (DS) sites had somewhat higher B-IBI values for fall and spring samples.

acceptable biological condition at the mine sites as B-IBI values within 8.2 units of the index value observed for reference sites.

The same level of sampling effort in headwater or high elevation streams often yields fewer individuals than similar sampling at lower elevations. Headwater streams are often nutrient poor with characteristically lower densities of invertebrates. An equivalent number of individuals from these types of streams, however, will often yield more taxa if human disturbance is absent or minimal. The current field sampling method for the Eagle River consistently failed to provide the specified target of 250 individuals. If the sampling protocol is changed to yield a larger number of individuals per sample, statistical precision will increase.

19.5 CONCLUSIONS

In the western United States, mining for heavy metals is a common source of degradation in mountain streams. Biological criteria based on B-IBI are appropriate for monitoring metal

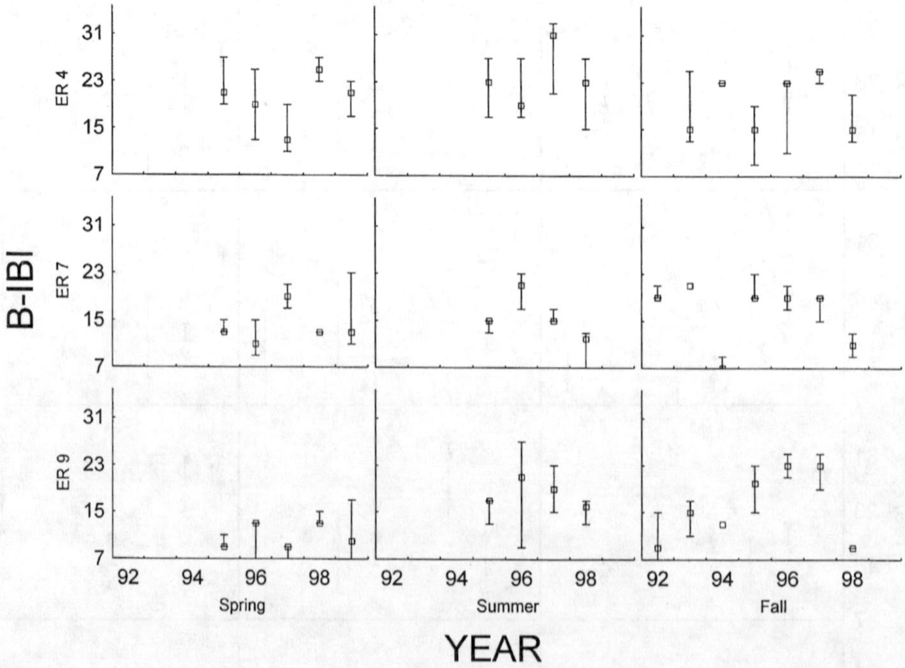

FIGURE 19.11 B-IBI values plotted by year for three Eagle River mine sites. B-IBI values for spring samples improved somewhat for ER4 and ER9 in recent years. B-IBI values for summer samples showed no change over time. Fall B-IBI values showed a steady increase at ER4 and ER9 over time until a sudden drop in 1998.

contamination and remediation efforts at these mine sites for three reasons. First, B-IBI was strongly correlated with metal concentration both on the Eagle River and in the larger ecoregion where streams with higher metal concentrations had significantly lower B-IBI values. In addition, the individual metrics included in the index were associated with mining disturbance at the regional scale; therefore, results reported here were not unique to the Eagle River. Second, the index had sufficient statistical power to detect 3.4 categories of biological condition based on the monitoring design currently used on the Eagle River. Therefore, a difference in B-IBI values greater than 8 points indicates a significant difference in biological condition. Third, when sites were compared spatially, B-IBI was consistently higher at upstream reference sites compared to mine sites on the Eagle River. For most years, index values at the upstream reference sites fell in the highest category of biological condition and index values for the two most downstream mine sites were in the middle or lowest category. The most upstream mining site was transitional, with B-IBI values similar to reference sites in some years and closer to mine sites in others. When site condition was compared through time, B-IBI values at two of the mine sites improved while the site closest to the mine showed no change. Thus, B-IBI can be used to compare the biological condition of sites within a single stream, across a region, or through time.

ACKNOWLEDGMENTS

Funding for this project was provided by the U.S. Environmental Protection Agency, Region 8, Denver, CO. Dames and Moore, Inc. collected the invertebrate and water chemistry samples. Chadwick and Associates, Inc. identified the invertebrates and S. Canton provided detailed information on laboratory protocols. J. Volosin, Parametrix, Inc. generously shared data files. R. Wisseman, Biological Associates, Inc., provided natural history information for invertebrates. J. Lazorchak, EPA, Cincinnati, OH provided background information for REMAP protocols.

W. Clements, Colorado State University, shared unpublished results from his analysis of the REMAP data. B. Wuerthele, EPA, Region 8, provided guidance and support. Discussions with W. Clements, L. Martin, D. Parachini, B. Quinlan, G. Taylor, J. Volosin, J. Woodling, and B. Wuerthele greatly improved this chapter.

REFERENCES

Barbour, M.T., J. Gerritsen, B.D. Snyder, and J. B. Stribling. 1999. *Rapid Bioassessment Protocols for Use in Wadeable Streams and Rivers: Periphyton, Benthic Macroinvertebrates, and Fish.* EPA 841-D-97–002. U.S. Environmental Protection Agency, Office of Water, Washington, D.C..

Beyers, D.W. 1998. Causal inference in environmental impact studies, *Journal of the North American Benthological Society,* 17, 367–373.

Carlisle, D.M. and W.H. Clements. 1999. Sensitivity and variability of metrics used in biological assessments of running waters, *Environmental Toxicology and Chemistry,* 18, 285–291.

Clements, W.H. and D. Carlisle. 1998. *Development and Validation of a Regional Biotic Index to Assess the Effects of Heavy Metals on Headwater Streams of the Southern Rocky Mountain Ecoregion.* U.S. Environmental Protection Agency, Region 8, Denver, CO.

Clements, W.H., D.M. Carlisle, J. M. Lazorchak, and P.C. Johnson. 2002. Heavy metals structure benthic communities in Colorado mountain streams. *Ecological Applications,* 10, 626–638.

Davis, W.S., B.D. Snyder, J.B. Stribling and C. Stoughton. 1996. *Summary of State Biological Assessment Programs for Streams and Rivers.* EPA 230-R-96–007.U.S. Environmental Protection Agency, Office of Policy, Planning, and Evaluation, Washington, D.C.

Doberstein, C.P., J. R. Karr, and L.L. Conquest. 2000. The effect of fixed-count subsampling on macroinvertebrate biomonitoring in small streams, *Freshwater Biology,* 44, 355–371.

Dunham, J.B. and G.L. Vinyard. 1997. Incorporating stream level variability into analyses of site level fish habitat relationships: some cautionary examples, *Transactions of the American Fisheries Society,* 126, 323–329.

Fore, L.S., J. R. Karr, and L.L. Conquest. 1994. Statistical properties of an index of biotic integrity used to evaluate water resources. *Canadian Journal of Fisheries and Aquatic Sciences,* 51, 1077–1087.

Fore, L.S., K. Paulsen, and K. O'Laughlin. 2001. Assessing the performance of volunteers in monitoring streams, *Freshwater Biology,* 46, 109–123.

Genter R.B. and R.M. Lehman, 2000. Metal toxicity inferred from algal population density, heterotrophic substrate use, and fatty acid profile in a small stream, *Environmental Toxicology and Chemistry,* 19, 869–878.

Heffner R.A., M.J. Butler IV, and C.K. Reilly. 1996. Pseudoreplication revisited, *Ecology,* 77, 2558–2562.

Hurlbert, S.H. 1984. Pseudoreplication and the design of ecological field experiments, *Ecological Monographs,* 54, 187–211.

Jessup, B. and J. Gerritsen. 2000. *Development of a Multimetric Index for Biological Assessment of Idaho Streams Using Benthic Macroinvertebrates.* Tetratech, Owings Mills, MD.

Karr, J.R. 1981. Assessment of biotic integrity using fish communities, *Fisheries,* 6(6), 21–27.

Karr, J.R. 1991. Biological integrity: a long-neglected aspect of water resource management, *Ecological Applications,* 1, 66–84.

Karr, J.R. 1998. Rivers as sentinels: using the biology of rivers to guide landscape management, in R.J. Naiman and R.E. Bilby (Eds.). *River Ecology and Management: Lessons from the Pacific Coastal Ecosystem.* Springer-Verlag, New York, 502–528.

Karr J.R., J.D. Allan, and A.C. Benke, 2000. River conservation in the United States and Canada: science, policy, and practice, in P.J. Boon, B.R. Davies, and G.E. Petts (Eds.). *River Conservation: Science, Policy, Practice.* J. Wiley & Sons, New York.

Karr, J.R. and E.W. Chu. 1999. *Restoring Life in Running Waters: Better Biological Monitoring.* Island Press, Washington, D.C.

Karr, J. R. and E.M. Rossano. 2001. Applying public health lessons to protect river health, *Ecology and Civil Engineering,* 4, 3–18.

Kuwabara, J.S. 1985. Phosphorus-zinc interactive effects on growth by *Selenastrum capricornutum* (Chlorophyta), *Environmental Science and Technology,* 19, 417–421.

Larsen, D.P. and A.T. Herlihy. 1998. The dilemma of sampling stream for macroinvertebrate richness, *Journal of the North American Benthological Society*, 17, 359–366.

Li, J., A. Herlihy, W. Gerth, P. Kaufmann, S. Gregory, S. Urquhart, and D.P. Larsen. 2001. Variability in stream macroinvertebrates at multiple spatial scales, *Freshwater Biology*, 46, 87–97.

Merritt, R.W. and K.W. Cummins (Eds.). 1996. *An Introduction to the Aquatic Insects of North America*, 3rd ed. Kendall/Hunt, Dubuque, IA.

Miltner, R.J. and E.T. Rankin. 1998. Primary nutrients and the biotic integrity of rivers and streams, *Freshwater Biology*, 40, 145–158.

Morley, S.A. and J.R. Karr. In press. Assessing the biological health of urban streams: tools for restoration and conservation, *Conservation Biology.*

Paulsen, S.G., R.M. Hughes, and D.P. Larsen. 1998. Critical elements in describing and understanding our nation's aquatic resources, *Journal of the American Water Resources Association*, 34, 995–1005.

Peterman, R.M. 1990. Statistical power analysis can improve fisheries research and management, *Canadian Journal of Fisheries and Aquatic Sciences*, 47, 2–15.

Ransel, K.P. 1995. The sleeping giant awakes: PUD No. 1 of Jefferson County v. Washington Department of Ecology, *Environmental Law*, 25, 255–283.

Stewart-Oaten, A., J.R. Bence, and C.W. Osenberg. 1992. Assessing effects of unreplicated perturbations: no simple solutions, *Ecology*, 73, 1396–1404.

U.S. Environmental Protection Agency. 1986. *Quality Criteria for Water.* EPA 440/5–86–001. U.S. Environmental Protection Agency, Office of Water, Washington, D.C.

Zar, J. H. 1984. *Biostatistical Analysis*, 2nd ed. Prentice-Hall, Englewood Cliffs, NJ.

Section V

Case Studies

20 Patterns in Water Quality and Fish Assemblages in Three Central Indiana Streams with Emphasis on Animal Feed Lot Operations

James R. Gammon, Wayne C. Faatz, and Thomas P. Simon

CONTENTS

0-8493-0905-0/03/$0.00+$1.50

20.1 INTRODUCTION

Concentration of animals on land causes direct and indirect effects on aquatic systems. Perhaps the most obvious impact is toxicity caused by increased runoff of animal waste into streams, especially ammonia and nitrite. These increases in ammonia and nitrite may be enhanced by changes in pH. These episodic events may cause degradation of water quality to the extent that fish kills are observed. However, such events are difficult to predict and may be due to a variety of causes including poor disposal practices, storm events, and differential densities of animal populations. Differences may also be attributed to the types of animals; however, this has not been well documented (Morris et al., Chapter 6, this volume). In addition, the land use associated with concentrations of animals typically results in channelization of streams and loss of riparian habitat. The increase of soil loss from animal movements and the sloughing of banks cause significant amounts of erodible soil movement through stream channels. This may often be a result of the need for access for animals but equally more likely is the encroachment of tillable acres.

Beginning as a comparison of 1960s to 1990s fish communities, this investigation quickly became an assessment of impacts from concentrations of farm animals after chemical testing revealed that contamination of west central Indiana streams extended far upstream into headwaters. A contemplated two-year study at approximately 70 locations scattered throughout three stream systems more than doubled in magnitude. The expectation that several examples of streams harboring excellent fish populations would be found was unrealized. Almost every stream was polluted to some extent at its source, usually by agricultural activities.

Regional land use is primarily agricultural. Corn and soybeans fields are prominent landscape features. Domesticated animals include beef cattle, hogs, and dairy cows, and minor numbers of sheep, horses, donkeys, poultry, elk, llamas, and bison. These animals are usually fenced or otherwise confined, and density varies considerably.

In recent years, the numbers and sizes of confined animal feed lot operations (CAFOs) have rapidly expanded. A few large factory farm operations exist, and CAFOs producing tens of thousands of hogs annually are not uncommon and are similar to the basic problems of human waste management. Often accompanying such operations are dozens of breeders pastured on land unsuitable for row crops, i.e., rough land adjacent to streams. Dairy farms present similar environmental problems although they are not as common as hog operations in this area of Indiana where hogs outnumber people.

Human society finds it imperative to treat our wastes, wherever we live, before they enter waterways. However, wastes from concentrated animals are treated only when the numbers are sufficiently high, and the treatment involves relatively simple, ineffective methods such as anaerobic lagoons. We face major problems in storing, handling, and disposing of wastes produced by concentrations of farm animals.

Most of the low order streams studied here are too small and numerous to include in routine monitoring programs. Nevertheless, they are inordinately important because they are the sites where pollutants and nutrients first enter the aquatic ecosystem and begin their journey to the sea. They are also the sites where efforts to reduce nutrient inputs must be applied. The "Dead Zone" in the Gulf of Mexico and the dissolved oxygen deficits and opacities of large rivers in the Corn Belt are directly traceable to the agricultural modifications of these tiny brooks.

20.2 METHODS

20.2.1 STUDY AREA

Big Raccoon, Big Walnut, and Deer Creek stream systems are located in central Indiana (Figure 20.1). Big Raccoon is a tributary of the Wabash River. Big Walnut and Deer Creek are tributaries of the Eel River drainage of the West Fork White River watershed, which drains into the Wabash River. These watersheds are located in the Eastern Corn Belt Plain and the Interior River Lowland ecoregions (Omernik and Gallant, 1988). A total of 91 sites located in these streams were sampled in 1993. Thirty-two sites were added the following summer, and 10 more sites were added in 1995 to better bracket feed lots and pastures during changing hydrologic conditions. Most sites were located near bridges for access. Some were intentionally located downstream from animal feed lots in an attempt to evaluate effects on the biotic community. It was necessary to add several new locations in upper headwaters during the course of the study because many small streams were found to have high ammonia concentrations, which eliminated them as reference sites.

Hydrologic conditions during the summer of 1993 were unusually wet and required organizing the collecting effort around rainstorm events and frequent periods of high water. Most of the smaller tributaries were sampled in early summer because they were the first to return to normal flows after a rainstorm event. Some stations on the mainstem of Big Walnut Creek were not sampled because of unfavorably high flows throughout most of the summer. All mainstem stations were, however, sampled in the summer of 1994 when weather patterns and stream flow were more normal. The summer of 1995, on the other hand, was unusually hot and dry and some headwater stations were completely dry in late summer.

Little Deer Creek originates in corn fields 13 km east of Greencastle and flows south and west approximately 9 km to join Deer Creek. Its headwaters are strongly affected by a large hog operation located on both sides of State Road 240. Four sampling sites were established on this creek in 1993 for fish and habitat assessment and water chemistry. Site LDC0, a small roadside pool directly fed by a field tile, was sampled to provide ambient water chemistry only. Site LDC1 was located about 0.4 km south of the hog operation at a bridge. Sites LDC2, LDC3, and LDC4 were located at bridges about 1.5, 4.8, and 7.7 km south of the hog operation, respectively. Site LDC2.5 was established at a bridge located between LDC2 and LDC3 during fall 1994. Another station was added in 1995 at a tributary that enters Deer Creek a short distance upstream from LDC2.5 after flowing through extensive cattle pastures to the north.

Deweese Creek flows southward from its two primary headwaters to join Deer Creek near Manhattan. Potential influences in the headwaters include discharges from the Lone Star cement plant where surface water passes through a series of three ponds before flowing into Deweese Creek (station DW2). In past years, it was reported that fish had difficulty surviving in the upper pond because of pollution, but recent reports indicate that fish are present. The eastern headwater passes through pastures (station DW1) and may be influenced negatively by an animal feed lot.

The headwaters of Owl Branch seep from agricultural fields just south of the eastern fringes of Greencastle. The stream soon enters an unusually well forested valley that has gradually undergone conversion to home development and ultimately flows into Deer Creek. Station OB1a was located on a small Order 0 tributary that drains an extensive hog and cattle operation a few miles southeast of Greencastle. This stream flows southward to join lower Owl Branch where two more sampling stations (OB1 and OB2) bracket the OB1a tributary. Station RI1 was located on a small intermittent tributary that enters Deer Creek about 0.8 km upstream from the mouth of Owl Branch. It was selected as an ambient comparison for station OB1a. It is very similar in terms of size of its watershed area (DBA = 0.5 mi^2) and physical characteristics to the OB1a tributary, but lacks inputs from animal feed lots.

A tributary of Big Raccoon Creek, the headwaters of North Ramp Creek, originate about 3 km south of Roachdale. The stream flows south-southwesterly through gently rolling terrain dominated

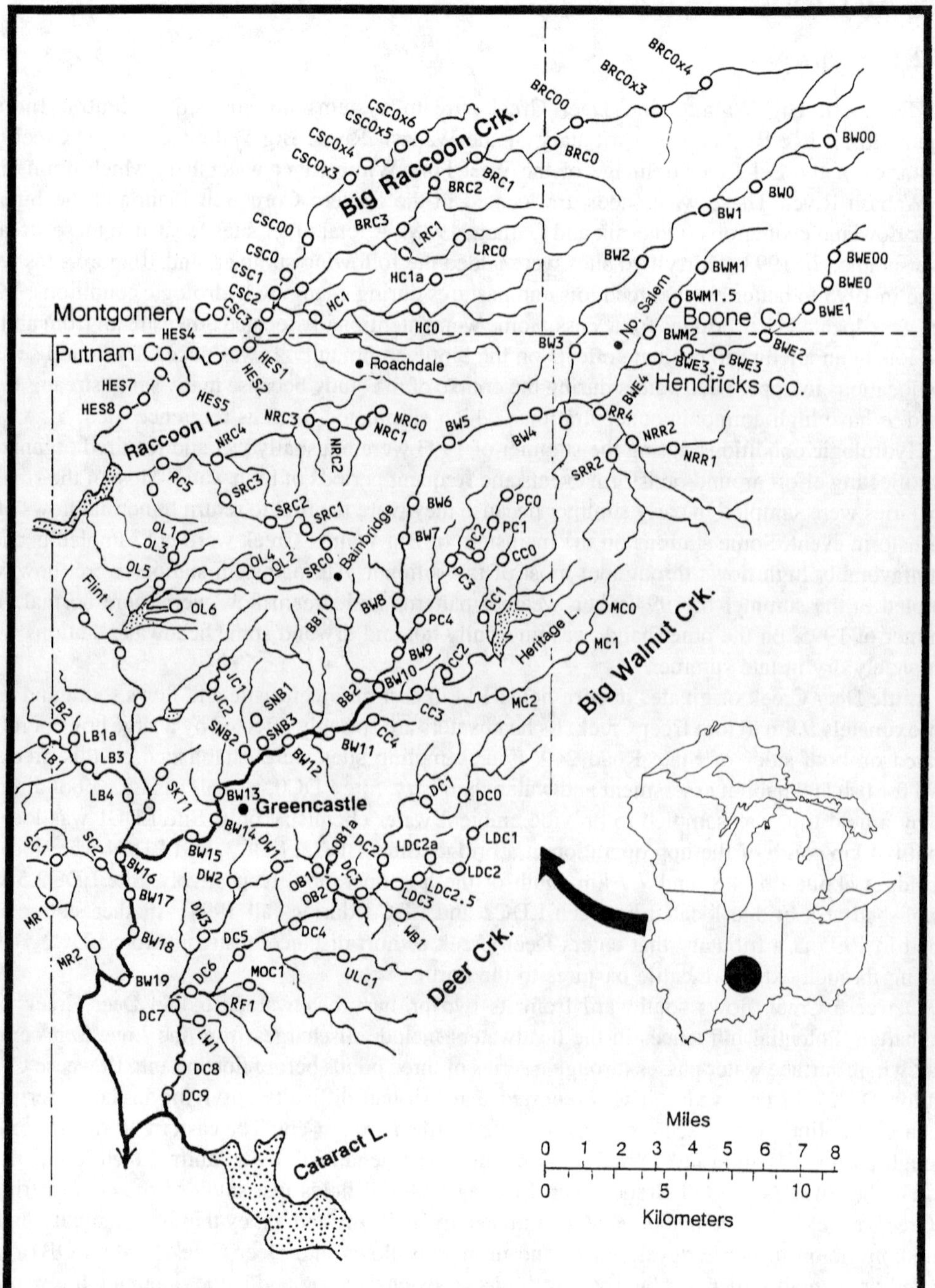

FIGURE 20.1 Locations of study streams and collecting sites.

by cropland and pastures and then enters a more highly dissected landscape. After flowing for approximately 12.9 km, it joins South Ramp Creek less than 1 km south of Fincastle. The initial focus on North Ramp Creek was an extensive, heavily overgrazed hog and cattle operation located in the southeast quarter of Section 15, T16N, R3W. Bridges immediately upstream and about 0.6 km downstream bracketed this feed lot (stations NRC2 above and NCR3 below). Two more stations (NRC1 and NRC0) were added upstream from NRC2 when another cattle feed lot and additional cattle pastures were discovered in the headwaters. Two downstream stations (NRC4 and NRC5) were also established. Station RC5 was located in lower Ramp Creek below the confluence of North and South Ramp Creeks. Ramp Creek then continues to flow westward to enter Big Raccoon Creek just above upper Hardin Lake.

The upper headwaters of South Ramp Creek originate in corn fields about 1.5 km northwest of Bainbridge. A large hog lot heavily impacted the southern branch and two smaller operations drain into the northern branch. Pastures further downstream support cattle and horses. This stream flows westward, joining with North Ramp Creek to form Ramp Creek, which continues flowing west, ultimately entering Big Raccoon Creek above Hardin Lake.

Cornstalk Creek is a long tributary of Big Raccoon Creek, which was a fine fishing stream 25 years ago. Its quality has declined significantly. The upper headwaters are channelized, flanked by corn fields, and devoid of riparian trees. The middle and lower sections are flanked by pastures. A heavily overgrazed pasture with holsteins occupied a section of the lower creek until 1995, when we also observed dredging in part of the lower section. Eight collecting sites were established on this stream.

Haw Creek and Little Raccoon Creek are also tributaries of Big Raccoon Creek. Little Raccoon Creek joins Big Raccoon Creek at Ladoga, and Haw Creek enters it about 9.5 km downstream.

Plum Creek is a tributary of Big Walnut Creek. It originates east of New Maysville, and flows south about 13 km before joining Big Walnut Creek. A small, open hog lot located about 1.6 km southwest of Groveland was initially the focus for chemical work. Access above and below the hog lot was good and 19 above–below water samples were taken and analyzed between March 3 and August 3, 1993.

Long Branch Creek joins lower Little Walnut Creek a short distance before Little Walnut enters Big Walnut Creek. Long Branch originates in well dissected sandstone hills near the Parke/Putnam county line near Vivalia and Keytsville. More forests cover the landscape here than in most Putnam County streams because of the hilly terrain, but agricultural activities near streams are perhaps more concentrated than elsewhere because flat land for rowcrop agriculture is less available. Five fish stations and one chemical sampling station were located immediately adjacent to a hog lot. The most severe negative influence from animal feed lots appeared to be in the extreme upper part of Long Branch, but more subtle effects may operate further downstream. Stream access is limited in the headwaters and one station (LB1 on the south fork) was used as an ambient site for both chemistry and fish. However, the assumption that station LB1 was unaffected by animal feed lots may not be valid. On at least two occasions, ammonia levels at LB1 exceeded 0.8 mg/l. Station LB0 was used only for water quality sampling since it receives direct runoff from a hog lot. Station LB2 was located about 1.9 km downstream from station LB0. Station LB2a, an intermittent stream, was sampled because it flows entirely within a forested landscape and therefore seemed to be a good headwater reference stream.

Bledsoe Branch originates in the vicinity of Bainbridge and flows south to enter Big Walnut Creek about 1 km above Wildwood Bridge. It has a much steeper gradient and better in-stream habitat than most streams in Putnam County, especially in the lower portion.

Clear Creek and Miller Creek, its major tributary, originate in the predominantly agricultural landscape east of Groveland and flow southwesterly to enter Big Walnut Creek about 1 km south of Wildwood Bridge, a few miles northeast of Greencastle.

The headwaters of another Owl Creek flow into Glenn Flint Lake from corn and soybean fields west of Bainbridge. Glenn Flint Lake was drained and its fish community was rotenoned by the Indiana Department of Natural Resources, Division of Fish and Wildlife in October 1995. Owl Creek was also partially subjected to rotenone then. Three sites were sampled on two main tributaries (OC1, OC3, OC5 and OC2, OC4, OC6). Sampling stations OC1 and OC2 were located in channelized pastures. Only OC2 was fenced. Station OC3 also flowed through open pasture.

Snake Creek and Maiden Run are located in adjacent watersheds north of Reelsville and flow into Big Walnut Creek. They were selected because their watersheds have more forests than most of the streams in the area and their fish communities were assumed to be relatively unaffected by the presence of animal feed lots. This supposition proved to be incorrect. A forested tract known as Fern Cliff, formerly the site of a sandstone quarry, flanks Snake Creek. It is currently under the jurisdiction of The Nature Conservancy. Residents of Thomas Lake in the upper watershed noted a problem that possibly had a negative influence on the fish communities found in 1993. Sometime earlier Thomas Lake was dredged and the sediment deposited behind a nearby cofferdam. Heavy rains washed out the cofferdam and much of the sediment washed into the headwaters of Snake Creek. Signs of sediment were observed in late April and early May at SC1.

Jones Creek is located about 5 mi northwest of Greencastle. It is one of the largest tributaries of Little Walnut Creek, originating a few miles west of Bainbridge and flowing southwesterly about 16 km. Jones Creek and its primary tributary, Falls Branch, were dammed about 20 years ago to create Madison Hills Lake. Station JC1 is located downstream from a wood lot cattle pasture. Sediment coated the streambed below the bridge in August 1995. Downstream from JC1 is a large hog operation located in the southeast quarter of Section 31, T15N R4W. This operation was responsible for fish kills on at least two occasions in the past two decades. The entire creek may be depressed because the Madison Hills Lake dam blocked repopulation from Little Walnut Creek after a fish kill. Another large eroding hog lot is located less than 1 km northeast on headwaters of an intermittent stream that enters Jones Creek between JC1 and JC2.

A variety of other smaller streams were included in the study to ensure adequate coverage of habitats. Presumably some area stream sites would support excellent fish communities. These sites would be variable in terms of size and drainage basin area (DBA) and would serve as reference communities to which all others could be compared. However, only a few scattered sites supported excellent fish communities. Because of this, a small stream system, Rattlesnake Creek, was added for comparative purposes. Rattlesnake Creek is mostly contained within the Owen–Putnam State Forest located a few miles south of the study area. Its fish communities were used as ambient comparative controls in an earlier study (Gammon et al., 1983).

The best fish communities in the area served as reference stations. They consisted of five sites on the Deer Creek mainstem (DC0 through DC4), four sites on Rattlesnake Creek (RS0 through RS3), and a variety of small headwaters (RF1, LWC1, DAY1, SKT1, and RI1).

The mainstem of Big Walnut Creek is formed by three main headwaters that unite near North Salem: the East Fork (BWE), the Middle Fork (BWM), and the West Fork. The East Fork and Middle Fork are treated here as tributaries, while the West Fork is considered the upper headwaters of the Big Walnut Creek mainstem (BW). The West Fork and Middle Fork flow through an agricultural landscape. However, the East Fork is currently undergoing a rapid transition from a mixture of agriculture and small woodlands to country suburban land use because of its proximity to I-74 and ready access to the western fringes of Indianapolis. The upper three Big Walnut Creek stations are channelized, as are the upper four stations of the East Fork.

The mainstem of Big Raccoon Creek begins with channelized headwaters (six stations: BRC0x4 to BRC2) and ends above Hardin Lake with eight heritage environmental services (HES) stations (HES1 through HES8). The HES sites have been used to monitor the HES landfill near Russellville annually since 1981. The landfill receives wastes from the HES industrial oil reprocessing operation from Indianapolis. No collecting stations were located below Hardin Lake, although Big Raccoon Creek continues on to enter the Wabash River south of Montezuma.

20.2.2 Fish Community Assessment

Representative fish assemblages were collected at each site using a Cybertronic backpack electrofisher over a timed interval (Hocutt and Stauffer, 1980). Sampling procedures were consistent with methodology specified by Simon (1991) for application of index of biotic integrity criteria. The duration depended upon the size of the stream. Small first order streams generally required only

15 to 20 min for adequate coverage, while larger third order streams required 40 to 50 min. Some mainstem sites were measured and timed.

Captured fish were placed in a floating live well until the collecting effort was completed. At the completion of the zone, all species collected were counted, measured, and weighed. The fish were returned to the stream with the exception of specimens of uncertain taxonomy that were placed in 5% formalin and returned to the laboratory for identification using standard keys such as Trautman (1981) and Pflieger (1975). Common and scientific names follow Robins et al. (1991).

20.2.2.1 Fish Assemblage Indicators: Index of Biotic Integrity and Index of Well-Being

Evaluation of the quality of the fish community at each collecting site was based on the computation of an index of biotic integrity (IBI) using metric criteria developed by Simon (1991). Simon modified Karr's original IBI (Karr, 1981; Karr et al., 1986) based on extensive collections from the Kankakee and Iroquois Rivers and their tributaries. These metrics are essentially those used by the state of Ohio (Ohio EPA, 1987, 1988) for the Eastern Corn Belt Plain ecoregion (Omernik and Gallant, 1988). Most study stations are located in that ecoregion. However, a few stations in the extreme southwest section of Big Walnut Creek are located outside the ecoregion.

Index of Biotic Integrity — The site numeric catch data (number of fish/km = number of fish/hr) were used to compute IBI values for each station. Simon (1991) developed different reference condition expectations, depending on the size of the drainage basin. Sites with drainage basin areas (DBAs) less than 20 mi^2 were designated headwater sites while sites in drainage basins with areas from 20 and 1000 mi^2 were called wadeable sites. A large number of collecting stations from this study were located on very small tributaries with DBAs smaller than 2 $mi.^2$ The wet weather experienced between June and July 1993 fostered successful documentation of resident fish communities from small Order I streams that normally contained little or no water.

Simon (1991) changed some of the original IBI metric categories in order to reflect variations in fish composition in streams differing in size. Since sunfish species are found mostly in larger streams, they were excluded from headwaters metrics where several assemblages of headwater minnows, lamprey, and darters dominated. In addition, headwater streams were often dominated by minnow species and rarely contained suckers and redhorses except perhaps during spring spawning. Headwaters usually lack sufficient depth for carnivores, which are generally larger species. The streams are episodic, drying in summer and flowing during wet periods. Several small pioneer species adapted to these conditions and are the first to reenter headwaters when flow resumed (Smith, 1971; Rankin and Simon, Chapter 10, this volume). Metrics common to both headwaters and wadeable IBI criteria include total number of species, number of sensitive species, percent individuals as tolerant species, percent individuals as omnivores, percent individuals as insectivores, percent individuals as simple lithophils that use gravel for spawning, catch per unit effort, and percent DELT (deformities, eroded fins, lesions, tumors) anomalies.

Simon (1991) developed a series of data plots to determine least impacted reference conditions for the IBI. All values for each individual metric were plotted against DBAs to assess the influence of stream size on that metric. The upper 95th percentile maximum species richness line was estimated (Fausch et al., 1984). This angle is then trisected into three equal portions. When the metric was stable between DBAs of about 20 to 300 mi^2, the trisecting lines were drawn horizontally. When a drainage area relationship was observed, an oblique line was drawn to compensate for lower expectations at smaller DBA sites. The three trisected data wedges were then rated as 1 (deviating strongly from the reference condition), 3 (deviating somewhat from the reference condition), and 5 (approximating the reference condition) for purposes of scoring. For each collection, site data for a single date were compared to metric scattergrams and the IBI criteria for headwater sites or wadeable sites (Simon, 1991). The site score is based on the summation of all 12 metric values into a single IBI value.

Index of Well-Being — Fish communities from mainstem collecting sites were also evaluated using a composite index of well-being (Iwb) (Gammon, 1976, 1980, and 1998):

$$Iwb = 0.5\,LnN + 0.5\,LnB + H(no) + H(wt)$$

where N = total number per kilometer electrofishing; B = total weight (kg) per kilometer electrofishing; H(no) = Shannon–Weiner index of diversity based on numbers; H(wt) = Shannon–Weiner index of diversity based on weights; Ln = natural logarithms.

This index was developed specifically for the Wabash River and has been used successfully on many smaller rivers in the midwest and elsewhere. It is not, however, a satisfactory composite index for use on streams smaller than about Order IV.

20.2.2.2 Habitat Assessment Protocols

Big Raccoon, Big Walnut, Little Walnut, and Deer Creek river systems were evaluated for the quality of their habitats in June and August of 1993. Most sites were easily accessible by bridge, and measurements were taken upstream or downstream depending on the side that provided the better quality riparian zone habitat.

Stream characteristics including stream gradient and sinuosity were determined from U.S. Geological Survey (USGS) topographic maps (1:24,000). The gradient of small streams was estimated by finding the altitude of the headwaters for that stream divided by the altitude at the mouth. For larger streams, the gradients at specific stations were determined using the distance between adjacent altitude lines. Sinuosity was measured by dividing the total distance of the stream from the headwater to the mouth by the straight line distance between these two points. Most determinations of drainage basin area (DBA) were obtained from Hoggatt (1975), although very small DBAs were directly measured from USGS maps using a Keuffel & Esser polar planimeter. Topographic maps were also utilized to determine stream orders. The smallest permanent streams were designated first order streams. All other stream orders followed procedures outlined by Horton (1945) and Strahler (1952).

Field measurements at each site were based on a 100-m section of stream subdivided into 10 equidistant transects. Stream width was recorded using field tape and depth was measured to the closest centimeter at intervals of approximately 0.7 m along each transect. The type of substrate was categorized at each depth interval as mud (M), sand (S), gravel (G), rock (R), bedrock (B), or clay (C). Measurements of bank height and high water marks were recorded for the left and right sides of the streams and estimates of riparian zone width and bank angle were measured for each side.

Two qualitative procedures were used to evaluate habitat: the qualitative habitat evaluation index (QHEI; Rankin, 1989) and the rapid bioassessment protocols (RBP; Plafkin et al., 1989). Both procedures are rapid assessment techniques. The QHEI highlights instream elements to a greater extent than the RBP, which emphasizes riparian elements.

20.2.3 Water Chemistry

Chemical sampling began in February 1993 and continued sporadically through summer 1995. Water samples were collected from each site with a polypropylene long-handled dipper. They were collected in glass containers, stored on ice, and delivered to the laboratory. Most analyses were conducted within 12 hr of sampling. Parameters measured in the field included water pH and temperature. Ammonia was determined with Chemettes, a LaMotte Model AQ-2 test kit, or standard methods. Turbidity was determined with a HF Scientific Model DRT15C turbidimeter. Beginning in 1994, a YSI Model 3800 datalogger on loan from the Indiana Department of Natural Resources' Division of Fish and Wildlife was used extensively for *in situ* determinations of temperature, pH, ammonia, dissolved oxygen, conductivity, turbidity, and salinity. Ammonia, dissolved oxygen, and pH probes were calibrated prior to use. In 1995, the datalogger was also used to obtain timed interval readings at a few sites over a 24-hr period.

20.3 RESULTS AND DISCUSSION

20.3.1 Patterns in Habitat and IBI Metric Values

Nearly 29,000 individuals and a total of 53 species of fish were collected from Big Raccoon Creek, Big Walnut Creek, Little Walnut Creek, and Deer Creek and their tributaries. The river chub (*Nocomis micropogon*) was absent from Big Walnut Creek and Deer Creek although it was quite common in the mainstem of Big Raccoon Creek. Mosquitofish (*Gambusia affinis*) were found only in Owl Branch of the Big Walnut Creek system.

IBI and QHEI values are summarized for all sites in the Big Raccoon Creek, Big Walnut Creek (including sites in Little Walnut Creek), Deer Creek, and Rattlesnake Creek systems in Appendix I. The distances of each site from beef cattle, dairy herds, or hog lots varied considerably because of the location of access sites at the closest bridges downstream. The number of animals at each concentration was usually not known and varied considerably. Collecting sites located in pastures (indicated by superscripted Ps) contained horses, beef cattle, sheep, or a combination of mammalian species. One common attribute was a sufficiently low animal density that allowed the existence of a grassy ground cover.

Sites in the headwaters of Big Raccoon and Big Walnut Creeks were located in streams that had been extensively channelized. Some were totally devoid of riparian vegetation (indicated by superscripted c0). Sites that had at least 50% of riparian wetlands consisting of trees and/or shrubs are indicated by superscripted CTs.

20.3.2 Ecological Assessment: Case Study of Area Streams

Both the IBI and Iwb indicate the magnitude and extent of negative impacts to fish assemblages (Hughes and Gammon, 1987; Yoder and Rankin, 1995; Yoder and Smith, 1999). We used this approach to establish a base of IBI values from the least impacted sites and then compared the fish communities of each individual stream to this reference standard.

The five Deer Creek mainstem sites (DC0 through DC4), four sites on Rattlesnake Creek (RS0 through RS3), and the other sites located on small streams (RF1, LWC1, DAY1, SKT1, and RI1) were considered least impacted or reference stream reaches. A log–log regression of IBI values from these reference stations on DBAs provided a comparative ceiling against which all other stream sites were compared. These reference fish communities represented the best available under present day land use, waste treatment, and agricultural practices. The communities do not exist under pristine conditions nor do they necessarily represent fish communities that are the best under current land use practices. These reference communities are certainly not the best that could have developed if best management practices (BMPs), including conservation tillage, were adopted wholesale by the agricultural community.

20.3.2.1 Little Deer Creek

Nutrient additions from headwaters corn fields and two hog lots promoted heavy algal growth in this open segment of Little Deer Creek, especially during periods of low flow. At least once, ammonia concentrations below one hog lot were sufficiently high to kill the profusion of algae that returned after ammonia levels declined to nontoxic levels. Ammonia levels at ambient LDC0 exceeded 1.0 mg/l twice from 1993 through 1996 (Figure 20.2). Below the main hog lot at LDC1, however, ammonia levels exceeded 1.0 mg/l on 5 of 19 occasions in 1993, 14 of 21 occasions in 1994, and 12 of 13 occasions in 1995. In contrast, ammonia levels in 1996 declined to background

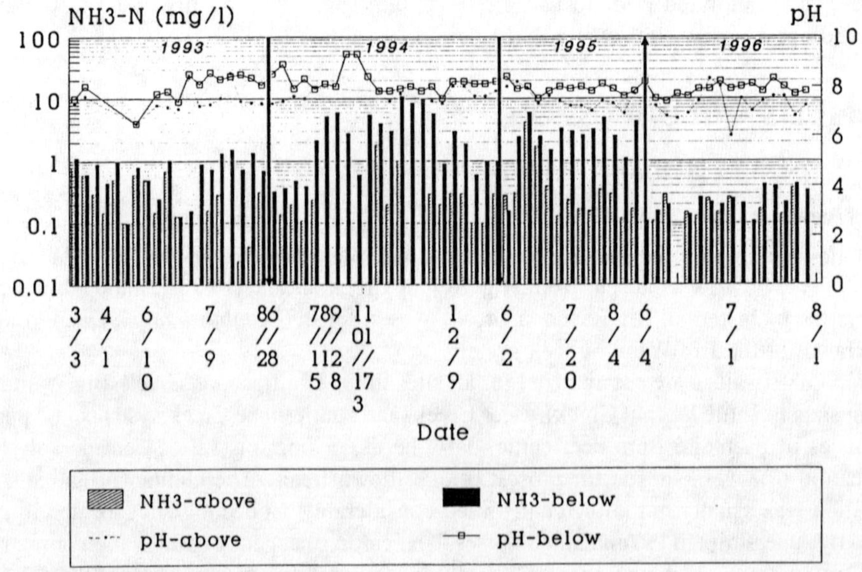

FIGURE 20.2 Total ammonia–nitrogen concentrations above (LDC0) and below (LDC1) a hog lot at the headwaters of Little Deer Creek.

levels after the Natural Resources Conservation Service (NRCS) and owners of the hog lots altered operation procedures slightly.

Ammonia–nitrogen levels increased sharply downstream from the hog lot in Summer 1993 and 1994, and higher concentrations were found in 1995 when stream flow was low. All sites had much higher ammonia concentrations in 1995 than they had in 1993. Longitudinal patterns in ammonia levels during 1994 were slightly higher than those found in 1993. However, during fall and early winter, ammonia concentrations immediately downstream from the hog lot were very high (mean NH_3-N = 4.48 mg/l). Ammonia levels at all other main stream sites were low and stable during this period, but the tributary flowing through cattle pastures exhibited ammonia concentrations exceeding about 1 mg/l (Figure 20.3).

Nitrate, conductivity, and turbidity values increased downstream of the hog lot. Nitrate concentrations averaged more than 11 mg/l at station LDC0 in summer 1995 (Appendix IIA). The levels increased further downstream from the hog lot, and then declined downstream thereafter (Figures 20.4, 20.5, and 20.6). Nitrate concentrations exceeded 3 mg/l even in the lower sites (LDC2.5 and LDC3). Conductivity values were also consistently elevated immediately downstream from the hog lot (Figure 20.5), but quickly declined to lower levels with increasing distance. Conductivity of water at tributary site LDC2a was also slightly elevated. Turbidity values increased from an average of 28 NTU at LDC0 to nearly 110 NTU at LDC1, but returned to ambient levels at LDC2 (Figure 20.6). Water turbidity at LDC1 was usually visually evident except when the ground was frozen.

The pH increased sharply between LDC0 and LDC1 and then leveled off (Figure 20.7). The pH below the hog lots was usually elevated considerably compared to ambient upstream sites as shown in Figure 20.7. Part of this shift in pH is no doubt due to photosynthesis during summer and fall months. Algae were present during the fall 1994 sampling, but temperatures were quite low (less than 5°C) and photosynthesis was probably low.

The fish communities at all sites on Little Deer Creek were affected negatively (Figure 20.8). The fish community below the hog lot (LDC1) received a poor rating (IBI = 26). The fish communities failed to reach their potential. The extensive cattle pastures along a tributary (LDC2a), which enters just above LDC3, provides a secondary depression due to high ammonia levels measured after rainstorms.

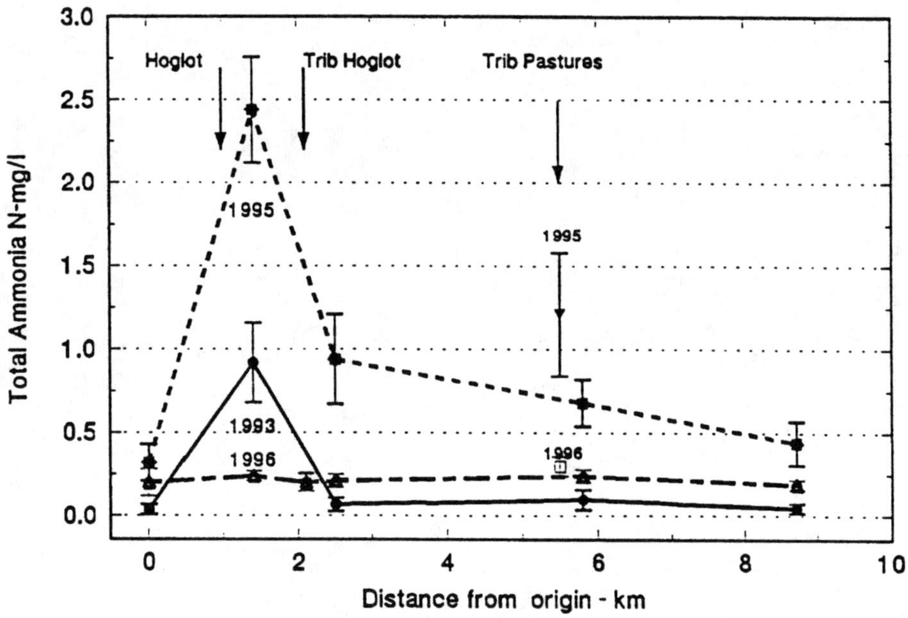

FIGURE 20.3 Mean total ammonia–nitrogen concentrations (mg/l) and S.E. of the mean at Little Deer Creek stations in 1993, 1995, and 1996.

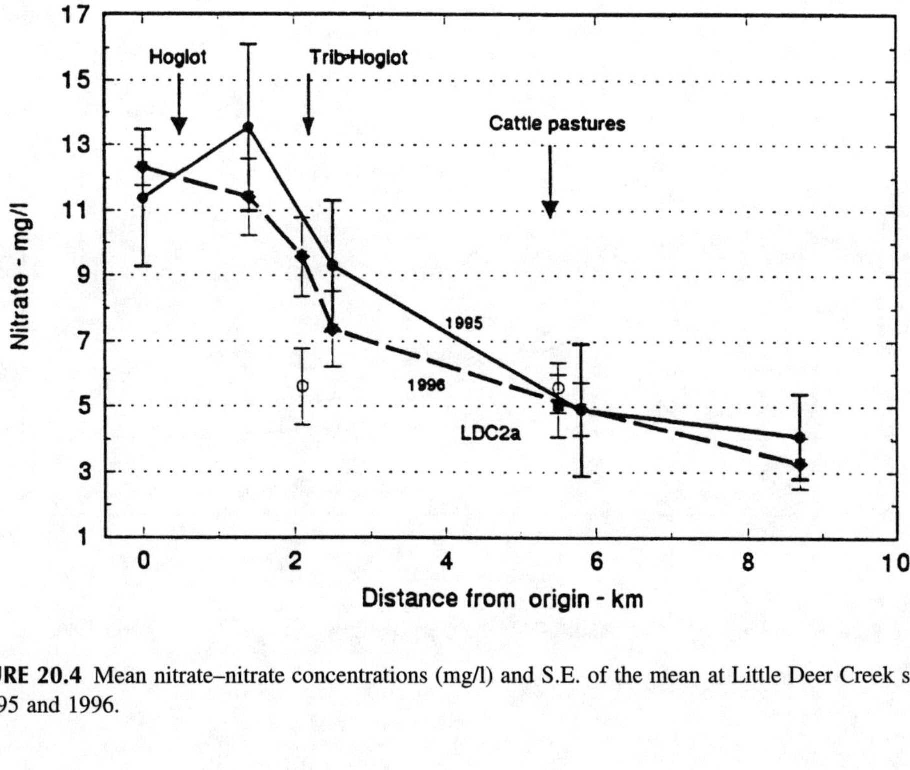

FIGURE 20.4 Mean nitrate–nitrate concentrations (mg/l) and S.E. of the mean at Little Deer Creek stations in 1995 and 1996.

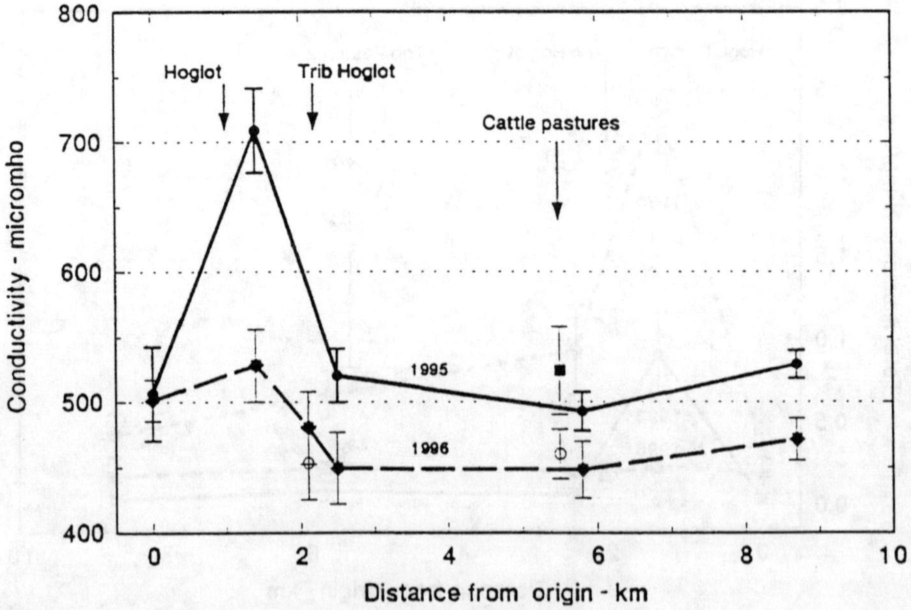

FIGURE 20.5 Mean conductivity (microhms) and S.E. of the mean at Little Deer Creek stations in 1995 and 1996.

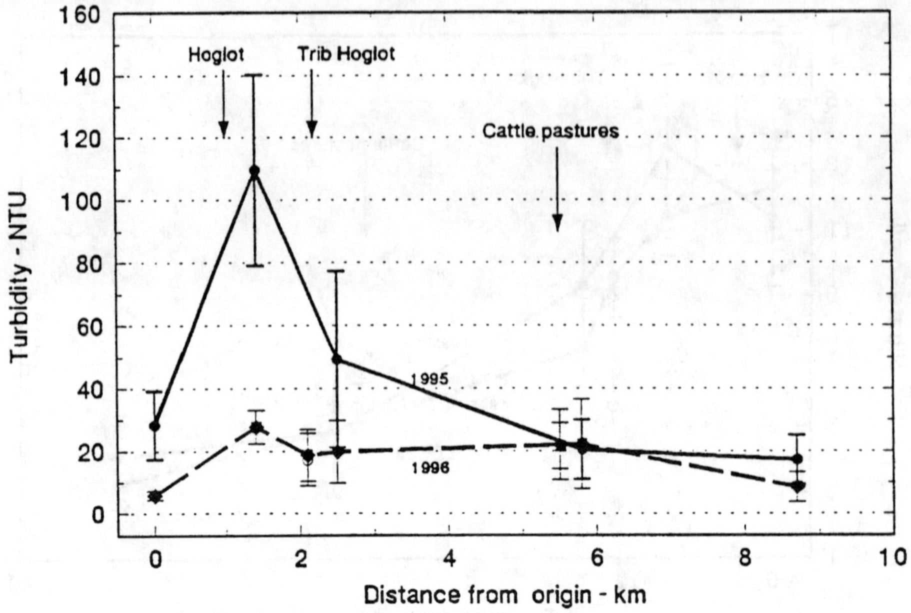

FIGURE 20.6 Mean turbidity (NTU) and S.E. of the mean at Little Deer Creek stations in 1995 and 1996.

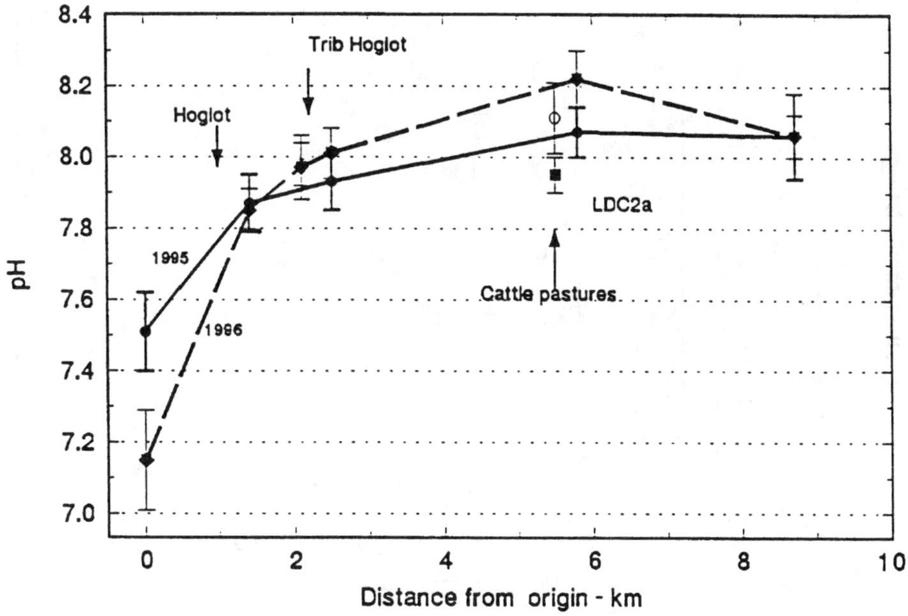

FIGURE 20.7 Mean pH and S.E. of the mean at Little Deer Creek Stations in 1995 and 1996.

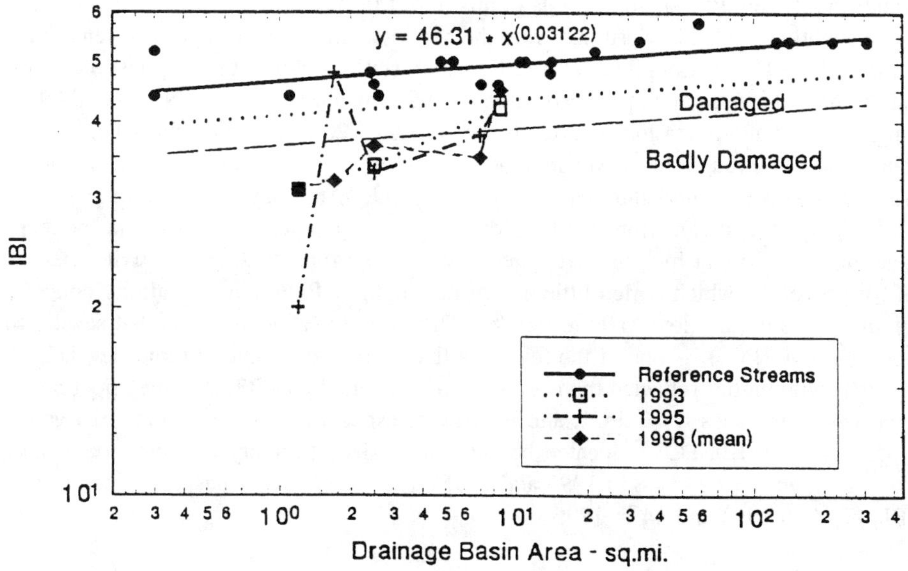

FIGURE 20.8 Comparison of fish community quality (IBI) of Little Deer Creek and reference stream sites in 1993 and 1995.

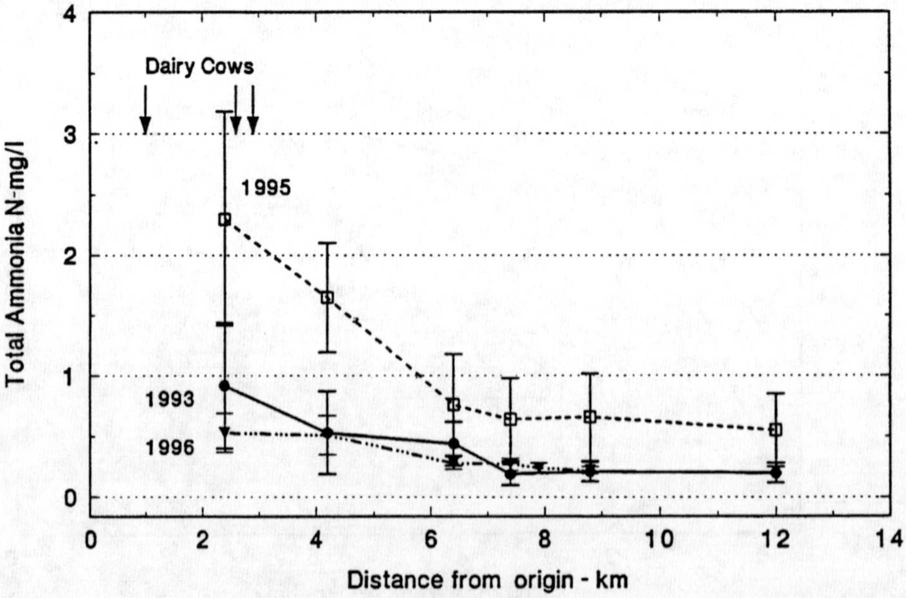

FIGURE 20.9 Mean and S.E. total ammonia concentrations (mg/l) at Plum Creek sampling sites in summers of 1993, 1995, and 1996.

20.3.2.2 Plum Creek

Dairy farms and cattle pastures occupy the headwaters of Plum Creek, a tributary of Big Walnut Creek and runoff from two small hog lots drains into it further downstream. Ammonia concentrations were slightly elevated downstream from the hog lots; ammonia levels were often higher above the hog lot than below it. Longitudinal profiles of six stations showed the highest mean ammonia concentrations in the extreme upper headwaters in 1993 (mean = 0.92 mg/l, N = 7) and 1995 (mean = 2.3 mg/l, N = 7) and much lower levels in 1996 (Figure 20.9 and Appendix IIB).

Concentrations throughout Plum Creek were about twice as high during the low-flow summer of 1995 than during the rainy summers of 1993 and 1996. Ammonia levels gradually declined to about 0.5 mg/l just upstream from the hog lots. The lots contained fewer hogs in the summer of 1995 than in 1993. Similar longitudinal patterns were noted for nitrate, conductivity, and turbidity, but not for pH levels, which varied little along the length of Plum Creek. Salinity concentrations at most sites were usually low (0.02 to 0.03%). They were considerably elevated several times at the uppermost site (PC0) — one of the few sites that exhibited elevated salinity levels.

The fish communities reflected the negative environmental conditions of the upper headwaters. The upper collecting sites (PC0, PC1, and PC2) were especially depressed; most IBI values were below 35 (Figure 20.10). PC3 is located below a good-sized tributary and supports a marginally good fish community (IBI of 48 in 1993 and 1995). PC4 in the lower basin declined to fair–good with IBIs of 46 in 1993 and 44 in 1995.

20.3.2.3 North Ramp Creek

Five sampling sites were located along this 12.9-km (8 mi) stream whose headwaters flow through several cattle pastures and a hog lot. The highest ammonia concentrations at this Big Raccoon Creek tributary were found at headwater stations NRC0 and NRC1 (Figure 20.11 and Appendix IIC). The uppermost station (NRC0) is immediately downstream from a wood lot occupied by cattle. The stream then flows through wooded terrain to NRC1. Ammonia levels declined precipitously from a high of about 1 mg/l to only 0.1 to 0.2 mg/l as the creek flowed through

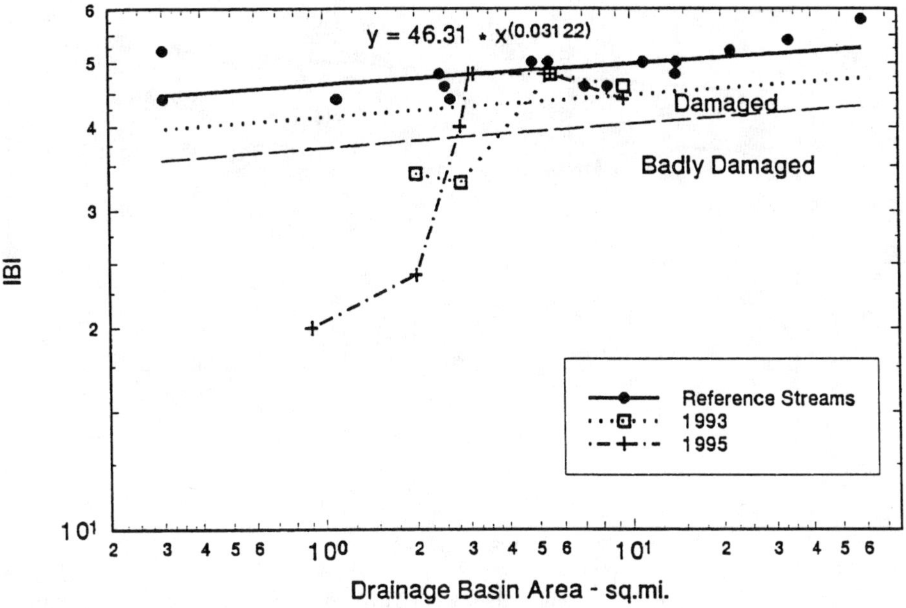

FIGURE 20.10 Comparison of fish community quality (IBI) of Plum Creek and reference stream sites in 1993 and 1995.

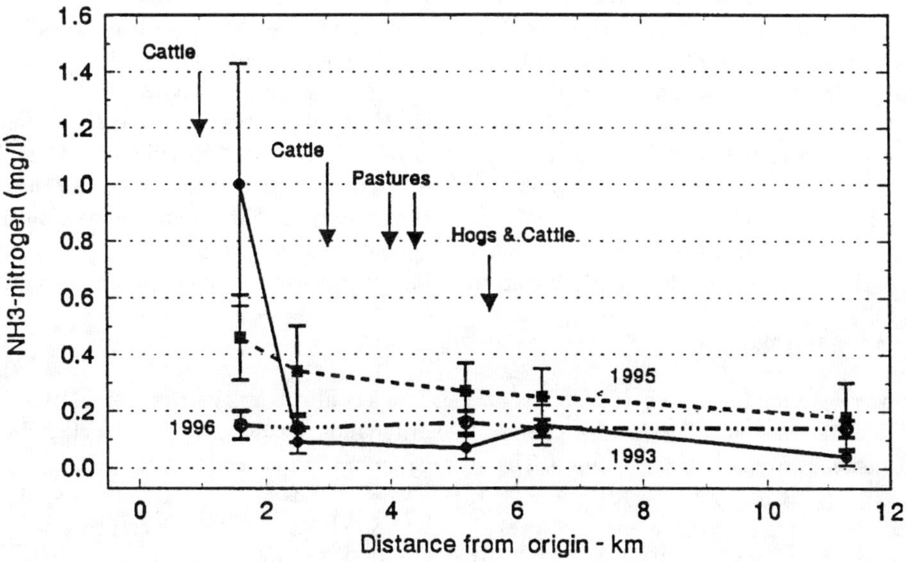

FIGURE 20.11 Mean and S.E. total ammonia concentrations (mg/l) at North Ramp Creek sampling sites in summers of 1993, 1995, and 1996.

extensive pastures and one cattle feed lot between NRC1 and NRC2. A cattle/hog operation between NRC2 and NRC3 caused a slight increase in ammonia in 1993 when rains were more frequent and animal density appeared higher than in 1995. Further declines in ammonia concentration were found between NRC3 and NRC4.

Longitudinal patterns of nitrate, conductivity, and turbidity were similar to the pattern of ammonia–nitrogen. Conductivity and turbidity increased downstream from the cattle/hog lot located between NRC2 and NRC3, but pH was unaffected.

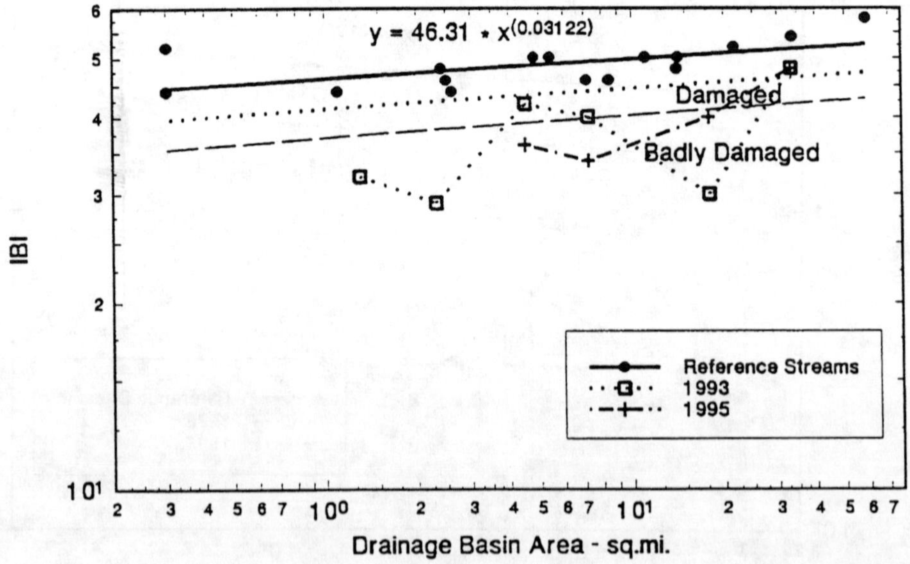

FIGURE 20.12 Comparison of fish community quality (IBI) of North Ramp Creek and reference stream sites in 1993 and 1995.

The fish communities of North Ramp Creek were poor at headwater sites NRC0 (IBI = 32) and NRC1 (IBI = 26) and included several small largemouth bass and bluegill species that were obviously out of place in small headwaters (Figure 20.12). Their presence probably resulted from frequent rains that washed small fish from several upstream ponds. These sites were dry or consisted of small isolated pools in summer of 1995. Some recovery occurred at NRC2 where fair to poor fish communities were found (IBI = 42 in 1993 and 36 in 1995). However, the fish community declined downstream from the hog and cattle lots at NRC3 (IBI = 38 in 1993 and 34 in 1995). The fish community at NRC4 was depressed in 1993 (IBI = 34), but improved to fair during the dry summer of 1995 (IBI = 40). The annual differences observed at NRC4 may indicate undetected sources of pollution that mainly affect the site during wet weather. The fish community improved to good (IBI = 48) only at RC5, downstream from the junction of North and South Ramp Creeks.

20.3.2.4 South Ramp Creek

The uppermost sampling station (SRC00) is northeast of a confined hog lot where field tiles drained corn fields and was only sampled for chemical determination. Station SRC0 is located about 0.5 km downstream from the hog operation. The average ammonia concentration at SRC0 in longitudinal surveys was below 1.0 mg/l (Figure 20.13). Levels were as high as 8.0 mg/l in spring 1993 and declined in the summer. Ammonia levels in lower South Ramp Creek averaged less than 0.4 mg/l during the summers of 1994 and 1995 (Appendix IIC). The nitrate profile was similar to that found at Plum Creek, with the highest levels at SRC00. Conductivity was occasionally elevated at SRC0; only SRC3 had a higher average. Turbidity was often high during the frequent rains of 1993. In 1995, turbidity was much higher downstream from the hog lot and pasture. The pH was also elevated downstream from the hog lot and at SRC3.

Trends in IBI values along the stream between 1993 and 1995 showed a minimum score of 12 (Figure 20.14). During 1993, only a single crayfish (no fish) was collected at station SRC0. Environmental conditions improved at the next station (SRC1) and the fish community was rated fair (IBI = 42). The IBI continued to rise downstream with values of 44 at SRC2 and 46 at SRC3. In 1995, no species except small young-of-the-year creek chubs were taken at SRC0 (IBI = 26).

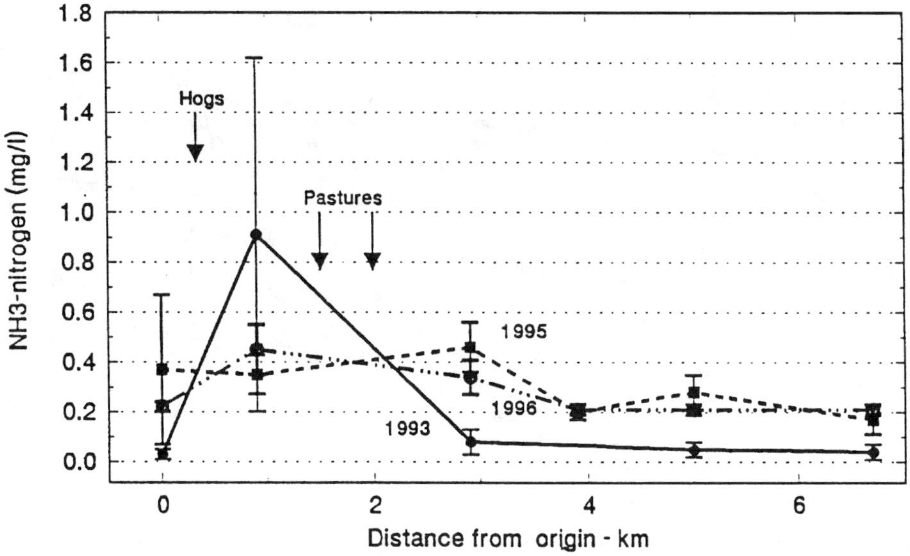

FIGURE 20.13 Mean total ammonia–nitrogen concentrations (mg/l) and S.E. of the mean at South Ramp Creek stations in 1993, 1995, and 1996.

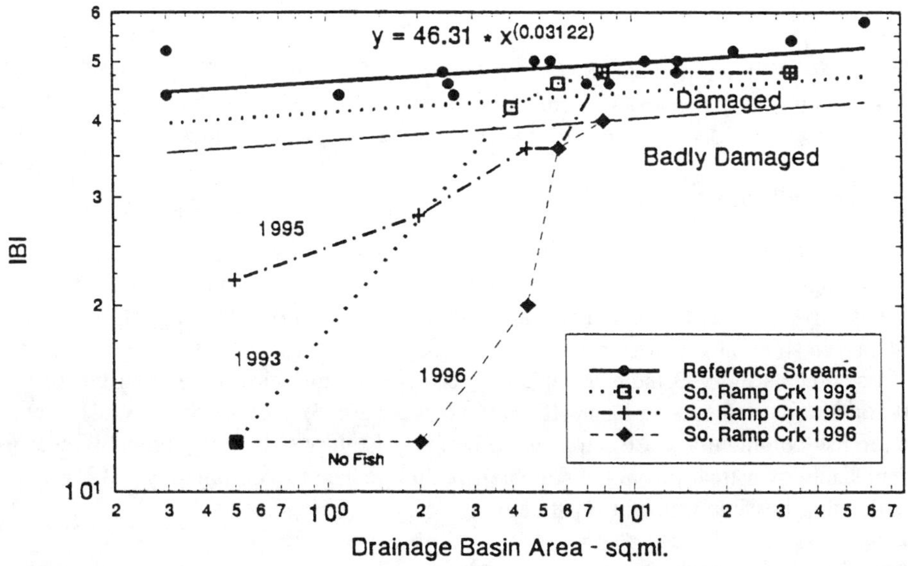

FIGURE 20.14 Comparison of fish community quality (IBI) of South Ramp Creek and reference stream sites in 1993, 1995, and 1996.

Fish communities improved slightly downstream, but only the lowermost station (SRC3) and RC5 harbored near-normal fish communities.

In the summer of 1994, routine chemical determinations revealed an unexpected environmental problem associated with farm animal production. On August 5, 1994 water discharge levels in South Ramp Creek were very low, as were levels in all other small streams in central Indiana. No water drained from the fields at station SRC00 and flows were sluggish at the other downstream stations. Water chemistry in the headwater branch stations (SRC0 and SRC1a) differed significantly, especially dissolved oxygen concentrations. Dissolved oxygen concentration at SRC0 was normal

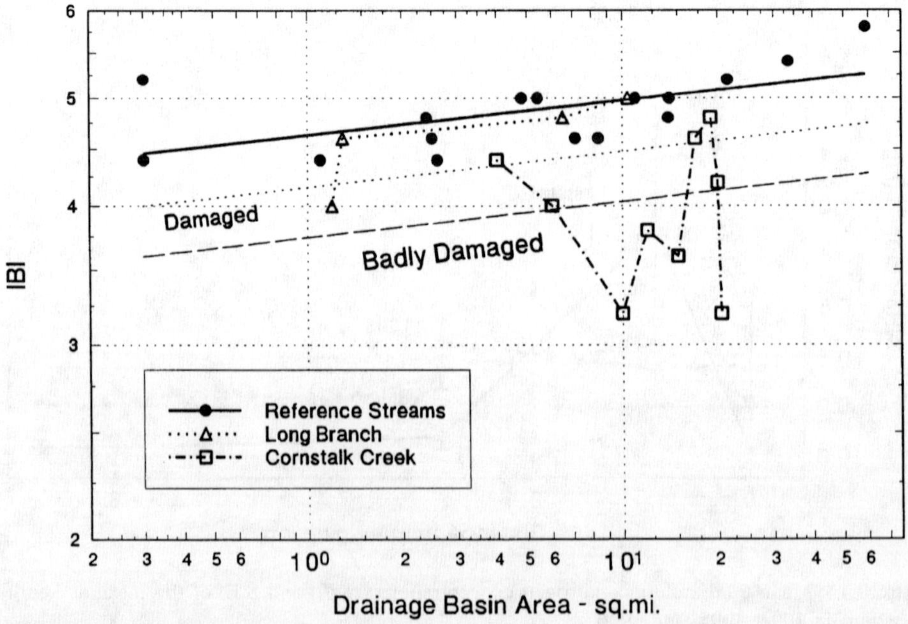

FIGURE 20.15 Comparison of fish community quality (IBI) of Long Branch and Cornstalk Creek and reference stream sites in 1993.

at 8.89 mg/l (97% saturation), but at SRC1, the level was 27.72 mg/l (367% saturation); T = 28.7°C, pH = 8.94, and turbidity = 33 NTU. Ammonia levels were relatively low: 0.61 mg/l at SRC0 and 0.20 mg/l at SRC1a. Supersaturation of dissolved oxygen of this magnitude is toxic to fish.

20.3.2.5 Long Branch

The ammonia concentration of runoff at the northernmost hog lot was never less than 1.0 mg/l and reached 90 mg/l on August 27, 1993 when runoff trickled into a black pool near the road. Nevertheless, live fish were observed within about 0.8 km (0.5 mi) of the hog lot. A larger hog lot a short distance south also drains into Long Branch, but no access was available for sampling.

IBI assessments in 1993 indicated fair to good fish communities throughout the stream with the best community in the lower basin (IBI = 50) (Figure 20.15). Despite the periodic flows of hog wastes, the fish community at LB2 appears to be fairly healthy (IBI = 46), probably because it is more than 1 mi downstream from the source of pollution. The fish community at LB1 had an IBI of 40, indicating possible problems upstream.

20.3.2.6 Cornstalk Creek

The upper headwaters of Cornstalk Creek flow in an open channelized stream that drain corn fields. Pastures flank the middle and lower sections. A heavily overgrazed pasture with a large herd of holsteins occupied a section of lower Cornstalk Creek until 1995. During the summer of 1995, we observed active cutting and bulldozing of the riparian corridor in the lower basin.

Nine fish collection and habitat evaluation sites were located at intervals of 2 to 4 km in 1994. The upper three stations (CSC0x4, CSC0x5, and CSC0x6) are channelized. Nevertheless, the upper two stations (CSC0x5 and CSC0x6) contained fair fish communities (IBI values of 40 and 44, respectively) (Figure 20.15). Community quality declined sharply at CSC0x4 (IBI = 32) and then increased at stations CSC0x3 and CSC00 (IBI = 38 and 36, respectively). IBI values continued to increase at CSC0 (IBI = 46), peaked at CSC1 (IBI = 48) above the dairy herd, and then sharply

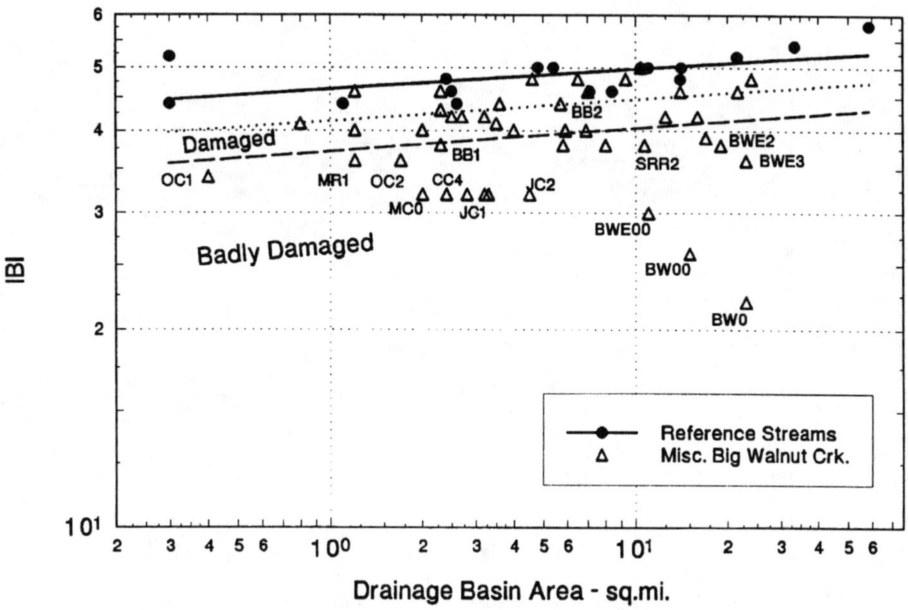

FIGURE 20.16 Comparison of fish community quality (IBI) of miscellaneous Big Walnut Creek tributaries and reference stream sites.

declined below the herd (IBI = 42 at CSC1.5 and 32 at CSC2). In the 1960s, Cornstalk Creek was an excellent fishing stream but it has deteriorated greatly since then.

20.3.2.7 Bledsoe Branch, Clear and Miller Creeks

During the late winter, spring, and early summer of a very wet and rainy period in 1993, the beef cattle near lower Bledsoe Branch were usually knee-deep in mud. Access upstream of the cattle was afforded only by a bridge (BB1), but good access was available downstream by road (BB2). Ammonia concentrations were less than 0.8 mg/l at both sites in March 1993 and declined even lower thereafter. Concentrations were usually but not always slightly higher downstream from the cattle.

Fish communities were only fair at both sites (IBI = 38 at BB1 and 44 at BB2) and slightly below expectations (Appendix IB). This small stream should be one of the best because it flows mostly through forested terrain and has a greater gradient than most area streams. Negative activities near the stream corridor between BB1 and BB2 may have an impact but they cannot be seen because of difficult access.

The upper part of Miller Creek (MC0) contains a degraded fish community (IBI = 32). Fish communities in the lower reaches (MC1 and MC2) are fair to good (IBI = 44 and 48, respectively) despite flowing through pastures (Figure 20.16). With the exception of station CC2, which is clearly impaired because it is downstream from the Heritage Lake dam, fish communities at other Clear Creek stations (CC0, CC1, and CC3) are near or slightly below expectations (Appendix IB) with IBIs ranging from 43 to 48.

20.3.2.8 Deweese Creek and Ramp Run

Deweese Creek enters the Indiana State Penal Farm in the upper basin. Station DW3 is at a bridge in the penal farm area. Fish communities at headwater stations DW1 and DW2 appeared slightly depressed. The community of DW3 was severely below expectations for unknown reasons (Figure 20.18).

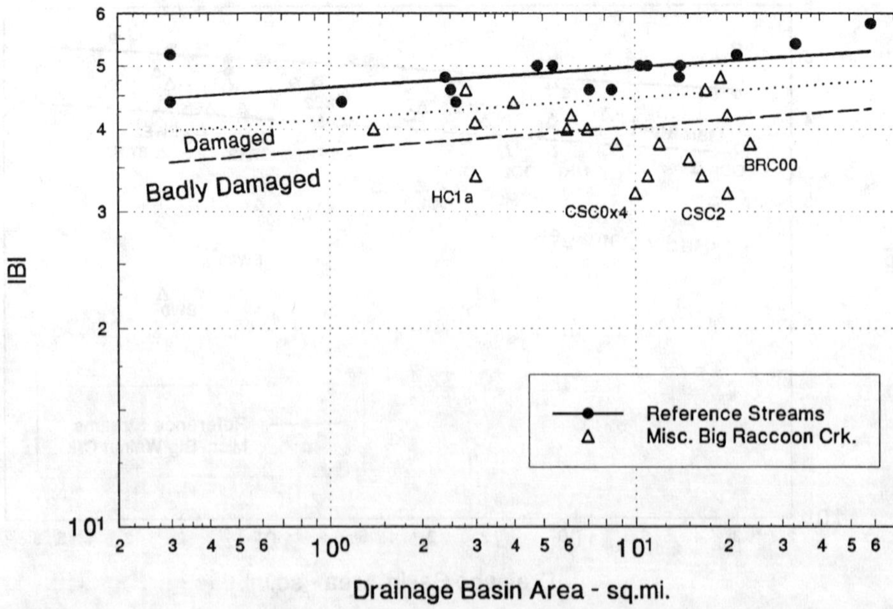

FIGURE 20.17 Comparison of fish community quality (IBI) of miscellaneous Big Raccoon Creek tributaries and reference stream sites.

Ramp Run is a small forked tributary of the east fork of Big Walnut Creek located south of North Salem. A single station was located on the south fork (SRR1) and two stations were sited on the north fork (NRR1 and NRR2). An additional station occupied the lower stream (RR3). The poorest fish community was found at SRR1 (IBI = 38), perhaps because of bridge construction upstream in 1993. The fish communities of North Ramp Run were somewhat better (IBI = 42 and 46), but the community at RR3 was depressed, possibly because of its location in a channelized section of open pasture.

20.3.2.9 Owl Creek (Little Walnut)

All stations except OC4 supported depressed fish communities (Figure 20.16). Station OC4 was located in a wooded corridor sufficiently remote from fields in the extreme headwaters. One unusual component of the Owl Creek fish communities was the presence of mosquitofish (*Gambusia affinis*) at five of the six stations.

20.3.2.10 Snake Creek, Maiden Run, and Jones Creek

Snake Creek station SC1 is adjacent to a beef cattle pasture on both sides connected by a narrow access corridor that crosses the creek. Several years ago, the landowner fenced the creek to minimize damage by cattle, the only such example observed in the entire area. Water chemistry was determined sporadically in 1993 and showed no unusual results.

The IBI values for Snake Creek indicated fair to good fish communities; IBI values for SC1 were 40 in 1993 and 1995. SC2 had an IBI of 50 in 1993 that dropped to 42 in 1995 (Appendix IB). Snake Creek contained both smallmouth bass (*Micropterus dolomieu*) and spotted bass (*Micropterus punctulatus*). With help, Snake Creek could become an excellent small stream.

On Maiden Run, a small hog lot was located upstream from MR1. Cattle occupied the wood lot flanking MR1. Both activities exerted negative impacts in the upper headwaters. MRI had an IBI score of 36, indicating poor to fair fish communities. The lack of significant improvement downstream at MR2 (IBI = 38) is probably indicative of other problems in the watershed (Figure 20.16).

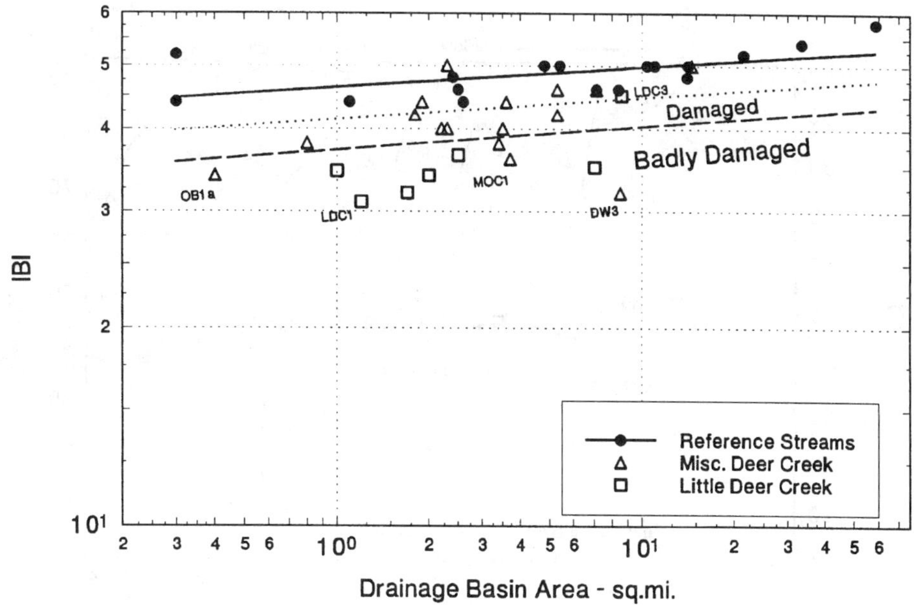

FIGURE 20.18 Comparison of fish community quality (IBI) of miscellaneous Deer Creek tributaries and reference stream sites.

Fish collection and water chemistry stations were established at the first bridges upstream and downstream from the hog operation on Jones Branch (JC1, upstream; JC2, downstream). Chemical tests indicated higher concentrations of ammonia in early July than later during the summer of 1993. Ammonia concentrations were higher downstream from the hog operation on four of six dates. Assessments of the fish communities at JC1 and JC2 indicate poor results at both sites (IBI at JC1 = 32; IBI at JC2 = 32) (Figure 20.16). In addition to these negative influences, other problems may exist upstream of JC1.

20.3.2.11 Haw Creek and Little Raccoon Creek

The fish community at Little Raccoon Creek station LRC0 was rated fair (IBI = 42), despite its channelization and location in a lightly grazed pasture (Appendix IA). The fish community at LRC1 (IBI = 38) is indicative of a problem affecting the two stations. After rain on August 5, 1995, an ammonia concentration of 0.14 mg/l was determined at LRC1.

The fish communities of Haw Creek also indicate multiple agricultural problems. Station HC1 is essentially an open algae-filled roadside ditch that underwent a sharp transition to a turbid stream in a large open pasture. The fish community was poor (IBI = 34) (Figure 20.17). Riparian forests were totally lacking. Stations HC0 and HC1 contained somewhat improved (fair) fish communities (IBI = 40 for both sites).

20.3.2.12 Big Walnut Mainstem

IBI and Iwb levels of the Big Walnut Creek mainstem collections were profiled in August 1993, shortly after the high summer flows returned to normal. The Iwb values were highly variable between sites and no obvious trend to increase or decrease along the length of the mainstem was noted (Figure 20.19). However, sampling effort ended at Greencastle because the river was too high for sampling success further downstream. Sampling was conducted again in 1994. The IBI profile was much more regular, with no evident trend over space. Fish communities at the channelized

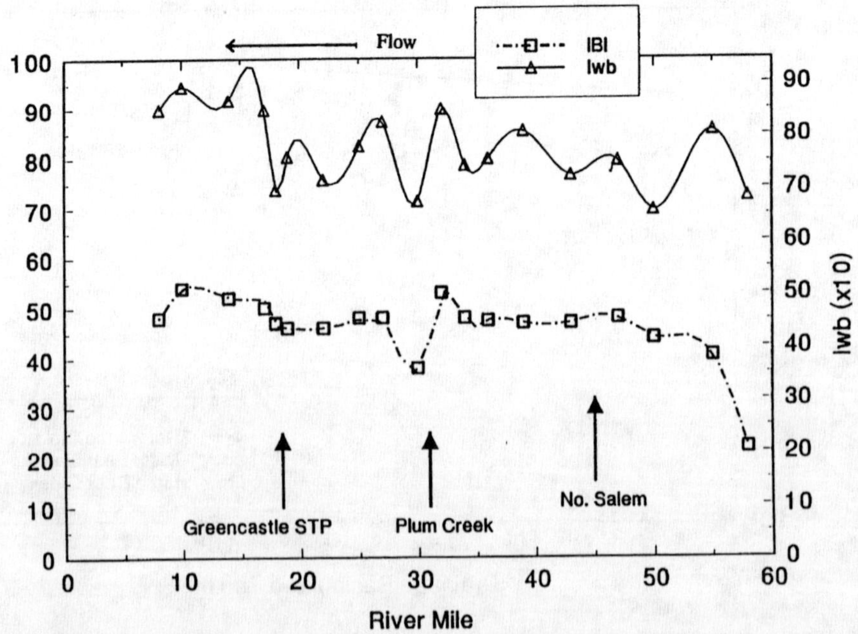

FIGURE 20.19 IBI and Iwb profiles of fish communities in the Big Walnut Creek mainstem based on mean 1993 and 1994 values.

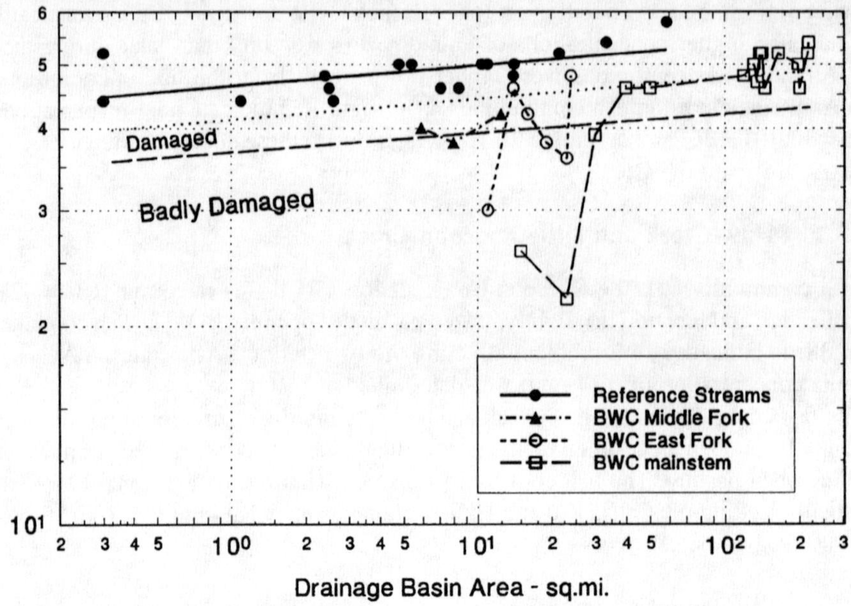

FIGURE 20.20 Comparison of fish community quality (IBI) of the mainstem, east fork, and middle fork of Big Walnut Creek and reference stream sites.

sites were clearly depressed (Figure 20.20). Those of the Middle Fork were less impacted but still substandard.

One significant depression indicated by both parameters occurred downstream from the mouth of Plum Creek. The proximity of this depression to Plum Creek suggested that Plum Creek might have exerted a negative impact on the fish communities of Big Walnut Creek. A closer examination of this reach indicated active lateral erosion, which is a more likely factor of degradation, especially during the high flows of 1993. The 1994 water levels were much more favorable for good sampling of the fish communities. Both the IBI and Iwb profiles indicated generally improving fish communities in a downstream direction. A small depression was still present below the mouth of Plum Creek, but its magnitude was no larger than depressions elsewhere. The lack of a depression downstream from North Salem and Greencastle is an encouraging sign that domestic and industrial wastes are treated adequately before they enter Big Walnut Creek. Several sections of river, however, are actively eroding because of inadequate riparian protection. In addition to the Plum Creek area, extensive erosion is in progress upstream from Wildwood Bridge, upstream from the mouth of Little Walnut Creek, and midway between Houk Covered Bridge and Reelsville.

20.3.2.13 Big Raccoon Creek Mainstem

Three of the upper collecting sites support fair fish communities (Figure 20.21); most contain poor communities. Only at the HES sites in a wooded valley do the communities suddenly improve to the good category. In the early 1980s, these sites contained fish communities that rated poor to fair (mean IBI = 38.5) and they lacked important components such as bass, sunfish, and darters. Gradual improvement since 1981 led to an average IBI value of 50.3 and a good rating in 1999 (J.R. Gammon, unpublished data). The 1993 profiles of IBI and Iwb values along the mainstem were similar (Figure 20.22). They both indicate depressed communities in upstream sections with a pronounced depression downstream from Ladoga, along a distance of 8 km. Recovery occurred at HES1 near the mouth of Cornstalk Creek. Problems with waste treatment were responsible for this depression and they were later rectified.

20.3.2.14 Deer Creek Mainstem

Only the upper watershed of Deer Creek mainstem was sampled chemically and biologically because of access difficulties in the lower watershed. The fish communities of four sites (DC1 through DC4) were included in the reference category because of high IBI values (50 to 58)

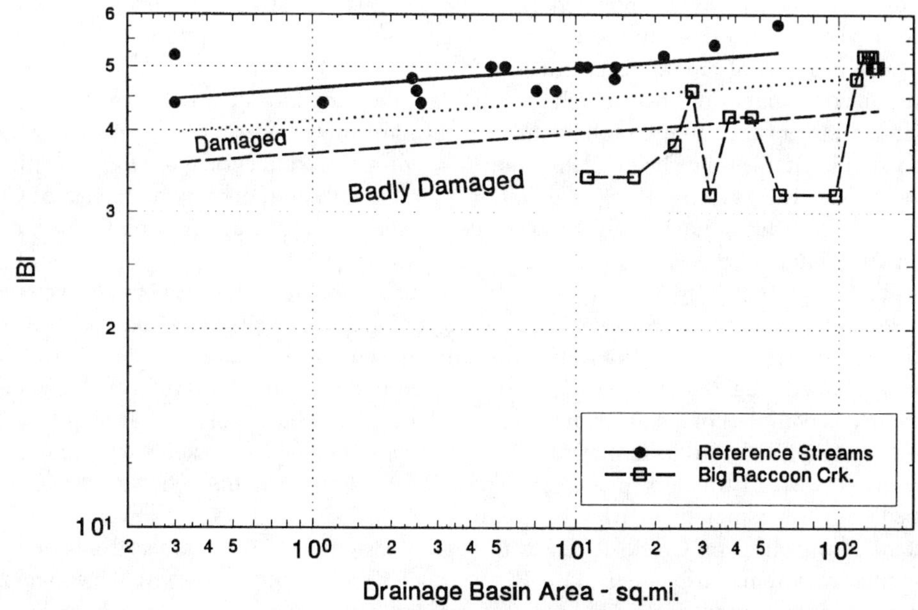

FIGURE 20.21 Comparison of fish community quality (IBI) of the mainstem, east fork, and middle fork of Big Raccoon Creek and reference stream sites.

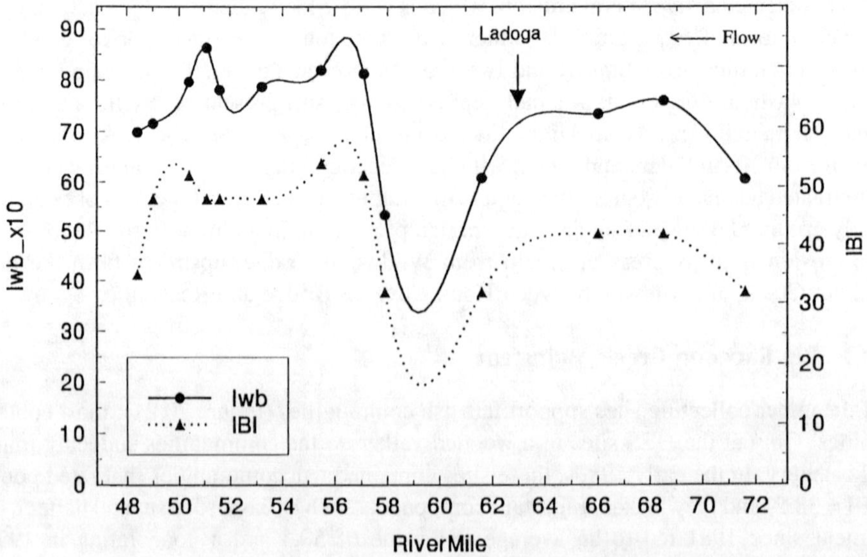

FIGURE 20.22 IBI and Iwb profiles along the Big Raccoon Creek mainstem in 1993.

(Appendix IC). Although several sections have been channelized, Deer Creek still contains the best fish communities among local streams.

20.3.2.15 Miscellaneous Streams

Ammonia concentrations were elevated (up to 3.5 mg/l) at Owl Branch (Deer Creek) at OB1 in the summers of 1993 and 1994, primarily after rain events. Determinations of the IBI above (OB1) and below (OB2) the polluted tributary indicated no negative effect once tributary water mixed with Owl Branch water (OB1 IBI = 42; OB2 IBI = 46) (Appendix IC). However, the low IBI value (34) of the fish community in the polluted tributary (OB1a) is indicative of degraded environmental conditions. The IBI for the nearby RI1 was 52, the highest value of the headwater stations and a higher value than one of the reference streams.

The fish communities of Owl Branch are rated fair; they are only slightly lower in quality than the reference streams. A more thorough study of this stream system should be made because of the probability that this watershed will continue to be converted from agriculture to residential use. The headwaters also may be subject to alteration from the industrial concentration east of Greencastle. It is not possible to predict with any confidence whether the stream communities will benefit or suffer from future changes.

Snyder Branch joins Big Walnut Creek slightly upstream from the Greencastle waterworks dam. The fish communities of Snyder Branch were sampled at two sites (SB1 and SB2) in 1993. IBI values of 40 were obtained at both sites, indicating the presence of fair fish communities. Single stations scattered throughout the area were also sampled to obtain additional information about the environmental conditions of small streams. Station JB1 is on a tributary of Big Raccoon Creek just west of Cornstalk Creek. It runs through an old pasture and contained a fair to good fish community (IBI = 46). The small tributary of Big Raccoon Creek that passes through New Ross (HR1), contained a fair fish community (IBI = 40). Peters Creek is a small tributary of lower Ramp Creek, a tributary of Big Raccoon Creek. It passes through a pasture at PTC1 and also contained a fair fish community. No small tributaries of the Big Raccoon Creek system possessed fish communities sufficiently good to warrant inclusion in the reference stream group.

Station RR4 is located on a small tributary of Big Walnut Creek just west of the Ramp Run system. It also contained a fair fish community (IBI = 41). Station CC4 was sited on another small tributary of Big Walnut Creek immediately south of lower Clear Creek. The stream passes through an open pasture at the collection site and contained a poor fish community (IBI = 32). Reference stream stations located in small tributaries of Big Walnut Creek were DAY1 and SKT1.

Three small tributaries of Deer Creek were examined: Wallace Branch (WB1), Upper Limestone Creek (ULC1), and Mosquito Creek (MOC1) (Figure 20.18). Mosquito Creek contained a poor to fair fish community (IBI = 38), perhaps because of activities in the basin in 1993 that caused considerable turbidity. The communities of Wallace Branch and Upper Limestone Creek were both fair (IBI = 40). Several other tributaries of Deer Creek had sufficiently good fish communities to warrant inclusion as reference sites (RF1, LWC1, and RI1). Rocky Fork Creek (RF1) was electrofished on May 25, 1995 when its waters flowed clear. However, in late July 1995 when RF1 was photographed, its waters were very turbid despite the fact that no rain had fallen for some time. Undoubtedly sediment was entering from the Kentucky Stone Company quarry upstream.

20.3.3 The Effects of Animal Feed Lots, Pastures, and Channelization

The effects of animal feed lots, pastures, and channelization on the fish communities were evaluated by a statistical analysis of the IBI values. Sites in each land use group were compared to all other sites (t-test), and a multivariate detrended correspondence analysis (DCA) of the catch rates (number of fish per hour) of the most common species using the CANOCO program was performed (Hill and Gauch, 1980; Ter Braak, 1987, 1989; Ter Braak and Prentice, 1988).

The IBI data for all sites with DBAs smaller than 20 mi^2 were partitioned into four groups: (1) sites located downstream and close to concentrations of farm animals; (2) sites located in pastures, usually with cattle, more rarely with horses, or a combination of both; (3) sites in sections of stream that had been channelized, and (4) all other sites not known to be directly influenced by animal feed lots, active pastures, or channelization.

IBI data for each land use category were compared statistically to IBI values of all other sites by means of a t-test using the MINITAB package. Descriptive statistics are summarized in Table 20.1 and shown graphically in Figure 20.23. Channelized sites having riparian borders of trees and shrubs were initially separated from open channelized sites. However, when IBI values from these groupings were compared using a t-test, they were statistically similar and therefore were combined in further analyses. A highly significant difference (T = 3.28, p = 0.0065) was found between the mean IBI of sites downstream from animal feed lots (IBI = 31.50) and the mean IBI of reference sites (IBI = 41.98) — more than a 10-point difference. The difference in IBI values of channelized sites (mean IBI = 37.21) compared to other sites (mean IBI = 41.98) was also highly significant statistically (T = 3.12, p = 0.0035). IBI values at sites in pastures did not differ statistically from reference sites by a narrow margin (T = 1.98, p = 0.066), although the average IBI value was 3.5 points lower.

TABLE 20.1
Descriptive Statistical Summary for Sites with Drainage Basin Areas Smaller than 20 mi.^{2}

Statistic	Other Sites	Below Feed lots	Pastures	Channelized
Number of sites	58	12	12	24
Mean IBI	42.0	31.5	38.5	37.2
Median IBI	42	33	38	38
Standard deviation	5.65	10.76	5.54	6.55
Standard error	0.74	3.11	1.60	1.34
Range	28–52	12–48	32–48	22–50
95% confidence interval	1.49	6.84	3.52	2.77

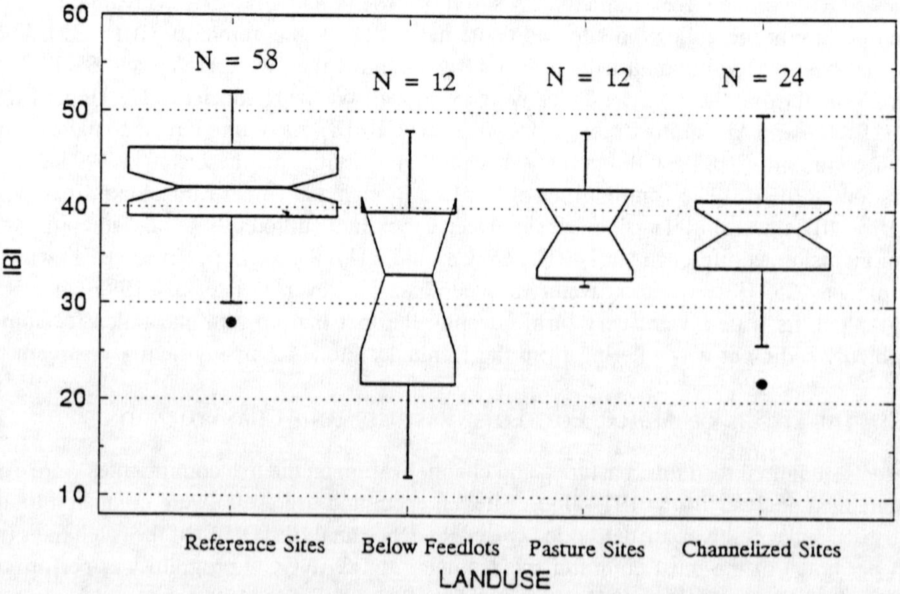

FIGURE 20.23 Boxplots of IBI data from sites with DBAs smaller than 20 mi^2 and located below animal feed lots, in pastures, in channelized sections of streams, and at other sites.

Detrended correspondence analysis (DCA) included catch data (number per hour) from all headwater sites with drainage basin areas smaller than 20 mi^2 (Figure 20.24). Thirty-eight collections of fish from the Big Raccoon Creek watershed, 54 from Big Walnut Creek, including Little Walnut Creek, 29 from Deer Creek, and 5 collections from Rattlesnake Creek were analyzed via DCA. There were no fish at site SRC0 in 1993. Therefore, in order to include it in the analysis, it was necessary to indicate a catch rate of 1.0 fish per hour for the most likely species, the creek chub (*Semotilus atromaculatus*). A few sites with slightly larger watersheds (CSC1, CSC1.5, and CSC2 located in Cornstalk Creek and DC4 located in Deer Creek) were included in the analysis.

The DCA showed compositional differences in the fish communities at the sites located downstream from animal concentrations compared to other sites. The DCA plot is subdivided into three main areas, (1) sites located near concentrations of farm animals (shown in black), (2) channelized sites, and (3) all other sites. All sites downstream from animal concentrations including sites more distant from those strongly and directly influenced by animal wastes in their extreme upper headwaters appear on the left side. Plum Creek (all sites downstream from PC0 and PC1) and South Ramp Creek (all sites downstream from SRC0) are included in this group. Many sites located higher up in the headwaters that are not directly impacted by farm animal concentrations are also included: Miller Creek (MC1 and MC2), Snake Creek (SC1), DeWeese Creek (DW1, DW2, and DW3), and many others. The "other sites" grouping includes most of the least impacted reference sites (Figure 20.25) and some sites downstream from a dairy herd on lower Cornstalk Creek. Five reference sites were included in the "below animal feed lots" grouping. They are mostly sites located in small tributaries that may periodically go dry.

Channelized sites are mostly scattered in the lower portion of the DCA plot (Figure 20.26). Nine channelized sites were mixed with the "below animal feed lots" section. Most sites affected by channelization are smaller headwater sites located in the upper more northerly portions of Big Raccoon Creek and Big Walnut Creek.

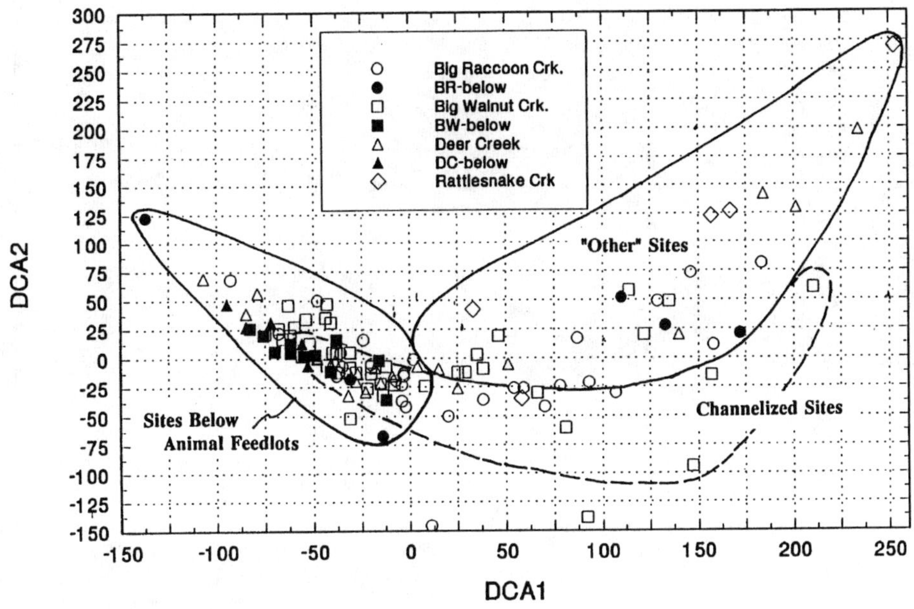

FIGURE 20.24 Detrended correspondence analysis of fish community compositions in headwater sites (DBA < 20 mi²) for Big Raccoon, Big Walnut, and Deer Creeks.

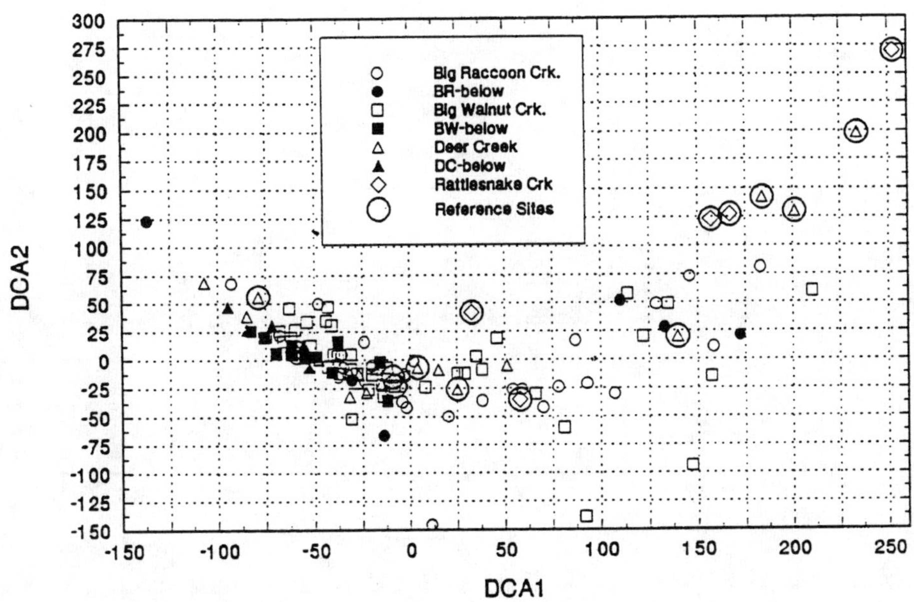

FIGURE 20.25 Positions of reference sites within the DCA array (large circles).

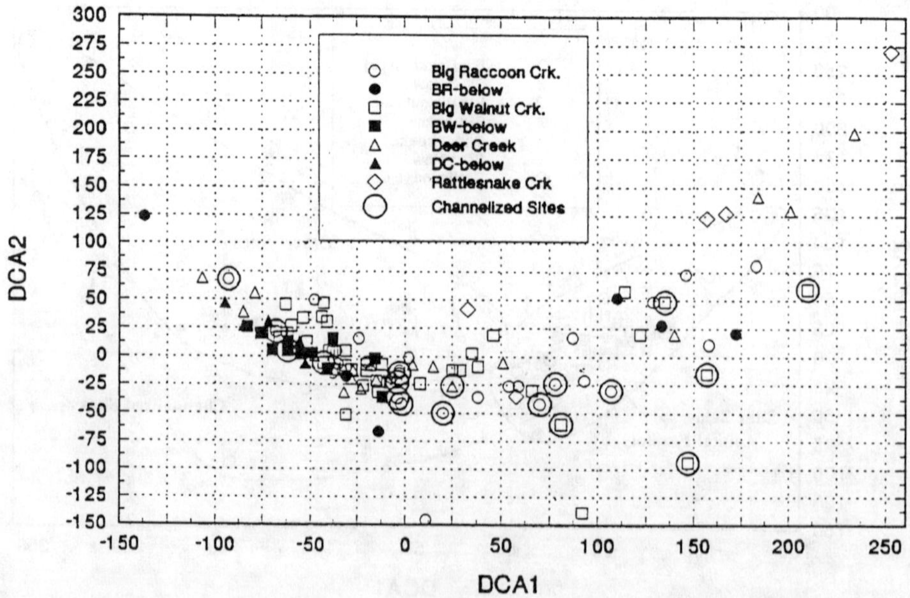

FIGURE 20.26 Positions of channelized sites within the DCA array (large circles).

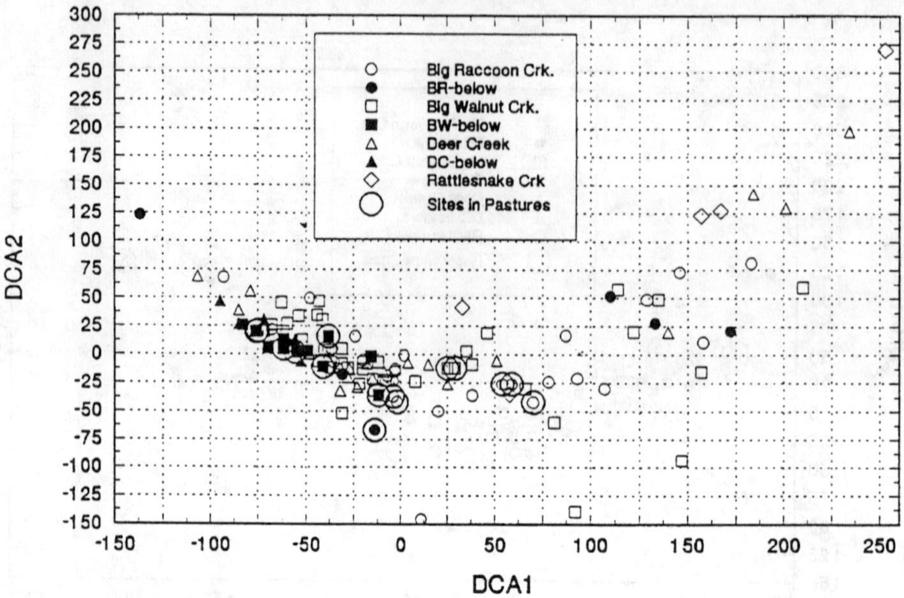

FIGURE 20.27 Positions of sites located in pastures within the DCA array (large circles).

Figure 20.27 shows the sites located in pastures. Only two sites are marginally in the "other site" cluster, three are in the "channelized" grouping, and the remaining sites are in the "below animal feed lots" group. Heavily pastured areas may exert the same negative effects on fish communities as hog lots.

20.3.3.1 Changes in Composition of Headwaters Fish Communities below Feed Lots

Species compositions of small polluted headwaters (DBA <5 mi^2) were compared to those of unpolluted headwaters by examining electrofishing catches of 11 least impacted sites (SC1, BB1, LB1, MC0, ULC1, LWC1, DAY1, SKT1, RI1, and RS0) and 6 sites subjected to significant animal feed lot runoff (LDC1, PC1, MR1, OB1a, SRC0, and JC2). All sites are in small watersheds with DBAs smaller than 5 mi^2. Included were data from sites that were electrofished during a wet summer (1993) and a dry one (1995).

Unpolluted headwaters were normally occupied by diverse fish communities that usually included creek chub (*Semotilus atromaculatus*), stoneroller (*Campostoma anomalum),* blacknose dace (*Rhinichthys atratulus),* bluntnose minnow (*Pimephales notatus),* white sucker (*Catostomus commersoni),* southern redbelly dace (*Phoxinus erythrogaster),* striped shiner (*Luxilus chrysocephalus),* johnny darter (*Etheostoma nigrum),* and orangethroat darter (*Etheostoma spectabile).* Many were designated by Ohio EPA (1989) as "headwater species" and were used to form the headwater metric developed by Simon (1991).

The fish communities less frequently included black redhorse (*Moxostoma duquesnei*), silverjaw minnow (*Ericymba buccata*), rainbow darter (*Etheostoma caeruleum*), greenside darter (*Etheostoma blennioides*), mottled sculpin (*Cottus bairdi*), green sunfish (*Lepomis cyanellus*), and longear sunfish (*Lepomis megalotis*). When small ponds were present, headwaters fish communities often included small bluegill (*Lepomis macrochirus*) and largemouth bass (*Micropterus salmoides*).

Other *Micropterus* bass species more typical of rivers and streams were also occasionally found. In one stream less influenced by agricultural activities than most others in the area (Snake Creek, SC1), both smallmouth bass (*Micropterus dolomieu*) and spotted bass (*Micropterus punctulatus*) were found.

Headwaters fish communities that were strongly influenced by feed lots sometimes lacked fish completely (SRC0–1993) or were represented only by young creek chub, presumably hatched from eggs deposited by spawning adults that subsequently moved downstream (SRC0–1995). More commonly, creek chub together with smaller numbers of bluntnose minnow, blacknose dace, and stoneroller comprised more than half the catch downstream from feed lots. These species were occasionally accompanied by orangethroat darter and southern redbelly dace.

The normal composition of a headwaters community, the mean number of species, and mean IBI values contrasted strongly with parameters for the depauperate fish community downstream from feed lots (Table 20.2). Only creek chub and bluntnose minnow were relatively more abundant in headwaters downstream from feed lots than they were in unpolluted headwaters. The orangethroat darter population was about equal in both types of headwaters although its relative abundance was quite low in both situations. Other species, if present, were much less common in the polluted headwaters, showing that environmental conditions were unsuitable or biologically stressful.

The reduction or elimination of the more sensitive species from the polluted headwaters led to a significant decrease in the total number of species present, an average of only 4.7 species compared to an average of 11.3 species in unpolluted headwaters. About the same magnitude of difference was found for median number of species. Similar disparities occurred for IBI values. All of these differences are statistically significant.

The abundance of fish in the natural headwaters was far greater (>3.51) than it was in headwaters receiving pollutive inputs of animal feed lots. The catch via electrofishing averaged 840 fish per hour in the natural headwaters compared to only 235 in headwaters downstream from animal feed lots and dairy herds.

TABLE 20.2
Percent Composition of Fish Communities in Unpolluted Headwaters and at Sites Downstream of Animal Feed Lots and Other Community Statistics

Species and Community Values	Unpolluted Headwaters	Headwaters below Feed Lots
Creek chub – *Semotilus atromaculatus*	28.7%	57.4%
Stoneroller – *Campostoma anomalum*	15.8%	6.5%
Blacknose dace – *Rhinichthys atratulus*	14.5%	8.4%
Bluntnose minnow – *Pimephales notatus*	6.6%	8.9%
White sucker – *Catostomus commersoni*	5.8%	1.8%
Southern redbelly dace – *Phoxinus erythrogaster*	5.8%	1.9%
Striped shiner – *Luxilus chrysocephalus*	4.4%	0.1%
Johnny darter – *Etheostoma nigrum*	3.1%	0.4%
Orangethroat darter – *Etheostoma spectabile*	3.0%	2.9%
Silverjaw minnow – *Ericymba buccata*	2.8%	0.2%
Number of collections	10	13
Mean number of species collected	11.3	4.7
Median number of species collected	12	4
Mean IBI	42.5	27.8
Median IBI	40.0	31.5
Catch per unit effort (number/hour)	840	235.0

Note: All sites are located in headwaters with DBAs <5 mi^2.

20.3.3.2 Changes in Compositions of Fish Communities in Channelized Headwaters

DCA analysis showed that few channelized sites supported normal fish communities. Great differences among channelized sites were often noted, as was considerable variability in instream habitats. These differences were not examined since our interest focused on confined feeding operations.

20.3.4 HABITAT

The physical characteristics of collecting sites in Big Raccoon Creek, Big Walnut Creek, Deer Creek, and Rattlesnake Creek were examined using the qualitative habitat evaluation index (QHEI) in 1993 (Rankin, 1989) and the rapid bioassessment protocols (RBPs) (Plafkin et al., 1989). The habitat conditions of headwater sites were estimated to be poorest in Big Raccoon Creek and about equal at sites on Big Walnut Creek and Deer Creek (Table 20.3). Only three wadeable sites were evaluated for Deer Creek, but the differences in conditions between sites on Big Raccoon and Big Walnut Creeks were about the same as those observed for headwater sites. Headwater sites generally rated lower than wadeable sites in all three stream systems (Figure 20.28).

The condition of the forested riparian wetlands was detailed by Gammon (1994). The vegetated riparian border along Big Raccoon Creek afforded the least protection, with 32% of the sites possessing riparian corridors less that 3 m wide. Riparian vegetation consisted primarily of saplings and shrubs. Big Raccoon Creek also had the most open canopy (60% sites, <50% shaded) and the greatest incidence of severe and moderate bank eroding (40%). Big Walnut Creek was better protected, with more than 50% of its riparian border measuring more than 9 m in width; most sites had >50% shading. About 30% of its banks are exposed to moderate to severe erosion and the vegetation covering its banks consisted primarily of saplings and shrubs, especially in headwater sites. Seventy percent of Deer Creek sites had good riparian protection (>9 m). About 80% of sites

TABLE 20.3
Summary of Statistics of Habitat Scores for Big Raccoon, Big Walnut, and Deer Creek Sites Using 1993 Habitat Criteria

Statistic	Big Raccoon Creek	Big Walnut Creek	Deer Creek
	Headwaters (DBA <20 mi²)		
Mean	60.08	67.34	66.43
Median	67.50	70.00	74.00
Standard error	4.16	2.28	3.91
Range	12–87	12–91	30–80
Number of sites	26	41	14
	Wadeable Sites (DBA >20 mi²)		
Mean	71.33	78.00	67.67
Median	72.00	80.00	67.00
Standard error	2.97	2.32	1.76
Range	53–92	46–97	65–71
Number of sites	15	23	3

were shaded, and <25% of banks were in a state of moderate to severe erosion. The primary vegetative cover for most headwater sites consisted of grass, indicative of the prevalence of pastures. The riparian wetlands of these stream systems are currently incapable of providing adequate protection from agricultural nonpoint source pollution. Furthermore, the degenerate riparian environment contributes to the negative aspects of animal feed lots.

A more detailed assessment of riparian condition was made from aerial photographs available in 1997. Only segments of the mainstem and tributaries within Putnam County were included. All of Deer Creek was included, but the upper channelized headwaters of Big Walnut and Big Raccoon Creeks were excluded. Classes of riparian condition included channelized, two bare banks, one bank bare and one vegetated, both banks vegetated (no trees), and well forested.

One third of the Big Raccoon Creek mainstem was well forested while bare banks occupied another 25%. Among its tributaries, 42% of Cornstalk Creek was channelized and 53% of Haw Creek was channelized or had at least one bare bank. Nearly half of the North Ramp Creek channel was forested, but only 17% of South Ramp Creek. Downstream from their junction, 53% of the Ramp Creek channel was forested.

The mainstem of Big Walnut Creek was 75% forested from the Putnam County line to U.S. 36, 50% forested from U.S. 36 to State Road 231, and only 32% forested from there to the mouth of Mill Creek. Forests flanked 33 to 51% of the eight tributaries examined. Long Branch and Little Walnut Creek were least affected by bare banks (15 to 17%); about 40% of Bledsoe, Snake, Owl, Jones, Snider, and Plum Creeks were either channelized or flanked by at least one bare bank.

Deer Creek differed from the two larger streams by having a greater percentage of channelization in the lower segment below Putnamville (14%), than above Putnamville (4%). This physical disturbance was particularly prevalent within the boundaries of the state farm prison. Forests in both segments flanked about 40% of the stream, however, and one or more bare banks flanked 38% of both segments. Only Little Deer Creek, Deweese Creek, and Owl Branch were assessed, but forests constituted the major riparian element (60 to 74%).

20.4 CONCLUSIONS

The effects of animal feed lots and pastures on water chemistry and fish communities were studied in the mainstems and tributaries of four creek systems in central Indiana between 1993 and 1995.

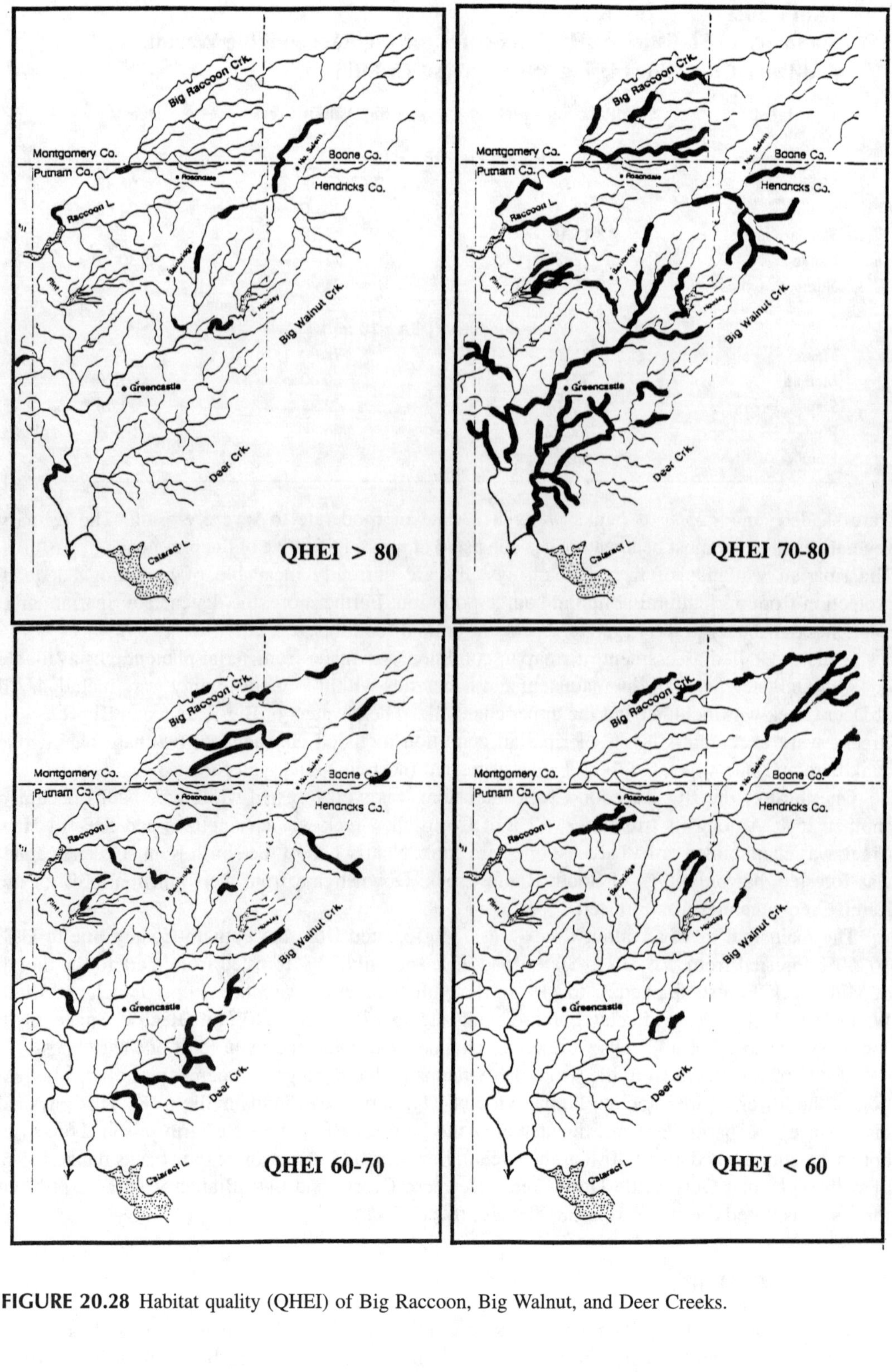

FIGURE 20.28 Habitat quality (QHEI) of Big Raccoon, Big Walnut, and Deer Creeks.

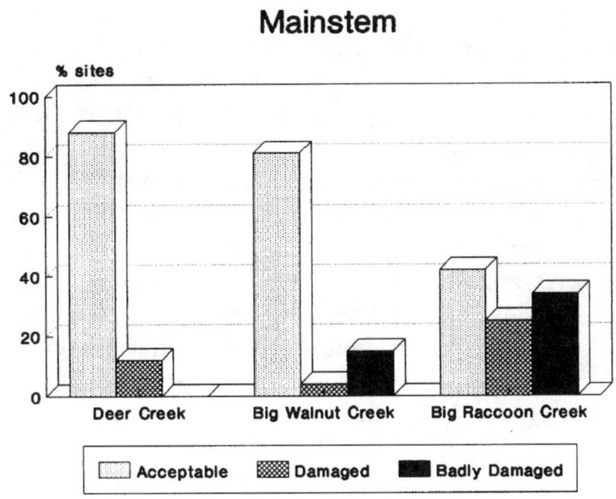

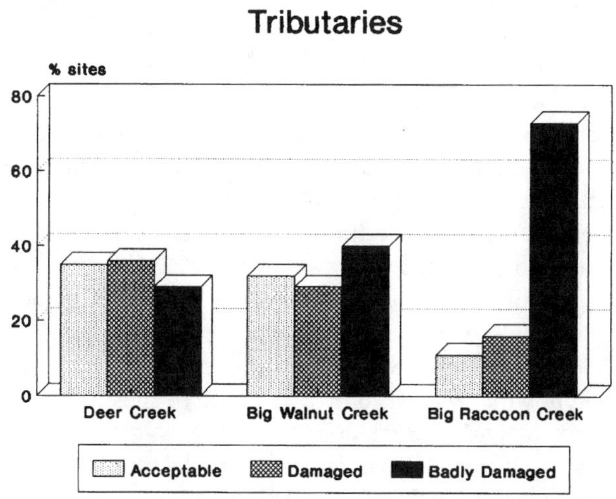

FIGURE 20.29 Overall assessment of fish community quality (IBI) of tributaries and mainstems of Big Raccoon, Big Walnut, and Deer Creeks.

Fish community composition, habitat characteristics, and water quality were sampled at 131 sites in Putnam County. The influence of animal feed lots was assessed by comparing sites downstream from feed lots with a group of least affected reference sites. Most of the animal feed lots were located near small headwaters and were primarily hog lots and dairy farms. Most pastures afforded ready access of cattle and horses to headwater streams, although fencing restricted access to a few streams.

Headwaters downstream from feed lots usually had elevated levels of ammonia, pH, turbidity, and conductivity. Ammonia concentrations were usually greater than 1.0 mg/l and exceeded 90 mg/l at one site. The pH was often elevated 0.2 to 0.4 units, turbidity was often quadrupled, and conductivity increased by 20 to 30% compared to upstream levels.

Fish communities downstream from animal feed lots were occasionally absent and usually degraded. Normal headwaters contained a diverse assemblage of fishes including creek chub (*Semotilus atromaculatus*), stoneroller (*Campostoma anomalum),* blacknose dace (*Rhinichthys atratulus),* bluntnose minnow (*Pimephales notatus),* white sucker (*Catostomus commersoni),* southern redbelly dace (*Phoxinus erythrogaster),* striped shiner (*Luxilus chrysocephalus),* johnny darter

(*Etheostoma nigrum*), and orangethroat darter (*Etheostoma spectabile*). Headwaters downstream from animal feed lots contained fewer fish and a limited number of species, predominantly creek chub and bluntnose minnow, in addition to a few of the other expected species. IBI values of fish communities located downstream of animal feed lots were statistically lower than communities of comparable streams lacking feed lots.

A toxic ammonia zone was also observed immediately downstream of heavily pastured lands. The distance affected by toxic ammonia was dependent upon rainfall frequency, intensity, and duration. Higher levels of ammonia were found in 1995 (a dry summer) than in 1993 (a wet summer). Stronger discharge and dilution may have reduced ammonia levels during wet periods. Heavily pastured lands also delivered pollutants intermittently to their contained streams, primarily as intermittent peaks during and after rainstorm events. Ammonia concentration, turbidity, and conductivity usually decreased with distance downstream from animal feed lots, The fish communities improved, but did not usually reach normal levels even when ammonia was reduced to background levels.

Ammonia is gradually oxidized to nitrite and nitrate as it is carried downstream from animal feed lots, but the nitrates, together with phosphates, form a potent fertilizer. In places where the riparian canopy of a stream was removed, sunlight penetrated to the bottoms of small streams. With plentiful nutrients, it stimulates the development of an algal carpet and secondary environmental problems. The subsequent diurnal changes in dissolved oxygen concentration and increased temperatures add to the overall deterioration of fish communities in small streams. Photosynthesis was shown to create supersaturated dissolved oxygen (>300%) levels during afternoon hours on sunny days. Respiration of the algal carpet at night along with concurrent decomposition of organic matter and oxidation of ammonia can cause a dissolved oxygen deficiency. Furthermore, elevated water temperatures may exceed the thermal capacities of some species of headwater fishes.

It is apparent that the conditions of riparian wetlands are closely linked to the effects of animal feed lots on small agriculturally-influenced streams. Only 40% of Big Raccoon Creek sites were 50% shaded, whereas about 65% of sites of Big Walnut Creek and about 75% of Deer Creek sites were similarly shaded. Approximately 33% of the sites on Big Raccoon Creek were buffered by riparian zones less than 3 m wide, whereas only about 15% of sites in Big Walnut Creek and Deer Creek possessed such narrow riparian corridors.

The cumulative negative effects on fish communities in the three streams differed in magnitude based on IBI criteria. Departures of eight or more IBI points from the reference regression constituted badly damaged communities and differences ranging from four to eight points indicated damaged communities. Appendix III summarizes tributary and mainstem data for all streams studied. Figure 20.29 shows overall effects. The quality of fish communities in tributaries of all three stream systems was considerably poorer than in their mainstems. The best communities were in Deer Creek and the poorest in Big Raccoon Creek; Big Walnut Creek fell somewhere between Deer Creek and Big Raccoon Creek. This suggests that improving habitat and water quality in tributaries would mutually benefit fish communities in tributaries and mainstems.

The negative impacts of animal feed lots could be reduced by implementing cost-effective best management practices recommended by several state and federal agencies familiar with the problems created by animal feed lots. The ammonia levels below the headwaters hog lot on Little Deer Creek quickly declined to background after moving lagoon wastes to fields more distant to the creek. Judicious use of fences along streams in order to restrict free access of farm animals to streams would also yield benefits. Establishing wetlands in headwater locations could provide a water treatment in suitable sites. The forested riparian wetlands that should flank every Indiana stream need to be restored. These critically important wetlands shade streams and keep them cool. They intercept eroded soil and nutrients from bottomland fields. In addition, these wetlands would also act as a buffer for feed lots and pastures.

The Natural Resources Conservation Service (NRCS) planted trees along nearly 40 mi of the Big Walnut Creek mainstem in 2001 (Fisher, personal communication), targeting eroding areas in particular. Future efforts should include tributaries as well.

ACKNOWLEDGMENTS

This project was supported in part by grants from Heritage Environmental Services, Indianapolis, Indiana, and the Indiana Department of Natural Resources with additional support from PSI Energy, Plainfield, Indiana, and Eli Lilly and Company, Indianapolis, Indiana. Fish collection assistance was provided by Bradley Garner, Matt Hall, John Baumann, John Riggs, Shawn Riggs, Mark Davis, and John Hecko. Lisa Portune, Julie Sear, Lee Schoenfeld, and Kim Hemmerlien performed habitat assessments. Chemical and microbiological sampling was done by Mutia Mateene, Becky May, and David Cruse. The Indiana Department of Natural Resources provided a YSI Model 3800 datalogger. Field assistance in 1995 was provided by John Garner, Gene DeClark, Simon Krughoff and Misty Robinson. In 1997, undergraduate students Eugene Chio and Kevin McKelvey examined riparian quality along the mainstems and major tributaries using aerial photographs of streams within Putnam County. The opinions expressed do not necessarily represent those of the U.S. Fish and Wildlife Service. No official endorsement by that agency should be inferred.

REFERENCES

Anonymous. 1985. *Standard Methods for the Examination of Water and Wastewater*, 16th ed. American Public Health Association, American Waterworks Association, and World Pollution Control Federation. Washington, D.C.

Anonymous. 2001. *Indiana Agricultural Statistics 2000–2001*. Purdue University, West Lafayette, IN.

Gammon, J.R. 1976. *The Fish Populations of the Middle 340 km of the Middle Wabash River.* Purdue University, West Lafayette, IN.

Gammon, J.R. 1980. The use of community parameters derived from electrofishing catches of river fish as indicators of environmental quality, *Seminar on Water Quality Trade-Offs*, U.S. Environmental Protection Agency, Washington, D.C. 335–363.

Gammon, J.R. 1994. The status of riparian wetlands in West-central Indiana Streams, *Proceedings of the Indiana Academy of Science,* 103, 195–213.

Gammon, J.R. 1998. *The Wabash River Ecosystem.* Indiana University Press, Bloomington.

Hill, M.O. and Gauch, H.G. 1980. Detrended correspondence analysis: an improved ordination technique, *Vegetatio,* 42, 47–58.

Hocutt, C.H. and J.R. Stauffer, Jr. 1980. *Biological Montoring of Fish.* Lexington Books, Lexington, MA.

Hoggatt, R.E. 1975. *Drainage Areas of Indiana Streams.*U.S. Department of the Interior, Washington, D.C.

Horton, R.E. 1945. Erosional development of streams and their drainage basins: hydrophysical approach to quantitative morphology, *Bulletin of the Geological Society of America*, 56, 275–370.

Hughes, R.M. and J.R. Gammon. 1987. Longitudinal changes in fish assemblages and water quality in the Willamette River, Oregon, *Transactions of the American Fisheries Society*, 116, 196–209.

Indiana Department of Environmental Management. 1992. Habitat Assessment and Physicochemical Parameters. Indianapolis, IN.

Karr, J.R. 1981. Assessment of biotic integrity using fish communities, *Fisheries,* 6, 21–27.

Karr, J.R., K.D. Fausch, P.L. Angermeier, P.R. Yant, and I.J. Schlosser. 1986. Assessing biological integrity in running waters: a method and its rationale. Illinois Natural History Survey Special Publication 5, Champaign, IL.

Omernik, J.M. and A.L. Gallant. 1988. Ecoregions of the Upper Midwest States. EPA 600/3-88/037. U.S. Environmental Protection Agency, Corvallis, OR.

Ohio Environmental Protection Agency. 1987. *Water Quality Implementation Manual*, 3rd update. Ohio Environmental Protection Agency, Columbus.

Ohio Environmental Protection Agency. 1988. *Biological Criteria for the Protection of Aquatic Life*. Vol. II. *Users' Manual for Biological Field Assessment of Ohio Surface Waters*. Columbus.

Ohio Environmental Protection Agency. 1989. *Biological Criteria for the Protection of Aquatic Life*. Vol. III. *Standardized Biological Field Sampling and Laboratory Methods for Assessing Fish and Macroinvertebrate Communities*. Div. Water Qual. Planning and Assessment.

Pflieger, W.L. 1975. *The Fishes of Missouri*. Missouri Department of Conservation, Columbia.

Plafkin, J.L., M.T. Barbour, K.D. Porter, S.K. Gross, and R.M. Hughes. 1989. Rapid bioassessment protocols for use in streams and rivers: benthic macroinvertebrates and fish. U.S. Environmental Protection Agency, Office of Water. EPA/444/4-89-001. May 1989.

Platts, W.S., W.F. Megahan, and G.W. Minshall. 1983. Methods for Evaluating Stream, Riparian, and Biotic Conditions. General Technical Report INT-138. U.S. Department of Agriculture, Ogden, UT.

Rankin, E.T. 1989. The Use of the Qualitative Habitat Evaluation Index for Use Attainability Studies in Streams and Rivers in Ohio. Ohio Environmental Protection Agency, Columbus.

Robins, C.R., R.M. Bailey, C.E. Bond, J.R. Brooker, E.A. Lachner, R.N. Lea, and W.B. Scott. *Common and Scientific Names of Fishes from the United States and Canada*. 5th ed. American Fisheries Society Special Publication 20.

Simon, T.P. 1991. Development of Index of Biotic Integrity Expectations for the Ecoregions of Indiana. I. Central Corn Belt Plain. EPA 905/9-91/025. U.S. Environmental Protection Agency, Chicago.

Smith, P.W. 1971. Illinois streams: a classification based on their fishes and an analysis of factors responsible for the disappearance of native species. *Illinois Natural History Survey Biological Notes* 76, Champaign, IL.

Strahler, A.N. 1952. Hypsometric (area-altitude) analysis of erosional topography, *Bulletin of the Geological Society of America*, 63, 1117–1142.

Ter Braak, C.J.F. 1987. The analysis of vegetation–environment relationships by canonical correspondence analysis, *Vegetatio*, 69, 69–77.

Ter Braak, C.J.F. 1989. CANOCO: an extension of DECORANA to analyze species–environment relationships, *Hydrobiologia*, 184, 169–270.

Ter Braak, C.J.F. and I.C. Prentice. 1988. A theory of gradient analysis, *Advances in Ecological Research*, 18, 271–317.

Trautman, M.B. 1981. *The Fishes of Ohio*. Ohio State University Press, Columbus.

Yoder, C.O. and E.T. Rankin. 1995. Biological criteria program development and implementation in Ohio, in Davis, W.S. and T.P. Simon, Eds., *Biological Assessment and Criteria: Tools for Water Resource Planning and Decision Making*. Lewis Publishers, Boca Raton, FL. 109–144.

Yoder, C.O. and M.A. Smith. 1999. Using fish assemblages in a state biological assessment and criteria program: essential concepts and considerations, in Simon, T.P., Ed., *Assessing the Sustainability and Biological Integrity of Water Resources Using Fish Communities,* CRC Press, Boca Raton, FL, 17–56.

APPENDIX IA
Summary of Habitat Evaluations (QHEIs) and Index of Biotic Integrity (IBI) Values for Big Raccoon Creek Sites, 1993 through 1996

Site	DBA (mi^2)	Index of Biotic Integrity				QHEI	1993 Hab. Score
		1993	1994	1995	1996		
BRC0x4[ct]	11.0			34		51	46
BRC0x3[ct]	16.5			34		64	45
BRC00[ct]	23.7			38		53	12
BRC0	27.5			46		61	22
BRC1[c0p]	32.0	32				74	80
BRC2[c0p]	38.0	42				59	54
BRC3	46.4	42				74	77
BRC4	59.5	32				78	65
BRC5	96.9	32				79	53
HES1	117	52	54	54	48	83	76
HES2	125	54	52	52	52	71	63
HES3	130	48	50	46	52	61	67
HES4	134	48	52	52	52	73	72
HES5	135	48	52	50	50	78	92
HES6	137	52	52	48	50	80	84
HES7	138	48	48	46	50	71	85
HES8	141	35	52	46	50	80	74
CSC0x6[c0]	4.0	44				39	23
CSC0x5[c0]	6.0	40				38	30
CSC0x4[ct]	10.0	32				63	47
CSC0x3[c0p]	12.0	38				67	87
CSC00	15.0	36				73	75
CSC0	17.0	46				70	59
CSC1	19.0	48				69	72
CSC1.5*	20.0	42				76	81
CSC2*	20.0	32				76	81
HC0	7.0	40				79	81
HC1	25.0	40				72	58
HC1a[c0p]	3.0	34				60	53
LRC0[p]	6.2	42				64	81
LRC1[p]	8.7	38				62	45
NRC0[p]	1.3	32		(dry)		44	48
NRC1	2.3	28		(dry)		50	65
NRC2	4.5	42		36		64	80
NRC3*	7.2	40		34		50	71
NRC4	17.8	30		40		77	74
RC5	33.1	48		49		74	70
SRC0*	0.5	12		23		66	79
SRC1[c0]	2.0	42		29		63	79
SRC1.5[p]	4.5			36			
SRC2	5.7	44		36		83	72
SRC3	8.0	46		48		69	70
PTC1[c0]	3.0	41				63	62
HR1	1.4	40				52	75
JB1[p]	2.8	46				65	53

Note: * = Downstream from an animal feed lot. [p] = Active cattle and/or horse pasture. [c0] = Channelized, open canopy. [ct] = Channeled, with 50% trees or shrubs as canopy.

APPENDIX IB
Summary of Habitat Evaluations (QHEIs) and Index of Biotic Integrity (IBI) Values for all Big Walnut Creek Sites, 1993 through 1996

Site	DBA (mi^2)	Index of Biotic Integrity 1993	1994	1995	1996	QHEI	1993 Hab. Score
BW00[c0]	15.0	26				59	72
BW0[c0]	23.0		22			60	81
BW1[c0]	30.0	42	39			55	66
BW2	40	42	46			83	90
BW3	50	50	46			81	87
BW4	119	46	48			78	82
BW5	131	44	50			83	82
BW6	135	48	47			79	84
BW7	138	44	52			78	85
BW8	141	54	52			76	69
BW9	145	30	46			70	63
BW10	164	44	52			76	75
BW11	197	46	50			75	76
BW12	200	46	46			76	80
BW13	216	40	52			80	93
BW14	218	40	54			77	86
BW15	220		50				
BW16	294		52				
BW17	304					73	46
BW18	306		54			79	76
BW19	308		48			83	97
BWM1	5.9	40				55	59
BWM1.5	8.0	38				64	49
BWM2	12.5	42				82	81
BWE00[c0]	11.0	30				34	72
BWE0[ct]	14.0	46				56	36
BWE1[c0]	16.0	42				56	67
BWE2[c0]	19.0	38				68	67
BWE3	23.0	36				77	77
BWE3.5	24.0	48				73	80
BWE4[c0]	25.0	50				71	80
BB1	2.3	38				77	78
BB2[*]	5.7	44				83	67
CC0	2.3	43				57	77
CC1	4.6	48				72	63
CC2	17.0	39				81	73
CC3	21.6	46				77	76
MC0	2.0	32				58	66
MC1	3.6	44				70	72
MC2[p]	9.3	48				71	52
MR1[*p]	1.2	36				75	66
MR2	5.8	38				72	75
PC0[*p]	0.9			20		48	66
PC1[*p]	2.0	34		24		74	47
PC2	2.8	33		40		65	83
PC2.3[*]	3.0			48			

APPENDIX IB (CONTINUED)
Summary of Habitat Evaluations (QHEIs) and Index of Biotic Integrity (IBI) Values for all Big Walnut Creek Sites, 1993 through 1996

Site	DBA (mi^2)	Index of Biotic Integrity 1993	1994	1995	1996	QHEI	1993 Hab. Score
PC2.6	3.1			48			
PC3	5.5	48		48		70	71
PC4	9.5	46		44		79	81
PC2a	2.8			32		47	
NRR1[p]	3.2	42				73	83
NRR2	7.0	46				64	71
RR3[c0p]	3.5	41				74	76
RR4	3.5	41				67	58
SRR2[ct]	10.7	38				76	76
SC1	3.1	40		40	42	64	68
SC2	5.4	50		42	44	76	67
SB1	2.0	40				62	62
SB2	4.0	40				71	78
RF1[c0]	2.5			46	42	75	
LWC1	2.6			44	42	79	
SKT1	1.1			44			
DAY1	2.4			48	42		
CC4[p]	2.4	32					
OC1[c0p]	0.4				34	63	
OC2[ct]	1.7				36	59	
OC3[p]	2.7				42	72	
OC4	2.3				46	72	
OC5	3.2				32	65	
OC6	6.9				40	72	
JC1[p]	3.3	32				73	66
JC2[*p]	4.5	32				65	50
LB1	1.2	40				81	91
LB2	1.2	46				73	70
LB2a	0.8	41				79	91
LB3	6.5	48				72	49
LB4	10.4	50				69	69

Note: * = Downstream from an animal feed lot. [p] = Active cattle and/or horse pasture. [c0] = Channelized, open canopy. [ct] = Channelized, with 50% trees or shrubs as canopy.

APPENDIX IC
Summary of Habitat Evaluations (QHEIs) and Index of Biotic Integrity (IBI) Values for Deer Creek Sites, 1993 through 1996

Site	DBA (mi^2)	Index of Biotic Integrity 1993	1994	1995	1996	QHEI	1993 Hab. Score
DC0000	1.9				36		
DC000	3.6				44		
DC00	3.7				44		
DC0	7.1	46				63	68
DC1	14.5	50				70	80
DC2	21.7	52				69	65
DC3	34.0	54				85	71
DC4	45.0	58	58			74	67
LDC1^{*c0p}	1.2	31		20	36\26	50	39
LDC1.5	1.7			48	30\34	66	
LDC2	2.5	34		34	33\40	73	74
LDC2.5	7.0			36	38\32	73	70
LDC3	8.6	42		40	44\46	70	75
LDC1.5a					38\30		
LDC2a^{p}	1.0			40	35\34	49	
OB1	5.3	42				73	68
OB1a*	0.4	34				74	76
OB2	5.3	46					
DW1^{p}	0.8	38				72	74
DW2^{ct}	2.2	40				71	74
DW3	8.5	32				74	74
MOC1	3.4	38				65	74
ULC1	3.5	40			40	61	59
WB1	1.8	42				70	65
DC2a*	2.3			50	40	49	
DC3a*	2.3			40			
RI1	0.3	52		44			

Note: * = Downstream from an animal feed lot. ^{p} = Active cattle and/or horse pasture. ^{c0} = Channelized, open canopy. ^{ct} = Channelized, with 50% trees or shrubs as canopy.

APPENDIX ID
Summary of Habitat Evaluations (QHEIs) and Index of Biotic Integrity (IBI) Values for Rattlesnake Creek Sites, 1993 through 1995

Site	DBA (mi^2)	Index of Biotic Integrity 1993	1994	1995	1996	QHEI	1993 Hab. Score
RAT0	4.8		50		42	Dry	59
RAT1	8.4		46		44	81	76
RAT2	11.0		50		44	81	81
RAT3	14.0		48		52	92	93
RAT4	19.0		42		50	77	68

APPENDIX IIA
Statistical Summary of Water Chemistry at Little Deer Creek Sites

Site	Km from Origin	Mean Concentration (S.E.) 1993	1995	1996
		NH_3-nitrogen (mg/l) (N_{93} = 5, N_{95} = 16, N_{96} = 15)		
LDC0	0.0	0.04 (0.03)	0.14 (0.11)	0.20 (0.08)
LDC1	1.4	0.92 (0.24)	2.44 (0.32)	0.24 (0.03)
LDC1.5	1.9	—	—	0.20 (0.05)
LDC2	2.5	0.07 (0.04)	0.94 (0.27)	0.21 (0.04)
LDC2.5	5.8	0.10 (0.06)	0.68 (0.14)	0.24 (0.04)
LDC3	8.7	0.21 (0.08)	0.44 (0.13)	0.19 (0.03)
LDC1.5a	Trib.	—	—	0.23 (0.03)
LDC2a	Trib.	—	1.21 (0.37)	0.30 (0.06)
		pH (N_{95} = 21, N_{96} = 15)		
LDC0	0.0	—	7.51 (0.11)	7.15 (0.14)
LDC1	1.4	—	7.87 (0.08)	7.85 (0.06)
LDC1.5	1.9	—	—	7.97 (0.09)
LDC2	2.5	—	7.93 (0.08)	8.01 (0.07)
LDC2.5	5.8	—	8.07 (0.07)	8.22 (0.08)
LDC3	8.7	—	8.09 (0.07)	8.09 (0.06)
LDC1.5a	Trib.			
LDC2a	Trib.		7.95 (0.05)	8.11 (0.10)
		NO_3-Nitrogen (mg/l) (N_{95} = 14, N_{96} = 15)		
LDC0	0.0	—	11.36 (2.10)	12.3 (0.55)
LDC1	1.4	—	13.54 (2.56)	11.36 (1.17)
LDC1.5	1.9	—	—	9.57 (1.21)
LDC2	2.5	—	9.29 (2.01)	7.36 (1.15)
LDC2.5	5.8		4.89 (2.03)	4.92 (0.80)
LDC3	8.7	—	4.09 (1.29)	3.27 (0.76)
LDC1.5a	Trib.	—	—	5.58 (1.17)
LDC2a	Trib.	—	5.02 (0.92)	5.57 (0.78)
		Conductivity (microhms) (N_{95} = 21, N_{96} = 15)		
LDC0	2.4	—	506 (36.1)	501 (15.9)
LDC1	4.2	—	709 (32.6)	528 (28.0)
LDC1.5	6.4	—	520 (20.8)	480.8 (27.8)
LDC2	7.4	—	493 (14.9)	448.7 (27.5)
LDC2.5	7.9	—	529 (10.7)	447.6 (21.5)
LDC3	8.8	—	524 (33.7)	471.1 (16.1)
LDC1.5a	Trib.		—	453.2 (28.4)
LDC2a	Trib.	—	524 (33.7)	459.7 (19.2)
		Turbidity (NTU) (N_{95} = 23, N_{96} = 15)		
LDC0	0.0	—	28.3 (10.9)	5.93 (1.45)
LDC1	1.4	—	109.7 (30.6)	27.7 (5.34)
LDC1.5	1.9	—	—	18.7 (8.32)
LDC2	2.5	—	49.2 (28.0)	19.7 (9.88)
LDC2.5	5.8	—	20.4 (9.5)	22.2 (14.1)
LDC3	8.7	—	17.1(7.9)	8.4 (4.73)
LDC1.5a	Trib.		—	7.3 (8.31)
LDC2a	Trib.	—	21.5 (7.3)	21.9 (11.2)

APPENDIX IIB
Statistical Summary of Water Chemistry at Plum Creek Sites

Site	Km from Origin	Mean Concentration (S.E.) 1993	1995	1996
		NH_3-nitrogen (mg/l) (N_{93} = 7, N_{95} = 7, N_{96} = 6)		
PC0	2.4	0.92 (0.52)	2.30 (0.88)	0.53 (0.16)
PC1	4.2	0.53 (0.34)	1.65 (0.45)	0.51 (0.16)
PC2	6.4	0.44 (0.18)	0.76 (0.42)	0.28 (0.05)
PC2.3	7.4	0.19 (0.09)	0.64 (0.34)	0.27 (0.05)
PC2.6	7.9	—	—	0.24 (0.04)
PC3	8.8	0.21 (0.08)	0.66 (0.36)	0.22 (0.03)
PC4	12.0	0.20 (0.08)	0.55 (0.30)	—
		pH (N_{95} = 7, N_{96} = 6)		
PC0	2.4	—	7.84 (0.08)	8.06 (0.15)
PC1	4.2	—	8.08 (0.09)	8.28 (0.21)
PC2	6.4	—	7.70 (0.13)	8.06 (0.08)
PC2.3	7.4	—	7.92 (0.12)	8.08 (0.07)
PC2.6	7.9	—	—	8.22 (0.06)
PC3	8.8	—	8.09 (0.07)	8.08 (0.06)
PC4	12.0	—	7.98 (0.09)	—
		NO_3-nitrogen (mg/l) (N_{95} = 5, N_{96} = 6)		
PC0	2.4	—	9.20 (3.48)	8.50 (1.75)
PC1	4.2	—	4.80 (1.88)	6.30 (1.52)
PC2	6.4	—	2.78 (1.46)	3.70 (1.12)
PC2.3	7.4	—	2.02 (0.99)	4.50 (1.35)
PC2.6	7.9	—	—	4.50 (1.06)
PC3	8.8	—	2.88 (2.28)	4.00 (1.51)
PC4	12.0	—	2.12 (1.23)	—
		Conductivity (microhms) (N_{95} = 12, N_{96} = 6)		
PC0	2.4	—	918 (202.5)	517 (35.2)
PC1	4.2	—	660 (35.5)	511 (37.3)
PC2	6.4	—	600 (30.3)	512 (56.5)
PC2.3	7.4	—	607 (15.4)	534 (45.8)
PC2.6	7.9	—	—	516 (52.7)
PC3	8.8	—	589 (15.6)	509 (26.2)
PC4	12.0	—	541 (17.8)	—
		Turbidity (NTU) (N_{95} = 10, N_{96} = 6)		
PC0	2.4	—	19.9 (6.55)	28.8 (14.35)
PC1	4.2	—	17.3 (4.37)	35.3 (19.04)
PC2	6.4	—	4.9 (1.78)	99.8 (81.24)
PC2.3	7.4	—	6.6 (2.40)	64.5 (39.8)
PC2.6	7.9	—	—	164.7 (158.9)
PC3	8.8	—	2.9 (1.43)	11.7 (8.65)
PC4	12.0	—	1.1(0.87)	—

APPENDIX IIC
Statistical Summary of Water Chemistry at North Ramp Creek Sites

Site	Km from Origin	Mean Concentration (S.E.) 1993	1995	1996
		NH_3-nitrogen (mg/l) (N_{93} = 6, N_{95} = 16, N_{96} = 15)		
NRC0	1.6	1.00 (0.43)	0.46 (0.15)	0.15 (0.05)
NRC1	2.5	0.09 (0.04)	0.34 (0.16)	0.14 (0.05)
NRC2	5.2	0.07 (0.04)	0.27 (0.10)	0.16 (0.04)
NRC3	6.4	0.15 (0.07)	0.25 (0.10)	0.14 (0.03)
NRC4	11.3	0.04 (0.03)	0.18 (0.12)	0.14 (0.03)
		pH (N_{95} = 5, N_{96} = 15)		
NRC0	1.6	—	8.01 (0.09)	7.92 (0.13)
NRC1	2.5	—	7.94 (0.07)	7.92 (0.15)
NRC2	5.2	—	8.07 (0.07)	8.06 (0.13)
NRC3	6.4	—	8.05 (0.09)	8.12 (0.14)
NRC4	11.3	—	8.04 (0.17)	8.15 (0.12)
		NO3-nitrogen (mg/l) (N_{95} = 1, N_{96} = 15)		
NRC0	1.6	—	12.0	10.2(1.11)
NRC1	2.5	—	16.0	9.4(1.60)
NRC2	5.2	—	6.0	6.0(1.68)
NRC3	6.4	—	4.0	6.0(1.82)
NRC4	11.3	—	3.0	4.4(1.36)
		Conductivity (microhms) (N_{95} = 5, N_{96} = 15)		
NRC0	1.6	—	772 (144.0)	479 (15.7)
NRC1	2.5	—	454 (47.0)	413 (40.1)
NRC2	5.2	—	466 (35.2)	478 (17.6)
NRC3	6.4	—	537 (14.3)	499 (14.2)
NRC4	11.3	—	568 (7.7)	511 (13.5)
		Turbidity (NTU) (N_{95} = 5, N_{96} = 15)		
NRC0	1.6	—	12.0 (6.9)	15.3 (7.0)
NRC1	2.5	—	8.0 (4.9)	37.9 (28.0)
NRC2	5.2	—	0.2 (0.2)	5.9 (2.6)
NRC3	6.4	—	2.6 (0.8)	6.9 (1.34)
NRC4	11.3	—	1.4 (0.6)	11.6 (9.1)

APPENDIX IID
Statistical Summary of Water Chemistry at South Ramp Creek Sites

Site	Km from Origin	Mean Concentration (S.E.) 1993	1995	1996
		NH_3-nitrogen (mg/l) (N_{93} = 6, N_{95} = 11, N_{96} = 15)		
SRC00	0,0	0.03 (0.02)	0.37 (0.30)	0.22 (0.02)
SRC0	0.9	0.91 (0.71)	0.35 (0.08)	0.45 (0.10)
SRC1	2.9	0.08 (0.05)	0.46 (0.10)	0.34 (0.07)
SRC1.5	3.9	—	0.20 (0.03)	0.21 (0.02)
SRC2	5.0	0.05 (0.03)	0.28 (0.07)	0.21 (0.02)
SRC3	6.7	0.04 (0.03)	0.17 (0.06)	0.21 (.0.02)
SRC1a	Trib.	—	0.35 (0.06)	0.24 (0.05)
		pH (N_{95} = 11, N_{96} = 15)		
SRC00	0,0	—	7.76 (0.27)	7.71 (0.11)
SRC0	0.9	—	7.87 (0.11)	8.02 (0.08)
SRC1	2.9	—	7.54 (0.12)	8.12 (0.13)
SRC1.5	3.9	—	7.84 (0.07)	7.84 (0.15)
SRC2	5.0	—	7.79 (0.06)	8.01 (0.08)
SRC3	6.7	—	7.96 (0.10)	7.92 (0.17)
SRC1a	Trib.	—	7.92 (0.10)	8.13 (0.15)
		NO3-nitrogen (mg/l) (N_{95} = 9, N_{96} = 15)		
SRC00	0,0	—	8.40 (1.17)	6.43 (1.02)
SRC0	0.9	—	3.64 (1.48)	6.86 (1.24)
SRC1	2.9	—	0.83 (0.21)	5.43 (1.02)
SRC1.5	3.9	—	0.83 (0.21)	3.07 (1.01)
SRC2	5.0	—	0.79 (0.21)	3.25 (0.98)
SRC3	6.7	—	0.90 (0.30)	3.18 (0.86)
SRC1a	Trib.	—	0.90 (0.30)	3.86 (0.83)
		Conductivity (microhms) (N_{95} = 14, N_{96} = 15)		
SRC00	0,0	—	543 (20.8)	481 (16.9)
SRC0	0.9	—	578 (16.2)	607 (111.2)
SRC1	2.9	—	736 (104.1)	513 (35.7)
SRC1.5	3.9	—	538 (9.2)	506 (22.8)
SRC2	5.0	—	530 (11.3)	493 (19.2)
SRC3	6.7	—	579 (10.3)	497 (18.8)
SRC1a	Trib.	—	519 (11.5)	507 (18.0)
		Turbidity (NTU) (N_{95} = 13, N_{96} = 15)		
SRC00	0,0	—	14.1 (4.3)	16.9 (8.1)
SRC0	0.9	—	60.3 (32.7)	141 (123.8)
SRC1	2.9	—	74.9 (68.1)	65.7 (57.6)
SRC1.5	3.9	—	0.5 (0.2)	4.6 (1.5)
SRC2	5.0	—	3.3 (1.6)	4.9 (1.7)
SRC3	6.7	—	2.1 (1.2)	7.4 (3.9)
SRC1a	Trib.	—	3.0 (1.3)	8.6 (3.6)

APPENDIX III
Stream Segment Lengths of Three Central Indiana Streams and Amounts Damaged by Animal Feed Lot Disturbances

Stream	DBA mi²	Stream Length (km)	Damaged (km)	Badly Damaged (km)
		Big Raccoon Creek		
Mainstem	141	53	13.0	18.0
Cornstalk Creek	20	21	4.0	16.0
Haw Creek	25	25	0	25.0
L. Raccoon Creek	9	13	6.1	6.9
Ramp Creek	26	6.2	0	0
North Ramp Creek	18	12	1.0	11.0
South Ramp Creek	8	7	1.0	6.0
Peters Creek	3	1	1.0	0
John's Branch	3	2	0	0
HR1	1.4	1	1.0	0
Total Length/% Length		88.5 km	14.2 km	64.9 km
			16.0%	73.3%
		Big Walnut Creek		
Mainstem	308	93.0	4.0	14.0
Middle Fork	13	9.7	9.7	0
East Fork	25	15.1	7.0	2.0
Bledsoe Branch	6	7.6	3.8	3.8
Clear Creek	22	18.4	4.0	0
Miller Creek	10	17.2	0	5.0
Maiden Run	6	2.7	0	2.7
Plum Creek	10	12.0	1.5	3.0
Ramp Run		2.6	0	2.6
North Ramp Run	11	5.0	2.5	0
South Ramp Run	11	7.9	0	7.9
Snake Creek	8	10.4	5.0	0
Rocky Fork Creek	3	2.9	0	0
Leatherwood Creek	3	3.6	3.6	0
Seketa Creek	1	2.0	0	0
Day Creek	3	7.8	0	0
Owl Creek	7	17.8	6.0	8.0
Jones Creek	5	12.5	0	12.5
Long Branch	11	10.5	2.0	0
Snider Creek	3	3.0	3.0	0
Total Length/% Length		168.7	48.1 km	39.5 km
			28.5%	39.5%
		Deer Creek		
Mainstem	60	26.0	3.2	0
Little Deer Creek	9	9.0	2.0	7.0
Owl Branch	5.5	6.3	3.3	3.0
Deweese Creek	10	9.4	4.4	5.0
Upper Limestone Creek	5	8.3	8.3	0
Wallace Branch	2.7	2.5	2.5	0
Riggs Creek	0.3	0.8	0	0
Mosquito Creek	4	4.1	0	4.1
Total Length/% Length		66.3 km	23.7 km	19.1 km
			35.7%	28.8%

21 Response Signatures of Four Biological Indicators to an Iron and Steel Industrial Landfill

Paul M. Stewart, Jason T. Butcher, and Thomas P. Simon

CONTENTS

21.1 INTRODUCTION

Response indicators of ecological change are used as estimates of biological integrity to assess community response to specific stressors (McKenzie et al., 1992; Stewart et al., 1999). Endpoint measurements include changes in species diversity, which generally decrease in response to disturbance from a variety of stressors (Suter and Bartell, 1993). Few studies have examined response signatures of different communities to complex stressors. Scientists tend to specialize in one

0-8493-0905-0/03/$0.00+$1.50

taxonomic group or one type of contaminant (Hansen and Johnson, 1999). An exception was the series of papers about ecological risk assessment for a large contaminated site, the Clinch River (Suter, 1999). Another example will be reported in this chapter, where the responses of four biological assemblages in relationship to an industrial landfill were examined.

One impairment to a water body's designated use is the presence of noxious algal blooms caused by eutrophication (Stewart, 1995). Algal communities are sensitive to water quality (Butcher, 1947; Cholnoky, 1960; Patrick, 1973; Rosen, 1995; Stewart, 1995) and have been used as indicators of pollution impacts (Patrick, 1949; Stewart et al., 1986; Patrick et al., 1992; Barbour et al., 1999). Changes in community structure may include a shift from a diverse community with abundant pollution-intolerant species to a community dominated by pollution-tolerant species (Stewart, 1995). Algal assemblages are ubiquitous in aquatic systems and are often more diverse than other aquatic organisms such as macroinvertebrates, making them ideal organisms to use in ecological assessments involving changes in water quality and increased pollution loads.

Aquatic plant species, populations, and communities are used as indicators of the aquatic environment, allowing detection of ecosystem response to different stressors (Stewart et al., 1999). Endpoints measured include the loss of sensitive species, decrease in species richness, trophic state changes (Ravera, 1983), plant index of biotic integrity (PIBI) (Simon et al., 2001), and shifts in growth habits (Mesters, 1995). Species diversity of an aquatic plant community may be a direct reflection of water quality. The loss or predominance of certain species may indicate the presence of pollutants.

Macroinvertebrates are integrally linked to aquatic habitats, and their abundance and community structure are related to both chemical and physical conditions, making them useful biological indicators (Hilsenhoff, 1987; Knorr and Fairchild, 1987; Rosenberg and Resh, 1993; Stewart et al., 2000). Macroinvertebrates are directly influenced by physical conditions such as substrate type, channel morphology, amount and type of detritus and aquatic vegetation, and canopy cover (Pedersen and Perkins, 1986; Niemi et al., 1990; Richards et al., 1993) and are indirectly affected by changing nutrient concentrations and shifts in primary productivity (Stewart and Robertson, 1992; Richards et al., 1996). Unlike water quality measurements, which only provide an instantaneous assessment of stream conditions, macroinvertebrate assemblages can be used to identify past disturbances and toxic effects that are not readily detected by chemical means.

Fish community assessment endpoints include species richness, species abundance, species diversity (Suter et al., 1999), and an index of biotic integrity (Karr, 1981; Simon, 1998; Simon and Stewart, 1998). Due to their widespread public appeal, fish are often used as indicators of aquatic biological integrity (Simon, 1999). Changes in community structural metrics may include a loss in species richness and diversity as well as a decrease in the number of sensitive or intolerant species. Functional changes include the loss or reduction in trophic guilds; for example, the loss of lithophilous species (those that lay their eggs among stones) in cases of extreme sedimentation. Indicators of fish condition include the proportion of fish with external anomalies such as lesions, fin erosion deformities, and tumors that can be found in severely degraded conditions.

One important environmental issue facing managers is the contamination of water, air, soil, and biota by persistent toxic substances. These include heavy metals (i.e., copper, zinc) and polycyclic aromatic hydrocarbons (PAHs). Lead concentrations in precipitation were higher in Northwest Indiana than in any other part of the Great Lakes region (Gatz et al., 1989). Indiana Dunes National Lakeshore (INDU) had the highest wet deposition levels of sulfate and nitrate of any monitored National Park in the country (NADP, 1995). Fish consumption advisories for Lake Michigan and many of its tributaries have been posted (Indiana State Department of Health, 1999; Zorn, 1999). These include the Grand Calumet River, which contains the largest lake and open water wetland complex in INDU, the Grand Calumet Lagoons. Former consumption advisories cited elevated concentrations of PCBs, chlordane, dieldrin, and DDT in fish, while current advisories cite concentrations of PCBs and mercury (Indiana State Department of Health, 1999).

Industrial landfills greatly modify surrounding areas by affecting chemical, physical, and biological integrity (Simon and Stewart, 1998; Stewart et al., 1999). Few data quantifying contaminant levels near landfills in sediments or in the organisms living near landfills exist. We examined several indicators of the aquatic community to determine whether a relationship existed between proximity to an industrial landfill and a decrease in biological integrity. The purpose was to determine patterns in community compositions and concentrations of contaminants in organisms and to assess the effects of contaminants on several trophic levels in the Grand Calumet Lagoons and adjacent ponds. In most aquatic systems, it is difficult to establish causal relationships between contaminants and ecosystem health due to the many ecological factors that can influence the responses of organisms and communities to particular stressors (Ham et al., 1997). This research will use a weight-of-evidence approach to confirm landfill effects on aquatic communities and sources of contaminants (Adams et al., 1999; Suter, 1999).

21.2 METHODS

21.2.1 Description of Study Area

The Grand Calumet Lagoons and adjacent ponds are in an area identified by the International Joint Commission as a Great Lakes Area of Concern (GLAOC). This area is one of 43 regions of the Great Lakes watershed having severe environmental contamination (Holowaty et al., 1992). The Grand Calumet River and Indiana Harbor Ship Canal (GCR/IHSC) constitute the only GLAOC with all 14 designated uses impaired. Much of the western section of the Grand Calumet Lagoons is surrounded by a large industrial landfill. Where the landfill forms the western border of the park, it directly impinges on the park's aquatic resources. Noteworthy are several hazardous waste sites, including a dumpsite that received 18.2 million kg per year of tar decanter waste for 20 years (Read, 1994).

The study area includes the watershed of a series of lagoons and adjacent ponds bordering the east side of Gary in Lake County, Indiana (Figure 21.1). The Grand Calumet Lagoons system is an area of land use extremes, including parks and recreational areas, residences, and light and heavy industry that includes a large industrial iron and steel landfill. These uses produce slag disposal, industrial storage, refuse dumping, scrap preparation, basic oxidation sludge processing, and coal, coke, and rail car storage, and hazardous waste. The 32.6-ha Grand Calumet Lagoons system drains a 3.5 km^2 (350 ha) watershed at the eastern edge of the area of concern.

The lagoons are divided into three sections of roughly similar size: the East, Middle, and West Lagoons. The East (7.9 ha) and Middle Lagoons (9.9 ha) are connected by a channel. The West Lagoon (14.8 ha) is slightly lower in elevation than the Middle Lagoon and is connected by a short, shallow stream. The Middle Lagoon contains three study site locations — two on opposite shores (ML1, ML2) and one upstream of the small stream (ML3). The eastern section (about a third) of the West Lagoon has two sites east of the landfill that represent a transition area between the more intact natural areas to the east and the industrial area further west (WL1, WL2). This landscape continues west for a short distance and finally yields to an industrial area with mixed uses. Two sites are located adjacent to this industrial area (WL3, WL4). The site located furthest west (WL5) is near a culvert that drains the West Lagoon and leads to the mouth of the Grand Calumet River. Sites in the Middle Lagoon are termed far-field sites and sites located in the West Lagoon are termed near-field sites.

The East and West Ponds are north of the lagoons in similar terrain and are separated by a high dune ridge (Figure 21.1). Aerial photographs indicate the ponds were formerly part of the Grand Calumet River and were probably outlets to Lake Michigan. Two sample sites were selected in each pond (EP1, EP2 and WP1, WP2) for all studies except for aquatic plant surveys, for which the two sites in each pond were combined. The landfill comprises the western edge of the West

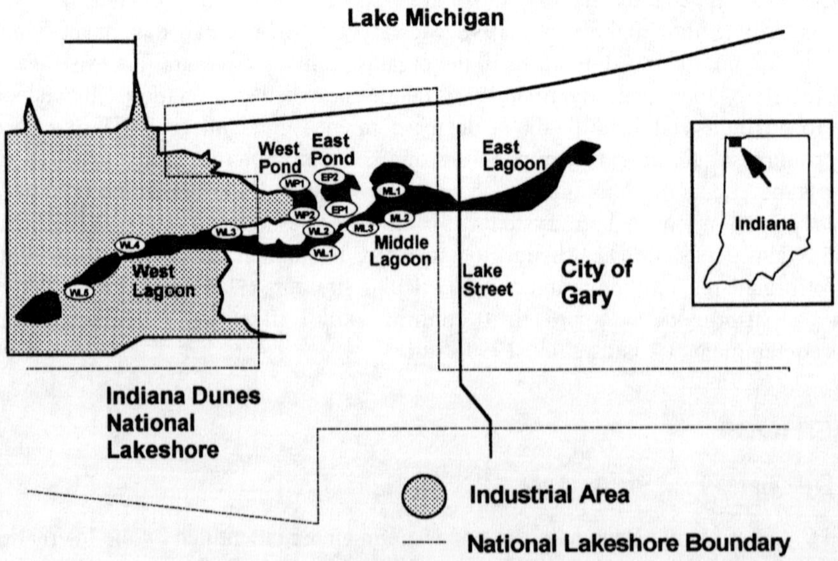

FIGURE 21.1 Map of the Grand Calumet Lagoons showing study sites and the relationship to Indiana Dunes National Lakeshore, the City of Gary, and industrial areas.

Pond, and from archived photographs, it is evident that a large portion of the West Pond has been filled by the industrial landfill. In addition, an electrical wire salvaging and burning area exists close to WP2.

21.2.2 Water Chemistry Analysis

Triplicate water samples were collected from 12 sites during placement (June 6, 1996) and retrieval (June 19, 1996) of periphytometers. Variables measured included dissolved oxygen, specific conductance, pH, total alkalinity, total hardness, and concentrations of sulfate, ammonia nitrogen, nitrite nitrogen, nitrate nitrogen, and total phosphate.

Dissolved oxygen concentration (mg l^{-1}) was measured in the field using a YSI Model 51B®oxygen meter* that was air calibrated. The pH (standard units) measurements were taken in the field using a Corning pH/°C Model 107® meter that was calibrated daily. One-liter water samples were collected in acid-washed (10% HCl) polyethylene bottles and transported to the laboratory on ice for analysis of additional variables (APHA, 1992). Specific conductance (μmhos/cm) was measured using the Fisher Scientific Accumet Model 50® pH/conductivity meter calibrated with standard conductance solutions (APHA, 1992). Total alkalinity (mg l^{-1} as $CaCO_3$), and total hardness (mg l^{-1} as $CaCO_3$) were measured by titration (APHA, 1992). Sulfate (mg l^{-1} SO_4^{2-}) was measured using the SulfaVer 4 method with a Hach DR2000® spectrophotometer. Nutrients, including total phosphate (mg l^{-1} as PO_4^{3-}), nitrite nitrogen (mg l^{-1} as N-NO_2-), nitrate nitrogen (mg l^{-1} as N-NO_3-), and ammonia nitrogen (mg l^{-1} as N-NH_3), were measured with a Hach DR2000® spectrophotometer using appropriate high and low range methods.

21.2.3 Sediment Chemistry Analysis

In field-collected sediments, PAHs almost always occur as a complex mixture of compounds (Swartz, 1999). Sediments were sampled and analyzed for contaminants at several lagoon sites. Sediments were collected in 1994 and 1995 for contaminant analysis by forcing a site-dedicated

* Use of brand names does not imply endorsement by the federal government.

decontaminated 7-cm PVC pipe into the top 20 to 30 cm. Sediments were transported on ice in clean airtight polyethylene jars. One sediment sample from each site was analyzed for texture, composition, and 22 contaminants by a contract laboratory using EPA-approved methods. Analyses included 15 acid-extractable organic contaminants and 78 base-neutral organic contaminants including PAHs. See Gillespie et al. (1998) for a further description of sampling methods and analysis.

21.2.4 Biological Analysis

21.2.4.1 Algal Assemblages

Periphyton samples were collected at 12 locations in the lagoons and ponds in 1996. Artificial substrates (periphytometers) were used for collecting algae in the lagoons and ponds (Patrick, 1949; Cairns, 1982). Triplicate periphytometers containing eight glass slides (25 mm × 75 mm) were placed at each site and incubated *in situ* for 2 weeks, with retrieval on August 3, 1993. Duplicate slides from each periphytometer were collected and placed in vials with site water and 3% Lugol's solution (APHA, 1992). All sites were successfully collected except ML1 where two periphytometers were vandalized. In this case, three slides collected from the same periphytometer were analyzed separately as triplicates.

Attached algae were removed from slides by thoroughly scraping both sides with the edge of a new glass slide. Several randomly selected slides were examined at 100× magnification to verify that all growth was removed from the slide. Scraped material was rinsed with distilled water into a 125-ml wide-mouthed jar. All residue and rinse water were allowed to concentrate by settling a minimum of 4 h per centimeter of depth of the final solution. After settling, the supernatant was carefully removed with a low-vacuum aspirator to avoid disturbance of the sediment. Samples of attached algal concentrate were adjusted to a known volume. The inverted microscope plate chamber was prepared by filling the chamber with distilled water. A known amount of distilled water was removed to match the volume of the sample to be added. Each sample was mixed by vigorously shaking the container 40 to 50 times. A measured aliquot of sample was extracted with a graduated pipette, placed into a counting chamber, mixed with the pipette, and allowed to settle for 4 h per centimeter of depth. Algae from a known area were identified and counted along a median strip across the bottom plate chamber at 1000× with a Zeiss Invertoscope D. The length of the strip counted was measured from the graduated mechanical stage. A minimum of 500 cells were counted per sample. Algae were enumerated as single cells; however, blue-green algal trichomes were counted as one cell equivalent to each 25-μ length. Counts were converted to quantitative cell densities as cells per square millimeter. Positive identification of problematic diatoms was made at 1000× from Hyrax® burn mounts prepared from at least two of the three replicates at each site.

21.2.4.2 Aquatic Plant Assemblages

Aquatic macrophytes were surveyed once during the late summer using a modified relevé sampling approach with a modification of the Braun–Blanquet cover abundance scale method of estimating percent cover (Mueller–Dombois and Ellenberg, 1974). One hundred-meter zones were surveyed along the near-shore, offshore, and littoral zones where facultative and obligate wetland species would be found. The level of effort ensured that the number of species found were in the upper limits of the species-area curve (Heck et al., 1975; de Caprariis et al., 1981; Smith et al., 1985).

The intent was to perform a representative qualitative survey, not an exhaustive census, and was targeted at estimates of biological diversity and relative abundance. All species of aquatic plants were identified, and an abundance rating (1 = observed, 2 = rare, 3 = rare/common, 4 = common, 5 = very common, 6 = abundant) was assigned to each species. Abundance categories were expanded from those used previously (Kershaw, 1964) and represent the number of individuals of a plant species; an "observed" rating was assigned when only one individual of a species was found. A "rare" rating was assigned when two to four individuals of a plant species were found. The "rare/common" rating was assigned when the plant species was represented by more than four individuals but was not a common component of the community. "Common" species were easily located and "very common" species comprised up to about 25% of the community. The "abundant" species dominated a site, comprising from 25 to almost, 100% of the plant community.

Identifications were made in the field and unknowns were identified using appropriate floristic manuals (Fasset, 1957; Winterringer and Lopinot, 1977; Gleason and Cronquist, 1991; Swink and Wilhelm, 1994). Scientific names followed Swink and Wilhelm (1994). Vouchered specimens were deposited at Purdue University's North Central herbarium (Stewart et al., 1999).

21.2.4.3 Macroinvertebrate Assemblages

Macroinvertebrates were collected at each site by sweep net samples in the littoral zones among the emergent aquatic vegetation. Sampling consisted of three sweep net samples composited at each site using a rectangular frame net measuring 0.46 × 0.23 m, with a mesh size of 600 µm. Each sweep measured 2 m in length (the overall length of the net and handle) for a total sampling surface area of 2.76 m^2 at each site. Samples were preserved in the field in polyethylene jars with 70% ethanol. Invertebrates were separated from debris in the laboratory, stored in glass jars with fresh 70% ethanol, and identified to family using standard taxonomic keys (Merritt and Cummins, 1984; Peckarsky et al., 1990). Taxonomy followed Merritt and Cummins (1984).

21.2.4.4 Fish Assemblages

Fish community structure and function were determined from daytime inventories at 12 sites collected in 1994 (Simon and Stewart, 1998). One hundred-meter zones were electroshocked along the near-shore, offshore, and littoral zones at each of the Middle and West Lagoon sites. All habitats within a sample area were surveyed relative to their frequency. We used a DC canoe-mounted electrofishing unit consisting of a T&J 1750-watt generator configuration with the net serving as the anode. Stunned fish were netted and placed into a live well for identification and enumeration. Fish were inspected for gross external deformities, eroded fins, lesions, and tumors (DELT) and released. Smaller species such as minnows, darters, madtoms, and sculpins were preserved in 10% formalin. Preserved specimens were identified in the laboratory using standard taxonomic references (Gerking, 1955; Smith, 1979; Becker, 1983), and scientific and common names follow Simon et al. (1992). The East and West ponds were sampled by repeated seining efforts with a 3.3-m minnow seine at each of the sites.

Stunned *Cyprinus carpio* (carp) were collected from several sites, wrapped in aluminum foil, placed on ice, and brought to the laboratory for freezing. Frozen fish were shipped overnight to the Great Lakes Science Center (U.S. Geological Survey, Ann Arbor, MI) for PAH analysis, after which 10 g of homogenized fish tissue composite was analyzed via gas chromographic mass spectrophotometry (van der Oost et al., 1994; Stein et al.,1995) with a Hewlett-Packard 5988A® research grade low resolution unit equipped with a HP5890® gas chromatograph and an HP7673A® autosampler. Detection limits were estimated at 5 to 30 ppb per compound.

21.2.5 Statistical Analyses

The nonparametric Mann–Whitney U-test was used to determine differences in periphyton, aquatic macrophyte, macroinvertebrate, and fish diversity between the Middle and West Lagoons and between the East and West Ponds at a significance level of $\alpha \leq 0.05$. Assemblage data were examined for trends and patterns using cluster analysis, both Bray–Curtis hierarchical agglomerative clustering (Primer 5© for Windows, Plymouth Marine Laboratories, U.K.; Clarke and Warwick, 1994) and a technique based on the coefficient of similarity (Pearson and Pinkham, 1992; Gonzales et al., 1993) were used to find natural groupings in the similarity data among assemblages found at all sites.

21.3 RESULTS

21.3.1 PATTERNS IN WATER QUALITY

Water quality data collected along with periphyton samples are reported for the two sampling dates corresponding to placement and retrieval of periphytometers (Table 21.1). Dissolved oxygen was significantly higher in the West Lagoon than the Middle Lagoon ($p = 0.015$) and was about 9.1 mg l^{-1} at WL1, decreasing westward with the exception of WL3. Dissolved oxygen measurements were significantly higher ($p = 0.015$) in the East Pond (over 7 mg l^{-1}) than the West Pond (less than 5 mg l^{-1}). Specific conductance was significantly higher in the West Lagoon than in the Middle Lagoon ($p < 0.001$). In the West Lagoon, specific conductance measurements tended to increase westward, with the highest readings taken at WL5. Specific conductance measurements in the East Pond were also significantly higher than those measured from the West Pond ($p = 0.016$). The pH was significantly higher in the Middle Lagoon than in the West Lagoon ($p < 0.001$), and decreased westward in the West Lagoon with the lowest measurements found at WL5. In the ponds, the pH was significantly higher in the East Pond than the West Pond ($p = 0.021$).

Alkalinity was about two times higher in the West Lagoon than in the Middle Lagoon ($p < 0.001$). Again, West Lagoon alkalinity readings increased from east to west, with the highest values found at WL5. Alkalinity measurements were significantly higher in the West Pond than in the East Pond ($p = 0.023$). Hardness was much higher in the West Lagoon than in the Middle Lagoon ($p < 0.001$). Total hardness in the West Lagoon also increased from east to west, with WL5 the highest. Total hardness concentrations were significantly higher in the East Pond than the West Pond ($p = 0.027$). Sulfate concentrations were highest in the West Lagoon, with the highest values found at the more westward sites. The East Pond had extremely low sulfate concentrations (< 2 mg l^{-1}) while the West Pond values (average = 37.5 mg l^{-1}) were much higher.

Ammonia nitrogen concentrations were higher ($p < 0.001$) in the West Lagoon (all > 0.65 mg l^{-1}) than in the East Lagoon (all at or near the detection limit, < 0.015 mg l^{-1}) and were the highest at WL5 (> 2.7 mg l^{-1}), thus increasing from east to west in the West Lagoon. Ammonia nitrogen concentrations in the West Pond tended to be about half those from the East Pond ($p = 0.018$). Nitrite nitrogen concentrations were significantly lower in the Middle Lagoon (all below the limit of detection of < 0.01 mg l^{-1}) than in the West Lagoon ($p < 0.001$). Nitrite nitrogen measurements from the East Pond and West Pond were also below detection limits. Nitrate nitrogen concentrations were significantly higher in the Middle Lagoon than the West Lagoon sites ($p = 0.035$), but at WL5, they tended to be about the same as the Middle Lagoon sites. There was no significant difference between the nitrate nitrogen concentrations found in the East and West Ponds ($p = 0.083$). Total phosphate was two to three times higher in the West Lagoon than the East Lagoon ($p < 0.001$) and was highest at WL5. Total phosphate concentrations were not significantly different between the East and West Ponds ($p = 0.059$).

21.3.2 SEDIMENT CONTAMINANT CHARACTERIZATION

Few contaminants were detected in the elutriate or the water samples analyzed. Sediment analyses revealed several heavy metals in the West Pond near an area of wire burning and recycling. The burning process released ash that contained metals and may have affected metal concentrations in the pond. In 1994, sediment organic contaminants were sampled and analyzed at several sites in the Grand Calumet Lagoons (Table 21.2). Total PAH at WL5 was 12.3% of the sediment, and of that, 77% was naphthalene.

TABLE 21.1
Means and Standard Deviations of Water Quality Variables Measured at 12 Sites in the Grand Calumet Lagoons and Adjacent Ponds at the Placement and Retrieval of Periphytometers

Site	DO (mg/l)	Conductance (μmhos/cm)	pH (s.u.)	Alkalinity (mg/l)	Hardness (mg/l)	Sulfate (mg/l)	Ammonia (mg/l)	Nitrite (mg/l)	Nitrate (mg/l)	Total Phos. (mg/l)
ML1	7.00 ± 0.09	439.3 ± 20.7	8.85	64.33 ± 2.94	120.00 ± 2.19	34.33 ± 1.63	<0.015	<0.01	0.408 ± 0.399	0.09 ± 0.04
ML2	7.26 ± 0.20	447.83 ± 8.0	8.78	72.67 ± 5.32	124.17 ± 4.40	34.83 ± 0.98	<0.015	<0.01	0.245 ± 0.210	0.08 ± 0.03
ML3	6.97 ± 0.12	446.8 ± 7.9	8.75	79.00 ± 7.67	125.67 ± 5.43	34.50 ± 1.64	<0.015	<0.01	0.225 ± 0.202	0.08 ± 0.02
WL1	9.10 ± 0.30	468.3 ± 30.1	8.64	116.17 ± 8.26	179.00 ± 35.52	47.50 ± 4.37	0.792 ± 0.404	0.029 ± 0.005	0.097 ± 0.063	0.22 ± 0.09
WL2	9.08 ± 0.23	469.2 ± 24.7	8.53	117.17 ± 5.15	184.33 ± 27.17	54.83 ± 2.40	0.677 ± 0.625	0.029 ± 0.003	0.115 ± 0.082	0.16 ± 0.05
WL3	9.28 ± 0.29	490.2 ± 20.7	8.53	129.00 ± 3.03	201.67 ± 15.87	63.00 ± 2.28	1.105 ± 0.568	0.044 ± 0.004	0.177 ± 0.135	0.21 ± 0.05
WL4	8.93 ± 0.12	490.8 ± 9.1	8.46	130.67 ± 4.84	201.00 ± 16.81	66.33 ± 2.80	1.325 ± 0.557	0.052 ± 0.002	0.242 ± 0.185	0.27 ± 0.07
WL5	6.45 ± 1.11	581.2 ± 6.6	7.97	151.67 ± 1.97	216.17 ± 13.36	75.83 ± 1.33	2.730 ± 0.457	0.068 ± 0.006	0.318 ± 0.258	0.37 ± 0.18
EP1	7.52 ± 0.07	407.0 ± 27.5	8.07	120.33 ± 5.13	194.33 ± 71.21	<2	0.882 ± 0.331	<0.01	0.317 ± 0.276	0.11 ± 0.02
EP2	7.08 ± 0.45	401.3 ± 23.8	8.05	124.00 ± 5.22	195.33 ± 70.89	<2	0.798 ± 0.334	<0.01	0.315 ± 0.292	0.10 ± 0.02
WP1	4.83 ± 0.55	374.5 ± 8.6	7.65	139.00 ± 7.13	176.67 ± 3.26	37.67 ± 0.82	0.435 ± 0.016	<0.01	0.265 ± 0.208	0.12 ± 0.02
WP2	4.76 ± 0.61	383.2 ± 5.9	7.61	138.33 ± 7.09	179.33 ± 4.50	37.00 ± 1.90	0.450 ± 0.017	<0.01	0.195 ± 0.115	0.12 ± 0.01

Note: N = 6 for each measurement; pH is reported as the median value.

TABLE 21.2
Concentrations of 17 PAHs from Sediments at Sites WL4 and WL5 of the Grand Calumet Lagoons

Contaminant	WL4	WL5
Naphthalene	BDL	95,455
Acenaphthene	BDL	5,666
2-Methylnaphthalene	BDL	4,800
Dibenzofuran	BDL	4,184
Phenanthrene	BDL	3,286
Fluorene	BDL	2,470
Fluoranthene	0.70	2,122
Pyrene	BDL	1,694
Anthracene	BDL	792
Benzoanthracene	BDL	523
Chrysene	BDL	453
Benzopyrene	BDL	444
Benzo(b) fluoranthene	0.80	440
Acenapthalene	BDL	420
Benzoperylene	BDL	300
Idenopyrene	BDL	221
Benzo(k) fluoranthene	1.00	206

Note: Concentrations shown in mg/kg wet weight. Detection limits for all contaminants were 0.70 mg/kg. BDL = below detection limit.

21.3.3 Biological Response Indicators and Measurement Endpoints

21.3.3.1 Algal Assemblages

A total of 116 algal species were found. The most numerous algal groups were the Chlorophyta (green algae, 47 species) and Bacillariophyta (diatoms, 45 species, Table 21.3). Significant differences existed in the number of species ($p < 0.001$), abundance ($p < 0.001$), and Shannon–Wiener diversity ($p < 0.001$) between the Middle and West Lagoons (Table 21.4). The number of species was higher in the Middle Lagoon than in the West Lagoon. Periphyton abundance was much lower in the Middle Lagoon (about one tenth) than in the West Lagoon, and abundance was highest at WL5. Shannon–Wiener diversity was higher in the Middle Lagoon than in the West Lagoon. The East Pond had a greater number of species than the West Pond ($p = 0.009$). Abundance (cells/mm^2) in the East Pond was less than half that of the West Pond ($p = 0.002$). Shannon–Wiener diversity in the East Pond was significantly greater ($p = 0.041$).

A cluster analysis was performed on the periphyton data from the Grand Calumet Lagoons and ponds (Figure 21.2). Periphyton assemblages from each lagoon and pond formed separate groupings, suggesting more similarity within than among lagoons and ponds. The samples with the highest degrees of similarity were collected from the West Lagoon (about 70 to 83% similarity).

21.3.3.2 Aquatic Plant Assemblages

Various indices, including the number of species, the coefficient of conservatism (Swink and Wilhelm, 1994), and the floristic quality index (Swink and Wilhelm, 1994), were used to analyze aquatic plant community structure (Table 21.5). The Middle Lagoon sites were not significantly

TABLE 21.3
Number of Species Found in Major Algal Groups from Periphyton Samples Collected from Grand Calumet Lagoons and Ponds

Water Body	Cyanophyta	Chlorophyta	Cryptophyta	Pyrrophyta	Chrysophyta	Bacillariophyta
Middle Lagoon	8	35	3	1	2	28
West Lagoon	6	23	2	1	1	22
East Pond	6	11	3	1	4	25
West Pond	6	8	2	0	3	19
Total	12	47	3	3	5	45

TABLE 21.4
Diversity Values (Mean of Triplicate Samples) from Periphyton Samples Collected from the Grand Calumet Lagoons and Ponds

Site	# Sp	Abundance	SW Div
ML-1	35.0	1.3	1.04
ML-2	28.0	0.4	0.69
ML-3	30.3	0.9	0.70
WL-1	25.7	6.4	0.52
WL-2	22.3	6.9	0.40
WL-3	20.0	9.2	0.37
WL-4	17.3	5.4	0.46
WL-5	24.7	14.4	0.31
EP-1	24.3	3.3	0.43
EP-2	24.3	3.0	0.43
WP-1	17.3	9.6	0.36
WP-2	13.7	7.9	0.31

Note: # Sp = number of species. Abundance values = cells/mm^2 (×1000). SW Div = Shannon–Wiener diversity.

different in number of species than those from the West Lagoon ($p = 0.071$). However, two of the three Middle Lagoon sites had the greatest numbers of aquatic plant species, and site ML3 had the greatest number of aquatic plant species of all sites. Site ML2 had only 11 species with no emergent representatives.

Sites WL3, WL4, and WL5 had the fewest aquatic plant species. The West Lagoon sites had a significantly lower mean coefficient of community ($p = 0.036$) and floristic quality index ($p = 0.036$) than the Middle Lagoon sites. Sites closest to the landfill had a lower mean coefficient of conservatism and lower floristic quality than sites located further from the landfill. The East and West Ponds had the same number of species. The coefficient of conservatism and the floristic quality index were higher for the West Pond than the East Pond.

The number of submergent plant species was reduced at most West Lagoon sites, and they were nonexistent at WL3, WL4, and WL5 (Table 21.6). Floating plant species were absent from the West Lagoon at the time of sampling although they were observed on previous trips to the lagoon (P.M. Stewart, personal observation). An exotic species, *Myriophyllum spicatum* (Eurasian watermilfoil), was found throughout the Middle Lagoon and at several sites in the West Lagoon earlier in the season.

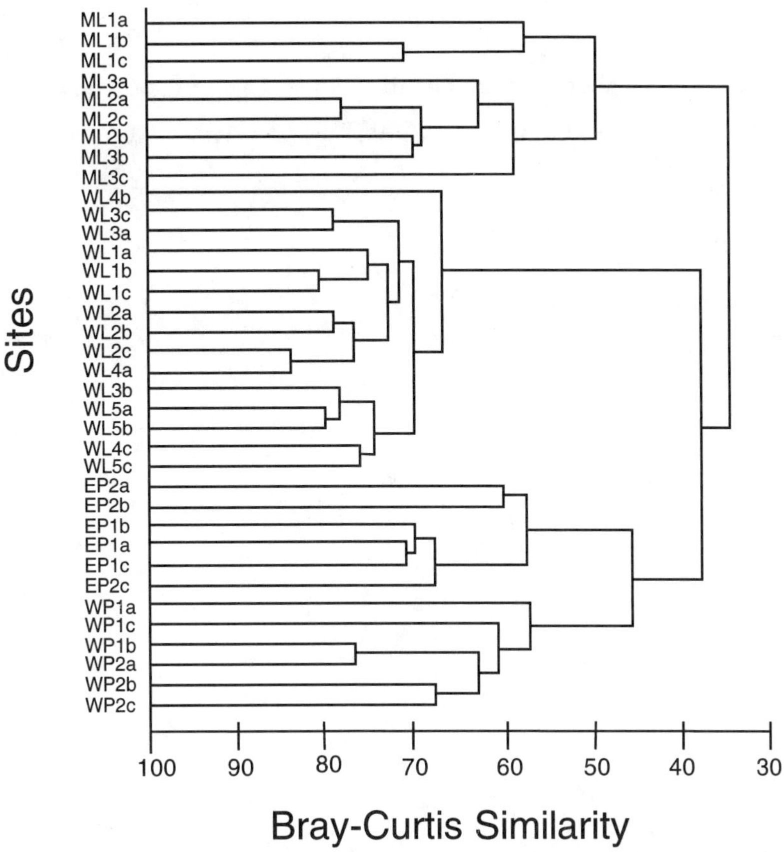

FIGURE 21.2 Cluster analysis of periphytic algal samples found in the Grand Calumet Lagoons and adjacent ponds.

An aquatic plant index of biotic integrity was developed and used to examine plant community health in the lagoons and ponds (Simon et al., 2001). This multimetric index, while experimental in scope, suggests that plant communities can be used in a multimetric manner, since results correctly indicated that communities found in the most polluted areas had the lowest PIBIs.

Cluster analysis based on the coefficient of similarity was performed to examine relationships among sites based on aquatic plant distributions. Results showed two main groupings (Figure 21.3); the Middle Lagoon sites and the East and West Ponds formed one cluster, and the five West Lagoon sites another. The West Pond and the East Pond plant assemblages clustered together and had the same number of species. The West Pond had a higher mean coefficient of conservatism and floristic quality index than the East Pond, which suggested that the West Pond was not affected by proximity to the industrial landfill. Lower mean coefficient of community and floristic quality index values are thought to be indicative of a plant community that is less representative of one that could be found in a natural area (i.e., an area displaying high biological integrity).

21.3.3.3 Macroinvertebrate Assemblages

There was no significant difference in macroinvertebrate diversity measurements (number of species, Shannon–Wiener diversity, evenness) between the Middle and West Lagoon sites or the East and West Pond sites ($p > 0.05$). A cluster analysis of the taxa at each site revealed four major groups (Figure 21.4). Sites WL3 through WL5 formed one cluster, the East Pond sites a second, and the West Pond sites a third. The fourth cluster included samples collected from the Middle

TABLE 21.5
Aquatic Plant Species Collected from Grand Calumet Lagoons and Ponds

Species	ML1	ML2	ML3	WL1	WL2	WL3	WL4	WL5	EP	WP
Asclepias incarnata							3	1		
Carex bebbii						1				
Carex comosa					4		4			4
Carex stricta				6	3					
Ceratophyllum demersum				2						
Chara globularis	4	5	5		2				6	
Cladium mariscoides			3						4	4
Eleocharis ovata					3		3	5		
Elodea canadensis	4		3	6						
Iris virginica v. shrevei	2		4		3	3	4	2		
Juncus acuminatus	6		4							
Juncus balticus v. littoralis			4	2	2					
Juncus brachycephalus										3
Juncus torreyi			4						3	
Lycopus virginicus				2	3	3	3	2		
Myriophyllum exalbescens				3						
Myriophyllum spicatum	3	4	6	3	4	6				
Najas flexilis	4	4	6						6	
Nitella sp.			3							
Nuphar advena	5	4	2							
Nymphaea tuberosa	4		3							
Panicum clandestinum									4	5
Phragmites australis					1	4	1	4		
Polygonum hydropiperoides			2							
Polygonum lapathifolium							3	5		
Pontederia cordata	6		3							
Potamogeton crispus		3								3
Potamogeton foliosus									1	
Potamogeton illinoensis	6	2	3						6	6
Potamogeton pectinatus	4	3	4	4	2				4	4
Potamogeton zosteriformis	3	4	2		2					2
Scirpus atrovirens									2	
Scirpus pungens	4	3	6		3	4	4		6	6
Scirpus validus v. creber	4								6	4
Triglochin maritima		2								
Typha angustifolia				6			3			
Typha latifolia				6	6	3	3	2	6	
Utricularia minor										5
Utricularia vulgaris				6						6
Vallisneria americana		4	5							
Number of species	14	11	19	11	13	7	10	7	12	12
$\overline{X}$ Coefficient of conservatism	6.3	6.9	6.6	5.3	5.5	4.5	4.1	4.3	5.5	7.2
Floristic quality index	21.9	19.4	26.3	16.8	18.1	11.0	13.0	11.3	18.1	23.8

Note: Numbers refer to an abundance scale: observed =1, rare = 2, rare/common = 3, common = 4, very common = 5, abundant = 6.

TABLE 21.6
Number of Aquatic Plant Species that Occupy Submergent, Floating, and Emergent Growth Habits at Each Site

Site	Submergent	Floating	Emergent	Total
ML1	7	2	5	14
ML2	8	1	2	11
ML3	9	2	8	19
WL1	6	0	5	11
WL2	4	0	9	13
WL3	0	0	7	7
WL4	0	0	10	10
WL5	0	0	7	7
EP	5	0	7	12
WP	6	1	5	12

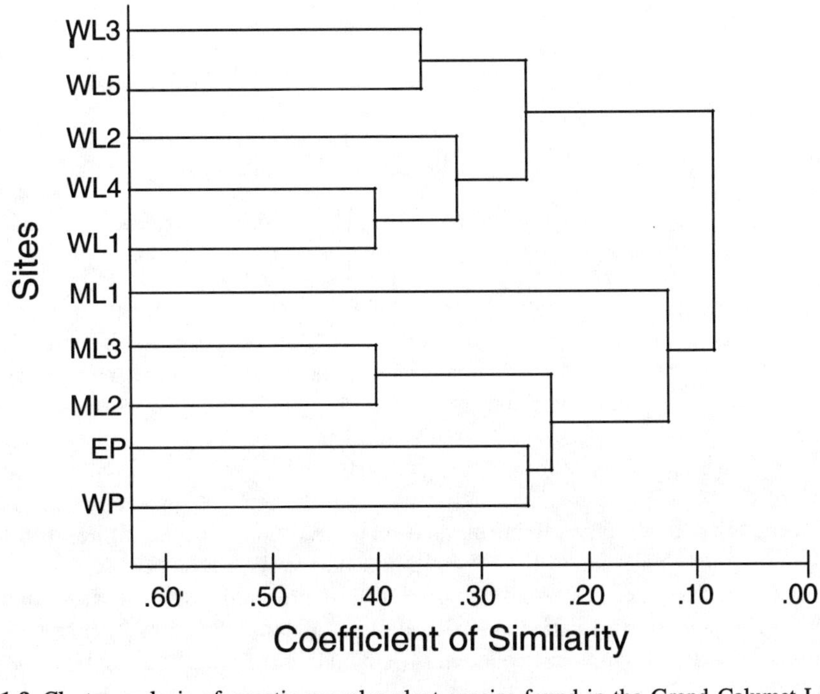

FIGURE 21.3 Cluster analysis of aquatic vascular plant species found in the Grand Calumet Lagoons and adjacent ponds.

Lagoon and WL1 and WL2. For the macroinvertebrate community living among the aquatic vegetation, there were differences between communities found at the far-western locations of the West Lagoon and those of the eastern section of the West Lagoon (WL1 and WL2). Samples from the eastern part of the West Lagoon (WL1 and WL2) were more similar to communities found in the Middle Lagoon. Shannon–Wiener macroinvertebrate diversity ($p = 0.036$) and evenness ($p = 0.036$) were significantly lower at the WL3 through WL5 sites than at the other sites (ML1 through ML3 and WL1 and WL2) when grouped and compared.

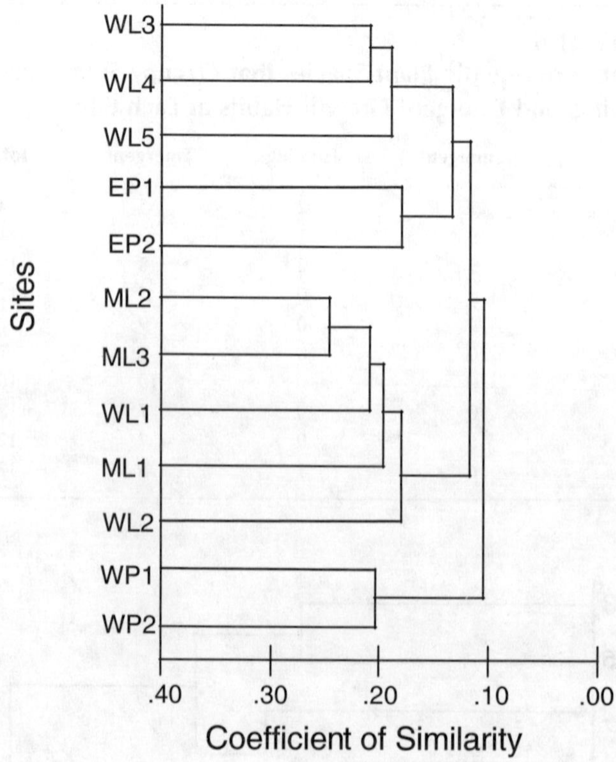

FIGURE 21.4 Cluster analysis of the macroinvertebrate community collected by sweep net samples from the Grand Calumet Lagoons and adjacent ponds.

Macroinvertebrate community data showed differences among dominants from ML1, ML2, ML3, WL1, WL2, and WL3 through WL5 (Figure 21.5). Talitrids (amphipods) were dominant at WL3 through WL5 (more than 95% of the organisms counted at WL5), were fewest at WL1 and WL2, and were present in moderate numbers at ML1 through ML3. Chironomids (midges) were most abundant at sites ML2, ML3, WL1, and WL2. They comprised a small percentage of the community at sites ML1 and WL3 through WL5. Coenagrionids (narrow-winged damselflies) were present at all of the Middle Lagoon sites and at WL2 but were absent from all other sites.

Macroinvertebrate community compositions from the East and West Pond sites differed greatly (Figure 21.6). The East Pond had 20 to 45% talitrids, and the West Pond had few at site WP2 and none at WP1. In contrast, the West Pond had 15 to 60% chironomids and the East Pond had very low numbers at EP1 and none at EP2. Planorbid mollusks (snails) were more abundant in the East Pond than in the West Pond. We found no asselids (isopods) in the West Pond and no coenagrionids in the East Pond.

21.3.3.4 Fish Assemblages

Seventeen species of fish were collected (Table 21.7). We found no significant difference between the Middle and West Lagoons in the number of fish species found ($p > 0.05$). Sites ML1, ML2, ML3, and WL3 each contained nine species of fish. Sites WL1, WL4, and WL5 had seven species each. There were differences in species composition between the West and Middle Lagoons. The lake chubsucker (*Erimyzon sucetta*) and the grass pickerel (*Esox americanus*) were found only in the Middle Lagoon. In contrast, the bluntnose minnow (*Pimephales notatus*) was found only in the West Lagoon.

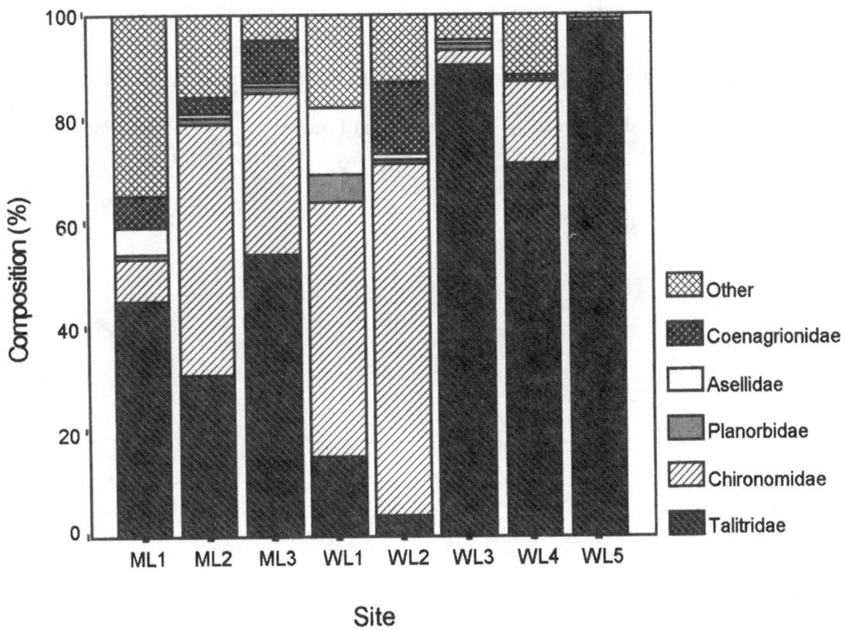

FIGURE 21.5 Dominant macroinvertebrate taxa collected by sweep net samples from the Grand Calumet Lagoons.

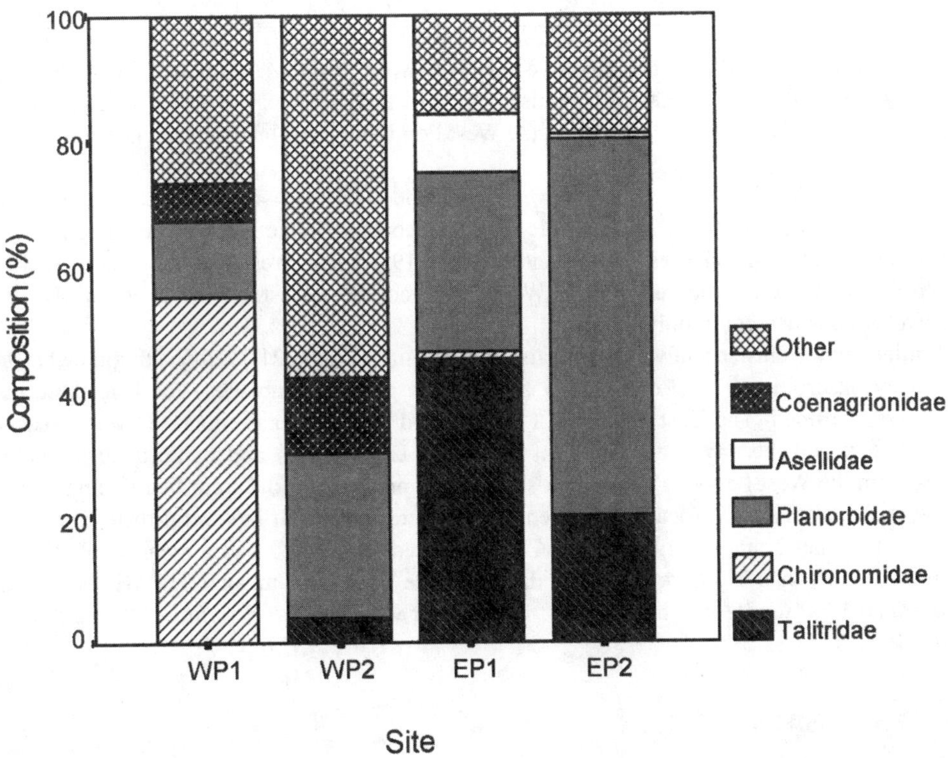

FIGURE 21.6 Dominant macroinvertebrate taxa collected by sweep net samples from the East Pond and West Pond.

TABLE 21.7
Fish Species and Numbers Collected from Grand Calumet Lagoons and Ponds

	Middle Lagoon			West Lagoon					East Pond		West Pond	
Fish Species	**1**	**2**	**3**	**1**	**2**	**3**	**4**	**5**	**1**	**2**	**1**	**2**
Ameiurus natalis			1									
Carassius auratus						1						
Cyprinus carpio	2			5	4	6	1	1				
Erimyzon sucetta	2		11								1	
Etheostoma exile	2	4	2			2	2					
Esox americanus	2	2	2						4	6		
Lepomis cyanellus	3	7	2	4	1	10						
Lepomis gibbosus	4	31	35	4	6	1	4	17	198	110	265	220
Lepomis gulosus		5	15		1	1						
Lepomis macrochirus	10	36	67	8	11	1	6	1				
Micropterus salmoides	6	15	5	21	7	25	21	20				
Notemigonus crysoleucas								2				
Noturus gyrinus		1										
Perca flavescens	1	5	3	11	5	3	4	2				
Pimephales notatus				10	3	23	2	13				
Pomoxis nigromaculatus	1											
Umbra limi		1										
Number of species	10	10	10	7	8	10	7	7	2	2	2	1
Number of individuals	33	107	143	63	38	73	40	56	202	116	266	220

The East and West Ponds had two species; both were heavily dominated by pumpkinseed sunfish (*Lepomis gibbosus*) (>300 individuals collected). The second species was the grass pickerel in the East Pond and the lake chubsucker in the West Pond (only one lake chubsucker was collected from the West Pond).

There was a significant difference between the Middle (mean = 42) and West Lagoon (mean = 33.4) in their IBI scores ($p = 0.036$). Biotic integrity scores from the lagoons indicated fair health in all of the Middle Lagoon sites (Simon and Stewart, 1998). The West Lagoon sites ranged from fair to poor at WL4 to poor to very poor at WL3. The other three West Lagoon sites had IBI scores indicative of poor fish community quality.

Results from a cluster analysis of the fish community (Figure 21.7) differed somewhat from those of the macroinvertebrate community and aquatic plant assemblages. For fish, three major clusters were formed. The East and West Pond formed one cluster dominated by pumpkinseed sunfish. The second cluster was comprised of the Middle Lagoon sites, and the third cluster included all sites from the West Lagoon. These data suggest the presence of different fish populations.

Polycyclic aromatic hydrocarbon concentrations were quite high in several fish samples from the Grand Calumet Lagoons (Table 21.8). Carp collected from WL3 had a mean total PAH concentration of more than 250 μg/kg in whole tissues, while WL5 had a mean total PAH concentration of more than 1000 mg/kg in whole tissues. No attempt was made to determine concentrations in separate tissues.

21.4 DISCUSSION

21.4.1 Water Chemistry

Clear differences in water quality existed between the Middle Lagoon and the West Lagoons for most of the variables measured. Most of the water chemistry variables, specifically dissolved

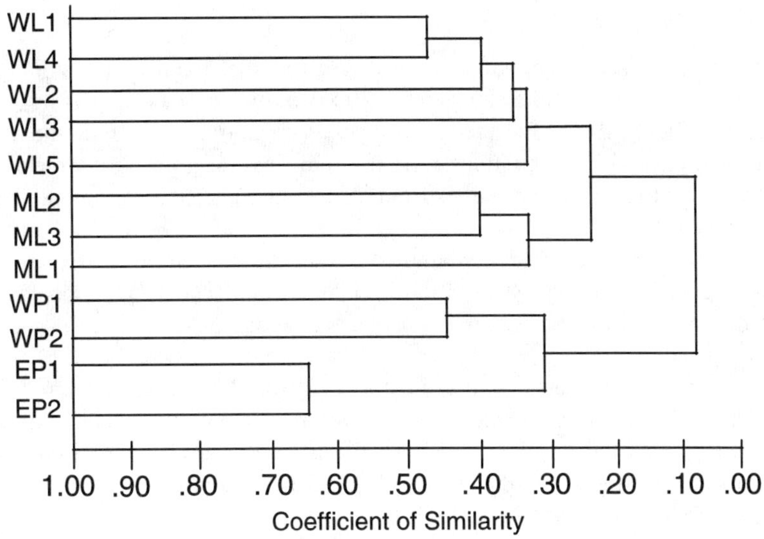

FIGURE 21.7 Cluster analysis of the fish community from the Grand Calumet Lagoons and adjacent ponds.

TABLE 21.8
Polycyclic Aromatic Hydrocarbon Concentrations in Fish Tissues from the Grand Calumet Lagoons

Analyte (μg/kg)	WL3-A	WL3-B	WL5-A	WL5-B
Naphthalene	29.5	29.6	212.1	145.9
Acenaphthene	80.9	34.3	450.9	446.6
Fluorene	14.0	11.1	138.6	105.5
Fluoranthene	11.6	36.5	84.5	18.5
Total PAHs	257.4	282.4	1413.4	807.3

oxygen, specific conductance, total alkalinity, total hardness, sulfate, and nutrients including ammonia nitrogen, nitrite nitrogen, and total phosphate were higher in the West Lagoon than in the Middle Lagoon. Higher specific conductance, sulfate, and nutrients suggest poorer water quality in the West Lagoon. In contrast, the Middle Lagoon was found to have higher pH and nitrate nitrogen than the West Lagoon. For specific conductance, total alkalinity, total hardness, sulfate, ammonia, nitrite, and total phosphate, we found a clear pattern of the values increasing from WL1 to WL5, which suggested a further decline in water quality from east to west, with WL5 having the poorest water quality in the lagoons. According to Trussell (1972), the toxic fraction of ammonia is 15% at a pH of 8.5 and 25°C. If the pH of the lagoons increased to 9.0, the percentage of the more toxic form of ammonia could exceed 35%, perhaps leading to increased toxicity.

The difference in water quality found in the ponds was more ambiguous, with higher dissolved oxygen, specific conductance, pH, hardness, and ammonia in the East Pond than in the West Pond. The West Pond had higher total alkalinity and sulfate than the East Pond. Large differences were found in dissolved oxygen, sulfate, and ammonia concentrations between the ponds. The location of the West Pond adjacent to the industrial landfill may suggest that water quality does not depend only on location of the landfill, but on other factors such as the nature of the landfill material or nature of the groundwater.

21.4.2 Sediment Quality

The extent of contaminant concentration in the Grand Calumet Lagoons is similar to that of the Grand Calumet River (MacDonald and Ingersoll, 2000) and increases from east to west (Gillespie et al., 1998; Stewart et al., 1999). Elevated heavy metal concentrations were found at several sites. In most cases, heavy metal concentrations were not indicative of gross heavy metal pollution but were comparable to sediment concentrations from other contaminated areas around the Great Lakes (U.S. Environmental Protection Agency, 1977). Gillespie et al. (1998) found that the average increase in metal and other sediment contaminants from the eastern portion of the West Lagoon was 2.5 times that of the reference sites in the Middle Lagoon. The average increase in sediment metal concentrations from the western reach of the West Lagoon was eight to nine times that of the reference sites.

Polycyclic aromatic hydrocarbons in sediments at WL5 (Table 21.2) were much higher than those reported from other industrialized rivers in the Great Lakes basin (Table 21.9). Clark and Jarvis (1990) determined the no-effect level for PAHs based on concentrations found in clean sediments around the Great Lakes (Table 21.10). These concentrations were well below those found in West Lagoon sediments. In addition, the PAH concentrations in the Grand Calumet Lagoons sediments were two to three orders of magnitude higher than the major biological effects level in sediments from several contaminated areas around the Great Lakes (Clark and Jarvis, 1990). These concentrations indicated a major PAH contamination problem and many of the PAHs were higher than sediment quality guidelines (Swartz, 1999).

Total PAH concentrations that did not cause biological effects in Great Lakes sediments were 3.8 mg/kg, while sites that showed major biological effects had total PAH concentrations of 6.8 mg/kg. Total PAHs in the sediment from WL5 were over 123 g/kg, comprising 12.3% of the sediment. Naphthalene comprised nearly 10% of the sediment at WL5. Total PAH concentrations at WL5 were more than 18,000 times that of Great Lake's sediments that cause major biological effects (Clark and Jarvis, 1990). Additionally, sediments from WL4 may be contaminated with lower levels of PAHs. This indicates the PAHs at WL5 are high enough to cause biological effects. Response signatures from this area should be related to PAH concentrations in the sediments and not heavy metal concentrations.

These data prompted additional investigations of contaminant chemistry in the Grand Calumet Lagoons and West Pond by the U.S. Army Corps of Engineers (ACOE) on July 16 and 17, 1996 (US ACOE, 1997). ACOE collected six sediment samples and two pore-water samples from the Grand Calumet Lagoons. The samples were analyzed for priority metals, volatiles, semivolatiles, pesticides, and additional variables. Contaminant data revealed high PAH concentrations at several locations in the lagoons and supported earlier findings. For the most part, these were a different group of contaminants than the ones reported in the Gillespie et al. (1998) study. Phenanthrene

TABLE 21.9
Concentrations of PAHs Found in Sediments from Industrialized Rivers in the Great Lakes Region and at Grand Calumet Lagoon Site WL5

Analyte (mg/kg)	Black River[1]	Cuyahoga River[1]	Fox River[1]	Hersey River[2]	WL5
Benz[a]anthracene	11.0	2.20	0.70	3.50	523
Benzo[a]pyrene	8.80	2.60	1.00	1.20	444
Ideno[1,2,3-cd]pyrene	6.40	1.40	BDL	—	221
Phenanthrene	—	—	—	4.10	3,286

[1]Baumann et al., 1991.
[2]Black et al., 1981.

TABLE 21.10
Comparison of Sediment PAH Concentrations from Clean Sites, Biologically Impaired Sites, and the Grand Calumet Lagoons

Analyte (mg/kg)	NEL/CLA[1]	MBE/COA[2]	ER-M[3]	Grand Calumet Lagoon (WL5)
Acenaphthene	—	2.5–7.5	0.65	5,666
Anthracene	—	1.0–15	0.96	792
Benzo[a]anthracene	—	2.0–25	1.60	523
Benzo[a]pyrene	0.03–0.05	2.5–20	2.50	444
Chrysene	0.075	3.0–30	2.80	453
Fluoranthene	0.08–1.00	7.0–35	3.60	2,122
Fluorene	—	2.0–15	0.64	2,470
Phenanthrene	0.03–0.07	5.0–100	1.38	3,286
Pyrene	0.05–0.10	5.0–50	2.20	1,644
Total PAHs	3.8	6.8	35.0	123,431

[1] NEL/CLA = No effects level/concentrations in clean sediments around the Great Lakes (Clark and Jarvis, 1990).
[2] MBE/COA = Major biological effects/concentrations in sediments from contaminated areas around the Great Lakes (Clark and Jarvis, 1990).
[3] ER-M = Effects range midway (Long and Morgan, 1990).

was found at concentrations as high as 18,000 mg/kg. As in our sampling, the sediments between WL4 and WL5 were loose, black, oily, and flocculent, with a strong hydrocarbon odor. Heavy metal contamination, while present, did not appear to be of great concern. Site WL5 had PAH concentrations much higher than levels showing major biological effects (Clark and Jarvis, 1990; Long and Morgan, 1990; Swartz, 1995, 1999).

MacDonald and Ingersoll (2000) used published literature to evaluate consensus-based probable effect quotients (PEC-Qs, MacDonald et al., 2000a), the published bioaccumulation-based sediment quality guidelines (SQGs) (NYDEC, 1994), the published toxicity thresholds for pore water (CCREM, 1987; USEPA, 2000b), and the published tissue residue guidelines (TRGs) for the protection of piscivorous fish (Newell et al., 1987). Those substances present in sediments within the various reaches of the lagoons at concentrations in excess of the chemical benchmarks were identified as contaminants of concern. Metals that exceeded the PEC-Qs and background levels were identified as priority substances (Ingersoll and MacDonald, 1999; MacDonald and Ingersoll, 2000). More specifically, metal levels were compared to the maximum background concentrations (mean plus four standard deviations) reported for lake and stream sediments in Northwest Indiana (Wente, 1994).

Sediment quality conditions sufficient to adversely affect sediment-dwelling organisms (i.e., as indicated by exceedances of effect-based PEC-Qs) were identified using a variety of measurement endpoints (MacDonald and Ingersoll, 2000). Examination of PEC-Qs in surficial sediments of the Grand Calumet Lagoons found that much of the Middle and East Lagoon sites were below concern levels. A reverse asymptotic curve existed, with highest PEC-Q levels in the far-western portion of the West Lagoon (Figure 21.8). Most of the sites with gross exceedances were in the far western sections of the West Lagoon.

21.4.3 Patterns in Biological Response Signatures

The number of species and Shannon–Wiener diversity are generally thought to increase with biological integrity (Stewart, 1995). In contrast, and depending on type of organism, organismal

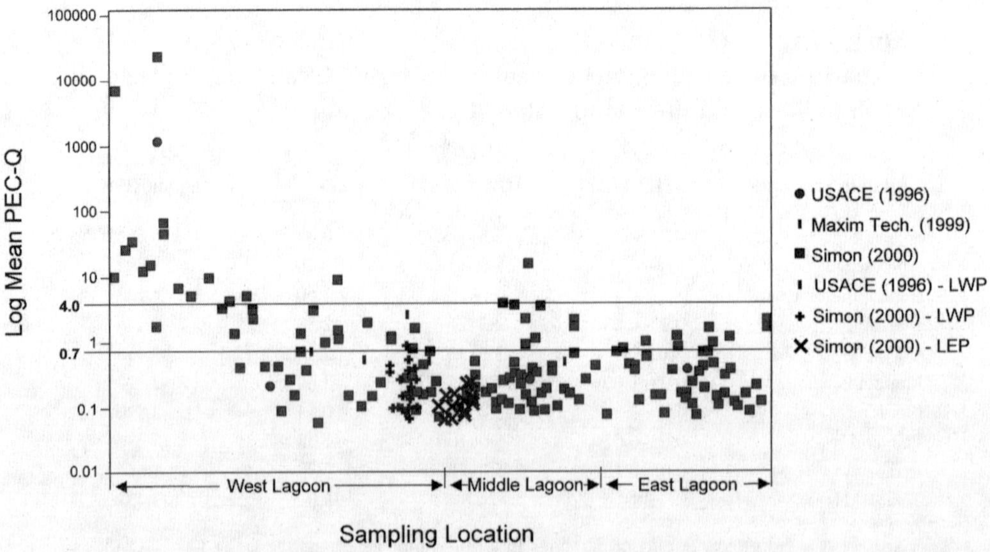

FIGURE 21.8 Spatial distribution of mean consensus-based probable effects quotients (PEC-Qs) in surficial sediments within the Grand Calumet Lagoons.

abundance usually increases with decreasing biological integrity. Patterns of these metrics suggested that the Middle Lagoon had higher algal biological integrity than the West Lagoon, and that the periphyton community of the East Pond also showed higher biological integrity than the assemblage in the West Pond. Algal data suggested differences between the lagoons and between the ponds with lower algal diversity found in close proximity to the landfill. Algal abundance was highest in the West Lagoon and West Pond, areas nearest the industrial landfill.

In the West Lagoon, aquatic plant assemblages showed a reduction in biological integrity (PIBI) in sites located nearest the industrial landfill. The West Lagoon sites showed a lower degree of fidelity to natural areas in the region. The aquatic plant assemblage from the West Ponds was an exception. In spite of its location next to the landfill, the community found at this site was more indicative of a natural area than the East Pond's community. Plant assemblage patterns from cluster analysis suggested differences in biological composition that could be related to proximity to the industrial landfill. However, causation is difficult to explain using the aquatic plant assemblages located here and the type of data collected. Ambiguity caused by differences in habitat and the high quality of the West Pond's plant assemblage, makes contaminants and proximity to the landfill difficult to blame. This observation may lead to the conclusion that the landfill has major spatial differences in contaminant amounts, an observation supported by the chemistry. Only at areas of extreme contamination were the plant communities severely depressed. At these areas, species found, such as *Phragmites communis*, were typical of other contaminated waste sites around the Great Lakes (Wilhelm et al., Chapter 14, this volume).

Macroinvertebrate community analysis based on family level taxonomic identification demonstrated differences in community composition relative to proximity to the industrial landfill. For example, WL5 contained nearly 95% Talitridae, more than any other site. Talitrids also dominated sites WL3 and WL4. Data suggested that macroinvertebrate communities at WL1 and WL2 were more closely comparable to the Middle Lagoon sites than those of the more westward West Lagoon sites. Macroinvertebrate communities collected from sites located near the landfill were different from those collected further away in the lagoons, but again results were more ambiguous in the ponds. The two ponds contained different macroinvertebrate communities, but cannot be compared to the lagoons. None of the pond sites had contaminant concentrations as high as the western area of the lagoons, and it is more difficult to link differences to landfill effects. Genus and species level taxonomic identifications would allow for finer discrimination, but with the clear differences shown by the macroinvertebrate collections at the family level in the lagoons, further discrimination was not needed for the purposes of this study. Further analysis and inventory should allow for species level identifications to form a baseline survey of the macroinvertebrate community in order to examine changes after future mitigation.

Fish communities sampled in the West Lagoon were different from those obtained from the East Lagoon. Several species were found exclusively in the Middle Lagoon or West Lagoon. It appeared that some fish species might have been separated from the Middle Lagoon by a small stream choked with cattails. This lack of migration may have allowed communities with lower biological integrity to develop at lagoon sites located closest to the landfill.

Carp are among the fish in Indiana waters subject to a do-not-eat-at-all advisory, and it is recommended that no fish in the Grand Calumet River be consumed (Zorn, 1999). Carp collected from WL5 had a mean total PAH concentration of more than 1000 μg/kg, which was similar to concentrations from some of the most contaminated sites in the Great Lakes area, including industrialized sites on the Cuyahoga River (Baumann et al., 1991) and the Hersey River (Black et al., 1981). These concentrations were also similar to values found in skin-on/scaleless carp fillets from the Grand Calumet River west of the lagoons (Jim Stahl, Indiana Department of Environmental Management, personal communication, March 29, 1999) and were at least an order of magnitude higher than other sites in Indiana. Lack of complete fish contaminant data at all sites make a gradient and linkage to landfill effects impossible.

21.5 CONCLUSIONS

Our analysis used a weight-of-evidence approach to analyze a complex field situation (Apple et al., 1986). We relied heavily on the biological assessment of several aquatic assemblages in our ecological work with a suite of indicators (Adams et al., 1999). An impaired aquatic community exists within and surrounding the Grand Calumet Lagoons and ponds. Concentrations of some elements in some areas of the Grand Calumet Lagoons and in the pond located near the landfill are elevated and PAHs elicit the greatest concern. Heavy metal concentrations are elevated at some sites, and this may be an additional cause for concern.

There were differences in several water quality variables between the West and Middle Lagoons, with lower water quality found in the West Lagoon. Major response signatures and trends are depicted for contaminants, algal, aquatic plant, macroinvertebrate, and fish assemblage indicators (Table 21.11). Since this is a correlative study, no causal relationship can be inferred. We merely examined general patterns and trends in response signatures. Contaminant concentrations increased and several response signatures decreased from the Middle to the West Lagoon with extremely high PAH concentrations and PEC-Qs in the far western region of the West Lagoon. Algal communities showed distinct differences depending on proximity to the landfill, especially in the number of species, algal diversity, and periphyton abundance. Aquatic plant communities also showed differences among the sites, with entire growth habits not appearing in the West Lagoon. These data must be interpreted with caution since habitat differences exist among the sites, but typically as a result of habitat disturbance from the landfill. Macroinvertebrate community data in the lagoons suggested differences among sites based on landfill proximity. In the West Lagoon, the biological integrity of the fish assemblage was depressed and there were differences in fish community composition between the Middle and West Lagoons. Carp collected from WL5 had PAH concentrations that exceeded 1,000 μg/kg dry weight and these are considered high levels in whole fish tissues.

The weight of evidence suggests that several aquatic communities in the lagoons were significantly impacted by proximity to the industrial landfill. In some cases, such as fish tissue PAH concentrations, we only had data at two sites. Severe impacts were found at WL5 and moderate

TABLE 21.11
Response Signatures and Trend Lines Showing General Increases Following the Directions of the Arrows

Indicator	Response Signature: Middle Lagoon ... West Lagoon
Contaminants	
Total PAH	————————→
PEC-Q	————————→
Algal indicators	
# of algal species	←————————
Periphyton abundance	————————→
Algal S-W diversity	←————————
Aquatic plants	
Coefficient of community	←————————
Floristic quality index	←————————
# of floating plants	←————————
Plant- IBI	←————————
Aquatic macroinvertebrates	
Percent talitrids	————————→
Percent coenagrionids	←————————
S-W diversity and evenness	----------changes--------←————
Fish assemblages	
Species composition	---------------changes---------------
IBI	←————————
Fish tissue PAH	————————→

impacts were found at WL4. Most communities and indices were clearly and significantly different between the lagoons and we did not need more qualitative estimates of damage (e.g., such as a 20% decrease in the metric; Jones, 1999). In other cases, it was difficult to blame proximity to the landfill, as in some cases the data set was incomplete.

The three most commonly documented assemblage-level response signatures to environmental stress are reduction in species richness, changes in kinds and abundance of species, and decrease in mean size of organisms (Gray, 1989). Others specific to different groups of organisms and specific stressors are the subject of this volume. The response signatures that occurred in the Grand Calumet Lagoons system with PAHs as the predominant contaminants were those commonly associated with a variety of impairment responses to PAH contamination. While no unique response signature was found to be indicative of PAH contamination, typical patterns such as reduction in biodiversity, increases in abundance, and reduction in biological integrity were often seen.

ACKNOWLEDGMENTS

This manuscript does not reflect the opinions of the U.S. Fish and Wildlife Service and no official endorsement should be inferred. This chapter is contribution 1159 of the USGS Great Lakes Science Center.

REFERENCES

Adams, S.M., M.S. Bevelhimer, M.S. Greeley, Jr., D.A. Levine, and S.J. The. 1999. Ecological risk assessment in a large river-reservoir: 6. Bioindicators of fish population health, *Environmental Toxicology and Chemistry*, 18(4), 628–640.

APHA. 1992. *Standard Methods for the Examination of Water and Wastewater*, 18th ed. Washington, D.C.

Apple, G.J., W.G. Hunter, and S. Bisgaard. 1986. Scientific data and environmental regulation, in M.H. DeGroot, S.E. Fienberg, and J.B. Kadane (Eds.). *Statistics and the Law.* John Wiley & Sons, New York.

Barbour, M.T., J. Gerritsen, B.D. Snyder, and J.B. Stribling. 1999. *Rapid Bioassessment Protocols for use in Streams and Wadeable Rivers: Periphyton, Benthic Macroinvertebrates, and Fish.* 2nd ed. EPA 841-B-99–002. U.S. Environmental Protection Agency, Office of Water, Washington, D.C.

Baumann, P.C., M.J. Mac, S.B. Smith, and J.C. Harshbarger. 1991. Tumor frequencies in Walleye (*Stizostedion vitreum*) and Brown bullhead (*Ictalurus nebulosus*) and sediment contaminants in tributaries of the Laurentian Great Lakes, *Canadian Journal of Fisheries and Aquatic Sciences*, 48, 1804–1810.

Becker, G.C. 1983. *The Fishes of Wisconsin.* University of Wisconsin Press, Madison.

Black, J.J., T.F. Hart, Jr., and E. Evans. 1981. HPLC studies of PAH pollution in a Michigan trout stream, in M. Cooke and A.J. Dennis (Eds.). *Chemical Analysis and Biological Fate: Polynuclear Aromatic Hydrocarbons.* Fifth International Symposium. Battelle Press, Columbus, OH, 343–355.

Bosserman, R.W. 1985. Distribution of heavy metals in aquatic macrophytes from Okefenokee Swamp. Heavy metals in water organisms, in *Symposia Biologica Hungarica*, Vol. 29. Akademiai Kiado, Budapest, Hungary, 31–39.

Butcher, R.W. 1947. Studies in the ecology of rivers. VII. The algae of organically enriched waters, *Journal of Ecology*, 35, 186–191.

Cairns, J., Jr. 1982. *Artificial Substrates.* Ann Arbor Science Publishers, Ann Arbor, MI.

Canadian Council of Resource and Environment Ministers. 1987. *Canadian Water Quality Guidelines.* Prepared by the CCME Task Force on Water Quality Guidelines, Ottawa, Ontario, Canada.

Cholnoky, B.J. 1960. The relationship between algae and the chemistry of natural waters, Council for Scientific and Industrial Research, Reprint R.W. No. 129:215–225.

Clark, J.U. and A.S. Jarvis. 1990. Regulatory evaluation of petroleum hydrocarbons in dredged material. Miscellaneous paper EL-90–11.U.S. Army Corps of Engineers Waterways Experimental Station, Vicksburg, Mississippi.

Conover, W.J. 1971. *Practical Nonparametric Statistics.* John Wiley & Sons, New York.

Cowgill, U.M. 1973. The determination of all detectable elements in the aquatic plants of Linsley Pond and Cedar Lake (North Bradford, Connecticut) by x-ray emission and optical emission spectroscopy, *Applied Spectroscopy*, 17, 5–9.

Custer, T.W., R.K. Hines, P.M. Stewart, M.J. Melancon, D.S. Henshel, and D.W. Sparks. 1998. Organochlorines, mercury, and selenium in great blue heron eggs from Indiana Dunes National Lakeshore, Indiana, *Journal of Great Lakes Research*, 24(1), 3–11.

de Caprariis, P., R. de Lindemann, and R. Haimes. 1981. A relationship between sample size and accuracy of species richness predictions, *Mathematical Geology*, 13, 351–355.

Fassett, N. 1957. *A Manual of Aquatic Plants.* University of Wisconsin Press, Madison, WI.

Gatz, D.F., V.C. Bowersox, and J. Su. 1989. Lead and cadmium loadings to the Great Lakes from precipitation, *Journal of Great Lakes Research*, 15, 246–264.

Gerking, S.D. 1955. Key to the fishes of Indiana, *Investigations of Indiana Lakes and Streams*, 4, 49–86.

Gillespie, R.B., J. Speelman, and P.M. Stewart. 1998. *An Evaluation of the Toxic Effects of Sediments in the Grand Calumet Lagoons near the Indiana Dunes National Lakeshore.* National Park Service, Indiana Dunes National Lakeshore, Porter, IN.

Gleason, H. and A. Cronquist. 1991. *Manual of Vascular Plants of Northeastern United States and Adjacent Canada.* D. Van Nostrand, New York.

Gonzales, D.A., J.G. Pearson, and C.F.A. Pinkham. 1993. *User's Manual BIOSIM1, Beta Version 1.0. A Program That Applies the Coefficient of Biotic Similarity to Complex Data Matrices.* EPA 600-R-93–219. U.S. Environmental Protection Agency, Environmental Monitoring and Support Laboratory, Las Vegas, NV.

Gray, J.S. 1979. Pollution induced changes in populations, *Philosophical Transactions of the Royal Society of London*, 286, 545–561.

Grootjans, A. and R. van Diggelen. 1995. Assessing the restoration prospects of degraded fens, in B.D. Wheeler, S.C. Shaw, W.J. Fojt and R.A. Robertson (Eds.). *Restoration of Temperate Wetlands*. John Wiley & Sons, Chichester, U.K, 73–89.

Ham, K.D., S.M. Adams, and M.J. Peterson. 1997. Application of multiple bioindicators to differentiate spatial and temporal variability from the effects of contaminant exposure on fish, *Ecotoxicology and Environmental Safety*, 37, 53–61.

Hanson, L.J. and M.L. Johnson. 1999. Conservation and toxicology: the need to integrate the disciplines, *Environmental Toxicology and Chemistry*, 18(10), 2121–2122.

Hardy, M.A. 1984. Chemical and Biological Quality of Streams at the Indiana Dunes National Lakeshore, Indiana, 1978–1980. U.S. Geological Survey, Water Resources Investigations 83–2408. Indianapolis, IN.

Heck, K.L., G. van Belle, and D. Simberloff. 1975. Explicit calculation of the rarefaction diversity measurement and calculation of sufficient sampling size, *Ecology*, 56, 1459–1461.

Hilsenhoff, W.L. 1987. An improved biotic index of organic stream pollution, *Great Lakes Entomologist*, 20(1), 31–39.

Hilsenhoff, W.L. 1988. Rapid field assessment of organic pollution with a family-level biotic index, *Journal of the North American Benthological Society*, 7, 65–68.

Holowaty, M.O., M. Reshkin, M.J. Mikulka, and R.D. Tolpa. 1992. Working toward a remedial action plan for the Grand Calumet River and Indiana Harbor and Ship Canal, in J.H. Hartig and M.A. Zarull (Eds.). *Under RAPs: Toward Grassroots Ecological Democracy in the Great Lakes Basin*. The University of Michigan Press, Ann Arbor, MI, 211–233.

Indiana State Department of Health. 1999. *1999 Indiana Fish Consumption Advisory*. Environmental Epidemiology Section, Indiana State Department of Health, Indianapolis, IN.

Ingersoll, C.G. and D.D. MacDonald. 1999. *An Assessment of Sediment Injury in the West Branch of the Grand Calumet River*, Vol. I. Environmental Enforcement Section, Environment and Natural Resources Division, U.S. Department of Justice, Washington, D.C.

Kershaw, K.A. 1964. *Quantitative and Dynamic Ecology*. Edward Arnold Publishing, London.

Knorr, D.F. and G.W. Fairchild. 1987. Periphyton, benthic invertebrates and fishes as biological indicators of water quality in the East Branch Brandywine Creek, *Proceedings of the Pennsylvania Academy of Science*, 61, 61–66.

Long, E.R. and L.G. Morgan. 1990. The Potential for Biological Effects of Sediment-Sorbed Contaminants Tested in the National Status and Trends Program. National Oceanic and Atmospheric Administration Technical Memorandum NOS OMA52. Seattle, WA.

McDonald, D.D. and C.G. Ingersoll. 2000. *An Assessment of Sediment Injury in the Grand Calumet River, Indiana Harbor Canal, Indiana Harbor, and the Nearshore Areas of Lake Michigan*. Vol. 1. U.S. Fish and Wildlife Service, Bloomington, IN.

McGroddy, S.E., J.W. Farrington, and P.M. Gschwend. 1996. Comparison of the *in situ* and desorption sediment-water partitioning of polycyclic aromatic hydrocarbons and polychlorinated biphenyls, *Environmental Science and Technology*, 30, 172–177.

McKenzie, H.H., D.E. Hyatt, and V.J. McDonald (Eds.). 1992. *Ecological Indicators*. Proceedings of the International Symposium on Ecological Indicators, October 16–19, Ft. Lauderdale, FL. Elsevier Science, Essex, U.K.

Merritt, R.W. and K.W. Cummins. 1984. *An Introduction to the Aquatic Insects of North America*, 2nd ed. Kendall/Hunt, Dubuque, IA.

Mesters, C.M.L. 1995. Shifts in macrophyte species composition as a result of eutrophication and pollution in Dutch transboundary streams over the past decades, *Journal of Aquatic Ecosystem Health and Management*, 4, 295–305.

Mueller-Dombois, D. and H. Ellenberg. 1974. *Aims and Methods of Vegetation Ecology*. John Wiley & Sons, New York.

NADP. 1995. National Atmospheric Deposition Program. Colorado State University, Fort Collins, CO.

Newell, A.J., D.W. Johnson, and L.K. Allen. 1987. *Niagara River Biota Contamination Project: Fish Flesh Criteria for Piscivorous Wildlife*. Technical Report 87–3. New York State Department of Environmental Conservation, Albany, N.Y.

New York State Department of Environmental Conservation. 1994. Technical Guidance for Screening Contaminated Sediments. Division of Fish and Wildlife, Division of Marine Resources, Albany, N.Y.

Niemi, G.J., P. Devore, N. Detenbeck, D. Taylor, A. Lima, J. Pastor, J.D. Yount, and R.J. Naiman. 1990. Overview of case studies on recovery of aquatic systems from disturbance, *Environmental Management*, 14, 571–588.

PAHLS (People Against Hazardous Landfills). 1993. *The Environment of Northwest Indiana. Contrasts and Dilemmas*. Sheffield Press, Hammond, IN.

Patrick, R. 1949. A proposed biological measure of stream conditions, based on a survey of the Conestoga basin, Lancaster County, Pennsylvania, *Proceedings of the Academy of Natural Sciences of Philadelphia*, 101, 277–341.

Patrick, R. 1973. Use of algae, especially diatoms, in the assessment of water quality, in *Biological Methods for the Assessment of Water Quality*. ASTM STP 528, American Society for Testing and Materials, Philadelphia, PA, 76–95.

Patrick, R., F. Douglass, D.M. Palavage, and P.M. Stewart. 1992. *Surface Water Quality: Have the Laws Been Successful*? Princeton University Press, Princeton, N.J.

Pearson, J.G. and C.F.A. Pinkham. 1992. Strategy for data analysis in environmental surveys emphasizing the index of biotic similarity and BIOSIM1, *Water and Environmental Research*, 64, 901–909.

Peckarsky, B.L., P.R. Fraissinet, M.A. Penton, and D.J. Conklin, Jr. 1990. *Freshwater Macroinvertebrates of Northeastern North America*. Comstock Publishing, Ithaca, N.Y.

Pedersen, E.R. and M.A. Perkins. 1986. The use of benthic invertebrate data for evaluating impacts of urban runoff, *Hydrobiologia*, 139, 13–22.

Plafkin, J.L., M.T. Barbour, K.D. Porter, S.K. Gross, and R.M. Hughes. 1989. *Rapid Bioassessment Protocols for Use in Streams and Rivers: Benthic Macroinvertebrates and Fish*. EPA/440/4–89/001.U.S. Environmental Protection Agency, Office of Water. Washington, D.C.

Ravera, O. 1983. Assessment of the trophic state of a water body, *Annals of Limnology*, 19, 229–234.

Richards, C., G.E. Host, and J.W. Arthur. 1993. Identification of predominant environmental factors structuring stream macroinvertebrate communities within a large agricultural catchment, *Freshwater Biology*, 29, 285–294.

Richards, C., L.B. Johnson, and G.E. Host. 1996. Landscape-scale influences on stream habitats and biota, *Canadian Journal of Fisheries and Aquatic Sciences*, 53, 295–311.

Rosenberg, D.M. and V.R. Resh. 1993. *Freshwater Biomonitoring and Benthic Macroinvertebrates*. Chapman & Hall, New York.

Simon, T.P. 1991. Development of Index of Biotic Integrity Expectations for the Ecoregions of Indiana, I. Central Corn Belt Plain. EPA 905–9–91–025.U.S. Environmental Protection Agency, Region 5, Environmental Sciences Division, Monitoring and Quality Assurance Branch: Ambient Monitoring Section, Chicago, IL.

Simon, T.P. 1998. Development of fish community reference conditions for dunal, palustrine wetlands along the southern shore of Lake Michigan, *Aquatic Ecosystem Health and Management*, 1, 57–70.

Simon, T.P. and P.M. Stewart. 1998. Development of fish community reference conditions for dunal, palustrine wetlands: an emphasis on assessment of non-point source landfill effects, *Aquatic Ecosystem Health and Management*, 1, 71–72.

Simon, T.P. and P.M. Stewart. 1999. Structure and function of fish communities in the southern Lake Michigan basin with emphasis on restoration of native fish communities, *Natural Areas Journal*, 19, 142–154.

Simon, T.P., P.M. Stewart, and P.E. Rothrock. 2001. Development of multimetric indices of biotic integrity for riverine and palustrine wetland plant communities along Southern Lake Michigan, *Aquatic Ecosystem Health and Management*, 4, 293–309.

Simon, T.P., J.O. Whitaker, J.S. Castrale, and S.A. Minton. 1992. Checklist of the vertebrates of Indiana, *Proceedings of the Indiana Academy of Science*, 101, 95–126.

Smith, E.P., P.M. Stewart, and J. Cairns, Jr. 1985. Similarities between rarefaction methods, *Hydrobiologia*, 120, 167–170.

Smith, P.W. 1979. *The Fishes of Illinois*. The University of Illinois Press, Champaign, IL.

St. Cyr, L. and P.G.C. Campbell. 1993. Trace metals in submerged plants of the St. Lawrence River, *Canadian Journal of Botany*, 72, 429–439.

Stein, J.E., T. Hom, T.K. Collier, D.W. Brown, and U. Varanasi. 1995. Contaminant exposure and biochemical effects in outmigrant juvenile chinook salmon from urban and nonurban estuaries of Puget Sound, Washington, *Environmental Toxicology and Chemistry*, 14, 1019–1029.

Stewart, P.M. 1995. Use of algae in aquatic pollution assessment, *Natural Areas Journal*, 15, 234–239.

Stewart, P.M., J.T. Butcher, and T.O. Swinford. 2000. Land use, habitat, and water quality effects on macroinvertebrate communities in three watersheds of a Lake Michigan associated marsh system, *Aquatic Ecosystem Health and Management*, 3, 179–189.

Stewart, P.M. and D.J. Robertson. 1992. Aquatic organisms as pollution indicators of water quality in suburban streams of the Lower Delaware River Region, USA, *Proceedings of the Pennsylvania Academy of Science*, 66, 135–141.

Stewart, P.M., K.G. Sappington, and J. Cairns, Jr. 1986. Community and population response of natural diatom communities to zinc stress in laboratory mesocosms, *Current Practices in Environmental Science and Engineering*, 2, 177–184.

Stewart, P.M., R.W. Scribailo, and T.P. Simon. 1999. The use of aquatic macrophytes in monitoring and in assessment of biological integrity, in A. Gerhardt (Ed.). *Biomonitoring of Polluted Water*, Environmental Research Forum, Trans Tech Publications, 275–302.

Suter, G.W., II. 1993. Defining the field, in G.W. Suter, II (Ed.). *Ecological Risk Assessment*, Lewis Publishers, Chelsea, MI, 3–20.

Suter, G.W., II. 1999. Lessons for small sites from assessments of large sites. *Environmental Toxicology and Chemistry*, 18(4), 579–580.

Suter, G.W., II, L.W. Barnthouse, R.A. Efroymson, and H. Jager. 1999. Ecological risk assessment in a large river-reservoir: 2. Fish community, *Environmental Toxicology and Chemistry*, 18(4), 589–598.

Suter, G.W., II and S. Bartell. 1993. Ecosystem-level effects, in G.W. Suter, II (Ed.). *Ecological Risk Assessment*, Lewis Publishers, Chelsea, MI, 275–308.

Swartz, R.C. 1999. Consensus sediment quality guidelines for polycyclic aromatic hydrocarbon mixtures, *Environmental Toxicology and Chemistry*, 18(4), 780-787.

Swartz, R.C., D.W. Schults, R.J. Ozretich, J.O. Lamberson, F.A. Cole, T.H. DeWitt, M.S. Redmond, and S.P. Ferraro. 1995. OAH: A model to predict the toxicity of polynuclear aromatic hydrocarbons in field-collected sediments, *Environmental Toxicology and Chemistry*, 14, 1977–1987.

Swink, F. and G. Wilhelm. 1994. *Plants of the Chicago Region.* 4th ed. Indiana Academy of Sciences, Indianapolis, IN.

Trussell, R.P. 1972. The percent of unionized ammonia in aqueous ammonia solutions at different pH levels and temperatures, *Journal of the Fisheries Research Board of Canada*, 29, 1505–1507.

U.S. Army Corps of Engineers. 1997. Grand Calumet River — Indiana Harbor Canal Sediment Cleanup and Restoration Alternatives Project. U.S. Army Corps of Engineers, Chicago District, Great Lakes and Ohio River Division, Chicago, IL.

U.S. Environmental Protection Agency. 1977. Guidelines for the Pollution Classification of Great Lakes Harbor Sediments. U.S. Environmental Protection Agency, Region V, Chicago, IL.

U.S. Environmental Protection Agency. 2000. Prediction of Sediment Toxicity Using Concensus-Based Freshwater Sediment Quality Guidelines. EPA 905/R-00/007. Great Lakes National Program Office, Chicago, IL.

van der Oost, R., F.J. Schooten, F. Ariese, H. Heida, K. Satumalay, and N.P.E. Vermeulen. 1994. Bioaccumulation, biotransformation and DNA binding of PAHs in feral eel (*Anguilla anguilla*) exposed to polluted sediments, *Environmental Toxicology and Chemistry*, 13, 859–870.

Vitousek, P.M., H.A. Mooney, J. Lubchenco, and J.M. Melillo. 1997. Human domination of earth's ecosystems, *Science*, 277, 494–499.

Weiss, J.C., R.E. Unsworth, and E. Ruder. 1997. *Assessment Plan for the Natural Resource Damage Assessment of the Grand Calumet River, Indiana Harbor Ship Canal, Indiana Harbor, and Associated Lake Michigan Environments.* Industrial Economics, Cambridge, MA.

Wente, S.A. 1994. Background Concentrations of Bioaccumulative and Sediment Concentrations of Contaminants for Indiana. Indiana Department of Environmental Management, Indianapolis, IN.

Wilhelm, G.S., T.P. Simon, and P.M. Stewart. 2002. Conservatism of confined disposal facilities based on the biological stability and integrity of plant communities: a case study in the Laurentian Great Lakes, Chapter 14, this volume.

Winterringer, G.S. and A.C. Lopinot. 1977. *Aquatic Plants of Illinois*. Illinois State Museum, Springfield, IL.

Zorn, T. 1999. State Puts Warning on Fish. *Post Tribune*, Gary, Indiana. March 20, 1999.

22 Response of Diatom Assemblages to Human Disturbance: Development and Testing of a Multimetric Index for the Mid-Atlantic Region (USA)

Leska S. Fore

CONTENTS

0-8493-0905-0/03/$0.00+$1.50

22.1 INTRODUCTION

Monitoring programs at the state level typically use fish or invertebrates rather than periphyton as biological indicators of stream condition (Rosen, 1995; Whitton and Kelly, 1995; Davis et al., 1996). This situation is rapidly changing as larger, federally funded projects include periphyton sampling as part of their routine site assessments. The results from these studies demonstrate the usefulness of periphyton as biological indicators (Leland, 1995; Charles, 1996; Cuffney et al., 1997; Carpenter and Waite, 2000; Hill et al., 2000). Recent studies have also shown that algae represent more than simple indicators of water chemistry and are often closely associated with direct measures of human disturbance such as land cover, land use, or riparian disturbance (Stewart, 1995; Kutka and Richards, 1996; Chessman et al., 1999; Pan, et al., 1999; Stewart et al., 1999; Carpenter and Waite, 2000; Leland and Porter, 2000).

Stream segments in the Mid-Atlantic region were sampled as part of the Environmental Monitoring and Assessment Program (EMAP; Herlihy et al., 2000) of the United States Environmental Protection Agency (USEPA). Sites were randomly selected and intensively sampled. Fish, invertebrates, and periphyton were collected and measurements were made of physical, chemical and hydrological conditions. Landscape scale measures were also assembled for each site's watershed. As part of this project, Pan et al. (1996 and 1999) demonstrated that the presence and abundance of specific diatom taxa were related to water chemistry variables and measures of human disturbance; although they did not test specific metrics. Hill et al. (2000) developed a periphyton index that included measures of periphyton metabolism, relative abundances of algal groups, and measures of the diatom assemblage.

This study differs from previous studies in this region in four ways: (1) a large set of potential metrics was evaluated and only those consistently associated with human disturbance were used to construct a multimetric index; (2) hypothesis testing based on correlation was used to select metrics rather than an exploratory analysis to look for patterns after metrics were selected; (3) only diatom data were used; and (4) diatom metrics were calculated at the species rather than genus level. The goal for this study was first to develop and select diatom metrics that were consistently associated with human disturbance measured at multiple spatial scales and across multiple years and, second, to define and test a multimetric index using an independent data set.

22.2 METHODS

Three steps were followed in developing and testing a multimetric index for diatoms. First, the various measures of site condition and human influence were evaluated and one measure at each of three spatial scales that best summarized human disturbance was selected. Second, candidate diatom metrics were tested for correlation with measures of human disturbance using two independent years of data. Metrics consistently associated with disturbance at multiple spatial scales were selected and combined into a multimetric index. Finally, the index was evaluated in terms of its association with disturbance using an independent data set, its correlation with natural, physical features, and its ability to detect change based on the statistical precision estimated from replicate samples.

TABLE 22.1
Number of Periphyton Samples Collected by Year and Habitat Type

Year	Pools	Riffles
1993	52 (5)	92 (6)
1994	47 (6)	92 (11)
1995	49 (4)	168 (16)
1996	12 (8)	30 (14)
Total	160 (23)	382 (47)

Note: Number of repeat visits collected within the same year are shown in parentheses

22.2.1 Study Area

The Mid-Atlantic region includes Pennsylvania, Maryland, Delaware, Virginia, West Virginia, and southeastern New York. It also includes the Blue Ridge, Central Appalachian Plateau, Central Appalachian Ridge and Valley, Northern Appalachian Plateau and Uplands, Western Allegheny Plateau, Piedmont, and Coastal Plain ecoregions (Omernik, 1987). Sites on wadeable streams were selected using a randomized sampling design (Herlihy et al., 2000). First, second, and third order streams were selected in approximate proportion to their occurrence. Sites ranged in elevation from 0 to 1400 m, with most sites (~95%) below 800 m. Over half the watershed areas upstream of the sample sites were less than 10 km^2, and 90% of the watersheds were less than 60 km^2.

22.2.2 Diatom Collection and Identification

Diatoms were collected from randomly selected positions (left, center, or right channel) along nine equally spaced transects within a site reach. Reach length was equal to 40 times the mean wetted width, not less than 150 m, and not greater than 500 m. Riffle habitat was sampled in most reaches; pools were sampled less frequently (Table 22.1). At each transect, periphyton were loosened from a 12-cm^2 area of the substrate using a toothbrush for cobble or simple agitation for finer substrate material, then suctioned with a PVC pipe. Pool and riffle samples were kept separate when both habitats were sampled in a reach. Scraped material from multiple transects was combined to obtain a single sample for each habitat type within a reach. Samples were preserved with formalin. Approximately 500 diatom valves from each sample were identified to species. For additional details see Hill et al. (2000).

22.2.3 Quantifying Human Disturbance

Over 100 variables related to channel dimensions, substrate and habitat condition, water chemistry, geographic features, and watershed land cover and use were measured for each site. Many approaches were possible for quantifying the intensities and types of human disturbance. To select among different measures before testing diatom metrics, the measures of site condition and human disturbance were first grouped into four categories: water chemistry, channel or instream condition, habitat condition, and watershed land cover and use. The four categories represented the influence of human disturbance at different spatial scales including the water flowing through the site from the upstream watershed, the channel condition at the reach scale, and the watershed upstream. Within each category, one measure was selected for testing diatom metrics that summarized similar measures made at the same scale and that, where possible, was least correlated with natural physical features such as stream size, watershed area, or slope. Disturbance measures that were least correlated with natural features were selected in order to isolate the influence of human disturbance on diatoms as much as possible.

For water chemistry, chloride was selected because Herlihy et al. (1998) found that it was the strongest indicator of human disturbance in the Mid-Atlantic region. Chloride was also significantly correlated with 13 other water chemistry measures, suggesting that it was representative of chemical condition. For channel condition, percent sand and fines were significantly correlated with 11 other measures of embeddedness, particle size, habitat type, and substrate type.

To summarize habitat condition, the mean of the rapid bioassessment protocol habitat measures (referred to subsequently as RH_XMET) was selected because it was significantly correlated with canopy cover, fish cover measures, and human disturbance in the riparian zone for this data set. RH_XMET averaged the scored values for eight channel measures: instream fish cover, epifaunal substrate, embeddedness, velocity/depth regime, channel alteration, sediment deposition, riffle frequency, and channel flow status; and four measures of riparian condition: bank condition, bank vegetative cover, grazing or other disruptive pressures, and riparian vegetation zone width (Barbour et al., 1999; Kaufmann et al., 1999). At the watershed scale, human population density was highly correlated with other measures of watershed condition, such as forested and urban land cover and road density, but was correlated with the *fewest* measures related to natural, physical features (e.g., watershed area, stream order, and slope).

While not part of the metric selection process, metric association with a risk index developed for a subset of 99 sites in the Mid-Atlantic region was also tested. The risk index summarized human influence in the riparian and upland areas using topographic maps, aerial photographs, and field information collected at the sites (Bryce et al., 1999). The risk index ranked sites in five categories according to the intensity of human disturbance.

22.2.4 Identifying Candidate Diatom Metrics

Based on other published sources, 23 attributes of the diatom assemblage related to tolerance and intolerance, autecological guild, community structure, and morphological guild were identified for testing in this study (Stevenson and Bahls, 1999). Because attributes, in most cases, could be expressed in terms of either number of species or percentage of valves, nearly twice as many candidate metrics (35) were tested and 19 were considered for possible inclusion in the diatom index because of their correlation with human disturbance. Nearly half of these were different measures of the same attribute (e.g., number of eutrophic species and percentage valves belonging to eutrophic species). For each pair, only one metric was selected leaving nine that were not redundant (Table 22.2).

Each of the candidate metrics described below represents a hypothesis test. Although some of the diatom metrics have been tested by other authors, many have only been suggested on the basis of species-level analyses. The following hypotheses were derived from other published sources and taxa lists before they were tested using the current data set. Some additional metrics were adapted from fish and invertebrate monitoring protocols.

22.2.4.1 Tolerance and Intolerance

Taxa were categorized as sensitive, tolerant, or very tolerant according to Bahls (1993), who modified initial assignments by Lange–Bertalot (1979) and Lowe (1974) to reflect diatom responses to disturbance in Montana. Bahls defined diatom species as generally tolerant to high nutrients (eutrophic), organics (polysaprobic), temperature (euthermal), salts (euhalobus), toxics, suspended solids, or unstable substrate (1993). The pollution tolerance index (PTI) was calculated as the sum over all species of the number of valves within each species multiplied by that species' tolerance value. This format is typical for many algal indices used in Europe (Whitton and Kelly, 1995).

22.2.4.2 Autecological Guilds

Diatom samples included species listed as tolerant to salt by Van Dam et al. (1994). Sources of salt include evaporation of irrigation water from agricultural fields, salt applied to roads during winter storms, and treated wastewater. Trophic state relates to the presence of inorganic nutrients such as nitrogen, phosphorus, silica and carbon; in contrast, saprobity refers to the presence of biodegradable organic matter and low oxygen concentrations (Van Dam et al., 1994). Eutrophic and polysaprobic diatoms were expected to increase if inorganic or organic nutrients were present in large amounts. Fertilizer would be one potential source of inorganic nutrient enrichment; livestock excrement and wastewater return could be sources of organic waste. In contrast, oligotrophic and oligosaprobic diatoms should decline with disturbances that increase inorganic or organic nutrient levels.

Diatoms in the genera *Epithemia* and *Rhopalodia* are nitrogen fixers because they harbor endosymbiotic cyanobacteria that allow them to convert atmospheric nitrogen into biologically useful forms such as ammonia (Mulholland, 1996). In contrast, diatoms classified as nitrogen heterotrophs can use amino acids created by other organisms as sources of carbon and nitrogen (Tuchman, 1996). Thus, nitrogen fixers should decline as they are replaced by nitrogen heterotrophs that can take advantage of the organic nitrogen associated with fertilization or other disturbances that increase nitrogen.

Many diatoms are known to be specifically sensitive to acidic or alkaline conditions. Construction or agriculture on alkaline soils can erode soils and may increase alkalinity. Irrigation and fertilization can also increase soil alkalinity. In contrast, acidophilic species may increase with acid rain.

22.2.4.3 Community Structure

Human activities that increase silt and sediment often reduce habitat complexity that may lead to decline of biodiversity and dominance by a few tolerant taxa. Percentage dominance was calculated in five different ways. The number of individual diatoms in the single most abundant species was divided by the total number of diatoms in the sample. Percentage dominance was also calculated for the two through five most abundant species. Each version of the dominance metric was tested separately. *Achnanthes minutissima*, is a common diatom species that is often the first to colonize a disturbed area. A high relative abundance of this species may indicate recent disturbances associated with scouring, metal contamination, or grazing (Medley and Clements, 1998; Stevenson and Bahls, 1999).

22.2.4.4 Morphological Structure Guilds

Motile diatoms tolerate silt because they can move across unstable substrates without being buried and are expected to increase as sediment increases. Diatom genera were assigned to three categories of motility (R.J. Stevenson, Michigan State University, personal communication, June 2000). Very motile genera included *Cymatopleura, Gyrosigma, Hantzschia, Nitzschia, Stenopterobia*, and *Surirella*; moderately motile genera included those with raphes except very motile genera; and a third group included nonmotile genera.

Algal mats are thought to follow a pattern of succession beginning with high spring flows that carry substrate-scouring sediment (McCormick, 1996; Peterson, 1996). The first algae that attach to the scoured surface attach along their length (prostrate); they are followed by algae that attach apically (adnate). Next, algae attach perpendicularly to the substrate (erect); and last are the stalked and filamentous algae that are typically taller and intolerant of fast current (Kutka and Richards, 1996). Diatom genera were assigned to these morphological guilds based on how cells attach to the substrate and to one another (Round et al., 1990; R. J. Stevenson, Michigan State University, personal communication, June 2000). Only percentages (and not taxon richness) were tested for

TABLE 22.2
Correlation of Candidate Diatom Metrics with Chloride, Percentage Sand and Fines, Habitat Condition (Average of Rapid Habitat Measures [RH_XMET]), and Human Population Density (Pop. Dens.)

		Chloride		% Sand & Fines		RH_XMET		Pop. Dens.	
Candidate metric	Selected?	1993 n = 86	1994 n = 81	1993 n = 47	1994 n = 47	1993 n = 47	1994 n = 46	1993 n = 86	1994 n = 69
Tolerance and Intolerance									
Intolerant species									
% intolerant	Yes	–0.53**	–0.39**		–0.36**		0.31*	–0.38**	–0.44**
Very tolerant species	Yes	0.53**	0.60**	0.46**	0.33*	–0.47**	–0.31*	0.52**	0.58**
% very tolerant	Yes	0.53**	0.65**	0.29*	0.3*	–0.4**	–0.32*	0.55**	0.60**
PTI		–0.42**			–0.27*			–0.25*	–0.29*
Salt-tolerant species	Yes	0.60**	0.56**	0.34**	0.35**		–0.42**	0.46**	0.55**
% salt tolerant	Yes	0.64**	0.62**	0.25*	0.4**		–0.61**	0.46**	0.61**
Species requiring high O_2		0.40**						0.25*	
% requiring high O_2			–0.41**				0.29*		–0.44**
Species tolerate low O_2	Yes	0.59**	0.68**	0.31*	0.47**	–0.27*	–0.51**	0.48**	0.68**
% tolerate low O_2	Yes	0.57**	0.71**	0.33*	0.44**	–0.40**	–0.58**	0.50**	0.70**
Autecological Guild									
Oligotrophic species							0.25*		
% oligotrophic			–0.27*	–0.28*			0.28*		
Eutrophic species	Yes	0.59**	0.65**	0.41**			–0.39**	0.45**	0.54**
% eutrophic	Yes	0.51**	0.63**	0.36**	0.27*	–0.32*	–0.5**	0.45**	0.58**

N heterotrophic species	Yes	0.50**	0.66**	0.50**		–0.37**	–0.37**	0.42**	0.63**
<u>% N heterotrophic</u>	Yes	0.54**	0.67**	0.41**	0.29*	–0.38**	–0.41**	0.50**	0.64**
Oligosaprobic species		0.42**						0.28*	
% oligosaprobic			–0.33*						–0.39**
Polysaprobic species	Yes	0.52**	0.61**	0.39**	0.26*	–0.42**	–0.30*	0.43**	0.54**
<u>% polysaprobic</u>	Yes	0.45**	0.64**		0.30*	–0.31*	–0.34**	0.34**	0.58**
Acidophilic species			–0.3*						
% acidophilic			–0.31*						
Alkaliphilic species	Yes	0.57**	0.54**	0.35**			–0.37**	0.41**	0.50**
<u>% alkaliphilic</u>	Yes	0.50**	0.49**	0.24*	0.26*	–0.26*	–0.34**	0.43**	0.50**
Community Structure									
Total taxa	Yes	0.48**	0.39**	0.32*	0.25*			0.29*	0.36**
% dominance (3 spp.)	Yes	–0.45**	–0.44**		–0.28*	0.27*		–0.32*	–0.39**
% *Achn. minutissima*			–0.38**		–0.34**		0.24*		–0.45**
Morphological Guild									
% prostrate							–0.38**		
% erect			–0.4**				0.42**	–0.3*	–0.32*
% stalked							0.26*		
% unattached						0.30*			
% non-motile							0.41**		
% moderately motile	Yes	0.57**	0.60**	0.34**	0.34**		–0.58**	0.50**	0.54**
<u>% very motile</u>	Yes	0.44**	0.63**	0.30*	0.41**	–0.38**	–0.49**	0.42**	0.66**

Note: Metrics marked "Yes" were considered for inclusion in the diatom index because six of eight tests were statistically significant (Spearman's *r*, one-sided test; for $n < 50$: * $p < 0.05$, ** $p < 0.01$; for $n > 50$: * $p < 0.01$, ** $p < 0.001$); insignificant correlations not shown. High values for chloride, percentage sand and fines, and population density indicate greater disturbance while high values for RH_XMET indicate less disturbance. The diatom index included the underlined metrics.

these candidate metrics because the physical structure of a diatom assemblage depends more on the percentage of valves of each type than the presence of a particular taxon.

22.2.5 Criteria for Metric Selection

Data collected from riffles in 1993 and 1994 were used to test candidate metrics. Metrics were selected for inclusion in a multimetric index if they satisfied four criteria: (1) they responded to disturbance in the predicted direction; (2) they were significantly correlated with at least three of the four measures of disturbance (Spearman's *r*); (3) they were significantly associated with disturbance in two different years (1993 and 1994); and (4) they were not redundant with other metrics. Hypothesis testing for the candidates metrics was based on one-sided tests. Correlation with disturbance had to be in the predicted direction to be considered significant. The predicted direction for each metric's association with disturbance was based primarily on Stevenson and Bahls (1999).

Sample sizes for each test varied because the data collected at each site varied. Because large sample sizes can yield statistically significant results for very small correlation coefficients, different *p*-values were selected for significance testing according to the sample size. For $n > 50$, p had to be less than 0.01 to be considered significant. For $n < 50$, a more standard values of p less than 0.05 was selected. Repeat visits to the same sites were excluded from testing to avoid inflating the significance of the correlation coefficient.

22.2.6 Testing and Evaluating the Diatom Index

The diatom index was tested with independent data collected from 1995 and 1996. Repeat visits to sites in the same year were again excluded. Three of the four measures used to select metrics for the 1993 to 1994 data were available for testing the diatom index in 1995 and 1996; these were chloride, RH_XMET, and population density. The fourth measure, percentage sand and fines, was not measured in 1995 or 1996. Other measures of human disturbance used to develop fish and invertebrate indexes for the Mid-Atlantic region were also used to test the diatom index developed in this study (McCormick et al., 2001; Klemm et al., 2002). These included Mn, Fe, acid neutralizing capacity, pH, NH_4, SO_4, total P, total N, turbidity, road density, percent urban area, percent forested area, and percent disturbed area (percentage urban, agricultural, and mining land cover).

To evaluate the influence of natural physical features, the diatom index was tested for correlation with watershed (drainage) area, stream order, stream temperature, dissolved oxygen, SiO_2, and stream reach elevation. Nonparametric tests (Spearman's *r*) were used to avoid concerns about the underlying distributions of the various measures.

From 1993 to 1996, a 27-site subset was visited two to six times. Repeat visits occurred one to four months after the initial visit. For each repeat site visit, diatoms were collected from riffles, pools, or both. ANOVA was used to estimate and partition the overall variance of the diatom index due to site differences, years, habitat type (pool vs. riffle), and repeat visits within the sampling season. Because habitat differences contributed little to the overall variance of the diatom index compared to other sources of variability, subsequent components of variance analysis were based on riffle samples only. Limiting the analysis to riffles also simplified a very unbalanced ANOVA design (i.e., many site–year combinations were missing) which can improve variance estimates.

To estimate the number of categories of biological condition that the diatom index could detect, the minimum detectable difference (MDD) was estimated and the range of the index was divided by that value. The MDD was calculated for a two-sample *t*-test with three replicates with alpha set equal to 0.05 and power (1 – beta) equal to 0.80 (Zar, 1984). Sample variance of the diatom index was estimated from repeat visits to the same site.

$$MDD = \sqrt{\frac{2s_p^2}{n}}(t_{\alpha,\,\nu} + t_{\beta(1),\,\nu})$$

where s^2 = mean squared error from ANOVA, $n = 3$, $\nu = 2\ (n - 1) = 4$, and t = Student's t for alpha = 0.05 (two-sided) and beta = 0.20 (one-sided). The use of normality-based statistical models is appropriate for multimetric indexes, and no transformation of the index was necessary (Fore et al., 1994).

Power calculations are model-dependent and the selection here of a t-test as a statistical model is somewhat arbitrary. Though not standard, the t-test is common enough and frequently used to illustrate a general case (Peterman, 1990; Carlisle and Clements, 1999); other authors used a regression model to estimate power (Hughes et al., 1998).

22.2.7 Identifying Metric Signatures

To test for specific metric "signatures" associated with different types of disturbance, three small groups of sites were selected whose watersheds were (1) dominated by agriculture (15 sites), (2) had high levels of urban land cover (21 sites), or (3) had high land covers associated with mining or had multiple toxic point sources of pollution (12 sites). The sites also tended to have very low incidences of other types of disturbance. The grouping variable for discriminant function analysis (DFA) was defined according to the three disturbance types and the nine diatom metrics were used as the independent variables for separating sites according to group membership.

All metrics were percentage measures and tended to be positively skewed in their distributions (long tail to the right). When the assumptions of multivariate normality are not properly met, solutions are less reliable and tests for significance must be interpreted with caution (Tabachnik and Fidell, 1989). One solution is a transformation of variables to approximate a normal distribution; however, when variables follow similar distributions (as in this case) the solution is not typically improved by transformation. Transformation also introduces additional problems including the necessity of interpreting results in terms of the *transformed* variables rather than the original measured variables. For this analysis, diatom metrics were not transformed because the patterns of association between metrics and site groups were of the greatest interest and testing for the statistical significance was not performed.

22.3 RESULTS

Of the 35 candidate metrics tested, 19 strongly and consistently correlated with human disturbance. Several of these were redundant, for example, taxa richness of alkaliphilic species and percentage of valves of alkaliphilic species. From these 19, nine metrics were selected that were correlated with disturbance in the predicted direction and were not conceptually redundant with other metrics. A multimetric diatom index was derived from these nine metrics; it was significantly associated with several measures of human disturbance and was not influenced by stream size or watershed area. Based on replicate sampling, the diatom index was less variable than its component metrics. In addition, most of the variance associated with the index was due to differences in site condition (80%) rather than replicate sampling (20%). Riffle and pool index values were similar as were values from repeat visits to the same sites. Specific metrics tended to be associated with different types of disturbance.

22.3.1 Metric Response to Disturbance

In general, if a diatom metric was significantly correlated with one measure of disturbance, it tended to be correlated with the other three measures as well (see Table 22.2). Metrics showed a slight tendency to be more consistently associated with chloride and population density than with percentage sand and fines or habitat condition. Typically, if an attribute was significantly associated with disturbance when measured as the total number of taxa, it was also significant when measured as the percentage of valves. Overall, diatom metrics associated with specific tolerances and ecological requirements were more closely associated with disturbance than measures related to community structure or morphological guild.

22.3.1.1 Tolerance and Intolerance

Although most metrics were significantly associated with disturbance whether measured as taxa richness or percentage of valves, intolerant diatom species were the exception; percentage intolerant valves were significantly associated with disturbance, but number of species was not. Diatoms belonging to very tolerant species, salt-tolerant species, and species that tolerate low oxygen were all significantly associated with disturbance whether measured as number of species or percentage of valves. Bahls' pollution tolerance index (PTI) was not consistently associated with disturbance and neither were diatoms belonging to species that require high oxygen.

22.3.1.2 Autecological Guilds

Diatoms that can process and take advantage of organic nutrients, inorganic nutrients, or amino acids (saprobic, eutrophic, and nitrogen heterotrophic species) all increased as disturbance increased. Diatoms that prefer alkaliphilic conditions also increased with disturbance. Diatoms that can survive with very low nutrients (oligotrophic and oligosaprobic) were not associated with disturbance. Acidophilic diatoms also did not increase with human disturbance.

22.3.1.3 Community Structure

Total taxa richness and percentage dominance were significantly associated with human disturbance, but not in the directions predicted (Stevenson and Bahls, 1999). The number of diatom species increased with disturbance and percent dominance decreased. In 1993, the percentage of *Achnanthes minutissima* valves present was not significantly correlated with disturbance; in 1994, the percentage *decreased* with disturbance rather than increasing as predicted.

22.3.1.4 Morphological Guilds

The percentages of valves belonging to both moderately motile and very motile genera increased with disturbance as predicted. Surprisingly, this metric was more highly correlated with chloride and population density than with percentage sand and fines. None of the other morphological measures (prostrate, erect, stalked, unattached, or non-motile) were consistently associated with human disturbance.

22.3.2 Constructing a Multimetric Index for Diatoms

Of the 19 metrics consistently associated with human disturbance, seven taxa richness metrics were eliminated because of redundancy with similar metrics based on percentage relative abundance; two metrics were eliminated because they were associated with disturbance in the opposite direction predicted; and one was eliminated because it was redundant conceptually. This left nine metrics for the final index: percentages of valves belonging to species that are intolerant, very tolerant, salt-tolerant, tolerant of low oxygen, eutrophic, nitrogen heterotrophic, polysaprobic, and alkaliphilic, and percentage of valves belonging to genera that are very motile. All measures increased with disturbance except for one, percentage of valves belonging to very intolerant species, which decreased.

Different measures of the same attribute were highly correlated and obviously redundant. Percentage of valves rather than number of species was selected for all metrics for two reasons. First, percentages of valves were more strongly associated with disturbance for some metrics. Second, for some metrics the number of species varied little across sites because very few species were found (typically 0 to 2 species). In contrast, the percentage of valves offered a broader range of potential values to distinguish among sites. Although some diatoms metrics included in the index were highly correlated, they were not necessarily calculated from the same species (Appendix I).

Some species were designated tolerant of low oxygen and polysaprobic, for example, and others were one or the other.

Total taxa richness and percent dominance satisfied the criteria for selection based on their correlation with disturbance, but the direction of association was opposite of the predicted direction (Stevenson and Bahls, 1999); therefore, they were not included in the index. Percentage very motile was selected rather than percentage moderately motile because it was more consistently associated with disturbance and conceptually more specific.

The nine metrics included in the diatom index were all significantly associated with a watershed risk index that ranked sites according to the intensity of human disturbance at the riparian and watershed scale (Figure 22.1; Bryce et al., 1999). None of the diatom metrics significantly correlated with stream order, watershed area, stream depth, stream width, or sinuosity (Table 22.3). All metrics were correlated with elevation measured either at the site or the highest point in the watershed. Most metrics were also correlated with slope measured at either the watershed or reach scale. Because the pattern of metric correlation with physical features was similar for each metric, the influence of elevation was evaluated for the multimetric index rather than individual metrics.

To combine metrics into a single multimetric index, metrics were rescaled: 5 indicated best biological condition; 3 indicated moderate condition; and 1 indicated degraded condition (Table 22.4). Approximately one third of the observed site values were assigned to each of the three scores. Exceptions to this approach occurred when more than a third of the sites had values between 0 and 1. A single scoring category was assigned to all values less than or equal to 1.

22.3.3 Index Performance

Based on independent data collected during 1995 and 1996, the diatom index was significantly correlated with four measures of human disturbance: chloride, habitat condition (RH_XMET), population density, and watershed disturbance category (Figure 22.2). The diatom index was not significantly correlated with natural physical and chemical features such as watershed area, stream order, stream temperature, dissolved oxygen, or silica (Table 22.5). The index was correlated with elevation, possibly because human development was more intense at lower elevations. After excluding high elevation sites (>600 m), the index was no longer correlated with elevation, but was still significantly correlated with measures of human influence related to water chemistry (ANC, total P, total N, chloride, and turbidity), habitat condition (RH_XMET), and land cover (road density, forested area, disturbed area [agriculture + urbanization + mining], and population density).

Overall, the variability associated with site differences was the largest contributor to the total variance of the index; this was also true for most metrics (Figure 22.3). A precise index should have high variability associated with site differences (the variable of interest) and low variability associated with natural differences such as year or time within year (nuisance variables). The next largest component was associated with different sampling times within a year, which may also be considered the measurement error associated with the sampling protocol. Site-by-year interaction and annual variability were near zero for most metrics and the index. Only percent intolerant, percent tolerant, and percent polysaprobic metrics had large components of their variance associated with site-by-year interaction. Overall, variability due to all sources other than site differences was smaller for the index than for any of its component metrics.

A larger ANOVA model was calculated that also included index values from pool samples. Neither habitat type (pool vs. riffle) nor the various interaction terms associated with habitat represented a significant source of variability to the index; variability associated with repeat visits still dominated other sources of natural variability (Table 22.6). Index values based on repeat visits to the same sites within a single year agreed moderately well ($r^2 = 0.67$; Figure 22.4).

To estimate the MDD for the diatom index, measurement error was estimated as the mean squared error from ANOVA. Measurement error associated with repeat visits to 27 riffle sites was 22.35 (standard deviation = 4.7) which translated into 2.5 categories of biological condition. Because less

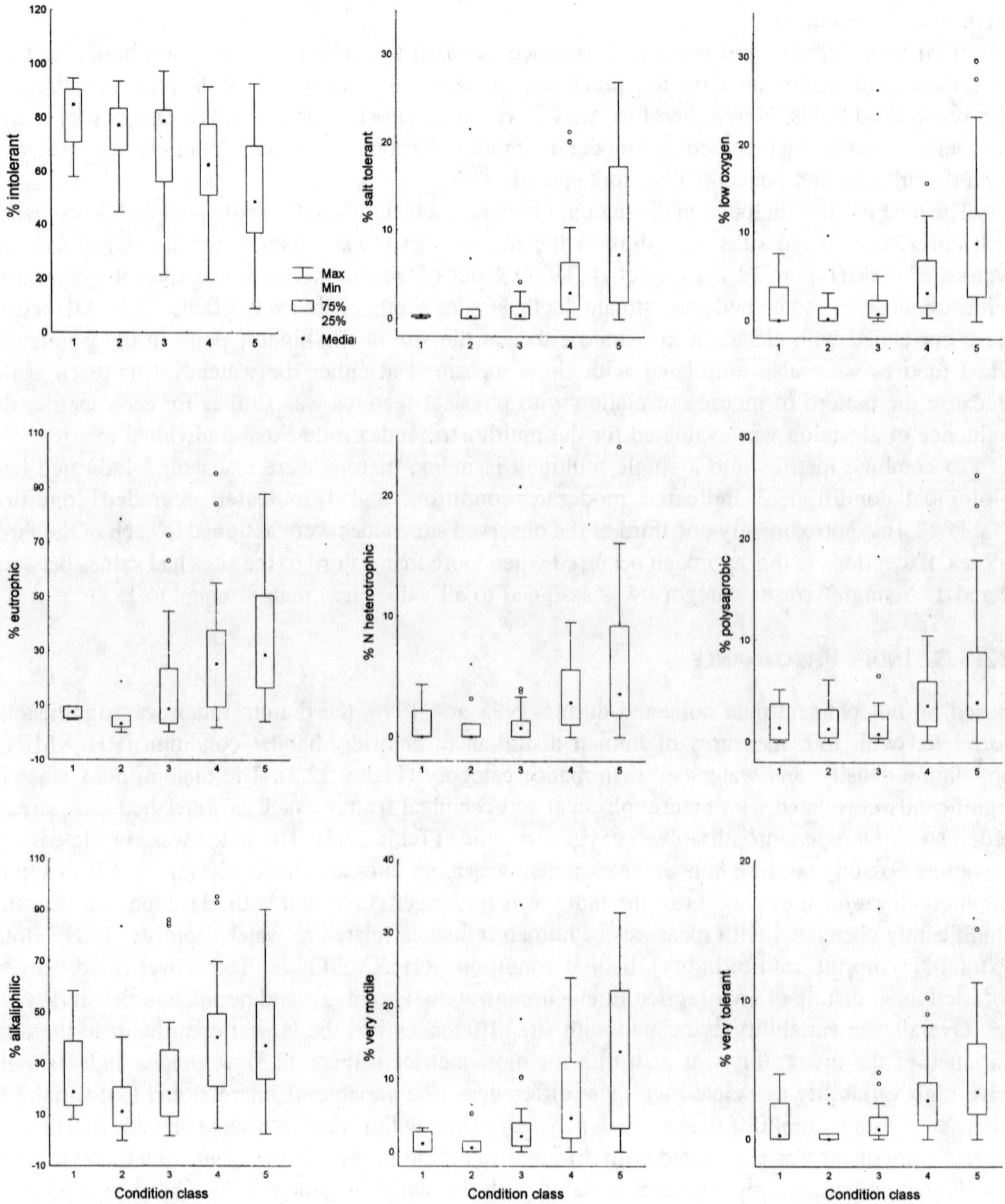

FIGURE 22.1 Diatom metrics were significantly associated with a watershed risk index (Bryce et al., 1999). Higher condition class represents greater human disturbance. Percent intolerant declined as disturbance increased; all other metrics increased with disturbance. Data are shown for riffle samples from 99 sites taken in 1993 and 1994.

disturbed sites were less variable, measurement error was also estimated for nine sites with the highest diatom index values (greater than 35). Measurement error was much smaller for these nine sites, 14.46 (standard deviation = 3.8), which translated into 3.1 categories of biological condition.

For riffle samples, the standard deviation associated with repeat sampling ranged from 1.2 to 8.6; for pool samples, the standard deviation ranged up to 14. Sites with the most variable index values also tended to have low index values — indicating more degraded condition — for both riffles and pools (Figure 22.5).

TABLE 22.3
Correlation of Diatom Metrics and Stream Order, Watershed Area Upstream of Sample Site, Stream Depth, Stream Width, Sinuosity, Site Elevation, Elevation at Highest Point in the Watershed, Watershed Slope, and Slope at the Stream Site

Metric	Stream Order n = 94	WS Area n = 94	Depth n = 93	Width n = 93	Sinuosity n = 94	Elevation n = 94	High Point n = 141	WS Slope n = 94	Slope n = 94
% intolerant						0.38**	0.4**	0.32**	0.24*
% very tolerant						–0.34**	–0.37**	–0.27**	–0.25*
% salt tolerant						–0.41**	–0.41**	–0.27**	–0.23*
% tolerant of low oxygen						–0.47**	–0.47**	–0.36**	–0.31**
% eutrophic						–0.31**	–0.3**	–0.21*	
% nitrogen heterotrophs						–0.36**	–0.35**	–0.33**	–0.29**
% polysaprobic						–0.3**	–0.33**	–0.25*	
% alkaliphilic						–0.34**	–0.39**		
% very motile						–0.43**	–0.48**	–0.32**	–0.25*

Note: Only significant correlations are shown (Spearman's *r*, two-sided test; * $p < 0.05$; ** $p < 0.01$). Sample size is shown for each test.

TABLE 22.4
Diatom Metrics Included in Multimetric Index, Their Responses to Disturbance, and Scoring Categories Used to Rescale Metric Values

Metric	Response	Scoring criteria 1	3	5
Tolerance and Intolerance				
% intolerant	Decrease	< 60	(60, 80)	> 80
% very tolerant	Increase	> 4	(1, 4)	<= 1
% salt tolerant	Increase	> 6	(1, 6)	<= 1
% tolerant of low oxygen	Increase	> 5	(1, 5)	<= 1
Autecological guild				
% eutrophic	Increase	> 30	(7, 30)	< 7
% nitrogen heterotrophs	Increase	> 4	(1, 4)	<= 1
% polysaprobic	Increase	> 4	(1, 4)	<= 1
% alkaliphilic	Increase	> 40	(15, 40)	< 15
Morphometric guild				
% very motile	Increase	> 6	(1, 6)	<= 1

Index values collected from pools and riffles at the same site were correlated, but were quite variable for same-day sampling (Figure 22.6). A subset of 14 sites had samples collected from both pools and riffles and were visited more than once (between two and five times). Index values for riffles and pools were averaged for repeat visits and again compared. Agreement improved suggesting that the biological condition of riffles and pools, as measured by the diatom index, was similar when samples were averaged through time.

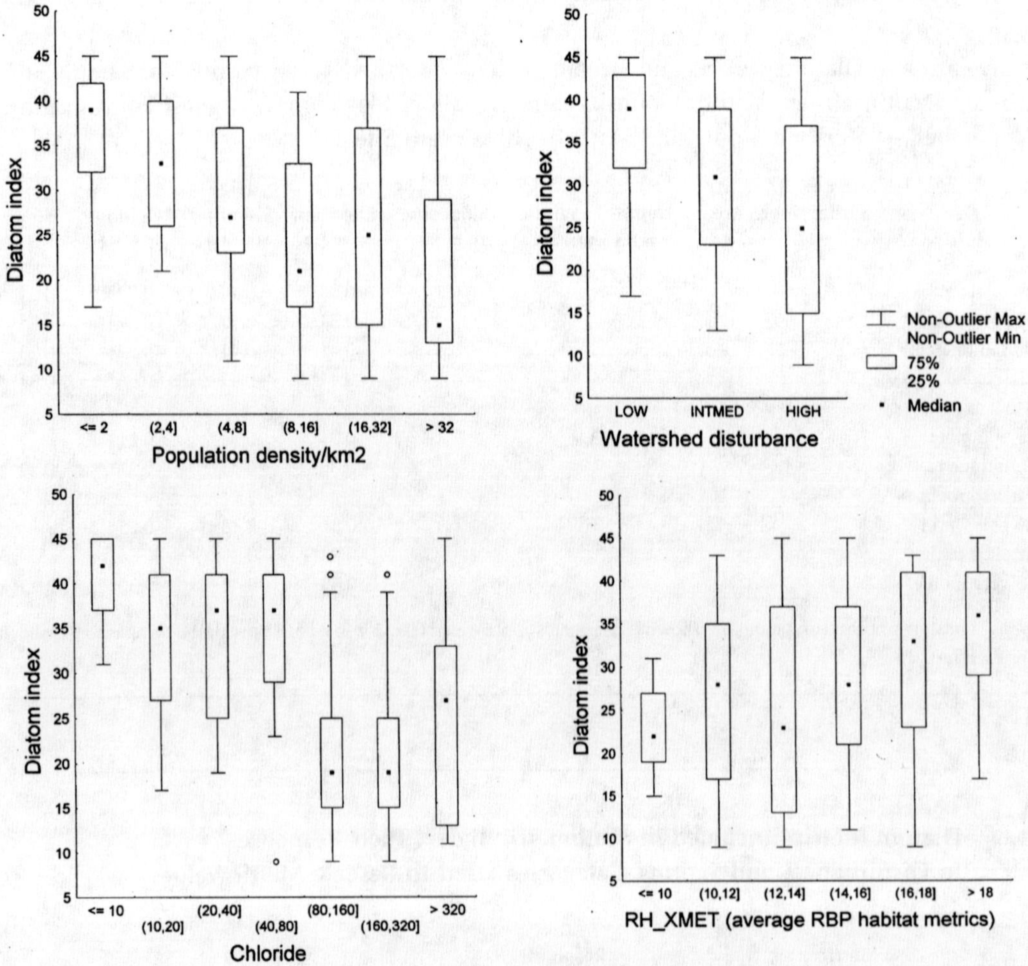

FIGURE 22.2 The diatom index was significantly associated with human disturbance measured as population density and watershed disturbance at the watershed scale, in terms of riparian condition near the site (RH_XMET) and as chemical condition of the stream (chloride). Data shown were collected in 1995 and 1996; they were not used to test or select metrics for the index.

22.3.4 Metric Signatures

Discriminant function analysis separated the three types of sites on the basis of differences in the diatom metric values (Figure 22.7). The first axis separated agricultural sites from urban and mining/toxic site groups. This axis was more highly correlated with the percentages of polysaprobic and very motile valves compared to the second axis (Table 22.7). The second axis separated most of the urban sites from the other two groups and was more highly correlated with the percentage of salt-tolerant valves. Thus, agricultural sites were characterized by a higher percentage of valves belonging to very motile genera that tolerate sediment. Urban sites were characterized by a higher percentage of valves belonging to salt-tolerant species. In addition, diatoms associated with organic sources of nutrients were more common at agricultural sites.

22.4 DISCUSSION

Much is known about the natural history and ecological requirements of individual diatom species (Stevenson and Bahls, 1999). This study builds on the work of other authors who tested and recorded

TABLE 22.5
Correlation of Diatom Index with Watershed Features and Measures of Human Disturbance

Measure	N	*r*	*p*-value	< 600 m
Physical Features				
Watershed area	151	0.03	0.68	No
Stream order	123	0.09	0.33	No
Stream temperature	119	–0.17	0.07	No
Dissolved oxygen	118	0.12	0.21	No
SiO_2	151	–0.11	0.18	No
Elevation	92	0.10	0.36	Yes
Disturbance Measures				
Mn	92	–0.03	0.77	Yes
Fe	92	–0.20	0.06	Yes
ANC	92	–0.56**	0.00	Yes
PH	151	–0.59**	0.00	No
NH_4	92	–0.24*	0.02	Yes
SO_4	92	–0.05	0.62	Yes
Total P	92	–0.48**	0.00	Yes
Total N	92	–0.56**	0.00	Yes
Chloride	92	–0.43**	0.00	Yes
Turbidity	92	–0.18	0.08	Yes
Average of habitat metrics (RH_XMET)	91	0.29*	0.01	Yes
Road density	92	–0.29**	0.00	Yes
% urban area	92	–0.14	0.18	Yes
% forested area	92	0.48**	0.00	Yes
% disturbed area (%ag + %urb + %mining)	92	–0.47**	0.00	Yes
Population density	92	–0.34**	0.00	Yes

Note: Sample size of the test (N), correlation coefficient (Spearman's r), significance (two-sided test; * $p < 0.05$; ** $p < 0.01$), and whether the sites included in the test were below 600 m are shown. Data are for riffle samples collected in 1995, first visits only.

their observations of diatom natural history (Van Dam et al., 1994; Bahls, 1993; Lange–Bertalot, 1979; Stevenson and Pan, 1999). Unlike fish or invertebrates, many diatom species are distributed globally; consequently, knowledge acquired about an individual species in Europe, for example, may be applied to that same species in North America. This study took what is known about diatoms at the species (or genus) level and summarized that information by grouping species according to autecological or morphological guilds. From a large set of biological attributes used to describe the diatom assemblage, a set of metrics were chosen that were strongly associated with human disturbance at multiple spatial scales and across multiple years. Following the approach developed for fish and invertebrates, metrics were combined into a multimetric index for diatoms (Karr et al., 1986; Kerans and Karr, 1994).

To be useful as a monitoring tool, a multimetric index should respond primarily to human disturbance rather than natural sources of variability. Designing an index with this objective in mind may be difficult when patterns of human land use follow natural features, for example, agriculture is more common on alluvial plains than at higher elevations (Rathert et al., 1999). The diatom index developed for this study was not associated with measures of stream size such as watershed area, stream order, or wetted width or depth. The index was correlated with elevation, possibly because human disturbance was less intense at higher elevations.

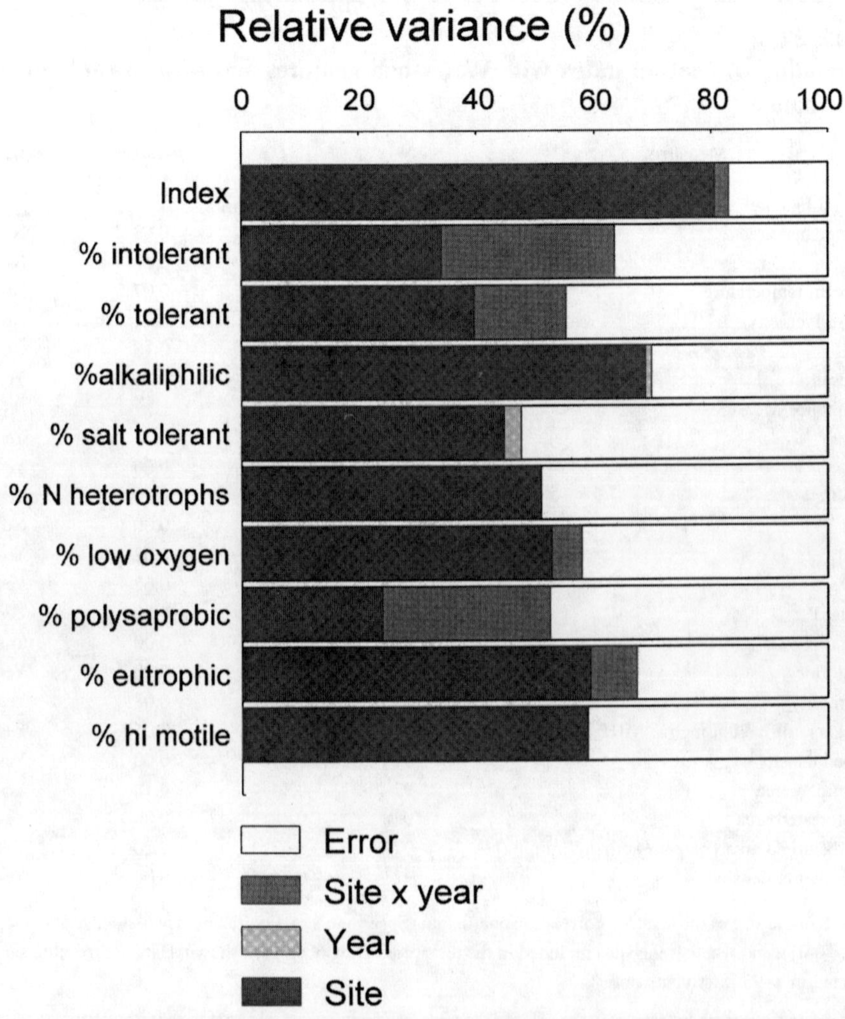

FIGURE 22.3 Components of variance estimates for the diatom index and its nine metrics. Variability associated with site differences was greatest followed by variance associated with repeat sampling (error); variances associated with yearly differences and site-by-year interaction were smallest. The index was more precise than its component metrics because yearly, site by year, and error variances were smallest. Data are for 27 riffle sites visited from 1993 to 1996.

A good monitoring tool should also be precise such that, in the absence of a change in human influence, it gives the same value from one visit to the next. For this study, least disturbed sites were less variable than more disturbed sites in terms of their diatom index values suggesting that the diatom index was sensitive to changes in human influence that occurred between sampling occasions. Finally, by including different aspects of diatom biology as metrics in the index, different types of human disturbance could be diagnosed according to the patterns, or signatures, of individual metric responses.

22.4.1 Diatoms and Human Disturbance

Of the 35 candidate metrics tested — representing 23 attributes of the diatom assemblage — nine were selected as metrics for the index on the basis of their strong association with human disturbance measured in terms of water chemistry, channel condition, habitat condition, and watershed land

TABLE 22.6
Components of Variance Estimates for the Diatom Index Due to Site, Year, Habitat (Riffle vs. Pool), and Interaction Terms for Each Factor

Source of Variance	DF	Variance	Percent of Total
Site	26	105.2	72
Year	3	–4.7	0
Habitat type	1	–2.3	0
Site by year	30	5.2	4
Site by habitat	13	3.5	2
Year by habitat	3	8.1	6
Error (repeat visits)	68	24. 5	17

Note: Measurement error was estimated for repeat visits to the sites during different months of the same year. Degrees of freedom, estimates of variance derived from Anova, and variance expressed as a percentage of total variance are shown.

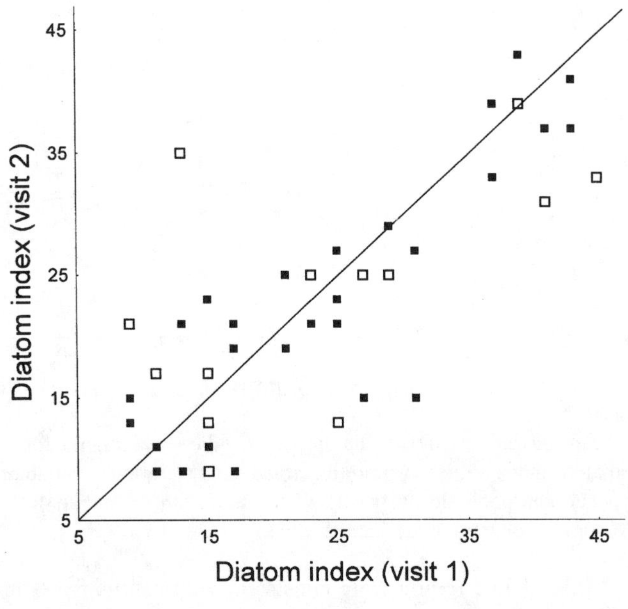

FIGURE 22.4 Diatom index values for riffle samples agreed for repeat visits collected within the same year (1993 through 1996), although they were somewhat variable ($r^2 = 0.67$). Sites sampled in 1996 (□) were somewhat more variable, possibly because they were sampled later in the year.

use. Measures in different categories were selected to capture human disturbance at different spatial scales. Selection of diatom metrics was a straightforward process because metrics associated with one measure of disturbance tended to be associated with most other measures. Significant correlations also tended to be consistent across years. In short, diatom metrics either worked or did not work. The initial rigorous screening for metric selection insured that the final index would also be strongly associated with disturbance when it was tested with an independent set of sites collected in subsequent years.

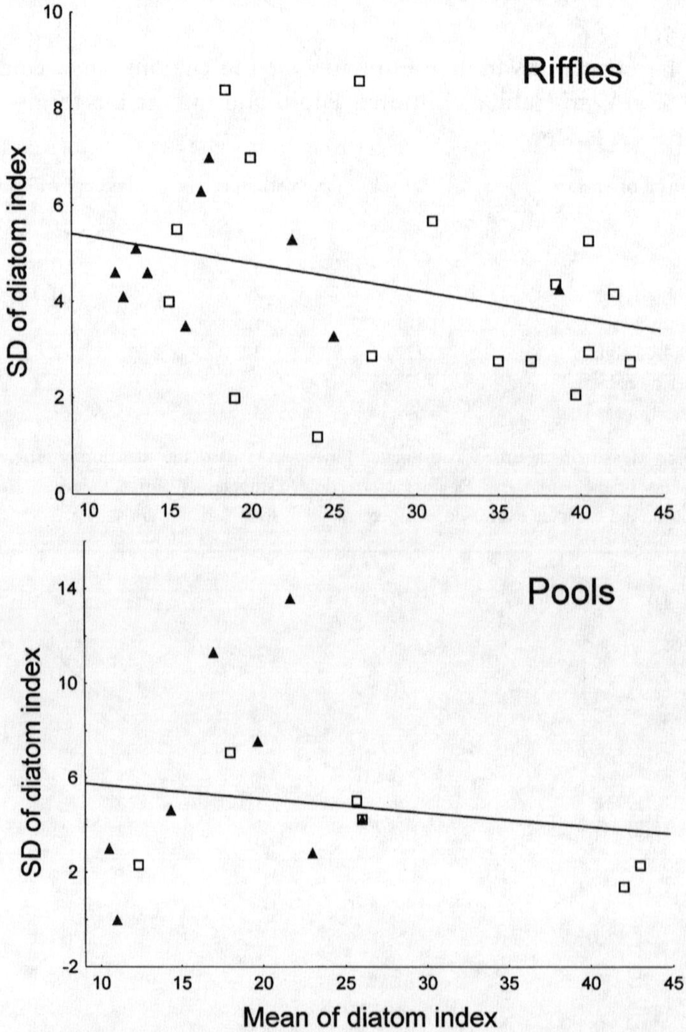

FIGURE 22.5 The standard deviation (SD) of diatom index values was higher for sites with lower index values indicating that diatom index values at more disturbed sites were more variable. Results were similar for both riffles (upper panel) and pools (lower panel). Watersheds with more than 40% of their land cover in agriculture, urban development, or mining are identified (▲).

Results from the Mid-Atlantic region were consistent with results from other regions. Diatom species defined as sensitive or tolerant to disturbance in other regions declined or increased with disturbance as predicted (Lange–Bertalot, 1979; Bahls, 1993; Fore and Grafe, in press). The salinity classification of Van Dam et al. (1994) was developed in relation to preference for marine (or brackish) vs. freshwater environments. The association of this metric with disturbance was something of a surprise. An increase in salt-tolerant or halophilic diatoms could be caused by the chloride associated with urban wastewater, salt used to remove road ice, or other industrial sources (Herlihy et al., 1998). Rott et al. (1998) noted an association between diatom species composition and chloride and Leland and Porter (2000) found that salt tolerant species were associated with human disturbances such as agriculture and urbanization.

The percentages of valves belonging to eutrophic, nitrogen heterotrophic, polysaprobic and alkaliphilic species all increased with human disturbance in the Mid-Atlantic region. Other studies report similar shifts in the diatom assemblage related to agriculture and urbanization (Leland, 1995;

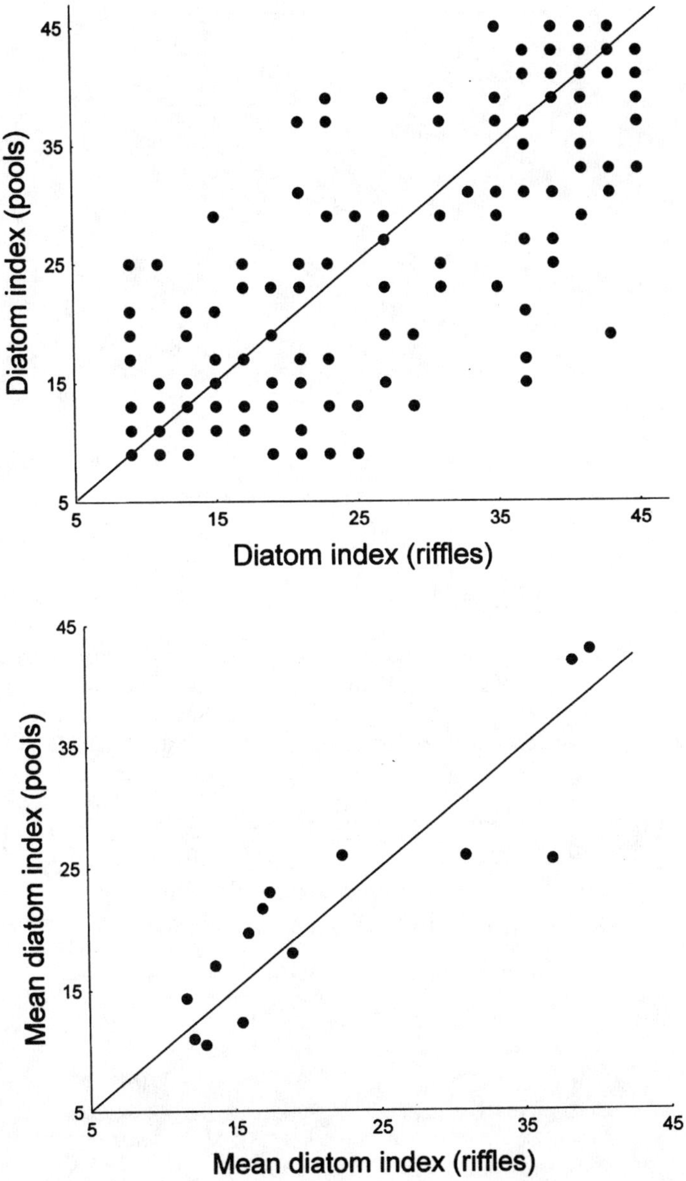

FIGURE 22.6 Diatom index values calculated for samples collected on the same day from riffle and pool habitats were similar in values but quite variable (upper panel). When index values were averaged separately for riffle and pool habitats through time, they showed better agreement (lower panel). The line of perfect agreement is shown for both graphs.

McCormick and O'Dell, 1996; Cuffney et al., 1997; Pan et al., 1999; Carpenter and Waite, 2000), alkalinity (Chessman et al., 1999), and organic pollution (Kelly et al., 1995; Rott et al., 1998; Leland and Porter, 2000). Often for these studies, changes in the diatom assemblage are described in terms of individual species rather than in terms of species guilds, or metrics, as was done for this study.

The response of diatom taxa richness, measured as the total number of species, to human disturbance seems to vary across studies, decreasing in some cases and increasing in others (Stewart, 1995; Medley and Clements, 1998; Chessman et al., 1999; see also review in Hill et al., 2001). For the Mid-Atlantic region, the number of diatom species increased with a variety of measures of

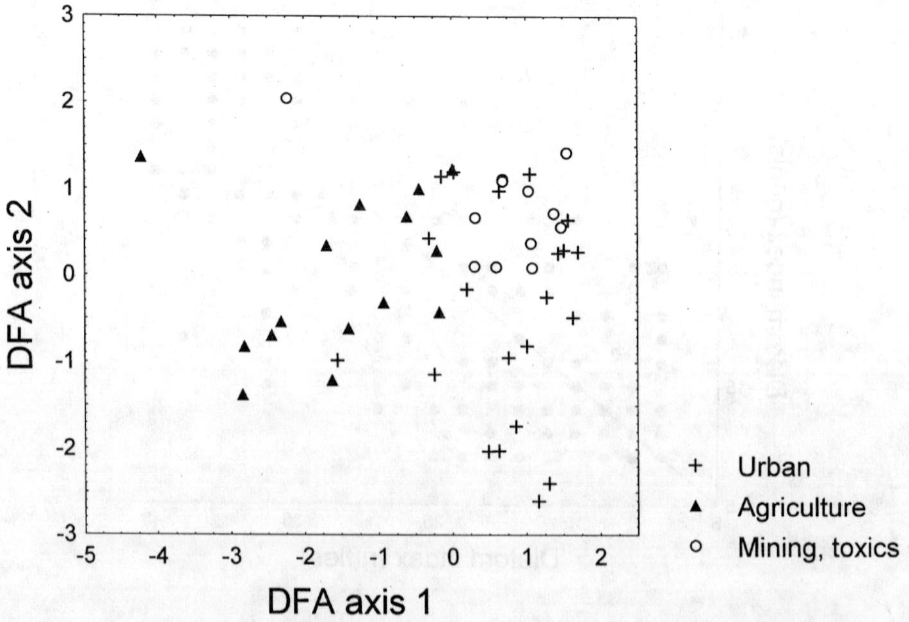

FIGURE 22.7 Discriminant function analysis based on nine diatom metrics-separated sites with watersheds dominated by urban development, agriculture, and mining or toxics. Urban sites were characterized by a larger percentage of diatoms that are tolerant of salt and inorganic nutrients (eutrophic). Agricultural sites were characterized by a larger percentage of motile diatoms (sediment tolerant). See Table 22.7 for coefficients.

TABLE 22.7
Correlation of Diatom Metrics with Axes Derived from Discriminant Function Analysis

Metric	Axis 1	Axis 2
% intolerant	0.37	0.24
% very tolerant	–0.53	–0.19
% salt tolerant	–0.06	**–0.69**
% tolerant of low oxygen	–0.45	–0.47
% eutrophic	–0.55	–0.72
% N heterotrophic	–0.42	–0.26
% polysaprobic	–0.48	–0.28
% alkaliphilic	–0.60	–0.46
% very motile	**–0.54**	–0.17

Note: Correlations unique to a particular axis and that may indicate a metric signature are bolded (see Figure 22.7).

human disturbance, contrary to expectations (Stevenson and Bahls, 1999). If sites are naturally nutrient-poor, an increase in species richness may be observed with nutrient enrichment caused by human disturbance. For the Mid-Atlantic region, an increase in taxa richness was observed for all types of disturbance. Because of the inconsistent response of this metric in other studies, the total number of species was not selected as a metric for the diatom index.

Many studies have noted an increase in motile diatoms associated with erosion and sedimentation which may be caused by a variety of human activities (Bahls, 1993; Kutka and Richards,

1996; Detenbeck et al., 2000). By moving across the substrate, these diatoms avoid being smothered by shifting sediments.

22.4.2 Comparison with Idaho Rivers

Seven of the nine metrics that showed a strong correlation with human disturbance in the Mid-Atlantic region for wadeable streams were also consistently associated with disturbance in Idaho for large rivers (Fore and Grafe, in press). Three differences between the results of the two studies were: (1) the percentage of valves belonging to species that require high oxygen was associated with disturbance only in Idaho; (2) the percentage of valves belonging to species that tolerate salt was only significant in the Mid-Atlantic region; and (3) the Idaho index included a metric that measures the percentage of deformed cells; such data were not available for the Mid-Atlantic region. The percentage of valves belonging to species that tolerate low oxygen was correlated with disturbance in both studies, but was not included in the Idaho index because it was redundant with other metrics.

The differences between these two studies are relatively small when one considers that 23 different diatom attributes (representing 35 candidate metrics) were tested and the results were nearly identical, both in terms of attributes significantly associated with human disturbance and attributes that were not. For both studies, attributes tended to be significantly correlated with disturbance whether measured as taxa richness or percentage of valves. The probability of selecting the same 7 of 23 attributes of the diatom assemblage due to chance alone is extremely small. Similarities are more striking when one considers the enormous differences in watershed size and dramatic ecoregional differences between these study areas.

Although the metrics selected were very similar, the range of values associated with minimal disturbance was strikingly different for several metrics. Percentages of valves belonging to very tolerant, nitrogen heterotrophic, polysaprobic, and very motile taxa were about three to seven times higher for Idaho rivers than for Mid-Atlantic streams. In contrast, percentage of species belonging to intolerant and alkaliphilic species defined a similar range for least disturbed sites in both regions.

The consistent responses of diatom metrics in these very different ecoregions suggest that diatom metrics may be particularly robust. Furthermore, the conceptual modifications and substitutions required to adapt fish, and to a lesser extent invertebrate, multimetric indices to new regions may not be as necessary for diatoms (Miller et al., 1988; Hughes and Oberdorff, 1999). The cosmopolitan nature of diatoms may represent a unique advantage for this assemblage. Biological monitoring protocols based on diatoms may be more readily applied to new regions or continents than protocols based on fish or invertebrates that must be adapted to the specific local fauna.

22.4.3 Diatoms and Natural Physical Features

Robust monitoring tools must distinguish human-caused changes in the biological assemblage from natural variability (Allen et al., 1999; Richards et al., 1996). Other studies based on multivariate analysis found that human disturbance had a greater influence on the diatom assemblages than natural features related to ecoregion or basin geology (Leland and Porter, 2000; Pan et al., 2000). For the current study, diatom metrics and the diatom index were not correlated with physical features such as stream or watershed size; nor were they associated with silica, the naturally occurring mineral that diatoms use to construct their shells or frustules. Dissolved oxygen and temperature also represent a potential source of natural variability but were not correlated with the diatom index.

The metrics and index were correlated with elevation and slope. Higher elevation sites had higher index values (indicating better biological condition). To test whether the index was simply measuring slope, high elevation sites (>600 m) were excluded. The correlation with elevation disappeared, but correlation with disturbance did not. Human disturbance was also associated with elevation with more urban and agricultural land use in lower elevations. Elevation and slope may also play a role in naturally flushing heavy nutrients and toxics from the system. If that is so, these streams may be able to tolerate a higher degree of pollution before the diatom community is measurably degraded. The diatom index could perhaps benefit from metric scoring that corrected for the influence of watershed slope. Similar corrections are often used for fish metrics to correct for the influence of watershed area (Hughes et al., 1998).

22.4.4 Index Variability and Precision

A robust monitoring tool should be sensitive to site differences associated with human disturbance and minimally affected by small differences in location or time of sampling (Fore et al., 1994; Barbour et al., 1999; Kaufmann et al., 1999; Fore et al., 2001). Most diatom index variability was associated with site differences which are the differences of greatest interest in the context of biological monitoring and assessment and should represent the largest component of variance. Variability associated with annual differences was smallest while most nuisance variability was associated with repeat sampling during the year. Based on repeat visits to riffle sites, the index could detect 2.5 categories of biological condition; if only less disturbed sites were considered, the index was more precise and could detect 3.1 categories. The variability of the index was somewhat high considering its strong association with human disturbance.

Index variability can be caused by measurement error associated with the way samples are collected or how they are analyzed in the laboratory, seasonal changes in the diatom assemblage, or changes in human disturbance during the year. Same-day replicate samples were not collected for this study, but same-day sampling for a similar study in Idaho showed that index values differed by only 3%. The field sampling protocol was similar for both studies, but the Idaho study was based on larger sample sizes which may increase precision (800 vs. 500; Fore and Grafe, in press). Variability could also be due to random changes in the diatom assemblage during the year.

Data were insufficient to test with confidence whether diatom index values changed seasonally because insufficient sites were sampled during different months. For sites visited in 1996, index values were more variable, possibly because the time between repeat visits was the longest and may have included more seasonal variability. If seasonality is a cause of variability, the issue could be resolved by defining a more narrow index period for sampling; the sampling period for this study ranged from April to August.

Changes in the diatom index may also track changes in human disturbance throughout the year. Sites with the most variable index values through time also had low index values, indicating greater human disturbance. If diatoms measure real changes in biological condition during the index period, this could be interpreted as either an asset or a liability depending on perspective. A quick response of diatoms means that they could be used to compare management practices that change within a year or season. This is not possible with invertebrates or fish. If on the other hand, a monitoring tool that integrates over a year is needed, diatoms may be too time-sensitive.

22.4.5 Metric Signatures

Multimetric indices summarize information from different levels of biological organization in order to be responsive to a variety of human disturbances and apply across a large geographic area (Angermeier and Karr, 1994; Karr and Chu, 1999). In addition, independent responses of the component metrics to different types of disturbance may define a "signature" for a particular disturbance (Yoder and Rankin, 1995). To address these issues, diatom metrics related to tolerance and intolerance, autecological guild, community structure, and morphological guild were selected for testing.

Metrics included in the diatom index primarily evaluated the diatom assemblage in terms of species groups. Community level metrics, such as total taxa richness and dominance, were not correlated with disturbance in the predicted direction and were not included in the index. At the other end of the spectrum, measurements made at the level of the individual valve were not recorded for these data. Other studies have found the percentage of deformed valves to be a strong indicator of metal contamination (McFarland et al., 1997). Future studies should consider additional metrics in these categories that would extend the current set of metrics to other levels of biological organization.

For this study, urban sites were characterized by a higher percentage of valves belonging to salt-tolerant species while agricultural sites had more sediment-tolerant (motile) valves. Salt-tolerant species may be better able to tolerate the chloride in treated urban wastewater. An increase in sediment-tolerant diatoms may be due to the erosion associated with plowing and harvesting of crops. Pan and Stevenson (1996) were also able to distinguish between different land uses (agriculture vs. mining) on the basis of the diatom assemblage.

Although polysaprobic and eutrophic diatoms both respond to enrichment, different diatoms may distinguish between organic and inorganic sources. A slight tendency was noted for diatoms that prefer organic nutrients to be found at agricultural sites and diatoms that prefer inorganic nutrients to be found at urban sites. Urban wastewater is typically treated before it is returned to the stream while livestock excrement is not; the diatoms may detect this difference. Other studies have noted a difference between diatom assemblages associated with organic and inorganic effluents (Kelly, 1998; Rott et al., 1998; Leland and Porter, 2000).

22.5 CONCLUSIONS

Two major conclusions emerge from this study. First, diatoms are influenced by more than water chemistry alone. The focus for algal assemblages has often been their association with chemical variables such as pH, phosphorus, and nitrogen (Carrick et al., 1988; Pan et al., 1996; Winter and Duthie, 2000). In some cases, water chemistry may represent a direct measure of human disturbance, for example, phosphorus levels associated with agricultural fertilizer (McCormick and O'Dell, 1996; Pan et al., 1996). In the Mid-Atlantic and other regions, diatoms are sensitive to other changes associated with human disturbance, such as increased sedimentation due to erosion. Although a proximate cause of change may be chemical, human disturbance at the watershed scale is strongly correlated with changes in the diatom assemblage.

Second, the unique sensitivities of individual diatom species justifies a move from one-dimensional indexes based on general tolerance (Prygiel and Coste, 1993; Kelly et al., 1995) to multimetric indexes that integrate across multiple dimensions of biological organization and sensitivity. Multiple metrics capture more of the biological complexity of the diatom assemblage and allow us to distinguish between different aspects of human disturbance on the basis of diatom sensitivities to salt, nutrients, or sediment. Thus, the associations of individual diatom species with specific types of disturbance noted by other authors are not merely anecdotal but can be summarized and tested in terms of an aggregate measure (e.g., percentage relative abundance). By grouping species in terms of guilds, changes associated with human disturbance can be more simply communicated to a broader audience unfamiliar with the specific sensitivities of individual diatom taxa. Advantages to the multimetric approach include a monitoring tool that is both sensitive to different types of human disturbance and capable of diagnosing between them based on metric signatures.

One goal of the USEPA's environmental monitoring and assessment project (EMAP) is to develop a blueprint for biological assessment that integrates information across multiple biological assemblages (Hughes et al., 2000). Fish, invertebrates, and diatoms represent different trophic levels that integrate environmental conditions over different temporal and spatial scales; therefore, we expect them to be affected differently by different types of disturbances (Allen et al., 1999). Interesting questions remain regarding which assemblages are most sensitive in different geographic settings to different human activities. The diatom index will likely contribute unique insights to this process.

ACKNOWLEDGMENTS

This study would not have been possible without the help and hard work of many people. Discussions with L. Bahls, J. Stevenson, and S. Porter helped ground this analysis in meaningful biology. A. Herlihy and B. Hughes helped place this study within the larger historical context and sampling design of the EMAP project. L. Bahls, A. Herlihy, B. Hughes, P. Kaufmann, P. Larsen, T. Simon, P. Stewart, and J. Stoddard provided many helpful comments that improved the manuscript. M. Arbogast, M. Cappaert, B. Hill and B. Subramanian developed the diatom database. Y. Pan identified the diatoms. The EMAP field and laboratory technicians collected and assembled the data. W. Davis provided project support. Funding was provided by the U.S. Environmental Protection Agency under U.S. Department of Commerce's Commerce Information Technical Solutions Contract 50-CMAA-900065 with Technology Planning and Management Corporation.

REFERENCES

Allen, A.P., T.R. Whittier, D.P. Larsen, P.R. Kaufmann, R.J. O'Connor, R.M. Hughes, R.S. Stemberger, S.S. Dixit, R.O. Brinkhurst, A.T. Herlihy and S.G. Paulsen. 1999. Concordance of taxonomic composition patterns across multiple lake assemblages: effects of scale, body size, and land use, *Canadian Journal of Fisheries and Aquatic Sciences*, 56, 2029–2040.

Angermeier, P.L. and J.R. Karr. 1994. Biological integrity versus biological diversity as policy directives, *Bioscience*, 44, 690–697.

Bahls, L.L. 1993. Periphyton Bioassessment Methods for Montana Streams. Water Quality Bureau, Department of Health and Environmental Science, Helena, MT.

Barbour, M.T., J. Gerritsen, B.D. Snyder, and J.B. Stribling. 1999. Rapid Bioassessment Protocols for Use in Streams and Wadeable Rivers: Periphyton, Benthic Macroinvertebrates, and Fish. 2nd ed. EPA 841-B-99-002. U.S. Environmental Protection Agency, Office of Water, Washington, D.C.

Bryce, S.A., D.P. Larsen, R.M. Hughes, and P.R. Kaufmann. 1999. Assessing relative risks to aquatic ecosystems: A mid-Appalachian case study, *Journal of the American Water Resource Association*, 35(1), 23–36.

Carlisle, D.M. and W.H. Clements. 1999. Sensitivity and variability of metrics used in biological assessments of running waters, *Environmental Toxicology and Chemistry*, 18, 285–291.

Carpenter, K.D. and I.R. Waite. 2000. Relations of habitat-specific algal assemblages to land use and water chemistry in the Willamette basin, Oregon, *Environmental Monitoring and Assessment*, 64, 247–257.

Carrick, H.J., R.L. Lowe, and J.T. Rotenberry. 1988. Guilds of benthic algae along nutrient gradients: relationships to algal community diversity, *Journal of the North American Benthological Society*, 7, 117–128.

Charles, D.F. 1996. Use of algae for monitoring rivers in the United States: some examples, in B. A. Whitton and E. Rott (Eds.). *Use of Algae for Monitoring Rivers II*. Institut für Botanik, AG Hydrobotanik, Universität Innsbruck. 109–117.

Chessman, B., I. Growns, J. Currey, and N. Plunkett-Cole. 1999. Predicting diatom communities at the genus level for the rapid biological assessment of rivers, *Freshwater Biology*, 41, 317–331.

Cuffney, T.F., M.R. Meador, S.D. Porter, and M.E. Gurtz. 1997. *Distribution of fish, benthic invertebrate, and algal communities in relation to physical and chemical conditions, Yakima River basin, Washington, 1990*. Water Resources Investigations Report 96–4280, U.S. Geological Survey, Raleigh, N.C.

Davis, W.S., B.D. Snyder, J. B. Stribling, and C. Stoughton. 1996. Summary of state biological assessment programs for streams and rivers. EPA 230-R-96–007. Office of Policy, Planning, and Evaluation, U.S. Environmental Protection Agency, Washington, D.C.

Detenbeck, N.E., S.L. Batterman, V.J. Brady, J. C. Brazner, V.M. Snarski, D.L. Taylor, J.A. Thompson, and J. W. Arthur. 2000. A test of watershed classification systems for ecological risk assessment, *Environmental Toxicology and Chemistry*, 19, 1174–1181.

Fore, L.S. and C. Grafe. In press. Using diatoms to assess the biological condition of large rivers in Idaho (USA). *Freshwater Biology*.

Fore, L.S., J. R. Karr, and L.L. Conquest. 1994. Statistical properties of an index of biotic integrity used to evaluate water resources, *Canadian Journal of Fisheries and Aquatic Sciences*, 51, 212–231.

Fore, L.S., K. Paulsen, and K. O'Laughlin. 2001. Assessing the performance of volunteers in monitoring streams, *Freshwater Biology*, 46, 109–123.

Herlihy, A.T., D.P. Larsen, S.G. Paulsen, N.S. Urquhart, and B.J. Rosenbaum. 2000. Designing a spatially balanced, randomized site selection process for regional stream surveys: the EMAP Mid-Atlantic pilot study, *Environmental Monitoring and Assessment*, 63, 95–113.

Herlihy, A.T., J.L. Stoddard, and C.B. Johnson. 1998. The relationship between stream chemistry and watershed land cover data in the Mid-Atlantic region, *U.S. Water, Air and Soil Pollution*, 105, 377–386.

Hill, B.H., A.T. Herlihy, P.R. Kaufmann, R.J. Stevenson, F.H. McCormick, and C.B. Johnson. 2000. The use of periphyton assemblage data as an index of biotic integrity, *Journal of the North American Benthological Society*, 19, 50–67.

Hill, B.H., R.J. Stevenson, Y. Pan, A.T. Herlihy, P.R. Kaufmann, and C.B. Johnson. 2001. Comparison of correlations between environmental characteristics and stream diatom assemblages characterized at genus and species levels, *Journal of the North American Benthological Society*, 20, 299–310.

Hughes, R.M., P.R. Kaufmann, A.T. Herlihy, T.M. Kincaid, L. Reynolds, and D.P. Larsen. 1998. A process for developing and evaluating indices of fish assemblage integrity, *Canadian Journal of Fisheries and Aquatic Sciences*, 55, 1618–1631.

Hughes, R.M. and T. Oberdorff. 1999. Applications of IBI concepts and metrics to waters outside the United States and Canada, in T.P. Simon (Ed.). *Assessing the Sustainability and Biological Integrity of Water Resources Using Fish Communities*, CRC Press, Boca Raton, FL. 79–93.

Hughes, R.M., S.G. Paulsen, and J.L. Stoddard. 2000. EMAP-surface waters: a multiassemblage probability survey of ecological integrity, *Hydrobiologia*, 442/443, 429–443.

Karr, J. R. and E.W. Chu. 1999. *Restoring Life in Running Waters: Better Biological Monitoring*. Island Press, Washington, D.C..

Karr, J.R., K.D. Fausch, P.L. Angermeier, P.R. Yant, and I.J. Schlosser. 1986. *Assessment of Biological Integrity in Running Water: A Method and Its Rationale*, Illinois Natural History Survey Special Publication 5, Champaign, IL.

Kaufmann, P.R., P. Levine, E.G. Robison, C. Seeliger, and D.V. Peck. 1999. *Quantifying Physical Habitat in Wadeable Streams*. EPA/620/R-99/003. U.S. Environmental Protection Agency, Washington, D.C.

Kelly, M.G. 1998. Use of the trophic diatom index to monitor eutrophication in rivers, *Water Research*, 32(1), 236–242.

Kelly, M.G., C.J. Penny, and B.A. Whitton. 1995. Comparative performance of benthic diatom indices used to assess river water quality, *Hydrobiologia*, 302, 179–188.

Kelly, M.G. and B.A. Whitton. 1998. Biological monitoring of eutrophication in rivers, *Hydrobiologia*, 384, 55–67.

Kerans, B.L. and J.R. Karr. 1994. A benthic index of biotic integrity (B-IBI) for rivers of the Tennessee Valley, *Ecological Applications*, 4, 768–785.

Klemm, D.J., K.A. Blocksom, W.T. Thoeny, F.A. Fulk, A.T. Herlihy, P.R. Kaufmann, and S.M. Cormier. 2002. Using macroinvertebrates as indicators of ecological conditions for streams in the Mid-Atlantic Highlands Region, *Environmental Monitoring and Assessment*.

Kutka, F.J. and C. Richards 1996. Relating diatom assemblage structure to stream habitat quality, *Journal of the North American Benthological Society*, 15, 469–480.

Lange-Bertalot, H. 1979. Pollution tolerance of diatoms as a criterion for water quality estimation, *Nova Hedwigia Beiheft*, 64, 285–304.

Leland, H.V. 1995. Distribution of phytobenthos in the Yakima River basin, Washington, in relation to geology, land use, and other environmental factors, *Canadian Journal of Fisheries and Aquatic Sciences*, 52, 1108–1129.

Leland, H.V. and S.D. Porter. 2000. Distribution of benthic algae in the upper Illinois River basin in relation to geology and land use, *Freshwater Biology*, 44, 279–301.

Lowe, R.L. 1974. Environmental requirements and pollution tolerance of freshwater diatoms. Environmental Monitoring Series 670/4–74–005. U.S. Environmental Protection Agency, Washington, D.C.

McCormick, F.H., R.M. Hughes, P.R. Kaufmann, A.T. Herlihy, Peck D.V., and J.L. Stoddard. 2001. Development of an index of biotic integrity for the Mid-Atlantic Highlands region, *Transactions of the American Fisheries Society*, 130, 857–877.

McCormick, P.V. 1996. Resource competition and species coexistence in freshwater benthic algal assemblages, in R.J. Stevenson, M.L. Bothwell, and R.L. Lowe (Eds.). *Algal Ecology: Freshwater Benthic Ecosystems.* Academic Press, San Diego, CA. 229–252.

McCormick, P.V. and M.B. O'Dell. 1996. Quantifying periphyton responses to phosphorus in the Florida Everglades: a synoptic-experimental approach, *Journal of the North American Benthological Society*, 15, 450–468.

McFarland, B.H., B.H. Hill, and W.T. Willingham. 1997. Abnormal *Fragilaria* spp. (Bacillariophyceae) in streams impacted by mine drainage, *Journal of Freshwater Ecology*, 12(1), 141–150.

Medley, C.N. and W.H. Clements. 1998. Responses of diatom communities to heavy metals in streams: the influence of longitudinal variation, *Ecological Applications*, 8, 631–644.

Miller, D.L., P.M. Leonard, R.M. Hughes, J.R. Karr, P.B. Moyle, L.H. Schreder, B.A. Thompson, R.A. Daniels, K.D. Fausch, G.A. Fitzhugh, J.R. Gammon, D.B. Haliwell, P.L. Angermeier, and D.J. Orth. 1988. Regional applications of an index of biotic integrity for use in water resource management, *Fisheries*, 13(5), 12–20.

Mulholland, P.J. 1996. Role in nutrient cycling in streams, in R.J. Stevenson, M.L. Bothwell, and R.L. Lowe (Eds.). *Algal Ecology: Freshwater Benthic Ecosystems.* Academic Press, San Diego, CA. 705–739.

Omernik, J.M. 1987. Aquatic ecoregions of the conterminous United States, *Annals of the Association of American Geographers*, 77, 118–125.

Pan, Y. and R.J. Stevenson. 1996. Gradient analysis of diatom assemblages in western Kentucky wetlands, *Journal of Phycology*, 32, 222–232.

Pan, Y., R.J. Stevenson, B.H. Hill, A.T. Herlihy. 2000. Ecoregions and benthic diatom assemblages in Mid-Atlantic Highlands streams, USA, *Journal of the North American Benthological Society*, 19, 518–540.

Pan, Y., R.J. Stevenson, B.H. Hill, A.T. Herlihy, and G.B. Collins. 1996. Using diatoms as indicators of ecological conditions in lotic systems: a regional assessment, *Journal of the North American Benthological Society*, 15, 481–495.

Pan, Y., R.J. Stevenson, B.H. Hill, P.R. Kaufmann, and A.T. Herlihy. 1999. Spatial patterns and ecological determinants of benthic algal assemblages in mid-Atlantic streams, USA, *Journal of Phycology*, 35, 460–468.

Peterman, R.M. 1990. Statistical power analysis can improve fisheries research and management, *Canadian Journal of Fisheries and Aquatic Sciences*, 47, 2–15.

Peterson, C.G. 1996. Response of benthic algal communities to natural physical disturbance, in R.J. Stevenson, M.L. Bothwell, and R.L. Lowe (Eds.). *Algal Ecology: Freshwater Benthic Ecosystems.* Academic Press, San Diego, CA. 375–402.

Prygiel, J. and M. Coste. 1993. The assessment of water-quality in the Artois-Picardie Water Basin (France) by the use of diatom indices, *Hydrobiologia*, 269, 343–349.

Rathert, D., D. White, J.C. Sifneos, and R.M. Hughes. 1999. Environmental correlates of species richness for native freshwater fish in Oregon, U.S.A., *Journal of Biogeography*, 26, 257–273.

Richards, C., L.B. Johnson, and G.E. Host. 1996. Landscape-scale influences on stream habitats and biota, *Canadian Journal of Fisheries and Aquatic Sciences*, 53(Supplement 1), 195–311.

Rosen, B.H. 1995. Use of periphyton in the development of biocriteria, in W.S. Davis and T.P. Simon (Eds.). *Biological Assessment and Criteria: Tools for Water Resource Planning and Decision Making.* Lewis Publishers, Boca Raton, FL. 209–215.

Rott, E., H.C. Duthie, and E. Pipp. 1998. Monitoring organic pollution and eutrophication in the Grand River, Ontario, by means of diatoms, *Canadian Journal of Fisheries and Aquatic Sciences*, 55, 1443–1453.

Round, F.E., R.M. Crawford, and D.G. Mann. 1990. *The Diatoms: Biology and Morphology of the Genera.* Cambridge University Press, Cambridge, U.K.

Stevenson, R.J. and L. Bahls. 1999. Periphyton protocols, in M.T. Barbour, J. Gerritsen, B.D. Snyder and J.B. Stribling (Eds.). *Rapid Bioassessment Protocols for Use in Streams and Wadeable Rivers: Periphyton, Benthic Macroinvertebrates, and Fish*, 2nd ed. EPA 841-B-99–002. U.S. Environmental Protection Agency, Office of Water, Washington, D.C.

Stevenson, R.J. and Y. Pan. 1999. Assessing environmental conditions in rivers and streams with diatoms, in E.F. Stoermer and J.P. Smol (Eds.). *The Diatoms: Applications for the Environmental and Earth Sciences.* Cambridge University Press, Cambridge, U.K. 11–40.

Stewart, P.M. 1995. Use of algae in aquatic pollution assessment, *Natural Areas Journal*, 15, 234–239.

Stewart, P.M., J.T. Butcher, and P.J. Gerovac. 1999. Diatom (Bacillariophyta) community response to water quality and land use, *Natural Areas Journal*, 19(2), 155–165.

Tabachnik, B.G. and L.S. Fidell. 1989. *Using Multivariate Statistics,* 2nd ed. HarperCollins Publishers, New York.

Tuchman, N.C. 1996. The role of heterotrophy in algae, in R.J. Stevenson, M.L. Bothwell, and R.L. Lowe (Eds.). *Algal Ecology: Freshwater Benthic Ecosystems.* Academic Press, San Diego, CA. 299–320.

Van Dam, H., A. Mertens, and J. Sinkeldam. 1994. A coded checklist and ecological indicator values of freshwater diatoms from The Netherlands, *Netherlands Journal of Aquatic Ecology*, 28(1), 117–133.

Winter, J.G. and H.C. Duthie, 2000. Epilithic diatoms as indicators of stream total N and total P concentration, *Journal of the North American Benthological Society*, 19, 32–49.

Whitton, B.A. and M.G. Kelly. 1995. Use of algae and other plants for monitoring rivers, *Australian Journal of Ecology*, 20, 45–56.

Yoder, C.O. and E.T. Rankin. 1995. Biological response signatures and the area of degradation values: new tools for interpreting multimetric data, in W.S. Davis and T.P. Simon (Eds.). *Biological Assessment and Criteria: Tools for Water Resource Planning and Decision Making.* Lewis Publishers, Boca Raton, FL. 263–286.

Zar, J.H. 1984. *Biostatistical Analysis*, 2nd ed. Prentice-Hall, Englewood Cliffs, N.J.

APPENDIX I
Natural History Attributes Used to Calculate Diatom Metrics for Selected Species

Taxon	Tolerant	Salt	Oxygen	Trophic	Nitrogen	Saprobity	pH	VD	Motility
Achnanthes biasolettiana							Alka.		
Achnanthes clevei							Alka.		
Achnanthes exigua							Alka.		
Achnanthes exilis							Alka.		
Achnanthes helvetica							Alka.		
Achnanthes hungarica			Low	Eutrophic			Alka.		
Achnanthes lanceolata				Eutrophic			Alka.		
Achnanthes lanceolata v. *robusta*							Alka.		
Achnanthes lanceolata v. *rostrata*				Eutrophic			Alka.		
Achnanthes minutissima v. *affinis*							Alka.		
Achnanthes minutissima v. *gracillima*							Alka.		
Achnanthes rosenstockii							Alka.		
Achnanthes thermalis		Tolerant							
Actinocyclus normanii		Tolerant		Eutrophic			Alka.		
Amphipleura pellucida						Polysaprobic	Alka.		
Amphora commutata		Tolerant		Eutrophic				NL	
Amphora montana				Eutrophic			Alka.		
Amphora ovalis				Eutrophic			Alka.		
Amphora pediculus				Eutrophic			Alka.		
Amphora veneta		Tolerant		Eutrophic		Polysaprobic	Alka.		
Anomoeoneis vitrea							Alka.		
Asterionella formosa							Alka.		
Aulacoseira ambigua				Eutrophic			Alka.		
Aulacoseira granulata				Eutrophic			Alka.		
Bacillaria paradoxa		Tolerant	Low	Eutrophic			Alka.		
Caloneis amphisbaena		Tolerant		Eutrophic			Alka.		
Caloneis bacillum							Alka.		
Caloneis schumanniana							Alka.		
Campylodiscus hibernicus				Eutrophic			Alka.		
Cocconeis disculus		Tolerant						NL	

Cocconeis pediculus		Tolerant		Eutrophic			Alka.		
Cocconeis placentula				Eutrophic			Alka.		
Cocconeis placentula v. *euglypta*				Eutrophic			Alka.		
Cocconeis placentula v. *lineata*				Eutrophic			Alka.		
Cocconeis scutellum							Alka.		
Cyclotella atomus		Tolerant		Eutrophic			Alka.		
Cyclotella distinguanda							Alka.		
Cyclotella meneghiniana		Tolerant	Low	Eutrophic	Heterotrophic	Polysaprobic	Alka.		
Cyclotella pseudostelligera				Eutrophic					
Cymatopleura solea				Eutrophic			Alka.		High
Cymatopleura solea v. *apiculata*								NL	High
Cymbella affinis				Eutrophic			Alka.		
Cymbella aspera							Alka.		
Cymbella brehmii							Alka.		
Cymbella cistula				Eutrophic			Alka.		
Cymbella ehrenbergii							Alka.		
Cymbella helvetica							Alka.		
Cymbella hustedtii Krasske							Alka.		
Cymbella mesiana							Alka.		
Cymbella microcephala							Alka.		
Cymbella naviculiformis				Eutrophic					
Cymbella prostrata				Eutrophic			Alka.		
Cymbella pusilla	Very							NL	
Cymbella tumida							Alka.		
Cymbella tumidula							Alka.		
Diatoma hyemalis							Alka.		
Diatoma tenue		Tolerant		Eutrophic			Alka.		
Diatoma vulgare							Alka.		
Diploneis elliptica							Alka.		
Diploneis oblongella							Alka.		
Diploneis puella							Alka.		
Diploneis smithii v. *dilatata*							Alka.		
Entomoneis ornata	Very								
Entomoneis paludosa (W. Smith) Reimer		Tolerant							
Epithemia adnata							Alka.		
Epithemia sorex				Eutrophic			Alka.		

APPENDIX I (CONTINUED)
Natural History Attributes Used to Calculate Diatom Metrics for Selected Species

Taxon	Tolerant	Salt	Oxygen	Trophic	Nitrogen	Saprobity	pH	VD	Motility
Epithemia turgida							Alka.		
Fragilaria arcus							Alka.		
Fragilaria bidens				Eutrophic			Alka.		
Fragilaria brevistriata							Alka.		
Fragilaria capucina v. *austriaca*							Alka.		
Fragilaria capucina v. *mesolepta*							Alka.		
Fragilaria capucina v. *vaucheriae*				Eutrophic			Alka.		
Fragilaria construens							Alka.		
Fragilaria construens v. *binodis*							Alka.		
Fragilaria construens v. *venter*							Alka.		
Fragilaria crotonensis							Alka.		
Fragilaria elliptica							Alka.		
Fragilaria fasciculata		Tolerant		Eutrophic			Alka.		
Fragilaria lapponica							Alka.		
Fragilaria leptostauron							Alka.		
Fragilaria parasitica							Alka.		
Fragilaria parasitica v. *subconstricta*							Alka.		
Fragilaria pinnata							Alka.		
Fragilaria vaucheriae				Eutrophic			Alka.		
Frustulia vulgaris							Alka.		
Gomphonema acuminatum				Eutrophic			Alka.		
Gomphonema affine							Alka.		
Gomphonema angustatum							Alka.		
Gomphonema angustum							Alka.		
Gomphonema auger							Alka.		
Gomphonema dichotomum							Alka.		
Gomphonema micropus				Eutrophic			Alka.		
Gomphonema minutum				Eutrophic					
Gomphonema olivaceum				Eutrophic			Alka.		
Gomphonema parvulum	Very		Low	Eutrophic	Heterotrophic	Polysaprobic			

Gomphonema parvulum v. *saprophilum*				Eutrophic		Polysaprobic			
Gomphonema psuedoauger				Eutrophic					
Gomphonema sarcophagus							Alka.		
Gomphonema truncatum							Alka.		
Gomphonema vibrio							Alka.		
Gyrosigma acuminatum				Eutrophic			Alka.		High
Gyrosigma attenuatum				Eutrophic			Alka.		High
Gyrosigma exile								NL	High
Gyrosigma nodiferum								NL	High
Gyrosigma obtusatum				Eutrophic			Alka.		High
Gyrosigma scalproides								NL	High
Gyrosigma spencerii								NL	High
Gyrosigma strigilis								NL	High
Hantzschia amphioxys									High
Mastogloia smithii		Tolerant					Alka.		
Melosira lineata		Tolerant		Eutrophic			Alka.		
Melosira varians				Eutrophic	Heterotrophic		Alka.		
Meridion circulare							Alka.		
Meridion circulare v. *constrictum*							Alka.		
Navicula accomoda	Very		Low	Eutrophic	Heterotrophic	Polysaprobic	Alka.		
Navicula agrestis	Very								
Navicula atomus	Very			Eutrophic	Heterotrophic	Polysaprobic	Alka.		
Navicula bacillum							Alka.		
Navicula capitata							Alka.		
Navicula capitata v. *lunebergensis*							Alka.		
Navicula capitatoradiata				Eutrophic			Alka.		
Navicula cincta				Eutrophic			Alka.		
Navicula clementis		Tolerant					Alka.		
Navicula cohnii		Tolerant		Eutrophic			Alka.		
Navicula confervacea		Tolerant		Eutrophic	Heterotrophic				
Navicula contenta							Alka.		
Navicula cryptotenella							Alka.		
Navicula cuspidata				Eutrophic			Alka.		
Navicula decussis							Alka.		
Navicula digitoradiata		Tolerant					Alka.		
Navicula elginensis				Eutrophic			Alka.		

APPENDIX I (CONTINUED)
Natural History Attributes Used to Calculate Diatom Metrics for Selected Species

Taxon	Tolerant	Salt	Oxygen	Trophic	Nitrogen	Saprobity	pH	VD	Motility
Navicula elginensis v. *cuneata*				Eutrophic			Alka.		
Navicula erifuga		Tolerant		Eutrophic			Alka.		
Navicula gastrum				Eutrophic			Alka.		
Navicula goeppertiana			Low	Eutrophic	Heterotrophic	Polysaprobic	Alka.		
Navicula gregaria		Tolerant	Low	Eutrophic			Alka.		
Navicula halophila		Tolerant		Eutrophic			Alka.		
Navicula halophilioides	Very							NL	
Navicula integra		Tolerant		Eutrophic					
Navicula lanceolata		Tolerant		Eutrophic			Alka.		
Navicula lenzii							Alka.		
Navicula menisculus				Eutrophic			Alka.		
Navicula menisculus v. *upsaliensis*							Alka.		
Navicula meniscus		Tolerant		Eutrophic			Alka.		
Navicula minima	Very		Low	Eutrophic	Heterotrophic	Polysaprobic	Alka.		
Navicula minunscula v. *muralis*				Eutrophic	Heterotrophic	Polysaprobic	Alka.		
Navicula minusculoides			Low	Eutrophic	Heterotrophic	Polysaprobic		NL	
Navicula modica							Alka.		
Navicula molestiformis v. *Hust.*			Low	Eutrophic	Heterotrophic	Polysaprobic	Alka.		
Navicula monoculata	Very			Eutrophic	Heterotrophic		Alka.		
Navicula mutica		Tolerant		Eutrophic					
Navicula pelliculosa	Very						Alka.		
Navicula placentula				Eutrophic			Alka.		
Navicula protracta		Tolerant		Eutrophic					
Navicula pseudoscutiformis							Alka.		
Navicula pseudoventralis							Alka.		
Navicula pupula v. *capitata*							Alka.		
Navicula pupula v. *elliptica*							Alka.		
Navicula pygmaea		Tolerant		Eutrophic	Heterotrophic		Alka.		
Navicula recens		Tolerant		Eutrophic			Alka.		
Navicula reinhardtii							Alka.		

Navicula rhynchocephala			Low				Alka.		
Navicula riparia				Eutrophic		Polysaprobic			
Navicula salinarum		Tolerant		Eutrophic					
Navicula saprophila			Low	Eutrophic	Heterotrophic	Polysaprobic			
Navicula schoenfeldii							Alka.		
Navicula schroeterii		Tolerant		Eutrophic			Alka.		
Navicula seminulum	Very		Low	Eutrophic	Heterotrophic	Polysaprobic			
Navicula stroemii							Alka.		
Navicula subrotundata							Alka.		
Navicula tenelloides	Very			Eutrophic			Alka.		
Navicula tenera	Very						Alka.		
Navicula tripunctata				Eutrophic			Alka.		
Navicula tripunctata v. *schizonemoides*		Tolerant		Eutrophic			Alka		
Navicula trivialis		Tolerant		Eutrophic			Alka.		
Navicula veneta	Very	Tolerant	Low	Eutrophic		Polysaprobic	Alka.		
Navicula viridula				Eutrophic			Alka.		
Navicula viridula v. *rostellata*				Eutrophic			Alka.		
Navicula vitabunda							Alka.		
Navicula vulpina							Alka.		
Nitzschia acicularis			Low	Eutrophic	Heterotrophic		Alka.		High
Nitzschia acidoclinata									High
Nitzschia agnita	Very	Tolerant						NL	High
Nitzschia amphibia				Eutrophic	Heterotrophic		Alka.		High
Nitzschia amphibia f frauenfeldii								NL	High
Nitzschia angustata									High
Nitzschia apiculata								NL	High
Nitzschia archibaldii				Eutrophic					High
Nitzschia aurariae	Very	Tolerant	Low		Heterotrophic		Alka.		High
Nitzschia bacillum							Alka.		High
Nitzschia bremensis							Alka.		High
Nitzschia brevissima		Tolerant		Eutrophic					High
Nitzschia calida		Tolerant		Eutrophic				NL	High
Nitzschia capitellata		Tolerant		Eutrophic		Polysaprobic	Alka.		High
Nitzschia clausii		Tolerant		Eutrophic			Alka.		High
Nitzschia communis	Very			Eutrophic	Heterotrophic	Polysaprobic	Alka.		High
Nitzschia commutata		Tolerant						NL	High
Nitzschia constricta		Tolerant		Eutrophic			Alka.		High
Nitzschia debilis							Alka.		High

APPENDIX I (CONTINUED)
Natural History Attributes Used to Calculate Diatom Metrics for Selected Species

Taxon	Tolerant	Salt	Oxygen	Trophic	Nitrogen	Saprobity	pH	VD	Motility
Nitzschia denticula								NL	High
Nitzschia dippelii								NL	High
Nitzschia dissipata							Alka.		High
Nitzschia dissipata v. *media*							Alka.		High
Nitzschia dubia		Tolerant		Eutrophic					High
Nitzschia fasciculata		Tolerant						NL	High
Nitzschia filiformis		Tolerant		Eutrophic	Heterotrophic		Alka.		High
Nitzschia filiformis v. *conferta*								NL	High
Nitzschia flexa									High
Nitzschia flexoides								NL	High
Nitzschia fonticola							Alka.		High
Nitzschia frustulum		Tolerant		Eutrophic	Heterotrophic		Alka.		High
Nitzschia frustulum v. *perpusilla*								NL	High
Nitzschia gessneri								NL	High
Nitzschia gracilis									High
Nitzschia hantzschiana									High
Nitzschia heufleriana							Alka.		High
Nitzschia homburgiensis									High
Nitzschia hungarica		Tolerant	Low	Eutrophic			Alka.		High
Nitzschia ignata								NL	High
Nitzschia incognita		Tolerant						NL	High
Nitzschia inconspicua		Tolerant		Eutrophic	Heterotrophic		Alka.		High
Nitzschia intermedia				Eutrophic					High
Nitzschia lanceolata		Tolerant		Eutrophic			Alka.		High
Nitzschia levidensis		Tolerant		Eutrophic			Alka.		High
Nitzschia levidensis v. *salinarum*								NL	High
Nitzschia levidensis v. *victoriae*		Tolerant		Eutrophic			Alka.		High
Nitzschia linearis							Alka.		High
Nitzschia linearis v. *subtilis*								NL	High
Nitzschia linearis v. *tenuis*									High
Nitzschia littoralis		Tolerant		Eutrophic			Alka.		High
Nitzschia lorenziana		Tolerant						NL	High

Nitzschia microcephala	Very			Eutrophic	Heterotrophic		Alka.		High
Nitzschia nana									High
Nitzschia nodosa								NL	High
Nitzschia normannii								NL	High
Nitzschia obtusa	Very							NL	High
Nitzschia palea	Very		Low	Eutrophic	Heterotrophic	Polysaprobic			High
Nitzschia palea v. *debiis*									High
Nitzschia palea v. *tenuirostris*									High
Nitzschia paleacea				Eutrophic	Heterotrophic		Alka.		High
Nitzschia paleaformis									High
Nitzschia pellucida								NL	High
Nitzschia pellucida								NL	High
Nitzschia perminuta							Alka.		High
Nitzschia perspicua	Very							NL	High
Nitzschia pumila								NL	High
Nitzschia pura								NL	High
Nitzschia pusilla	Very								High
Nitzschia recta							Alka.		High
Nitzschia reversa								NL	High
Nitzschia rosenstockii								NL	High
Nitzschia scalpelliformis	Very							NL	High
Nitzschia sigma		Tolerant		Eutrophic			Alka.		High
Nitzschia sigmoidea				Eutrophic			Alka.		High
Nitzschia sinuata v. *tabellaria*							Alka.		High
Nitzschia sociabilis				Eutrophic					High
Nitzschia solita		Tolerant		Eutrophic				NL	High
Nitzschia subacicularis							Alka.		High
Nitzschia subtilioides								NL	High
Nitzschia terrestris									High
Nitzschia tropica								NL	High
Nitzschia trybionella v. *victoriae*								NL	High
Nitzschia tubicola	Very							NL	High
Nitzschia umbonata	Very		Low	Eutrophic	Heterotrophic	Polysaprobic			High
Nitzschia vermicularis							Alka.		High
Nitzschia wuellerstorffii								NL	High
Pleurosigma salinarum		Tolerant					Alka.		
Pleurosira laevis		Tolerant		Eutrophic			Alka.		
Rhoicosphenia curvata				Eutrophic			Alka.		

APPENDIX I (CONTINUED)
Natural History Attributes Used to Calculate Diatom Metrics for Selected Species

Taxon	Tolerant	Salt	Oxygen	Trophic	Nitrogen	Saprobity	pH	VD	Motility
Rhopalodia gibberula		Tolerant					Alka.		
Stauroneis producta		Tolerant					Alka.		
Stauroneis smithii							Alka.		
Stenopterobia curvula									High
Stenopterobia delicatissma									High
Stenopterobia densestriata									High
Stephanodiscus hantzschii			Low	Eutrophic	Heterotrophic	Polysaprobic	Alka.		
Stephanodiscus minutus				Eutrophic			Alka.		
Stephanodiscus parvus				Eutrophic			Alka.		
Surirella amphioxis		Tolerant		Eutrophic			Alka.		High
Surirella angusta	Very			Eutrophic			Alka.		High
Surirella biseriata				Eutrophic			Alka.		High
Surirella bohemica								NL	High
Surirella Brébissonii		Tolerant					Alka.		High
Surirella Brébissonii v. *kuetzingii*				Eutrophic			Alka.		High
Surirella elegans									High
Surirella lineraris v. *helvetica*									High
Surirella minuta				Eutrophic			Alka.		High
Surirella ovalis		Tolerant	Low	Eutrophic			Alka.		High
Surirella roba									High
Surirella robusta									High
Surirella splendida							Alka.		High
Surirella spp.								NL	High
Surirella subsalsa		Tolerant						NL	High
Surirella tenera				Eutrophic			Alka.		High
Synedra parasitica							Alka.		
Thalassiosira weissflogii		Tolerant		Eutrophic	Heterotrophic		Alka.		

Note: Hundreds more species without specific attributes are not shown. Diatom species were described as very tolerant by Bahls (1993). Diatom species tolerant of salt, tolerant of low oxygen, eutrophic, nitrogen heterotrophic, polysaprobic, and alkaliphilic were described by Van Dam et al. (1994). Species not listed by Van Dam et al. are cited as "NL" in the VD column. Highly motile genera were defined by R.J. Stevenson (Michigan State University, June 2000).

23 Response Patterns of Great River Fish Assemblage Metrics to Outfall Effects from Point Source Discharges

Erich B. Emery, Frank H. McCormick, and Thomas P. Simon

CONTENTS

23.1 INTRODUCTION

Human disturbance alters key attributes of aquatic ecosystems such as water quality, habitat structure, hydrological regime, energy flow, and biological interactions (Karr and Dudley, 1981; Ward and Stanford, 1989; Sparks, 1995). In great rivers, this is particularly evident because they are disproportionately degraded (Karr et al., 1985a; Simon and Sanders, 1999; Gammon and Simon, 2000) by habitat alteration (Ward and Stanford, 1995; Poff et al., 1997) and industrial and municipal discharges (Pearson and Krumholz, 1984; Simon and Stahl, 1998). Water quality degradation as a result of point and nonpoint source pollution further impacts the ecological integrity of large rivers such as the Ohio River (Sparks et al., 1990; Bayley, 1995). By examining patterns in the responses of fish assemblages to potential stressors associated with point source discharges, it may be possible

0-8493-0905-0/03/$0.00+$1.50

to assess the extent to which pollution alters water quality and affects biotic integrity (Karr and Dudley, 1981; Bayley, 1995; Yoder and Rankin, 1995a).

The index of biological integrity (IBI) assesses the conditions of water bodies by direct evaluation of biological attributes (Karr, 1981; Karr et al., 1986). It integrates structural, ecological, trophic, and reproductive attributes of fish assemblages at multiple levels of organization (Fausch et al., 1990). The IBI was originally developed for assessment of Midwestern warmwater streams and was modified for use in other regions and waters (Miller et al., 1988; Simon, 1992; Simon and Lyons, 1995; Hughes and Oberdorff, 1999; Simon and Stahl, 1998), including the upper Ohio River basin (Simon and Emery, 1995; Emery et al., 1999; Simon and Sanders, 1999; Emery et al., in review).

Emery and Thomas (Chapter 9, this volume) found that point source effects on biological communities of the Ohio River are limited to the immediate influence of the outfall. Typically, studies of the impacts of point source discharges to aquatic ecosystems have been limited to comparisons of the impacted area to an upstream, unimpaired reference condition. They described an approach of incrementally sampling outfalls that was intended to detect gradients of fish assemblage responses to effluents. This traveling zone (T-zone) approach was based on the computation of an IBI based on fish assemblage metrics from ten continuous 100-m segments. Data can be aggregated and metrics calculated to show incremental changes in response to the effects from point source discharges. These metrics can be evaluated individually or combined to form a multimetric index of biological integrity for the Ohio River. The purpose of this paper is to compare the responses of select metrics to three types of industrial and municipal wastewater discharges using data collected by the T-zone approach.

23.2 METHODS

23.2.1 Study Area

The Ohio River begins at the confluence of the Monongahela and Allegheny Rivers at Pittsburgh, PA (Rkm 0) and flows southwesterly to the confluence with the Mississippi River near Cairo, IL (1578.4 km) (Figure 23.1). The Ohio River crosses four ecoregions: the Western Allegheny Plateau, Interior Plateau, Interior River Lowland and Mississippi Alluvial Plain (Omernik, 1987). Nearly 10% of the nation's population (more than 25 million people) reside in the Ohio River basin. Over 600 permitted discharges reach its waters from industries, power generating facilities, and municipalities. Twenty navigational dams provide a 2.75-m minimum depth for commercial navigation that transports approximately 250 million tons of cargo annually.

23.2.2 Sample Collection and Comparison of Outfall and Control Sites

Field collections were conducted at 11 outfall sites in 1999 by night electrofishing from boats starting in early July through late October when the river is at stable low to moderate flow. We selected large point source discharges with effluent plumes discharged at or near the surface. The discharge locations were relatively unaffected by other anthropogenic disturbance. We measured habitat characteristics at each outfall site and selected upstream control sites with similar conditions. Control sites were upstream of the discharge plume, were least disturbed by human activities, and possessed similar habitat type and composition. Night electrofishing was conducted at outfall and control locations along a continuous 1000 m of shoreline, using the T-zone approach described by Emery and Thomas (Chapter 9, this volume). This approach provides a spatial resolution (at 100-m increments) of the responses of fish assemblages to discharges and the equivalent of two contiguous 500-m electrofishing zones. Each outfall location was divided into ten 100-m samples, which provided six 500-m T-zones (Table 23.1).

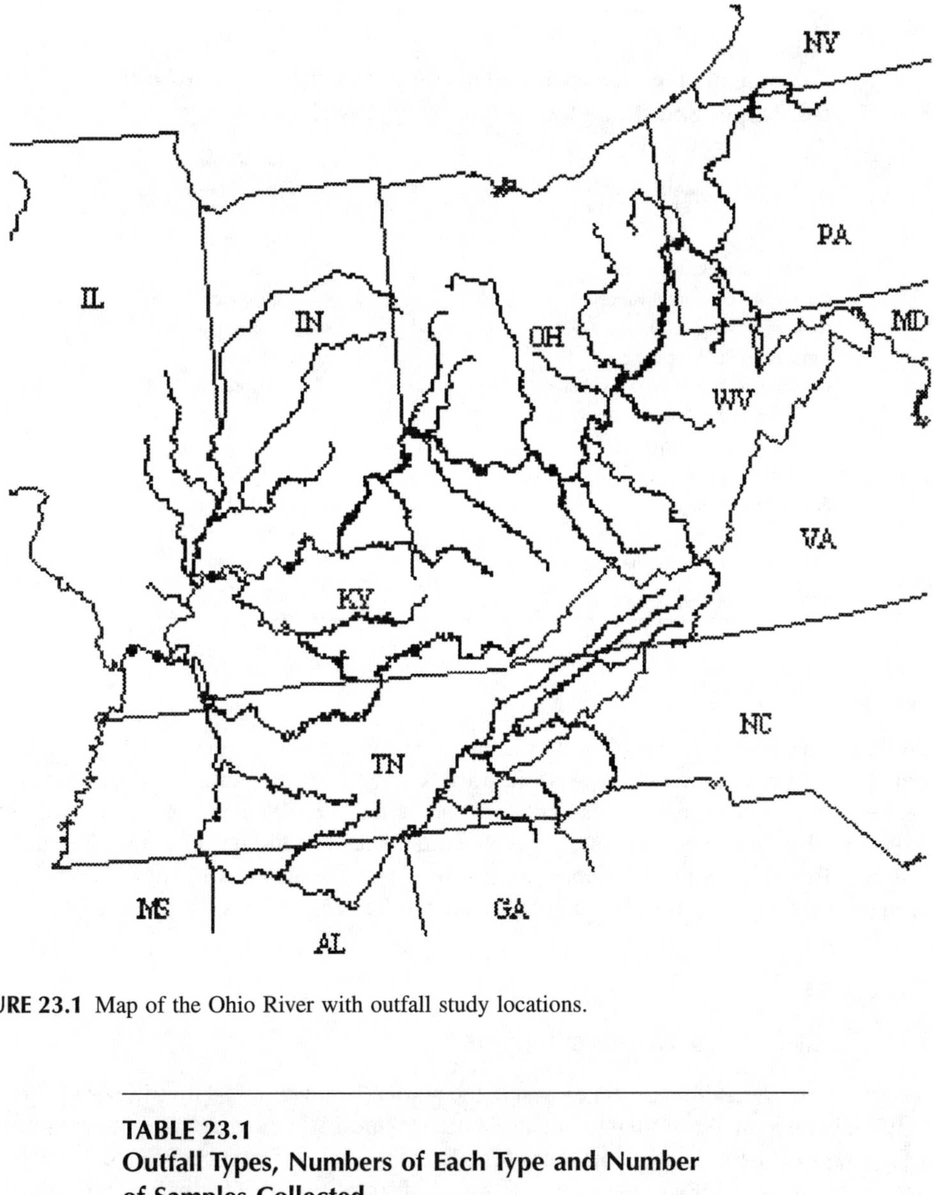

FIGURE 23.1 Map of the Ohio River with outfall study locations.

TABLE 23.1
Outfall Types, Numbers of Each Type and Number of Samples Collected

Outfall Type	Number of Sites	Number of Events
Chemical	4	12
Thermal	4	12
Wastewater	3	9
Controls	11	31

Note: Control sites were paired with outfalls at each outfall location. Two control sites could not be sampled during one of the rounds of sampling.

TABLE 23.2
Fish Assemblage Metrics Tested for Responsiveness to Point Source Discharges and Predicted Responses to Disturbance

Metric	Expected Response to Disturbance
Native species richness	Decrease
Number of sucker species	Decrease
Number of centrarchid species	Decrease
Number of great river species	Decrease
Number of intolerant species	Decrease
Percentage of tolerant individuals	Increase
Percentage of simple lithophils	Decrease
Percentage of invertivores	Decrease
Percentage of detritivores	Increase
Percentage of piscivores	Decrease
Percentage of nonindigenous species	Increase
Number of DELT anomalies	Increase
CPUE	Decrease

23.2.3 Data Analysis

We evaluated 13 metrics for each T-zone for their responses to disturbance (Table 23.2). We used the ANOVA GLM procedure in SAS (SAS Institute, Cary, NC) to calculate the least squares means differences of metric and index scores of control and outfall sites. We compared control and outfall sites (standard 500-m zones) and control versus outfall sites for differences among T-zones. Only probabilities associated with preplanned comparisons were used. Plots of mean scores for each metric in each T-zone were used to graphically depict differences between control and outfall sites.

23.3 RESULTS

23.3.1 Gradient Patterns among T-Zones

Nine metrics show initial impairment closest to the outfall (i.e., within 700 m) followed by recovery to levels approximating those found at control sites (Figure 23.2). Four metrics (number of native species, number of sucker species, number of intolerant species, and number of deformities, eroded fins, lesions and tumors (DELT anomalies)) showed little or no response to discharges. Eight metrics exhibited u-shaped responses, indicating immediate responses at the discharge points with higher expectations immediately falling off and then recovering over the remainder of the zone.

23.3.2 Differentiating Control Condition and Outfall Effects

We found significant ($p < 0.05$) differences between control and outfall sites for 8 of the 13 metrics (Table 23.3). The metric scores were significantly higher ($p < 0.0001$) at control sites. Two metrics did not respond as predicted (Table 23.3). The percent individuals as nonindigenous species and as tolerant species were greater at control sites with higher quality habitats. When comparing the first 500-m sample reach immediately below an outfall to the second contiguous 500 m at the same location, no significant differences were observed for any of the metrics.

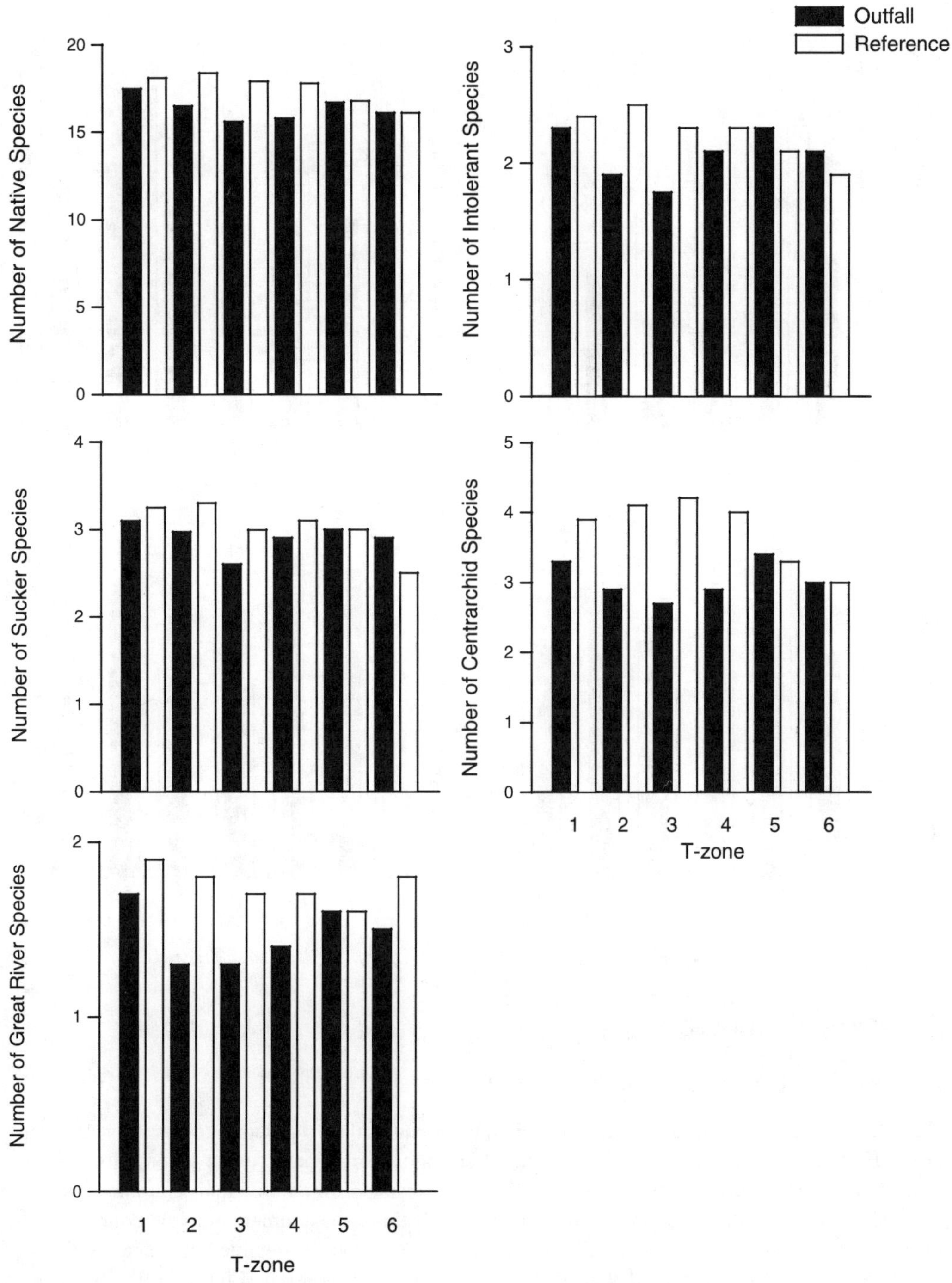

FIGURE 23.2 Response of ORFIn metric and index values by traveling zone.

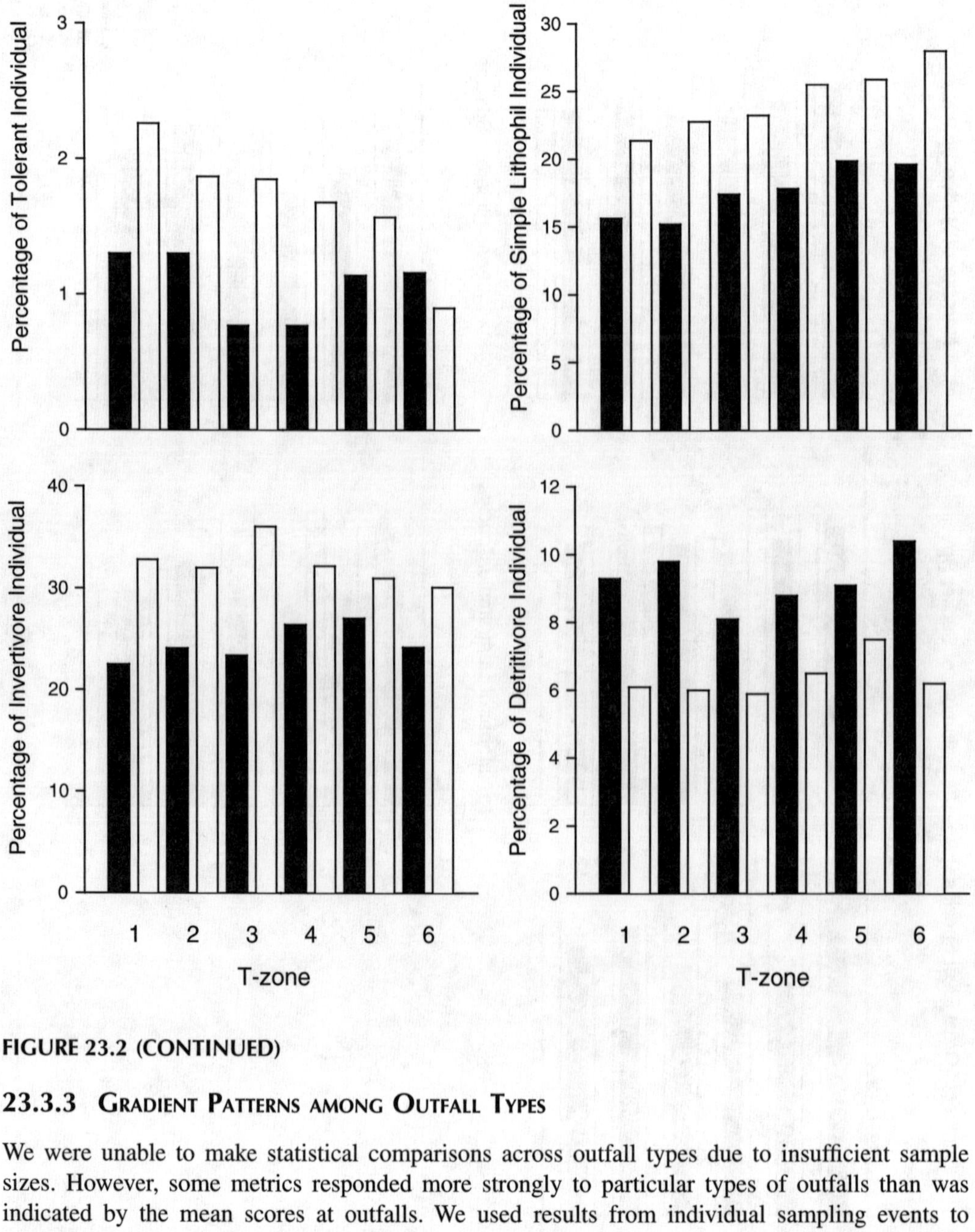

FIGURE 23.2 (CONTINUED)

23.3.3 Gradient Patterns among Outfall Types

We were unable to make statistical comparisons across outfall types due to insufficient sample sizes. However, some metrics responded more strongly to particular types of outfalls than was indicated by the mean scores at outfalls. We used results from individual sampling events to graphically display these response signatures and distinguish summer and fall collections. At chemical outfall sites, the percent individuals as invertivore species declined between the first and second transects, then increased with distance away from the outfall (Figure 23.3) — a pattern reflected in both summer and fall samples. The percentage of individuals as lithophilous spawning species did not recover until the most downstream zone, and were absent during the summer sampling period. At sites with thermal discharges, the number of species and catch-per-unit-of-effort (CPUE) increased with distance from outfalls (Figure 23.3) in both the summer and fall samples; summer expectations were much lower. Sites affected by wastewater effluent showed no change in values for CPUE or percent individuals as tolerant species or a decrease with distance from outfall (Figure 23.3).

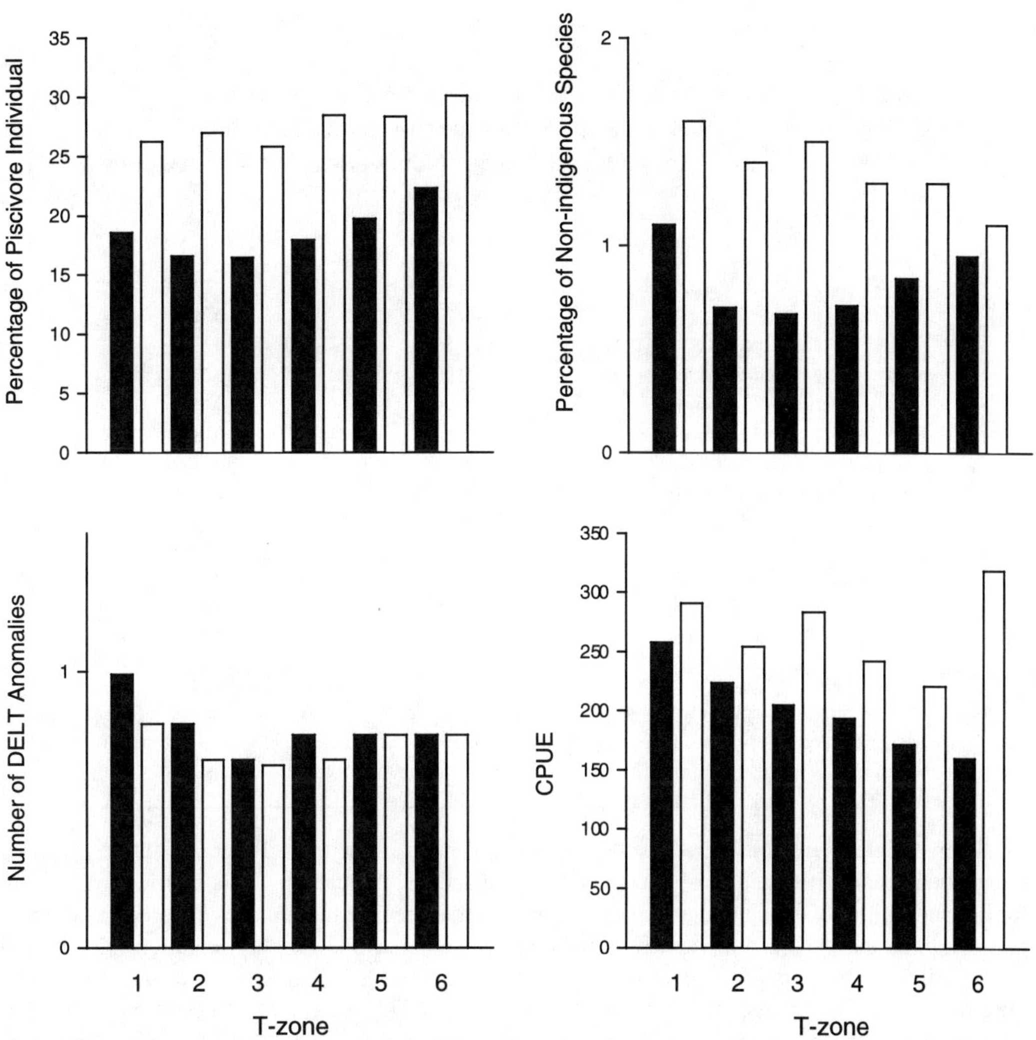

FIGURE 23.2 (CONTINUED)

Habitat quality in the vicinity of outfalls has an effect on biological integrity. For example, habitat quality may mitigate the deleterious effects of outfalls. Habitats with coarse (good) substrates tend to have greater diversity, even at outfalls, than shallow (poor) habitats with fine substrates (Figure 23.4).

23.4 DISCUSSION

23.4.1 Differentiating Control Condition and Outfall Effects

Outfalls on the Ohio River have definable effects on fish communities. Eight of the 13 metrics detect significant differences between control sites and outfall sites. Some metrics did not respond as expected for several reasons. The nonindigenous species metric was intended to track influence of invasive aquatic species on fish assemblages in the Ohio River, not track pollution. Less than 100 years ago, the common carp was the only species considered an exotic or nonnative species (Fuller et al., 1999).

TABLE 23.3
Least Square Mean Probability Values Based on Comparisons of Control and Outfall Data for Point Source Outfalls Sampled in 1999 Using the T-Zone Design and for Nonoverlapping 500-m Zones

Metric	Control vs. Outfall	Upper vs. Lower 500-m Outfall
Native species richness	0.01	0.30
Number of sucker species	0.36	0.55
Number of centrarchid species	0.007	0.52
Number of great river species	0.006	0.51
Number of intolerant species	0.42	0.64
Percentage of tolerant individuals	**0.0001**	0.76
Percentage of simple lithophils	0.0001	0.33
Percentage of invertivores	0.004	0.79
Percentage of detritivores	0.01	0.70
Percentage of piscivores	0.0001	0.37
Percentage of nonindigenous species	**0.001**	0.68
Number of DELT anomalies	0.62	0.42
CPUE	0.008	0.10

Note: Bold values indicate metric response values contrary to our *a priori* predictions (i.e., higher at control sites). Upper vs. lower outfall comparisons represent nonoverlapping 500-m zones at outfall sites.

Twelve species are considered exotic or nonindigenous in the Ohio River. As exotic and nonindigenous species increase at a site, the biological integrity decreases. This metric is important for measuring improvements in the conservation of native species. Fish species classified as tolerant that comprise the percent individuals as tolerant species metric are highly pollution-tolerant and reflect water quality conditions that prevailed before the 1980s. As conditions in the Ohio River improved after the passage of the Clean Water Act of 1972, Emery et al. (1999) reported that tolerant species have become increasingly scarce as impacts are more localized. The number of sucker species, number of intolerant species, and number of deformities, eroded fins, lesions, and tumor (DELT) anomalies metrics all responded predictably although not significantly so. Not all of the metrics in a multimetric index must respond at the same time in order to distinguish impacted and nonimpacted conditions. The IBI was developed to respond to a number of environmental disturbances. Point source impacts are only one of the many types of perturbations to which the index responds.

23.4.2 Gradient Patterns among T-Zones

The T-zone approach detected gradients at the outfalls that were not evident in the two sequential (upper and lower) 500-m zones. Most metrics showed distinct differences between control and outfall sites, even among sequential T-zones. The ability to detect a response gradient and indicate community recovery is essential to establishing cause-and-effect relationships, recommending future actions, and monitoring the success of pollution reduction efforts.

Most of the metrics indicate a community response within the first 700 m. This is a relatively short distance for community recovery compared to the areas of impact seen in smaller streams and rivers (Karr et al., 1985b; Simon, 1992; Simon et al., Chapter 22; Dufour et al., Chapter 24). Some metrics recover more quickly than others and some show little or no response. Seven of the

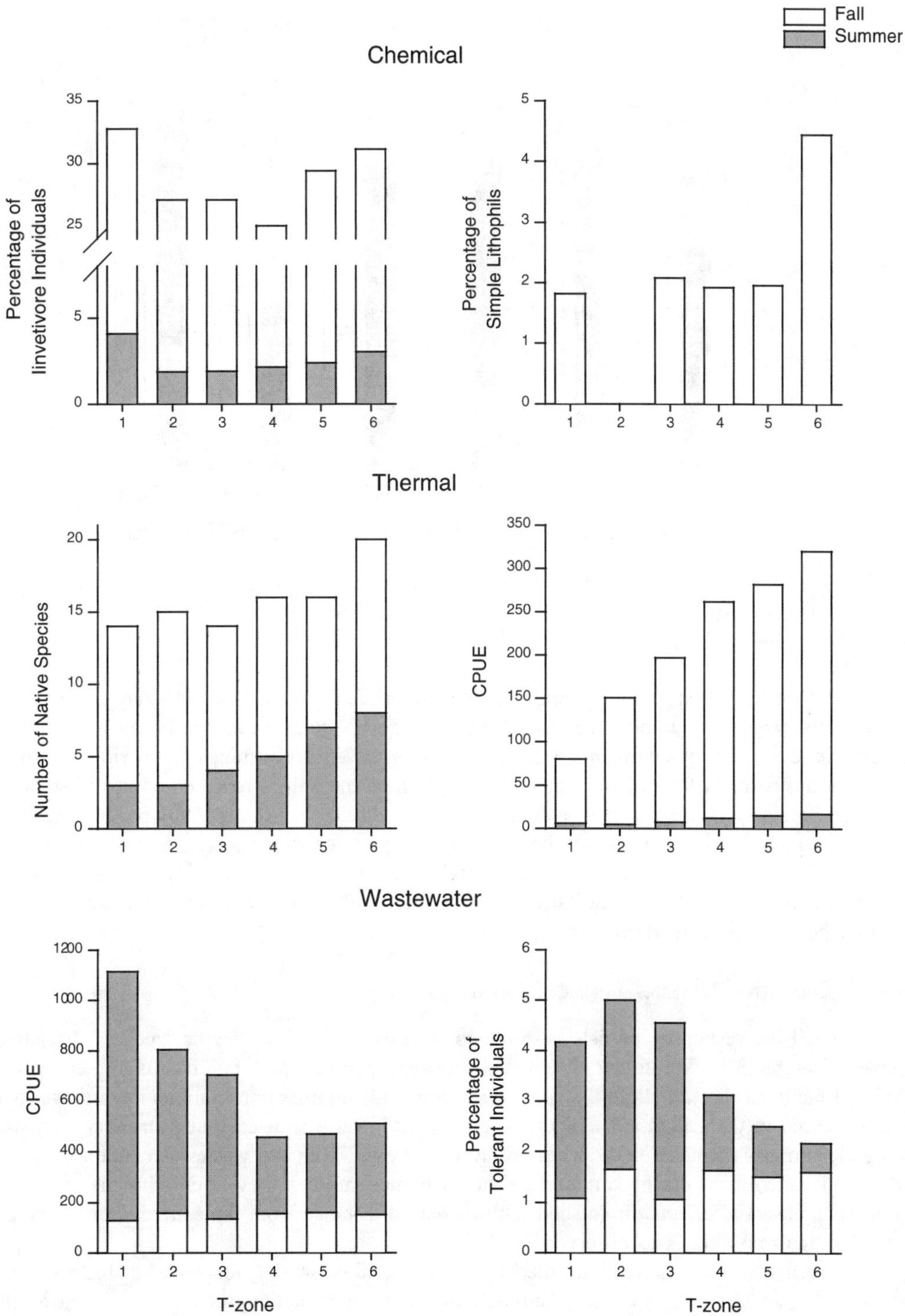

FIGURE 23.3 Responses of selected ORFIn metrics by traveling zone at a chemical, thermal and wastewater discharge.

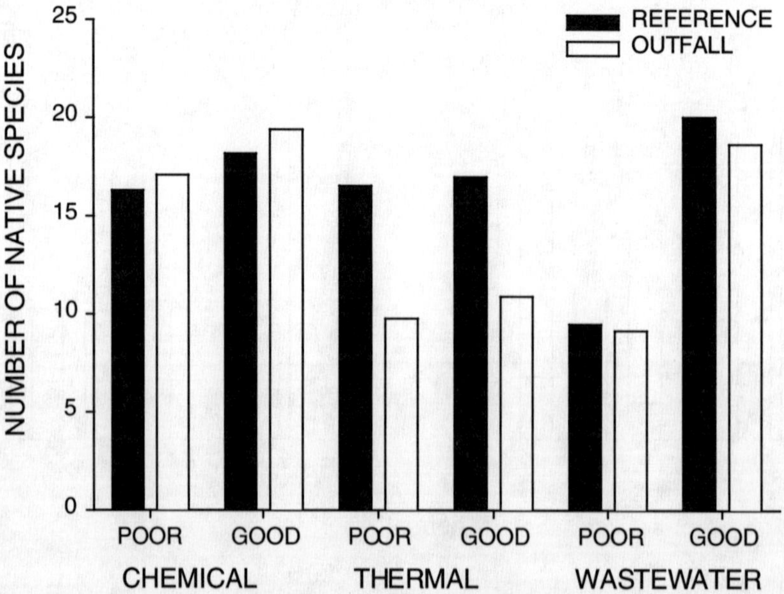

FIGURE 23.4 Responses of native species at reference and outfall sites for three types of discharges. Sites with poor habitat quality are shallow and have sandy substrates. Sites with good habitat quality are deep and have coarse substrates.

metrics (Figure 23.2) displayed u-shaped responses, indicated by slightly inflated values at the points of discharge that rapidly decreased over the next 100 to 200 m, then began to recover to background levels. This phenomenon may be due to the movement of transient individuals into the zone from upstream of the effluent that artificially inflate the values represented at the points of discharge. Similarly, in areas of poor habitat quality, the discharge structure may provide attractive fish cover for some species. Fish may be drawn to the area due to the increased flow or modified habitats typical of outfalls. Some type of bank stabilization usually in the form of rip-rap (cobble- to boulder-sized rock) or the outfall structure itself may offer some type of cover otherwise not found in the vicinity of the discharge.

23.4.3 Gradient Patterns among Outfall Types

We examined the responses of fish assemblage metrics at the three major types of discharges sampled (Figure 23.3). We found differences between summer and fall results for each type. Chemical facilities showed slight u-shaped responses with summer expectations much lower than those observed in the fall months. Thermal effluents typically showed much stronger responses during the summer months due to increased thermal stress. Municipal wastewater treatment plants showed opposite effects during summer and fall. Summer samples showed enrichment nearest the outfall. The observed effects diminished with increased distance from the source. During the fall months, little or no effect was observed.

Outfall effects are sometimes masked by habitat quality so that response to disturbance is mitigated (Figure 23.4). Higher quality outfall sites may have greater expectations than control sites with lower quality habitat. Figure 23.4 shows that thermal effects are equal across habitats, while chemical effects are similar across habitats but cause greater responses at the higher quality substrate sites. Wastewater effects are dramatically different in poor and good quality habitats, probably because of shifts in macroinvertebrate assemblage structure and function caused by soft sediments.

The response resolutions of most fish assemblage metrics to discharge effects at finer scales suggests that deleterious impacts are restricted to a few hundred meters. The T-zone method diagnoses response to a stressor, provides a more robust sampling approach, and identifies responses that may have otherwise been overlooked. We view this paper as a preliminary effort to initially test candidate Ohio River fish index metrics responses to particular point source discharges. We do not have sufficient data to adequately test the statistical significance of each outfall type to individual IBI metrics.

23.5 CONCLUSIONS

The T-zone approach is similar to the area degradation value (ADV) of Yoder and Rankin (1995b) because both techniques are designed to measure the decline or recovery of a community immediately downstream of a discharge. Both approaches are successful in determining the extent and magnitude of impacts from point source discharges. However, the T-zone approach allows the dissection of specific impacts within large rivers. We developed a technique for evaluating fish community response that is applicable to situations in which the zone of impairment is too small to be adequately represented by a standard-sized boat electrofishing zone. By collecting data in 100-m increments along a continuous 1000 m, we were able to construct traveling zones or T-zones 500 m in length and incrementally 100 m further from the point of impact. This technique requires the sampling effort of two standard- sized boat electrofishing zones and provides the equivalent of six standard-sized zones. This overlapping technique provides a 100-m resolution, increasing the researcher's ability to document community responses usually missed by standard 500-m zones.

We examined the responsiveness of select metrics to changes in water quality associated with point source discharges. We conducted night electrofishing at sites immediately downstream of point source discharges and at upstream control sites, maintaining uniform habitat conditions at test and control locations. We used an electrofishing method based on overlapping sampling zones to reveal indicator response along a gradient of human disturbance. Our results showed that 11 of 13 metrics responded to disturbance in a predictable manner. We were able to differentiate high-quality fish assemblages at control sites from assemblages with lower biotic integrity along disturbance gradients.

ACKNOWLEDGMENTS

We thank Matt Wooten, Jim Hawkes, and Robert Row for helping to collect the data used to analyze the effectiveness of the traveling zones. The opinions expressed here do not necessarily represent those of the U.S. Fish and Wildlife Service. No official endorsement by that agency should be inferred.

REFERENCES

Bayley, P.B. 1995. Understanding large river-floodplain ecosystems, *BioScience,* 45, 153–158.

Courtenay, W.R., Jr. and J.R. Stauffer, Jr. 1984. *Distribution, Biology, and Management of Exotic Fishes,* Johns Hopkins Press, Baltimore, MD.

Emery, E.B. and J.A. Thomas. 2002. A method for assessing outfall effects on Great River fish populations: the traveling zone approach, Chapter 9, this volume.

Emery, E.B., T.P. Simon, and R. Ovies. 1999. Influence of the family *Catostomidae* on the metrics developed for a great rivers index of biotic integrity, in T.P. Simon (Ed.). *Assessing the Sustainability and Biological Integrity of Water Resources Using Fish Communities,* CRC Press, Boca Raton, FL, 203–224

Emery, E.B., T.P. Simon, F.H. McCormick, P.L. Angermeier, C.O. Yoder, J.E. DeShon, R.E. Sanders, W.D. Pearson, G.D. Hickman, R.J. Reash, M. Miller, and J.G. Shulte. In review. Development of a multimetric index of biological condition for assessment of the Ohio River, *Transactions of the American Fisheries Society.*

Fausch, K.D., J. Lyons, J.R. Karr and P.L. Angermeier. 1990. Fish communities as indicators of environmental degradation. in S.M. Adams (Ed.). *Biological Indicators of Stress in Fish,* American Fisheries Society, Symposium 8, Bethesda, MD, 123–144

Fuller, P.L., L.G. Nico, and J.D. Williams. 1999. *Non-indigenous Fishes Introduced into the Inland Waters of the United States*. American Fisheries Society, Special Publication 27, Bethesda, MD.

Gammon, J.R. and T.P. Simon. 2000. Variation in a great river index of biotic integrity over a 20-year period, *Hydrobiologia,* 422/423, 291–304.

Hughes, R.M. and T. Oberdorff. 1999. Applications of IBI concepts and metrics to waters outside the United States and Canada, in T.P. Simon (Ed.). *Assessing the Sustainability and Biological Integrity of Water Resources Using Fish Communities*, CRC Press, Boca Raton, FL, 79–93.

Karr, J.R. 1981. Assessment of biological integrity using fish communities. *Fisheries*, 6(6), 21–27.

Karr, J.R. and D.R. Dudley. 1981. Ecological perspective on water quality goals, *Environmental Management*, 5, 55–68.

Karr, J.R., L.A. Toth, and D.R. Dudley. 1985a. Fish communities of Midwestern rivers: a history of degradation, *Bioscience,* 35, 90–95.

Karr, J.R., R.C. Heidinger, and E.H. Helmer. 1985b. Sensitivity of the index of biotic integrity to changes in chlorine and ammonia levels from wastewater treatment facilities, *Journal of the Water Pollution Control Federation,* 57, 912–915.

Karr, J.R., K.D. Fausch, P.L. Angermeir, P.R. Yant, and I.J. Schlosser. 1986. *Assessing Biological Integrity in Running Waters: A Method and Its Rationale*, Illinois Natural History Survey Special Publication 5.

Miller, D.L., P.M. Leonard, R.M. Hughes, J.R. Karr, P.B. Moyle, L.H. Schrader, B.A. Thompson, R.A. Daniels, K.D. Fausch, G.A. Fitzhugh, J.R. Gammon, D.B. Halliwell, P.L. Angermeier, and D.J. Orth.. 1988. Regional applications of an index of biotic integrity for use in water resource management, *Fisheries,* 13(5), 12–20.

Ohio River Valley Water Sanitation Commission. 2000. Quality assurance project plan for the collection of fish population samples as part of the fish community biocriteria development program, ORSANCO, Cincinnati, OH.

Ohio River Valley Water Sanitation Commission. 1999. Guidelines for Delineating Mixing Zones for Ohio River Discharges: Part I: Calculation of Mixing and Review of State Policies, Limno-Tech, Ann Arbor, MI.

Omernik, J.M. 1987. Ecoregions of the conterminous United States, *Annals of the Association of American Geographers*, 77, 179–190.

Pearson, W.D. and L.A. Krumholz. 1984. Distribution and Status of Ohio River fishes, ORNL/Sub/79–7831/1.U.S. Department of Energy, Oak Ridge National Laboratory, Oak Ridge, TN.

Poff, N.L., J.D. Allen, M.B. Bain, J.R. Karr, K.L. Prestegaard, B.D. Richter, R.E. Sparks, and J.C. Stromberg. 1997. The natural flow regime: a paradigm for river conservation. *BioScience*, 47, 769–784.

Simon, T.P. 1992. Development of Biological Criteria for Large Rivers with an Emphasis on an Assessment of the White River Drainage, Indiana, EPA 905/R-92/006, U.S. Environmental Protection Agency, Region 5, Chicago, IL.

Simon, T.P. and E.B. Emery. 1995. Modification and assessment of an index of biotic integrity to quantify water resource quality in great rivers, *Regulated Rivers: Research and Management*, 11, 283–298.

Simon, T.P. and J. Lyons. 1995. Application of the index of biotic integrity to evaluate water resource integrity in freshwater ecosystems, in W.S. Davis and T.P. Simon (Eds.). *Biological Assessment and Criteria: Tools for Water Resource Planning and Decision Making,* Lewis Publishers, Boca Raton, FL, 245–262

Simon, T.P. and R.E. Sanders. 1999. Applying an index of biotic integrity based on great river fish communities: considerations in sampling and interpretation, in T.P. Simon (Ed.). *Assessing the Sustainability and Biological Integrity of Water Resources Using Fish Communities,* CRC Press, Boca Raton, FL, 475–506.

Simon, T.P. and J.R. Stahl. 1998. Development of Index of Biotic Integrity Expectations for the Wabash River, EPA 905/R-96/026, USEPA, Water Division, Watershed and Non-Point Source Branch, Chicago, IL.

Sparks, R.E. 1995. Need for ecosystem management of large rivers and their floodplains, *BioScience*, 45, 169–182.

Sparks, R.E., P.B. Bayley, S.L. Kohler, and L.L. Osborne. 1990. Disturbance and recovery of large floodplain rivers, *Environmental Management,* 14, 699–709.

Statistical Application Software. 1996. PROC GLM module. SAS Institute, Cary, N.C.

Ward, J.V. and J.A. Stanford. 1989. Riverine ecosystems: the influence of man on catchment dynamics and fish ecology, Canadian Special Publications in Fisheries and Aquatic Sciences, 106, 56–64.

Ward, J.V. and J.A. Stanford. 1995. Ecological connectivity in alluvial river systems and its disruption by flow regulation, *Regulated Rivers*: *Research and Management,* 11, 105–119.

Yoder, C.O. and E.T. Rankin. 1995a. Biological criteria program development and implementation in Ohio, *in* W.S. Davis and T.P. Simon (Eds.). *Biological Assessment and Criteria: Tools for Water Resource Planning and Decision Making*. Lewis Publishers, Boca Raton, FL, 109–144.

Yoder, C.O. and E.T. Rankin. 1995b. Biological response signatures and the area of degradation value: new tools for interpreting multimetric data, in W.S. Davis and T.P Simon (Eds.). *Biological Assessment and Criteria: Tools for Water Resource Planning and Decision Making*. Lewis Publishers, Boca Raton, FL, 263–286.

24 Evaluating the Effects of Thermal Discharges on Aquatic Life: Patterns in Multimetric Indices from Three Case Studies in Large and Great Rivers of the Midwestern United States

Ronda L. Dufour, Thomas P. Simon, and Steven A. Newhouse

CONTENTS

0-8493-0905-0/03/$0.00+$1.50

24.1 INTRODUCTION

Water is used in many industrial applications to cool machinery and produce condensed steam. The use of water for cooling and its discharge to the environment generally increase downstream water temperatures from electric generating facilities (Majewski and Miller, 1979; Coutant, 1997; Bulleit, 2000). These heated water discharges are regulated by the Clean Water Act, Section 316(a), which prohibits the release of heated water in amounts that would cause adverse effects on aquatic life using the normative standard that is used to judge harm based on the requirement for protection of a balanced indigenous population. The standard has been interpreted as a balanced ecological community of mostly native species (Coutant, 2000). This concept is in agreement with the primary goal of the Clean Water Act: to restore and maintain biological integrity.

Karr and Dudley (1985) described biological integrity as "the capability of supporting and maintaining a balanced, integrated, adaptive community of organisms having a species composition, diversity, and functional organization comparable to that of natural habitats of the region." Biological criteria are "narrative and numerical expressions that describe the reference [least-impacted] biological integrity of aquatic communities inhabiting waters of a given designated aquatic life use" (U.S. Environmental Protection Agency [USEPA], 1990). Simon (2000) suggested that in the broadest sense, biological criteria include narrative and numerical expressions that can be based on such measurements as diversity indices, univariate indices, population and stock assessment indices, and multimetric indices of biological integrity and sustainability.

As a result of the 316(a) regulation, many studies have been conducted to assess the impacts of heated water on biological assemblages. Virtually all biological indicators (zooplankton, phytoplankton, macroinvertebrares, and fish) were studied to determine their responses to heated effluents (Brown, 1976; Hokanson and Biesinger, 1980; EPRI, 1981; Raney and Menzel, 1969). Several laboratory studies have been conducted with these taxonomic groups to determine lethal minimum or maximum temperatures and individual species responses.

The paradox in studying heated effluents is that although many datasets characterize the structures of biological assemblages at population levels in the vicinity of heated discharges, only a few studies have attempted to assess the structures and functions of biological indicators at a community level. Gammon (1973, 1976, 1983) studied phytoplankton, aquatic macroinvertebrates, and fish assemblages of the middle Wabash River in the vicinity of two thermal generating facilities for over two decades.

Gammon and Simon (2000) evaluated changes in multimetric indices in the vicinity of these discharges to determine patterns in the index of biotic integrity based on hydrologic cycle, recruitment, and gear effects. Simon (1992) evaluated fish assemblages in the lower White River using a modified index of biotic integrity for large rivers. In addition, power industries conduct annual site-specific studies that describe the structures of aquatic assemblages in the vicinities of their stations, but few calculate multimetric indices or describe the effects of these discharges on downstream biological integrity.

The intent of this chapter is to describe patterns in aquatic macroinvertebrate and fish assemblage structures and functions near thermal generating stations and describe the responses of large and great river indices of biotic integrity upstream and downstream of these discharges. In addition, evaluations of patterns in fish assemblage structure and function at electric generating stations on the Wabash River, White River, and Ohio Rivers were conducted to determine multimetric responses.

24.1.1 ENVIRONMENTAL IMPACTS FROM THERMAL DISCHARGES AND COOLING WATER INTAKES

The largest industrial user of cooling water in the U.S. is the steam electric power industry, which reports that 44% of steam electric generating capacity utilizes once-through cooling systems (EEI, 1996). A once-through cooling system draws water from the ambient environment to cool the heat-producing components of energy production. The water is then discharged into the environment after a single use. About 50% of generating plants use closed-cycle-cooling systems in which water is withdrawn from a cooling tower, basin, cooling pond, or cooling lake, pumped to the condenser, then returned to the tower, basin, pond, or lake. An additional 5% of generating capacity uses a combination of once-through cooling and closed-cycle systems or operates in mixed-cycle mode.

Once-through cooling systems generally affect the environment through the discharge of heated water, which produces two primary effects; avoidance of the heated water or acute toxicity from rapid changes as a result of physiological response. Presumably aquatic life exposed to heated water would avoid the area by evacuating to safer water or die as a result of rapid changes in somatic tissue temperature.

The discharge of heated water to a receiving system and the associated thermodynamics are well documented. A simple mass–balance relationship requires one British thermal unit (BTU) of heat to raise the temperature of one pound of water one degree Fahrenheit. If one knows the mass and temperature of the discharge water, the final receiving water temperature can be calculated.

24.1.2 Adaptive Risk Assessment and Management Approach to 316(a)

Coutant (2000) described a variety of issues concerning flow, distribution of organisms, mortality rates, and finally the consideration of balanced indigenous communities that can create erroneous conclusions in assessing 316(b) issues. The balanced indigenous communities were thought to be on an adverse scale between normative and disturbed. The issues of a balanced indigenous community and development of a response scale depicting normative and disturbed states are probably the most important and controversial issues for assessing the in-stream effects of heated waters.

In the revised legislation, Section 316(a) mandates that baseline biological analyses be conducted. Coutant (2000) suggests that this baseline analysis approach has merit for Sections 316(a) and 316(b) since the combination of effects from a power station will or will not cause changes in the community in ways detrimental for human ecosystem values and we agree with this assertion. However, we disagree with Coutant's suggestion that a biotic community in a large water body is never "balanced." Coutant (2000) speaks about populations, while the intent of the section clearly focuses on evaluation of community levels of organization.The aspects of the community that he addresses are the dynamic processes of large rivers over cyclical seasonal changes. By limiting index periods for sampling and establishing reference conditions and levels of expectation for each measurable response characteristic for any specific water body using appropriate methods, the dynamic cyclical processes become mute issues (Simon and Sanders, 1999; Seegert, 2000a and b; Emery et al., in review). Since the baseline condition is calibrated and validated using multimetric approaches specific to a certain water body (reference condition) and has specific time periods for application (index period), changes outside the index period can only be evaluated within the sampling period to which the reference condition has been calibrated (Davis and Simon, 1995; Simon, 1999). The response measurements in the Wabash River even over severe hydrologic cycles (e.g., drought and flood conditions) resulted in changes of only 4 to 8 IBI points (Gammon and Simon, 2000).

Coutant (2000) recommends the development of a conceptual and numerical adverse scale for the normative and disturbed standards as a method to evaluate trends. By definition, the calibration of the multimetric index should consider variations in biological characteristics of the indicator over space within a specific time. Thus, differences in scoring are based on ranges that are interpreted as similar to reference conditions (score of 5), deviating slightly from reference conditions (score of 3), or deviating strongly from reference conditions (score of 1). Emery et al. (in review) added a 0 score for the Ohio River fish index (ORFIn) that is used when none of the attributes of a specific metric are present in the sampled reach. This extension of the numerical adverse scale between the normative and disturbed scores should limit variations in low-end IBI scores sometimes referred to as "noise."

Coutant (2000) strongly recommends careful consideration of the biological sustainabilities of some systems in light of the blend of natural and cultural attributes of aquatic ecosystems. The impoundment of rivers, flood control devices, diked shorelines, and other modifications will compromise some features of natural aquatic systems. The goal of biological criteria and reference conditions should not be the return to pre-Columbian settings (i.e., the absence of all disturbance). The detection of changes must be realistic and responsive in that certain variations may be defined as adverse for evaluating conditions and regulatory purposes, but ultimately will be based on cultural requirements and not impose an undue burden on industry to meet unrealistic expectations. A baseline should reflect the reference conditions for the entire watershed. It should not be so site-specific that expectations are based only on an upstream comparison to downstream. Reference conditions for ecoregions, pools, and within-pool habitats for large rivers should be compared and considered for biological sustainability and integrity (Simon and Sanders, 1999; Reash, 1999; see Emery et al., Chapter 23).

USEPA recommends and has provided guidance in regulating mixing; it suggested that situations be defined on a case-by-case basis after the assimilative capacity of the receiving system has been considered. It also suggested that mixing zones should be applied to prevent adverse impacts on immobile species, such as benthic communities (1994). Further guidance on mixing zones (USEPA, 1991) recommends that they be of size and physical quality that will not block or prevent the free movement of fish species for migration, spawning and other purposes. The mixing zone guidelines promulgated for Indiana water quality standards state: "The mixing zone should be limited to no more than one-fourth of the cross-sectional area and/or volume of flow of the stream, leaving at least three-fourths free as a zone of passage for aquatic biota, nor should it extend over one-half of the width of the stream" (327 IAC 2-1-4).

Some permits for electric generating facilities permit mixing zones measured in miles versus the much more conservative recommended guidance. Veil and Moses (1996) explained the economic and environmental impacts that would be imposed upon the power industry by limiting thermal mixing zones to 1000 feet. It was suggested that most plants affected by the proposed changes would retrofit cooling towers and some would retrofit diffusers. The installation of cooling tower diffusers would exert only a 1.0 to 5.8% energy penalty on plant output. As suggested by the CWA, the mixing zone should define the extent of any thermal impact on existing biota; the extent or size of the mixing zone is critical for protecting the aquatic community of the receiving waters.

24.1.3 Overview of Avoidance Temperatures

Field and laboratory thermal preference studies were completed in the late 1970s as some of the Section 316 demonstrations required by the CWA. It was anticipated that species preferring temperatures below 29°C would disappear from the vicinities of mixing zones. The absence of cool water species (e.g., redhorse, sauger, walleye, northern pike) was anticipated based on thermal preferences. Species thermal tolerances derived from controlled laboratory experiments were used to estimate predicted fish assemblages in the vicinities of thermal discharges (Brungs and Jones, 1977; EPRI, 1981; Gammon, 1983; Simon, 1992). Gammon examined the ambient thermal preferences of common Wabash River species based on laboratory data. Close agreements were observed between the thermal preferences of predicted and observed species. Minor variances were attributed to differences in life stages, since most species-specific testing is conducted with juvenile fish.

Gammon (1983) used intake control samples to evaluate several river models along a spatial scale. All cases assumed instantaneous mixing; based on temperature, the changes in community composition observed during start-up of a power station were attributed to temperature. Gammon had three hypotheses that assumed the thermal regime, using mean river discharge and overall mean monthly ambient temperatures, was under maximum thermal loading by the generating station. The second hypothesis assumed complete mixing of temperatures, using mean river discharge and extreme high mean monthly temperatures. A final hypothesis assumed that the entire flow of the

Wabash River passed through the generating station when it operated at maximum capacity. Superimposed on these three cases were the thermal preferences of common species.

The results of the first scenario predicted that temperature would exceed the thermal preferences of coolwater species, e.g., redhorse, sauger, and walleye, for much of the summer. In addition, temperature would exceed the preferences of sensitive thermophilic warm water species, e.g., smallmouth bass, goldeye, mooneye, and *Pimephales* spp., for a single month. Assuming the second scenario, all cool water species would be eliminated, as would white crappie, skipjack herring, and shiners for perhaps a single month each summer. Changes attributed to removing the entire flow of the river and passing it through the generating facilities, which is an extreme hypothesis and not necessarily legal, eliminated even heat-tolerant channel catfish and flathead catfish when resultant temperatures were above 30°C and also eliminated all cool water and sensitive thermophilic warm-water species. Changes in the thermal regime affect compositional integrity, reproduction, competition, and the trophic dynamics of the assemblage. These diffuse or direct competitive interactions cannot be adequately modeled, but exist as possible effects of thermal discharge.

24.2 METHODS

24.2.1 Study Area

Since the primary siting criterion for power generating facilities is the abundance of water, many plants are located on large and great rivers. We evaluated the responses of fish and macroinvertebrate assemblages in the vicinity of thermal stations on the Wabash, White, and Ohio Rivers in Indiana (Figure 24.1). Ample data were available for facilities chosen for study. The facilities were included in an Ohio River research program sponsored by the power consortium or were studied extensively by the authors over the past decade. The plants on the Wabash River are located at RM 252, near Cayuga in Vermillion County and RM 220 near Terre Haute in Vigo County. Both plants were well studied over the last 30 years by DePauw University, the power industry, and their consultants. The power stations on the White River are from the West Fork, RM 268.7, near Indianapolis, Hamilton County; West Fork, River, RM 196.23, near Martinsville, Morgan County; and on the lower White River, near Petersburg, Pike and Knox Counties. Thermal generating facilities on the Ohio River include single facilities in the Smithland, Newburgh, and Markland Pools.

24.2.2 Sampling Considerations

We followed the basic premise of large river sampling established by USEPA (1988), Ohio EPA (1989), Simon and Sanders (1999), and Emery et al. (in review) so that a representative sample of the aquatic assemblage was obtained from a 500-m reach of river. Collection of fish community data was based on boat-mounted electrofishing methods. A pulsed DC generator was carried on board a modified V-hull Jon boat or Coleman sport canoe. Electricity was applied to the anodes through an electrosphere or a hand-held electrified dipnet. Each procedure required a minimum of 1800 s to a maximum of 3600 s sampling time in the zone where each habitat type was targeted. Species were netted and placed into a live well until the completion of work in the zone. They were then identified, counted, weighed, measured, and inspected for deformities, eroded fins, lesions, and tumor (DELT) anomalies. Voucher specimens, small minnows, darters, and madtoms that required further inspection in the laboratory were preserved.

24.2.3 Reference Conditions

A reference condition is a baseline or standard that evaluates a representative assemblage collected at a site. The reference condition for this study was based on Simon (1992) for the White River, Simon and Stahl (1998) for the Wabash River, and Emery et al. (in press) for the Ohio River. Sites used to establish the reference condition were least impacted and considered representative of the

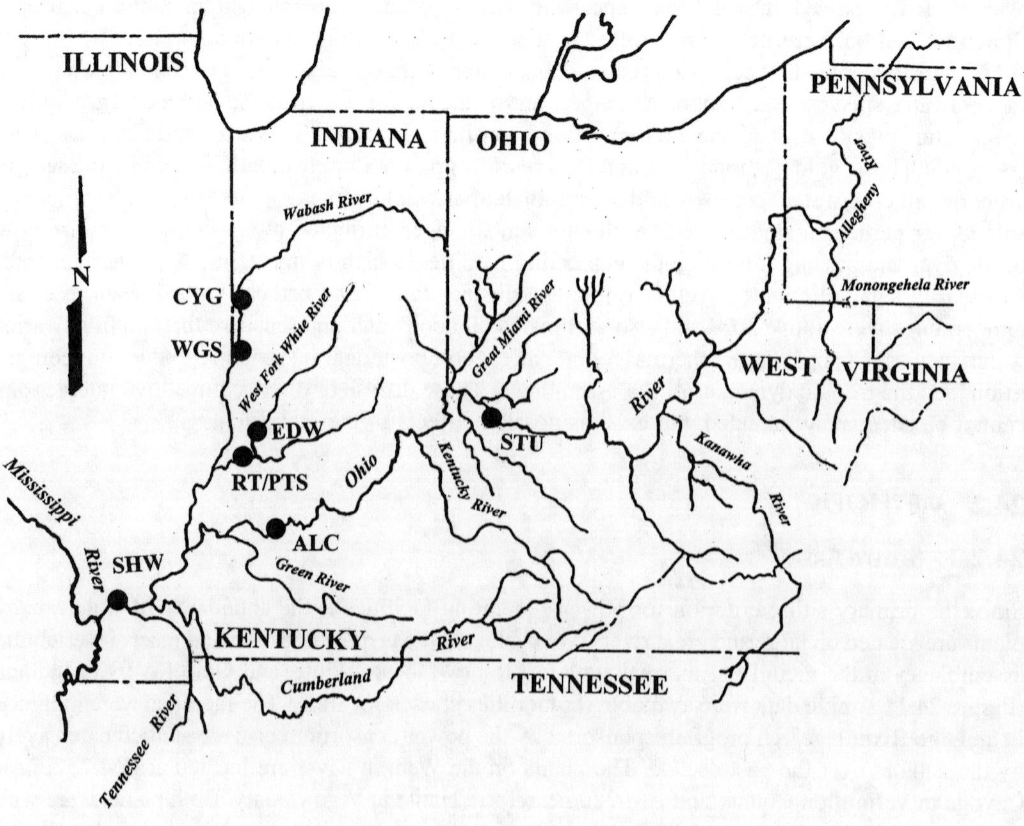

FIGURE 24.1 Study area showing the locations of power generating facilities utilized for a comparison of metric responses for large and great rivers in the Wabash, White, and Ohio Rivers. CYG = Cayuga Generating Station, WGS = Wabash Generating Station, EDW = Edwardsport Generating Station, RT = Ratts Generating Station, PTS = Petersburg Generating Station, SHW = Shawnee Generating Station, ALC = ALCOA thermal outfall, and STU = Stuart Station.

reach of the river. Sites were selected also for ease of transporting equipment. Simon (1992) sampled at 45 sites in the White River drainage. Simon and Stahl (1998) worked with a multiagency team of biologists to collect fish assemblage data at 28 sites in the middle to lower Wabash River. Emery et al. (in press) sampled 450 sites to calibrate the Ohio River fish index (ORFIn) for use on the entire Ohio River.

24.2.4 Other Datasets

Most electric generating facilities possess a wealth of information on the fish and macroinvertebrate assemblages upstream and downstream of their heated effluents. We evaluated a 5-year dataset for the middle Wabash River in possession of the Cayuga and Wabash River generating stations between 1992 and 1996. Both facilities were well studied but never at the multimetric level scale (Lewis and Seegert, 2000). The benefit of using generating station data is that the facilities conducted multiple sampling events over several years at the same locations. Both Wabash River plants contained an equivalent number of sampling stations upstream and downstream of their heated outfalls and included data collected for fish and macroinvertebrate assemblages.

Ambient background temperature data were based on a model established by the U.S. Geological Survey (USGS) that depicts the changes in water correlated with latitude and longitude (USGS, 1979) and on fixed station temperature data collected from the Wabash River between Lafayette and Vincennes from 1990 through 2001 by the Indiana Department of Environmental Management data management system (AIMS). This temperature data represented a conservative evaluation of water temperature because data were based on grab samples taken from the river at discrete intervals. Historical grab temperature measurements of rivers used in these calibrations were typically made at mid-day. *In situ* monitors indicated that distinct diurnal temperature fluctuations of as much as 5°F were observed in the early evening. Diurnal fluctuation should be considered in an updated thermal model of river latitudinal relationships of temperatures.

24.2.5 Statistics

Comparisons of patterns among electric generating facilities were tested using Kruskal–Wallis analysis of variation, histograms, and box and whisker box plots (Statistica, 1999). Patterns in responses of individual metrics were compared using the hypothesis that scoring at sites downstream of power generating facilities should not deviate from the reference condition if no impacts result from operation. Sites that show statistical differences from reference values were considered significant departures from the calibrated references for the respective rivers.

24.3 RESULTS

24.3.1 Seasonal Patterns in Fish and Aquatic Macroinvertebrate Assemblages in Large Rivers

Surveys of thermal generating stations typically are based on monthly sampling from May until October. Few studies are conducted during the late fall, winter or early spring. Biological data collected from the Cayuga and Wabash River generating facilities (Lewis et al., 1993a, b; 1995; 1996a, b, c; 1997a, b; 1998a, b) were graphed to evaluate the monthly responses of the great river metrics for macroinvertebrate and fish assemblages (Figures 24.2 through 24.4).

Macroinvertebrate Assemblages — Invertebrate communities generally shows declining trends in ICI scores with increasing distance downstream of thermal inputs (Figure 24.2a). The decline in ICI below the Cayuga generating station at the zone immediately upstream of the discharge (C3) was not outside the eddy currents and should be considered a near-field effect station. Likewise, the Wabash River generating station zone W4 was outside the thermal discharge plume, on the opposite bank, and should be considered an upstream control station. Otherwise, ICI scores showed initial declines and recoveries downstream with increasing distance from the heated effluent. The ICI classification group membership showed increasing scores during February and October, with steep declines during March and recovery from April through September (Figure 24.2b). Thus, the fall appears to be the best sampling period to obtain maximum species diversity and biological integrity scores.

Individual metrics did not show significantly different responses for the total number of mayfly taxa, total number of caddisfly taxa, total number of diptera taxa, and number of EPT taxa (Figure 24.3). Differences in metric scores were observed for percent individuals as mayflies, percent individuals as caddisflies, percent individuals as tribe tanytarsini midges, and percent individuals as other Diptera and noninsects (Figure 24.3). This is in agreement with the predictions of Patrick (1950) concerning declines in sensitive algal, macroinvertebrate, and fish assemblages with increased degradation.

Fish Assemblages — Metrics generally respond with an increasing metric relationship between May and October (Figure 24.4). The number of species, percent individuals as large river species, number of sensitive species, percent individuals as tolerant species, and percent individuals as simple lithophilous spawners all showed increasing trends. The number of sunfish species and percent individuals as carnivores showed declining trends as months passed. The remaining metric

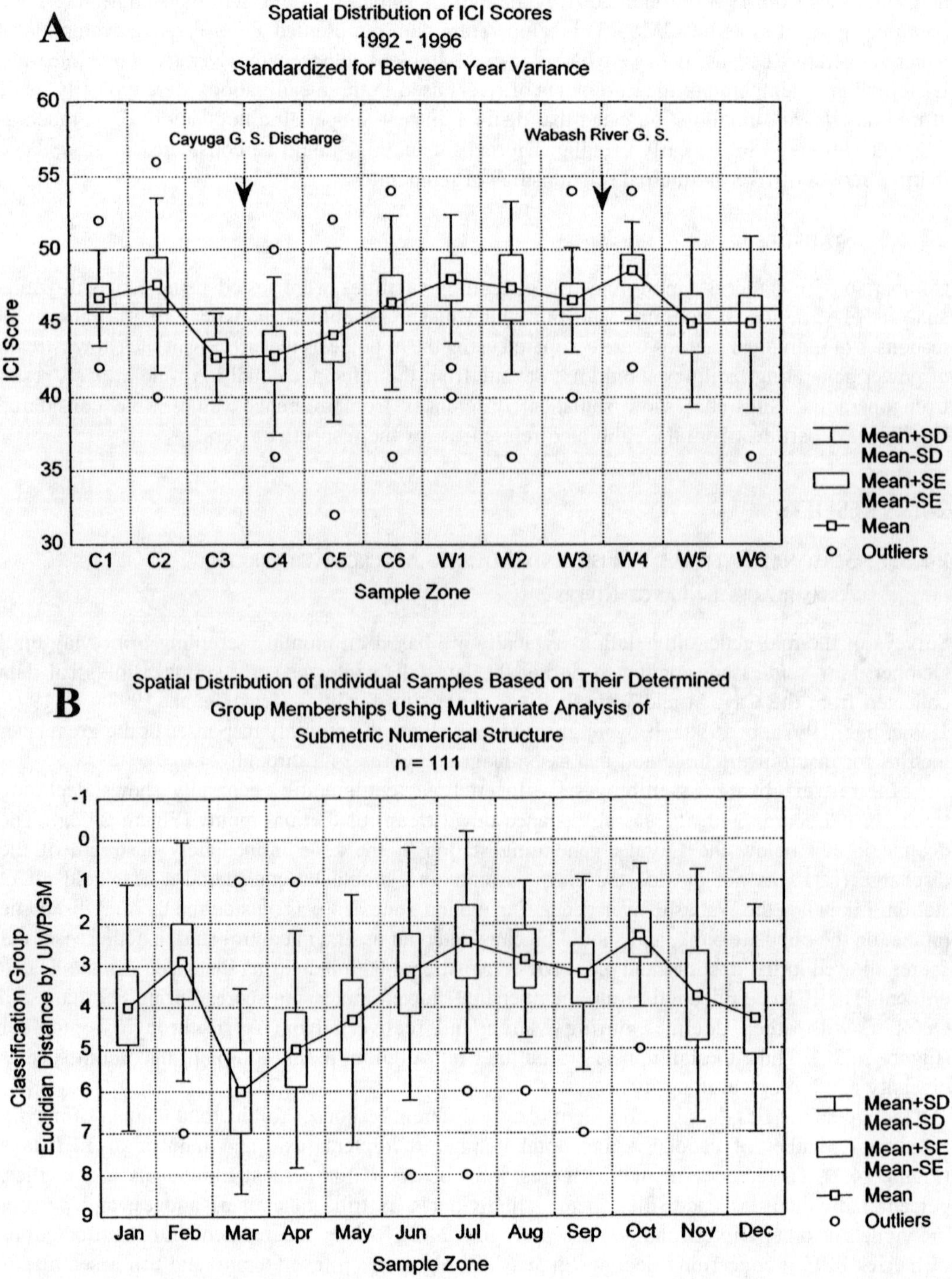

FIGURE 24.2 Box and whisker plots of mean, standard deviation, and standard error estimates of seasonal responses of macroinvertebrate assemblage metrics used to evaluate large and great rivers based on data from two facilities on the Wabash River. A: Total ICI score based on upstream and downstream effects from two thermal effluents. B: Monthly classification groupings based on ICI scores for all data from upstream and downstream zones from two thermal generating stations.

scores, percent individuals as omnivores, percent individuals as insectivores, and number of round-bodied sucker species were relatively stable and did not show patterns with changes in season.

Metric score responses showed that summer collections (June through August) were typically the lowest over the 6 months studied. Early fall (September and October) tended to show the highest metric scores. We believe several metrics were not adequately scored by the data and may reveal additional trends after proper assessments are conducted. For example, the DELT anomalies metric is typically given a score of 5 even though DELT information may not have been collected (Randy Lewis, Cinergy, personal communication). At a minimum, this metric should receive a score no higher than 3 when information is insufficient or absent. In addition, problems occur when black spot, parasite, and other non-DELT disease information is included in the calculation of this metric. As accurate DELT information becomes available, patterns in DELT anomalies may increase our knowledge of emaciation, lesions, and eroded fin conditions caused from exposure to heated effluents. Another concern was the low relative abundance reported by the regulated community. It is unclear whether this was an effect of thermal effluent or a result of less effort within the sampling zone. We assumed the effect arose from gear and personnel used, not from thermal discharge.

24.3.2 Comparison of Upstream and Downstream Changes in Assemblage Structures and Functions of Large and Great Rivers

Many of the power generating facilities we reviewed shared similar study designs. Both Wabash River facilities have three stations above the heated outfalls and three stations at various distances downstream of the heated outfalls. This is the same design used for the Ohio River research program. Typically, representatives of the thermal generating facilities determine where the sampling stations will be located. We noted that the sites immediately upstream and immediately downstream of these facilities alternated between being more or less degraded than expected. However, upon reviewing the selection criteria for these two locations, we found that the sites immediately downstream of the heated effluent were in proximity to discharge plumes and often located on opposite banks away from the immediate influence of the thermal discharge. The site immediately upstream was affected by habitat modification from the plant.

In order to evaluate the influence of upstream and downstream responses to heated effluents we selected several sensitive metrics that should have shown differences in response to heated effluent input. For example, round body suckers are among species that are the most sensitive to heat. The redhorse (*Moxostoma*), blue sucker (*Cycleptus*), chubsucker (*Erimyzon*), and spotted sucker (*Minytrema*) are considered intolerant to increased heat (Gammon, 1976; Emery et al., 1999; Reash et al., 2000). Simon (1992) also found that application of critical thermal maximums obtained from species exposure data from the laboratory tended to correlate well with observations in the ambient environment (Gammon, 1983). However, Reash et al. (2000) modified and raised the short-term upper thermal tolerances for redhorse species, based on results from species collected in the vicinity of two Ohio generating facilities on the Muskingum River. After review of this study, we found significant departures from standard procedures for holding specimens that resulted in heavy losses of fish kept in outdoor hydroponic vats while awaiting testing. This may have biased the results of this study and we do not consider the results valid.

Results for upstream sites (facility sites 1 through 3) for Cayuga and Wabash stations were graphically depicted using box and whisker box plots with standard deviation and standard error measures against sites downstream (facility sites 4 through 6) from the heated effluent (Figure 24.5). We did not modify or eliminate sites that may have been biased due to siting from this study and used all information available between 1992 and 1996. The input of heated effluent and habitat modification in the vicinity of the heated discharge dropped the mean scores to statistically significant levels ($p < 0.01$) below upstream conditions. In the case of the Cayuga station, zone 6 showed recovery, but the zone is 7.4 river miles downstream of the facility. The Wabash River generating

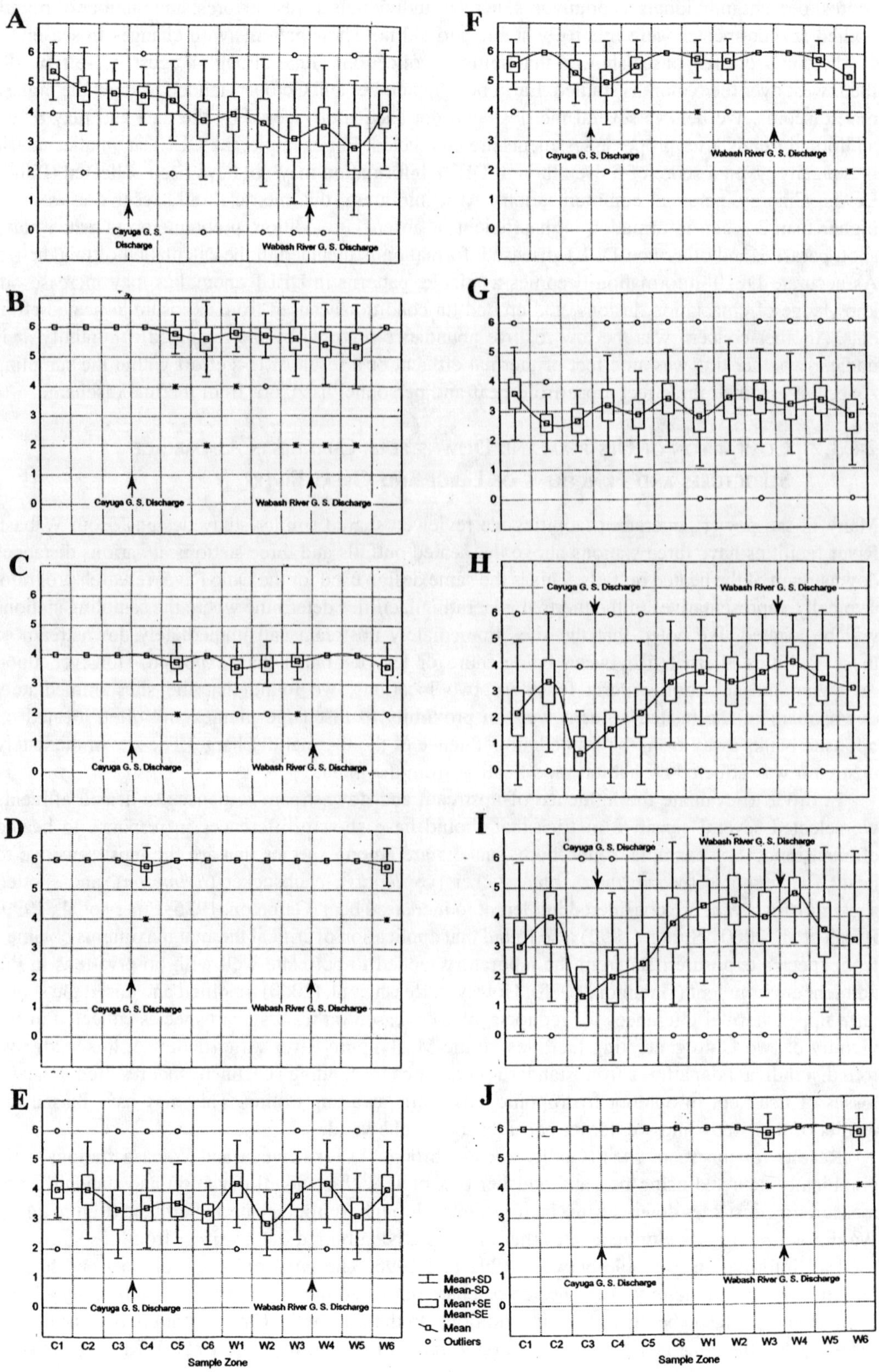
A
B
C
D
E
F
G
H
I
J
Cayuga G. S. Discharge
Wabash River G. S. Discharge
C1 C2 C3 C4 C5 C6 W1 W2 W3 W4 W5 W6
Sample Zone
Mean+SD
Mean-SD
Mean+SE
Mean-SE
Mean
Outliers

station does not show recovery downstream of the heated input within the area measured, which is 2 river miles downstream.

24.3.3 Comparison of Downstream and Reference Condition Effects

An alternate consideration for evaluating the influence of heated effluents on the downstream aquatic communities is to evaluate the response compared to the reference condition. Fish assemblage data collected in 1993 by a multiagency team of scientists were used to calibrate a great river IBI for the Wabash River (Simon and Stahl, 1998). This information was compared to data collected by the power industry during similar sampling years.

The reference condition is considered a preliminary calibration of the IBI. We graphically show the results of two metrics to compare the results of the reference site condition to those upstream and downstream of the facility discharge. The number of individuals in the sample (Figure 24.6a and b) and the number of sensitive species (Figure 24.7a and b) were compared to show that even upstream sites did not exhibit high biological integrity. A comparison of reference and nonreference sites from the multiagency study showed a mean of 396 fish per 500-m zone at reference or least-impacted sites, and 180 fish per 500-m zone at nonreference sites (Figure 24.6a). Compared to the upstream and downstream studies conducted by the facilities, the mean number of individuals collected was less than 50 fish per zone. It is not apparent whether this is a result of effort or the number of fish was lower in the vicinity of the generating facilities (Figure 24.6b). However, we suspect that the information content of biological integrity is strengthened by increased numbers of individuals sampled.

The number of sensitive species likewise showed similar results with upstream and downstream locations collected near thermal generating facilities. Both Wabash River facilities showed significantly fewer sensitive species than expected from the reference condition (Figure 24.7a and b). The number of sensitive species collected from least-impacted sites had a mean of about ten species, while the nonreference site mean was about four. The number of sensitive species upstream of the thermal discharge was slightly better than at nonreference sites, but significantly (t-test, $p = < 0.01$) different from the least-impacted condition.

24.4 DISCUSSION

24.4.1 Patterns in Fish Assemblage Metrics

Data collected by the authors, ORSANCO (unpublished data), and the power industry were included in a larger dataset to evaluate trends in great and large river IBI scores for a variety of thermal stations on the White, Wabash, and Ohio Rivers (Table 24.1). Only sites located downstream of the thermal generating facility heated discharge collected during the same index period were included to evaluate patterns in responses among generating stations. Limited information was available for aquatic macroinvertebrate assemblages so we will discuss data only for fish assemblages.

Patterns in metric responses downstream of heated effluents showed that several metrics did not show a statistically significant difference (t-test, $p < 0.01$) between reference condition expectations and observed values and metric scores downstream of heated effluents (Table 24.1). Metrics that did not show any relationship included total number of species, number of centrarchid species,

FIGURE 24.3 (previous page) Box and whisker plots of mean, standard deviation, and standard error estimates of individual metrics of the ICIs upstream and downstream of two thermal generating facilities. A: Total number of taxa. B: Total number of mayfly taxa. C: Total number of caddisfly taxa. D: Total number of Diptera taxa. E: Percent individuals as mayfly taxa. F: Percent individuals as caddisfly taxa. G: Percent individuals as tribe Tanytarsini midge taxa. H: Percent individuals as other Diptera and noninsect taxa. I: Percent individuals as tolerant organisms. J: Total number of EPT taxa.

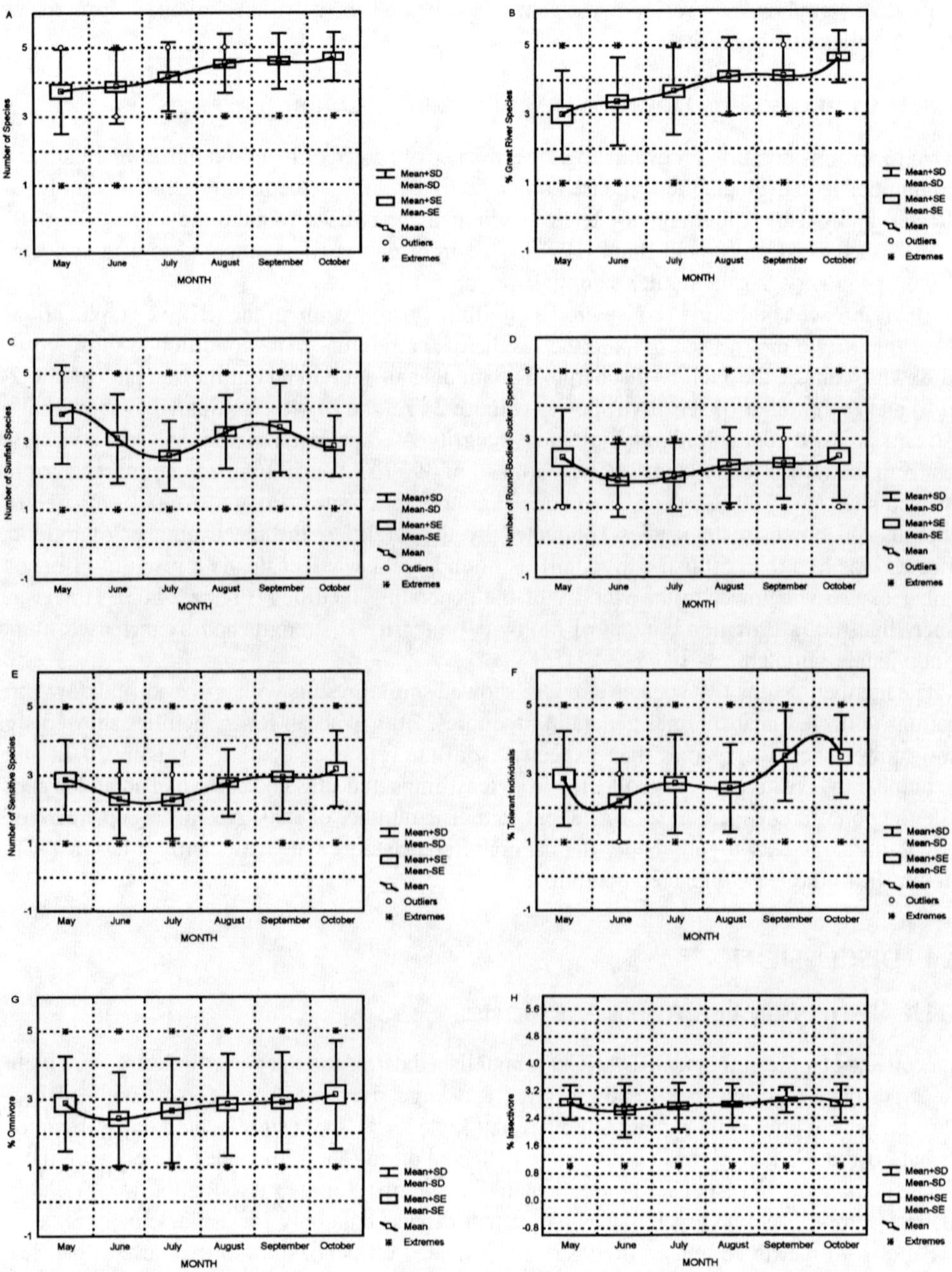

FIGURE 24.4 Box and whisker plots of mean, standard deviation, and standard error estimates of seasonal responses of fish assemblage metrics used to evaluate large and great rivers based on data from two facilities on the Wabash River.

number of sensitive and intolerant species, and percent individuals as large river species. Large river metrics that showed a response to heated effluents included number of round-bodied sucker species (t-test, $p < 0.001$, $n = 40$, 100% of sites), number of sensitive species (t-test, $p < 0.001$), $n = 40$, 87.5% of sites), and relative abundance (t-test, $p < 0.001$, $n = 40$, 80% of sites). In addition, the percent individuals as tolerant species (t-test, $p < 0.001$, $n = 40$, 80% of sites), omnivore species (t-test, $p < 0.001$, $n = 40$, 77.5% of sites), insectivore (t-test, $p < 0.001$, $n = 40$, 80% of sites),

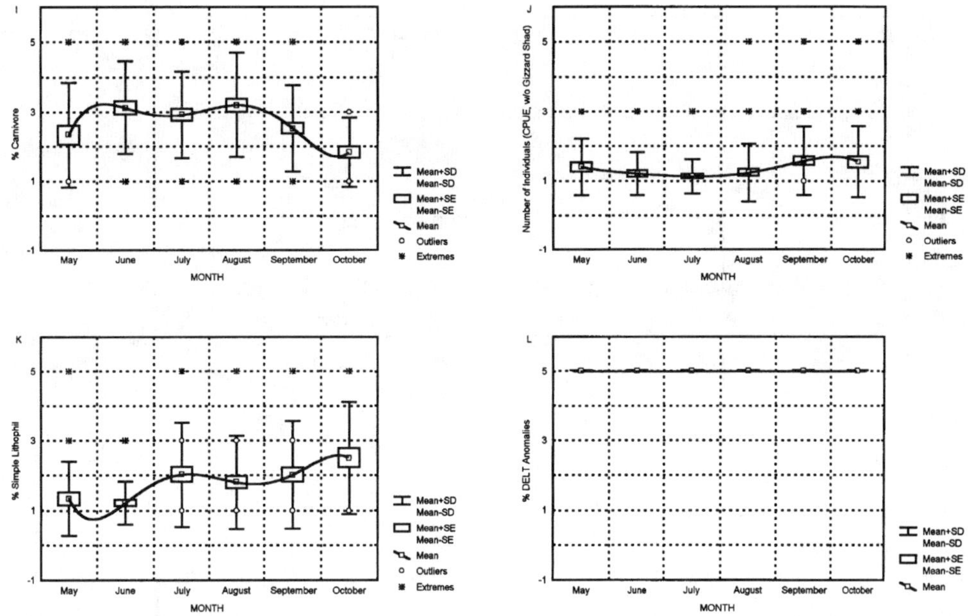

FIGURE 24.4 (CONTINUED)

carnivore (t-test, $p < 0.001$, $n = 40$, 67.5% of sites), and simple lithophils (t-test, $p < 0.001$, $n = 40$, 90% of sites) were statistically different from expected reference conditions.

Great River metrics showed similar responses, but the biological sustainability of the Ohio River should be carefully considered in light of the blend of natural and cultural attributes of the aquatic system (Table 24.1). Statistically significant differences were observed in the total number of species (t-test, $p < 0.001$, $n = 20$, 95% of sites), number of great river species (t-test, $p < 0.001$, $n = 20$, 100% of sites), number of suckers (t-test, $p < 0.001$, $n = 20$, 95% of sites), number of centrarchid species (t-test, $p < 0.001$, $n = 20$, 100% of sites), number of intolerant species (t-test, $p < 0.001$, $n = 20$, 70% of sites), number of exotic species (t-test, $p < 0.001$, $n = 20$, 80% of sites), and relative abundance (t-test, $p < 0.01$, $n = 20$, 70% of sites).

The percent individuals as tolerant species was not statistically different from expected reference conditions, while the percent individuals as detritivores ($p < 0.001$, $n = 20$, 85% of sites), invertivores ($p < 0.001$, $n = 20$, 75% of sites), top piscivores (t-test, $p < 0.001$, $n = 20$, 85% of sites), simple lithophils (t-test, $p < 0.001$, $n = 20$, 95% of sites), and relative number of DELT anomalies (t-test, $p < 0.001$, $n = 20$, 60% of sites) were statistically significant. We consider the decline in biological integrity attributed to the number of great river species to be indicative of a much larger problem in the Ohio River; however, the impact of heated effluents may be contributing to the overall decline.

24.4.2 Influence of Heated Effluents on Fish Assemblage Structure and Function

As the structure and function of fish communities become simplified beneath heated effluent discharges, the stability or standard of normative of the community is affected in ways that show declines in biological integrity. For the most part, the distance over which these effects are observed is considered short — less than 700 m on the Ohio River (Emery et al., Chapter 23, this volume). For other rivers this is not the case and greater distances may be affected, e.g., Cayuga facility, Wabash River near Cayuga. As declines in the total number of species, percent individuals as round-body suckers, and percent individuals as great river species, number of sensitive species, percent

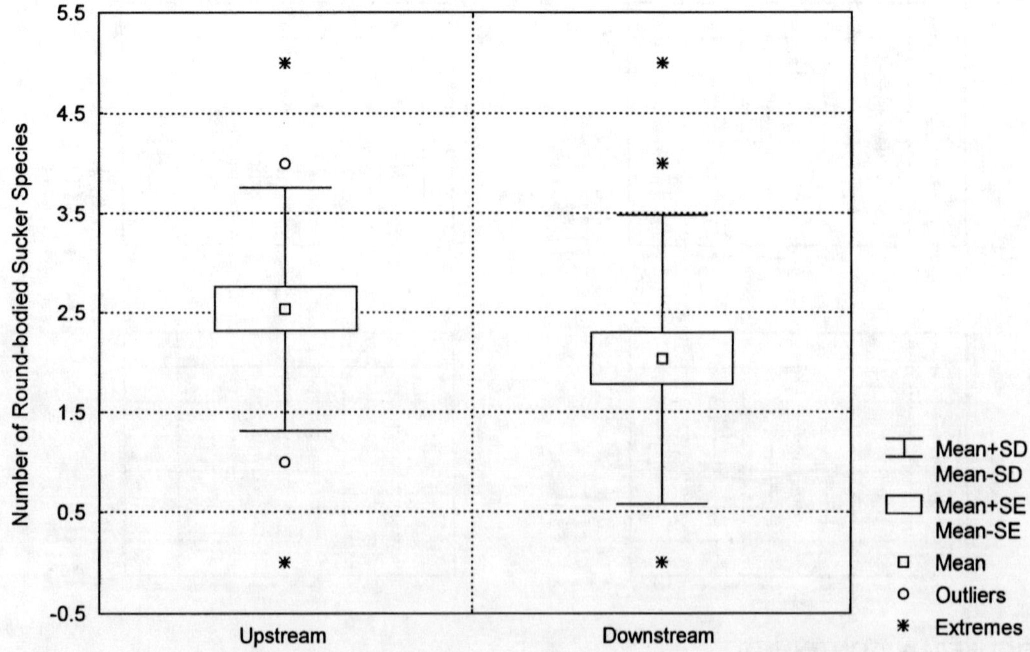

FIGURE 24.5 Comparisons of upstream to downstream responses of the percent individuals as round-body suckers metric for two power generating facilities on the Wabash River. Sampling zones 1 through 3 are upstream of heated discharge input and zones 4 through 6 are downstream of heated inputs.

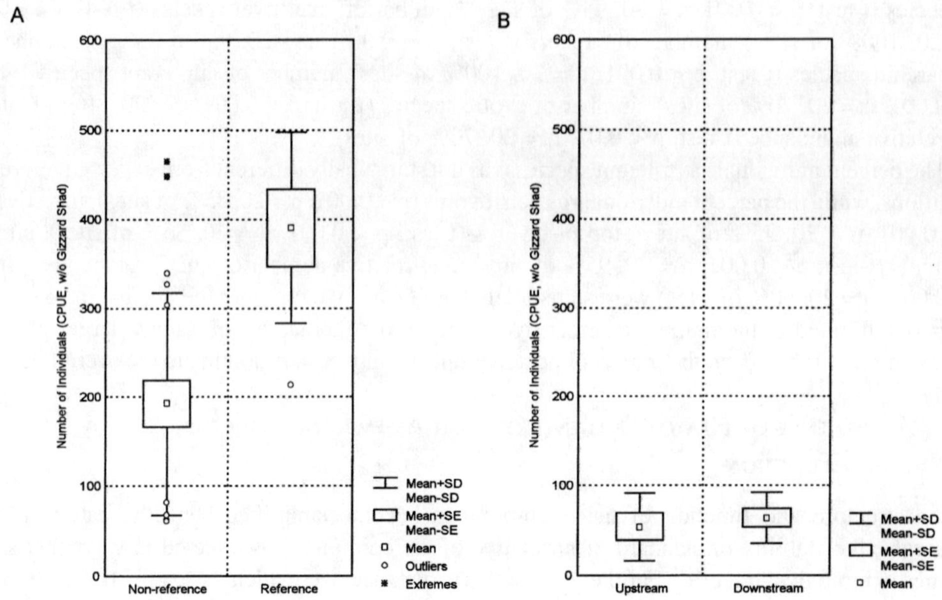

FIGURE 24.6 Comparison of results for catch-per-unit-of-effort metric between reference and nonreference sites sampled during the multi-agency study of the Wabash River conducted in 1993 (A) and upstream and downstream data from two Wabash River thermal generating facilities collected in fall 1993 (B).

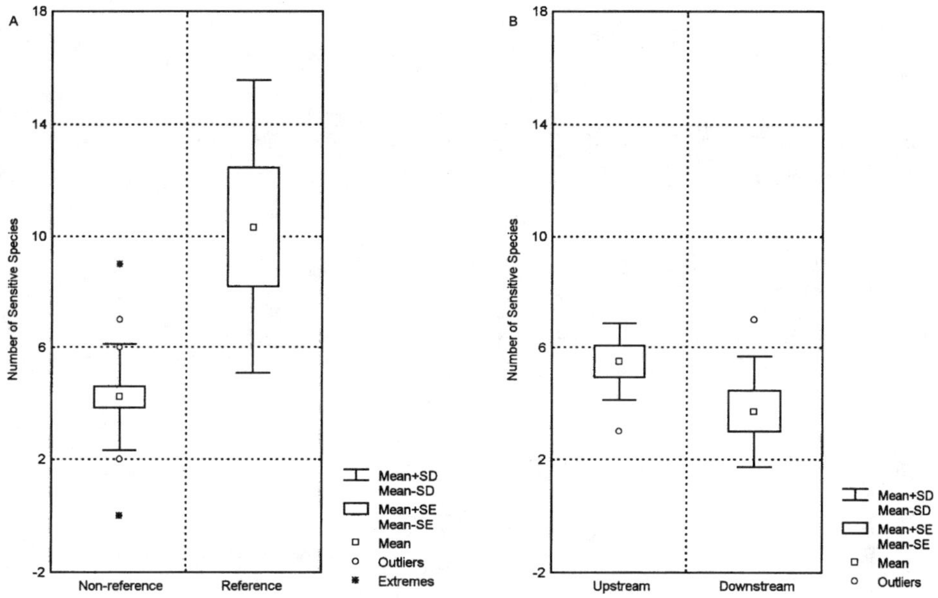

FIGURE 24.7 Comparison of results for the number of sensitive species metric between reference and nonreference sites sampled during the multi-agency study of the Wabash River conducted in 1993 (A) and upstream and downstream data from two Wabash River thermal generating facilities collected in fall 1993 (B).

individuals as simple lithophils, and catch per unit of effort occur, significant changes in the biological integrity of the rivers are seen near thermal generating stations.

Although the loss of integrity may be seasonal, the mistaken assumption is that the fish go elsewhere and return when conditions allow. It can be argued that avoidance is a biological impairment. First, resources such as habitat, food, and individual condition cannot be affected for long periods without effects in populations. The warm discharges that are attractive nuisances for species during winter months have not been well studied and may present an even greater risk for fish assemblage structure with the loss of fitness, increase in DELT anomalies, and decreased reproductive potential as a result of metabolic activity not permitting a resting or recovery phase for species. As further studies are conducted to evaluate downstream recovery (Emery and Thomas, Chapter 9; Emery et al., Chapter 23, this volume) the impacts of thermal generating stations on fish assemblage structure and function may be further understood. However, it is clear from this preliminary study that the effects of thermal generating facilities are detected by the great and large river calibrations for index of biotic integrity.

24.5 CONCLUSIONS

Electric generating facilities cause declines in fish community structure and function downstream of electric generating facilities. As the argument concerning the standard of normative and biological integrity increases it becomes important in the regulatory agencies that we define culturally what is desired for large and great river aquatic communities. A study of conditions on the Wabash River showed that fish and aquatic macroinvertebrate assemblages could measure impacts downstream of power facilities. The degree of impact was observed in specific metrics that recognized shifts in assemblage structure and function. The apparent decline of individuals in the vicinity of electric generating facilities may have been a result of the heated effluents but possibly was a lack of information content as a result of lower sampling effort in the dataset evaluated.

TABLE 24.1
Responses of Individual Metric Values and Scores (in Parentheses) for Great and Large River IBIs for Eight Power Generating Facilities in the Wabash, White, and Ohio River Drainages for August or September Collections for Designated Years

Watershed/ Site		Number of Species								Percent Individuals as						
		Total	Great River	Sucker	Centrarchid	Rnd Bodied	Sensitive[1]	Exotic	CPUE	Lg River Species	Tolerant	Omnivores[1]	Insectivores[1]	Carnivores[1]	Simple Lithophil	DELT*
Wabash River	Cayuga GS															
C4	1992	18(3)	—	—	5(5)	1(1)	3(3)	—	113(1)	33.6(3)	49.6(3)	49.6(1)	21.2(1)	22.1(5)	0.9(1)	— (3)
C5	1992	18(3)	—	—	3(3)	2(3)	3(3)	—	91(1)	53.8(5)	42.9(3)	27.5(3)	48.4(3)	14.3(3)	2.2(1)	— (3)
C6	1992	21(5)	—	—	2(3)	3(3)	6(3)	—	74(1)	48.6(3)	24.3(3)	13.5(5)	64.9(3)	13.5(3)	27.0(3)	— (3)
C4	1993	22(5)	—	—	6(5)	3(3)	8(5)	—	100(1)	30.0(3)	71.0(1)	59.0(1)	17.0(1)	13.0(3)	10.0(1)	— (3)
C5	1993	18(3)	—	—	3(3)	1(3)	3(3)	—	50(1)	62.0(5)	70.0(1)	26.0(3)	34.0(3)	20.0(5)	14.0(1)	— (3)
C6	1993	20(3)	—	—	4(3)	2(3)	5(3)	—	53(1)	43.4(3)	49.1(3)	17.0(3)	58.5(3)	9.4(1)	13.2(1)	— (3)
C4	1994	23(5)	—	—	5(5)	1(1)	4(3)	—	139(1)	33.8(3)	63.3(1)	48.9(1)	33.1(3)	14.4(3)	4.3(1)	— (3)
C5	1994	17(3)	—	—	2(3)	2(3)	4(3)	—	87(1)	54.0(5)	66.7(1)	34.4(1)	31.0(3)	28.7(5)	11.5(1)	— (3)
C6	1994	24(5)	—	—	4(3)	2(3)	6(3)	—	64(1)	48.4(3)	39.1(3)	12.5(5)	48.4(3)	26.6(5)	31.3(5)	— (3)
C4	1995	22(5)	—	—	6(5)	0(1)	4(3)	—	122(1)	54.1(5)	28.7(3)	54.1(1)	31.1(3)	14.8(3)	5.7(1)	— (3)
C5	1995	13(3)	—	—	2(3)	0(1)	1(1)	—	73(1)	67.1(5)	57.5(1)	24.7(3)	37.0(3)	37.0(3)	0(1)	— (3)
C6	1995	17(3)	—	—	3(3)	2(3)	3(3)	—	41(1)	68.3(5)	53.7(1)	14.6(5)	53.7(5)	26.8(5)	4.9(1)	— (3)
C4	1996	18(3)	—	—	3(3)	0(1)	1(1)	—	146(1)	51.4(5)	37.7(3)	65.8(1)	26.7(3)	6.9(1)	41.8(5)	— (3)
C5	1996	13(3)	—	—	1(1)	1(1)	3(3)	—	50(1)	70.0(5)	68.0(1)	18.0(3)	34.0(3)	28.0(5)	8.0(1)	— (3)
C6	1996	18(3)	—	—	2(3)	2(3)	5(3)	—	67(1)	40.3(3)	29.9(3)	13.4(5)	74.6(5)	4.5(1)	44.8(5)	— (3)
	Wabash River GS															
W5	1992	14(3)	—	—	4(3)	1(1)	3(3)	—	80(1)	45.0(3)	62.5(1)	21.3(3)	26.3(3)	38.8(3)	1.3(1)	— (3)
W6	1992	16(5)	—	—	3(3)	1(1)	2(1)	—	48(1)	58.3(5)	56.3(1)	27.1(1)	29.2(1)	27.1(1)	12.5(1)	— (3)
W5	1993	13(3)	—	—	2(3)	1(1)	3(3)	—	63(1)	71.4(5)	41.3(3)	17.5(3)	42.9(3)	34.9(3)	1.6(1)	— (3)
W6	1993	15(3)	—	—	1(1)	1(1)	4(3)	—	38(1)	57.9(5)	42.1(3)	23.7(1)	52.6(1)	18.4(1)	7.9(1)	— (3)
W5	1994	14(3)	—	—	4(3)	0(1)	2(1)	—	62(1)	50.0(3)	54.8(1)	8.1(5)	37.1(3)	50.0(1)	1.6(1)	— (3)
W6	1994	17(3)	—	—	3(3)	2(3)	3(3)	—	67(1)	68.7(5)	29.9(3)	20.9(3)	61.2(3)	11.9(3)	23.9(3)	— (3)
W5	1995	12(3)	—	-	4(3)	0(1)	2(1)	—	35(1)	57.1(5)	25.7(3)	11.4(5)	54.3(1)	28.6(1)	2.9(1)	— (3)
W6	1995	15(3)	—	—	1(1)	0(1)	0(1)	—	51(1)	68.6(5)	41.2(3)	37.3(1)	35.3(3)	21.6(5)	19.6(3)	— (3)

W5	1996	16(3)	—	—	3(3)	0(1)	2(1)	—	79(1)	48.1(3)	51.9(1)	27.8(3)	48.1(3)	21.5(5)	13.9(1)	— (3)
W6	1996	16(3)	—	—	1(1)	2(3)	3(3)	—	105(1)	78.1(5)	20.0(5)	69.5(1)	19.0(1)	7.6(1)	66.7(5)	— (3)
White River	Edwardsport															
Outfall	1990	34(5)	—	—	8(5)	1(1)	6(3)	—	335(1)	29.6(5)	62.1(1)	31.0(1)	38.8(1)	17.0(5)	4.2(1)	— (3)
d/s 0.75 miles	1990	36(5)	—	—	6(5)	2(3)	5(3)	—	1022(5)	25.7(3)	19.9(3)	35.7(1)	59.4(3)	1.4(1)	8.1(1)	— (3)
d/s 2.5 miles	1990	30(5)	—	—	5(5)	0(1)	2(3)	—	629(3)	17.2(3)	18.9(1)	21.9(1)	73.0(5)	2.5(3)	0.5(1)	— (3)
Outfall	1992	34(5)	—	—	5(5)	3(3)	9(5)	—	446(1)	52.0(5)	42.2(1)	49.3(1)	30.3(1)	7.4(5)	9.4(1)	— (3)
d/s 0.75 miles	1992	29(5)	—	—	4(3)	1(1)	4(3)	—	1520(5)	12.4(1)	5.4(5)	11.6(5)	84.9(5)	0.7(1)	2.6(1)	— (3)
d/s 2.5 miles	1992	32(5)	—	—	4(3)	1(1)	5(3)	—	1220(5)	7.2(1)	11.0(5)	10.5(5)	86.4(5)	1.6(1)	1.5(1)	— (3)
Outfall	1993	25(5)	—	—	6(5)	2(3)	6(3)	—	182(1)	35.2(5)	50.0(1)	35.7(1)	41.8(3)	14.8(5)	1.6(1)	— (3)
d/s 0.75 miles	1993	25(5)	—	—	3(3)	1(1)	7(3)	—	642(3)	17.6(3)	9.3(5)	16.8(3)	76.2(5)	2.8(3)	3.4(1)	— (3)
d/s 2.5 miles	1993	28(5)	—	—	3(3)	1(1)	4(3)	—	1406(5)	27.5(3)	10.5(5)	31.6(1)	61.5(3)	1.4(1)	4.6(1)	— (3)
Outfall	1994	35(5)	—	—	7(5)	1(1)	9(5)	—	494(3)	38.1(5)	43.9(1)	39.3(1)	34.4(1)	13.8(5)	3.4(1)	— (3)
d/s 0.75 miles	1994	35(5)	—	—	6(5)	1(1)	9(5)	—	1439(5)	37.0(5)	7.2(5)	40.2(1)	56.4(3)	2.4(3)	0.8(1)	— (3)
d/s 2.5 miles	1994	38(5)	—	—	4(3)	4(3)	9(5)	—	1712(5)	20.9(3)	13.0(5)	26.8(3)	69.7(5)	1.9(1)	1.2(1)	— (3)
	Petersburg															
SR 61	1990	13(1)	—	—	5(3)	0(1)	1(1)	—	147(1)	21.8(3)	51.7(1)	57.1(1)	33.3(1)	9.5(5)	0(1)	0(5)
SR 61	1991	17(3)	—	—	4(2)	0(1)	4(1)	—	1501(5)	7.1(1)	3.7(5)	8.7(5)	89.7(5)	0.9(1)	0.7(1)	0(5)
	Ratts															
d/s outfall	1991	28(5)	—	—	7(5)	0(1)	6(1)	—	1152(5)	30.8(3)	10.7(1)	27.0(1)	66.1(3)	3.6(1)	5.4(1)	0(5)
Ohio River	Shawnee Plant															
	1999	17(3)	1(1)	3(3)	1(1)	—	1(1)	0(5)	357(5)	—	0.0(5)	2.0(5)	54.0(5)	5.0(1)	0(0)	2(3)
	1999	15(3)	2(3)	4(3)	0(0)	—	1(1)	2(1)	86(1)	—	2.0(1)	24.0(1)	4.0(1)	30.0(1)	0(0)	0(5)
	1999	9(1)	1(1)	3(3)	0(0)	—	1(1)	5(1)	61(1)	—	5.0(3)	27.0(1)	3.0(1)	2.0(1)	0(0)	4(1)
	1999	12(3)	1(1)	3(3)	2(1)	—	0(0)	4(1)	55(1)	—	0(5)	22.0(1)	6.0(1)	4.0(1)	0(0)	4(1)
	1999	9(1)	0(0)	2(1)	0(0)	—	0(0)	10(3)	115(3)	—	7.0(1)	18.0(1)	0(0)	3.0(1)	0(0)	2(3)
	ALCOA															
1999	13(3)	2(3)	3(3)	1(1)	—	0(0)	0(5)	74(1)	—	0(5)	39.0(1)	13.0(1)	8.0(1)	3.0(1)	4(1)	
1999	15(3)	1(1)	3(3)	4(3)	—	0(0)	0(5)	271(5)	—	1.0(5)	10.0(3)	2.0(1)	5.0(1)	0(0)	11(1)	
1999	15(3)	1(1)	3(3)	1(1)	—	1(0)	0(5)	1365(5)	—	0(5)	4.0(5)	37.0(5)	1.0(1)	1.0(1)	3(3)	

TABLE 24.1 (CONTINUED)
Responses of Individual Metric Values and Scores (in Parentheses) for Great and Large River IBIs for Eight Power Generating Facilities in the Wabash, White, and Ohio River Drainages for August or September Collections for Designated Years

Watershed/ Site		Number of Species									Percent Individuals as					
		Total	Great River	Sucker	Centrarchid	Rnd Bodied	Sensitive[1]	Exotic	CPUE	Lg River Species	Tolerant	Omnivores[1]	Insectivores[1]	Carnivores[1]	Simple Lithophil	DELT*
Stuart Plant																
Outfall	2000	3(1)	0(0)	2(1)	0(0)	—	0(0)	0(0)	7(1)	—	0(5)	100(1)	0(0)	0(0)	0(0)	2(3)
500 m	2000	8(1)	1(1)	2(1)	0(0)	—	0(0)	1(1)	42(1)	—	0(5)	67.0(1)	3(1)	10(1)	0(0)	1(5)
Outfall	2000	7(1)	1(1)	1(1)	1(1)	—	0(0)	0(0)	58(1)	—	0(5)	21(1)	0(0)	42(1)	9.0(1)	4(1)
500 m	2000	13(3)	1(1)	3(3)	0(0)	—	1(5)	1(1)	95(1)	—	0(5)	22(1)	1(1)	28(3)	23.0(1)	0(5)
Outfall	2000	13(3)	2(3)	5(3)	1(1)	—	8(5)	0(0)	105(1)	—	0(5)	12(3)	2(1)	75(5)	35.0(3)	2(3)
500 m	2000	20(5)	2(3)	6(5)	0(0)	—	1(5)	1(1)	336(5)	—	2.0(5)	37.0(1)	5(1)	52(5)	29.0(3)	2(3)
Outfall	2000	10(3)	0(0)	3(3)	2(1)	—	6(1)	0(0)	45(1)	—	0(0)	56.0(1)	6(1)	19(1)	13.0(1)	3(3)
500 m	2000	10(3)	2(3)	3(3)	3(3)	—	6(1)	0(0)	41(1)	—	6.0(1)	44.0(1)	0(0)	17(1)	6.0(1)	0(5)
1000 m	2000	12(3)	2(3)	3(3)	0(0)	—	0(0)	1(0)	86(1)	—	0(0)	17.0(1)	9(1)	52(5)	26.0(1)	0(5)
Outfall	2000	11(3)	1(1)	3(3)	0(0)	—	7(5)	1(1)	77(1)	—	0(5)	14(3)	46.0(5)	11(1)	14(1)	0(5)
500 m	2000	9(1)	0(1)	4(3)	0(0)	—	3(5)	1(1)	244(3)	—	0(5)	2.0(5)	64.0(5)	12(1)	16(1)	0(5)
1000 m	2000	18(3)	3(3)	5(3)	2(1)	—	4(5)	2(1)	97(1)	—	0(5)	22.0(1)	30.0(3)	23(3)	43(5)	0(5)

Note: [1] = Percent individuals as omnivore species is replaced by percent individuals as detritivore species; number of sensitive species is replaced by number of intolerant species; percent individuals as insectivores and carnivores is replaced by percent invertivores and top piscivores, respectively, for Great River metric application, * = DELT metric not measured in the field and given an average score of 3. Reference conditions are based on Simon (1992), Simon and Stahl (1998), and Emery et al. (in review).

Based on seasonal data collected by the utilities, collections conducted in the fall showed the highest diversity and biological integrity at the test sites. Reductions in biological integrity downstream of these facilities compared to reference conditions within designated index periods and expectations raised questions whether the production of electricity and protection of native aquatic communities could coexist without increased offstream cooling or increased mixing zone protection. A comparison of facilities on the Ohio, Wabash, and White Rivers showed that declines in metrics from the late summer were observed at all plants evaluated. This suggests a much greater impact than perceived and the problem may be more persistent than originally thought since species are forced to move to cooler waters. With increases in thermal outputs and requests for additional variances, these results will only be more conservative and the impacts associated with power production enhanced.

ACKNOWLEDGMENTS

The authors wish to express their gratitude to the Electric Power Research Institute (EPRI) for enabling TPS to attend several conferences concerning 316(a) issues including the national conference in Atlanta, GA in 1999. Cinergy generously provided the data and information used for this analysis. The opinions and results presented do not necessarily represent those of the Indiana Department of Environmental Management or the U.S. Fish and Wildlife Service, although this study may have been funded by those agencies.

REFERENCES

Brown, H.W. 1976. *Handbook of the Effects of Temperature on Some North American Fishes*, American Electric Power Corporation, Canton, OH.

Brungs, W.A. and B.R. Jones. 1977. *Temperature Criteria for Freshwater Fish: Protocol and Procedures*, EPA 600/3–77/061. U.S. Environmental Protection Agency, Environmental Research Laboratory, Duluth, MN.

Bulleit, K.A.N. 2000. Achieving sustainable development: a review of the environmental balancing act in two key clean water act regulatory programs, *Environmental Science and Policy*, 3, xv-xx.

Coutant, C.C. 1997. Thermal pollution, in *Encyclopedia of Chemical Technology*. Vol. 23, 4th ed., John Wiley & Sons, New York. 963–984.

Coutant, C.C. 2000. What is "normative" at cooling water intakes? Defining normalcy before judging adverse, *Environmental Science and Policy*, 3, 37–42.

Davis, W.S. and T.P. Simon (Eds.). 1995. *Biological Assessment and Criteria: Tools for Water Resource Planning and Decision Making*, Lewis Publishers, Boca Raton, FL.

Edison Electric Institute. 1996. *Environmental Directory of U.S. Power Plants – 1996*, Edison Electric Institute, Washington, D.C.

Emery, E.B., T.P. Simon, and R. Ovies. 1999. Influence of the Family Catostomidae on the metrics developed for a Great River Index of Biotic Integrity, in T.P. Simon (Ed.). *Assessing the Sustainability and Biological Integrity of Water Resources Using Fish Communities*, CRC Press, Boca Raton, FL, 203–224.

Emery, E.B., T.P. Simon, F.H. McCormick, P.L. Angermeier, C.O. Yoder, G.D. Hickman, J.E. DeShon, R.E. Sanders, M. Miller, R.J. Reash, W.D. Pearson, J.G. Shulte. In review. Development of an index of biotic condition for assessing the Ohio River. *Transactions of the American Fisheries Society*.

Emery, E.B. and J.A. Thomas. 2002. A method for assessing outfall effects on Great River fish populations: the traveling zone approach, in T.P. Simon (Ed.). *Biological Response Signatures: Patterns in Biological Integrity Using Freshwater Aquatic Assemblages*, CRC Press, Boca Raton, FL.

Emery, E.B, F.H. McCormick, and T.P. Simon. 2002. Response patterns of Great River fish assemblage metrics to outfall effects from point source discharges, in T.P. Simon (Ed.). *Biological Response Signatures: Patterns in Biological Integrity Using Freshwater Aquatic Assemblages*, CRC Press, Boca Raton, FL.

Electric Power Research Institute. 1981. *Literature Review: Response of Fish to Thermal Discharges*. EPRI EEA-1840/ORN/EIS-193. EPRI, Oak Ridge National Laboratory, Oak Ridge, TN.

Gammon, J.R. 1973. The effects of thermal input on the populations of fish and macroinvertebrates in the Wabash River. Purdue University Water Resources Research Center Technical Report 21.

Gammon, J.R. 1976. The fish populations of the middle 340 km of the Wabash River. Purdue University Water Resources Research Center Technical Report 86, Lafayette, IN.

Gammon, J.R. 1983. Changes in the fish community of the Wabash River following power plant start-up: projected and observed, in W.E. Bishop, R.D. Cardwell, and B.B. Heidolph (Eds.). *Aquatic Toxicology and Hazard Assessment: Sixth Symposium.* ASTM STP 802, 350–366.

Gammon, J.R. and T.P. Simon. 2000. Variation in a great river index of biotic integrity over a 20-year period, *Hydrobiologia*, 422/423, 291–304.

Hokanson, K.E.F. and K.E. Biesinger. 1980. *A National Compendium of Freshwater Fish and Water Temperature Data.* Vol. II. *Field Validation of Freshwater Fish Temperature Requirements.* U.S. Environmental Protection Agency, Environmental Research Laboratory, Duluth, MN.

Karr, J.R. and D.R. Dudley. 1985. Ecological perspectives on water quality goals, *Environmental Management,* 5, 55–68.

Lewis, R.B., M.A. Turner, J.E. Pike, J.S. Palla. 1993a. Fish and Benthic Macroinvertebrate Community Analysis near the Cayuga Generating Station May-October 1992. PSI Energy Technical Report EPW-046, Plainfield, IN.

Lewis, R.B., M.A. Turner, J.E. Pike, and J.S. Palla. 1993b. Fish and Benthic Macroinvertebrate Community Analysis near the Wabash River Generating Station May-October 1992. EPW-048. PSI Energy Technical Report, Plainfield, IN.

Lewis, R.B., J.S. Palla, and J.E. Pike. 1995. Fish and Benthic Macroinvertebrate Community Analysis near the Wabash River Generating Station 1993. EPW-052, PSI Energy Technical Report, Plainfield, IN.

Lewis, R.B., J.S. Palla, and J.E. Pike. 1996a. Fish and Benthic Macroinvertebrate Community Analysis near the Cayuga Generating Station, 1993. PSI Energy Technical Report EPW-053, Plainfield, IN.

Lewis, R.B., J.E. Pike, and D. Weston. 1996b. Fish and Benthic Macroinvertebrate Community Analysis near the Cayuga Generating Station 1994. PSI Energy Technical Report EPW-057, Plainfield, IN.

Lewis, R.B., J.E. Pike, and D. Weston. 1996c. Fish and Benthic Macroinvertebrate Community Analysis near the Wabash River Generating Station 1994. PSI Energy Technical Report EPW-056, Plainfield, IN.

Lewis, R.B., J.E. Pike, and K.E. Richard. 1997a. Fish and Benthic Macroinvertebrate Community Analysis near the Cayuga Generating Station 1995. PSI Energy Technical Report EPW-060, Plainfield, IN.

Lewis, R.B., J.E. Pike, K.E. Richard. 1997b. Fish and Benthic Macroinvertebrate Community Analysis near the Wabash River Generating Station 1995. PSI Energy Technical Report EPW-058, Plainfield, IN.

Lewis, R.B., J.E. Pike, and A.M. Edwards. 1998a. Fish and Benthic Macroinvertebrate Community Analysis near the Cayuga Generating Station 1996. PSI Energy Technical Report EPW-062, Plainfield, IN.

Lewis, R.B., J.E. Pike, and A.M. Edwards. 1998b. Fish and Benthic Macroinvertebrate Community Analysis near the Wabash River Generating Station 1996. PSI Energy Technical Report EPW-063, Plainfield, IN.

Lewis, R.B. and G. Seegert. 2000. Entrainment and impingement studies at two power plants on the Wabash River in Indiana, *Environmental Science and Policy,* 3, S303-S312.

Majewski, W. and D.C. Miller. 1979. Predicting effects of power plant once-through cooling on aquatic systems. *Technical Papers in Hydrology,* 20, UNESCO, Paris.

Ohio Environmental Protection Agency. 1989. *Biological Criteria for the Protection of Aquatic Life,* Volume II. *Users Manual for Biological Field Assessment of Ohio Surface Water.* Ohio EPA, Division of Water Quality Planning and Assessment, Ecological Assessment Section, Columbus, OH.

Patrick, R. 1950. Biological measure of stream conditions, *Sewage and Industrial Wastes,* 22, 926–938.

Raney, E.C. and B.W. Menzel. 1969. *Heated Effluents and Effects on Aquatic Life with Emphasis on Fishes. A Bibliography.* Water Resource Information Center Bulletin 2, Washington, D.C.

Reash, R.J. 1999. Considerations for characterizng midwestern large-river habitats, in T.P. Simon (Ed.), *Assessing the Sustainability and Biological Integrity of Water Resources using Fish Communities.* CRC Press, Boca Raton, FL, 475–505.

Reash, R.J., G.L. Seegert, and W.L. Goodfellow. 2000. Experimentally-derived upper thermal tolerances for redhorse suckers: revised 316(a) variance conditions at two generating facilities in Ohio, *Environmental Science and Policy,* 3, S191-S196.

Seegert, G. 2000a. The development, use, and misuse of biocriteria with an emphasis on the index of biotic integrity, *Environmental Science and Policy,* 3, S51-S58.

Seegert, G. 2000b. Considerations regarding development of index of biotic integrity metrics for large rivers, *Environmental Science and Policy*, 3, S99-S106.

Simon, T.P. 1992. Development of biological criteria for large rivers with an emphasis on an assessment of the White River Drainage, Indiana. EPA 905-R-92–026. U.S. Environmental Protection Agency, Chicago, IL.

Simon, T.P. (Ed.). 1999. *Assessing the Sustainability and Biological Integrity of Water Resources using Fish Communities*, CRC Press, Boca Raton, FL.

Simon, T.P. 2000. The use of biological criteria as a tool for water resource management, *Environmental Science and Policy*, 3, 43–49.

Simon, T.P. and R.E. Sanders. 1999. Applying an index of biotic integrity based on great river fish communities: considerations in sampling and interpretation. in T.P. Simon (Ed.). *Assessing the Sustainability and Biological Integrity of Water Resources using Fish Communities*. CRC Press, Boca Raton, FL. 475–505.

Simon, T.P. and J.R. Stahl. 1998. Development of index of biotic integrity expectations for the Wabash River. EPA 905-R-96–005. U.S. Environmental Protection Agency, Chicago, IL.

Simon, T.P. and J.R. Stahl. 2001. Clarifying Statement for the Report Entitled, *Development of Index of Biotic Integrity Expectations for the Wabash River*. U.S. Environmental Protection Agency, Chicago, IL.

Statisica. 1999. General Conventions and Statistics I. StatSoft, Inc., http://www.statsoft.com, Tulsa, OK.

U.S. Environmental Protection Agency. 1988. Standard Operating Procedures for Conducting Rapid Assessments of Ambient Surface Water Quality Using Fish. USEPA, Region 5, Central Regional Laboratory, Chicago, IL.

U.S. Environmental Protection Agency. 1990. Biological Criteria: National Program Guidance for State Managers. EPA 440–4–90–010. Office of Water, Washington, D.C.

U.S. Environmental Protection Agency. 1991. Technical Support Document for Water Quality-Based Toxics Control. EPA/505/2–90–001 PB91–127415. USEPA, Office of Water, Washington, D.C.

U.S. Environmental Protection Agency. 1994. *Water Quality Standards Handbook*, 2nd ed. EPA-823-B-94–005a. USEPA, Office of Water, Washington, D.C.

Veil, J. A. and D.O. Moses. 1996. Consequences of proposed changes to Clean Water Act thermal discharge requirements. International Clean Water Conference: Clean Water: Factors That Influence Its Availability, Quality and Its Use, La Jolla, CA 28–30 November 1995, *Water, Air and Soil Pollution*, 90, 41–52.

25 Assessing the Ecological Integrity of the East Branch of the Grand Calumet River: Responses of Four Biological Indicators

Thomas P. Simon, Scott A. Sobiech, Daniel W. Sparks, and Krysztof M. Jop

CONTENTS

0-8493-0905-0/03/$0.00+$1.50

25.1 INTRODUCTION

As the complexity of water resource problems increases, the need for more sophisticated approaches grows. A biological system is healthy and has ecological integrity when its inherent potential is realized, its condition is stable, its capacity for self-repair is maintained, and external support for maintenance is minimal (Karr et al., 1986; Karr, 1993). The policy of independent application (IA) of the United States Environmental Protection Agency (USEPA) suggests that all environmental data be weighted equally for evaluation. It used the analogy of a three-legged stool to support different monitoring approaches, i.e., water quality parameters, whole effluent toxicity testing, and ambient biological surveys.

Karr (1993) challenged this concept on the basis that it was too rigid and inadequate to address changing environmental conditions. He proposed the analogy of a tripod supporting a spotting scope. In order to see a distant object, such as a designated use, the three legs of the tripod must be adjusted to accommodate the terrain, and this is the nature of the water resource problem. IA is considered controversial by many since the biological data that are direct measures of aquatic life designated uses can usually only affect management decisions in a unilateral manner. Under IA, biological data can only affect management decisions when both the whole effluent toxicity and water chemistry data indicate that no problem exists; thus, assessments can only cause more stringent environmental management decisions.

The International Joint Commission designated several specific areas of the Great Lakes as predominantly contributing to basinwide degradation (Hartig and Zarull, 1992). These areas of concern are being evaluated through a series of remedial action plans (RAPs) that identify impaired designated uses. The development and implementation of RAPs constitute a two-stage process: (1) identification and implementation of remedial actions necessary to resolve immediate problems in the near-term, and (2) investigation and continued planning to identify and implement remedial actions to fully restore all impaired beneficial uses (International Joint Commission, 1989). The RAP process consists of three components that identify, assess, and implement management measures to remediate conditions and provide important milestones for documenting success. The loss of aesthetics, decline in biological integrity and biological diversity, loss of habitat, and increase of contaminants are common attributes of areas of concern.

The Grand Calumet River is urban and industrialized. It suffered severe degradation that has been documented since the mid 1960s (U.S. Department of the Interior, 1966, 1967; USEPA, 1985; Simon et al., 1989). Spills, municipal and industrial waste-water discharges, and combined sewer overflows contributed to declines in water and sediment quality. Simon et al. (1989) suggested that the East Branch of the Grand Calumet River has the greatest potential for improvement and select areas could marginally support biological integrity goals of the remedial action plan. Sediment and water quality have been identified as significant impediments to the establishment of stable biological communities in the East Branch (U.S. Department of the Interior, 1966, 1967; USEPA, 1985; Simon et al., 1989; Simon, 1989).

The purpose of our study was to compare the ecological impacts downstream of a major iron and steel manufacturer in the East Branch, using four biological response indicators chosen by the International Joint Commission. We evaluated sediment toxicity, fish community, macroinvertebrate community, and habitat indicators as measurements of disturbance. Limited evaluation of environmental indicators was conducted to substantiate initial perceptions of impaired designated uses. Hartig and Law (1994) suggested basinwide evaluation should not unnecessarily focus on areas of concern for each of the Great Lakes; rather, it should reflect goals and objectives.

25.2 STUDY AREA AND METHODS

25.2.1 Study Area Description

The Grand Calumet River occurs in a low relief area occupying the bed of glacial Lake Chicago. Its basin is located in northwestern Indiana and encompasses 43,242 acres contained almost entirely within Lake County, Indiana (USEPA, 1985). The river is a small watershed, approximately 34 km in length. The general flow is sluggish and westward in the East Branch, eastward or westward in the West Branch, depending on Lake Michigan levels, and northward in the Indiana Harbor Canal, an artificial connection to Lake Michigan. Land use disturbance is extensive. Modifications include ditching, channelization, flow modification, development of urban centers, and development of one of the most concentrated steel and petrochemical industrial complexes in the United States.

Historically, the East Branch of the Grand Calumet River was dredged and iron was mined from the dredge spoil and processed through the steel plant (W.L. Redmond, personal communication). The iron and steel manufacturer provides the most substantial discharge to the East Branch. Average flow from 14 permitted outfalls is 400 ft^3/s, which is essentially (67%) noncontact cooling water (Crawford and Wangsness, 1987). Seven outfalls contribute most of the wastewater that has an average river discharge of 127 ft^3/s upstream and 491 ft^3/s downstream of the plant (USEPA, 1985). Typical effluent chemistry and receiving water quality are shown in Table 25.1 (Crawford and Wangsness, 1987; T.P. Simon, unpublished data). Severe sediment contamination has been documented for the upper 5 mi of the East Branch including polycyclic aromatic hydrocarbons (PAHs), metals, and oil and grease (Hoke et al., 1993).

25.2.2 Community Collection and Reach Selection

Fish and macroinvertebrate communities were sampled in river reaches bracketing point source discharge outfalls, so that outfalls that constitute a large portion of the load to the East Branch could be evaluated. Sample reaches were selected based on previous water and sediment sampling both above and below the mill in the East Branch (Crawford and Wangsness, 1987; Hoke et al., 1993). Five reaches were evaluated to assess environmental impacts (Figure 25.1). All reaches in this study are located on the iron and steel manufacturer's property. Reach 1, upstream of Tennessee Street (adjacent to the coke batteries), is the furthest upstream location below outfall 2 (RM 13.5); Reach 2 is downstream of Broadway Avenue (below oil booms, RM 13.2); Reach 3 is upstream of the Interstate 90 exit ramps which are below Outfall 7 (RM 12.3); Reach 4 is downstream of Bridge Street (RM 10.0); and Reach 5 is upstream of the Wabash Railroad trestle (immediately upstream of the Gary Sanitary District outfalls) and is the lowermost zone (RM 9.0) in the study area incorporating all remaining steel mill outfalls.

Benthic macroinvertebrate communities were collected using Hester–Dendy multiplate samplers and qualitative dip netting methods (Plafkin et al., 1989; Ohio Environmental Protection Agency, 1989). Aquatic macroinvertebrates were sampled with modified Hester–Dendy multiplate artificial substrate samplers at each sampling reach. Multiplate samplers were made from 0.3175-cm tempered hardboard circular discs. Each sampler consisted of five circular discs with a total area of 0.02468 m^2. Five replicate samplers were deployed at each sampling location for 6 wks. Qualitative dip netting using 500-µ mesh netting was conducted for each site for all available habitats in each reach for a minimum of 15 min.

All invertebrates were preserved in 70% ethyl alcohol and labeled with the sampled location information. Invertebrates were returned to the laboratory for sorting and identification to family level. We assessed biological integrity using a modified version of the invertebrate community index (ICI) metrics (Ohio Environmental Protection Agency, 1989). Because suitable reference conditions for nonwadeable streams in Indiana do not exist for the study area, the Ohio Environmental Protection Agency's (1989) reference database served as reference conditions for this study.

TABLE 25.1
Representative Characteristics of East Branch Effluent and Water Quality (Fall, Low Flow)

Parameter	Upstream Mill Effluent	Mill Effluent	Near-Field	Far-Field
Dissolved oxygen (ppm)	5.5–8.2	5.3–8.0	7.4	7.5
pH	7.6–8.1	6.6–8.1	7.9	6.6
Specific conductance (μS/cm)	356–367	335–420	359	735
Suspended solids (mg/L)	3–10	2–5	4	2
Total dissolved solids (mg/L)	173–203	162–523	186	523
Chloride (mg/L)	13–25	13–190	18	190
Fluoride (mg/L)	0.2	0.1–1.3	0.3	0.2
Sulfate (mg/L)	26–33	26–47	31	47
Hardness (mg/L $CaCO_3$)	110	110–130	110	280
BOD_5 (mg/L)	2.0–3.8	1.0–2.4	1.8	12.0
Total phenols (μg/L)	<1–17	<1–67	2	67
Total cyanide (mg/L)	0.04–0.05	<0.01–0.06	0.02	<0.01
Total nitrogen (mg/L as N)	0.1–0.5	<0.1–0.2	0.1	0.4
Total ammonia (mg/L as N)	0.58–1.5	0.22–1.70	0.8	0.22
Total nitrite (mg/L as N)	0.02–0.05	0.01–0.03	0.06	0.18
Total nitrate (mg/L as N)	0.21	0.18–0.21	0.21	0.11
Total phosphorus (mg/L)	0.02–0.04	<0.01–0.04	0.03	0.03
Total chromium (μg/L)	<1–3	<1–1	2	1
Total hexavalent chromium (μg/L)	<1	<1	<1	<1
Total copper (μg/L)	<1–2	<1–1	4	<1
Total iron (μg/L)	380–1500	250–1100	810	1100
Total lead (μg/L)	1–7	<1–20	6	<1
Total mercury (μg/L)	0.2–0.5	0.3–1.0	0.2	0.9
Total nickel (μg/L)	4–5	5–11	8	8
Total zinc (μg/L)	20–30	20–30	40	30

The Ohio EPA has shown that ICI score expectations do not vary among ecoregions and we determined they could be used as suitable references for our study. We calculated a family biotic index (FBI) that provides a measure of organic degradation in a stream based on the pollution tolerances of the invertebrates collected (Hilsenhoff, 1988) and an ICI for each location. Functional feeding guild metrics were based on designations in Merritt and Cummins (1984), Cummins and Wilzbach (1985), and Thorp and Covich (1991).

Fish species composition and relative abundance data were gathered by electrofishing surveys of river reaches using a model 6A Smith–Root boat-mounted electrofisher. Electrofishing surveys included systematic sampling within reaches of bank areas, usually for a distance of 500 m for a minimum of 1000 s. Captured fish were placed in an onboard holding tank until sampling was completed. Data recorded for each survey included the identification, weight, and enumeration of captured fish and sample and habitat conditions.

Species level fish population data were gathered on all fish and included total length (mm) range, total weight (g), and results of external examination for disease and anomalies. Relative abundance was expressed as catch-per-unit-of-effort (CPUE) or the number of fish in the total catch per 500 m for each reach. Biological integrity of the Each Branch was assessed using the index of biotic integrity (Karr et al., 1986). Regional reference conditions representing least disturbed conditions within the region (Davis and Simon, 1995), were based on the established biological expectations for the Lake Michigan division, Lake Michigan section of the Central Corn Belt Plain Ecoregion (Simon, 1991). The reference condition is a compilation of sites (Simon, 1991) used as

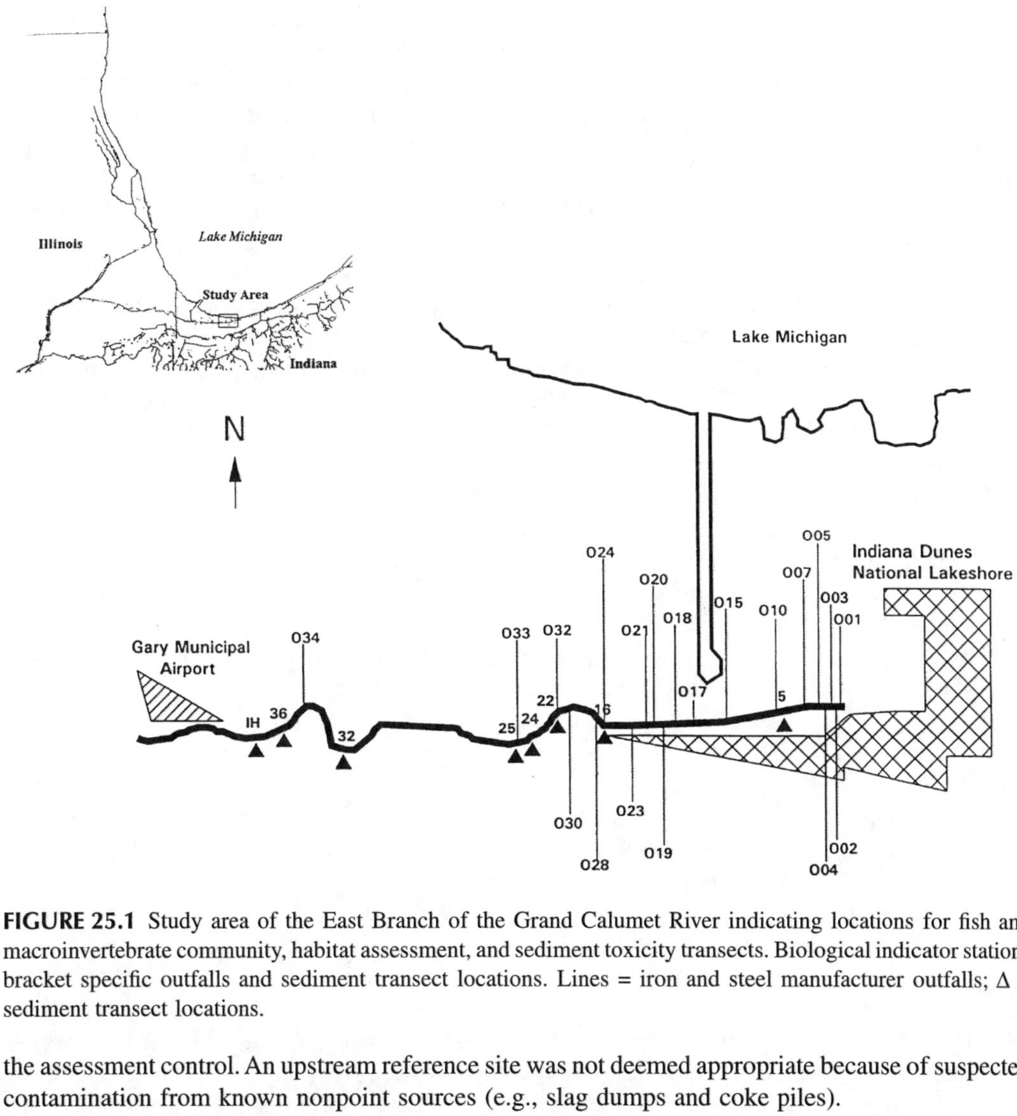

FIGURE 25.1 Study area of the East Branch of the Grand Calumet River indicating locations for fish and macroinvertebrate community, habitat assessment, and sediment toxicity transects. Biological indicator stations bracket specific outfalls and sediment transect locations. Lines = iron and steel manufacturer outfalls; Δ = sediment transect locations.

the assessment control. An upstream reference site was not deemed appropriate because of suspected contamination from known nonpoint sources (e.g., slag dumps and coke piles).

25.2.3 Habitat Assessments

We conducted habitat evaluations for the entire 34-km section of the East Branch to identify major land use types and principal habitat characteristics. Site-specific habitat assessments were conducted using the qualitative habitat evaluation index (QHEI) to assess microhabitat and other variables following Rankin (1995). The QHEI systematically evaluates habitat features including substrate types, instream cover, channel morphology, bank erosion, riparian corridor quality, pool and riffle development, and channel modifications. Habitat quality was determined for each of the five sites within the study reach by compositing habitat category scores; a maximum score of 100 points indicated the highest quality.

25.2.4 Sediment Toxicity Testing

We collected sediment samples from the five community assessment reaches in the East Branch. Long Lake, an interdunal palustrine wetland in the Indiana Dunes National Lakeshore (IDNL), was used as a reference for sediment comparison. Grab samples of surface sediment (targeting the upper 0 to 150 mm) were collected with a stainless steel petite ponar dredge that was decontaminated after every use. Approximately 2 L of sediment were collected from each location, stored on ice at 4°C and shipped via overnight delivery for analysis. Samples were immediately tested upon receipt or refrigerated at 4°C, then warmed to room temperature prior to testing. Species were cultured and tested using the same water recommended for bulk and elutriate tests.

Prior to testing, each sediment sample was strained through a 2-mm stainless steel sieve to remove rocks, debris, and large clumps of sediment. Samples were homogenized in a clean 1000-mL glass beaker by stirring for 10 min before division into replicate test vessel aliquots. A total of 200 mL of sediment per vessel was used for each of the five test replicates. We used 1000-mL beakers as test vessels and a sediment depth of 2 cm. The 10-day exposure for each test was conducted in a temperature-controlled water bath capable of maintaining test solutions at 20 ± 1°C. Overlying well water (800 mL), was gently added to each replicate and had a total hardness of 39 to 40 mg/L $CaCO_3$, alkalinity of 26 mg/L $CaCO_3$, pH 7.2 to 7.3, total organic carbon of 0.56 mg/L, and specific conductance between 130 and 140 μohms/cm. Well water filtered through an Amberlite XAD-7 resin column to remove potential organic contaminants was used as a diluent for short-term elutriate chronic tests. The dilution water characteristics included total hardness between 46 and 48 mg/L, total alkalinity of 27 to 28 mg/L, pH of 7.4 to 7.6, and specific conductance range of 150 to 180 μohms/cm. Photoperiods for both tests were 16 h of light and 8 h of darkness, with a light intensity range between 30 and 50 ft-candles.

The acute toxicity of bulk sediment samples was measured using the epibenthic amphipod, *Hyalella azteca*. Toxicity test procedures followed the standards described in the American Society for Testing Materials' "Guidelines for conducting sediment toxicity tests with freshwater invertebrates" (1992). *Hyallela azteca*, less than 1 wk old, were reared in culture water that had total hardness of 60 to 80 mg/L $CaCO_3$, pH of 7.3 to 7.8, temperature 22 to 24°C, and specific conductance between 100 and 200 μohms/cm. The test initiated when 20 amphipods were added to each of six replicate exposure vessels. Test vessels were covered with plastic wrap and aerated to reduce evaporation and ensure adequate levels of dissolved oxygen.

Test solutions were renewed three times weekly by carefully siphoning 75% (approximately 600 mL) of the existing overlying water and gently replacing it with fresh overlying water. Amphipods were fed daily a suspension of flake fish food (1000 mg/mL) at a rate of 300 μL per test vessel. At completion, survival of amphipods was determined by sieving the sediment from each replicate exposure test vessel and counting the live organisms. Measurements of test solutions and biological observations were recorded for each test vessel at the initiation (0 h), mid-test (168 h), and termination (240 h) and on other days in alternate test vessels. Dissolved oxygen, pH, temperature, hardness, alkalinity, and specific conductance were measured for new and old renewal water.

The chronic toxicity of contaminants released from the mixing of the sediment–water interface was measured by exposing *Ceriodaphnia dubia* to elutriated sediment. The elutriate test is a simulation of potential contaminant release from dredged material under the hydraulic conditions existing during open water disposal events. The short-term chronic tests using *C. dubia* followed standard procedures described by the U.S. Environmental Protection Agency (Method 1002.0, 1989). Four parts sediment and one part water were thoroughly mixed for 10 min by shaking. The resulting pore water from each sediment reach was used for test solutions at varying dilutions. These tests were static renewal exposures conducted in 30- mL disposable plastic cups containing 15 mL of elutriate solution and a single daphnid. Plastic cups were covered with plastic wrap and placed in a styrofoam (2.5-cm insulation board) rack in a water bath. Tests were conducted at 25 ± 1C. The test included four concentrations (100, 50, 25, and 10%) and a dilution water control (soft reconstituted water). Each treatment consisted of ten replicates of single organisms less than 25 h old hatched within the same 8-h period. Test solutions were renewed daily by transferring adult organisms to the appropriate prepared test solution. During the renewal process, each cup was examined and the number of offspring produced over the preceding 24-h period was recorded.

The mean number of offspring in a treatment was the total number of offspring produced during the exposure to treatment divided by the total number of females in the treatment.

25.2.5 Statistics

Statistical analyses were carried out by standard methods (Zar, 1984). Differences between reaches were examined by performing single factor ANOVA, followed by Dunnett's test for comparison of upstream, near-, and far-field stations if significant differences were observed. Student t-tests were used to evaluate the longitudinal effects from the mill effluent. Correlation and regression analyses were performed using SAS and tested for significance. Chronic toxicity values were determined using standard procedures (USEPA, 1989). We used Wilcoxon's rank sum test to determine significance of *Ceriodaphnia* reproduction at a level of $p < 0.05$.

25.3 RESULTS

25.3.1 Habitat Analysis

Our assessment suggests that habitat alterations related to human modification reduced the quality of the East Branch of the Grand Calumet River (Table 25.2). Relatively little difference existed in the quality of microhabitats within the study area, but two habitat zones could be recognized: sites above and including RM 13.5 (upstream) and those below RM 13.2 (near-field and far-field). Habitats in the East Branch above and including RM 13.5 included a constrained channelized ditch possessing little or no instream cover, shallow depths, slow flow, extensive siltation and soft bottom substrates. The riparian corridor consisted of coal piles and coke batteries of the iron and steel manufacturer and a steep-graded grass bank prone to erosion and sloughing into the stream. Habitat was principally run habitat of constant uniform depth with no riffle or pool habitat occurring in this stream reach. This upper zone scored the poorest of the study segments, possessing a total habitat score of 22.

A statistically significant difference (Student t-test, $p \geq 0.05$) in QHEI scores occurred between RM 13.5 and the near-field (RM 13.2 to 12.3) and all subsequent far-field stations downstream (RM 10.0 to 9.0). Station microhabitat scores improved dramatically below RM 13.2 as the numbers and types of substrates improved. Substrates below RM 12.2 included a variety of materials including silt, clay, cobble, sand, and gravel. The number of habitat cycles also improved with run and pool habitats occurring at most sites. However, the quality of the riparian corridor, morphology of the stream channel, and erosional characteristics of the stream still reflected degraded conditions reflective of bank modification. The far-field stations (RM 10.0 to 9.0) possessed additional types of instream cover including woody debris, emergent and submergent aquatic macrophytes, and less modified riparian corridors than the near-field stations.

25.3.2 Benthic Macroinvertebrate Community Assessment

25.3.2.1 Taxa Distribution and Relative Abundance

Generally, the number of taxa collected increased with distance from the mill in the near-field and far-field stations. Total numbers of individuals and density varied at all locations sampled and did not follow any trend. Number of taxa, total number of individuals collected, and density were greatest at the far-field station furthest from the mill. A single taxon constituted as much as 89% of the total invertebrates collected within a reach. Differences in invertebrate taxa diversity and density were not attributed to habitat differences since the QHEI scores were not substantially different for near- and far-field stations, although differences in assemblages were observed between stations.

TABLE 25.2
Habitat Characteristics of the East Branch Watershed

Reach	Habitat Type	Channel Morphology	Substrate	Instream Cover	Other Observations
Upstream Mill Discharges (above RM 13.5)	Run 100%	Confined channel within steep banks	Silt, muck coal fines	None	Erosion and sloughing of steep grass banks; low flow; shallow, uniform depths; no aquatic macrophytes or periphyton; heavily disturbed land use associated with coal fines and runoff particulates
Near-field (RM 13.2-12.3)	Run 50% Pool 50%	Occasionally confined channel within steep banks	Clay, silt, muck, cobble, large gravel	Submerged aquatic macrophytes, artificial structures, and few deep pools	Erosion of steep banks often severe, pilings and pipe supports provide some cover and scouring of substrates; moderate variety of depth; significant input of discharge from outfalls including oil and grease; riparian corridor unnatural
Far-field (RM 10.0-9.0)	Run 20% Pool 80%	Channel seldom confined within steep banks	Muck, silt, sand	Emergent and submerged aquatic macrophytes; woody debris common; more frequent deep pools; shallows along meandering channel	Riparian corridors often with hardwood forest; little erosion or sloughing of banks; significant differences in depths of pools; less significant input of outfall dscharge; channel morphology heavily modified

25.3.2.2 Community Structure and Function

The invertebrate communities were indicative of severe environmental degradation. We did not collect any sensitive taxa from upstream, near-field, or far-field stream segments. Dominant taxa were Dipteran and noninsect invertebrates that are tolerant of environmental perturbations. With the exception of Reach 4, invertebrate communities were dominated (60.5 to 89.2%) by a single taxon. FBI's ranged from 6.23 to 8.63, showing general improvement in invertebrate communities in the near- and far-field reaches.

Gathering-collectors (75.4%) and scrapers (20.7%), dominated the East Branch invertebrate community. We observed a few predacious invertebrates during the survey; shredding and filter-feeding invertebrates were represented by single individuals. The invertebrate community trophic structure was variable at all locations sampled. The unbalanced trophic structure of the invertebrate community is typical of poor environmental conditions.

25.3.3 Fish Community Assessment

25.3.3.1 Species Distribution and Relative Abundance

Species distribution and relative abundance, as estimated by CPUE, varied significantly (Student t-test, $p \geq 0.05$) among reaches in the upstream, near-field, and far-field stations. The most disturbed area was the upstream station (RM 13.5) that had the fewest species and lowest relative abundance of any of the zones. The near-field and far-field stations contained 10 species and had similar relative abundances. Although not significant, the near-field stations actually showed slightly better CPUE values. This initial increase in relative abundance may be a result of the increased flow velocities and initial increase in habitat heterogeneity providing habitat opportunities for the depauperate fish fauna.

25.3.3.2 Community Structure and Function

Community structure in all three stream zones showed aspects of extreme environmental degradation. Dominant species collected in all stream reaches were tolerant of severe environmental degradation and toxins. Tolerant species, including common carp *Cyprinus carpio*, goldfish *Carassius auratus*, and green sunfish *Lepomis cyanellus*, possess a broad niche breadth and are adaptable to extremes in environmental conditions. No sensitive species, headwater species, benthic insectivores, and few or no simple lithophilic spawning species were collected from upstream, near-field, or far-field segments.

Far-field stations differed (Chi-squared test, $p \geq 0.05$) from near-field and upstream stations, showing improvements in the percentage of insectivorous species, fewer hybrids, and fewer pioneer species. Species considered insectivorous, e.g., pumpkinseed *Lepomis gibbosus*, green sunfish, and golden shiner *Notemigonus crysoleucas*, are all mid-water column species that can feed on terrestrial or surface-dwelling aquatic insects. Thus, changes in these species' relative abundance may be inaccurate indicators of improvement. The percentage of hybrids was significantly different (Chi-squared test, $p \geq 0.05$) between the near-field (4.0 to 13.5%) and the far-field (0 to 1.3%) stations. Likewise, pioneer species showed increased relative abundance in disturbed systems or were the first to recolonize sites after removal of impacts in water quantity or quality. Pioneer species were significantly (Chi-squared test, $p \geq 0.01$) more abundant in upstream zones (100%) and less significant (Chi-squared test, $p \geq 0.05$) for near-field zones (15.5 to 7.1%) than far-field zones (2.7 to 14.6%).

The number of minnow species showed a significant increase (Chi-squared test, $p \geq 0.01$) in the near- and far-field stations. We collected five minnow species at the near- and far-field stations as compared to a single species at the upstream station. The minnow species we collected were all nonindigenous or exotic species that successfully colonized these degraded habitats in the absence of competition. Species such as the Eurasian rudd *Scardinius erythrophthalmus*, goldfish, common carp, and the tolerant native bluntnose minnow *Pimephales notatus* have been able to increase. We consider this phenomenon contrary to the original intention of the metric as originally proposed by Simon (1991). We regard the increase of these exotic species to be a negative reflection of environmental quality.

25.3.3.3 Occurrence of Deformities, Eroded Fins, Lesions, and Tumors

Deformities, eroded fins, lesions, and tumors (DELT anomalies) are considered indicators of degraded conditions contributing to a decline of individual fitness and health (Karr et al., 1986; Davis and Simon, 1995). Without anthropogenic input of toxins, individual fish frequently show few or no individuals with DELT anomalies (Davis and Simon, 1995; Sanders et al., 1999). For instance, individuals associated with agricultural land uses seldom exhibit DELT anomalies (T.P. Simon, unpublished data).

None of the few fish species collected upstream of the mill outfalls had DELT anomalies. The highest percentage of DELT anomalies was found in the near-field zone downstream of the coke piles. We observed a reduction in DELT anomalies with increased downstream distance from the iron and steel industry's outfalls. The most frequently encountered DELT anomalies were lesions observed in most near- and far-field stations.

25.3.3.4 Biological Integrity

The invertebrate community index (ICI) and index of biotic integrity (IBI) are multimetric indices used to evaluate the biological conditions of invertebrate and fish communities with structure and function attributes (Karr et al., 1986; Ohio EPA, 1989; Davis and Simon, 1995). The IBI was calibrated for use in northwestern Indiana (Simon, 1991) by using modifications from Karr's original index (Table 25.3).

TABLE 25.3
Modified IBI Characteristics Used to Evaluate the Lake Michigan Division, Central Corn Belt Plain Ecoregion for Least Impacted Regional Reference Conditions for Drainage Areas Less than 20 mi² (after Simon, 1991)

Attribute	Scoring 1 (Worst)	3	5 (Best)
Species Composition			
Total number of species	<6	6–12	>12
Number of darter/sculpin/madtom species	<1	1–2	>2
Number of sunfish species	<2	2–4	>4
Number of minnow species	<2	2–4	>4
Number of sensitive species	<2	2–4	>4
Percent tolerant species	>50%	25–50%	<25%
Trophic Composition			
Percent omnivorous feeding species	<38.7%	19.3–38.7%	<19.3%
Percent insectivorous feeding species	<25%	25–50%	>50%
Percent pioneer species	<24.7%	24.7–49.4%	>49.4%
Fish Condition			
Catch per unit effort	<100	100–275	>275
Percent simple lithophilous spawners	<16.5%	16.5–33.9%	>34%
Percent DELT anomalies	>1.3%	0.1–1.3%	<0.1%

The condition of the East Branch's invertebrate and fish communities were severely impaired. ICI scores were <2 at all stations sampled. ICI integrity classifications ranged from very poor in the upstream and near-field stations to poor in the far-field stations. Furthest upstream fish communities scored the poorest with an IBI assessment rating of very poor (IBI score of 12). Many of the percentage metrics scored as strongly deviating from the reference condition because fewer than 25 individuals were present in the 500-m station reach. All other near- and far-field stations were assessed as very poor to poor (IBI of 22).

25.3.4 Sediment Toxicity Assessment

25.3.4.1 *Hyallela* Acute Whole Sediment Toxicity

Acute whole sediment toxicity tests showed severe toxicity for *H. azteca* ranging from 1.0 to 1.9 toxic units (TUs) (Table 25.4). The control sediment from Long Lake showed replicate survival ranging from 97 to 99%. All ten sediment samples showed significant deviation from the control sediment. Replicates were not significantly different among sites, with the exception of Transect 21 (ANOVA, $p \geq 0.05$). Percent survival of amphipods was low (0 to 4%) for all reaches sampled. Reaches 1, 3, 4, and 5 showed 100% mortality, while Reach 2 had 96% mortality. No correlations were observed between sites with higher survival and recorded water quality parameters (Table 25.5).

Hyallela azteca exhibited differential trajectories between sediment transects on the East Branch of the Grand Calumet River (Figure 25.2). Three distinct response trajectories were observed on Day 0. The lowest response trajectory split into two paths after 48 h. Three of the response trajectories had similar final responses with nearly 100% mortality. The lowest trajectory from Transect 21 was the only one that showed moderate toxicity (52% mortality). Reach 1 exhibited the fastest toxicity response with 100% mortality within 24 h (occurring within the initial 30 min). Transects 2 and 4 showed the slowest response after 24 h. These reaches responded with only 5% mortality after 24 h and 20% mortality at 48 h.

TABLE 25.4
Percent Survival of *Hyallela azteca* for Five Replicates at Ten Localities in the East Branch of the Grand Calumet River

Sediment Transect Locality	Reach	Percent Survival A	B	C	D	E	$\bar{X}$[d]	$\bar{X}$[e]
Long Lake (INDL)[a]	—	100	100	100	95	100	99	
GCR-6	1	0	0	0	0	0	0[c]	0
GCR-11	2	35	25	10	40	30	28[c]	
GCR-18	2	0	0	20	0	0	4[c]	16
GCR-21	3	55	50	50	55	30	48[c]	
GCR-24	3	0	0	0	0	0	0[c]	
Long Lake (INDL)[b]	—	100	95	100	90	100	97	
GCR-28	3	5	0	0	0	0	1[c]	16.3
GCR-32	4	0	0	0	0	0	0[c]	
GCR-34	4	20	15	20	15	10	16[c]	8
GCR-35	5	0	0	0	0	0	0[c]	
GCR-36	5	0	0	0	0	0	0[c]	0

[a] Control for East Branch Sediment Transects 6, 11, 18, 21, and 24 tested July 8 through 18, 1994.
[c] Mean concentration mortality is significantly different from control ($p < 0.01$).
[b] Control for East Branch Sediment Transects 28, 32, 34, 35, and 36 tested July 6 through 16, 1994.
[d] Value represents concentration mean for specific transect.
[e] Mean value represents composite of transects within each community sampling reach.

TABLE 25.5
Chemical Characteristics of Sediment Exposure Water during 10-Day Acute Static Renewal Tests with *Hyallela azteca* at Ten Localities in the East Branch of the Grand Calumet River

Sediment Sample	Dissolved Oxygen (mg/L)	pH	Temperature (°C)	Hardness (mg/L $CaCO_3$)	Alkalinity (mg/L $CaCO_3$)	Specific Conductance
Long Lake (INDL)[a]	7.2–8.9	6.9–7.9	21	48–56	30–36	170–180
GCR-6	6.1–8.4	7.0–8.0	21	48	48	220
GCR-11	6.5–8.9	7.0–7.9	21	48–84	48–68	220–250
GCR-18	6.4–8.9	7.0–8.3	21	60–96	60–86	220–250
GCR-21	6.7–8.9	7.0–8.1	21	68–104	64–112	200–250
GCR-24	6.6–8.8	7.0–8.0	21	56–104	60–88	215–240
Long Lake (INDL)[b]	7.1–8.9	7.0–8.5	20–21	40–44	32	170–200
GCR-28	5.2–8.9	7.3–8.4	20–21	46–66	46–66	180–200
GCR-32	5.5–8.7	7.1–8.2	20–21	44–64	48–72	175–200
GCR-34	4.2–8.9	7.3–8.2	20–21	48–72	44–56	180–200
GCR-35	5.7–8.6	7.1–7.6	20–21	48–60	40–46	180–200
GCR-36	5.6–8.5	7.2–7.9	20–21	44–64	40–50	180–200

[a] Control for East Branch Sediment Transects 6, 11, 18, 21, and 24 tested July 8 through 18, 1994.
[b] Control for East Branch Sediment Transects 28, 32, 34, 35, and 36 tested July 6 through 16, 1994.

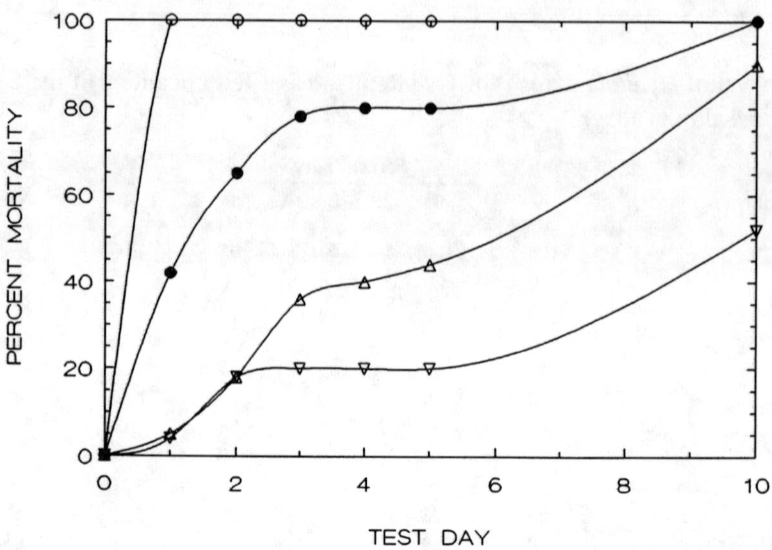

FIGURE 25.2 Response trajectories of *Hyallela azteca* mortality from 10 sediment transects in the East Branch of the Grand Calumet River. o = Transect 6; • = Transects 24 and 36; Δ = Transects 11, 18, 28, 32, 34, and 35; ∇ = Transect 21.

TABLE 25.6
Preliminary Survival and Mean Number of Young Produced per Female for Sediment Elutriate Toxicity Dilution Assessment Using Survival and Reproduction Endpoints in Chronic Exposures of *Ceriodaphnia dubia*

Station/Dilution	Mean Percent Survival	Mean Number of Young per Female ± SE
Long Lake	100	49.8 ± 25.1
GCR-6		
10%	100	37.0 ± 23.4
25%	100	34.2 ± 22.5
50%	100	30.2 ± 16.5
100%	100	5.3 ± 4.1
GCR-35		
10%	100	37.7 ± 22.1
25%	100	35.2 ± 22.3
50%	100	37.3 ± 25.0
100%	80	22.8 ± 13.4
GCR-36		
10%	100	34.8 ± 22.0
25%	100	39.5 ± 25.7
50%	90	35.3 ± 23.0
100%	100	35.7 ± 22.9

25.3.4.2 *Ceriodaphnia* Chronic Sediment–Water Interface Assessment

Elutriate toxicity tests were conducted on the three most contaminated sediment reaches. We used the daphnid *C. dubia* endpoints for survival and reproduction and a dilution series of 0.5 (Table 25.6). Mean survival of *C. dubia* ranged from 80 to 100%. We did not observe significant differences in survival between the control and any of the reaches. *Ceriodaphnia* females exposed to 100% elutriate samples produced 5.3 to 35.7 mean numbers of young. Reproduction showed a significant dose response decline from the control ($P \geq 0.01$) in elutriate samples from Reach 1. We found a no-observed-effect concentration (NOEC) of 50% elutriate and lowest-observed-effect concentration (LOEC) of 100% elutriate for Reach 1 (Transect 6). Neither of the transect samples from Reach 5 (Transects 35 and 36) showed significant reproduction differences from controls. Higher specific conductance in overlying water (450 μohms/cm) of transect 6 sediment was significantly different from transect 35 (260 to 270 μohms/cm) and 36 (250 to 290 μohms/cm).

25.4 DISCUSSION

25.4.1 Habitat Quality

The Grand Calumet River is a significant ecological resource because its corridor contains the largest amounts of dune and swale wetlands in northwestern Indiana. Approximately 11% of the study area is considered dune and swale wetlands (U.S. Fish and Wildlife Service, unpublished data). In addition, the wetlands provide an important flyway for resting and feeding of migrating birds, and serve as habitats for nesting and raising young. Coastal wetlands along the Great Lakes are important as spawning and nursery habitats for many fish species, yet are considered the rarest habitat resources (Goodyear et al., 1982).

The mill virtually eliminated the pre-existing habitat typical of the dune and swale wetlands that existed along the East Branch of the Grand Calumet River. The quality of habitat improved in the far-field stations as a result of a less constrained channel and the presence of more emergent and submergent wetland macrophytes along the stream margin. The best habitat we observed was near RM 10. This site possessed woody debris and a slight meander to the stream channel that enabled the formation of deep pool and run habitats and development of macrophyte beds. The riparian corridor possessed forested bottomland habitat on one side of the channel. However, even with habitat improvement, little increase in biological integrity was observed. Rankin (1995) found that sites capable of supporting a balanced and integrated biological community needed to score a minimum of 65 QHEI points to be comparable to reference conditions.

25.4.2 Benthic Macroinvertebrate Community Assessment

Benthic macroinvertebrate communities in the East Branch were similar to those at degraded sites (Plafkin et al., 1989; Hilsenhoff, 1988; Merrtt and Cummins, 1984). We noted low numbers of invertebrate taxa at all sites during the survey, with the exception of Reach 5 where the invertebrate community was uneven and dominated (89.2%) by a single taxon. The severe degradation in the East Branch caused the absence of sensitive taxa, high percentages of tolerant (95.5%) species, and FBI scores ranging from poor to very poor. No sensitive invertebrate taxa were collected from the East Branch. The low numbers of individuals collected from Reaches 1 through 4 and the dominance of a single taxon indicate the extent of degradation.

Degraded environmental conditions in the East Branch adversely affected invertebrate trophic composition. Gathering collectors (75.4%) and scrapers (20.7%) dominated the East Branch. The trophic structure dominance of gathering collectors and the low number of predacious, shredding, and filter-collector feeding insects are considered irregular and unbalanced. Trophic imbalance is usually a response to an overabundance or lack of particular food sources (Klemm et al., 1990) or may be indicative of a particular impact. The near-absence of shredders may be a result of poor riparian conditions because of a lack of coarse particulate organic matter, a lack of microbial flora or the absorption of toxicants to organic matter (Plafkin et al., 1989). Additionally, the near-absence of filter-feeding invertebrates suggests that water-borne toxicants adversely affect the East Branch (Plafkin et al., 1989).

25.4.3 Fish Community

We found slight increases in species distribution, diversity, and relative abundance at RM 13.2 and RM 10.0, compared to historical conditions (Simon et al., 1989). Similar species dominated in 1994 as among near-field and far-field reaches in normal (1985) and low (1988) water years and during periods when the mill did not operate (1986). Thus, species composition may be limited by habitat parameters, persistent discharge of toxics, and toxic constituents accumulated in the sediments.

The lack of sensitive species, headwater species, benthic insectivores, and few or no simple lithophilic spawning species from any of the upstream, near-field, or far-field stations are indicators of extreme degraded conditions (Karr et al., 1986; Simon, 1991; Davis and Simon, 1995). These characteristics of fish communities were lacking for the entire Grand Calumet River watershed. Benthic insectivorous darter species, such as logperch *Percina caprodes* and Iowa darter *Etheostoma exile*, were found in the Grand Calumet River (Simon et al., 1989; T.P. Simon and P.M. Stewart, unpublished data) and Iowa darter still exist in some of the higher quality portions of the watershed.

Our expectations for this watershed are based on over 300 collections in the tributaries of southern Lake Michigan that represent least impacted conditions that were used to modify the IBI

(Simon et al., 1989; Simon, 1991). Based on these observations, we expect headwater species, such as blacknose dace *Rhinichthys atratulus*, to be collected since they are present in the system and have been collected in all other major tributaries to Lake Michigan (Gerking, 1945; Simon and Stewart, 1999; T.P. Simon, unpublished data). We also expected a higher proportion of simple lithophilous spawning species, such as blacknose dace, white sucker *Catostomus commersoni*, and other sucker species belonging to such genera as *Carpiodes, Moxostoma,* and *Ictiobus*. We did not collect any of these anticipated species from the East Branch, which may suggest that toxic contamination of sediments and suffocation of eggs and developing embryos are results of siltation, bank sloughing, and discharge of suspended solids.

Fish communities of high biological integrity in the Lake Michigan drainage possess a greater percentage of insectivorous species, fewer hybrids, and fewer pioneer species (Simon, 1991). Trophic guilds are functional response indicators of disruption in fish communities. The difficulty in assessing trophic guild response is that a number of insectivorous species can switch between omnivorous and specialized feeding in response to differential water quality and habitat. Karr et al. (1986) considered only insectivorous cyprinids for evaluating community structure. Simon (1991), however, considered the historical distribution of minnows in the Lake Michigan drainage and replaced this metric because too few species are possible. Simon (1991) did not exclude exotic minnow species from enumeration. The increase in number of exotic species in the Grand Calumet River will bias the overall intention of this metric. An inadequate macroinvertebrate food base in the East Branch can cause some species, such as golden shiners, to feed as omnivores rather than insectivores.

Many authors have not found a correlation between the presence of hybrids and degradation. Simon (1991) rejected the percentage of hybrids metric substituting the percentage of simple lithophilous species since hybrids occur in high quality systems, usually in low proportions. Disruption of spawning events because of conditions, e.g., flow, turbidity or because of mechanical indiscriminate behavior, have usually resulted in sterile or noncontributing individuals to localized demes because of deleterious mutations. We do not consider the high proportion of hybrids in the East Branch in the near-field areas of impact a coincidence. Further study of increased numbers of hybrids needs to be pursued.

Simon (1991) used a stream recovery metric in the IBI for northwest Indiana that is based on Smith (1971). Smith coined the phrase "pioneer species" to suggest a guild relationship for certain species that are the first to invade or recolonize an area after sufficient recovery (Rankin and Simon, Chapter 10, this volume). Pioneer species are usually unable to compete successfully in more stable communities and attain highest levels in disturbed or severely impacted habitats (Smith, 1971; Simon, 1991). Pioneer species are often the only taxa able to persist under extreme conditions in the Great Lakes basin (Simon, 1991).

Few studies correlated *in situ* DELT anomalies with known whole effluent pollution sources and impacts (Reash and Berra, 1989; Davis and Simon, 1995; Sanders et al., 1999), except for studies that evaluated the influence of single compounds or effluent types, such as pulp mill effluents (Swanson et al., 1994; Hodson et al., 1992) and sediment (Black, 1983; West et al., 1988; Baumann et al., 1991). No studies other than that of Baumann et al. (1987) evaluated the influence of iron and steel mill effluents on incidence of DELT anomalies. Our studies found an increased frequency of lesions and cutaneous abrasions in the vicinity of the iron and steel mill effluents. We suggest that the absence of DELT anomalies in fish from the upstream site was a result of reduced exposure. Species that exhibit increased incidence of DELT are benthic omnivores such as common carp and white sucker. Baumann et al. (1987) used bullheads and white suckers as indicator organisms because of increased tumor frequency in the vicinity of a steel mill with high PAH levels. Reash and Berra (1989) found benthic species comprising a variety of trophic guilds that developed DELT in an Ohio stream receiving treated municipal and industrial effluents.

25.4.4 Biological Integrity

The biological integrity of invertebrate and fish communities was rated as poor or very poor (Table 25.7). The macroinvertebrate and fish communities in the East Branch possessed low numbers of individuals and species diversity. We collected few or no indicator species such as mayflies, darters, sunfish, and salmon. No darter, caddisflies or mayfly species, which are sensitive or intolerant of low dissolved oxygen, siltation, and toxicants, were collected from the East Branch (Karr et al., 1986; Simon, 1991). Low numbers of sunfish species suggest relatively poor substrate conditions in pools and an inadequate macroinvertebrate food base (Karr et al., 1986; Simon, 1991). Salmon wedre stocked routinely in Lake Michigan tributaries (Simon, 1991; Simon and Stewart, 1999); however, no salmon were collected from any of the near- or far-field stations even though they were present several kilometers downstream. Thermal changes as a result of noncontact cooling water and effluent pollutants may act as barriers to fish movements.

The East Branch fish community is the epitome of the very poor community described by Karr et al. (1986) at sites with total IBI scores below 22. These sites possess few fish, are dominated by tolerant species, commonly contain hybrids, and frequently have high proportions of DELT anomalies. We observed skewed trophic function. Absence of carnivorous fish may cause cascading trophic level responses. High percentages of omnivores and the abundance of golden shiners are symptomatic of declining environmental quality (Simon, 1991). CPUEs were low, ranging from 3 to 103 fish per 500 m sampled. Four white suckers collected from three sampling reaches were the only simple lithophils collected. A high occurrence of external anomalies is a sensitive IBI metric for estimating biological integrity (Karr et al., 1986).

Simon (1991) found that modifications were necessary for IBI scoring when low numbers of fish were collected. He (1991) suggested that a minimum of 25 individuals should be collected from sites with drainage areas of 20 mi.[2] If fewer than 25 fish are collected, erroneous interpretations of percentage metrics could result since metric response would be unpredictable. When fewer than 25 fish are collected, all percentage metrics are low-end scored to avoid interpretation bias. The

TABLE 25.7
Summary of Response Indicators for Reaches in the East Branch of the Grand Calumet River

	Reach				
Indicator	**1**	**2**	**3**	**4**	**5**
Invertebrates					
Total number of invertebrate taxa	5	7	7	11	14
Total number of organisms collected	96	199	99	87	488
Density (organisms/m^2)	—	673	308	486	3160
FBI score	6.59	7.23	7.47	8.63	6.23
ICI score	1	1	0	0	2
ICI integrity classification	Very poor	Very poor	Very poor	Very poor	poor
Fish					
Total number of fish species	2	7	9	10	6
Total number of fish collected	3	103	99	89	74
Catch per unit of effort (fish/500m)	3	103	99	89	74
IBI Score	12	18	22	22	22
IBI integrity classification	Very poor	Very poor	Very poor	Very poor	Very poor
Habitat					
QHEI score	22	41	46	51	48
Sediment toxicity					
% amphipod survival	0	4	0	0	0

furthest downstream site draining the East Branch is 18.13 mi^2 (Scott Morloch, U.S. Geological Survey, personal communication); thus, all sampling sites in our study are considered headwater streams. Although East Branch drainage is relatively small, the quantity of water added to the river from the steel mill's outfalls makes the river resemble a much larger river than its drainage size would indicate; thus its biological integrity may be overestimated.

25.4.5 Sediment Toxicity Assessment

The use of several test organisms to quantify responses from effluent (Cairns, 1985; Birge et al., 1986) or sediment toxicity (Burton et al., 1989; Geisy et al., 1988; Simon et al., 1990) is preferred to single species tests due to differing modes of action and metabolic processes. Ecosystem sensitivity is influenced by a number of factors, such as indigenous species sensitivity, physiochemical alterations of toxicity (due to natural or anthropogenic factors), seasonal effects, and food web interactions. The sensitivity of species varies with site and contaminant types (Burton et al., 1989). Simon et al. (1990) found that iron and steel manufacturing effluents typically showed significant results for *Ceriodaphnia* survival (correlation coefficient = 0.781) for subchronic toxicity endpoints. Simon (1989), during previous testing of iron and steel discharges, found a significant difference in three endpoints. Fathead minnow hatchability, survival, and teratogenicity were significantly different from control results in the summer of 1986.

Hoke et al. (1993) evaluated three areas in the East Branch of the Grand Calumet River using the burrowing midge *Chironomus tentans* near Reaches 1 and 3 and Transect 5. We noted similar responses between *Chironomus* and *Hyallela*. Transects 6, 35, and 36 exhibited 100% mortality during the exposure period. Burton et al. (1989) used several test organisms to evaluate toxicity to whole sediment exposures. The response of *Hyallela* was more sensitive to Indiana Harbor Canal sediments than *Daphnia magna* but was significantly (t-test, $P \geq 0.01$) less sensitive than *C. dubia*. Our results showed that *Hyallela* exhibited more increased sensitivity to bulk sediments than *C. dubia* did to elutriate samples from overlying water.

Our *Ceriodaphnia* toxicity test results did not find any toxic effects on survival or young production, with the exception of Transect 6 (Reach 1). *Hyallela* bulk sediment survival was significantly different from chronic elutriate tests. The interaction between the sediment and overlying water may be overestimated in laboratory testing since the oxidized microzone does not represent ambient conditions. Laboratory test conditions never showed reducing conditions.

25.4.6 Evaluation of Four Biological Endpoints

The environmental condition of the East Branch required adjustment of our tripod to reflect a stronger emphasis on biological community indicators (Table 25.7). Severe sediment degradation in the East Branch would not be able to predict aquatic life quality using bulk and elutriate sediment toxicity tests. The sensitivity of the survival and reproduction chronic elutriate endpoints could not discriminate between changes in contaminated sediment to allow modeling of sediment contaminant affects as a result of transport through the system. Response trajectories could serve as an alternate endpoint for discriminating between magnitude and severity at heavily contaminated sites (Figure 25.2). In our study, it was apparent that the sediment toxicity response, when initially low after the first 48 h, usually resulted in lower final toxicity units. The elutriate tests we conducted on overlying water from Transects 6, 35, and 36 did not find significant results for survival and reproduction. Burton et al. (1989) found *Ceriodaphnia* to be the species most sensitive to contaminated Great Lakes harbor sediments, but based on the reproductive endpoint, significant adverse effects could not be determined from some of the worst contaminated sites in the East Branch.

The response signature determined from the iron and steel manufacturer shows a reduction in total number of species, loss of benthic specialists, loss of sensitive species, skewed trophic guilds to omnivores and generalist feeders, loss of habitat specialists, an increase in tolerant, exotic, and pioneer species, and an increase in lesions. Increases in ICI and IBI scores reflecting the presence of high quality macroinvertebrate and fish communities are better able to evaluate the severity and extent of contamination that could provide targets for remediation. An all-or-none response using sediment toxicity tests is not effective in evaluating remediation alternatives unless it can discriminate among levels of contamination. We recommend using survival results from bulk sediment and reproduction endpoints from elutriate tests to evaluate terminal endpoints for remediation and restoration of severely contaminanted sediments. Sediment toxicity tests may not be used to determine the magnitude and extent of sediment remediation, but can serve as measures of final sediment remediation success.

We compared our findings to those from various freshwater site studies of ambient toxicity and biological community response (Mount et al., 1984; Mount et al., 1985; Mount and Norberg-King, 1986; Norberg-King and Mount, 1986; Mount et al., 1986). Mount et al. (1986) found that as river size increased, fish species avoidance or complete absence of select taxa would be necessary to show a response. Studies on the Kanawha and Ottawa Rivers showed a significant correlation ($P \leq 0.005$) between macroinvertebrate community response and *Ceriodaphnia* chronic toxicity (Mount et al., 1984; Mount and Norberg-King, 1986), however, we did not observe similar results for benthic macroinvertebrate community responses or the results from *Ceriodaphnia* elutriate toxicity tests.

Our studies of the East Branch did not show significant toxicity as a result of survival or reproduction differences from the control even though benthic macroinvertebrate communities were severely degraded. Results of toxicity testing on Skeleton Creek found that impact at stream stations was correlated with toxicity and the number of lost species (Norberg-King and Mount, 1986). We found *Hyallela* survival after exposure to bulk sediment toxicity to be correlated with loss of species richness but not to *Ceriodaphnia* elutriate survival or reproduction. The results of community response and toxicity from Five Mile Creek downstream of a coke plant found that no single indicator could predict the impacts at different stations (Mount and Norberg-King, 1985). Predictions using acute toxicity tests found that usually half or fewer of the species are present at those sites. We found that species richness increased away from the coke plant on the East Branch and that the *Hyallela* survival response trajectory was the most reliable predictor of sediment toxicity (Figure 25.2).

25.5 CONCLUSIONS

The Grand Calumet River and Indiana Harbor Canal is a small, industrial watershed with unique and complex problems. It has been designated an area of concern by the International Joint Commission. The Grand Calumet River is the only area of concern impaired for all 14 designated uses. Bulk and elutriate sediment toxicity tests, fish and macroinvertebrate community evaluations, and habitat assessments were conducted to evaluate status of endpoints used for Stage II remediation. Habitat was degraded and upstream ratings were significantly different from near-field and far-field habitats. Upstream habitats had slower flows, soft substrates, more silt deposition, a confined channel, and a single habitat type. The amphipod bulk sediment test had significantly different survival from the control for all 10 transects tested. *Ceriodaphnia* elutriate test results did not show significant survival differences from the control; the reproduction endpoint showed significant dose response differences for a single reach. Severely degraded fish and macroinvertebrate communities were observed at upstream, near-field, and far-field zones in the East Branch. Hilsenhoff FBI values ranged from 6.23 to 8.63. All stations were classified as possessing very poor or poor fish and invertebrate biotic integrity. The assessment endpoints selected by the International Joint Commission confirm that the East Branch of the Grand Calumet River is severely impaired for the biological endpoints measured. The sediment toxicity evaluations confirm our suspicion that, in addition to poor physical habitat conditions, the toxicity of the sediments further prevents invertebrates and fish communities from colonizing.

ACKNOWLEDGMENTS

The authors want to thank Aaron Spicer for field and laboratory assistance. We appreciate the input of various colleagues including Ronald Kovach, Thomas Bramscher, and Dave Hudak. Paul M. Stewart improved a previous version of this chapter. Although the U.S. Environmental Protection Agency, Region 5, Compliance and Enforcement Branch in Chicago and the U.S. Fish and Wildlife Service funded this project, the opinions expressed by the authors do not necessarily represent those of the agencies.

REFERENCES

American Society for Testing and Materials. 1992. *Standard Guide for Conducting Sediment Toxicity Test with Freshwater Invertebrates.* American Society for Testing and Materials, Philadelphia, Pennsylvania. Volume 11.04, Designation E 1383–92, 1116–1138.

Baumann, P.C., W.D. Smith, and W.K. Parland. 1987. Tumor frequencies and contaminant concentrations in brown bullhead from an industrialized river and a recreational lake, *Transactions of the American Fisheries Society,* 116, 79–86.

Baumann, P.C., M.J. Mac, S.B. Smith, and J.C. Harshbarger. 1991. Tumor frequencies in walleye (*Stizostedion vitreum*) and brown bullhead (*Ictalurus nebulosus*) and sediment contaminants in tributaries of the Laurentian Great Lakes, *Canadian Journal of Fsheries and Aquatic Sciences*, 48, 1804–1810.

Birge, W.J., J.A. Black, and B.A. Ramey. 1986. Evaluation of effluent biomonitoring systems, in H.L. Bergman, R.A. Kimerle, A.W. Maki (Eds.). *Environmental Hazard Assessment of Effluents.* SETAC, Special Publication Series, Pergamon Press, Elmsford, N.Y, 66–80.

Black, J.J. 1983. Field and laboratory studies of environmental carcinogenesis in Niagara River fish, *Journal of Great Lakes Research*, 9, 326–334.

Burton, G.A., Jr., B.L. Stemmer, K.L. Wnks, P.E. Ross, and L.C. Burnett. 1989. A multitrophic level evaluation of sediment toxicity in Waukegan and Indiana Harbors, *Environmental Toxicology and Chemistry,* 8, 245–258.

Cairns, J., Jr. 1985. *Multispecies Toxicity Testing.* SETAC Special Publication, Pergamon Press, New York.

Crawford, C.G. and D.J. Wangness. 1987. Streamflow and water quality of the Grand Calumet River, Lake County, Indiana, and Cook County, Illinois, October 1984. U.S. Geological Survey, Water Resources Investigation report 86–4208.

Cummins, K.W. and M.A. Wilzbach. 1985. Field Procedures for Analysis of Functional Feeding Groups of Stream Macronvertebrates. Contribution 1611. University of Maryland, Forstburg.

Davis, W.S. and T.P. Simon (Eds.). 1995. *Biological Assessment and Criteria: Tools for Water Resource Planning and Decision Making.* Lewis Publishers, Boca Raton, FL.

Geisy, J.P., R.I. Graney, J.L. Newsted, C.J. Rosiu, A. Benda, R.G. Kreis, Jr., and F.J. Horvath. 1988. Comparison of three sediment bioassay methods using Detroit River sediments, *Environmental Toxicology and Chemistry,* 7, 483–498.

Gerking, S.D. 1945. Distribution of Indiana fishes, *Investigations of Indiana Lakes and Streams*, 4, 1–137.

Goodyear, C.D., T.A. Edsall, D.M. Ormsby Dempsey, G.D. Moss, and P.E. Polanski. 1982. *Atlas of Spawning and Nursery Areas of the Great Lakes Fishes.* FWS/OBS-82/52. U.S. Fish and Wildlife Service, Washington, D.C.

Hartig, J.H. and N.L. Law. 1994. Progress in Great Lakes Remedial Action Plans: Implementing the Ecosystem Approach in Great Lakes Areas of Concern. EPA 905-R-24–020. U.S. Environmental Protection Agency and Environment Canada. 210 pp.

Hartig, J.H. and M.A. Zarull (Eds.). 1992. *Under RAPs: Towards Grassroots Ecological Democracy in the Great Lakes Basin.* University of Michigan Press, Ann Arbor, MI.

Hilsenhoff, W.L. 1988. Rapid field assessment of organic pollution with a famly-level biotic index, *Journal of the North American Benthological Society,* 7, 65–68.

Hodson, P.V., M. McWhirter, K. Ralph, B. Gray, D. Thivierge, J.C. Carey, G. Van Der Kraak, D.M. Whittle, and M.C. Levesque. 1992. Effects of bleached kraft mill effluent on fish in the St. Maurice River, Quebec, *Environmental Toxicology and Chemistry,* 11, 1635–1651.

Hoke, R.A., J.P. Giesy, M. Zabi, and M. Unger. 1993. Toxicity of sediments and sediment pore waters from the Grand Calumet River – Indiana Harbor, Indiana area of concern, *Ecotoxicology and Environmental Safety*, 26, 86–112.

International Joint Commission. 1989. *Report on the Great Lakes Water Quality.* Great Lakes Water Quality Board, Windsor, Ontario, Canada.

Karr, J.R. 1993. Defining and assessing ecological integrity: beyond water quality, *Environmental Toxicology and Chemistry*, 12, 1521–1531.

Karr, J.R., K.D. Fausch, P.L. Angermeier, P.R. Yant, and I.J. Schlosser. 1986. *Assessment of Biological Integrity in Running Waters: A Method and its Rationale*. Illinois Natural History Survey, Special Publication 5, Champaign, IL.

Klemm, D.J., P.A. Lewis, F. Fulk, and J.M. Lazorchak. 1990. *Macroinvertebrate Field and Laboratory Methods for Evaluating the Biological Integrity of Surface Waters.* EPA 600/4–90/030. U.S. Environmental Protection Agency, Cincinnati, OH.

Merritt, R.W. and K.W. Cummins (Eds.). 1984. *An Introduction to the Aquatic Insects of North America.* 2nd ed. Kendall/Hunt Publishing Company, Dubuque, Iowa.

Mount, D.I. and T. Norberg-King (Eds.). 1986. *Validity of Ambient Toxicity Tests for Predicting Biological Impact, Kanawha River, Charleston, West Virginia.* EPA 600/3–86/006. U.S. Environmental Protection Agency, Office of Research and Development, Duluth, MN.

Mount, D.I. and T. Norberg-King (Eds.). 1986. *Validity of Ambient Toxicity Tests for Predicting Biological Impact, Kanawha River, Charleston, West Virginia.* EPA 600/3–86/006. U.S. Environmental Protection Agency, Office of Research and Development, Duluth, MN.

Mount, D.I., A.E. Steen, and T. Norberg-King (Eds). 1985. *Validity of Ambient Toxicity Tests for Predicting Biological Impact on Five Mile Creek, Birmingham, Alabama.* EPA 600/8–85/015. U.S. Environmental Protection Agency, Office of Research and Development, Duluth, MN.

Mount, D.I., A.E. Steen, and T. Norberg-King (Eds). 1986. *Validity of Ambient Toxicity Tests for Predicting Biological Impact, Ohio River, near Wheeling, West Virginia.* EPA 600/3–85/071. U.S. Environmental Protection Agency, Office of Research and Development, Duluth, MN.

Mount, D.I., N.A. Thomas, T.J. Norberg, M.T. Barbour, T.H. Roush, and W.F. Brandes. 1984. *Effluent and Ambient Toxicity Testing and Instream Community Response on the Ottawa River, Lima, Ohio.* EPA 600/3–84/080. U.S. Environmental Protection Agency, Office of Research and Development, Duluth, MN.

Norberg-King, T.J. and D.I. Mount (Eds.). 1986. *Validity of Ambient Toxicity Tests for Predicting Biologcal Impact, Skeleton Creek, Enid, Oklahoma.* EPA 600/8–86/002. U.S. Environmental Protection Agency, Office of Research and Development, Duluth, MN.

Ohio Environmental Protection Agency. 1989. *Biological Criteria for the Protection of Aquatic Life.* Vol. III. Standardized Biological Field Sampling and Laboratory Methods for Assessing Fish and Macroinvertebrate Communities. Ohio Environmental Protection Agency, Division of water Quality Monitoring and Assessment, Ecological Assessment Section, Columbus, OH.

Plafkin, J.L., M.T. Barbour, K.D. Porter, S.K. Gross, and R.M. Hughes. 1989. *Rapid Bioassessment Protocols for Use in Streams and Rivers: Benthic Macroinvertebrates and Fish.* EPA 440/4–89/001. U.S. Environmental Protection Agency, Assessment and Watershed Protection Division, Washington, D.C.

Rankin, E.T. 1995. Habitat indices in water resource quality assessments, in W.S. Davis and T.P. Simon (Eds.). *Biological Assessment and Criteria: Tools for Water Resource Planning and Decision Making*. Lewis Publishers, Boca Raton, FL, 181–208.

Reash, R.J. and T.M. Berra. 1989. Incidence of fin erosion and anomalous fishes in a polluted stream and a nearby clean stream, *Water, Air, and Soil Pollution*, 47, 47–63.

Sanders, R.E., R.J. Miltner, C.O. Yoder, and E.T. Rankin. 1998. The use of external deformities, erosion, lesions, and tumors (DELT anomalies) in fish assemblages for characterizing aquatic resources: a case study of seven Ohio streams. in T.P. Simon (Ed.). *Assessing the Sustainability and Biological Integrity of Water Resources Using Fish Communities.* CRC Press, Boca Raton, FL, 225–246.

Simon, T.P. 1989. Sub-chronic toxicity evaluations of major point source dischargers in the Grand Calumet River and Indiana Harbor Canal, Indiana, using the embryo-larval survival and teratogenicity test, *Proceedings of the Indiana Academy of Science,* 98, 241–255.

Simon, T.P. 1991. Development of Index of Biotic Integrity Expectations for the Ecoregions of Indiana. I. Central Corn Belt Plain. EPA 905/9–91/025. U.S. Environmental Protection Agency, Region 5, Environmental Sciences Division, Monitoring and Quality Assurance Branch: Ambient Monitoring Section, Chicago, IL.

Simon, T.P. and P.M. Stewart. 1999. Structure and function of fish communities in the southern Lake Michigan basin with emphasis on restoration of native fish communities, *Natural Areas Journal*, 19, 142–154.

Simon, T.P., A. Lubin, and F. LeCurieux. 1990. Predictive abilities of Environmental Protection Agency subchronic toxicity test endpoints for complex effluents, *Proceedings of the Indiana Academy of Science,* 99, 29–38.

Simon, T.P., G.R. Bright, J. Rud, and J. Stahl. 1989. Water quality characterization of the Grand Calumet River basin using the Index of Biotic Integrity, *Proceedings of the Indiana Academy of Science*, 98, 257–265.

Smith, P.W. 1971. Illinois streams: a classification based on their fishes and an analysis of factors responsible for the disappearance of native species, Biological Notes 76. Illinois Natural History Survey, Champaign, IL.

Swanson, S.M., R. Schryer, R. Shelast, P.J. Kloepper-Sams, and J.W. Owens. 1994. Exposure of fish to biologically treated bleached-kraft mill effluent. 3. Fish habitat and population assessment, *Environmental Toxicology and Chemistry*, 13, 1497–1507.

Thorp, J.H. and A.P. Covich. 1991. *Ecology and Classification of North American Freshwater Invertebrates.* Academic Press, San Diego, CA.

U.S. Department of the Interior. 1966. Report of the Water Quality of Lower Lake Michigan, Calumet River, Little Calumet River and Wolf Lake by Department of the Interior for the Period January 1966 through June 1966, Illinois-Indiana. Federal Water Pollution Control Administration, Great Lakes Region, U.S. Department of the Interior, Chicago, IL.

U.S. Department of the Interior. 1967. Report of the water quality of lower Lake Michigan, Calumet River, Little Calumet River and Wolf Lake by Department of the Interior for the Period July 1966 through December 1966, Illinois-Indiana. Federal Water Pollution Control Administration, Great Lakes Region, U.S. Department of the Interior, Chicago, IL.

U.S. Environmental Protection Agency. 1985. Master Plan for Improving Water Quality in the Grand Calumet River/Indiana Harbor Canal. EPA 905–9–84–003C. U.S. EPA, Water Division, Region 5, Chicago, IL.

U.S. Environmental Protection Agency. 1989. Methods for Estimating the Chronic Toxicity of Effluent and Receiving Waters to Freshwater Organisms. EPA 600–4–89–001. U.S. Environmental Protection Agency, Cincinnati, OH.

West, W.R., P.A. Smith, G.M. Booth, and M.L. Lee. 1988. Isolation and detection of genotoxic components in a Black River sediment, *Environmental Science and Technology*, 22, 224–228.

Zar, J.H. 1984. *Biostatistical Analysis.* Prentice Hall, Englewood Cliffs, N.J.

26 Variable Assemblage Responses to Multiple Disturbance Gradients: Case Studies in Oregon and Appalachia, USA

Sandra A. Bryce and Robert M. Hughes

CONTENTS

26.1 INTRODUCTION

In the past 20 years, assessing biological condition through biological monitoring has assumed a more prominent position alongside chemical assessments of water quality. There has been a gradual evolution within federal and state water resource agencies to consider degradation to include physical habitat loss, adverse biological changes, and chemical alteration. As a result, much recent effort has been directed toward the development of multimetric indices of biotic integrity (IBIs) for various aquatic and riparian indicator taxa, particularly fish (Karr et al., 1986; Simon, 1999),

0-8493-0905-0/03/$0.00+$1.50

benthic macroinvertebrates (Kerans and Karr, 1994; Fore et al., 1996), algae (Hill et al., 2000; Fore, Chapter 22, this volume), and riparian birds and mammals (Brooks et al., 1991; O'Connell et al., 1998). Models predicting particular species from known environments compared against observed species also have been used for benthic macroinvertebrates (Wright, 1995; Norris and Hawkins, 2000).

Few investigators have examined multiple assemblages across a sample to compare their differing responses, sensitivities, and clarity of signal in indicating the condition of aquatic systems. Brooks et al. (1991) found that the differences between minimally disturbed and disturbed watersheds in Pennsylvania were more clearly revealed by avian indicators in the riparian corridor than fish and macroinvertebrates in the aquatic environment. In Michigan, fish IBI scores were more strongly related to nearby land use/land cover than to local habitat structure, while macroinvertebrate measures were more strongly correlated to reach habitat conditions (Lammert and Allan, 1999). Allen et al. (1999) examined concordant patterns of taxonomic composition among benthic macroinvertebrates, riparian birds, sedimentary diatoms, fish, and pelagic zooplankton assemblages in 186 northeastern U.S. lakes. Birds, fish, and macroinvertebrates correlated most strongly with broad scale factors such as climate and forest composition; diatoms and zooplankton correlated highly with water chemistry (pH) and lake depth. Unlike the Michigan results, in the lakes study, the macroinvertebrates joined the larger bodied organisms in responding to broad scale factors, possibly because the geographic variables used were at a coarser regional scale than the watershed scale variables used in the Michigan study. In any case, these results suggest that disparate assemblages operating in different media and at various scales are likely to have varying sensitivities and thresholds in response to a wide range of disturbances. Better recognition of the strengths and weaknesses of various indicator assemblages may help us choose different indicator taxa for water column, substrate, or riparian zone assessments.

In addition to choice of indicators, a successful evaluation of the extent of impairment to aquatic ecosystems hinges on our ability to recognize and measure human-caused alterations of stream ecosystems. An advantage of biological indicators over strictly chemical measurements of water quality is that they integrate the effects of stressors in their surroundings. This is also a disadvantage, because it causes problems in interpretation. When interpreting biological response, we must disentangle the complex interactions between natural variation in the environment and the effects of human activity on the landscape.

Developers of multimetric indices use their ecological knowledge to choose attributes of an indicator assemblage that they expect to vary systematically across a gradient of human disturbance. When presented graphically, these attributes may respond to increasing disturbance as expected, but the signal may be diluted by variability due to inconsistent individual metric responses and the influence of a number of interacting stressors (Wiley et al., Chapter 12, this volume). Presumably, the responses would be more dramatic (and more likely to satisfy those who expect statistical significance) if one could isolate a human activity on the landscape and evaluate biological response along a gradient of that single disturbance (Yoder and Rankin, 1995; Karr and Chu, 1999).

We compared index results in two different regions: a small sample of 13 stream reaches in the Willamette Valley of western Oregon, and a larger sample of 102 stream reaches in the mid-Appalachian region of Maryland, Pennsylvania, Virginia, and West Virginia (Figures 26.1A and B). We used indices and metrics for fish (Hughes et al., 1998; McCormick et al., 2001), benthic macroinvertebrates (Li et al., 2001; Klemm et al., in review), diatoms (Fore, Chapter 22, this volume), and riparian birds (Bryce et al., in press) for both geographical areas that were previously evaluated using a similar measure of disturbance (Bryce et al., 1999; USEPA, 2000). A common measure of disturbance simplified our ability to look across assemblages and refine the disturbance gradient as needed. We also examined outliers among the index and metric scores. Outliers provided important evidence to aid in interpreting the pathways of disturbance and biological response (Karr and Chu, 1999). They provided insight into situations that increased or limited quality for a representative class of streams. Thus, evidence from outliers could be used in the development of best management practices for streams.

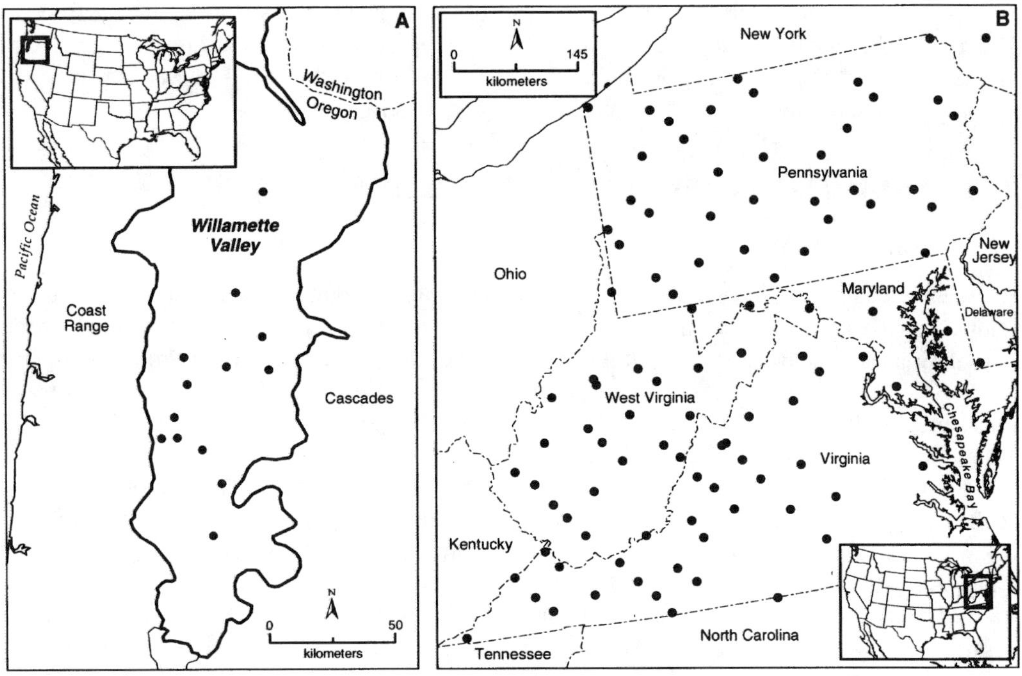

FIGURE 26.1 Spatial distribution of (A) 13 Willamette Valley and (B) 102 mid-Appalachian stream reaches included in this study.

Our objectives in this study were (1) to compare the responses of multiple aquatic and riparian assemblages to a general disturbance gradient; (2) to examine assemblage responses along more refined disturbance gradients representing individual disturbance types; and (3) to seek examples of individual metrics that were most clearly diagnostic of particular disturbances. We aimed to present a process suggesting ways to evaluate probable causes of variability in both predictor and response variables.

26.2 DEFINING THE DISTURBANCE GRADIENT

To develop the initial disturbance gradient in the mid-Appalachian region, we screened the sample sites and their watersheds using an iterative process of map analysis, aerial photo interpretation, and physical habitat information (Bryce et al., 1999). We recorded and ranked the number, proximity, intensity, and extent of all human alterations to riparian and upland areas that were detectable using these data. Once we catalogued the stressors affecting the stream reaches and watersheds, we classified the watersheds into a scoring class based on the accumulating severity of stressors. The scores ranged from 1 to 5, minimal to high disturbance. Scores of 2 denoted a relative lack of human influence, but had one or more disqualifying factors, such as a road paralleling the stream. A score of 3 denoted watersheds with intermediate risk of impairment where, though human alteration was dominant, mitigating factors tended to preserve stream quality. Watersheds that scored 4 also had attributes mitigating the major impacts of intensive farming, mining, or urban development that placed others in the highest disturbance category (score of 5).

From an aquatic indicator perspective, the disturbance index incorporated some measures of stream channel condition, such as fish cover, amount of woody debris, shading, substrate, and visual sedimentation estimates (general water clarity and embedddness), but it did not include other water column effects such as nutrient concentration, toxic pollutants, or suspended sediments. The resulting disturbance index scores were consistent with measures of stream condition based on water chemistry and benthic macroinvertebrates, neither of which had been used in the scoring process. The method was applied again in the Willamette Valley of Oregon, to test bird response metrics in the development of a bird integrity index (Bryce et al., in press). Thus, 115 watersheds in two similarly ranked regions were used to draw our indicator comparisons.

26.3 WILLAMETTE VALLEY, OREGON, CASE STUDY

26.3.1 Background

Between 1992 and 1996, physical habitat, riparian birds, fish, and benthic macroinvertebrates were sampled on 13 stream reaches in the Willamette Valley as part of a pilot project for the Environmental Monitoring and Assessment Program (EMAP) of the U.S. Enviromental Protection Agency (USEPA). The objectives of the study were to advance the development of aquatic and riparian indicators and develop field methods for stream sampling in the western states. The Willamette Valley lies between the Coast Range and the Cascade Mountains in western Oregon. It includes the floodplain and terraces of the Willamette River, as well as surrounding foothills. Before settlement by European immigrants in the mid-19th Century, the valley was mostly savanna with oak woodlands, wetlands, and an 8- to 13-km wide gallery forest in the Willamette River floodplain (Hughes et al., 1998). These areas were largely converted to cropland by 1900.

Today, land use in the valley is predominantly agricultural and urban; the low-gradient, meandering streams are often channelized in developed areas and on farms to maximize arable acreage. While winters are generally wet, seasonal drought and irrigation withdrawals cause many streams to dry or become intermittent in the summer months (Omernik and Gallant, 1986). Eight of the stream reaches were randomly selected to ensure that they were representative of first to third order perennial streams (Herlihy et al., 1997). Five additional streams were hand-selected to increase the range of disturbance types and intensities in developed areas. The Willamette Valley is a relatively homogeneous ecological region, and the agriculture/urban disturbance gradient is simpler than the multiple disturbance gradients in the diverse mountainous terrain of the mid-Appalachian region. We sampled each site, which simplified the ranking of sites to construct the disturbance gradient and aided in interpreting the results.

26.3.2 Comparison of Fish, Bird, and Benthic Macroinvertebrate Responses

26.3.2.1 Assemblage Index Responses to an Agricultural/Urban Disturbance Gradient

To examine assemblage responses across a disturbance gradient, we compared two measures of biotic responses developed during the Willamette Valley pilot study: a bird integrity index (BII) and a fish index of biotic integrity (IBI) (Bryce et al., in press; Hughes et al., 1998). Both fish and birds clearly responded to the disturbance gradient (Figures 26.2A and B). Of the two indicator groups, the bird assemblage response was the most closely associated with the disturbance gradient (Figure 26.2A). This was partly because the initial set of 62 candidate bird metrics were tested for response against this disturbance gradient as a first screening before undergoing redundancy and variability testing. The high correlation also suggests that the augmented probability sample comprised a complete and representative disturbance gradient. The agricultural and urban land uses represented in the gradient directly affect the quality and the very existence of birds' riparian habitats.

While index scores for birds and fish were in a similar range at both ends of the disturbance gradient, fish IBI scores were generally higher (>55) than bird index scores (<55) for a group of agricultural streams in the middle of the distribution (stream reaches 4A through 4D). This pattern

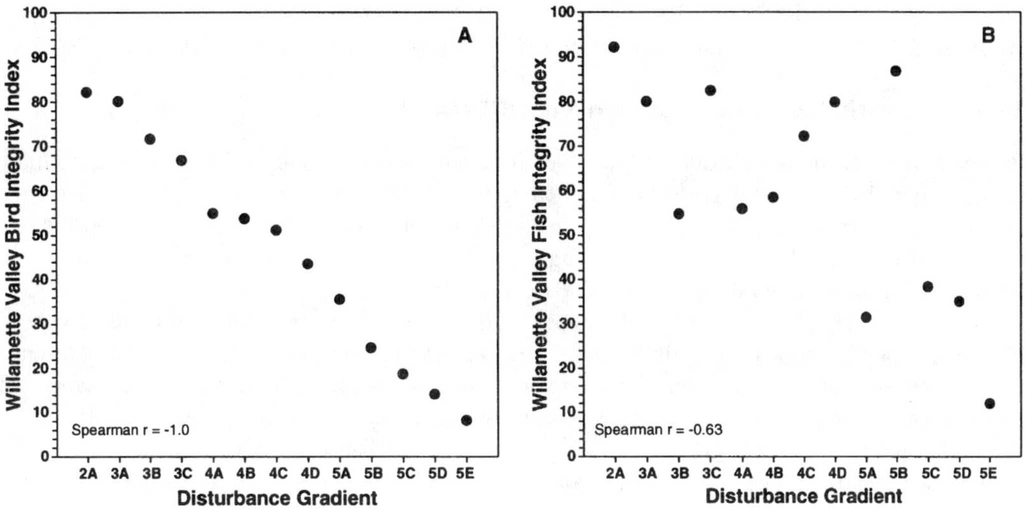

FIGURE 26.2 Relationship of (A) bird IBI and (B) fish IBI to an agriculture/urban disturbance gradient in the Willamette Valley of Oregon.

might suggest that birds are more sensitive to the impact gradient than fish. However, the disturbance index was developed as a general measure of human activity across the landscape. As a result, it may not capture all aspects necessary to characterize the response of a particular assemblage. While it does contain stream channel measures, the disturbance index was composed predominantly of riparian and watershed elements. Riparian cover is important to fish for shading and cover, but riparian structure, quality, and extent are vital to birds for food, shelter, and reproduction.

The fish index scores fell into three groups (Figure 26.2B): those for reaches in close proximity to the Willamette Valley foothills (scores between 82 and 92), agricultural reaches with wooded riparian buffers (scores between 56 and 62), and reaches without a riparian buffer (one agricultural ditch and 4 urban reaches with scores between 12 and 37). Three streams had scores that were higher or lower than expected. Reach 5B was on an urban stream with a very high fish index score of 86.7; the same location received a bird index score of 24.8 that was more typical of an urban riparian zone. This anomalous stream reach was spring-fed with a high volume of cold water that apparently compensated for the loss of riparian cover. As a result, it supported anadromous and resident salmonids that caused increased fish metric scores. Stream Reach 4D also scored higher than expected in the fish IBI. Here almost a kilometer of the reach length was in ungrazed woodlot with large old trees and riparian wetland. However, the watershed above the reach had characteristics that increased its disturbance index score; it was largely cleared, with several channelized tributaries. Yet the reach fish sample yielded high numbers of native, benthic, and hider species (including 25-cm cutthroat trout), and no alien or tolerant fish. One cannot generalize from one stream reach, but it does raise interesting questions about the nature and optimum length of stream refugia in disturbed areas. In agricultural watersheds in southwestern Minnesota, Marsh and Luey (1982) concluded that such oases were instrumental in mitigating the impacts associated with upstream channelization and drainage because they reduced levels of turbidity and sedimentation.

Finally, the fish IBI score for Reach 3B was lower than expected. The reach is in a U.S. Fish and Wildlife Service National Wildlife Refuge and two thirds of the watershed extend into the forested Willamette Valley foothills. Reach 3B had high sinuosity, a wide riparian zone, riparian wetlands, and an abundance of woody debris in the channel; it was also turbid, as were many streams on the valley floor. Its index score was low because, out of eleven fish species collected, three were aliens and six were tolerants (defined as species that tolerate warmer water, greater turbidity, higher nutrients, and lower dissolved oxygen). The wildlife refuge had been a past recipient of fish introductions as well as alluvial plain management for waterfowl forage.

26.3.2.2 Metric Responses to an Agricultural/Urban Disturbance Gradient

In the previous section, we examined the overall responses of fish and bird assemblages to three recognizable disturbance categories: (1) limited agriculture/wooded headwaters; (2) intensive agriculture/wooded riparian buffer; and (3) agriculture/urban/no riparian buffer. This section focuses on some of the individual metrics comprising the indices. We will examine a pair of metrics each for benthic macroinvertebrates (Li et al., 2001), fish, and birds.

The macroinvertebrate metric, EPT richness (Figure 26.3A), is the number of Ephemeroptera, Plecoptera, and Trichoptera (mayfly, stonefly, and caddisfly) taxa found at each site. The presence of these often-sensitive macroinvertebrates is commonly understood to indicate high quality stream reaches having clear water and coarse substrate with little sedimentation (Plafkin et al., 1989). While this metric distinguished fair from highly disturbed urban reaches (disturbance scores 4A through 5E), it showed high variability in scores for the less disturbed foothills (Reaches 2A through 3C). EPT richness was lower than expected at Reach 3B, the same reach that had a low fish IBI score, probably because of fine substrates and high sediment load. Stream Reach 3C, located entirely within the wooded foothills of the Willamette Valley, had a very high EPT score. Although the watershed was heavily logged and the alluvial plain cleared for agriculture (grazing) and rural residences, the stream reach had a wooded riparian buffer zone; and its relatively clear, cold water, and coarse substrate provided better EPT habitat than a typical Willamette Valley stream.

The second macroinvertebrate metric, observed/expected taxa (O/E), (Figure 26.3B), is based on the RIVPACS approach (Wright, 1995). This metric compares the number of macroinvertebrate taxa observed (O) with the number expected (E) in similar habitats largely unaffected by humans. In this case, expected values were derived from the six least disturbed stream reaches in the sample. This metric showed a steady decline in scores with increasing disturbance. It represented a general response across the entire sample to a gradual loss of habitat and increasing sedimentation. Of the two macroinvertebrate metrics, the EPT metric (Figure 26.3A) was more diagnostic, based on the uniformly low scores at the urban stream reaches.

The outlier points raised questions and provided additional corroborating evidence. Why did Stream Reach 3B at the minimal to moderately disturbed end of the disturbance gradient share an EPT score with the most highly disturbed reaches? The factors that 3B shared with the more highly disturbed streams were sedimentation, turbidity, and lack of coarse substrate. Why did Reach 5B receive a low EPT score and consistently high (outlier) fish scores? Reach 5B was along an urban stream that had been channelized, it was spring fed, and had good fish cover, undercut banks and shading from blackberry thickets. The stream channel and substrate were not as suitable for EPT habitat because of minimal riffle/pool development along the straight incised channel. While some coarse substrate was present, embeddedness and fines measured 59 and 30%, respectively.

A commonly used metric for assessing the integrity of fish assemblages is native species richness. In this case, native fish species response to the agriculture/urban gradient was convex rather than linear (Figure 26.4A). Species numbers increased at the mid-range of disturbance and then declined at the more highly disturbed agricultural and urban reaches. This pattern has been called the intermediate disturbance concept or perturbation theory (Odum et al., 1979; Odum, 1985). Odum et al. (1979) noted that human and natural disturbances often result in increased biotic diversity at low impact levels, followed by a decline in diversity with more intense disturbance. In our study, higher fish scores occurred at stream reaches with narrow riparian buffers in mid-valley farm fields. Many of these valley streams originate in the foothills of the Willamette Valley; as the cool water warms when it reaches the valley floor, warm water fish species join the species pool (Li et al., 1987; Rahel and Hubert, 1991). However, fish species richness is also covariate with stream size and/or watershed area (Fausch et al., 1984; Karr et al., 1986).

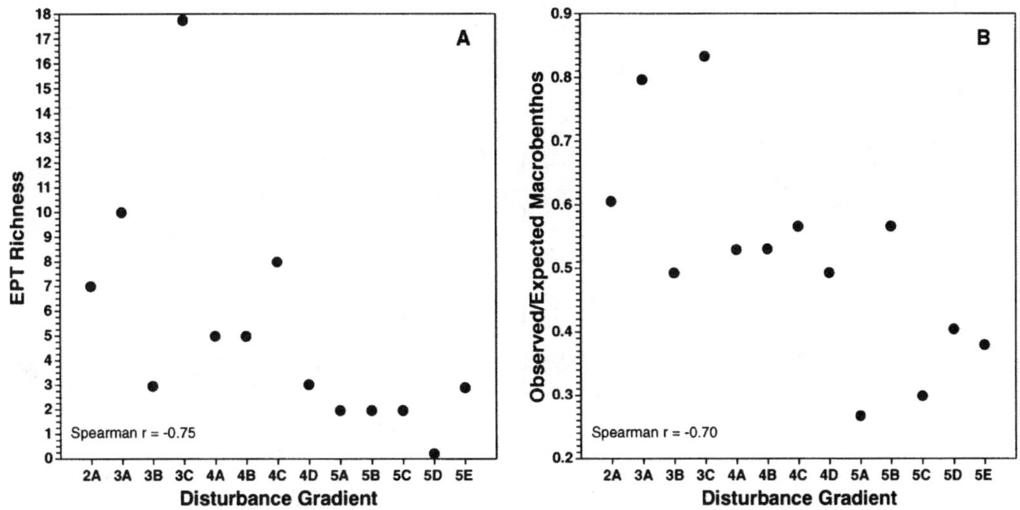

FIGURE 26.3 Relationship of two benthic macroinvertebrate metrics, (A) Ephemeroptera–Plecoptera–Trichoptera (EPT) richness and (B) observed number of taxa/expected number of taxa (O/E), to a Willamette Valley agriculture/urban disturbance gradient.

When we corrected for watershed area, the apparent effect of intermediate disturbance on the number of native species disappeared, except for the most highly disturbed urban stream reaches. Another confounding factor is the location of the stream in the drainage network. Osborne et al. (1992) found that species richness was higher in streams with a direct outlet to a main channel stream (order 5 or higher) than in streams of the same size situated in the headwaters or upper half of the watershed. Presumably, the proximity of main channels allowed greater fish immigration and recruitment to main channel tributaries than headwater tributaries. Eight of 13 sample stream reaches in our study were less than 2 km from a confluence with the Willamette River, and of those, four had higher than expected native species richness. Reaches near the river's main channel that did not have higher native species richness had more alien and tolerant species. Thus, while the native fish species richness metric clearly indicated general disturbance, its diagnostic power was confounded by other factors.

The hider species richness metric (Figure 26.4B) appeared more indicative of a specific disturbance than native species richness, showing less variability at both ends of the disturbance gradient. Hiders include species such as cutthroat trout (that require coarse substrates, undercut banks, overhanging vegetation, and woody debris for cover), sandrollers (that select fine roots, submerged brush and aquatic plants), and juvenile lamprey (that require stable patches of coarse sand). We did not correct this metric for watershed area, even though it was a richness measure, partly because we were dealing with a limited number of species and also because of the step pattern of response. The foothills and riparian buffered stream reaches had an average of three hider species, dropping to none in the agricultural ditch (5A) and the urban reaches (5C through 5E). The high scoring urban reach (5B) was the anomalous spring-fed urban stream discussed above. Hider species responded to changes in physical habitat, in this case the loss of cover and the simplification and sedimentation of stream channels in developed areas.

All 13 bird metrics showed clear responses to the agricultural/urban gradient. However, two of the metrics, percent bark-gleaning species and number of intolerant individuals, were particularly strong in the agricultural to urban transition. The percent bark-gleaning species metric (Figure 26.5A) produced an on–off signature similar to the fish hider species metric. Neither the data nor the reach descriptions suggested reasons for the two outliers, Reaches 4A and 4D. Clearly bark gleaning is a foraging strategy that depends on the presence of trees, and, for some bird

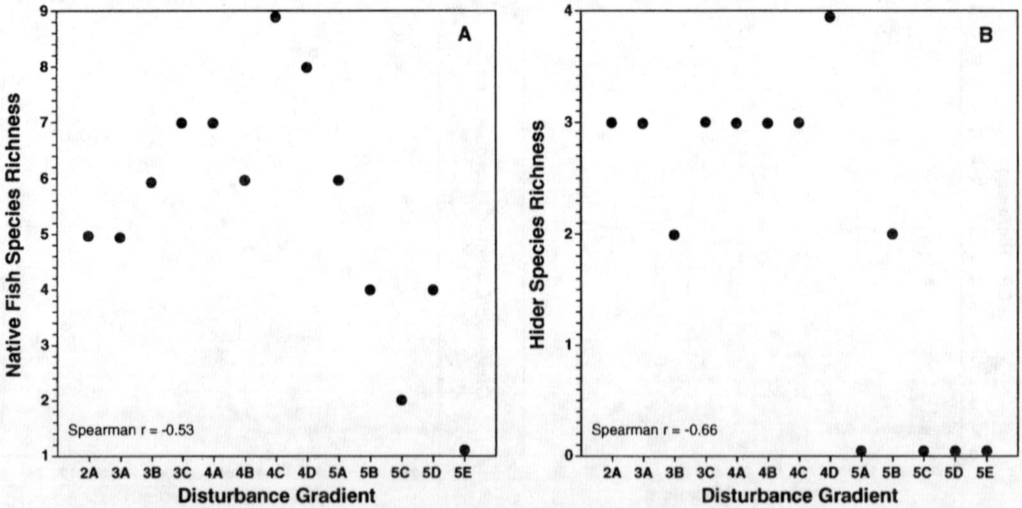

FIGURE 26.4 Relationship of two fish IBI metrics, (A) native fish species richness and (B) hider species richness, to a Willamette Valley agriculture/urban disturbance gradient.

species, trees of adequate size. During the breeding season, bark-gleaning species such as the hairy woodpecker and red-breasted nuthatch require larger diameter trees for nesting. A typical narrow agricultural buffer strip or urban tree planting often does not provide adequate habitat for both foraging and nesting. The other bird metric, number of intolerant individuals (Figure 26.5B), declined steadily to zero at the urban reaches (5C through 5E). This metric represents the response of sensitive bird species to a combination of stressors related to industrial farming and urbanization, such as riparian woodland habitat loss, increased exposure to nest predators and parasites, pet predation, and competition from alien species.

In summary, in the Willamette Valley study, we found that avian assemblages were particularly well-suited for assessing riparian integrity, that wooded riparian zones improved scores for all three assemblages over streams with no riparian buffer, and that macroinvertebrates responded strongly to substrate type and sedimentation. The convex pattern in the native fish species metric suggested some of the factors confounding clear responses from fish assemblages. The interpretation of outliers suggested that riparian oases in agricultural areas and high quality water resources in urban areas can produce marked improvements in fish assemblages. We now move to a more diverse region to determine whether similar relationships exist between stressors and biological responses.

26.4 MID-APPALACHIAN REGION CASE STUDY

26.4.1 BACKGROUND

Between 1993 and 1996, the USEPA conducted a survey of wadeable streams in the mid-Appalachian region as part of EMAP. Physical habitat, water chemistry, periphyton, benthic macroinvertebrates, and fish were sampled at 309 first through third order streams identified as solid blue lines on 1:100,000 scale maps. Klemm and Lazorchak (1994a and 1994b) described the sample collection and processing methods for all indicators. A region-wide assessment has been completed (USEPA, 2000), and several authors reported on indicator response and index development for fish (McCormick et al., 2001), macroinvertebrates (Klemm et al., in review), and periphyton (Hill et al., 2000; Fore, Chapter 22, this volume).

Agriculture, mining, acidic deposition, and logging are the main anthropogenic disturbances of the mid-Appalachian region (USEPA, 2000). Settlers began farming the fertile valleys and

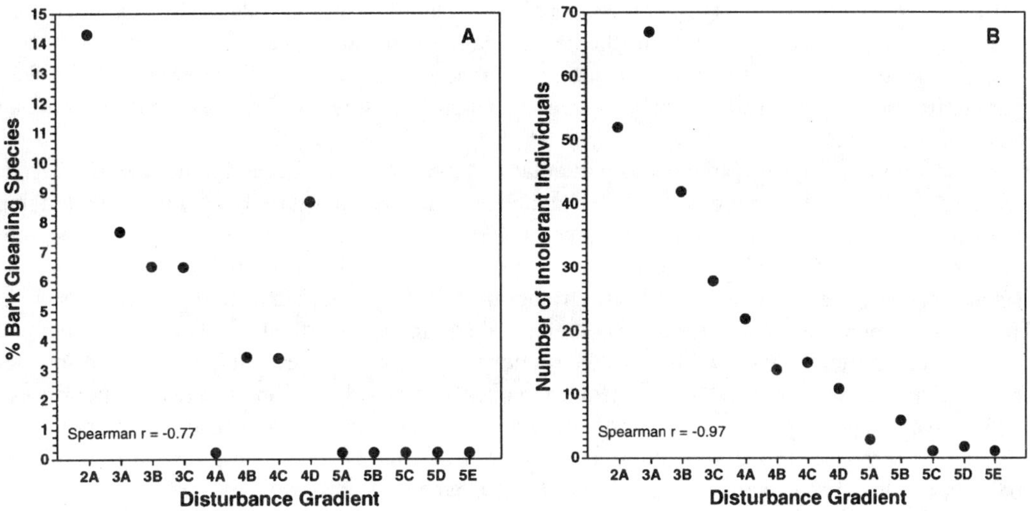

FIGURE 26.5 Relationship of two bird IBI metrics, (A) percent bark-gleaning species and (B) number of intolerant individuals, to a Willamette Valley agriculture/urban disturbance gradient.

woodland slopes in the late 1700s; timber removal was complete by the 1920s; and mining, both under- and above-ground, followed episodic economic cycles and advancements in extraction technology since the early 19th Century (Caudill, 1963). Woodland agriculture has declined considerably and reforestation has occurred over much of the region, but agriculture, logging, and mining still contribute high levels of sediment to streams (Hart,1991). In-stream habitat loss continues through the loss of pools, channelization, absence of large woody debris, and destruction of headwater streams by filling with mountain top mining spoil. In the valleys, the growth of concentrated animal feeding operations (e.g., chickens, turkeys, hogs, and cattle) has led to excessive nutrient levels in streams.

Most of the sampled watersheds in the mid-Appalachian database were moderately to seriously disturbed by human activities. Of the 102 watersheds evaluated, only 14 fit the criteria of the minimally disturbed (or reference) category. Although a number were intensively farmed or mined watersheds, few were urban or industrial reaches. Without a sufficient number of stream reaches at both ends of the disturbance gradient, it was difficult to demonstrate indicator response, particularly when organisms responded to a heterogeneous array of disturbances that resulted in a homogeneous severity of impairment.

26.4.2 Comparisons of Fish, Benthic Macroinvertebrate, and Diatom Responses

26.4.2.1 Assemblage Index Responses to General Disturbance Gradient

We began by comparing the responses of benthic macroinvertebrates, diatoms, and fish across the entire sample of 102 watersheds. We plotted index scores for each of the three assemblages against the disturbance index scores (Figure 26.6 A-C). Though the index scores for all three assemblages declined with increasing disturbance, the corresponding r-values suggested that benthic macroinvertebrates gave the clearest signal followed by diatoms and fish. However, the clarity of response for all three assemblages was reduced by high variability in scores, particularly in the highest disturbance classes. Some sources of natural variability had been considered in that macroinvertebrate and fish metrics were adjusted for stream size, using watershed area. Stream size is correlated with natural gradients such as stream slope, temperature, and geographic position (e.g., headwater vs. valley) (McCormick el al., 2001; Klemm et al., in review). In addition to natural sources of variability, indicator assemblages were responding to a highly variable pattern of landscape disturbance. Pooling all the watersheds in the sample may demonstrate the ability of each indicator assemblage to recognize generalized disturbance across a geographically diverse region, but it does not clarify the strengths and weaknesses of each indicator or their abilities to diagnose particular types of disturbance.

To address biological responses to particular disturbances, we reduced the number of reaches from 102 to 39 in four categories: farmed, channelized, mined, and forested with little to no logging pressure. Boxplots of fish, diatom, and macroinvertebrate index scores associated with the four categories of disturbance were inconclusive. Correlation strengths were reduced by placing the stream reaches into disturbance categories, rather than along a continuous gradient (Hughes et al., 1998). To achieve a clearer result, we created a continuous gradient for two major regional disturbances, agriculture and mining. In the sections that follow, we present benthic macroinvertebrate (Klemm et al., in review), diatom (Fore, Chapter 22, this volume), and fish (McCormick et al., 2001) assemblage responses to the agriculture and mining gradients; and macroinvertebrate and fish metric responses to the agriculture and mining gradients. Diatom metric responses to disturbance are covered elsewhere in this volume (Fore, Chapter 22, this volume).

26.4.2.2 Development of Agricultural and Mining Gradients

We developed agricultural and mining gradients by ranking farmed and mined stream reaches within each class of the disturbance index. For example, agricultural reaches within disturbance classes 4 and 5 were ranked by increasing disturbance 4A, 4B, 4C, etc. We compared assemblage scores at these reaches with those in predominantly forested watersheds with little to no logging pressure (Table 26.1). Of course, most disturbed landscapes are affected by multiple stressors. In our mid-Appalachian sample, moderately disturbed agricultural watersheds were characterized by a simple forest/agriculture mosaic. With increasing disturbance, as a greater proportion of the watersheds were cleared for agriculture, additional stressors, such as channelization, mining, oil drilling, and urbanization appeared in the agricultural landscape. Thus, simply ranking reaches by percent agricultural area, using classified thematic map satellite imagery, as is common in geographic analysis, did not accurately reflect the accumulation of disturbances.

We turned to the initial screening process used in the development of the disturbance index (Bryce et al., 1999), and refined the ranking by considering the intensity, extent, and stream proximity of each additional stressor, in addition to percent agriculture or percent mining in the watershed (Table 26.2). For example, agricultural stream Reach 5F clearly had less watershed area in agriculture than reaches 5B through 5D; however, on the aerial photos, stream banks on 5F appeared raw and unvegetated, and field forms stated that the stream was dredged every few years. In this case, continued channel disturbance was considered more damaging than a single episode and similar in severity to streams experiencing industrialization or urbanization. Thus, the agricultural gradient began with six minimally disturbed forested watersheds (1A through 2D) (Table 26.1), followed by 12 watersheds with increasing intensities of agriculture and other related disturbances (3A through 5G, Table 26.2). The proportion of watershed in agriculture did not increase gradually, but started at 50% watershed area.

The mined watersheds (Table 26.3) were generally larger than those included in the agricultural gradient. Six of 10 were >6000 ha^2 in contrast to the agricultural watersheds, where only one was >5000 ha^2. More of the mined watersheds were forested than farmed, so they were somewhat more amenable to ranking by percent area mined. We used the percent watershed area mined, recorded through aerial photo interpretation, rather than data acquired from classified thematic map satellite imagery from the same time (Table 26.3).

Mining effects differed across stream samples. Streams in the north central Appalachians and parts of the western Allegheny Plateau are more susceptible to acidification from mine drainage because of the prevalence of sandstone and other resistant base-poor rocks. In the south central

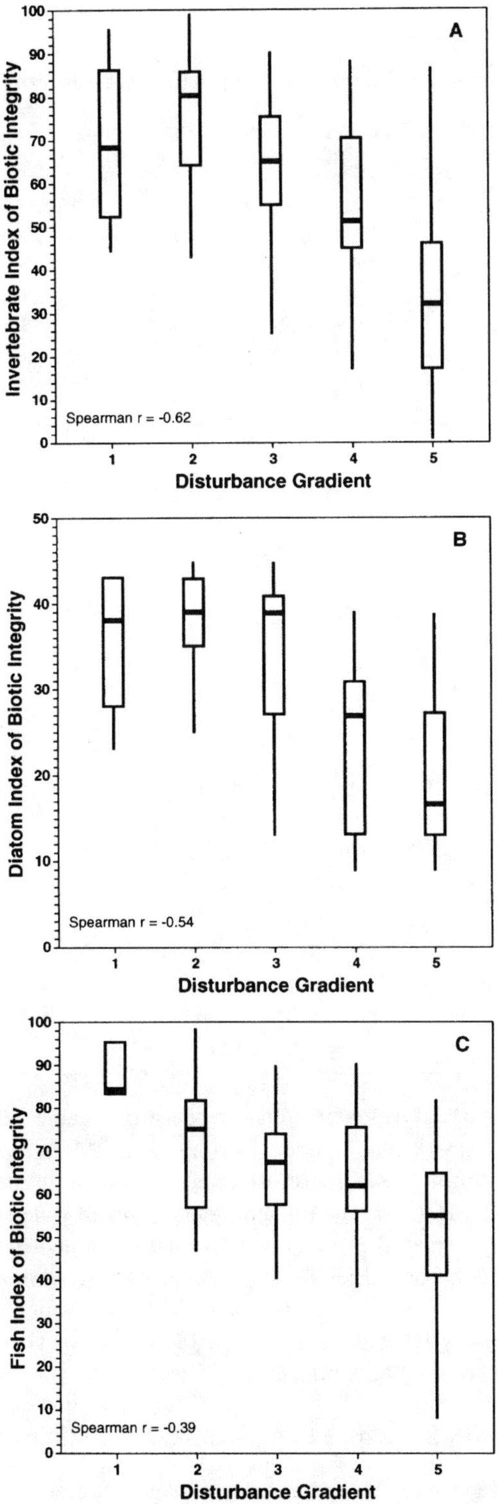

FIGURE 26.6 Relationship of (A) macroinvertebrate, (B) diatom, and (C) fish IBIs to a general disturbance gradient in the mid-Appalachian region. Boxes show ranges, quartiles, and medians.

TABLE 26.1
Forested Reference Streams for Agricultural and Mining Gradients

Stream ID	Watershed Area (ha^2)	Elevation Meters (ft)	Road Density (m/ha)	Stress Type
1A	141.11	1012.2 (3320)	0	No logging
1B	176.66	487.8 (1600)	5.64	National park
2A	4375.96	396.3 (1300)	5.58	Wilderness area
2B	1019.36	719.5 (2360)	5.06	No logging
2C	794.66	490.9 (1610)	3.77	Jeep trail parallels stream
2D	374.50	463.4 (1520)	11.04	Road follows stream; 30% logged

TABLE 26.2
Characteristics of Streams Included in the Agricultural Gradient

Stream ID	Watershed Area (ha^2)	Elevation Meters (ft)	Road Density	TM % Ag	Stress Type[a]
3A	72.8	250 (820)	16.27	54.62	A
4A	705	497 (1630)	15.20	65.25	A, L
4B	493	12.2 (40)	11.06	81.04	A, C, S
4C	3298	323.2 (1060)	24.77	83.66	A, C, R
4D	205.8	622 (2040)	42.08	52.38	A, C, L, R
5A	5836.8	115.9 (380)	19.15	49.23	A, C, L,
5B	3339.5	122 (400)	17.19	84.11	A, C, L, S
5C	2514	170.7 (560)	23.04	79	A, L, E, S
5D	2392.7	128 (420)	20.90	88.37	A, S, R
5E	2798	146.3 (480)	21.08	60.01	A, S, E, I
5F	1998	9.8 (32)	12.44	50.16	A, C, dredged
5G	5160	79.3 (260)	26.97	84.88	A, C, S, urbanization (15%)

[a] In addition to percent agriculture, we refined the rankings by considering the intensity, extent, and stream proximity of each additional stressor. A = agriculture. L = livestock. C = channelization. S = sedimentation. R = residential. E = bank erosion. I = industrialization.

Appalachians, in the Cumberland Mountains, the presence of weatherable overburden neutralizes acid drainage and reduces heavy metal toxicity (Herlihy et al., 1990). Reclamation of mined areas in the north often includes liming to increase pH levels. In our sample, we had direct knowledge of liming at one stream (5E, pH 7.89). The remaining eight mined watersheds had reach pH values in the low to mid-8 range (median 8.2); three were in the north and their pH levels suggest that they may have been limed (5B, 5C, and 5F, Table 26.3). In contrast, all minimally disturbed forested reaches had pH values in the mid-7 range (median 7.25). The mining gradient compared the minimally-disturbed forested watersheds (1A through 2D) (Table 26.1) with the mined watersheds, ranging from 10 to 70% watershed area mined (Table 26.3).

26.4.2.3 Assemblage Index Responses to Agricultural and Mining Gradients

Agriculture — Benthic macroinvertebrates and more than fish, diatoms were associated with the agricultural disturbance gradient from minimal to high impact (Figures 26.7A through C). There was a visible threshold on the plots for both macroinvertebrates and diatoms between the highest scoring minimally disturbed reaches (upper left quadrant) and the lower scoring disturbed reaches (lower right quadrant) (Figures 26.7A and B). The pattern differed for the fish IBI scores

TABLE 26.3
Characteristics of Streams Included in the Mining Gradient

Stream ID	Watershed Area (ha^2)	Elevation Meters (ft)	Road Density (m/ha)	TM % Mined	TM % Agriculture	Stress Type[a]
3	2408.58	440.5 (1445)	20.53	0	6.0	5% mined; oil/gas well grid across watershed
3A	3621.9	390.2 (1280)	4.54	0.16	0	No logging; 0% mines, road up stream
3B	8303.4	487.8 (1600)	13.12	5.54	1.66	20% mined on watershed boundary; 80% forested
5A	11,654.5	500 (1640)	17.08	8.83	0.29	10% mined near stream channel; R, C, S
5B	7347	228.7 (750)	22.93	0.59	63.75	10% mined on stream channel; A, R, C; fish stocking
5C	4570.9	420.7 (1380)	20.54	1.10	29.39	20% mined; fish stocking; R, A, S
5D	7725.2	225.6 (740)	12.15	0	0	5% mined; high S; construction on reach
5E	684.23	436 (1430)	24.07	6.62	34.82	35% mined; A; limed
5F	2178.3	277.4 (910)	20.90	19.55	28.57	35% mined; A, S, R, C
5G	6577.3	292.7 (960)	10.46	5.01	0	50% mined; S, T, C, LF
5H	3554.4	292.7 (960)	9.73	2.62	0	70% mined; T, S, C, R

Note: While the stream ID sequence is similar (3A, 3B, etc.), these streams are different from those included in the agricultural gradient. Percent area mined in text taken from aerial photo interpretation (listed under Stress Type) rather than thematic map satellite imagery percentages (which we determined was inaccurate).

[a] In addition to percent mined, we refined the rankings by considering the intensity, extent, and stream proximity of each additional stressor. R = residential. C = channelization. S = sedimentation. A = agriculture. T = trash. LF = landfill.

(Figure 26.7C). While the general response for fish was similar to that for macroinvertebrates and diatoms (Spearman's $r = -0.63$), most of the points clustered in the upper half of the plot. Fish IBI scores for several highly disturbed sites were higher than expected. As in the Willamette Valley, proximity to large rivers may have elevated fish scores. The highest scoring eight (of 13) agricultural stream reaches were within 8 to10 km of major river confluences.

Mining — Macroinvertebrate scores showed stronger responses to mining disturbances than diatom and fish scores (Figures 26.8A through C). Macroinvertebrate index scores were unexpectedly high at Reaches 5E, 5G, and 5H (Figure 26.8A). Diatom scores showed a general decline with increased mining in the watershed (Figure 26.8B), but the response was more variable than diatom response to the agricultural gradient (Figure 26.7B). That is, fewer highly disturbed mined watersheds received low diatom scores than did highly disturbed agricultural watersheds. Two heavily mined watersheds, 5E and 5G, with 35 and 50% mined areas, respectively, had unexpectedly high diatom scores. Fish IBI scores declined with increasing mining disturbance, but most clustered between 60 and 69, as they did for the agriculture gradient (Figure 26.8C). Like macroinvertebrates, fish had unexpectedly high scores at reaches 5E, 5G, and 5H. Reach 5E was limed, which may have stimulated biota or at least counteracted some mining effects. Reaches 5G and 5H had no agriculture and low road densities. This suggests that they were only affected by mining rather than mining plus agriculture. The greater percentage of forest in the watersheds may have mitigated the effects of mining.

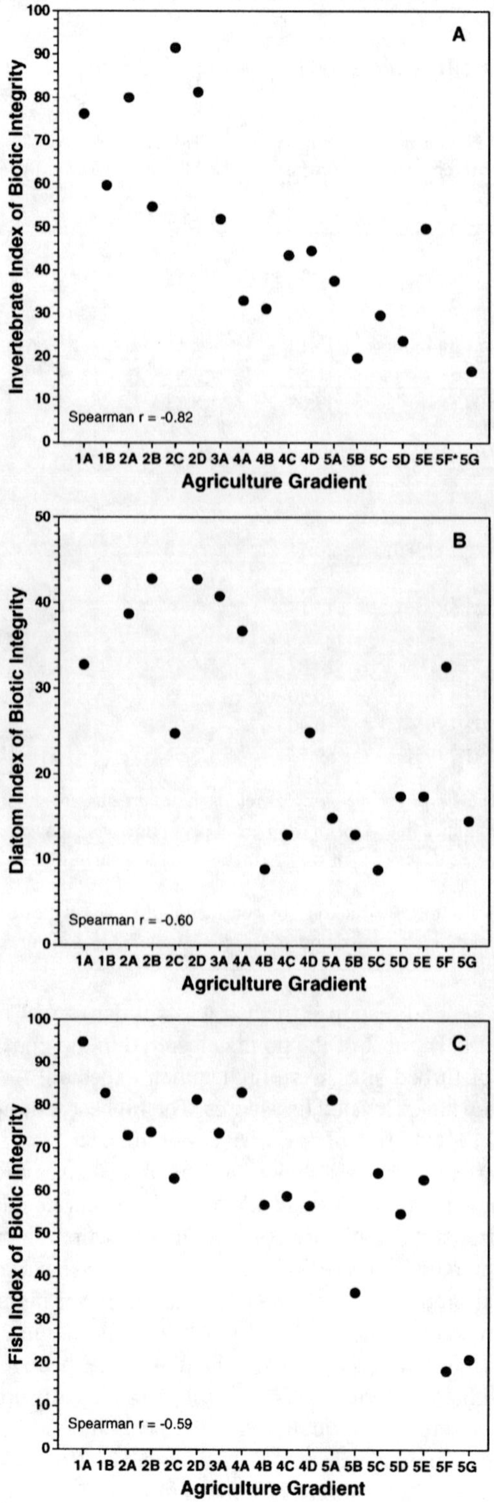

FIGURE 26.7 Relationship of (A) macroinvertebrate, (B) diatom, and (C) fish IBIs to an agricultural disturbance gradient in the mid-Appalachians; * = one sample inadequate for invertebrate evaluation (5F).

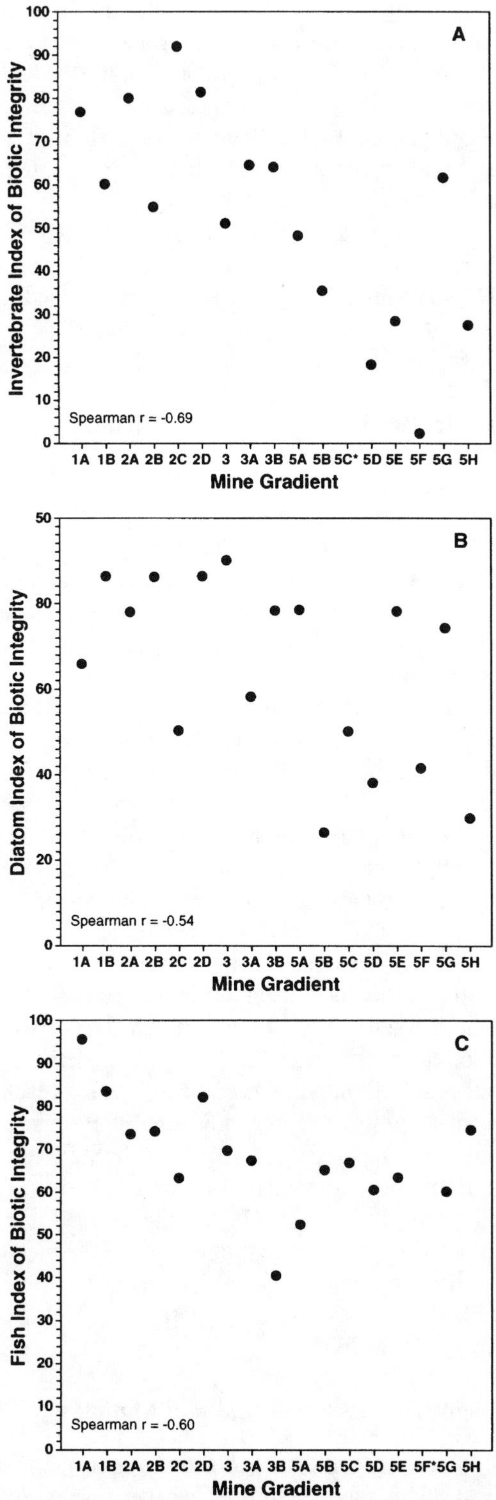

FIGURE 26.8 Relationship of (A) macroinvertebrate, (B) diatom, and (C) fish IBIs to a mining disturbance gradient in the mid-Appalachians; * = one sample inadequate for invertebrate evaluation (5C) and one with no fish collected (5F).

In summary, all three assemblages showed negative responses to both increased agriculture and increased mining, though the responses were stronger for agriculture than for mining. Macroinvertebrates were more strongly correlated with impact than diatoms or fish, possibly because benthic macroinvertebrates, dependent on substrate interstices, were most strongly affected by increased sediments from agriculture and mining. Both fish and motile diatoms can better occupy space above or on a sedimented substrate. Finally, fish scores may have been elevated by stream reach proximity to large river confluences.

The following two sections examine macroinvertebrate and fish metric responses to the agriculture and mining disturbance gradients. Diatom metric responses are discussed elsewhere in this volume (Fore, Chapter 22, this volume). As in the Willamette Valley study, we expected the diagnostic capabilities of various indicator assemblages to emerge more clearly through the examination of metric responses.

26.4.2.4 Macroinvertebrate Metric Responses to Agricultural and Mining Gradients

Several of the most responsive macroinvertebrate metrics related to the numbers or percentages of Ephemeroptera, Plecoptera, or Tricoptera taxa (EPT) in the sample. The Plecopteran richness metric was the most responsive of the EPT metrics to the agricultural gradient (Figure 26.9A). Except for one outlier (5A) that had a wooded riparian zone (and possibly cooler water), the Plecopteran richness score remained at 2 or below for all of the agricultural streams. The response for Trichopteran richness (not shown) was more erratic across the agricultural gradient, possibly because filter feeding caddisflies may have responded to the presence of suspended organic matter in agricultural streams, particularly if sedimentation was not too severe. Trichopterans are also less susceptible to warming than Plecopterans. For the mining gradient, both the Ephemeropteran (Figure 26.9B) and Trichopteran richness metrics gave clear responses with a number of zero scores, while Plecopteran richness showed some scatter in scores over the mined stream reaches. Stream Reaches 5B and 5D received zero scores even though only 10 and 5% of their areas were mined. However, 5B included strip mining above and parallel to the stream channel, and 5D had active construction on the sampled reach. Conversely, Reaches 5G and 5H scored higher than expected, possibly because a high percentage of forest cover in the watersheds mitigated the effects of mining.

Another macroinvertebrate metric, the Hilsenhoff biotic index (HBI), is an indicator of organic enrichment, including both human sewage and animal manure as sources. Macroinvertebrate species are assigned organic pollution tolerance values of 0 to 10. The HBI score for a stream reach is an average of the tolerance values for all individuals collected from a reach. HBI scores are scaled to conform with the other metrics, with low scores indicating high organic enrichment. The nutrient connection was clear when we compared HBI plots for the agricultural and mining gradients (Figures 26.9C and D). The HBI scores dropped in a stepwise pattern and stayed low with increasing agriculture and residential impacts (Table 26.2). On the other hand (Table 26.3), the mining gradient had more scatter in the HBI scores for the more highly disturbed watersheds. One reason could be that the lowest scoring disturbed stream reaches (5B, 5E, and 5F) all had >28% agriculture in their watersheds, while the disturbed reaches with higher HBI scores (5A, 5G, and 5H) had little to no agriculture in their watersheds (Table 26.3).

26.4.2.5 Fish Metric Responses to Agricultural and Mining Gradients

Of the nine fish metrics, the percent cottid metric showed a strong response. The percent piscivore/invertivores and percent aliens metrics showed weaker responses to the agricultural and/or mining gradients. The others showed little or no relationship to either gradient. The percent cottid metric displayed a stepwise pattern in the plot and declined to zero in the more highly disturbed watersheds of both the agricultural and the mining gradients. The agricultural gradient plot is shown

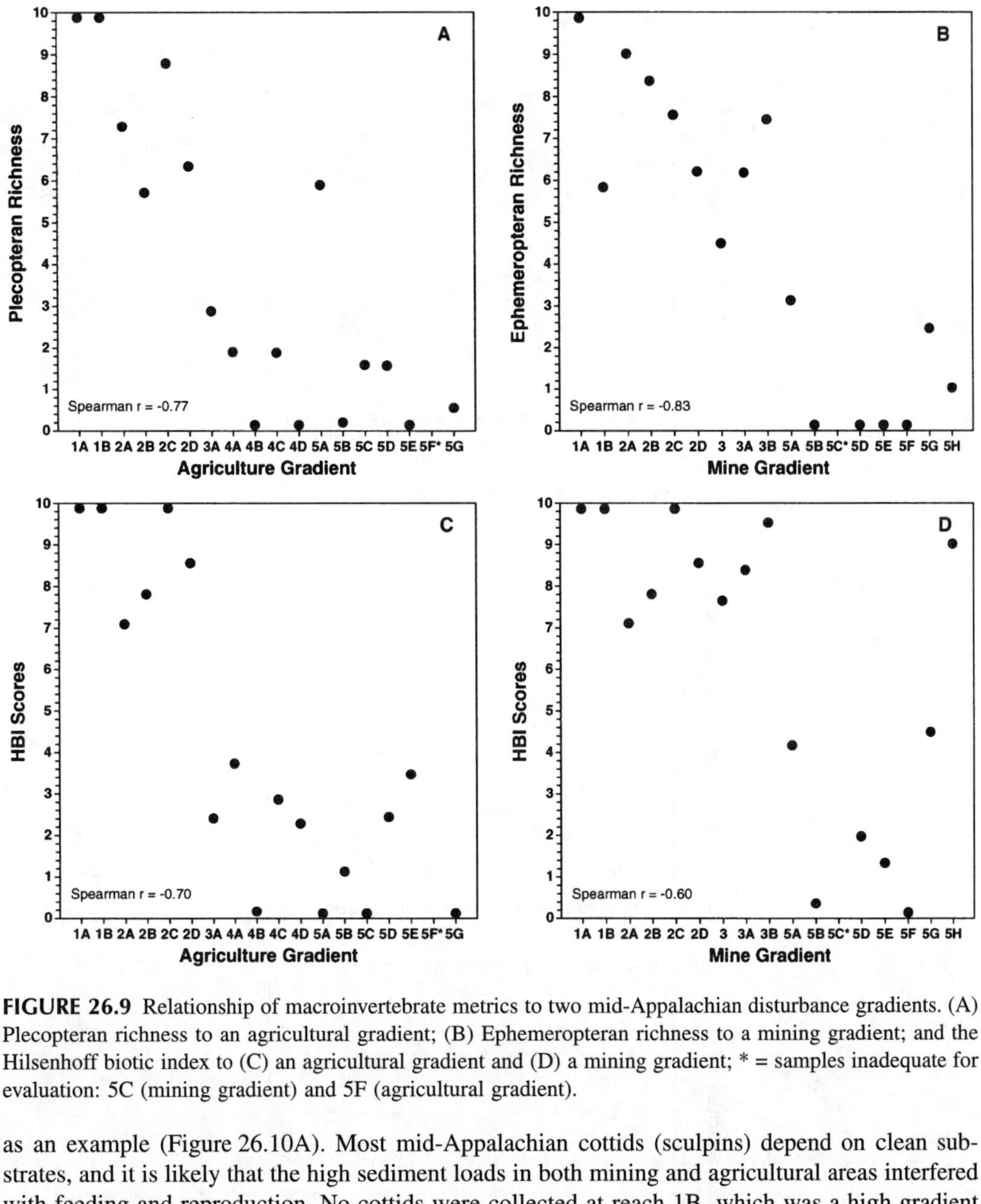

FIGURE 26.9 Relationship of macroinvertebrate metrics to two mid-Appalachian disturbance gradients. (A) Plecopteran richness to an agricultural gradient; (B) Ephemeropteran richness to a mining gradient; and the Hilsenhoff biotic index to (C) an agricultural gradient and (D) a mining gradient; * = samples inadequate for evaluation: 5C (mining gradient) and 5F (agricultural gradient).

as an example (Figure 26.10A). Most mid-Appalachian cottids (sculpins) depend on clean substrates, and it is likely that the high sediment loads in both mining and agricultural areas interfered with feeding and reproduction. No cottids were collected at reach 1B, which was a high gradient stream (9.1% slope), that precluded sculpin colonization and persistence. We found no explanation for the unexpectedly high number of sculpins at reach 5B, which was highly disturbed.

Piscivores/invertivores responded similarly to both the agricultural and the mining gradients (Figure 26.10B). These fish are sight feeders and the turbidity caused by mining and agriculture presumably hindered their foraging ability. The percent alien metric included introduced game fish (e.g., salmonids, centrarchids), bait fish, and Asian minnows like carp and goldfish. This metric was more responsive to the agricultural gradient (Figure 26.10C) than the mining gradient. Unexpectedly low scores occurred at forested reference reaches 2A and 2C because of the presence of introduced trout. Trout introductions often occur in high quality waters, making percent aliens more a direct indicator of biological disturbance than water pollution or habitat disturbance.

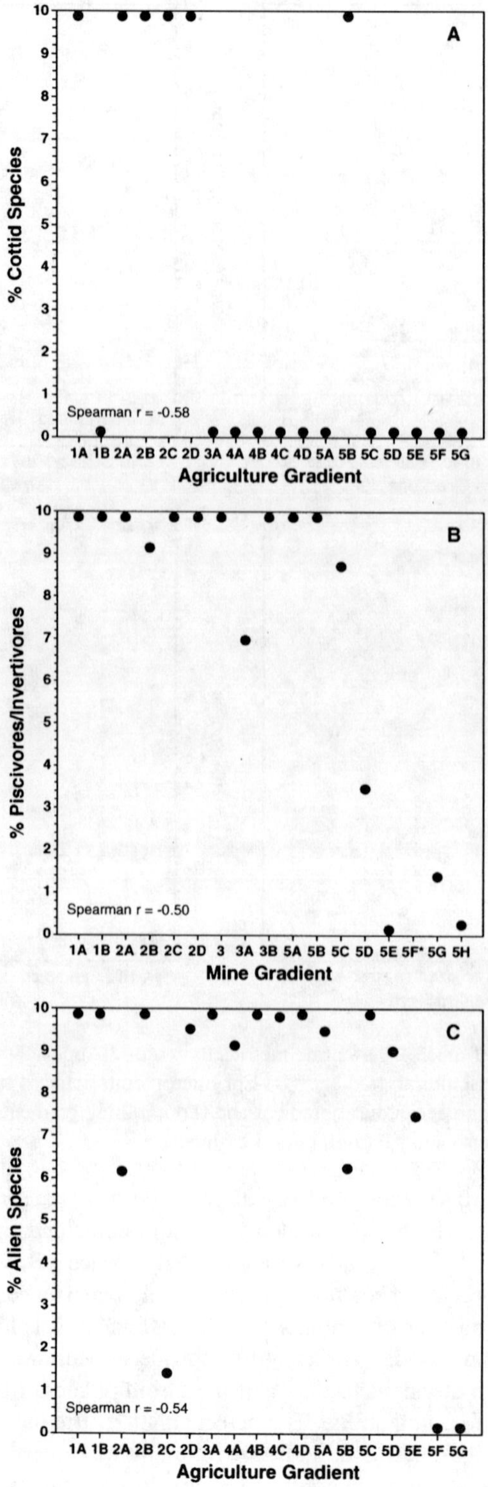

FIGURE 26.10 Relationship of fish metrics to two mid-Appalachian disturbance gradients: (A) percent cottid species to an agricultural gradient, (B) percent piscivore/invertivore species to a mining gradient, and (C) percent alien species to an agricultural gradient; * = one sample with no fish collected (5F, mining gradient).

Fish are the aquatic organisms that elicit the highest concern from the public and water resource managers, but fish populations have been manipulated throughout the history of human settlement in the U.S. Fish assemblage responses may be unexpected due to intermediate disturbance, watershed area, proximity to big rivers, alien introductions, and put-and-take fisheries, all of which confound the development of a responsive fish index. A related reason that fish assemblages may perform more poorly than macroinvertebrates is that species distributions have been increasingly homogenized (Scott and Helfman, 2001). Scott and Helfman propose that the accumulation of farming, logging, and mining disturbances in the Appalachians has transformed many cool highland streams into streams that function more like low elevation streams because of immigration of generalist species (i.e., first native and then introduced) that are tolerant of wider temperature extremes and increased sedimentation.

Native generalists take the role of native invaders in that they expand their ranges into highland habitats, and share the habitat with highland species that can persist in suboptimal conditions. The presence of the native invaders may inflate species richness metrics and other positive-scoring guild measures (such as percentage of piscivores/invertivores and gravel spawners). Scott and Helfman cautioned that measuring native diversity without accounting for such deviations from expected natural distribution patterns may give a misleading interpretation of regional integrity. Karr (1981), in his original fish IBI paper, also cautioned against using species richness metrics without considering the identities of the species included in those metrics.

26.5 CONCLUSIONS

The patterns observed in this chapter may be best understood if they are presented as a series of observations.

1. Regional examples demonstrate the value of using multiple assemblages to assess stream and riparian habitats. In each geographic region, the three assemblage indices agreed on the general level of disturbance. However, individual streams scored differently depending on specific indicator responses to substrate, water column, or channel and riparian habitat.
2. IBIs for all the indicators showed responses to general disturbances in each region. The examination of response at the metric level highlighted assemblage response to particular stressor gradients and each indicator's capability of conveying a strong signal of that disturbance. Bird proportionate abundances were directly related to changes in the structure of riparian habitats, and benthic macroinvertebrates responded strongly to substrate type and sedimentation. The absence of patterns in some of the fish metrics helped suggest confounding factors that may have diluted response to the disturbance gradients. The most highly diagnostic metrics focused on habitat specialists. For example, the plots for percent hider species (fish) and percent bark gleaners (birds) in Oregon and percent cottid fish species in the mid-Appalachians all showed an on–off or step pattern with increasing disturbance.
3. We suspect that the associations for all of the assemblages with disturbances would have been clearer if a larger number of minimally disturbed (and completely disturbed) stream reaches were available for comparison. The disturbance gradients for both geographical areas were truncated at the minimally disturbed end due to a shortage of reference streams. It is important to build an IBI around a model of expected condition. If adequate field reference streams do not exist, a conceptual model of reference condition may be constructed using historical information about minimally disturbed systems and associated assemblage distributions. In the Willamette Valley study, both the bird and fish species lists were augmented with extirpated species to counter the inflation of index scores based solely on present day field data collected from least disturbed (rather than

minimally disturbed) areas (Hughes et al., 1998; Bryce et al., in press). In a similar manner, fish response in the mid-Appalachian study might be clarified by incorporating expected pre-disturbance fish distribution patterns into metrics as suggested by Scott and Helfman (2001). Creating a model of historic fish distribution patterns would help distinguish present day species maxima from those expected under minimally disturbed conditions.

4. It is important to determine why biological indicators respond in unexpected ways to disturbance. In this study, outliers offered valuable insights about the diagnostic capabilities of indicator assemblages and metrics. However, besides being a function of noise in the biological response, unexpected results may also be due to a disconnection between landscape-scale and reach-scale disturbance indicators. Karr and Chu (2000) argued that land use estimates are often inaccurate or calculated at the wrong scale. Norris and Hawkins (2000) based prediction of biological response on reach level disturbance indicators simply because instream conditions are more directly associated with aquatic biota. The challenge for applied biologists is to meaningfully link both scales of information to biological responses and human management decisions (Hughes et al., 2000). Aquatic and riparian indicators, analyzed together with reach and watershed landscape information, offer a clearer multiscale picture of stream condition than any single indicator. We can capitalize on the differential responses of various indicators along a disturbance gradient to focus our monitoring efforts and tailor our interpretations of aquatic ecosystem responses.
5. The major advantage of a probability-based sample survey design, as opposed to hand-picked reaches, is that it provides a means of inferring results from the sampled stream reaches to make rigorous estimates of regional condition. A key disadvantage to a large probabilistic survey is that the clarity of biological response is reduced by high variability in biological scores across multiple disturbance gradients. To clarify assemblage response, it is necessary to examine a smaller number of stream reaches along clearer disturbance or stressor gradients. We need both large probabilistic surveys and smaller gradient studies to become familiar with the linkages of human land use, instream stressors, and biological responses. In so doing, we will be better able to link analysis with land management options.

ACKNOWLEDGMENTS

We are very grateful to Alan Herlihy and Leska Fore for sharing and discussing metric and index files. Thanks to Randy Hjort who provided prompt and patient assistance with many iterations of graphics and to Suzanne Pierson for making them print-ready. Thom Whittier and Ian Waite made invaluable suggestions to help clarify and organize the manuscript. This chapter was produced outside of normal work hours and was not subjected to the institution's review or clearance processes.

REFERENCES

Allen, A.P., T.R. Whittier, D.P. Larsen, P.R. Kaufmann, R.J. O'Connor, R.M. Hughes, R.S. Stemberger, S.S. Dixit, R.O. Brinkhurst, A.T. Herlihy, and S.G. Paulsen. 1999. Concordance of taxonomic composition patterns across multiple lake assemblages: effects of scale, body size, and land use, *Canadian Journal of Fisheries and Aquatic Sciences*, 56, 2029–2040.

Brooks, R.P., M.J. Croonquist, E.T. D'Silva, and J.E. Gallagher. 1991. Selection of biological indicators for integrating assessments of wetland, stream, and riparian habitats. *Proceedings of Biological Criteria: Research and Regulation.* December 1990, Arlington, Virginia, U.S. Environmental Protection Agency, Office of Water, Washington, D.C.

Bryce, S.A., D.P. Larsen, R.M. Hughes, and P.R. Kaufmann. 1999. Assessing relative risks to aquatic ecosystems: A mid-Appalachian case study, *Journal of the American Water Resources Association,* 35(1), 23–36.

Bryce, S.A., R.M. Hughes, and P.R. Kaufmann. In press. Development of a bird integrity index: using bird assemblages as indicators of riparian condition. *Environmental Management.*

Caudill, H.M. 1963. *Night Comes to the Cumberlands.* Little, Brown, Boston, MA.

Fausch, K.D., J.R. Karr, and P.R. Yant. 1984. Regional application of an index of biotic integrity based on stream fish communities, *Transactions of the American Fisheries Society,* 113, 39–55.

Fore, L.S., J.R. Karr, and R.W. Wisseman. 1996. Assessing invertebrate responses to human activities: Evaluating alternative approaches, *Journal of the North American Benthological Society,* 15, 212–231.

Fore, L.S. 2002. Response of diatom assemblages to human disturbance: Development and testing of a multimetric index for the mid-Atlantic region (USA), Chapter 22, this volume.

Hart, J.F. 1991. *The Land that Feeds Us.* W.W. Norton, New York.

Herlihy, A.T., P.R. Kaufmann, and M.E. Mitch. 1990. Regional estimates of acid mine drainage impact on streams in the mid-Atlantic and southeastern United States, *Water, Air, and Soil Pollution*, 50, 91–107.

Herlihy, A.T., P.R. Kaufmann, L. Reynolds, J. Li, and G. Robison. 1997. Developing indicators of ecological condition in the Willamette basin: an overview of the Oregon prepilot study for EPA's EMAP program, in A. Laenen and D.A. Dunnette. (Eds.). *River Quality, Dynamics and Restoration.* Lewis Publishers, CRC Press, Boca Raton, FL, 275–282.

Hill, B.H., A.T. Herlihy, P.R. Kaufmann, R.J. Stevenson, F.H. McCormick, and C. Burch Johnson. 2000. Use of periphyton assemblage data as an index of biotic integrity, *Journal of the North American Benthological Society,* 19(1), 50–67.

Hughes, R.M., P.R. Kaufmann, A.T. Herlihy, T.M. Kincaid, L. Reynolds, and D.P. Larsen. 1998. A process for developing and evaluating indices of fish assemblage integrity, *Canadian Journal of Fisheries and Aquatic Sciences,* 55, 1618–1631.

Hughes, R.M., S.G. Paulsen, and J.L. Stoddard. 2000. EMAP-Surface Waters: a national, multiassemblage probability survey of ecological integrity, *Hydrobiologia*, 423, 429–443.

Karr, J.R. 1981. Assessment of biotic integrity using fish communities, *Fisheries,* 6(6), 21–27.

Karr, J.R., K.D. Fausch, P.L. Angermeier, P.R. Yant, and I.J. Schlosser. 1986. *Assessment of Biological Integrity in Running Water: A Method and Its Rationale.* Illinois Natural History Survey Special Publication 5, Champaign, IL.

Karr, J.R. and E.W. Chu. 1999. *Restoring Life in Running Waters: Better Biological Monitoring.* Island Press, Covelo, CA.

Karr, J.R. and E.W. Chu. 2000. Sustaining living rivers, *Hydrobiologia,* 422, 1–14.

Kerans, B.L. and J.R. Karr. 1994. A benthic index of biointegrity (B-IBI) for rivers of the Tennessee Valley, *Ecological Applications*, 4, 768–785.

Klemm, D.J. and J.M. Lazorchak (Eds.). 1994a. *Environmental Monitoring and Assessment Program: Surface Waters and Region 3 Regional Environmental Monitoring and Assessment Program 1994 Pilot Laboratory Methods Manual for Streams.* EPA/620/R-94/003. Environmental Protection Agency, Environmental Monitoring Systems Laboratory, Cincinnati, OH.

Klemm, D.J. and J.M. Lazorchak (Eds.). 1994b. *Environmental Monitoring and Assessment Program: Surface Waters and Region 3 Regional Environmental Monitoring and Assessment Program 1994 Pilot Field Operations and Methods Manual for Streams.* EPA/620/R-94/004. Environmental Protection Agency, Environmental Monitoring Systems Laboratory, Cincinnati, OH.

Klemm, D.J., K.A. Blocksom, F.A. Fulk, A.T. Herlihy, R.M. Hughes, P.R. Kaufmann, D.V. Peck, J.L. Stoddard, W.T. Thoeny, M.B. Griffith, and W.S. Davis. In review. A macroinvertebrate biotic integrity index (MBII) for regionally assessing mid-Atlantic highlands streams, *Journal of the North American Benthological Society.*

Lammert, M. and J.D. Allan. 1999. Assessing biotic integrity of streams: effects of scale in measuring the influence of land use/cover and habitat structure on fish and macroinvertebrates, *Environmental Management*, 23(2), 257–270.

Li, H.W., C.B. Schreck, C.E. Bond, and E. Rexstad. 1987. Factors influencing changes in fish assemblages of Pacific Northwest streams, in W.J. Matthews and D.C. Heins (Eds.). *Community and Evolutionary Ecology of North American Stream Fishes.* University of Oklahoma Press, Norman, OK. 193–202.

Li, J., A.T. Herlihy, W. Gerth, P.R. Kaufmann, S.V. Gregory, N.S. Urquhart, and D.P. Larsen. 2001. Quantifying variation in stream macroinvertebrate assemblages at multiple spatial scales: the relative influence of sample size and spatial distribution, *Freshwater Biology,* 46, 87–97.

Marsh, P.C. and J.E. Luey. 1982. Oases for aquatic life within agricultural watersheds, *Fisheries*, 7(6), 16–24.

McCormick, F.H., R.M. Hughes, P.R. Kaufmann, D.V. Peck, and J.L. Stoddard. 2001. Development of an index of biotic integrity for the mid-Atlantic highlands region, *Transactions of the American Fisheries Society,* 130, 857–877.

Norris, R.H. and C.P. Hawkins. 2000. Monitoring river health, *Hydrobiologia*, 435, 5–17.

O'Connell, T.J., L.E. Jackson, and R.P. Brooks. 1998. A bird community index of biotic integrity for the mid-Atlantic highlands, *Environmental Monitoring and Assessment,* 51(1–2), 145–156.

Odum, E.P., J.T. Finn, and E.H. Franz. 1979. Perturbation theory and the subsidy-stress gradient, *Bioscience,* 29(6), 349–352.

Odum, E.P. 1985. Trends expected in stressed ecosystems, *Bioscience*, 35(7), 419–422.

Omernik, J.M. and A.L. Gallant. 1986. *Ecoregions of the Pacific Northwest.* EPA/600/3–86/033. U.S. Environmental Protection Agency, Environmental Research Laboratory, Corvallis, OR.

Osborne, L.L., S.L. Kohler, P.B. Bayley, D.M. Day, W.A. Bertrand, M.J. Wiley, and R. Sauer. 1992. Influence of stream location in a drainage network on the index of biointegrity, *Transactions of the American Fisheries Society,* 121, 635–643.

Plafkin, J.L., M.T. Barbour, K.D. Porter, S.K. Gross, and R.M. Hughes. 1989. *Rapid Assessment Protocols for Use in Streams and Rivers: Benthic Macroinvertebrates and Fish.* EPA/444/4–89–001. U.S. Environmental Protection Agency, Washington, D.C.

Rahel, F.J. and W.A. Hubert. 1991. Fish assemblages and habitat gradients in a Rocky Mountain-Great Plains stream: biotic zonation and additive patterns of community change, *Transactions of the American Fisheries Society,* 130, 319–332.

Scott, M.C. and G.S. Helfman. 2001. Native invasions, homogenization, and the mismeasure of integrity of fish assemblages, *Fisheries,* 26(11), 6–15.

Simon, T.P. (Ed.). 1999. *Assessing the Sustainability and Biological Integrity of Water Resources Using Fish Communities.* CRC Press, Boca Raton, FL.

U.S. Environmental Protection Agency. 2000. *Mid-Atlantic Highlands Streams Assessment.* EPA/903/R-00/015. U.S. Environmental Protection Agency Region 3. Philadelphia, PA.

Wiley, M.J., P.W. Seelbach, K. Wehrly, and J.S. Martin. 2002. Regional ecological normalization using linear models: a meta-method for scaling stream assessment indicators, Chapter 12, this volume.

Wright, J.F. 1995. Development and use of a system for predicting the macroinvertebrate fauna in flowing waters, *Australian Journal of Ecology,* 20, 181–197.

Yoder, C.O. and E.T. Rankin. 1995. Biological response signatures and the area of degradation value: new tools for interpreting multimetric data, in W.S. Davis and T.P. Simon. (Eds.). *Biological Assessment and Criteria: Tools for Water Resource Planning and Decision Making.* Lewis Publishers, Boca Raton, FL, 263–286.

Index

A

B

C

D

F

G

H

I

J

K

L

M

N

O

P

V

W